ELSEVIER'S
DICTIONARY OF
MAMMALS

ELSEVIER'S DICTIONARY OF MAMMALS

in
Latin, English, German, French and Italian

compiled by

MURRAY WROBEL
London, United Kingdom

ELSEVIER

Amsterdam • Boston • Heidelberg • London • New York • Oxford
Paris • San Diego • San Francisco • Singapore • Sydney • Tokyo

Elsevier
Radarweg 29, PO Box 211, 1000 AE Amsterdam, The Netherlands
The Boulevard, Langford Lane, Kidlington, Oxford OX5 1GB, UK

First edition 2007

Library of Congress Cataloging-in-Publication Data
A catalog record for this book is available from the Library of Congress

British Library Cataloguing in Publication Data
A catalogue record for this book is available from the British Library

ISBN-13: 978-0-444-51877-4
ISBN-10: 0-444-51877-0

For information on all Elsevier publications
visit our website at books.elsevier.com

Printed and bound in The Netherlands

07 08 09 10 11   10 9 8 7 6 5 4 3 2 1

Preface

This dictionary has been compiled with the aim of giving an overview of the English, German, French and Italian names of mammals. The Basic Table contains, in alphabetical order, the scientific names of families, genera, species and some sub-species and synonyms with, again in alphabetical order, the identified names in English, German, French and Italian. These are given in the singular for species and sub-species and in the plural for other terms.

The relevant order and family are given for each term. I have followed Duff & Lawson's "Mammals of the World - A Checklist" for orders, families and genera. The synonyms and subspecies are those that I have come across in other sources and are, of course, not a complete listing.

I have shown a number of alternative spellings of vernacular names which, although not always "correct", appear to have gained common currency.

My thanks go to Dott. Bruno Grazioli of Royal Holloway University of London for checking the Italian names and to Katayou Asadollahzadeh-Sarab, who prepared this work for publication.

Murray Wrobel

Contents

Abbreviations

The following abbreviations denote a general, but not necessarily exclusive, use in the various areas.

(ANZ)   -   Australia and New Zealand

(CSA)   -   Central and Southern Africa

(NA)   -   North America

(Qué)   -   Quebec

Basic Table

A

1 *Abditomys*
Rodentia - Muridae
e Luzon Broad-toothed Rats
d Luzon-Breitzahnratten

2 *Abditomys latidens*
Rodentia - Muridae
e Luzon Broad-toothed Rat
d Luzon-Breitzahnratte

3 *Abrawayaomys*
Rodentia - Muridae
e Ruschi's Rats
d Ruschi-Ratten

4 *Abrawayaomys ruschii*
Rodentia - Muridae
e Ruschi's Rat
d Ruschi-Ratte

5 *Abrocoma*
Rodentia - Abrocomidae
e Chinchilla Rats; Rat Chinchillas
d Chinchillaratten
f Rats-chinchillas

6 *Abrocoma bennetti*
Rodentia - Abrocomidae
e Bennett's Chinchilla Rat
d Chilenische Chinchillaratte

7 *Abrocoma boliviensis*
Rodentia - Abrocomidae
e Bolivian Chinchilla Rat
d Bolivische Chinchillaratte

8 *Abrocoma budini*
Rodentia - Abrocomidae
e Budin's Chinchilla Rat; Sierra di
Ambato Chinchilla Rat

9 *Abrocoma cinerea*
Rodentia - Abrocomidae
e Ashy Chinchilla Rat
d Chinchillaratte
f Rat-chinchilla cendré

10 *Abrocoma famatina*
Rodentia - Abrocomidae
e Famatina Chinchilla Rat; Sierra di
Famatina Chinchilla Rat

11 *Abrocoma schistacea*
Rodentia - Abrocomidae
e Tontal Chinchilla Rat; Sierra di
Tontal Chinchilla Rat

12 *Abrocoma uspallata*
Rodentia - Abrocomidae
e Uspallata Chinchilla Rat

13 *Abrocoma vaccarum*
Rodentia - Abrocomidae
e Vacas Chinchilla Rat

14 Abrocomidae
Rodentia
e Chinchilla Rats; Rat Chinchillas
d Chinchillaratten
f Abrocomidés; Rats-chinchillas
i Abrocomidi

15 *Abrothrix*
Rodentia - Muridae
e Grass Mice
d Anden-Feldmäuse

16 *Abrothrix illuteus*
Rodentia - Muridae
e Grey Grass Mouse

17   **Abrothrix lanosus**
     Rodentia - Muridae
 e   Woolly Grass Mouse

18   **Abrothrix longipilis**
     Rodentia - Muridae
 e   Long-haired Grass Mouse; South
     American Field Mouse

19   **Abrothrix mansoensis**
     Rodentia - Muridae
 e   Manso Grass Mouse

20   **Abrothrix sanborni**
     Rodentia - Muridae
 e   Sanborn's Grass Mouse

21   **Abrothrix xanthorhinus**
     Rodentia - Muridae
 e   Yellow-nosed Grass Mouse

     Acanthion syn. *Hystrix* q.v.

22   **Acerodon**
     Chiroptera - Pteropodidae
 e   Flying Foxes; Island Fruit Bats;
     Acerodons
 d   Große Flughunde; Riesenflughunde
 f   Renards volants; Roussettes
 i   Volpi volanti

     Acerodon argentatus syn. *A.*
     celebensis q.v.

23   **Acerodon celebensis**
     Chiroptera - Pteropodidae
 e   Celebes Flying Fox; Sulawesi Flying
     Fox; Sulawesi Fruit Bat; Signal-
     winged Acerodon
 d   Celebes-Flugfuchs
 f   Renard volant des Célèbes

24   **Acerodon humilis**
     Chiroptera - Pteropodidae

 e   Talaud Flying Fox; Talaud Fruit Bat;
     Talaud Acerodon

25   **Acerodon jubatus**
     Chiroptera - Pteropodidae
 e   Golden-capped Fruit Bat; Golden-
     crowned Flying Fox; Golden-capped
     Acerodon
 d   Goldkronenflughund; Luzon-
     Flughund
 f   Renard volant des Philippines

26   **Acerodon lucifer**
     Chiroptera - Pteropodidae
 e   Panay Golden-capped Fruit Bat;
     Panay Flying Fox; Panay Giant Fruit
     Bat; Panay Golden-capped Acerodon
 d   Panay-Riesenflughund
 f   Renard volant de l'île Panay;
     Roussette de Panay

27   **Acerodon mackloti**
     Chiroptera - Pteropodidae
 e   Sunda Flying Fox; Sunda Fruit Bat;
     Sunda Acerodon

     Acerodon turquatus syn. *A.*
     celebensis q.v.

28   **Acinonyx**
     Carnivora - Felidae
 e   Hunting Leopards; Cheetahs
 d   Jagdleoparden; Geparde
 f   Guépards
 i   Ghepardi

29   **Acinonyx jubatus**
     Carnivora - Felidae
 e   Hunting Leopard; Cheetah; Ounce
 d   Jagdleopard; Gepard
 f   Guépard
 i   Ghepardo; Ghepardo africano

30   **Acinonyx jubatus guttatus**
     Carnivora - Felidae
 e   North African Cheetah

31   *Acinonyx jubatus hecki*
Carnivora - Felidae
e   Northwest African Cheetah
d   Nordwest Afrikanischer Gepard;
       Nordafrikanischer Gepard

32   *Acinonyx jubatus jubatus*
Carnivora - Felidae
e   South African Cheetah; African
       Cheetah
d   Südafrikanischer Gepard
f   Guépard commun

33   *Acinonyx jubatus ngorongorensis*
Carnivora - Felidae
e   Eastern East African Cheetah
d   Tansania-Gepard

34   *Acinonyx jubatus raddei*
Carnivora - Felidae
e   Asiatic Cheetah; Turkestan Cheetah
d   Kaspischer Gepard

35   *Acinonyx jubatus rex*
Carnivora - Felidae
e   King Cheetah
d   Königsgepard
f   Guépard royal
i   Ghepardo reale

36   *Acinonyx jubatus soemmeringi*
Carnivora - Felidae
e   Sudan Cheetah
d   Sudan-Gepard

37   *Acinonyx jubatus velox*
Carnivora - Felidae
e   Western East Africa Cheetah
d   Kenia Gepard

38   *Acinonyx jubatus venaticus*
Carnivora - Felidae
e   Indian Cheetah
d   Indischer Gepard

f   Guépard d'Iran
i   Ghepardo asiatico

39   *Acomys*
Rodentia - Muridae
e   Porcupine Mice; Spiny Mice; African
       Spiny Mice
d   Stachelmäuse
f   Souris épineuses; Rats épineux
i   Topi spinosi

40   *Acomys cahirinus*
Rodentia - Muridae
e   Egyptian Spiny Mouse
d   Stachelmaus; Stachelratte;
       Ägyptische Stachelmaus
f   Souris épineuse; Rat épineux; Souris
       épineuse égyptienne
i   Topo spinoso egiziano

41   *Acomys cahirinus cahirinus*
Rodentia - Muridae
e   Cairo Spiny Mouse; Dusky Egyptian
       Spiny Mouse
d   Nil-Stachelmaus; Dunkle Nil-
       Stachelmaus; Schwarze Nil-
       Stachelmaus
f   Souris épineuse égyptienne

42   *Acomys cahirinus dimmidiatus*
Rodentia - Muridae
e   Arabian Spiny Mouse
d   Sinai-Stachelmaus; Palästina-
       Stachelmaus
f   Souris épineuse arabe; Souris
       épineuse du Caire

43   *Acomys cahirinus hunteri*
Rodentia - Muridae
e   Abu Simbel Spiny Mouse
d   Abu-Simbel-Stachelmaus

44   *Acomys cahirinus seurati*
Rodentia - Muridae
e   Tibesti Spiny Mouse
d   Tibesti-Stachelmaus

45   *Acomys cilicicus*
Rodentia - Muridae
e   Asia Minor Spiny Mouse; Turkey
Spiny Mouse
d   Türkei-Stachelmaus

46   *Acomys cineraceus*
Rodentia - Muridae
e   Grey Spiny Mouse
d   Graue Stachelmaus

47   *Acomys ignitus*
Rodentia - Muridae
e   Fiery Spiny Mouse
d   Feuerrote Stachelmaus

48   *Acomys kempi*
Rodentia - Muridae
e   Kemp's Spiny Mouse
d   Kemp-Stachelmaus

49   *Acomys louisae*
Rodentia - Muridae
e   Louise's Spiny Mouse
d   Louise-Stachelmaus
f   Souris épineuse de Somalie

50   *Acomys minous*
Rodentia - Muridae
e   Crete Spiny Mouse
d   Kreta-Stachelmaus
f   Rat épineux de Crète; Souris
épineuse de Crète
i   Topo spinoso cretese; Topo spinoso

51   *Acomys mullah*
Rodentia - Muridae
e   Mullah Spiny Mouse
d   Mullah-Stachelmaus

52   *Acomys nesiotes*
Rodentia - Muridae
e   Cyprus Spiny Mouse
d   Zypern-Stachelmaus

i   Topo spinoso di Cipro

53   *Acomys percivali*
Rodentia - Muridae
e   Percival's Spiny Mouse
d   Percival-Stachelmaus

54   *Acomys russatus*
Rodentia - Muridae
e   Golden Spiny Mouse
d   Goldstachelmaus
f   Souris épineuse rousse; Souris
épineuse dorée

55   *Acomys russatus lewisi*
Rodentia - Muridae
e   Jordan Spiny Mouse
d   Jordan-Stachelmaus

56   *Acomys spinosissimus*
Rodentia - Muridae
e   Spiny Mouse; Common Spiny Mouse
d   Zwergstachelmaus

57   *Acomys subspinosus*
Rodentia - Muridae
e   Cape Spiny Mouse
d   Kap-Stachelmaus

58   *Acomys wilsoni*
Rodentia - Muridae
e   Wilson's Spiny Mouse; Kenya Spiny
Mouse
d   Wilson-Stachelmaus; Kenia-
Stachelmaus
f   Souris épineuse de Wilson

Aconaemys syn. *Pithanotomys* q.v.

59   *Acrobates*
Diprotodontia - Acrobatidae
e   Pygmy Flying Phalangers; Pygmy
Flying Possums; Pygmy Gliding
Possums; Pygmy Gliders; Feathertail
Gliders (ANZ)

d   Zwerggleitbeutler ; Australien-
      Mausflugbeutler ; Mausflugbeutler
f   Acrobates
i   Acrobati

60   *Acrobates pygmaeus*
      Diprotodontia - Acrobatidae
e   Pygmy Flying Phalanger; Pygmy
      Flying Possum; Pygmy Gliding
      Possum; Pygmy Glider; Feathertail
      Glider (ANZ) ; Flying Mouse;
      Feather-tailed Pygmy Glider
d   Australischer Zwerggleitbeutler;
      Australien-Mausflugbeutler;
      Zwerggleitbeutler;
      Federschwanzgleitbeutler
f   Acrobate pygmée; Acrobate pygmée
      d'Australie
i   Acrobata pigmeo; Planatore dalla
      coda a penna

61   Acrobatidae
      Diprotodontia
e   Feathertail Gliders and Possums;
      Gliders; Feather-tailed Gliders and
      Feather-tailed Possums
d   Zwerggleitbeutler
f   Acrobatidés
i   Acrobatidi

62   *Addax*
      Artiodactyla - Bovidae
e   Addaxes
d   Mendes-Antilopen
f   Addax

63   *Addax nasomaculatus*
      Artiodactyla - Bovidae
e   Addax
d   Mendes-Antilope; Addax
f   Addax à nez tacheté; Addax;
      Antilope à nez tacheté
i   Antilope addax; Addax; Antilope di
      Mendes

Adenota syn. *Kobus* q.v.

Aello syn. *Mormoops* q.v.

64   *Aepeomys*
      Rodentia - Muridae
e   Montane Mice
d   Anden-Mäuse

65   *Aepeomys fuscatus*
      Rodentia - Muridae
e   Dusky Montane Mouse

66   *Aepeomys lugens*
      Rodentia - Muridae
e   Olive Montane Mouse

67   *Aepeomys reigi*
      Rodentia - Muridae
e   Reig's Montane Mouse

68   *Aepyceros*
      Artiodactyla - Bovidae
e   Impalas
d   Schwarzfersenantilopen; Impalas
f   Impalas
i   Impala

69   *Aepyceros melampus*
      Artiodactyla - Bovidae
e   Impala
d   Schwarzfersenantilope; Impala
f   Impala
i   Impala; Gazella dai piedi neri

70   *Aepyceros melampus melampus*
      Artiodactyla - Bovidae
e   Southern Impala; Angolan Impala
d   Transvaal-Impala
f   Impala

71   *Aepyceros melampus petersi*
      Artiodactyla - Bovidae
e   Black-faced Impala; Peters's Impala
d   Schwarznasenimpala
f   Impala à muffle noir;

Impala à face noire
i   Impala dalla faccia nera

72   *Aepyceros melampus rendilis*
Artiodactyla - Bovidae
e   East African Impala
d   Kenia-Impala
f   Gazelle aux pieds noirs

73   *Aepyprymnus*
Diprotodontia - Macropodidae
e   Large Rat Kangaroos; Rufous Rat
Kangaroos; Rufous Bettongs (ANZ)
d   Großrattenkängurus; Rote
Rattenkängurus
f   Rats-kangourous

74   *Aepyprymnus rufescens*
Diprotodontia - Macropodidae
e   Rufous Rat Kangaroo; Rufous Jerboa
Kangaroo; Rufous Bettong (ANZ)
d   Rotes Rattenkänguru;
Großrattenkänguru; Großes
Rattenkänguru
f   Rat-kangourou rougeâtre

75   *Aeretes*
Rodentia - Sciuridae
e   North Chinese Flying Squirrels;
Chinese Flying Squirrels; Groove-
toothed Flying Squirrels
d   Furchenzahngleithörnchen
i   Scoiattol volanti cinesi

76   *Aeretes melanopterus*
Rodentia - Sciuridae
e   North Chinese Flying Squirrel;
Chinese Flying Squirrel; Groove-
toothed Flying Squirrel
d   Furchenzahngleithörnchen
i   Scoiattolo volante cinese del nord

77   *Aeromys*
Rodentia - Sciuridae
e   Giant Black Flying Squirrels; Large
Black Flying Squirrels

d   Schwarze Gleithörnchen

Aeromys phaeomelas syn. A.
tephromelas q.v.

78   *Aeromys tephromelas*
Rodentia - Sciuridae
e   Large Black Flying Squirrel; Black
Flying Squirrel
d   Schwarzes Gleithörnchen
i   Scoiattolo volante della Malesia;
Scoiattolo volante del Borneo;
Scoiattolo volante di Sumatra

79   *Aeromys thomasi*
Rodentia - Sciuridae
e   Thomas's Flying Squirrel
d   Thomas-Gleithörnchen

80   *Aethalops*
Chiroptera - Pteropodidae
e   Pygmy Fruit Bats

81   *Aethalops alecto*
Chiroptera - Pteropodidae
e   Pygmy Fruit Bat; Smaller Tailless
Fruit Bat

Aethechinus syn. Erinaceus q.v.

82   *Aethomys*
Rodentia - Muridae
e   Etho Mice; African Rock Rats; Bush
Rats
d   Buschfelsenratten; Afrikanische
Buschratten
f   Rats des rochers

83   *Aethomys bocagei*
Rodentia - Muridae
e   Bocage's Rock Rat
d   Bocage-Buschratte

84   *Aethomys chrysophilus*
Rodentia - Muridae

e   Red Veld Rat; Red Rock Rat
d   Südafrikanische Veldratte; Rote
      Buschratte

85   *Aethomys granti*
      Rodentia - Muridae
e   Grant's Rock Mouse; Grant's Rock
      Rat
d   Grant-Buschratte

86   *Aethomys hindei*
      Rodentia - Muridae
e   Hinde's Rock Rat
d   Hinde-Buschratte

87   *Aethomys kaiseri*
      Rodentia - Muridae
e   Kaiser's Rock Rat
d   Kaiser-Buschratte

88   *Aethomys namaquensis*
      Rodentia - Muridae
e   Namaqua Rock Mouse; Namaqua
      Rock Rat
d   Namaqua-Buschratte; Namaqua-
      Felsenmaus

89   *Aethomys nyikae*
      Rodentia - Muridae
e   Nyika Veld Rat; Nyika Rock Rat
d   Nyika-Buschratte

Aethomys paeduleus syn. *Thallomys
paeduleus* q.v.

90   *Aethomys silindensis*
      Rodentia - Muridae
e   Silinda Rat; Silinda Rock Rat
d   Silinda-Buschratte

91   *Aethomys stannarius*
      Rodentia - Muridae
e   Tinfields Rock Rat
d   Zinnfeld-Buschratte

92   *Aethomys thomasi*
      Rodentia - Muridae
e   Thomas's Rock Rat
d   Thomas-Buschratte

Aethosciurus syn. *Paraxerus* q.v.

Afganomys syn. *Ellobius* q.v.

Agouti syn. *Cuniculus* q.v.

93   Agoutidae
      Rodentia
e   Pacas; Agoutis and Pacas
d   Pakas
f   Agoutidés
i   Agoutidi

Agricola syn. *Microtus* q.v.

94   *Ailuropoda*
      Carnivora - Ursidae
e   Giant Pandas
d   Ailuropes; Große Pandas;
      Bambusbären
f   Pandas; Pandas géants; Ailuridés
i   Panda giganti

95   *Ailuropoda melanoleuca*
      Carnivora - Ursidae
e   Giant Panda; Great Panda; Bamboo
      Bear
d   Großer Panda; Bambusbär;
      Riesenpanda
f   Panda; Panda géant; Grand panda;
      Ailuridé
i   Panda maggiore; Panda gigante; Orso
      del bambù

96   *Ailurops*
      Diprotodontidae - Phalangeridae
e   Bear Phalangers; Bear Cuscuses
d   Bärenkuskuse
f   Couscous des Célèbes
i   Cusci ursini

97   *Ailurops ursinus*
Diprotodontidae - Phalangeridae
e   Bear Phalanger; Bear Cuscus;
Sulawesi Bear Cuscus
d   Bärenkuskus; Eigentlicher
Bärenkuskus
f   Couscous des Célèbes
i   Cusco ursino

98   *Ailurus*
Carnivora - Ursidae
e   Lesser Pandas; Red Pandas; Red Cat
Bears; Himalajan Raccoons
d   Kleine Pandas; Katzenbären
f   Petits pandas
i   Panda minori

99   *Ailurus fulgens*
Carnivora - Ursidae
e   Lesser Panda; Cat Bear
d   Kleiner Panda; Katzenbär
f   Petit panda
i   Panda minore; Panda rosso

100   *Ailurus fulgens fulgens*
Carnivora - Ursidae
e   Red Panda; Western Cat Bear
d   Westlicher Katzenbär
f   Petit panda de l'Inde

101   *Ailurus fulgens styani*
Carnivora - Ursidae
e   Styan's Panda; Styan's Cat Bear
d   Styans Katzenbär
f   Petit panda de Styan

102   *Akodon*
Rodentia - Muridae
e   Grass Mice; South American Field
Mice
d   Südamerikanische Feldmäuse
f   Souris des champs

103   *Akodon aerosus*
Rodentia - Muridae

e   Highland Grass Mouse
d   Hochlandgrasmaus

104   *Akodon affinis*
Rodentia - Muridae
e   Colombian Grass Mouse
d   Kolumbianische Grasmaus

Akodon akodontius syn.
Oxymycterus akodontius q.v.

105   *Akodon albiventer*
Rodentia - Muridae
e   White-bellied Grass Mouse
d   Weißbauchfeldmaus

106   *Akodon aliquantulus*
Rodentia - Muridae
e   Tucuman Grass Mouse

Akodon angularis syn. *Oxymycterus*
angularis q.v.

107   *Akodon azarai*
Rodentia - Muridae
e   Azara's Grass Mouse
d   Azara-Feldmaus; Azara-Grasmaus

108   *Akodon bogotensis*
Rodentia - Muridae
e   Bogota Grass Mouse
d   Bogota-Grasmaus

109   *Akodon boliviensis*
Rodentia - Muridae
e   Bolivian Grass Mouse
d   Bolivianische Feldmaus

110   *Akodon budini*
Rodentia - Muridae
e   Budin's Grass Mouse
d   Budin-Grasmaus

111 *Akodon cursor*
Rodentia - Muridae
e Cursor Grass Mouse; Cursorial Grass Mouse
d Cursor-Grasmaus

112 *Akodon dayi*
Rodentia - Muridae
e Day's Grass Mouse
d Day-Grasmaus

Akodon delator syn. *Oxymycterus delator* q.v.

113 *Akodon dolores*
Rodentia - Muridae
e Dolorous Grass Mouse
d Trauergrasmaus

114 *Akodon fumeus*
Rodentia - Muridae
e Smoky Grass Mouse
d Rauchgraue Grasmaus

115 *Akodon hershkovitzi*
Rodentia - Muridae
e Hershkovitz's Grass Mouse

Akodon hispidus syn. *Oxymycterus hispidus* q.v.

Akodon iheringi syn. *Brucepattesonius iheringi* q.v.

Akodon illuteus syn. *Abrothrix illuterus* q.v.

Akodon inca syn. *Oxymycterus inca* q.v.

116 *Akodon iniscatus*
Rodentia - Muridae
e Intelligent Grass Mouse
d Kluge Grasmaus

117 *Akodon juninensis*
Rodentia - Muridae
e Junin Grass Mouse
d Junin-Grasmaus

118 *Akodon kempi*
Rodentia - Muridae
e Kemp's Grass Mouse
d Kemp-Grasmaus

119 *Akodon kofordi*
Rodentia - Muridae
e Koford's Grass Mouse
d Koford-Grasmaus

Akodon lanosus syn. *Abrothrix lanosus* q.v.

120 *Akodon latebricola*
Rodentia - Muridae
e Ecuadorean Grass Mouse

121 *Akodon lindberghi*
Rodentia - Muridae
e Lindbergh's Grass Mouse
d Lindbergh-Grasmaus

Akodon longipilis syn. *Abrothrix longipilis* q.v.

Akodon mansoensis syn. *Abrothrix mansoensis* q.v.

122 *Akodon markhami*
Rodentia - Muridae
e Markham's Grass Mouse
d Markham-Grasmaus

123 *Akodon mimus*
Rodentia - Muridae
e Thespian Grass Mouse

124 *Akodon molinae*
Rodentia - Muridae

e Molina's Grass Mouse
d Molina-Grasmaus

125 *Akodon mollis*
Rodentia - Muridae
e Soft Grass Mouse
d Weiche Grasmaus

126 *Akodon mystax*
Rodentia - Muridae
e Caparao Grass Mouse

127 *Akodon neocenus*
Rodentia - Muridae
e Neuquen Grass Mouse
d Neuquen-Grasmaus

128 *Akodon nigrita*
Rodentia - Muridae
e Blackish Grass Mouse
d Schwärzliche Grasmaus

129 *Akodon oenos*
Rodentia - Muridae
e Wine Grass Mouse

130 *Akodon olivaceus*
Rodentia - Muridae
e Olive Grass Mouse

131 *Akodon orophilus*
Rodentia - Muridae
e El Dorado Grass Mouse
d El-Dorado-Grasmaus

132 *Akodon paranaensis*
Rodentia - Muridae
e Parana Grass Mouse

133 *Akodon pervalens*
Rodentia - Muridae
e Robust Grass Mouse

134 *Akodon philipmyersi*

Rodentia - Muridae
e Myers's Grass Mouse
d Myers Grasmaus

135 *Akodon puer*
Rodentia - Muridae
e Altiplano Grass Mouse
d Altiplano-Feldmaus

Akodon roberti syn. *Oxymycterus roberti* q.v.

Akodon rufus syn. *Oxymycterus rufus* q.v.

Akodon sanborni syn. *Abrothrix sanborni* q.v.

136 *Akodon sanctipaulensis*
Rodentia - Muridae
e Sao Paolo Grass Mouse
d Sao-Paulo-Grasmaus

137 *Akodon serrensis*
Rodentia - Muridae
e Serra do Mar Grass Mouse
d Serrado-Mar-Grasmaus

138 *Akodon siberiae*
Rodentia - Muridae
e Cochabamba Grass Mouse
d Cochabamba-Grasmaus

139 *Akodon simulator*
Rodentia - Muridae
e Grey-bellied Grass Mouse
d Graubauchgrasmaus

140 *Akodon spegazzinii*
Rodentia - Muridae
e Spegazzini's Grass Mouse
d Spegazzini-Grasmaus

141 *Akodon subfuscus*
Rodentia - Muridae

e   Puno Grass Mouse
d   Puno-Grasmaus

142   *Akodon surdus*
       Rodentia - Muridae
e   Silent Grass Mouse
d   Stille Grasmaus

143   *Akodon sylvanus*
       Rodentia - Muridae
e   Forest Grass Mouse
d   Waldgrasmaus

144   *Akodon toba*
       Rodentia - Muridae
e   Chaco Grass Mouse
d   Chaco-Grasmaus

145   *Akodon torques*
       Rodentia - Muridae
e   Cloud Forest Grass Mouse
d   Wolkenwaldgrasmaus

146   *Akodon urichi*
       Rodentia - Muridae
e   Northern Grass Mouse

147   *Akodon varius*
       Rodentia - Muridae
e   Variable Grass Mouse
d   Variable Grasmaus

       Akodon xanthorhinus syn.
       Abrothrix xanthorhinus q.v.

148   *Alcelaphus*
       Artiodactyla - Bovidae
e   Hartebeests; Kongonis
d   Kuhantilopen
f   Bubales
i   Alcelafi

149   *Alcelaphus buselaphus*
       Artiodactyla - Bovidae

e   Hartebeest; Kongoni; Red
       Hartebeest; Cape Hartebeest; Coke's
       Hartebeest
d   Kuhantilope
f   Bubale; Kongoni; Bubale
i   Alcelafo

150   *Alcelaphus buselaphus buselaphus*
       Artiodactyla - Bovidae
e   Bubal Hartebeest; Northern
       Hartebeest; Northern Bubal;
       Northern Red Hartebeest
d   Nordafrikanische Kuhantilope;
       Nordafrikanisches Hartebeest
f   Bubale du Nord

151   *Alcelaphus buselaphus caama*
       Artiodactyla - Bovidae
e   Cape Red Hartebeest; Red
       Hartebeest; Caama Hartebeest
d   Kaama
f   Bubale caama
i   Alcelafo Caama

152   *Alcelaphus buselaphus cokii*
       Artiodactyla - Bovidae
e   Kongoni; Kongoni Hartebeest;
       Coke's Hartebeest
d   Kongoni
f   Bubale de Cole
i   Alcelafo di Coke

153   *Alcelaphus buselaphus jacksoni*
       Artiodactyla - Bovidae
e   Jackson's Hartebeest
f   Bubale de Jackson

154   *Alcelaphus buselaphus lelwel*
       Artiodactyla - Bovidae
e   Kenya Highlands Hartebeest;
       Lelwel's Hartebeest
d   Lelwel-Hartebeest
f   Bubal de Lelwel

155   *Alcelaphus buselaphus major*
       Artiodactyla - Bovidae

e Western Hartebeest
d Westafrikanische Kuhantilope
f Bubale major

156 *Alcelaphus buselaphus swaynei*
Artiodactyla - Bovidae
e Swayne Hartebeest
d Somali-Kuhantilope
f Bubale de Swayne

157 *Alcelaphus buselaphus tora*
Artiodactyla - Bovidae
e Tora Hartebeest
d Tora
f Bubale tora

158 *Alces*
Artiodactyla - Cervidae
e Moose (NA); Elk ; Elks
d Elche
f Élans
i Alci

159 *Alces alces*
Artiodactyla - Cervidae
e European Elk ; Elk; European
   Moose; Moose (NA); Elk Deer;
   Moose Deer; European Elk Deer;
   European Moose Deer

d Europäischer Elch; Westlicher Elch;
   Elen; Elentier; Elch
f Élan; Élan à crinière; Élan original
i Alce

160 *Alces alces alces*
Artiodactyla - Cervidae
e Common Elk
d Nordelch
f Élan européen

161 *Alces alces americana*
Artiodactyla - Cervidae
e Moose (NA)
d Ostamerikanischer Elch

f Élan de l'Est

162 *Alces alces andersoni*
Artiodactyla - Cervidae
e Manitoba Moose; Northwestern
   Moose
d Westkanadischer Elch; Manitoba-
   Elch
f Élan de Manitoba

163 *Alces alces cameloides*
Artiodactyla - Cervidae
e Ussuri Moose; Siberian Moose
d Zwergelch; Ussuri-Elch; Amur-Elch
f Élan de Sibérie

164 *Alces alces caucasicus*
Artiodactyla - Cervidae
e Caucasian Moose; Caucasian Elk
d Kaukasus-Elch; Kakasus-Elen
f Élan du Caucase

165 *Alces alces gigas*
Artiodactyla - Cervidae
e Alaskan Moose; Tundra Moose;
   Yukon Moose
d Alaska-Elch
f Élan d'Alaska

166 *Alces alces shirasi*
Artiodactyla - Cervidae
e Yellowstone Moose; Shiras Moose;
   Wyoming Moose
d Yellowstone-Elch
f Élan de Shiraz

Aletesciurus syn. *Sundasciurus* q.v.

Alexandromys syn. *Microtus* q.v.

167 *Alionycteris*
Chiroptera - Pteropodidae
e Mindanao Pygmy Fruit Bats
d Kurznasenflughunde

168   *Alionycteris paucidentata*
          Chiroptera - Pteropodidae
e   Mindanao Pygmy Fruit Bat

169   *Allactaga*
          Rodentia - Dipodidae
e   Four-and-five-toed Jerboas; Earth
          Hares; Five-toed Jerboas
d   Erdhasen; Pferdespringer ;
          Wüstenspringmäuse
f   Gerboises
i   Gerboa

170   *Allactaga balikunica*
          Rodentia - Dipodidae
e   Balikun Jerboa
d   Balikun-Springmaus; Balikun-Jerboa

171   *Allactaga bullata*
          Rodentia - Dipodidae
e   Gobi Jerboa
d   Gobi-Springmaus; Gobi-Jerboa
f   Gerboise du Gobi

172   *Allactaga elater*
          Rodentia - Dipodidae
e   Little Jerboa; Small Five-toed Jerboa;
          Five-toed Jerboa
d   Zwergpferdespringer; Kleine
          Springmaus; Kleiner Pferdespringer;
          Springer
f   Petite jerboa; Petite gerboise

173   *Allactaga euphratica*
          Rodentia - Dipodidae
e   Euphrates Jerboa; Williams's Jerboa
d   Euphrat-Springmaus; Williams-
          Springmaus

174   *Allactaga firouzi*
          Rodentia - Dipodidae
e   Iranian Jerboa
d   Iranischer Pferdespringer

175   *Allactaga hotsoni*

          Rodentia - Dipodidae
e   Hotson's Jerboa; Hotson's Five-toed
          Jerboa
d   Hotson-Springmaus

176   *Allactaga major*
          Rodentia - Dipodidae
e   Greater Jerboa; Great Jerboa
d   Jerboa; Erdhase; Großer
          Pferdespringer
f   Jerboa alactaga sauteur; Gerboise
          allactaga sauteur
i   Gerboa maggiore

177   *Allactaga severtzovi*
          Rodentia - Dipodidae
e   Severtzov's Jerboa
d   Sewerzows Pferdespringer;
          Sewertzow-Springmaus

178   *Allactaga sibirica*
          Rodentia - Dipodidae
e   Siberian Jerboa; Mongolian Five-
          toed Jerboa; Siberian Five-toed
          Gerboa
d   Sibirische Springmaus; Sibirischer
          Pferdespringer; Eigentlicher
          Pferdespringer; Mongolischer
          Pferdespringer
f   Gerboise de Sibérie
i   Gerboa della Siberia

179   *Allactaga tetradactyla*
          Rodentia - Dipodidae
e   Four-toed Jerboa
d   Vierzehenpferdespringer;
          Vierzehenspringmaus
f   Gerboise de Vinogradov; Gerboise à
          quatre doigts
i   Gerboa dalle quattro dita

180   *Allactaga vinogradovi*
          Rodentia - Dipodidae
e   Vinogradov's Jerboa
d   Vinogradov-Jerboa; Winogradow-
          Jerboa; Vingradov-Jerboa

Allactaga williamsi syn. *A. euphratica* q.v.

Allactagulus syn. *Pygeretmus* q.v.

181   *Allactodipus*
Rodentia - Dipodidae
e   Bobrinski's Jerboas
d   Bobrinski-Sprimgmäuse
f   Gerboises de Bobrinski

182   *Allactodipus bobrinskii*
Rodentia - Dipodidae
e   Bobrinski's Jerboa
d   Bobrinskis Springer; Bobrinskis Pferdespringer; Bobrinski-Springmaus
f   Gerboise de Bobrinski

183   *Allenopithecus*
Primates - Cercopithecidae
e   Allen's Swamp Monkeys
d   Schwarzgrüne Meerkatzen; Sumpfmeerkatzen
f   Cercopithèques noir-et-verts
i   Cercopithechi di palude

184   *Allenopithecus nigroviridis*
Primates - Cercopithecidae
e   Swamp Guenon; Blackish-green Guenon; Allen's Swamp Monkey; Swamp Monkey
d   Schwarzgrüne Meerkatze; Sumpfmeerkatze
f   Cercopithèque noir-et-vert
i   Cercopiteco di palude; Cercopiteco di Allen; Cercopiteco verde-nero

185   *Allocebus*
Primates - Cheirogaleidae
e   Hairy-eared Dwarf Lemurs
d   Büschelohrige Katzenmakis; Büschelohr-Katzenmakis
f   Chirogales
i   Chirogali dalle orecchie pelose

186   *Allocebus trichotis*
Primates - Cheirogaleidae
e   Hairy-eared Dwarf Lemur; Hairy-eared Mouse Lemur
d   Büschelohriger Katzenmaki; Kleiner Katzenmaki; Büschelohrkatzenmaki
f   Chirogale aux oreilles velues; Allocèbe
i   Chirogale dalle orecchie pelose

187   *Allocricetelus*
Rodentia - Muridae
e   Mongolian Hamsters
d   Zwerghamster ; Mittelgroße Zwerghamster
f   Hamsters nains
i   Criceti mongoli

188   *Allocricetulus curtatus*
Rodentia - Muridae
e   Mongolian Hamster; Chinese Hamster
d   Mongolischer Zwerghamster
f   Hamster de Mongolie
i   Criceto mongolo

189   *Allocricetulus eversmanni*
Rodentia - Muridae
e   Eversmann's Hamster; Short Dwarf Hamster
d   Eversmann-Zwerghamster
f   Hamster d'Eversmann; Hamster kazakh
i   Criceto di Eversman
Alopex syn. Vulpes q.v.

190   *Alouatta*
Primates - Atelidae
e   Howler Monkeys; Howlers
d   Brüllaffen
f   Hurleurs
i   Scimmie urlatrici; Aluatte

191   *Alouatta belzebul*
Primates - Atelidae
e   Rufous-handed Howler; Rufous-

handed Howler Monkey; Black-and-
red Howler; Black-and-red Howler
Monkey; Red-handed Howler; Devil-
handed Howler; Red-handed Howler
Monkey
d   Rothandbrüllaffe
f   Hurleur aux mains rousses
i   Aluatta dalle mani rosse

192   *Alouatta caraya*
Primates - Atelidae
e   Black Howler; Black Howler
Monkey
d   Schwarzer Brüllaffe
f   Hurleur moir
i   Aluatta nera; Micete nero

193   *Alouatta coibensis*
Primates - Atelidae
e   Coiba Howler Monkey; Coiba Island
Howler; Coiba Island Howler
Monkey
d   Coiba-Brüllaffe
f   Singe hurleur de l'île Coiba
i   Aluatta dell'isola di Coiba

Alouatta fusca syn. *A. guariba* q.v.

194   *Alouatta guariba*
Primates - Atelidae
e   Brown Howler; Brown Howler
Monkey; Ursine Howler; Ursine
Howler Monkey; Guariba
d   Brauner Brüllaffe
i   Aluatta bruna

195   *Alouatta macconnelli*
Primates - Atelidae
e   Guyanan Howler Monkey
d   Bolivianischer Roter Brüllaffe

196   *Alouatta nigerrima*
Primates - Atelidae
e   Amazon Black Howler Monkey

197   *Alouatta palliata*

Primates - Atelidae
e   Mantled Howler; Mantled Howler
Monkey; Mantled Black Howler
Monkey
d   Mantelbrüllaffe
f   Hurleur à manteau; Singe hurleur à
manteau
i   Aluatta col mantello; Aluatta dal
mantello

198   *Alouatta pigra*
Primates - Atelidae
e   Mexican Black Howler Monkey;
Guatemalan Howler; Guatemalam
Mantled Howler Monkey;
Guatemalan Howler Monkey;
Yucatan Black Howler Monkey
d   Guatemala-Brüllaffe; Mexikanischer
Brüllaffe; Mexikanischer Schwarzer
Brüllaffe
f   Hurleur de Guatemala
i   Aluatta del Guatemala; Scimmia
urlatrice messicana; Scimmia
urlatrice nera del Guatemalo

199   *Alouatta sara*
Primates - Atelidae
e   Bolivian Red Howler Monkey;
Bolivian Red Howler; Bolivian
Howler Monkey
d   Bolivien-Brüllaffe
f   Hurleur de Bolivie
i   Aluatta rossa della Bolivia

200   *Alouatta seniculus*
Primates - Atelidae
e   Red Howler; Red Howler Monkey;
Venezuelan Red Howler Monkey
d   Roter Brüllaffe
f   Hurleur roux
i   Aluatta rossa; Scimmia urlatrice
rossa

Alouatta villosa syn. *A. pigra* q.v.

Alsomys syn. *Apodemus* q.v.

201 **Alticola**
Rodentia - Muridae
e Mountain Voles; High Mountain Voles
d Gebirgsmäuse; Gebirgswühlmäuse
f Campagnols des montagnes

202 **Alticola albicauda**
Rodentia - Muridae
e White-tailed Mountain Vole
d Weißschwanzgebirgswühlmaus

203 **Alticola argentatus**
Rodentia - Muridae
e Silver Mountain Vole
d Silberne Gebirgswühlmaus

204 **Alticola barakshin**
Rodentia - Muridae
e Gobi Altai Mountain Vole; Barakshin's Mountain Vole
d Gobi-Altai-Gebirgswühlmaus

205 **Alticola lemminus**
Rodentia - Muridae
e Lemming Vole
d Lemmingwühlmaus

206 **Alticola macrotis**
Rodentia - Muridae
e Large-eared Vole
d Transbaikalische Gebirgswühlmaus; Mongolische Gebirgsmaus; Großohrwühlmaus

207 **Alticola montosa**
Rodentia - Muridae
e Central Kashmir Vole
d Zentralkaschmir-Wühlmaus

208 **Alticola roylei**
Rodentia - Muridae
e Royle's Mountain Vole; Royle's High Mountain Vole
d Royles Gebirgswühlmaus

f Campagnol de Royle

209 **Alticola semicanus**
Rodentia - Muridae
e Mongolian Silver Vole
d Mongolische Gebirgsmaus; Mongolische Gebirgswühlmaus

210 **Alticola semicanus alleni**
Rodentia - Muridae
e Grey Mountain Vole
d Silbergraue Bergwühlmaus

211 **Alticola stoliczkanus**
Rodentia - Muridae
e Stoliczka's Mountain Vole; Stoliczka's High Mountain Vole
d Stoliczkas Gebirgswühlmaus

212 **Alticola stracheyi**
Rodentia - Muridae
e Strachey's Mountain Vole

213 **Alticola strelzowi**
Rodentia - Muridae
e Flat-headed Mountain Vole
d Gebirgsmaus; Flachkopfwühlmaus

214 **Alticola tuvinicus**
Rodentia - Muridae
e Tuva Silver Vole
d Tuwa-Silberwühlmaus

215 **Amblonyx**
Carnivora - Mustelidae
e Oriental Small-clawed Otters
d Zwergotter
f Loutres cendrées
i Lontre nane

216 **Amblonyx cinereus**
Carnivora - Mustelidae
e Oriental Small-clawed Otter; Clawless Otter; Calcutta Otter; Indian Otter; Short-clawed Otter;

Asian Short-clawed Otter
d Zwergotter; Indischer
Kurzkrallenotter
f Loutre cendrée
i Lontra nana; Lontra senza unghie

217 Amblysomus
Chrysochloridea - Chrysochloridae
e Golden Moles; African Golden
Moles; South African Golden Moles
d Kupfergoldmulle; Afrikanische
Goldmulle; Kufermulle
f Taupes dorées

218 Amblysomus corriae
Chrysochloridea - Chrysochloridae
e Western Cape Golden Mole

219 Amblysomus gunningi
Chrysochloridea - Chrysochloridae
e Gunning's Golden Mole

220 Amblysomus hottentotus
Chrysochloridea - Chrysochloridae
e Hottentot Golden Mole
d Hottentotten-Goldmull

221 Amblysomus hottentotus iris
Chrysochloridea - Chrysochloridae
e Zulu Golden Mole

222 Amblysomus julianae
Chrysochloridea - Chrysochloridae
e Juliana's Golden Mole

223 Amblysomus obtusirostris
Chrysochloridea - Chrysochloridae
e Yellow Golden Mole

224 Amblysomus robustus
Chrysochloridea - Chrysochloridae
e Robust Golden Mole

225 Amblysomus septentrionalis
Chrysochloridea - Chrysochloridae

e Highveld Golden Mole

226 Ametrida
Chiroptera - Phyllostomidae
e Lesser Wrinkle-faced Bats; Little
White-shouldered Bats

227 Ametrida centurio
Chiroptera - Phyllostomidae
e Lesser Wrinkle-faced Bat; Little
White-shouldered Bat; Wrinkle-faced
Bat

228 Ammodillus
Rodentia - Muridae
e Somali Gerbils; Ammodiles
d Somali-Rennmäuse
f Gerbilles

229 Ammodillus imbellis
Rodentia - Muridae
e Ammodile; Somali Gerbil; Walo
d Somali-Rennmaus

230 Ammodorcas
Artiodactyla - Bovidae
e Clarke's Gazelles; Dibatags
d Lamagazellen; Stelzengazellen;
Dibatag-Antilopen; Dibatags
f Dibatags
i Gazzelle di Clarke

231 Ammodorcas clarkei
Artiodactyla - Bovidae
e Clarke's Gazelle; Dibatag
d Lamagazelle; Stelzengazelle;
Dibatag-Antilope
f Dibatag
i Dibatag; Gazella di Clarke

232 Ammospermophilus
Rodentia - Sciuridae
e Antelope Squirrels; Ground Antelope
Squirrels; Chipmunks; White-tailed
Chipmunks

d Antilopenziesel
f Écureuils antilopes
i Scoiattolini del deserto; Scoiattoli di antilope a terra

233 *Ammospermophilus harrisii*
Rodentia - Sciuridae
e Harris's Antelope Squirrel
d Harris-Antilopenziesel
f Écureuil antilope d'Harris
i Scoiattolo di antilope del Harris

234 *Ammospermophilus insularis*
Rodentia - Sciuridae
e Espiritu Santo Island Antelope Squirrel
d Espiritu-Santo-Antilopenziesel
i Ardilla; Scoiattolo di antilope insulare

235 *Ammospermophilus interpres*
Rodentia - Sciuridae
e Texas Antelope Squirrel
d Texas-Antilopenziesel
f Écureuil antilope du Texas
i Scoiattolo di antilope del Texas

236 *Ammospermophilus leucurus*
Rodentia - Sciuridae
e White-tailed Antelope Squirrel
d Weißchwanzantilopenziesel
f Écureuil antilope à queue blanche
i Scoiattolo di antilope bianco-unito

237 *Ammospermophilus nelsoni*
Rodentia - Sciuridae
e Nelson's Antelope Squirrel; San Joaquin Antelope Squirrel; Nelson's Antelope Ground Squirrel; San Joaquin Antelope Ground Squirrel
d Nelson-Antilopenziesel
f Écureuil antilope de San Joaquin
i Scoiattolo di antilope di San Joaquin; Scoiattolo di antilope di Nelson

238 *Ammotragus*
Artiodactyla - Bovidae
e Barbary Sheep ; Aodads; Aoudads; Arruis
d Mähnenschafe; Mähnenspringer ; Böcke
f Aoudads
i Ammotragi

239 *Ammotragus lervia*
Artiodactyla - Bovidae
e Barbary Sheep; Ruffled Sheep of Barbary; Aoudad; Arrui; Maned Mouflon; Ruffled Mouflon
d Mähnenspringer; Mähnenschaf; Afrikanischer Tur
f Aoudad; Mouflon aux manchettes
i Ammotrago; Capra crinita; Pecora crinita

240 *Ammotragus lervia angusi*
Artiodactyla - Bovidae
e Aïr Barbary Sheep; Aïr Arrui
d Aïr-Mähnenspringer

241 *Ammotragus lervia blainei*
Artiodactyla - Bovidae
e Sudan Barbary Sheep
d Kordofan-Mähnenspringer

242 *Ammotragus lervia fassini*
Artiodactyla - Bovidae
e Libyan Barbary Sheep
d Tripolis-Mähnenspringer

243 *Ammotragus lervia lervia*
Artiodactyla - Bovidae
e North African Barbary Sheep
d Mauretanischer Mähnenspringer

244 *Ammotragus lervia ornata*
Artiodactyla - Bovidae
e Egyptian Barbary Sheep; Egyptian Aoudad; Egyptian Aodad; Egyptian Arui
d Ägyptischer Mähnenspringer ;

Ägyptisches Mähnenschaf;
Ägyptischer Tur; Ägyptische
Mähnenziege

f   Mouflon aux manchettes d'Égypte;
Aoudad d'Égypte

245 ***Ammotragus lervia saharensis***
Artiodactyla - Bovidae

e   Saharan Arrui

d   Sahara-Mähnenspringer

246 ***Amorphochilus***
Chiroptera - Furipteridae

e   Smoky Bats

247 ***Amorphochilus schnablii***
Chiroptera - Furipteridae

e   Smoky Bat; Smoky Thumbless Bat

248 ***Amphinectomys***
Rodentia - Muridae

e   Amphibious Rats

d   Wasserratten

249 ***Amphinectomys savamis***
Rodentia - Muridae

e   Amphibious Rat

d   Wasserratte

250 ***Anathana***
Scandentia - Tupaiidae

e   Indian Tree Shrews

d   Elliots Tupajas; Indische Tupajas

f   Tupaïas d'Eliot

i   Tupaie di Elliot

251 ***Anathana ellioti***
Scandentia - Tupaiidae

e   Indian Tree Shrew; Madras Tree
Shrew

d   Elliots Tupaja; Indisches Tupaja;
Madras Spitzhörnchen; Indisches
Spitzhörnchen

f   Tupaïa d'Elliot

Andalgalomys syn. Graomys q.v.

252 ***Andinomys***
Rodentia - Muridae

e   Andean Mice

d   Anden-Mäuse

i   Topi andini

253 ***Andinomys edax***
Rodentia - Muridae

e   Andean Mouse

d   Anden-Maus

i   Topo andino

Anguistodontus syn. *Salpingotus* q.v.

254 ***Anisomys***
Rodentia - Muridae

e   New Guinea Giant Rats; Powerful-
toothed Rats; Squirrel-toothed Rats

d   Hörnchenzahnratten

255 ***Anisomys imitator***
Rodentia - Muridae

e   New Guinea Giant Rat; Powerful-
toothed Rat; Squirrel-toothed Rat;
Uneven-toothed Rat; Narrow-toothed
Giant Rat

d   Hörnchenzahnratte

256 ***Anoa***
Artiodactyla - Bovidae

e   Anoas

d   Anoas

f   Anoas

i   Anoa

257 ***Anoa depressicornis***
Artiodactyla - Bovidae

e   Anoa; Lowland Anoa

d   Anoa; Gemsbüffel; Tiefland-Anoa

f   Anoa des plaines

i   Bufalo pigmeo di pianura; Anoa;
Buffalo pigmeo

258   *Anoa mindorensis*
      Artiodactyla - Bovidae
  e   Tamarau; Tamaraw
  d   Tamarau
  f   Tamarau
  i   Tamaru; Bufalo di Mindoro

259   *Anoa quarlesi*
      Artiodactyla - Bovidae
  e   Mountain Anoa
  d   Berganoa; Hochlandanoa
  f   Anoa des montagnes
  i   Anoa di montagna

260   **Anomaluridae**
      Rodentia
  e   African Flying Squirrels; Scaly-tailed
      Flying Squirrels; Scaly-tailed
      Squirrels
  d   Dornschwanzhörnchen
  f   Anomaluridés
  i   Scoiattoli volanti; Scoiattoli di
      volata; Anumaluridi

      Anomalurops syn. *Anomalurus* q.v.

261   *Anomalurus*
      Rodentia - Anomaluridae
  e   Scaly-tailed Flying Squirrels; Brush-
      tailed Flying Squirrels; Large Scaly-
      tailed Squirrels; Flying Foxes;
      Anomalures
  d   Dornschwanzhörnchen
  f   Anomalures
  i   Anomaluri

262   *Anomalurus beecrofti*
      Rodentia - Anomaluridae
  e   Beecroft's Flying Squirrel; Beecroft's
      Scaly-tailed Squirrel; Beecroft's
      Scaly-tailed Flying Squirrel
  d   Beecroft-Dornschwanzhörnchen
  f   Anomalure de Beecroft; Écureuil
      volant de Beecroft
  i   Anomaluro dal ventre rosso

263   *Anomalurus derbianus*
      Rodentia - Anomaluridae
  e   Lord Derby's Scaly-tailed Squirrel;
      Lord Derby's Anomalure; Lord
      Derby's Flying Squirrel
  d   Lord Derby-Dornschwanzhörnchen
  f   Anomalure volant de Derby
  i   Anomaluro derbiano

264   *Anomalurus pelii*
      Rodentia - Anomaluridae
  e   Pel's Flying Squirrel; Pel's Scaly-
      tailed Squirrel; Pel's Scaly-tailed
      Flying Squirrel
  d   Pel-Dornschwanzhörnchen
  f   Anomalure de Pel
  i   Anomaluro di Pel

265   *Anomalurus pusillus*
      Rodentia - Anomaluridae
  e   Little Flying Squirrel; Pygmy Scaly-
      tailed Squirrel; Dwarf Scaly-tailed
      Squirrel
  d   Zwergdornschwanzhörnchen
  f   Anomalure nain; Anomalure pygmée

266   *Anonymomys*
      Rodentia - Muridae
  e   Mindoro Rats
  d   Mindoro-Ratten

267   *Anonymomys mindorensis*
      Rodentia - Muridae
  e   Mindoro Rat; Mindoro Climbing Rat
  d   Mindoro-Ratte

268   *Anotomys*
      Rodentia - Muridae
  e   Ecuador Fish-eating Rats
  d   Ecuador-Fischratten

269   *Anotomys leander*
      Rodentia - Muridae
  e   Ecuador Fish-eating Rat
  d   Ecuador-Fischratte

270   *Anoura*
         Chiroptera - Phyllostomidae
e   Tailless Bats; Long-nosed Bats;
         Geoffroy's Long-nosed Bats
i   Pipistrelli senza coda

         Anoura brevirostrum syn. *A.*
         cultrata q.v.

271   *Anoura caudifera*
         Chiroptera - Phyllostomidae
e   Tailed Tailless Bat

272   *Anoura cultrata*
         Chiroptera - Phyllostomidae
e   Handley's Tailless Bat; Handley's
         Hairy-legged Bat; Handley's Long-
         tongued Bat

273   *Anoura geoffroyi*
         Chiroptera - Phyllostomidae
e   Geoffroy's Tailless Bat; Geoffroy's
         Hairy-legged Bat
f   Chauve-souris de Geoffroy
i   Pipistrello senza coda di Geoffroy

274   *Anoura latidens*
         Chiroptera - Phyllostomidae
e   Broad-toothed Tailless Bat

275   *Anoura luismanueli*
         Chiroptera - Phyllostomidae
e   Luis Manuel's Tailless Bat

         Anoura werckleae syn. *A. cultrata*
         q.v.

276   *Anourosorex*
         Soricomorpha - Soricidae
e   Szechuan Burrowing Shrews; Mole
         Shrews
d   Stummelschwanzspitzmäuse

277   *Anourosorex squamipes*
         Soricomorpha - Soricidae
e   Szechuan Burrowing Shrew; Mole

         Shrew
d   Stummelschwanzspitzmaus

         Antechinomys syn. *Sminthopsis* q.v.

278   *Antechinus*
         Dasyuromorphia - Dasyuridae
e   Marsupial Mice; Broad-footed
         Marsupial Mice; Antechinuses
d   Springbeutelmäuse;
         Beutelspringmäuse;
         Breitfußbeutelmäuse
f   Rats marsupiaux; Petites gerboises
i   Antechini

279   *Antechinus adustus*
         Dasyuromorphia - Dasyuridae
e   Rusty Antechinus

280   *Antechinus agilis*
         Dasyuromorphia - Dasyuridae
e   Agile Antechinus
d   Bewegliches Antechinus
f   Antéchinus agile
i   Antechino agile

281   *Antechinus apicalis*
         Dasyuromorphia - Dasyuridae
e   Southern Dibbler (ANZ); Freckled
         Marsupial Mouse; Freckled
         Antechinus
d   Sprenkelbeutelmaus; Südliche
         Sprenkelbeutelmaus; Süddibbler
i   Dibbler

282   *Antechinus bellus*
         Dasyuromorphia - Dasyuridae
e   Fawn Antechinus; Fawn Marsupial
         Mouse
d   Augenfleckbreitfußbeutelmaus; Kitz-
         Antechinus

283   *Antechinus bilarni*
         Dasyuromorphia - Dasyuridae
e   Harney's Marsupial Mouse;
         Sandstone Dibbler; Northern Dibbler

(ANZ); Sandstone Antechinus;
Sandstone Marsupial Mouse
d Nördliche Sprenkelbeutelmaus;
Sandsteindibbler

284 *Antechinus flavipes*
Dasyuromorphia - Dasyuridae
e Yellow-footed Marsupial Mouse;
Yellow-footed Dunnart; Yellow-
footed Antechinus (ANZ); Broad-
footed Marsupial Mouse
d Gelbfußbeutelmaus; Gelbfüssiges
Antechinus
i Antechino giallo

285 *Antechinus godmani*
Dasyuromorphia - Dasyuridae
e Godman's Antechinus; Yellow-
footed Antechinus; Atherton
Antechinus
d Queensland-Breitfußbeutelmäuse;
Atherton-Antechinus
i Antechino di Atherton

286 *Antechinus leo*
Dasyuromorphia - Dasyuridae
e Cinnamon Antechinus
d Zimtbreitfußbeutelmaus;
Zimtantechinus
f Lion de canelle
i Antechino canella

287 *Antechinus macdonnellensis*
Dasyuromorphia - Dasyuridae
e Fat-tailed Marsupial Mouse; Fat-
tailed Antechinus; Fat-tailed
Pseudantechinus
d Fettschwanzbreitfußbeutelmaus;
Fettschwänzige Breitfußbeutlmaus
i Antechino dalle orecchie rosse;
Antechino dalla coda grassa

288 *Antechinus melanurus*
Dasyuromorphia - Dasyuridae
e Black-tailed Antechinus; Black-tailed
Dasyure
d Samt-Breitfußbeutelmaus

289 *Antechinus mimulus*
Dasyuromorphia - Dasyuridae
e Carpentarian Pseudantechinus

290 *Antechinus minimus*
Dasyuromorphia - Dasyuridae
e Little Tasmanian Marsupial Mouse;
Swamp Antechinus (ANZ)
d Hummock-Breitfußbeutelmaus;
Sumpfantechinus
i Antechino di palude

291 *Antechinus naso*
Dasyuromorphia - Dasyuridae
e Long-nosed Antechinus
d Seiden-Breitfußbeutelmaus

292 *Antechinus ningbing*
Dasyuromorphia - Dasyuridae
e Ningbing Pseudantechinus; Ningbing
Antechinus
d Ningbing-Antechinus

293 *Antechinus roryi*
Dasyuromorphia - Dasyuridae
e Tan Pseudantechinus
f Petite gerboise rousse

294 *Antechinus rosamondae*
Dasyuromorphia - Dasyuridae
e Little Red-tailed Marsupial Mouse;
Little Red Kaluta (ANZ); Red-eared
Antechinus
d Rote Breitfußbeutelmaus; Kleine
Rote Kaluta

295 *Antechinus stuartii*
Dasyuromorphia - Dasyuridae
e Brown Antechinus; Brown Marsupial
Mouse; Stuart's Antechinus;
Macleay's Marsupial Mouse
d Stuart-Breitfußbeutelmaus
i Antechino marrone

296 *Antechinus subtropicus*
Dasyuromorphia - Dasyuridae
e Subtropical Antechinus

297 *Antechinus swainsonii*
Dasyuromorphia - Dasyuridae
e Swainson's Marsupial Mouse;
Swainson's Antechinus; Swainson's
Phascogale; Dusky Antechinus
(ANZ); Dusky Marsupial Mouse
d Swainson-Breitfußbeutelmaus;
Düsteres Antechinus

298 *Antechinus wilhelmina*
Dasyuromorphia - Dasyuridae
e Lesser Antechinus
d Wilhelmina-Breitfußbeutelmaus;
Kleinantechinus
f Petite gerboise marsupiale

299 *Antechinus woolleyae*
Dasyuromorphia - Dasyuridae
e Woolley's Antechinus; Woolley's
Pseudantechinus
d Wooleys Antechinus

300 *Anthops*
Chiroptera - Rhinolophidae
e Flower-faced Bats
d Blumennasenfledermäuse

301 *Anthops ornatus*
Chiroptera - Rhinolophidae
e Flower-faced Bat
d Blumennasenfledermaus

Anthorhina syn. *Mimon* q.v.

302 *Antidorcas*
Artiodactyla - Bovidae
e Springboks; Springbocs
d Springböcke
f Antidorcas
i Antidorcadi

303 *Antidorcas marsupialis*
Artiodactyla - Bovidae
e Springbock; Springboc
d Springbock
f Antidorcas; Springbok
i Antilope eucore; Antilope saltante;
Springbok; Antidorcas

304 *Antilocapra*
Artiodactyla - Antilocapridae
e Pronghorns; Pronghorn Antelopes
d Gabelhornantilopen; Gabelböcke
f Antilopes aux cornes fourchues
i Antilocapre

305 *Antilocapra americana*
Artiodactyla - Antilocapridae
e Pronghorn; Pronghorn Antelope;
North American Prongbuck; Mexican
Pronghorn Antelope
d Gabelhornantilope; Gabelbock;
Niederkalifornischer Gabelbock
f Antilope d'Amérique; Antilope
d'Amérique aux cornes fourchues;
Chèvre américaine; Antilope à
fourche du Mexique
i Antilocapra; Fantasma della prateria;
Antilocapra americana

306 *Antilocapra americana peninsularis*
Artiodactyla - Antilocapridae
e Peninsular Pronghorn
d Baja-California-Gabelbock
i Antilocapra della Bassa California

307 *Antilocapra americana sonoriensis*
Artiodactyla - Antilocapridae
e Sonoran Pronghorn Antelope
d Sonora-Gabelbock
f Antilope du Sonora
i Antilocapra di Sonora

308 **Antilocapridae**
Artiodactyla
e Pronghorns; Pronghorn Antelopes
d Gabelhornträger; Gabelhorntiere

f Antilocapridés
i Antilocapridi

309 *Antilope*
  Artiodactyla - Bovidae
e Blackbucks; Indian Antelopes
d Hirschziegenantilopen
f Antilopes; Antilopes cercicapres
i Antilopi cervicapra

310 *Antilope cervicapra*
  Artiodactyla - Bovidae
e Blackbuck; Indian Antelope;
  Blackbuck Antelope
d Hirschziegenantilope
f Antilope cervicapre
i Antilope cervicapra

311 **Antrozoidae**
  Chiroptera
e Antrozoid Bats
d Blasse Fledermäuse
f Antrozoidés
i Antrozoidi

312 *Antrozous*
  Chiroptera - Antrozoidae
e Pallid Bats; Pale Bats; Desert Bats
d Wüstenfledermäuse
f Oreillards pâles

313 *Antrozous pallidus*
  Chiroptera - Antrozoidae
e Pale Bat; Desert Bat; Pallid Bat;
  Desert Pallid Bat
d Blasse Fledermaus;
  Wüstenfledermaus
f Oreillard pâle; Chauve-souris blonde

314 *Aonyx*
  Carnivora - Mustelidae
e Clawless Otters
d Eigentliche Fingerotter ;
  Kleinkrallenotter

f Loutres aux joues blanches
i Lontre; Lontre senza unghie

315 *Aonyx capensis*
  Carnivora - Mustelidae
e African Clawless Otter; Cape
  Clawless Otter
d Kap-Otter; Afrikanischer
  Kurzkrallenotter; Weißwangenotter
f Loutre aux joues blanches du Cap
i Lontra dalle guance bianche; Lontra
  capense; Aonice capense; Lontra
  senza unghie dal Capo

  Aonyx cinereus syn. *Amblonyx
  cinereus* q.v.

316 *Aonyx congicus*
  Carnivora - Mustelidae
e Zaire Clawless Otter; Small-toothed
  Clawless Zaire Otter; Congo
  Clawless Otter; Lower Congo
  Clawless Otter; Swamp Otter
d Kongo-Fingerotter; Kongo-
  Weißwangenotter; Fleckenotter;
  Kongo-Kleinkrallenotter
f Loutre aux joues blanches du Congo
i Lontra senza unghie del Congo;
  Lontra dalle guance bianche del
  Congo

317 *Aonyx congicus microdon*
  Carnivora - Mustelidae
e Cameroon Clawless Otter; West
  African Clawless Otter; Small-
  toothed Clawless Otter
f Loutre aux joues blanches du
  Cameroun
i Lontra dalle guance bianche del
  Camerun

318 *Aotus*
  Primates - Atelidae
e Night Monkeys; Douroucoulis; Owl
  Monkeys
d Mirikinas; Nachtaffen; Eulenaffen;
  Douroucoulis

f   Nyctipithèques; Singes des nuits;
     Douroucoulis; Singes nocturnes
i   Aoti

Aotus azarae syn. *A. azarai* q.v.

319 Aotus azarai
     Primates - Atelidae
e   Azara's Night Monkey; Southern
     Night Monkey; Southern Owl
     Monkey
i   Aoto di Azara; Mirichina

320 Aotus azarai infulatus
     Primates - Atelidae
e   Feline Night Monkey; Kuhl's Night
     Monkey
i   Aoto felino

Aotus brumbacki syn. *A. lemorinus
     brumbacki* q.v.

321 Aotus hershkovitzi
     Primates - Atelidae
e   Hershkovitz's Night Monkey
d   Hershkovitzs Nachtaffe
i   Aoto di Hershkovitz
     Aotus infulatus syn. A. azarai
     infulatus q.v.

322 Aotus lemurinus
     Primates - Atelidae
e   Lemurine Night Monkey; Western
     Night Monkey
d   Nördlicher Nachtaffe
f   Douroucouli aux pattes grises; Singe
     de nuit
i   Aoto lemurino; Dourmoucouli;
     Scimmia gufo

323 Aotus lemurinus brumbacki
     Primates - Atelidae
e   Brumbach's Night Monkey
i   Aoto di Brumbach

324 Aotus miconax

     Primates - Atelidae
e   Andean Night Monkey; Thomas's
     Night Monkey; Peruvian Night
     Monkey
f   Singe de nuit
i   Aoto delle Ande

325 Aotus nancymaae
     Primates - Atelidae
e   Ma's Night Monkey; Nancy Ma's
     Night Monkey; Peruvian Red-necked
     Owl Monkey
f   Singe de nuit
i   Aoto di Nancy Ma

326 Aotus nigriceps
     Primates - Atelidae
e   Black-headed Night Monkey
d   Musmuqui
f   Singe de nuit
i   Aoto dalla testa nera

327 Aotus trivirgatus
     Primates - Atelidae
e   Three-banded Night Monkey;
     Northern Night Monkey;
     Douroucouli; Owl Monkey;
     Humboldt's Night Monkey
d   Mirikina; Nachtaffe
f   Singe de nuit; Douroucouli commun;
     Douroucouli
i   Aoto dalle tre striscie

328 Aotus vociferans
     Primates - Atelidae
e   Noisy Night Monkey; Spix's Night
     Monkey
f   Singe de nuit
i   Aoto vocifero

329 Aplodontia
     Rodentia - Apolodontidae
e   Mountain Beavers; Sewellels
d   Stummelschwanzhörnchen
f   Castors de montagne
i   Castori di Montagna

330 *Aplodontia rufa*
Rodentia - Apolodontidae
e Mountain Beaver; Sewellel
d Stummelschwanzhörnchen;
Biberhörnchen
f Castor de montagne
i Castoro di montagna

331 *Aplodontia rufa nigra*
Rodentia - Apolodontidae
e Point Arena Mountain Beaver

332 *Aplodontia rufa phaea*
Rodentia - Apolodontidae
e Point Reyes Mountain Beaver

333 *Aplodontia rufa raineri*
Rodentia - Apolodontidae
e Cascade Mountain Beaver

334 **Aplodontidae**
Rodentia
e Mountain Beavers; Sewellels
d Stummelschwanzhörnchen
f Aplodontidés; Castors de montagne
i Aplodontidi; Castori della montagna

335 *Apodemus*
Rodentia - Muridae
e Old World Wood and Field Mice;
Wood Mice; Field Mice; Long-tailed
Field Mice
d Brandmäuse; Waldmäuse; Wald- und
Feldmäuse
f Mulots; Souris des bois
i Topi selvatici

336 *Apodemus agrarius*
Rodentia - Muridae
e Striped Field Mouse; Black-striped
Field Mouse
d Brandmaus
f Mulot rayé; Souris agraire; Souris
des champs; Souris rayée; Souris
rousse; Rat des champs; Rat à bandes

i Topo selvatico dorso-striato; Sorcio
campagnolo

337 *Apodemus alpicola*
Rodentia - Muridae
e Alpine Field Mouse; Alpine Mouse
d Alpenwaldmaus
f Mulot alpestre
i Topo selvatico alpino

338 *Apodemus argenteus*
Rodentia - Muridae
e Small Japanese Field Mouse
d Kleine Japanische Waldmaus

339 *Apodemus arianus*
Rodentia - Muridae
e Persian Field Mouse; Kohrud Wood
Mouse
d Iran-Waldmaus; Persische Waldmaus

340 *Apodemus chevrieri*
Rodentia - Muridae
e Chevrier's Field Mouse
d Chevrier-Waldmaus

341 *Apodemus draco*
Rodentia - Muridae
e Formosan Field Mouse; South China
Field Mouse
d Südchinesische Waldmaus

342 *Apodemus flavicollis*
Rodentia - Muridae
e Yellow-necked Field Mouse;
Yellow-necked Mouse; South China
Field Mouse; Yellow-necked Wood
Mouse; Tell
d Große Waldmaus; Gelbhalsmaus;
Halsbandmaus; Springmaus
f Mulot à collier; Mulot fauve; Mulot à
collier roux
i Topo selvatico collo-giallo; Topo
selvatico dal collo giallo

343   *Apodemus fulvipectus*
    Rodentia - Muridae
e   Yellow-breasted Field Mouse
d   Gelbbrustmaus

344   *Apodemus gurkha*
    Rodentia - Muridae
e   Himalajan Field Mouse
d   Himalaja-Waldmaus

345   *Apodemus hermonensis*
    Rodentia - Muridae
e   Mount Hermon Field Mouse
d   Hermon-Waldmaus

346   *Apodemus hyrcanicus*
    Rodentia - Muridae
e   Caucasus Field Mouse; Caucasian
    Field Mouse
d   Kaukasus-Waldmaus

    Apodemus krkensis syn. *A.
    sylvaticus* q.v.

347   *Apodemus latronum*
    Rodentia - Muridae
e   Sichuan Field Mouse
d   Sichuan-Waldmaus

    Apodemus microps syn. *A. uralensis*
    q.v.

348   *Apodemus mystacinus*
    Rodentia - Muridae
e   Broad-toothed Mouse; Broad-toothed
    Field Mouse; Rock Mouse
d   Schnurrbartmaus; Kleinasiatische
    Felsenmaus; Felsenmaus
f   Mulot rupestre
i   Topo selvatico rupestre; Topo
    selvatico greco

349   *Apodemus peninsulae*
    Rodentia - Muridae
e   Korean Field Mouse

d   Koreanische Waldmaus

350   *Apodemus ponticus*
    Rodentia - Muridae
e   Black Sea Field Mouse
d   Schwarzmeer-Waldmaus

351   *Apodemus rusiges*
    Rodentia - Muridae
e   Kashmir Field Mouse
d   Kaschmir-Waldmaus

352   *Apodemus semotus*
    Rodentia - Muridae
e   Taiwan Field Mouse
d   Taiwanische Waldmaus

353   *Apodemus speciosus*
    Rodentia - Muridae
e   Large Japanese Field Mouse
d   Asiatische Waldmaus; Große
    Japanische Feldmaus; Große
    Japanische Waldmaus

354   *Apodemus sylvaticus*
    Rodentia - Muridae
e   Field Mouse; Common Field Mouse;
    Long-tailed Field Mouse; Wood
    Mouse; European Wood Mouse; Krk
    Mouse
d   Kleine Waldmaus; Europäische
    Waldmaus; Westliche Waldmaus;
    Feld-Waldmaus; Gartenwaldmaus;
    Langschwänzige Feldmaus;
    Gelbbraune Maus; Waldmaus
f   Souris de terre; Mulot sylvestre;
    Souris des bois; Rat-mulot; Rat
    sauteur; Rat sauteur d'Islande; Mulot
    ordinaire; Mulot gris
i   Topo selvatico

355   *Apodemus sylvaticus hebridensis*
    Rodentia - Muridae
e   Hebridean Mouse

356 **Apodemus sylvaticus hirtensis**
Rodentia - Muridae
e St. Kilda Field Mouse
d St. Kilda Feldmaus

357 **Apodemus uralensis**
Rodentia - Muridae
e Ural Field Mouse; Pygmy Field Mouse
d Ural-Waldmaus
f Mulot pygmée
i Topo selvatico pigmeo

358 **Apodemus wardi**
Rodentia - Muridae
e Ward's Field Mouse
d Ward-Waldmaus

359 **Apomys**
Rodentia - Muridae
e Philippine Forest Mice
d Philippinische Ratten

360 **Apomys abrae**
Rodentia - Muridae
e Luzon Cordillera Forest Mouse

361 **Apomys datae**
Rodentia - Muridae
e Luzon Montane Forest Mouse
d Mount-Data-Ratte

362 **Apomys gracilirostris**
Rodentia - Muridae
e Large Mindoro Forest Mouse

363 **Apomys hylocoetes**
Rodentia - Muridae
e Mount Apo Forest Mouse; Mindanao Mossy Forest Mouse

364 **Apomys insignis**
Rodentia - Muridae
e Mindanao Montane Forest Mouse

365 **Apomys littoralis**
Rodentia - Muridae
e Mindanao Lowland Forest Mouse; Mindanao Lowland Mouse

366 **Apomys microdon**
Rodentia - Muridae
e Small Luzon Forest Mouse

367 **Apomys musculus**
Rodentia - Muridae
e Least Forest Mouse; Least Philippine Forest Mouse

Apomys petraeus syn. *A. hylocoetes* q.v.

368 **Apomys sacobianus**
Rodentia - Muridae
e Long-nosed Luzon Forest Mouse

Aporodon syn. *Reithrodontomys* q.v.

369 **Aproteles**
Chiroptera - Pteropodidae
e Bulmer's Fruit Bats
d Bulmers Fruchtfledermäuse

370 **Aproteles bulmerae**
Chiroptera - Pteropodidae
e Bulmer's Fruit Bat
d Bulmers Fruchtfledermaus

Arborimus syn. *Phenacomys* q.v.

371 **Archboldomys**
Rodentia - Muridae
e Mount Isarog Shrewmice
d Spitzmausratten

372 **Archboldomys luzonensis**
Rodentia - Muridae
e Mount Isarog Shrewmouse; Isarog Shrewmouse; Mount Isarog

Striped Rat
d   Mount-Isarog-Spitzmausratte

373   *Archboldomys musseri*
Rodentia - Muridae
e   Palanan Shrewmouse
d   Palanan-Spitzmaus

374   *Arctitis*
Carnivora - Viverridae
e   Binturongs
d   Binturongs
f   Binturongs
i   Binturong

375   *Arctitis binturong*
Carnivora - Viverridae
e   Binturong; Bearcat
d   Binturong
f   Binturong
i   Binturong

***Arctitis whitei syn*. A. binturong q.v.**
376   *Arctocebus*
Primates - Loridae
e   Golden Pottos; Calabar Pottos;
        Angwantibos
d   Bärenmakis; Angwantibos
f   Pottos de Calabar; Arctocèbes;
        Agwantibos
i   Arctocebi

377   *Arctocebus aureus*
Primates - Loridae
e   Golden Angwantibo; Golden Potto
d   Goldener Bärenmaki
f   Potto doré de Calabar; Agwantibo du
        Sud
i   Artocebo dorato

378   *Arctocebus calabarensis*
Primates - Loridae
e   Calabar Angwantibo; Calabar Potto;
        Agwantibo
d   Bärenmaki; Angwantibo; Calabar-

Bärenmaki
f   Potto de Calabar; Angwantibo du
        Nord; Perodictique agwantibo
i   Artocebo del Calabar; Potto dorato;
        Anguantibo

379   *Arctocephalus*
Carnivora - Otariidae
e   Southern Fur Seals
d   Seebären; Südliche Seebären
f   Otaries; Otaries à fourrure australes
i   Arctocefali; Leoni marini

380   *Arctocephalus australis*
Carnivora - Otariidae
e   Falkland Fur Seal; South American
        Fur Seal
d   Südamerikanischer Seebär
f   Otarie à fourrure australe
i   Foca dell'Uruguay; Arctocefalo del
        Sudamerica

381   *Arctocephalus australis australis*
Carnivora - Otariidae
e   Falkland Islands Fur Seal
d   Falkland-Seebär
f   Otarie à fourrure d'Amérique du Sud

382   *Arctocephalus forsteri*
Carnivora - Otariidae
e   New Zealand Fur Seal (ANZ);
        Antipodean Fur Seal; Western
        Australian Fur Seal; Southern Fur
        Seal (ANZ)
d   Neuseeländischer Seebär
f   Otarie d'Australie
i   Arctocefalo di Forster

383   *Arctocephalus forsteri doriferos*
Carnivora - Otariidae
e   South Australian Fur Seal
d   Südaustralischer Seebär

384   *Arctocephalus galapagoensis*
Carnivora - Otariidae

e Galapagos Fur Seal; Galapagos
Islands Fur Seal
d Galapagos-Seebär
f Otarie des Galapagos
i Arctocefalo delle Galapagos

385 *Arctocephalus gazella*
Carnivora - Otariidae
e Antarctic Fur Seal; Kerguelen Fur
Seal
d Kerguelen-Seebär; Antarktischer
Seebär
f Otarie de Kerguelen
i Lupo di mare antartico; Arctocefalo
delle Kerguellen

386 *Arctocephalus philippii*
Carnivora - Otariidae
e Juan Fernandez Fur Seal; Chilean Fur
Seal; Philippi's Fur Seal
d Juan-Fernandez-Seebär
f Otarie de Chile
i Arctocefalo di Juan Fernandez

387 *Arctocephalus pusillus*
Carnivora - Otariidae
e Southern Fur Seal (CSA); Afro-
Australian Fur Seal; Cape Fur Seal;
Giant Fur Seal; Brown Fur Seal;
Tasmanian Fur Seal; Tasman Seal
d Südafrikanischer Seebär;
Zwergseebär
f Otarie à fourure d'Afrique; Otarie à
fourrure du Sud; Otarie à fourrure du
Cap
i Otaria da pelliccia del Capo;
Arctocefalo sudafricano; Foca del
Sudafrica; Leone marino del
Sudafrica; Otaria del Capo

388 *Arctocephalus pusillus pusillus*
Carnivora - Otariidae
e South African Fur Seal
d Südafrikanische Pelzrobbe
f Otarie à fourrure d'Afrique du Sud

Arctocephalus tasmanicus syn. A.

pusillus q.v.

389 *Arctocephalus townsendi*
Carnivora - Otariidae
e Guadaloupe Fur Seal
d Guadeloupe-Seebär
f Otarie de Townsend
i Arctocefalo della Guadalupa; Foca
della Guadalupa

390 *Arctocephalus tropicalis*
Carnivora - Otariidae
e Subantarctic Fur Seal; Amsterdam
Island Fur Seal
d Subantarktischer Seebär
f Otarie à fourrure de l'île Amsterdam;
Otarie à fourrure subantarctique;
Arctocephale d'Australie; Otarie à
fourrure de l'Atlantique tropical;
Otarie à fourrure d'Amsterdam
i Lupo di mare subantartico;
Arctocefalo subantartico

391 *Arctogalidia*
Carnivora - Viverridae
e Small-toothed Civets; Small-toothed
Three-striped Palm Civets; Small-
toothed Palm Civets
d Streifenroller
f Civettes palmistes à trois bandes

392 *Arctogalidia trivirgata*
Carnivora - Viverridae
e Small-toothed Civet; Small-toothed
Three-striped Palm Civet; Three-
striped Palm Civet; Small-toothed
Palm Civet
d Streifenroller
f Civette palmiste à trois bandes

393 *Arctogalidia trivirgata bilineata*
Carnivora - Viverridae
e Java Palm Civet

394 *Arctonyx*
Carnivora - Mustelidae

e   Hog Badgers
d   Schweinsdachse
f   Arctonyx ; Blaireaux à gorge blanche
i   Tassi

395   *Arctonyx collaris*
       Carnivora - Mustelidae
e   Hog Badger
d   Schweinsdachs; Riesendachs
f   Blaireau à gorge blanche
i   Tasso naso di porco; Tasso porcino

396   *Ardops*
       Chiroptera - Phyllostomidae
e   Tree Bats
d   Baumfledermäuse
f   Chauve-souris arboricoles

397   *Ardops nichollsi*
       Chiroptera - Phyllostomidae
e   Dominican Tree Bat; Lesser
       Antillean Tree Bat; Lesser Antilles
       Tree Bat; Tree Bat
f   Ardops des Petites Antilles

398   *Ardops nichollsi annectens*
       Chiroptera - Phyllostomidae
e   Marie-Galante Tree Bat

       Arielulus syn. *Pipistrellus* q.v.

399   *Ariteus*
       Chiroptera - Phyllostomidae
e   Fig-eating Bats

400   *Ariteus flavescens*
       Chiroptera - Phyllostomidae
e   Jamaican Fig-eating Bat

401   *Artibeus*
       Chiroptera - Phyllostomidae
e   American Fruit-eating Bats;
       Neotropical Fruiteating Bats;
       American Fruit Bats; Neotropical

       Fruit Bats; Fruit-eating Bats; Fruit
       Bats
f   Artibées

402   *Artibeus amplus*
       Chiroptera - Phyllostomidae
e   Large Fruit-eating Bat

403   *Artibeus anderseni*
       Chiroptera - Phyllostomidae
e   Andersen's Fruit-eating Bat

404   *Artibeus aztecus*
       Chiroptera - Phyllostomidae
e   Highland Fruit-eating Bat; Aztec
       Fruit-eating Bat

405   *Artibeus cinereus*
       Chiroptera - Phyllostomidae
e   Pygmy Fruit-eating Bat; Gervais's
       Fruit-eating Bat; Little Brazilian
       Fruit-eating Bat

406   *Artibeus concolor*
       Chiroptera - Phyllostomidae
e   Surinam Fruit Bat; Brown Fruit-
       eating Bat

407   *Artibeus fimbriatus*
       Chiroptera - Phyllostomidae
e   Fringed Fruit-eating Bat

408   *Artibeus fraterculus*
       Chiroptera - Phyllostomidae
e   Fraternal Fruit-eating Bat

409   *Artibeus glaucus*
       Chiroptera - Phyllostomidae
e   Silver Fruit-eating Bat

410   *Artibeus hartii*
       Chiroptera - Phyllostomidae
e   Velvety Fruit-eating Bat; Hart's Little
       Fruit Bat; Little Fruit-eating Bat
d   Schokoladen-Fruchtzwerg

411 **Artibeus hirsutus**
Chiroptera - Phyllostomidae
e Hairy Fruit-eating Bat

412 **Artibeus incomitatus**
Chiroptera - Phyllostomidae
e Solitary Fruit-eating Bat

413 **Artibeus inopinatus**
Chiroptera - Phyllostomidae
e Honduran Fruit-eating Bat; Honduras Fruit-eating Bat

414 **Artibeus intermedius**
Chiroptera - Phyllostomidae
e Intermediate Fruit-eating Bat

415 **Artibeus jamaicensis**
Chiroptera - Phyllostomidae
e Jamaican Fruit-eating Bat; Mexican Fruit-eating Bat; Lesser Trinidadian Fruit Bat; Lesser Trinidad Fruit Bat; Common Fruit Bat; Mexican Fruit Bat; Jamaican Bat
d Jamaika-Fruchtfledermaus; Jamaika-Fruchtvampir
f Artibée de la Jamaïque

416 **Artibeus jamaicensis parvipes**
Chiroptera - Phyllostomidae
e Caribbean Fruit Bat

417 **Artibeus lituratus**
Chiroptera - Phyllostomidae
e Great Fruit-eating Bat
f Grande Artibée

418 **Artibeus obscurus**
Chiroptera - Phyllostomidae
e Dark Fruit-eating Bat

419 **Artibeus phaeotis**
Chiroptera - Phyllostomidae
e Dwarf Fruit-eating Bat

d Zwergfruchtfledermaus

420 **Artibeus planirostris**
Chiroptera - Phyllostomidae
e Flat-faced Fruit-eating Bat

421 **Artibeus toltecus**
Chiroptera - Phyllostomidae
e Lowland Fruit-eating Bat; Toltec Fruit-eating Bat

422 **Artibeus watsoni**
Chiroptera - Phyllostomidae
e Thomas's Fruit-eating Bat

423 **Arvicanthis**
Rodentia - Muridae
e Grass Mice; Striped Mice; Grass Rats; Nile Rats; Nile Grass Rats; African Grass Rats; Kusu Rats
d Grasratten; Kusu-Grasratten
f Rats roussards

424 **Arvicanthis abyssinicus**
Rodentia - Muridae
e Abyssinian Grass Rat
d Äthiopische Kusu-Grasratte

425 **Arvicanthis blicki**
Rodentia - Muridae
e Blick's Grass Rat

426 **Arvicanthis nairobae**
Rodentia - Muridae
e Nairobi Grass Rat

427 **Arvicanthis neumanni**
Rodentia - Muridae
e Neumann's Grass Rat

428 **Arvicanthis niloticus**
Rodentia - Muridae
e Grass Mouse; Striped Mouse; Grass Rat; Nile Rat; Nile Grass Rat; African Grass Rat; Kusu Rat

d Nil-Grasratte
f Rat du Nil; Rat roussard du Nil

429 *Arvicola*
Rodentia - Muridae
e Water Voles; Bank Voles
d Wühlmäuse; Wasserratten; Große
Wühlmäuse; Schernäuse
f Arvicolas; Rats d'eau; Campagnol
d'eau
i Arvicole

Arvicola amphibius syn. *A. terrestris*
q.v.

430 *Arvicola sapidus*
Rodentia - Muridae
e Southwestern Water Vole
d Westschermaus
f Arvicola occidentale; Campagnol
amphibie; Campagnol aquatique
gallo-ibérique
i Arvicola d'acqua

431 *Arvicola terrestris*
Rodentia - Muridae
e Ground Vole; Water Vole; European
Water Vole; Vole Rat; Water Rat;
Northern Water Vole
d Schermaus; Ostschermaus; Große
Wühlmaus; Große Europäische
Wühlmaus; Große Erdmaus; Große
Feldmaus; Mollmaus; Wühlratte;
Erdratte; Hamstermaus
f Arvicola terrestre; Campagnol
terrestre; Rat taupier
i Arvicola terrestre; Ratto d'acqua;
Arvicola d'acqua; Arvicola acquatica;
Topo d'acqua; Topo russo

432 *Arvicola terrestris exitus*
Rodentia - Muridae
e Alpine Ground Vole
d Alpenschermaus
i Arvicola terrestre alpina

Aschizomys syn. Eothenomys q.v.

433 *Asellia*
Chiroptera - Rhinolophidae
e Trident Bats; Leaf-nosed Bats; North
African Trident Bats; Trident Leaf-
nosed Bats
d Dreizackblattnasen
f Asellias

434 *Asellia patrizii*
Chiroptera - Rhinolophidae
e Patrizi's Trident Bat; Patrizi's Trident
Leaf-nosed Bat
f Asellia de Patrizi

435 *Asellia tridens*
Chiroptera - Rhinolophidae
e Trident Bat; Trident Leaf-nosed Bat
d Dreizackblattnase
f Asellia à trois endentures; Trident

436 *Aselliscus*
Chiroptera - Rhinolophidae
e Tate's Trident-nosed Bats

437 *Aselliscus stoliczkanus*
Chiroptera - Rhinolophidae
e Stoliczka's Trident Bat

438 *Aselliscus tricuspidatus*
Chiroptera - Rhinolophidae
e Dobson's Trident Bat; Temminck's
Trident Bat; Trident Horseshoe Bat

Atelerix syn. *Erinaceus* q.v.

439 *Ateles*
Primates - Atlidae
e Spider Monkeys
d Klammeraffen
f Singes-araignés; Coatas; Atèles
i Scimmie ragno

440 *Ateles belzebuth*
Primates - Atlidae
e Long-haired Spider Monkey; White-
bellied Spider Monkey; Marimonda

d Goldstirnklammeraffe
f Singe-araigné à ventre blanc; Atèle à ventre blanc
i Atele belzebù; Scimmia ragno; Atele dal ventro bianco

441 *Ateles chamek*
Primates - Atlidae
e Chamek Spider Monkey; Black-faced Spider Monkey; Western Black Spider Monkey; Peruvian Spider Monkey
d Schwarzgesichtklammeraffe
f Singe-araigné a tête noire; Atèle a tête noire
i Atele nero

442 *Ateles fusciceps*
Primates - Atlidae
e Brown-headed Spider Monkey; Black-headed Spider Monkey
d Braunkopfklammeraffe
f Singe-araigné à tête brune; Atèle à tête brune
i Atele dalla testa bruna

443 *Ateles geoffroyi*
Primates - Atlidae
e Miriki; Central American Spider Monkey; Geoffroy's Spider Monkey
d Geoffroy-Klammeraffe
f Singe-araigné de Geoffroy; Atèle de Geoffroy
i Atele dalle mani nere; Atele di Geoffroy; Scimmia ragno dalle mani nere

444 *Ateles geoffroyi frontatus*
Primates - Atlidae
e Black-handed Spider Monkey
d Schwarzbrauen-Geoffroy-Klammeraffe
f Singe-araigné aux mains noirs; Atèle aux mains noirs

445 *Ateles geoffroyi panamensis*
Primates - Atlidae

e Panama Spider Monkey
d Panama-Klammeraffe
f Singe-araigné aux mains noires de Panama; Atèle aux mains noires de Panama
i Atele del Panama

446 *Ateles hybridus*
Primates - Atlidae
e Variegated Spider Monkey; Brown Spider Monkey
d Goldstirnklammeraffe
f Singe-araigné varié; Atèle varié
i Atele variegato

447 *Ateles marginatus*
Primates - Atlidae
e White-whiskered Spider Monkey; White-bellied Spider Monkey
d Weißwangenklammeraffe
f Singe-araigné aux joues blanches; Atèle aux joues blanches
i Atele dalle basette bianche

448 *Ateles paniscus*
Primates - Atlidae
e Black Spider Monkey; Coaita
d Schwarzer Klammeraffe
f Singe-araigné noir; Atèle noir; Coata noir
i Atele dalla fascia rossa; Atele nero

449 *Ateles paniscus paniscus*
Primates - Atlidae
e Red-faced Spider Monkey
d Rotgesicht Schwarzer Klammeraffe
f Singe-araigné noir; Atèle noir à visage rouge
i Atele nero

450 **Atelidae**
Primates
e New World Monkeys
d Neuwelt Affen
f Atélidés
i Scimmie del nuovo mondo

451   *Atelocynus*
Carnivora - Canidae
e   Short-eared Dogs
d   Kurzohrfüchse
f   Chiens des buissons
i   Cani dalle orecchie corte

452   *Atelocynus microtis*
Carnivora - Canidae
e   Short-eared Dog; Short-eared Bush
Dog; Small-eared Dog; Small-eared
Zorro
d   Kurzohrfuchs
f   Chien des buissons aux oreilles
courtes; Renard aux petites oreilles
i   Cane dalle orecchie corte

453   *Atherurus*
Rodentia - Hystricidae
e   Brush-tailed Porcupines
d   Quastenstachler
f   Athérures; Porcs-épics africaines
i   Aterura

454   *Atherurus africanus*
Rodentia - Hystricidae
e   African Brush-tailed Porcupine; West
African Brush-tailed Porcupine
d   Afrikanischer Quastenstachler;
Westafrikanischer Quastenstachler
f   Athérure africain
i   Aterurua ; Istrice arborea

455   *Atherurus macrourus*
Rodentia - Hystricidae
e   Malayan Brush-tailed Porcupine;
Asiatic Brush-tailed Porcupine;
Brush-tailed Porcupine; Asiatic
Bush-tailed Porcupine
d   Langschwanzquastenstachler;
Langschwänziges Stachelschwein
f   Athérure à longue queue; Athérure
malais
i   Aterura macrura

456   *Atilax*
Carnivora - Herpestidae
e   Marsh Mongooses; Water
Mongooses
d   Sumpfmangusten
f   Mangoustes des marais
i   Manguste della palude africanoa

457   *Atilax paludinosus*
Carnivora - Herpestidae
e   Marsh Mongoose; Water Mongoose;
West African Water Mongoose;
African Water Mongoose
d   Sumpfmanguste; Wassermanguste;
Sumpfichneumon
f   Mangouste des marais
i   Mangusta della palude africana

458   *Atlantoxerus*
Rodentia - Sciuridae
e   Barbary Ground Squirrels
d   Borstenhörnchen
f   Écureuils de Barbarie
i   Xeri del Nordafrica

459   *Atlantoxerus getulus*
Rodentia - Sciuridae
e   Barbary Ground Squirrel; North
African Ground Squirrel; Maghreb
Ground Squirrel
d   Nordafrikanisches Erdhörnchen;
Atlas-Hörnchen
f   Écureuil de Barbarie
i   Xero del Nordafrica; Xero
dell'Atlante

Atopogale syn. *Solenodon* q.v.

Aulacomys syn. *Microtus* q.v.

460   *Auliscomys*
Rodentia - Muridae
e   Big-eared Mice
d   Südamerikanische Blattohrmäuse

461   *Auliscomys boliviensis*
          Rodentia - Muridae
   e   Bolivian Big-eared Mouse

462   *Auliscomys pictus*
          Rodentia - Muridae
   e   Painted Big-eared Mouse

463   *Auliscomys sublimis*
          Rodentia - Muridae
   e   Andean Big-eared Mouse

464   *Australophocoena*
          Cete - Phocoenidae
   e   Spectacled Porpoises
   d   Brillenschweinswale
   f   Marsouins
   i   Focene dagli occhiali

465   *Australophocoena dioptrica*
          Cete - Phocoenidae
   e   Spectacled Porpoise; South Atlantic
          Porpoise; Southern Harbour Porpoise
   d   Brillenschweinswal; Brillentümmler
   f   Marsouin à lunettes; Marsouin de
          Lahille
   i   Focena dagli occhiali

466   *Avahi*
          Primates - Indriide
   e   Woolly Lemurs; Avahis; Woolly
          Indris
   d   Wollmakis
   f   Avahis
   i   Licanoti

467   *Avahi laniger*
          Primates - Indriide
   e   Eastern Woolly Lemur; Eastern
          Avahi; Woolly Indri; Avahi Lemur;
          Woolly Lemur; Avahi
   d   Wollmaki; Vliesmaki; Avahi
   f   Avahi laineux oriental
   i   Licanoto lanoso; Avahi

468   *Avahi occidentalis*
          Primates - Indriide
   e   Western Woolly Lemur; Western
          Avahi
   d   Westlicher Wollmaki
   f   Avahi laineux occidental
   i   Licanoto occidentale

469   *Avahi unicolor*
          Primates - Indriide
   e   Unicoloured Avahi
   d   Einfarbiger Wollmaki;
          Nordwestlicher Wollmaki;
          Sambirano-Wollmaki
   f   Avahi laineux unicolore

470   *Axis*
          Artiodactyla - Cervidae
   e   Axis Deer
   d   Fleckenhirsche; Tüpfelhirsche
   f   Cerfs axis
   i   Cervi porcini

471   *Axis axis*
          Artiodactyla - Cervidae
   e   Chital; Indian Spotted Deer; Spotted
          Deer; Axis Deer
   d   Axishirsch; Tüpfelhirsch; Chital
   f   Axis; Cerf axis
   i   Cervo pomellato; Cervo axis

472   *Axis axis axis*
          Artiodactyla - Cervidae
   e   Indian Axis Deer
   d   Indischer Axishirsch
   f   Cerf axis de l'Inde
   i   Cervo axis

473   *Axis axis ceylonensis*
          Artiodactyla - Cervidae
   e   Ceylonese Axis Deer; Ceylon
          Spotted Deer
   d   Ceylon-Axishirsch
   f   Cerf axis de Ceylan

474   *Axis calamianensis*
         Artiodactyla - Cervidae
e   Calamian Deer; Calamian Hog-Deer
d   Calamian-Schweinshirsch; Calamian-
         Hirsch
f   Cerf calamian
i   Cervo porcino di Calamian

475   *Axis kuhli*
         Artiodactyla - Cervidae
e   Bawean Deer; Kuhl's Deer
d   Bawean-Schweinshirsch; Bawean-
         Hirsch; Kuhl-Hirsch
f   Cerf de Kuhl
i   Cervo porcino di Kuhl; Cervo
         porcino di Bawean

476   *Axis porcinus*
         Artiodactyla - Cervidae
e   Hog Deer
d   Schweinshirsch
f   Cerf-cochon
i   Cervo porcino

477   *Axis porcinus annamiticus*
         Artiodactyla - Cervidae
e   Indochinese Hog Deer; Ganges Hog
         Deer; Thai Hog Deer
d   Hinterindischer Schweinshirsch
f   Cerf-cochon du Gange
i   Cervo porcino dell'Indocina

478   *Axis porcinus porcinus*
         Artiodactyla - Cervidae
e   Common Hog Deer
d   Vorderindischer Schweinshirsch
f   Cerf-cochon commun
i   Cervo porcino comune

B

479 **Babyrousa**
Artiodactyla - Suidae
e Babiroussas; Babirussas
d Babirussas; Hirscheber
f Babiroussas
i Babirussa

480 **Babyrousa babyrussa**
Artiodactyla - Suidae
e Babiroussa; Babirussa; Indian Hog;
Hog Deer; Deer Hog; Pig Deer;
Horned Hog
d Hirscheber; Babirussa
f Babiroussa
i Babirussa; Porco cervo

481 **Babyrousa babyrussa celebensis**
Artiodactyla - Suidae
e Mainland Babiroussa; Sulawesi
Babirussa
d Sulawesi-Hirscheber
f Babiroussa des Célèbes

482 **Babyrousa babyrussa togeanensis**
Artiodactyla - Suidae
e Togian Babiroussa
f Babiroussa de l'île Togian

Baeodon syn. *Rhogeessa* q.v.

483 **Baiomys**
Rodentia - Muridae
e Pygmy Mice; American Pygmy Mice
d Amerikanische Zwergmäuse
f Souris pygmées

484 **Baiomys musculus**
Rodentia - Muridae
e Southern Pygmy Mouse
d Südliche Amerikanische Zwergmaus

485 **Baiomys taylori**
Rodentia - Muridae
e Northern Pygmy Mouse
d Nördliche Amerikanische
Zwergmaus
f Souris pygmée d'Amérique; Souris
pygmée de Taylor

Baiyankamys syn. *Hydromys* q.v

486 **Balaena**
Cete - Balaenidae
e Right Whales; Bow-headed
Greenland Whales
d Grönland-Wale; Riesenglattwale;
Nordwale
f Baleines de Groenland
i Balene

487 **Balaena australis**
Cete - Balaenidae
e Southern Right Whale
d Südkaper; Südlicher Glattwal
f Baleine franche australe
i Balena franca australe

489 **Balaena glacialis**
Cete - Balaenidae
e Atlantic Right Whale; North Atlantic
Right Whale; Northern Right Whale;
Right Whale; Black Right Whale;
Great Right Whale
d Nordkaper; Biskayer Wal;
Atlantischer Nordkaper; Biskaya-
Wal
f Baleine des Basques; Baleine noire;
Baleine franche noire; Baleine
franche des Basques; Baleine franche
i Balena nera; Balena franca boreale;
Balena franca nordatlantica; Balena
franca settentrionale; Balena dei
Baschi; Balena glaciale

488   *Balaena japonica*
Cete - Balaenidae
e   North Pacific Right Whale
d   Nordpazifik-Glattwal; Pazifischer
      Nordkaper; Glattwal; Nordkap-Wal
f   Baleine noire du Pacific Nord
i   Balena franca nordpacifica

490   *Balaena mysticetus*
Cete - Balaenidae
e   Arctic Whale; Greenland Whale;
      Common Whale; Bow-headed
      Whale; Great Polar Whale; Arctic
      Right Whale; Common Right Whale;
      Greenland Right Whale; Bowhead
      Whale; Bowhead
d   Grönland-Wal; Polar-Wal; Nordwal;
      Grönlandischer Walfisch
f   Baleine de Groenland; Baleine
      boréale; Baleine franche; Baleine du
      Nord; Baleine vraie; Baleine franche
      du Groenland
i   Balena boreale; Balena della
      Groenlandia; Balena franca della
      Groenlandia

491   Balaenidae
Cete
e   Right Whales; Bow-headed Whales
d   Glattwale
f   Baleines franches; Balénidés
i   Balene; Balenidi

492   *Balaenoptera*
Cete - Balaenopteridae
e   Rorquals; Finback Whales; Finny
      Whales
d   Finnwale
f   Baleinoptères; Rorquals
i   Balenottere

493   *Balaenoptera acutorostrata*
Cete - Balaenopteridae
e   Minke; Minke Whale; Little Piked
      Whale; Piked Whale; Pikehead
      Whale; Sharp-headed Whale; Lesser
      Rorqual; Lesser Finnback; Sharp-

headed Finnback; Common Minke
Whale; Northern Minke Whale
d   Zwergwal; Schnabelwal;
      Sommerwal; Zwergfinnfisch;
      Zwergfinnwal
f   Rorqual rostré; Rorqual à rostre;
      Rorqual à museau pointu; Petit
      rorqual; Baleinoptère rostré;
      Finbeque; Rorqual de Minke
i   Balenottera rostrata; Balenottera
      minore

**494   *Balaenoptera acutorostrata
        acutorostrata***
Cete - Balaenopteridae
e   Atlantic Minke Whale
f   Rorqual à beque; Baleine d'été;
      Gibard

**495   *Balaenoptera acutorostrata
        scammoni***
Cete - Balaenopteridae
e   North Pacific Minke Whale
f   Rorqual du Pacifique Nord

496   *Balaenoptera bonaerensis*
Cete - Balaenopteridae
e   Antarctic Minke Whale
d   Südlicher Zwergwal
f   Rorqual de l'Antarctique;
      Baleinoptère à museau pointu de
      l'Antarctique; Rorqual à museau
      pointu de l'Antarctique
i   Balenottera minore antartica

497   *Balaenoptera borealis*
Cete - Balaenopteridae
e   Sei Whale; Sardine Whale; Coalfish
      Whale; Rudolphi's Rorqual; Pollack
      Whale; Japan Finner
d   Rudolphi-Finnwal; Seiwal
f   Baleinoptère boréal; Rorqual boréal;
      Rorqual du Nord; Rorqual de
      Rudolphi; Rorqual de Minke arctique
i   Balenottera boreale

498   *Balaenoptera borealis borealis*
Cete - Balaenopteridae

e Northern Sei Whale; Northern
Rorqual

f Rorqual du Nord

499 *Balaenoptera borealis schlegeli*
Cete - Balaenopteridae

e Southern Rorqual

Balaenoptera brevicaudata syn. *B.
musculus* q.v

500 *Balaenoptera brydei*
Cete - Balaenopteridae

e Bryde's Whale; Eden's Whale

d Eden-Wal; Bryde-Wal; Brydes Wal

f Baleinoptère de Bryde; Rorqual de
Bryde; Rorqual tropical

i Balenottera di Eden; Balenottera di
Bryde

Balaenoptera edeni syn. *B. brydei*
q.v.

501 *Balaenoptera musculus*
Cete - Balaenopteridae

e Blue Whale; Great Whale; Great
Blue Whale; Sibbald's Rorqual; Blue
Rorqual; Great Northern Rorqual

d Riesenwal; Großmäuliger Finnfisch;
Blauwal; Großer Nördlicher
Furchenwal; Sibbalds Furchenwal

f Baleinoptère bleu; Baleine bleue;
Grande baleine bleue; Rorqual bleu;
Rorqual de Sibbald; Grand Rorqual

i Balenottera azzura

502 *Balaenoptera musculus brevicauda*
Cete - Balaenopteridae

e Pygmy Blue Whale

d Zwergblauwal

f Baleine bleue pygmée

503 *Balaenoptera musculus indica*
Cete - Balaenopteridae

e Indian Ocean Blue Whale

504 *Balaenoptera musculus intermedia*
Cete - Balaenopteridae

e Southern Blue Whale

505 *Balaenoptera musculus musculus*
Cete - Balaenopteridae

e Northern Blue Whale

f Baleine bleue de Biscaye; Baleine
bleue des Basques

506 *Balaenoptera physalus*
Cete - Balaenopteridae

e Common Fin Whale; Fin Whale;
Finar Whale; Herring Whale;
Finback Whale; Common Finback
Whale; Common Rorqual; Razorback

d Gemeiner Finnwal; Finnwal;
Finnfisch; Gemeiner Furchenwal

f Rorqual rostré; Rorqual de la
Méditerranée

i Balenottera comune; Dorso a rasio;
Balenottera fisalo; Rorqualo; Rorqual
comune; Rorcuallo; Capidoglio

507 *Balaenoptera physalus physalus*
Cete - Balaenopteridae

e Northern Fin Whale

f Rorqual commun; Baleinoptère
commune

508 *Balaenoptera physalus quoyi*
Cete - Balaenopteridae

e Southern Fin Whale

509 Balaenopteridae
Cete

e Rorquals; Finback Whales; Finny
Whales; Fin Whales

d Finnwale; Furchenwale

f Balénoptèridés; Rorquals;
Baleinoptères

i Balenottere

510 *Balantiopteryx*
Chiroptera - Emballonuridae

e Least Sac-winged Bats

511   *Balantiopteryx infusca*
Chiroptera - Emballonuridae
e   Ecuadorian Sac-winged Bat

512   *Balantiopteryx io*
Chiroptera - Emballonuridae
e   Thomas's Sac-winged Bat; Thomas's
Least Sac-winged Bat; Least Sac-
winged Bat

513   *Balantiopteryx plicata*
Chiroptera - Emballonuridae
e   Peters's Sac-winged Bat; Peters's Bat;
Grey Sac-winged Bat

514   *Balionycteris*
Chiroptera - Pteropodidae
e   Spotted-wings Fruit Bats; Spot-
winged Fruit Bats

515   *Balionycteris maculata*
Chiroptera - Pteropodidae
e   Spotted-wings Fruit Bat; Spot-
winged Fruit Bat

516   *Bandicota*
Rodentia - Muridae
e   Bandicoot Rats; Molerats;
Bandicoots; Asian Bandicoot Rats
d   Bandikutratten
f   Rats bandicoots
i   Ratti della talpa

517   *Bandicota bengalensis*
Rodentia - Muridae
e   Black Bandicoot Rat; Bengali
Bandicoot Rat; Lesser Bandicoot
Rat; Indian Molerat; Lesser
Bandicoot; Sind Rice Rat; Bengal
Bandicoot
d   Kleine Bandikutratte; Indische
Maulwurfsratte
f   Rat bandicoot du Bengale

518   *Bandicota bengalensis gracilis*
Rodentia - Muridae

e   Ceylon Mole Rat; Lesser Bandicoot
Rat
d   Indische Pestratte

519   *Bandicota bengalensis insularis*
Rodentia - Muridae
e   Northern Ceylon Mole Rat
d   Inselbandikutratte

520   *Bandicota indica*
Rodentia - Muridae
e   Large Bandicoot; Large Bandicoot
Rat; Greater Bandicoot; Greater
Bandicoot Rat
d   Bandikutratte; Große Indische
Bandikutratte; Große Bandikutratte
f   Rat pipistrelle d'Inde; Grand
bandicoot

521   *Bandicota indica nemorivaga*
Rodentia - Muridae
e   Mole Rat

522   *Bandicota maxima*
Rodentia - Muridae
e   Giant Bandicoot Rat
f   Rat bandicoot géant
i   Bandicota gigantea

523   *Bandicota savilei*
Rodentia - Muridae
e   Saville's Bandicoot Rat; Burmese
Bandicoot Rat

524   *Barbastella*
Chiroptera - Vespertilionidae
e   Barbastelles
d   Breitohrfledermäuse;
Mopsfledermäuse
f   Barbastelles
i   Barbastelli

525   *Barbastella barbastellus*
Chiroptera - Vespertilionidae
e   Common Barbastelle; Western
Barbastelle; Barbastelle;

Barbastelle Bat

d   Breitohrfledermaus; Breitohr;
Kurzmaul Gemeine Mopsfledermaus;
Mopsfledermaus; Kurzmaul;
Gemeine Mopsfledermaus;
Breitohrige Fledermaus
f   Barbastelle d'Europe; Barbastelle
i   Barbastello; Barbastello comune

526   ***Barbastella leucomelas***
Chiroptera - Vespertilionidae
e   Asian Barbastelle; Eastern
Barbastelle; Asiatic Wide-eared Bat
d   Östliche Mopsfledermaus
f   Barbastelle d'Asie

527   ***Barbastella leucomelas***
darjelingensis
Chiroptera - Vespertilionidae
e   Brown Long-eared Bat

528   ***Bassaricyon***
Carnivora - Procyonidae
e   Olingos
d   Makibären
f   Bassaricyons; Olingos

529   ***Bassaricyon alleni***
Carnivora - Procyonidae
e   Allen's Olingo
d   Allens Olingo

530   ***Bassaricyon beddardi***
Carnivora - Procyonidae
e   Beddard's Olingo
d   Beddards Olingo

531   ***Bassaricyon gabbii***
Carnivora - Procyonidae
e   Bushy-tailed Olingo
d   Schlankbär
i   Bassaricione di Gabb; Olingo dalla
coda pelosa

532   ***Bassaricyon lasius***

Carnivora - Procyonidae
e   Harris's Olingo
d   Harris-Olingo

533   ***Bassaricyon pauli***
Carnivora - Procyonidae
e   Chiriqui Olingo
d   Chriqui-Olingo

534   ***Bassariscus***
Carnivora - Procyonidae
e   Ringtails; Ring-tailed Cats;
Cacomistles
d   Katzenfretten
f   Basssaris; Bassarides
i   Bassarischi del Centroamerica

535   ***Bassariscus astutus***
Carnivora - Procyonidae
e   North American Cacomistle;
Ringtail; Ring-tailed Cat; Civet;
Civet Cat; Miner's Cat
d   Nordamerikanisches Katzenfrette
i   Ringtail; Gatto dalla coda ad anelli;
Bassarisco

536   ***Bassariscus sumichrasti***
Carnivora - Procyonidae
e   Central American Cacomistle;
Cacomistle; Ringtail Cacomistle;
Central American Ring-tailed Cat
d   Mittelamerikanisches Katzenfrette
f   Bassaride rusé; Bassari rusée
i   Bassarisco del Centroamerica

537   **Bathyergidae**
Rodentia
e   Molerats; African Molerats;
Blesmols
d   Sandgräber
f   Rats-taupes africains; Bathyergidés
i   Batiergidi

538   ***Bathyergus***
Rodentia - Bathyergidae

e Cape Sand Moles; Cape Molerats;
Cape Dune Molerats; Dune Molerats
d Strandgräber ; Sandgräber
f Rats-taupes
i Batierghi

539 ***Bathyergus janetta***
Rodentia - Bathyergidae
e Namaqua Dune Molerat; Dune
Molerat
d Namaqua-Strandgräber

540 ***Bathyergus suillus***
Rodentia - Bathyergidae
e Cape Dune Molerat; Cape Sand Mole
d Kap-Strandgräber; Kap-Sandgräber
f Fouisseur du Cap
i Batiergo marittimo

541 ***Batomys***
Rodentia - Muridae
e Luzon Forest Rats; Philippine Forest
Rats; Hairy-tailed Rats
d Waldratten; Philippinen-
Haarschwanzratten

542 ***Batomys dentatus***
Rodentia - Muridae
e Large-toothed Hairy-tailed Rat
d Luzon-Waldratte

543 ***Batomys granti***
Rodentia - Muridae
e Luzon Hairy-tailed Rat
d Grants Waldratte

544 ***Batomys russatus***
Rodentia - Muridae
e Dinagat Hairy-tailed Rat
d Dinagat-Haarschwanzratte

545 ***Batomys salomonseni***
Rodentia - Muridae
e Mindanao Rat; Mindanao Hairy-
tailed Rat

d Mindanao-Waldratte

546 ***Bauerus***
Chiroptera - Antrozoidae
e Van Gelder's Bats

547 ***Bauerus dubiaquercus***
Chiroptera - Antrozoidae
e Van Gelder's Bat; Cuban Bat

548 ***Bdeogale***
Carnivora - Herpestidae
e Black-legged Mongooses; Bushy-
tailed Mongooses
d Dickschwänzige Hundemangusten
f Mangoustes aux pieds noires
i Manguste

549 ***Bdeogale crassicauda***
Carnivora - Herpestidae
e Four-toed Mongoose; Bushy-tailed
Mongoose; Sokoke Bushy-tailed
Mongoose
d Dickschwänzige Hundemanguste;
Buschschwanzmanguste
i Mangusta dalla coda folta

550 ***Bdeogale jacksoni***
Carnivora - Herpestidae
e Jackson's Mongoose
d Jackson-Manguste
i Mangusta di Jackson

551 ***Bdeogale nigripes***
Carnivora - Herpestidae
e Black-footed Mongoose; Black-
legged Mongoose
d Schwarzfußichneumon;
Schwarzfußmanguste
f Mangouste aux pieds noires;
Mangouste aux pattes noires
i Mangusta dai piedi neri

552 ***Beamys***
Rodentia - Muridae
e Long-tailed Pouched Rats; East

African Long-tailed Pouched Rats
d   Kleine Hamsterratten

553   *Beamys hindei*
Rodentia - Muridae
e   Long-tailed Pouched Rat; East
African Long-tailed Pouched Rat;
Lesser Pouched Rat
d   Kleine Hamsterratte; Kleine
Ostafrikanische Hamsterratte;
Kleinste Hamsterratte

554   *Beamys major*
Rodentia - Muridae
e   Greater Long-tailed Pouched Rat;
Greater Hamster Rat

Beatragus syn. *Damaliscus* q.v.

555   *Belomys*
Rodentia - Sciuridae
e   Hairy-footed Flying Squirrels;
Tufted-ears Flying Squirrels
d   Haarfuß-Gleithörnchen
i   Scoiattoli pelosi

556   *Belomys pearsoni*
Rodentia - Sciuridae
e   Hairy-footed Flying Squirrel; Tufted-
ears Flying Squirrels; Hair-foot
Flying Squirrel
d   Haarfuß-Gleithörnchen

557   *Belomys pearsoni kaleensis*
Rodentia - Sciuridae
e   Taiwan Hairy-footed Squirrel

558   *Berardius*
Cete - Hyperoodontidae
e   Pacific Beaked Whales; Giant Bottle-
nosed Whales; Giant Beaked Whales
d   Schwarzwale
f   Bérardiens; Grans baleines à bec;
Bérardies
i   Berardi

559   *Berardius arnuxii*
Cete - Hyperoodontidae
e   Arnoux's Beaked Whale; New
Zealand Beaked Whale; Southern
Beaked Whale; Arnoux's Whale;
Smaller Whale; Ziphiid Whale;
Southern Four-toothed Whale;
Southern Giant Bottle-nosed Whale
d   Südlicher Schwarzwal
f   Baleine d'Arnoux; Bérardie australe
i   Berardio di Arnoux; Berardio
australe

560   *Berardius bairdii*
Cete - Hyperoodontidae
e   Baird's Whale; Baird's Beaked
Whale; North Pacific Giant Bottle-
nosed Whale; Giant Bottle-nosed
Whale; Northern Four-toothed
Whale; Northern Beaked Whale;
North Pacific Bottle-nosed Whale
d   Nordischer Schwarzwal; Baird-Wal
f   Baleine de Baird; Bérardie boréale
i   Berardio boreale; Berardio di Baird

561   *Berylmys*
Rodentia - Muridae
e   White-toothed Rats
d   Weißzahnratten

562   *Berylmys berdmorei*
Rodentia - Muridae
e   Grey Rat; Small White-toothed Rat;
Lesser White-toothed Rat
d   Kleine Weißzahnratte

563   *Berylmys bowersi*
Rodentia - Muridae
e   Bowers's Rat; Bower's White-toothed
Rat
d   Bowers Ratte

564   *Berylmys mackenziei*
Rodentia - Muridae
e   Kenneth's White-toothed Rat
d   Kenneths Weißzahnratte

565   *Berylmys manipulus*
Rodentia - Muridae
e   Manipur Rat; Manipur White-toothed
Rat
d   Manipur-Weißzahnrattte

566   *Bettongia*
Diprotodontia - Macropodidae
e   Short-nosed Rat Kangaroos; Jerboa
Kangaroos; Bettongs (ANZ); Rat
Kangaroos
d   Bürstenkängurus
f   Rats-kangarous à nez court;
Bettongies; Bettongs
i   Bettonge

Bettongia cuniculus syn. *B.*
gaimardi q.v.

567   *Bettongia gaimardi*
Diprotodontia - Macropodidae
e   Eastern Bettong; Gaimard's Rat
Kangaroo; Tasmanian Rat Kangaroo;
Tasmanian Bettong (ANZ);
Gaimard's Tasmanian Rat-Kangaroo
d   Festland-Bürstenkängaruh;
Tasmanisches Bettong; Gaimards
Bürstenrattenkänguru
f   Rat-kangourou de Gaimard; Rat-
kangourou à nez court de Gaimard;
Rat-kangourou à nez court
d'Australie; Bettongie à nez court de
Gaimard; Bettongie de Gaimard

568   *Bettongia lesueur*
Diprotodontia - Macropodidae
e   Lesueur' Rat Kangaroo; Short-nosed
Rat Kangaroo; Boodie; Burrowing
Bettong; Boodie (ANZ)
d   Lesueur-Bürstenkänguru
f   Rat-kangourou de Lesueur

569   *Bettongia penicillata*
Diprotodontia - Macropodidae
e   Brush-tailed Bettong (ANZ); Woylie;
Northern Bettong (ANZ); Brush-
tailed Rat Kangaroo; Tropical
Bettong

d   Bürstenschwanz-Rattenkänguru
f   Rat-kangourou à queue touffue; Rat-
kangourou à queue en brosse
i   Bettongia dalla coda a spazzola

570   *Bettongia tropica*
Diprotodontia - Macropodidae
e   Northern Bettong
d   Nord-Bettong
f   Bettong nordique

571   *Bibimys*
Rodentia - Muridae
e   Crimson-nosed Rats
d   Südamerikanische Rotnasenratten

572   *Bibimys chacoensis*
Rodentia - Muridae
e   Chaco Crimson-nosed Rat

573   *Bibimys torresi*
Rodentia - Muridae
e   Torres's Crimson-nosed Rat

Bibos syn. *Bos* q.v.

574   *Bison*
Artiodactyla - Bovidae
e   Bisons
d   Wisente; Bisons
f   Bisons
i   Bisonti

575   *Bison bison*
Artiodactyla - Bovidae
e   Buffalo (NA); American Bison;
North American Buffalo
d   Bison; Amerikanischer Bison
f   Bison américain; Bison; Bison
d'Amérique
i   Bisonte americano; Bufalo
americano; Bufalo

576   *Bison bison athabascae*
Artiodactyla - Bovidae

e   Wood Bison

d   Waldbison

f   Bison des bois; Bison des forêts

i   Bisonte dei boschi

577   *Bison bison bison*
      Artiodactyla - Bovidae

e   Plains Bison; Southern Prairie Bison; Southern Prairie Buffalo; Southern Great Plains Buffalo; Southern Great Plains Bison

d   Prairiebison; Mähnenbison; Südlicher Prairiebison

f   Bison des plaines du Sud

i   Bufalo

578   *Bison bison oreganus*
      Artiodactyla - Bovidae

e   Oregon Bison

579   *Bison bison pennsylvanicus*
      Artiodactyla - Bovidae

e   Eastern Woodland Bison

i   Bisonte dell'Est

580   *Bison bonasus*
      Artiodactyla - Bovidae

e   Bison; European Bison; European Wisent; Wisent

d   EuropäischerWisent; Flachlandwisent; Wisent

f   Bison d'Europe; Wisent

i   Bisonte europeo

581   *Bison bonasus caucasicus*
      Artiodactyla - Bovidae

e   Caucasian Bison

d   Bergwisent; Kaukasus-Wisent

f   Bison du Caucase

i   Bisonte del Caucaso

582   *Bison bonasus montanus*
      Artiodactyla - Bovidae

e   Highland Bison

583   *Biswamoyopterus*
      Rodentia - Sciuridae

e   Namdapha Flying Squirrels

d   Namdapha-Gleithörnchen

584   *Biswamoyopterus biswasi*
      Rodentia - Sciuridae

e   Namdapha Flying Squirrel

d   Namdapha-Gleithörnchen

585   *Blanfordimys*
      Rodentia - Muridae

e   Afghan Voles

d   Afghanische Wühlmäuse

586   *Blanfordimys afghanus*
      Rodentia - Muridae

e   Afghan Vole

587   *Blanfordimys bucharicus*
      Rodentia - Muridae

e   Bukharian Vole

588   *Blarina*
      Soricomorpha - Soricidae

e   Short-tailed Shrews; American Short-tailed Shrews; Greater North American Short-tailed Shrews

d   Nordamerikanische Spitzschwanzspitzmäuse; Kurzschwanzspitzmäuse

f   Musaraignes d'Amérique

i   Toporagni americani

589   *Blarina brevicauda*
      Soricomorpha - Soricidae

e   Northern Short-tailed Shrew; Short-tailed Shrew (NA); Short-tailed Mole Shrew; American Short-tailed Shrew; Greater North American Short-tailed Shrew

d   Nördliche Kurzschwanzspitzmaus; Kurzschwanzspitzmaus

f   Grande musaraigne; Grande musaraigne à queue courte

i   Toporagno americano dalla coda
      corta

590 Blarina brevicauda telmalestes
      Soricomorpha - Soricidae
e   Dismal Swamp Short-tailed Shrew

591 Blarina carolinensis
      Soricomorpha - Soricidae
e   Southern Short-tailed Shrew; Swamp
      Short-tailed Shrew
d   Kleine Wasserspitzmaus;
      Sumpfspitzmaus; Südliche
      Kurzschwanzspitzmaus

592 Blarina hylophaga
      Soricomorpha - Soricidae
e   Elliot's Short-tailed Shrew
d   Elliot-Kurzschwanzspitzmaus

593 Blarinella
      Soricomorpha - Soricidae
e   Asiatic Short-tailed Shrews
d   Asiatische Kurzschwanzspitzmäuse
i   Toporagni asiatici

594 Blarinella quadricauda
      Soricomorpha - Soricidae
e   Short-tailed Moupin Shrew; Asiatic
      Short-tailed Shrew; Chinese Short-
      tailed Shrew; Sichuan Short-tailed
      Shrew
d   Asiatische Kurzschwanzspitzmaus

595 Blarinella wardi
      Soricomorpha - Soricidae
e   Ward's Short-tailed Shrew; Burmese
      Short-tailed Shrew

596 Blarinomys
      Rodentia - Muridae
e   Brazilian Shrewmice
d   Brasilianische Spitzmausratten
i   Toporagni brasiliani

597 Blarinomys breviceps

      Rodentia - Muridae
e   Brazilian Shrewmouse
d   Wühlhamster; Brasilianische
      Spitzmausratte
i   Toporagno brasiliano

598 Blastocerus
      Artiodactyla - Cervidae
e   Marsh Deer ; Swamp Deer
d   Sumpfhirsche
f   Cerfs marécageux
i   Cervi delle paludi

599 Blastocerus dichotomus
      Artiodactyla - Cervidae
e   Marsh Deer; Swamp Deer
d   Sumpfhirsch
f   Cerf marécageux; Cerf des marais
i   Cervo delle paludi

600 Bolomys
      Rodentia - Muridae
e   Bolo Mice
d   Bolomäuse
f   Souris

601 Bolomys amoenus
      Rodentia - Muridae
e   Pleasant Bolo Mouse

      Bolomys arviculoides syn. *B.
      lasiurus* q.v.

602 Bolomys lactens
      Rodentia - Muridae
e   Rufous-bellied Bolo Mouse

603 Bolomys lasiurus
      Rodentia - Muridae
e   Hairy-tailed Bolo Mouse; South
      American Field Mouse

604 Bolomys obscurus
      Rodentia - Muridae
e   Dark Bolo Mouse

605 **_Bolomys punctulatus_**
Rodentia - Muridae
e   Spotted Bolo Mouse

606 **_Bolomys temchuki_**
Rodentia - Muridae
e   Temchuk's Bolo Mouse

607 **_Boneia_**
Chiroptera - Pteropodidae
e   Manado Fruit Bats

608 **_Boneia bidens_**
Chiroptera - Pteropodidae
e   Manado Fruit Bat; Greater Sulawesi
Rousette

609 **_Boocercus_**
Artiodactyla - Bovidae
e   Bongos
d   Bongos
f   Antilopes bongo
i   Bongo

610 **_Boocercus eurycerus_**
Artiodactyla - Bovidae
e   Bongo
d   Zentralafrikanischer Bongo;
Westafrikanischer Bongo; Bongo
f   Bongo; Antilope bongo

611 **_Boocercus eurycerus eurycerus_**
Artiodactyla - Bovidae
e   Western Bongo
d   Westlicher Bongo
f   Bongo de l'Ouest

612 **_Boocercus eurycerus isaaci_**
Artiodactyla - Bovidae
e   Eastern Bongo; Kenya Bongo;
Mountain Bongo
d   Gebirge Bongo; Östlicher Bongo
f   Bongo de l'Est

613 **_Bos_**

Artiodactyla - Bovidae
e   True Cattle; Oxen
d   Echte Rinder; Eigentliche Rinder
f   Boeufs
i   Bovini

614 **_Bos frontalis_**
Artiodactyla - Bovidae
e   Gaur; Seladang; Gayal
d   Gaur
f   Gayal
i   Gaur

615 **_Bos frontalis frontalis_**
Artiodactyla - Bovidae
e   Domestic Gayal
d   Gayal

616 **_Bos frontalis gaurus_**
Artiodactyla - Bovidae
e   Indian Gaur
d   Vorderindien-Gaur
f   Gaur
i   Gaur ; Bisonte indiano; Seladang

617 **_Bos frontalis hubbacki_**
Artiodactyla - Bovidae
e   Malayan Gaur
d   Malaya-Gaur

618 **_Bos frontalis readei_**
Artiodactyla - Bovidae
e   Indochinese Gaur
d   Hinterindien-Gaur

619 **_Bos grunniens_**
Artiodactyla - Bovidae
e   Domestic Yak
d   Grunzochse; Yak
f   Yack
i   Yak selvatico

620 **_Bos grunniens grunniens_**
Artiodactyla - Bovidae

e   Yak
d   Hausyak
f   Yack domestique

621   *Bos indicus*
     Artiodactyla - Bovidae
e   Zebu; Humped Cattle
d   Indischer Zebu; Buckelrind;
     Indisches Hausrind
f   Zébu
i   Zebù

622   *Bos javanicus*
     Artiodactyla - Bovidae
e   Banteng; Bali Cattle
d   Banteng; Rotrind; Bali-Rind
f   Banteng
i   Banteng

623   *Bos javanicus birmanius*
     Artiodactyla - Bovidae
e   Burmese Banteng
d   Birma-Banteng

624   *Bos javanicus domesticus*
     Artiodactyla - Bovidae
e   Domestic Bali Cattle
d   Bali-Hausrind
f   Banteng domestique
i   Banteng domestico

625   *Bos javanicus javanicus*
     Artiodactyla - Bovidae
e   Javan Banteng
d   Java-Banteng
i   Banteng di Giava

626   *Bos javanicus lowi*
     Artiodactyla - Bovidae
e   Borneo Banteng
d   Borneo-Banteng
i   Banteng di Borneo

627   *Bos mutus*

     Artiodactyla - Bovidae
e   Wild Yak
d   Wildyak; Wildjak
f   Yack sauvage
i   Yak

628   *Bos sauveli*
     Artiodactyla - Bovidae
e   Cambodian Forest Ox; Kouprey
d   Graurind; Kouprey
f   Kouprey
i   Couprey

629   *Bos taurus*
     Artiodactyla - Bovidae
e   Domestic Cattle; Cattle
d   Ur; Auerochse; Hausrind
f   Aurochs
i   Bue domestico

630   *Boselaphus*
     Artiodactyla - Bovidae
e   Nilgai ; Nylghau ; Bush Cows;
     Bluebucks; Blue Bulls ; Nilghau
d   Nilgau-Antilopen
f   Nilgauts
i   Tori blu

631   *Boselaphus tragocamelus*
     Artiodactyla - Bovidae
e   Nilgai; Nilghau; Bush Cow;
     Bluebuck; Blue Bull ; Nylghau
d   Blaubock; Nilgau-Anantilope
f   Nilgaut
i   Toro blu; Nilgau

632   Bovidae
     Artiodactyla
e   Hollow-horned Ruminants; Horned
     Ungulates; Cattle, Antelopes etc.;
     Cattle, Antelopes, Sheep and Goats
d   Horntiere; Hornträger; Hohlhörner;
     Rinder; Rinderartige
f   Bovidés; Boeufs
i   Antilopi, Bufali, Caprini etc.; Bovidi

633 **Brachiones**
Rodentia - Muridae
e Przewalski's Gerbils
d Przewalski-Wüstenmäuse;
Przewalski-Rennmäuse
f Gerbilles de Przewalski
i Gerbilli di Przewalski

634 **Brachiones przewalskii**
Rodentia - Muridae
e Przewalski's Gerbil; Gobi Short-
eared Gerbil
d Przewalski-Wüstenmaus; Przewalski-
Rennmaus; Przewalski-Renratte
f Gerbille de Przewalski
i Gerbil di Przewalski

635 **Brachylagus**
Lagomorpha - Leporide
e Pygmy Rabbits
d Zwergkaninchen
f Lapins pygmées de l'Idaho
i Conigli pimei

636 **Brachylagus idahoensis**
Lagomorpha - Leporide
e Pygmy Rabbit
d Zwergkaninchen
f Lapin pygmée de l'Idaho
i Coniglio pigmeo

637 **Brachyphylla**
Chiroptera - Phyllostomidae
e West Indian Fruit-eating Bats
d Antillen-Fruchtvampire
f Brachyphylles

638 **Brachyphylla cavernarum**
Chiroptera - Phyllostomidae
e St. Vincent Fruit-eating Bat;
Antillean Fruit-eating Bat; Lesser
Antillean Fruit-eating Bat
d Antillen-Fruchtvampir
f Brachyphylle des cavernes; Chauve-
souris à tête de cochon; Guimbo

639 **Brachyphylla nana**
Chiroptera - Phyllostomidae
e Cuban Fruit-eating Bat; Greater
Antillean Fruit-eating Bat; Antillean
Nectar Bat

Brachyphylla pumila syn. *B. nana*
q.v.

640 **Brachytarsomys**
Rodentia - Muridae
e White-tailed Rats
d Kurzfußinselratten
f Souris aux pattes courtes

641 **Brachytarsomys albicauda**
Rodentia - Muridae
e White-tailed Rat
d Kurzfußinselratte
f Souris aux pattes courtes

642 **Brachyteles**
Primates - Atelidae
e Woolly Spider Monkeys; Muriquis
d Spinnenaffen
f Brachytèles
i Muriki

643 **Brachyteles arachnoides**
Primates - Atelidae
e Woolly Spider Monkey; Southern
Muriqui
d Spinnenafffe; Muriki
f Brachytèle araigné; Atèle araigné;
Singe-araigné laineux; Muriqui
i Murichi; Murichi meridionale;
Scimmia ragno lanosa; Muriqui
meridionale

644 **Brachyteles hypoxanthus**
Primates - Atelidae
e Northern Muriqui
f Atèle arachnoïde
i Murichi settentrionale; Muriquí ;
Muriqui settentrionale

645   Brachyuromys
Rodentia - Muridae
e   Short-tailed Rats; Malagasy Voles
d   Madagaskar-Inselratten;
Madagaskar-Kurzschwanzratten

646   Brachyuromys betsileoensis
Rodentia - Muridae
e   Betsileo Short-tailed Rat

647   Brachyuromys ramirohitra
Rodentia - Muridae
e   Gregarious Short-tailed Rat

648   Bradypodidae
Pilosa
e   Three-toed Sloths; Tree Sloths
d   Dreifingerfaultiere;
Dreizehenfaultiere; Faultiere
f   Bradypodidés
i   Bradipodidi

649   Bradypus
Pilosa - Bradypodidae
e   Three-toed Sloths; Maned Sloths
d   Dreizehenfaultiere;
Dreifingerfaultiere
f   Bradypes; Aïs
i   Bradipi

Bradypus boliviensis syn. *B. variegatus* q.v.

Bradypus griseus syn. *B. variegatus* q.v.

Bradypus infuscatus syn. *B. variegatus* q.v.

650   Bradypus pygmaeus
Pilosa - Bradypodidae
e   Pygmy Three-toed Sloth
i   Bradipo pigmeo

651   Bradypus torquatus

Pilosa - Bradypodidae
e   Necklace Sloth; Maned Sloth; Maned Three-toed Sloth; Collared Sloth
d   Kragenfaultier
f   Bradype à collier
i   Bradypo dal collare

652   Bradypus tridactylus
Pilosa - Bradypodidae
e   Ai; Three-toed Sloth; Pale-throated Sloth; Pale-throated Three-toed Sloth
d   Dreizehenfaultier; Dreifingerfaultier
f   Aï
i   Bradipo tridattilo

653   Bradypus variegatus
Pilosa - Bradypodidae
e   Brown-throated Sloth; Grey Three-toed Sloth; Brown-throated Three-toed Sloth; Bolivian Three-toed Sloth
d   Bolivianisches Dreizehenfaultier; Braunkehldreifingerfaultier
f   Aï paresseux; Aï de Bolivie; Tridactyle de Bolivie
i   Bradipo boliviano; Bradipo variegato; Bradipo della Bolivia

654   Brucepattersonius
Rodentia - Muridae
e   Brucies
d   Brucies
i   Brucies

655   Brucepattersonius albinasus
Rodentia - Muridae
e   White-nosed Brucie

656   Brucepattersonius griserufescens
Rodentia - Muridae
e   Grey-bellied Brucie

657   Brucepattersonius guarani
Rodentia - Muridae
e   Guarani Brucie

658 **Brucepattersonius igniventris**
Rodentia - Muridae
e Red-bellied Brucie

659 **Brucepattersonius iheringi**
Rodentia - Muridae
e Ihering's Brucie; Iherings Hocicudo

660 **Brucepattersonius misiones**
Rodentia - Muridae
e Misiones Brucie

661 **Brucepattersonius paradisus**
Rodentia - Muridae
e Beautiful Brucie

662 **Brucepattersonius soricinus**
Rodentia - Muridae
e Soricine Brucie

663 **Bubalus**
Artiodactyla - Bovidae
e Asiatic Buffalos; Water Buffalos
d Asiatische Wasserbüffel ;
Wasserbüffel ; Büffel ; Asiatische
Büffel
f Buffles d'Asie
i Bufali

664 **Bubalus arnee**
Artiodactyla - Bovidae
e Wild Water Buffalo; Arna
d Wilder Wasserbüffel
f Buffle domestique
i Bufalo indiano; Bufalo asiatico;
Bufalo d'acqua; Bufalo acquatico

665 **Bubalus bubalis**
Artiodactyla - Bovidae
e Water Buffalo; Asiatic Water
Buffalo; Indian Buffalo; Asian Water
Buffalo; Domestic Water Buffalo
d Asiatischer Büffel; Wasserbüffel
f Buffle indien; Buffle de l'Inde; Buffle
asiatique; Buffle d'Asie

i Bufalo domestico

666 **Bubalus bubalis bubalis**
Artiodactyla - Bovidae
e Swamp Buffalo; Wild Buffalo;
Indian Water Buffalo
d Hausbüffel
f Buffle des marais

667 **Bubalus bubalis carabanensis**
Artiodactyla - Bovidae
e River Buffalo
f Buffle des rivières

668 **Bubalus bubalis hosei**
Artiodactyla - Bovidae
e Bornean Buffalo
d Borneo-Wasserbüffel
f Buffle de Bornéo

669 **Bubalus bubalis migona**
Artiodactyla - Bovidae
e Sri Lanka Water Buffalo
d Ceylon-Wasserbüffel
f Buffle de Ceylan

670 **Budorcas**
Artiodactyla - Bovidae
e Takins
d Gnuziegen; Takine; Rindergemsen
f Takins
i Takin

671 **Budorcas taxicolor**
Artiodactyla - Bovidae
e Takin
d Takin
f Takin
i Takin

672 **Budorcas taxicolor bedfordi**
Artiodactyla - Bovidae
e Shensi Takin; Golden Takin

d Schensi-Takin
f Takin doré

673 *Budorcas taxicolor taxicolor*
Artiodactyla - Bovidae
e Mishmi Takin
d Mishmi-Takin
f Takin mishmi

674 *Budorcas taxicolor tibetani*
Artiodactyla - Bovidae
e Tibetan Takin; Sichuan Takin
d Szetschuan-Takin
f Takin asiatique; Takin de Sichuan;
Takin du Tibet

675 *Bullimus*
Rodentia - Muridae
e Philippine Rats; Moss Mice
d Philippinen-Moosmäuse

676 *Bullimus bagobus*
Rodentia - Muridae
e Bagobo Rat; Large Mindanao Forest
Rat

677 *Bullimus gamay*
Rodentia - Muridae
e Camiguin Forest Rat

678 *Bullimus luzonicus*
Rodentia - Muridae
e Luzon Forest Rat; Large Luzon
Forest Rat

679 *Bunolagus*
Lagomorpha - Leporidae
e Riverine Rabbits
d Buschmannhasen
f Lapins hottentots
i Lepri di fiume; Conigli di fiume

680 *Bunolagus monticularis*
Lagomorpha - Leporidae
e Bushman Hare; Riverine Rabbit

d Buschmannhase
f Lapin hottentot
i Lepre di fiume; Coniglio di fiume

681 *Bunomys*
Rodentia - Muridae
e Hill Rats
d Sulawesi-Bergratten

Bunomys adspersus syn. *B. andrewsi*
q.v.

682 *Bunomys andrewsi*
Rodentia - Muridae
e Andrews's Hill Rat; Andrews's
Bunomys; Andrews's Shrew Rat

683 *Bunomys chrysocomus*
Rodentia - Muridae
e Yellow-haired Hill Rat; Golden-
capped Bunomys

684 *Bunomys coelestis*
Rodentia - Muridae
e Heavenly Hill Rat

685 *Bunomys fratrorum*
Rodentia - Muridae
e Fraternal Hill Rat; Lesser Hill Rat

686 *Bunomys heinrichi*
Rodentia - Muridae
e Heinrich's Hill Rat; Heinrich's
Bunomys
d Heinrichs Mausratte; Heinrichs Ratte

687 *Bunomys penitus*
Rodentia - Muridae
e Inland Hill Rat; Summit Shrewrat

688 *Bunomys prolatus*
Rodentia - Muridae
e Long-headed Hill Rat

Bunopithecus syn. *Hylobates* q.v.

Burmeisteria syn. *Chlamyphorus* q.v.

689 Burramyidae
Diprotodontia
e Pygmy Possums
d Bilchbeutler; Schlafbeutler; Bilch-
und Zwerggleitbeutler
f Burramyidés; Possums pygmées
i Opossum pigmei; Burramidi; Possum
pigmei

690 *Burramys*
Diprotodontia - Burramyidae
e Mountain Pygmy Possums
d Bergbilchbeutler
f Opossums nains des montagnes
i Opossum delle montagne; Possum
delle montagne

691 *Burramys parvus*
Diprotodontia - Burramyidae
e Mountain Pygmy Possum; Broom's
Pygmy Possum; Pygmy Possum
(ANZ)
d Bergbilchbeutler
f Opossum nain des montagnes
i Opossum pigmeo delle montagne ;
Possum pigmeo delle montagne

C

692 Cabassous
Cingulata - Dasypodidae
e Eleven-banded Armadillos; Naked-
tailed Armadillos
d Nacktschwanzgürteltiere
f Tatous à onze bandes
i Armadilli dalla coda nuda

693 Cabassous centralis
Cingulata - Dasypodidae
e Naked-tailed Armadillo; Northern
Naked-tail Armadillo; Five-toed
Armadillo; Central American Five-
toed Armadillo
d Mittelamerikanisches
Nacktschwanzgürteltier; Nördliches
Nacktschwanzgürteltier
f Tatou épineux; Tatou à queue nue du
Nord
i Armadillo dalla coda nuda del nord;
Armadillo dalla coda molle
settentrionale

694 Cabassous chacoensis
Cingulata - Dasypodidae
e Chacoan Naked-tailed Armadillo
d Kleines Nacktschwanzgürteltier;
Chaco-Nacktschwanzgürteltier
f Tatou à queue nue du Chaco; Tatou
de Paraguay

695 Cabassous tatouay
Cingulata - Dasypodidae
e Greater Naked-tailed Armadillo;
Yellow Armadillo; Six-banded
Armadillo
d Großes Nacktschwanzgürteltier
f Tatou à queue nue
i Grande armadillo dalla coda nuda;

Tatouay

696 Cabassous unicinctus
Cingulata - Dasypodidae
e Eleven-banded Armadillo; Twelve-
banded Armadillo; Broad-banded
Armadillo; Southern Naked-tailed
Armadillo
d Nacktschwanzgürteltier
f Tatou à onze bandes

Cabreramops syn. *Molossops* q.v.

697 Cacajao
Primates - Atelidae
e Uakaris; Wakaris ; Ouakaris
d Kurzschwanzaffen; Uakaris
f Ouakaris
i Uacari; Uakari

698 Cacajao calvus
Primates - Atelidae
e Bald Uakari; Bald-headed Uakari;
White Uakari
d Scharlachgesicht; Rotes Uakari;
Uakari
f Ouakari chauve
i Uacari calvo; Cacajao calvo; Uakari
calvo

699 Cacajao calvus calvus
Primates - Atelidae
e White Bald-headed Uakari; White
Uakari
d Weißer Uakari
i Uacari bianco; Uakari bianco

700 Cacajao calvus novaesi
Primates - Atelidae
e Novaes's Bald-headed Uakari
i Uacari dorato; Uakari dorato

701 Cacajao calvus rubicundus
Primates - Atelidae
e Red Uakari; Red Bald-headed Uakari

d Roter Uakari; Golduakari

f Ouakari rubicond

i Macaco rubicondo; Uacari rosso; Uakari rosso

702 *Cacajao melanocephalus*
Primates - Atelidae

e Black-headed Uakari; Black Uakari; Golden-headed Uakari; Humboldt's Black-headed Uakari; Black-headed Uacari; Black Uacari; Golden-headed Uacari; Humboldt's Black-headed Uacari

d Schwarzkopfuakari; Schwarzer Uakari

f Ouakari à tète noire

i Uacari dalla faccia nera

703 *Cacajao melanocephalus ouakari*
Primates - Atelidae

e Golden-backed Uakari

704 *Caenolestes*
Paucituberculata - Caenolestidae

e Ecuadorian Rat Opossums

d Ekuador-Opossummäuse; Mausopossums

705 *Caenolestes caniventer*
Paucituberculata - Caenolestidae

e Northern Shrew Opossum; Grey-bellied Shrew Opossum

d Graubäuchige Opossummaus

706 *Caenolestes condorensis*
Paucituberculata - Caenolestidae

e Condor Shrew Opossum

707 *Caenolestes convelatus*
Paucituberculata - Caenolestidae

e Blackish Shrew Opossum

d Schwarze Opossummaus

708 *Caenolestes fuliginosus*
Paucituberculata - Caenolestidae

e Ecuadorian Rat Opossum; Silky

Shrew Opossum

d Ekuador-Opossummaus

709 Caenolestidae
Paucituberculata

e Shrew Opossums; Opossum Rats; Caenolestids; Rat Opossums

d Opossummäuse; Mausopossums

f Cénolestidés

i Cenolestidi

Calcochloris syn. *Amblysomus* q.v.

710 *Callicebus*
Primates - Atelidae

e Titi Monkeys; Titis

d Springaffen; Titis

f Titis; Callicèbes; Singes titis

i Callicebi

711 *Callicebus baptista*
Primates - Atelidae

e Baptista Lake Titi

i Callicebo del lago Baptista

712 *Callicebus barbarabrownae*
Primates - Atelidae

e Blond Titi

713 *Callicebus bernhardi*
Primates - Atelidae

e Prince Bernhard's Titi

d Prinz Bernard-Springaffe

f Singe titi de Prince Bernhard

i Callicebo del principe Bernardo

714 *Callicebus brunneus*
Primates - Atelidae

e Brown Titi; Brown-masked Titi Monkey; Brown Titi Monkey

f Titi brun

i Callicebo bruno; Callicebo marrone

715   *Callicebus caligatus*
          Primates - Atelidae
   e   Booted Titi; Red-bellied Titi
          Monkey; Booted Titi Monkey
   i   Callicebo dal ventre rosso

716   *Callicebus cinerascens*
          Primates - Atelidae
   e   Ashy Titi; Ashy Titi Monkey; Ashy-
          black Titi; Ashy-grey Titi
   i   Callicebo cenerino

717   *Callicebus coimbrai*
          Primates - Atelidae
   e   Coimbra-Filho's Titi

718   *Callicebus cupreus*
          Primates - Atelidae
   e   Coppery Titi; Coppery Titi Monkey
   d   Roter Springaffe; Kupferfarbener
          Springaffe
   f   Titi cuivreux; Callicèbe rouge
   i   Callicebo rosso

719   *Callicebus cupreus cupreus*
          Primates - Atelidae
   e   Red Titi Monkey
   d   Sumpfspringaffe
   f   Callicèbe rouge

720   *Callicebus discolor*
          Primates - Atelidae
   e   Double-browed Titi

721   *Callicebus donacophilus*
          Primates - Atelidae
   e   Bolivian Titi; Bolivian Titi Monkey;
          White-eared Titi
   i   Callicebo della Bolivia

722   *Callicebus dubius*
          Primates - Atelidae
   e   Dubious Titi; Dubious Titi Monkey;
          Herschkovitz's Titi
   i   Callicebo misterioso

723   *Callicebus hoffmannsi*
          Primates - Atelidae
   e   Hoffmanns's Titi; Hoffmanns's Titi
          Monkey
   d   Hoffmanns-Springaffe
   i   Callicebo di Hoffmanns

724   *Callicebus lucifer*
          Primates - Atelidae
   e   Lucifer Titi

725   *Callicebus lugens*
          Primates - Atelidae
   e   Mourning Titi

726   *Callicebus medemi*
          Primates - Atelidae
   e   Medem's Titi

727   *Callicebus melanochir*
          Primates - Atelidae
   e   Northern Masked Titi; Black-handed
          Titi

728   *Callicebus modestus*
          Primates - Atelidae
   e   Modest Titi; Lönnberg's Titi
          Monkey; Modest Titi Monkey; Rio
          Beni Titi; Lönnberg's Tamarin
   f   Tamarin de Lönnberg
   i   Callicebo di Lonnberg; Callicebo del
          Rio Beni

729   *Callicebus moloch*
          Primates - Atelidae
   e   Dusky Titi; Orabassu Titi; Dusky Titi
          Monkey; Orabazu
   d   Grauer Springaffe; Graues
          Springäffchen
   f   Callicèbe orabussu; Titi molok;
          Callicèbe gris
   i   Callicebo grigio

730   *Callicebus nigrifrons*
          Primates - Atelidae
   e   Black-fronted Titi

731 *Callicebus oenanthe*
Primates - Atelidae
e Andean Titi; Andean Titi Monkey; Peruvian Mountain Titi; Rio Mayo Titi
d Anden-Springaffe
i Callicebo delle Ande; Callicebo del Rio Mayo

732 *Callicebus olallae*
Primates - Atelidae
e Olalla's Titi; Beni Titi Monkey; Olalla's Titi Monkey; Olalla Brothers Titi
i Callicebo del Rio Beni

733 *Callicebus pallescens*
Primates - Atelidae
e Pallid Titi

734 *Callicebus personatus*
Primates - Atelidae
e Masked Titi; Masked Titi Monkey
d Schwarzköpfiger Springaffe; Maskenspringaffe; Maskenspringäffchen
f Callicèbe à masque; Titi à masque
i Callicebo mascherato; Callicebo atlantico

735 *Callicebus purinus*
Primates - Atelidae
e Red-crowned Titi

736 *Callicebus regulus*
Primates - Atelidae
e Kinglet Titi

737 *Callicebus stephennashi*
Primates - Atelidae
e Stephen Nash's Titi
f Singe titi de Stephen Nash; Singe de Stephen Nash

738 *Callicebus torquatus*
Primates - Atelidae
e Yellow-handed Titi; Widow Monkey; White-handed Titi; Yellow-handed Titi Monkey; Widow Titi; Collared Titi
d Witwenaffe
f Callicèbe à fraise; Titi à collier
i Callicebo dal collare

739 *Callimico*
Primates - Callitrichidae
e Goeldi's Marmosets; Goeldi's Monkeys; Goeldi's Tamarins; Spring Tamarins
d Springtamarins
f Tamarins de Goeldi
i Callimico

740 *Callimico goeldii*
Primates - Callitrichidae
e Goeldi's Marmoset; Goeldi's Monkey; Goeldi's Tamarin; Spring Tamarin; Callimico
d Springtamarin; Brüllaffe; Goeldi-Tamarin
f Tamarin de Goeldi
i Callimico di Goeldi; Calimico

741 *Callithrix*
Primates - Callitrichidae
e Marmosets; True Marmosets; Short-tusked Marmosets; Ouistitis
d Marmorsetten; Büschelaffen
f Ouistitis; Callitriches

742 *Callithrix acariensis*
Primates - Callitrichidae
e Rio Acari Marmoset

743 *Callithrix argentata*
Primates - Callitrichidae
e Bare-eared Marmoset; Bare Ear Marmoset; Silvery Marmoset
d Silberäffchen; Schwarzschwänziges Silberäffchen; Silberaffe
f Ouistiti argenté
i Uistitì argentato; Apale argentata

744   *Callithrix aurita*
Primates - Callitrichidae
- *e*   White-eared Marmoset; White-shouldered Marmoset; Santarem Marmoset; Buffy Tufted-ear Marmoset; Buffy-tufted Marmoset
- *d*   Weißohrseidenäffchen; Weißohrseidenaffe
- *f*   Ouistiti oreillard
- *i*   Uistitì dalle orecchie bianche

745   *Callithrix chrysoleuca*
Primates - Callitrichidae
- *e*   Gold-and-white Marmoset; Golden-white Tassel-eared Marmoset
- *i*   Uistitì dalla coda gialla

746   *Callithrix emiliae*
Primates - Callitrichidae
- *e*   Snethlage's Marmoset; Emilia's Marmoset
- *i*   Uistsiti di Snethlage

747   *Callithrix flaviceps*
Primates - Callitrichidae
- *e*   Buffy-headed Marmoset; Buff-headed Marmoset; White-headed Marmoset
- *d*   Gelbkopfbüscheläffchen; Gelbkopfbuschelafffe
- *f*   Ouistiti à tête jaune
- *i*   Uistitì dalla testa gialla; Uistitì testa gialla

748   *Callithrix geoffroyi*
Primates - Callitrichidae
- *e*   Geoffroy's Marmoset; Geoffroy's Tufted-ear Marmoset; White-fronted Marmoset; White-faced Marmoset
- *d*   Weißgesichtseidenaffe
- *f*   Ouistiti de Geoffroy
- *i*   Uistitì di Geoffroy; Uistiti a fronte bianca

749   *Callithrix humeralifer*
Primates - Callitrichidae
- *e*   Tassel-eared Marmoset; Tasselear Marmoset; Yellow-legged Marmoset; Santarem Marmoset; Golden Marmoset; Silky Marmoset
- *d*   Weißschulterseidenäffchen; Langohrseidenäffchen; Weißschulterseidenaffe; Gelbfußäffchen; Gelbschwanzäffchen
- *f*   Ouistiti à camail; Ouistiti aux pieds jaunes
- *i*   Uistitì dalle spalle bianche; Uistitì di Santarem

750   *Callithrix humilis*
Primates - Callitrichidae
- *e*   Black-crowned Dwarf Marmoset; Dwarf Marmoset
- *d*   Krallenäffchen
- *i*   Uistitì nano; Uistitì nano dalla corona nera

751   *Callithrix intermedia*
Primates - Callitrichidae
- *e*   Aripuanà Marmoset; Hershkovitz's Marmoset
- *i*   Uistitì dell'Aripuanà

752   *Callithrix jacchus*
Primates - Callitrichidae
- *e*   White-tufted Marmoset; Common Marmoset; Ouistiti; Wistit; Tufted-ear Marmoset; Tuft-eared Marmoset
- *d*   Weißbüscheläffchen; Weißbüschelaffe
- *f*   Ouistiti sagouin; Ouistiti à toupet blanc
- *i*   Uistitì dai pennacchi bianchi; Scimmia grigia

753   *Callithrix kuhli*
Primates - Callitrichidae
- *e*   Wied's Black Tufted-ear Marmoset; Wied's Marmoset; Wied's Black-tufted Marmoset; Wied's Tufted-ear Marmoset
- *d*   Schwarzohrbüschelaffe
- *i*   Uistitì di Wied

754 *Callithrix leucippe*
Primates - Callitrichidae
e White Marmoset

755 *Callithrix manicorensis*
Primates - Callitrichidae
e Rio Manicore Marmoset; Manicore Marmoset

756 *Callithrix marcai*
Primates - Callitrichidae
e Marca's Marmoset
i Uistitì di Marca

757 *Callithrix mauesi*
Primates - Callitrichidae
e Maues Marmoset; Rio Maues Marmoset
i Uistitì del fiume Maués

758 *Callithrix melanura*
Primates - Callitrichidae
e Black-tailed Marmoset
f Ouistiti mélanure
i Uistitì dalla coda nera

759 *Callithrix nigriceps*
Primates - Callitrichidae
e Black-headed Marmoset
i Uistitì dalla testa nera

760 *Callithrix penicillata*
Primates - Callitrichidae
e Black-pencilled Marmoset; Black-eared Marmoset; Black-tufted Marmoset; Black Tufted-ear Marmoset; Black-plumed Marmoset
d Schwarzäffchen; Schwarzpinselaffe; Schwarzbüscheläffchen
f Ouistiti à pinceau noir
i Uistitì dai pennacchi neri

761 *Callithrix pygmaea*
Primates - Callitrichidae
e Pygmy Marmoset

d Zwergseidenäffchen
f Ouistiti nain; Ouistiti mignon
i Uistitì pigmeo

762 *Callithrix pygmaea niveiventris*
Primates - Callitrichidae
e White-bellied Pygmy Marmoset
d Weißbauchzwergseidenaffe

763 *Callithrix pygmea pygmea*
Primates - Callitrichidae
e Western Pygmy Marmoset
d Westliches Zwergseidenäffchen

Callithrix santaremensis syn. *C. humeralifer* q.v.

764 *Callithrix sateri*
Primates - Callitrichidae
e Sateré Marmoset; Sateré-Maues Marmoset
i Uistitì di Sateré

765 **Callitrichidae**
Primates
e Marmosets
d Krallenaffen; Krallenäffchen
f Callitrichidés
i Scimmie orso; Uistitì; Tamarini; Callitricidi

766 *Callorhinus*
Carnivora - Otariidae
e Northern Fur Seals
d Nordische Seebären
f Otaries à fourrure du Nord
i Callorini

767 *Callorhinus ursinus*
Carnivora - Otariidae
e Sea Bear; Sea Cat; Northern Fur Seal; Pribilof Fur Seal; Alaska Fur Seal
d Nördlicher Seebär; Bärenrobbe; Seebär; Nördliche Bärenrobbe

f   Otarie à fourrure du Nord
i   Callorino dell'Alaska; Foca orsina;
       Otaria orsino; Callorhino

768   *Callosciurus*
         Rodentia - Sciuridae
e   Beautiful Squirrels; Common
       Oriental Squirrels; Oriental Tree
       Squirrels; Tricoloured Squirrels
d   Schönhörnchen; Eigentliche
       Schönhörnchen
f   Écureuils de Bornéo

769   *Callosciurus adamsi*
         Rodentia - Sciuridae
e   Ear-spotted Squirrel; Ear-spot
       Squirrel; Bornean Ear-spot Squirrel
d   Ohrfleckhörnchen

770   *Callosciurus albescens*
         Rodentia - Sciuridae
e   Kloss's Squirrel; Pale Squirrel
d   Kloss-Hörnchen

       Callosciurus albicauda syn.
       Sundasciurus moellendorffi q.v.

771   *Callosciurus baluensis*
         Rodentia - Sciuridae
e   Kinabalu Squirrel
d   Kinabalu-Hörnchen

       Callosciurus brookei syn.
       Sundasciurus brookei q.v.

772   *Callosciurus caniceps*
         Rodentia - Sciuridae
e   Golden-backed Squirrel; Grey-
       bellied Squirrel
d   Graubauchhörnchen

773   *Callosciurus erythraeus*
         Rodentia - Sciuridae
e   Pallas's Squirrel; Belly-banded
       Squirrel; Yellow-handed Squirrel;
       Red-bellied Squirrel

d   Pallas-Hörnchen; Pallas-
       Schönhörnchen
f   Écureuil à ventre rouge
i   Scoiattolo dal ventre rosso

       Callosciurus ferrugineus syn. *C.*
       finlaysoni q.v.

774   *Callosciurus finlaysoni*
         Rodentia - Sciuridae
e   Finlayson's Squirrel; Variable
       Squirrel
d   Finlayson-Schönhörnchen;
       Finlayson-Hörnchen
f   Écureil d'Indochine; Écureuil de
       Finlayson
i   Scoiattolo di Finlayson; Scoiattolo
       variabile; Scoiattolo variabile
       tailandese

       Callosciurus flavimanus syn. *C.*
       erythraeus q.v.

       Callosciurus hippurus syn.
       Sundasciurus hippurus q.v.

       Callosciurus hoogstraali syn.
       Sundasciurus hoogstraali q.v.

775   *Callosciurus inornatus*
         Rodentia - Sciuridae
e   Inornate Squirrel; Plain Squirrel;
       Laotian Squirrel
d   Einfarbiges Schönhörnchen

       Callosciurus jentinki syn.
       Sundasciurus jentinki q.v.

       Callosciurus juvencus syn.
       Sundasciurus juvencus q.v.

       Callosciurus lowii syn. *Sundasciurus*
       lowii q.v.

       Callosciurus macclellandi syn.
       Tamiops macclellandi q.v.

Callosciurus maritimus syn.
Tamiops maritimus q.v.

776 *Callosciurus melanogaster*
Rodentia - Sciuridae
e Mentawai Squirrel; Loga Squirrel
d Mentawai-Hörnchen

Callosciurus mindanenensis syn.
Sundasciurus mindanensis q.v.

Callosciurus moellendorffi syn.
Sundasciurus moellendorffi q.v.

777 *Callosciurus nigrovittatus*
Rodentia - Sciuridae
e Black-banded Squirrel; Malayan
Squirrel; Black-striped Squirrel
d Schwarzstreifenschönhörnchen

778 *Callosciurus notatus*
Rodentia - Sciuridae
e Tri-coloured Squirrel; Plantain
Squirrel
d Dreifarbenhörnchen;
Bananenhörnchen;
Plantagenhörnchen
f Écureuil à trois couleurs
i Scoiattolo marrone

779 *Callosciurus orestes*
Rodentia - Sciuridae
e Borneo Black-banded Squirrel;
Bornean Black-banded Squirrel
d Borneo-Schwarzbindenhörnchen

780 *Callosciurus phayrei*
Rodentia - Sciuridae
e Phayre's Squirrel
d Phayre-Hörnchen

Callosciurus philippinensis syn.
Sundasciurus philippinensis q.v.

781 *Callosciurus prevosti*
Rodentia - Sciuridae

e Prevost's Squirrel; Tri-coloured
Squirrel
d Prevost-Hörnchen; Flaggenhörnchen
f Écureuil à trois couleurs
i Scoiattolo di Prevost

782 *Callosciurus prevosti borneoensis*
Rodentia - Sciuridae
e Pontianak Prevost's Squirrel
d Pontianak-Prevost-Hörnchen

783 *Callosciurus prevosti carimonensis*
Rodentia - Sciuridae
e Carimon Prevost's Squirrel
d Karimon-Prevost-Hörnchen

784 *Callosciurus prevosti prevosti*
Rodentia - Sciuridae
e Malacca Prevost's Squirrel
d Malakka-Prevost-Hörnchen

785 *Callosciurus prevosti rafflesi*
Rodentia - Sciuridae
e Sumatran Prevost's Squirrel
d Sumatra-Prevost-Hörnchen

786 *Callosciurus pygerythrus*
Rodentia - Sciuridae
e Irrawaddy Squirrel; Hoary-bellied
Himalajan Squirrel
d Irawadi-Hörnchen

787 *Callosciurus quinquestriatus*
Rodentia - Sciuridae
e Anderson's Squirrel; Five-striped
Squirrel
d Anderson-Hörnchen

Callosciurus rabori syn.
Sundasciurus rabori q.v.

Callosciurus rodolphei syn. *Tamiops
rodolphei* q.v.

Callosciurus rubriventer syn.
Rubrisciurus rubriventer q.v.

Callosciurus samarensis syn.
Sundasciurus samarensis q.v.

Callosciurus simus syn. *Glyphotes*
simus q.v.

Callosciurus steeri syn.
Sundasciurus steerei q.v.

Callosciurus swinhoei syn. *Tamiops*
swinhoei

Callosciurus tenuis syn.
Sundasciurus tenuis q.v.

Callospermophilus syn.
Spermophilus q.v.

788 *Calomys*
  Rodentia - Muridae
 e Vesper Mice
 d Verspermäuse; Abendmäuse
 f Souris des champs

789 *Calomys boliviae*
  Rodentia - Muridae
 e Bolivian Vesper Mouse

790 *Calomys callidus*
  Rodentia - Muridae
 e Crafty Vesper Mouse

791 *Calomys callosus*
  Rodentia - Muridae
 e Large Vesper Mouse

792 *Calomys expulsus*
  Rodentia - Muridae
 e Rejected Vesper Mouse

793 *Calomys hummellincki*
  Rodentia - Muridae
 e Hummelinck's Vesper Mouse

794 *Calomys laucha*
  Rodentia - Muridae
 e Small Vesper Mouse

795 *Calomys lepidus*
  Rodentia - Muridae
 e Andean Vesper Mouse

796 *Calomys musculinus*
  Rodentia - Muridae
 e Drylands Vesper Mouse
 d Brasilianische Vespermaus

797 *Calomys sorellus*
  Rodentia - Muridae
 e Peruvian Vesper Mouse

798 *Calomys tener*
  Rodentia - Muridae
 e Delicate Vesper Mouse

799 *Calomys venustus*
  Rodentia - Muridae
 e Pretty Vesper Mouse

800 *Calomyscus*
  Rodentia - Muridae
 e Mouse-like Hamsters
 d Mäusehamster ; Maushamster ;
    Mausartiger Zwerghamster
 f Calomysques; Hamster souriciforme

801 *Calomyscus bailwardi*
  Rodentia - Muridae
 e Iranian Mouse-like Hamster
 d Iran-Maushamster
 f Calomysque

802 *Calomyscus baluchi*
  Rodentia - Muridae
 e Baluchi Mouse-like Hamster;
    Baluchistan Mouse-like Hamster
 d Belutschistan-Maushamster

803 **Calomyscus hotsoni**
  Rodentia - Muridae
e Hotson's Mouse-like Hamster
d Hotsons Maushamster

804 **Calomyscus mystax**
  Rodentia - Muridae
e Afghan Mouse-like Hamster
d Türkmenischer Maushamster

805 **Calomyscus tsolovi**
  Rodentia - Muridae
e Tsolov's Mouse-like Hamster
d Syrischer Maushamster

806 **Calomyscus urartensis**
  Rodentia - Muridae
e Urartsk Mouse-like Hamster
d Kaukasus-Maushamster

807 **Caloprymnus**
  Diprotodontia - Macropodidae
e Desert Rat Kangaroos; Plains Rat Kangaroos
d Nacktbrustkängurus
f Rats-kangourous du désert; Caloprymnes
i Ratti canguro campestre

808 **Caloprymnus campestris**
  Diprotodontia - Macropodidae
e Desert Rat Kangaroo; Plains Rat Kangaroo; Buff-nosed Rat Kangaroo; Plain-loving Jerboa Kangaroo; Rat Kangaroo (ANZ); Plains Kangaroo
d Nacktbrustkänguru; Nacktbrustrattenkänguru; Wüstenkänguru
f Rat-kangourou du désert
i Ratto canguro campestre

809 **Caluromys**
  Didelphimorphia - Didelphidae
e Woolly Opossums
d Wollbeutelratten; Wollige Opossums

f Opossums laineux
i Opossum lanosi; Possum lanosi

810 **Caluromys derbianus**
  Didelphimorphia - Didelphidae
e Central American Woolly Opossum; Derby's Woolly Opossum; Woolly Opossum
d Bindenwollbeutelratte; Derbys Wolliges Opossum
f Opossum laineux de Derby

811 **Caluromys lanatus**
  Didelphimorphia - Didelphidae
e Western Woolly Opossum; Ecuadorian Woolly Opossum; Common Woolly Opossum
d Ecuadorisches Wolliges Opossum
i Opossum lanoso; Possum lanoso

812 **Caluromys philander**
  Didelphimorphia - Didelphidae
e Philander Opossum; Bare-tailed Woolly Opossum; Lower Amazonian Opossum
d Gelbwollbeutelratte; Gelbe Wollbeutelratte; Bloßschwanz Opossum
i Opossum lanoso giallo; Possum lanoso giallo

813 **Caluromysiops**
  Didelphimorphia - Didelphidae
e Black-shouldered Opossums
d Wollbeutelratten; Bindenwollbeutelratten
f Opossums aux épaules noires

814 **Caluromysiops irrupta**
  Didelphimorphia - Didelphidae
e Black-shouldered Opossum
d Schwarzgeschultertes Opossum; Schwarzschulterbeutelratte
f Opossum aux épaules noires

Calyptophractus syn. *Chlamyphorus* q.v.

815 *Camelidae*
Artiodactyla
- *e* Camels; Camels and Llamas; Camels and Relatives; Camels and allies
- *d* Kamele; Kamelartige
- *f* Camélidés; Chameaux
- *i* Cammelli; Camelidi

816 *Camelus*
Artiodactyla - Camelidae
- *e* Camels
- *d* Kamele; Altwelt-Kamele
- *f* Chameaux
- *i* Cammelli

817 *Camelus bactrianus*
Artiodactyla - Camelidae
- *e* Two-humped Camel; Bactrian Camel
- *d* Zweihöckriges Kamel
- *f* Chameau de Bactriane; Chameau à deux bosses
- *i* Cammello a due gobbe; Cammello; Ca\mello con due gobbi

818 *Camelus bactrianus bactrianus*
Artiodactyla - Camelidae
- *e* Domestic Camel
- *d* Hauskamel
- *f* Chameau domestique
- *i* Cammello della Battriana

819 *Camelus dromedarius*
Artiodactyla - Camelidae
- *e* Dromedary; Arabian Camel; One-humped Camel; Dromedary Camel
- *d* Dromedar; Einhöckriges Kamel
- *f* Dromadaire
- *i* Dromedario

820 *Camelus ferus*
Artiodactyla - Camelidae
- *e* Wild Camel; Wild Two-humped Camel; Wild Bactrian Camel
- *d* Wildkamel; Trampeltier
- *f* Chameau sauvage de Tartarie;

Chameau sauvage à deux bosses; Chameau de Bactriane
- *i* Cammello selvaggio

821 **Canidae**
Carnivora
- *e* Canids; Dogs and Foxes; Dogs; Dogs, Wolves and Foxes
- *d* Hunde; Hundeartige; Hundeartige Raubtiere
- *f* Canidés
- *i* Canidi

822 *Canis*
Carnivora - Canidae
- *e* Dogs and Wolves; Dogs, Wolves, Coyotes and Jackals
- *d* Hunde und Wölfe; Wolfs-und Schakalartige
- *f* Chiens
- *i* Cani

823 *Canis adustus*
Carnivora - Canidae
- *e* Side-striped Jackal
- *d* Streifenschakal
- *f* Chacal rayé; Chacal aux flancs rayés
- *i* Sciacallo striato

824 *Canis aureus*
Carnivora - Canidae
- *e* Northern Jackal; Common Jackal; Golden Jackal; Jackal
- *d* Goldschakal; Goldwolf; Wolfsschakal; Schakal; Gemeiner Schakal; Turkestanischer Schakal
- *f* Chacal; Chacal commun; Chacal doré; Loup doré
- *i* Sciacallo dorato; Sciacallo

825 *Canis aureus indicus*
Carnivora - Canidae
- *e* Asiatic Jackal; Oriental Jackal
- *d* Himalaja-Goldschakal

826   *Canis aureus lanka*
Carnivora - Canidae
e   Ceylon Jackal
d   Ceylon-Goldschakal

827   *Canis familiaris*
Carnivora - Canidae
e   Domestic Dog; Feral Dog
d   Haushund
f   Chien domestique
i   Cane domestico

828   *Canis latrans*
Carnivora - Canidae
e   Prairie Wolf; Brush Wolf
d   Heuwolf; Koyote; Kojote
f   Coyote
i   Coyote; Lupo della prateria

829   *Canis latrans clepticus*
Carnivora - Canidae
e   Coyote

830   *Canis latrans nebracenis*
Carnivora - Canidae
e   Plains Coyote

831   *Canis lupus*
Carnivora - Canidae
e   Common Wolf; Grey Wolf; Wolf;
Timber Wolf
d   Wolf; Schwarzer Wolf; Europäisch-
asiatischer Wolf; Mittelrussischer
Waldwolf; Lobowolf; Tundrawolf
f   Loup commun; Loup vulgaire; Loup;
Loup gris
i   Lupo; Lupo comune

832   *Canis lupus albus*
Carnivora - Canidae
e   Eurasian Tundra Wolf; Russian
Tundra Wolf; White Tundra Wolf;
Tundra Wolf; White Wolf
d   Polarwolf; Weißer Wolf
f   Loup de la toundra eurasienne; Loup

de la toundra; Loup de Sibérie; Loup
blanc
i   Lupo della tundra dell'Eurasia

833   *Canis lupus alces*
Carnivora - Canidae
e   Kenai Peninsula Wolf; Kenai
Peninsula Grey Wolf
d   Kenai Halbinsel- Wolf; Kenai-Wolf
f   Loup de la presqu'île Kenai; Loup
commun de la presqu'île Kenai; Loup
vulgaire de la presqu'île Kenai; Loup
de Kenai

834   *Canis lupus arabs*
Carnivora - Canidae
e   Arabian Wolf
d   Arabischer Wolf
f   Loup d'Arabie
i   Lupo della penisola Arabica

835   *Canis lupus arctos*
Carnivora - Canidae
e   Melville Island Wolf; Arctic Wolf;
White Wolf
d   Melville-Island-Wolf; Arktischer
Wolf
f   Loup arctique; Loup de l'Arctique
i   Lupo bianco

836   *Canis lupus baileyi*
Carnivora - Canidae
e   Mexican Wolf; Mexican Grey Wolf
d   Mexikanischer Wolf
f   Loup du Mexique
i   Lupo del Messico

837   *Canis lupus beothucus*
Carnivora - Canidae
e   Newfoundland Wolf; Common
Newfoundland Wolf; Newfoundland
White Wolf; Newfoundland Grey
Wolf
d   Neufundland-Wolf
f   Loup de Terre-neuve; Loup commun
de Terre-neuve; Loup vulgaire de

Terre-neuve

i   Lupo di Terranova

838   *Canis lupus bernardi*
Carnivora - Canidae

e   Banks Island Tundra Wolf; Bernard's Wolf; Banks Island Wolf

d   Banks Island Tundrawolf; Bernards Wolf

f   Loup de la Terre Victoria; Loup de Bernard

839   *Canis lupus campestris*
Carnivora - Canidae

e   Steppe Wolf

d   Steppenwolf; Steppewolf

f   Loup des steppes

i   Lupo delle steppe

840   *Canis lupus chanco*
Carnivora - Canidae

e   Mongolian Wolf

d   Mongolischer Wolf

f   Loup de Chine

841   *Canis lupus columbianus*
Carnivora - Canidae

e   British Columbia Wolf

d   British-Columbia-Wolf

f   Loup de Colombie

842   *Canis lupus communis*
Carnivora - Canidae

e   Grey Wolf

d   Grauwolf

843   *Canis lupus crassodon*
Carnivora - Canidae

e   Vancouver Island Wolf

d   Vancouver-Insel-Wolf

f   Loup de Vancouver

844   *Canis lupus cubanensis*
Carnivora - Canidae

e   Caspian Sea Wolf

d   Kaspischer Seewolf

f   Loup du Caucase

845   *Canis lupus deitanus*
Carnivora - Canidae

e   Spanish Wolf

d   Spanischer Wolf

f   Loup de Murcie

846   *Canis lupus dingo*
Carnivora - Canidae

e   Dingo; Australian Native Dog

d   Dingo

i   Dingo

847   *Canis lupus fuscus*
Carnivora - Canidae

e   Cascade Mountains Wolf; Cascade Mountains Grey Wolf; Common Cascade Mountains Wolf; Cascade Brown Wolf; Cascade Mountains Brown Wolf; Brown Wolf; Cascade Common Wolf

d   Kaskadengebirgs-Wolf; Cascade Mountains-Wolf

f   Loup des Montagnes Cascades; Loup commun des Montagnes Cascades; Loup vulgaire des Montagnes Cascades

848   *Canis lupus griseoalbus*
Carnivora - Canidae

e   Manitoba Wolf; Grey-white Wolf

d   Grauweißer Wolf

f   Loup des prairies

849   *Canis lupus hallstromi*
Carnivora - Canidae

e   New Guinea Singing Dog

d   Neuguinea-Dingo; Neuguinea-Hund; Urwalddingo; Hallstrom-Dingo

850   *Canis lupus hattai*
Carnivora - Canidae

e   Kishida Wolf; Hokkaido Wolf

d	Hokkaido-Wolf
f	Loup de Hokkaido

851 **_Canis lupus hodophilax_**
Carnivora - Canidae
e Japanese Wolf; Honshu Wolf; Japanese Grey Wolf; Common Japanese Wolf; Shamanu
d Japanischer Wolf
f Loup du Japon; Loup commun du Japon; Loup vulgaire du Japon
i Lupo giapponese

852 **_Canis lupus hudsonicus_**
Carnivora - Canidae
e Hudson Bay Wolf
d Hudson-Bay-Wolf; Hudson-Wolf
f Loup de l'Hudson

853 **_Canis lupus irremotus_**
Carnivora - Canidae
e Northern Rocky Mountains Wolf; Northern Rockies Wolf
d Felsengebirge-Wolf
f Loup des Rocheuses septentrionales

854 **_Canis lupus italicus_**
Carnivora - Canidae
e Italian Wolf
d Italienischer Wolf; Apenninischer Wolf
f Loup d'Italie
i Lupo italiano; Lupo appenninico

855 **_Canis lupus labradorius_**
Carnivora - Canidae
e Labrador Wolf
d Labrador-Wolf
f Loup du Labrador

856 **_Canis lupus laniger_**
Carnivora - Canidae
e Tibetan Wolf
d Tibetanischer Wolf
f Loup de Tibet

i Lupo cinese; Lupo tibetano

857 **_Canis lupus ligoni_**
Carnivora - Canidae
e Alexander Archipelago Wolf
d Alexander Archipelago-Wolf
f Loup de l'archipel Alexandre

858 **_Canis lupus lupus_**
Carnivora - Canidae
e Wolf; European Wolf; Eurasian Wolf; Common Grey Wolf
d Europäischer Wolf; Gemeiner Wolf; Gemeiner Grauer Wolf
f Loup commun; Loup d'Europe
i Lupo europeo

859 **_Canis lupus lyacon_**
Carnivora - Canidae
e Eastern Timber Wolf; Timber Wolf (NA)
d Östlicher Bauholz-Wolf
f Loup de l'Est; Loup du Canada
i Lupo dei boschi; Lupo del Canada; Lipo solitario; Lupo americano

860 **_Canis lupus mackenzii_**
Carnivora - Canidae
e Northwest Territories Wolf; Mackenzie Tundra Wolf
d Mackenzie-Tundrawolf
f Loup de Mackenzie

861 **_Canis lupus manningi_**
Carnivora - Canidae
e Baffin Island Tundra Wolf; Baffin Island Wolf
d Baffin-Insel-Wolf
f Loup de la Terre de Baffin

862 **_Canis lupus minor_**
Carnivora - Canidae
e Austro-Hungarian Wolf; Reed Wolf; Common Reed Wolf; Grey Reed Wolf
d Austro-Ungarischer Wolf; Rohrwolf

f   Loup de la canne; Loup commun de
     la canne; Loup vulgaire de la canne;
     Loup austro-hongrois

i   Lupo dei canneti

863   *Canis lupus mogollensis*
        Carnivora - Canidae

e   Mogollon Mountain Wolf;
      Southwestern Wolf; Mogollon Wolf;
      Arizona Wolf; Mogollon Mountains
      Grey Wolf; Common Mongollon
      Mountains Wolf; New Mexico Wolf;
      New Mexican Grey Wolf

d   Südwestlicher Wolf

f   Loup mogollon

864   *Canis lupus monstrabilis*
        Carnivora - Canidae

e   Texas Grey Wolf; Texas Wolf;
      Common Texas Wolf

d   Texas-Wolf; Texas-Grauwolf;
      Osttexas-Wolf; Osttexas-Grauwolf

f   Loup de Texas; Loup commun de
      Texas; Loup vulgaire de Texas

865   *Canis lupus nubilus*
        Carnivora - Canidae

e   Buffalo Wolf; Great Plains Wolf;
      Plains Wolf; Nebraska Grey Wolf;
      Great Plains Grey Wolf; Great Plains
      Lobo Wolf; Buffalo Runner; Loafer
      Wolf; Lobo

d   Nebraska-Wolf; Prairiewolf;
      Büffelwolf; Mittelwesten-Wolf

f   Loup des plaines; Loup commun du
      Nebraska; Loup vulgaire du
      Nebraska; Loup des bisons

i   Lupo del Nebraska

866   *Canis lupus occidentalis*
        Carnivora - Canidae

e   Mackenzie Valley Wolf; Western
      Wolf; Rocky Mountains Wolf;
      Northwest Territories Wolf

d   Mackenzie-Wolf; Mackenzie-Senke-
      Wolf

f   Loup d'Alberta

i   Lupo canadese

867   *Canis lupus orion*
        Carnivora - Canidae

e   Greenland Wolf

d   Grönland-Wolf

f   Loup de Groenland

868   *Canis lupus pallipes*
        Carnivora - Canidae

e   Iranian Wolf; Indian Wolf

d   Indischer Wolf; Pallipes-Wolf

f   Loup des Indes

i   Lupo indiano

869   *Canis lupus pambasileus*
        Carnivora - Canidae

e   Interior Alaskan Wolf; Alaskan Wolf

d   Alaska-Wolf

f   Loup d'Alaska

870   *Canis lupus signatus*
        Carnivora - Canidae

e   Iberian Wolf

d   Iberischer Wolf

f   Loup de l'Espagne

i   Lupo della penisola iberica

871   *Canis lupus tundrarum*
        Carnivora - Canidae

e   Alaska Tundra Wolf; Tundra Wolf

d   Alaska-Tundrawolf

f   Loup de la toundra d'Alaska

i   Lupo dell'Alaska

872   *Canis lupus youngi*
        Carnivora - Canidae

e   Southern Rocky Mountains Wolf;
      Southern Rockies Wolf; Southern
      Rocky Mountains Grey Wolf;
      Southern Rocky Mountains Common
      Wolf

d   Südlicher Felsenbebirge-Wolf

f   Loup des Montagnes Rocheuses du
      Sud; Loup commun des Montagnes

Rocheuses du Sud; Loup vulgaire des
Montagnes Rocheuses du Sud; Loup
des Rocheuses méridionales

873   ***Canis mesomelas***
Carnivora - Canidae
e   Black-backed Jackal; Silver-backed
Jackal
d   Schabrackenschakal
f   Chacal à chabraque; Chacal à dos
noir
i   Lupo dalla Gualdrappa; Sciacallo
dalla Gualdrappa; Sciacallo dal dorso
argentato

Canis niger syn. ***C. rufus*** q.v.

874   ***Canis rufus***
Carnivora - Canidae
e   Red Wolf
d   Rotwolf
f   Loup rouge
i   Lupo rosso

875   ***Canis rufus floridanus***
Carnivora - Canidae
e   Florida Red Wolf; Eastern Red Wolf;
Florida Black wolf; Florida Wolf;
Black Wolf
d   Florida-Rotwolf
f   Loup rouge de Floride

876   ***Canis rufus gregoryi***
Carnivora - Canidae
e   Gregory's Red Wolf
d   Sumpfwolf; Mississippi-Rotwolf
f   Loup de Gregory

877   ***Canis rufus rufus***
Carnivora - Canidae
e   Common Red Wolf
d   Gemeiner Rotwolf
f   Loup rouge commun

878   ***Canis simensis***
Carnivora - Canidae

e   Simien Fox; Simien Jackal;
Ethiopian Wolf; Abyssinian Wolf;
Simien Wolf; Ethiopian Red Wolf
d   Abessinien-Fuchs; Abessinischer
Fuchs; Kaberu; Kontsal; Walke;
Äthiopischer Wolf; Äthiopischer
Schakal; Asiatisch-Indischer Wolf;
Simien-Schakal
f   Loup d'Abyssinie
i   Lupo del Simien; Lupo d'Etiopia

879   ***Cannomys***
Rodentia - Muridae
e   Lesser Bamboo Rats
d   Kleine Bambusratten
f   Petits rats des bambous
i   Piccoli ratti del bambù

880   ***Cannomys badius***
Rodentia - Muridae
e   Lesser Bamboo Rat; Bay Bamboo
Rat; Assam Bamboo Rat
d   Kleine Bambusratte; Kleine Assam-
Bambusratte; Kleine Bambusratte aus
Assam
f   Petit rat des bambous
i   Piccolo ratto del bambù

881   ***Cansumys***
Rodentia - Muridae
e   Gansu Hamsters
d   Kansu-Zwerghamster
f   Hamsters

882   ***Cansumys canus***
Rodentia - Muridae
e   Gansu Hamster
d   Kansu-Zwerghamster

883   ***Caperea***
Cete - Balaenidae
e   Pygmy Right Whales
d   Zwergglattwale
f   Baleines pygmées
i   Capreree

884   *Caperea marginata*
          Cete - Balaenidae
e   Pygmy Right Whale; Dwarf Right
    Whale
d   Zwergglattwal
f   Baleine pygmée; Baleine franche
    naine
i   Caperea

885   *Capra*
          Artiodactyla - Bovidae
e   Goats
d   Wildziegen; Steinböcke; Ziegen
f   Chèvres véritables
i   Capre

886   *Capra aegagrus*
          Artiodactyla - Bovidae
e   Wild Goat; Balkan Wild Goat; Ibex;
    Grecian Pasang; Feral Goat
d   Paseng; Steinbock; Wildziege
f   Chèvre égagre; Chèvre de
    Bélouchistan; Chèvre sauvage;
    Chèvre véritable; Panseng; Égagre;
    Chèvre à bézoard
i   Capra di Montecristo; Capra
    selvatica; Egagro; Capra del Bezoar;
    Markhor dell'Afghanistan

887   *Capra aegagrus aegagreus*
          Artiodactyla - Bovidae
e   Bezoar Ibex
d   Bezoarbock; Bezoarziege;
    Kleinasiatische Bezoarziege;
    Vorderasiatische Bezoarziege;
    Vorderindische Bezoarziege
f   Bézoar de Pasang

888   *Capra aegagrus blythi*
          Artiodactyla - Bovidae
e   Sindh Ibex
d   Sind-Bezoarziege
i   Capra selvatica del Sind

889   *Capra aegagrus cretica*
          Artiodactyla - Bovidae

e   Kri-kri; Agrimi; Cretan Goat; Cretan
    Ibex
d   Kretische Bezoarziege; Kretische
    Wildziege
f   Chèvre sauvage de Crète; Chèvre de
    Crète
i   Agrimia di Creta

890   *Capra aegagrus turcmenica*
          Artiodactyla - Bovidae
e   Bearded Goat
d   Turkmenische Bezoarziege; Sind-
    Ziege

891   *Capra caucasica*
          Artiodactyla - Bovidae
e   Caucasian Tur; West Caucasian Tur
d   Westkaukasischer Tur;
    Westkaukasischer Steinbock;
    Kubanischer Tur; Kuban-Tur
f   Tour du Caucase; Bouquetin de
    Caucase occidental
i   Stambecco del Caucaso

892   *Capra cylindricornis*
          Artiodactyla - Bovidae
e   East Caucasian Tur
d   Ostkaukasischer Tur; Dagestanischer
    Steinbock; Dagestanischer Tur;
    Ostkaukasischer Steinbock
f   Tour du Caucase oriental
i   Tar

893   *Capra falconeri*
          Artiodactyla - Bovidae
e   Markhor
d   Schraubenziege; Markhor; Markhur
f   Markhor
i   Capra di Falconer; Markhor

894   *Capra falconeri cashmiriensis*
          Artiodactyla - Bovidae
e   Kashmir Markhor; Pir Panjal
    Markhor
d   Kaschmir-Markhorn

895   *Capra falconeri chialtanensis*
Artiodactyla - Bovidae
e   Chialtan Markhor; Chialtan Goat;
Chialtan Wild Goat
d   Chialtan-Markhor

896   *Capra falconeri falconeri*
Artiodactyla - Bovidae
e   Astor Markhor; Flare-horned
Markhor; Flat-horned Markhor
d   Astor-Schraubenziege; Astor-
Markhor

897   *Capra falconeri heptneri*
Artiodactyla - Bovidae
e   Tajik Markhor
d   Bucharische Schraubenziege

898   *Capra falconeri jerdoni*
Artiodactyla - Bovidae
e   Suleiman Markhor; Straight-horned
Markhor
d   Suleiman-Schraubenziege; Suleiman-
Markhor
f   Markhor de Suleiman

899   *Capra falconeri megaceros*
Artiodactyla - Bovidae
e   Straight-horned Markhor; Kabul
Markhor
d   Kabul-Markhor

900   *Capra hircus*
Artiodactyla - Bovidae
e   Domestic Goat
d   Hausziege
f   Chèvre domestique
i   Capra domestica

901   *Capra ibex*
Artiodactyla - Bovidae
e   Ibex; Alpine Ibex; Ibex of the Alps;
Siberian Ibex; Himalajan Ibex
d   Alpensteinbock; Bergsteinbock;
Steinbock der Alpen; Europäischer
Steinbock; Steinwild

f   Bouc astain; Bouc des rochers; Bouc
sauvage; Chèvre sauvage; Bouquetin
des Alpes; Bouquetin
i   Stambecco alpino; Stambecco delle
alpi; Stambecco

902   *Capra nubiana*
Artiodactyla - Bovidae
e   Nubian Ibex; Nubian Goat
d   Nubischer Steinbock
f   Bouquetin de Nubie
i   Stambecco nubiano

903   *Capra pyrenaica*
Artiodactyla - Bovidae
e   Spanish Ibex; Spanish Tur
d   Iberien-Steinbock; Iberischer
Bergsteinbock
f   Bouquetin d'Espagne; Bouquetin des
Pyrénées; Bouquetin
i   Stambecco iberico

904   *Capra pyrenaica hispanica*
Artiodactyla - Bovidae
e   Beceite Ibex; Southeastern Spanish
Ibex; Ronda Ibex
d   Sierra-Nevada-Steinbock

905   *Capra pyrenaica pyrenaica*
Artiodactyla - Bovidae
e   Pyrenean Ibex
d   Pyrenäen-Steinbock
f   Bouquetin des Pyrénées; Tur
d'Espagne
i   Stambecco dei Pirenei

906   *Capra pyrenaica victoriae*
Artiodactyla - Bovidae
e   Gredos Ibex
d   Gredos-Steinbock
f   Bouquetin de Gredos

907   *Capra sibirica*
Artiodactyla - Bovidae
e   Siberian Ibex
d   Sibirischer Steinbock

f Yanghir; Bouquetin de Sibérie
i Stambecco siberiano

908 ***Capra walie***
Artiodactyla - Bovidae
e Walia Ibex; Abyssinian Ibex; Simien Ibex
d Abessinischer Steinbock; Äthiopischer Steinbock
f Bouquetin d'Abyssinie
i Stambecco del Simien

909 ***Capreolus***
Artiodactyla - Cervidae
e Roe Deer ; Roebucks
d Rehe
f Chevreuils
i Caprioli

Capreolus capraea syn. *C. capreolus* q.v.

910 ***Capreolus capreolus***
Artiodactyla - Cervidae
e European Roe Deer; Roe Deer; Roebuck; Roe; Western Roe Deer
d Europäisches Reh; Reh
f Chevreuil; Cerf chevreuil; Chevreuil commun; Chevreuil d'Europe; Chevreuil ordinaire; Chevreuil européen
i Capriolo

911 ***Capreolus capreolus bedfordi***
Artiodactyla - Cervidae
e Chinese Roe Deer
d Chinesisches Reh
f Chevreuil chinois
i Capriolo cinese

912 ***Capreolus pygargus***
Artiodactyla - Cervidae
e Eastern Roe Deer
d Sibirisches Reh
f Chevreuil de Sibérie; Chevreuil d'Asie

i Capriolo siberiano

913 ***Capricornis***
Artiodactyla - Bovidae
e Serows
d Seraue
f Serows
i Capricorni

914 ***Capricornis crispus***
Artiodactyla - Bovidae
e Japanese Serow
d Japanischer Serau
f Serow japonais
i Capricorno del Giappone

915 ***Capricornis sumatraensis***
Artiodactyla - Bovidae
e Serow; Sumatran Serow; White-maned Serow; Southern Serow; Mainland Serow
d Sumatra-Serau; Waldziegenantilope
f Serow de Sumatra
i Capricorno di Sumatra; Seran

916 ***Capricornis swinhoei***
Artiodactyla - Bovidae
e Taiwan Serow; Formosa Serow
d Taiwan-Serau
f Serow de Formosa

917 ***Caprolagus***
Lagomorpha - Leporidae
e Bristly Rabbits; Assam Rabbits; Harsh-furred Hares; Hispid Hares; Hispid Rabbits
d Rauhkaninchen ; Borstenkaninchen
f Lapins asiatiques; Lapins de l'Assam
i Caprolaghi

918 ***Caprolagus hispidus***
Lagomorpha - Leporidae
e Bristly Rabbit; Assam Rabbit; Hispid Rabbit; Hispid Hare
d Rauhkaninchen; Borstenkaninchen

---

 f Lapin asiatique; Lapin de l'Assam
 i Caprolago ispido

919 **Capromyidae**
  Rodentia
 e Hutias; Hutias and Coypus
 d Baum-und-Ferkelratten;
  Baumrattenverwandte
 f Capromyidés
 i Capromidi

920 ***Capromys***
  Rodentia - Capromyidae
 e Desmarest's Hutias; Long-tailed
  Hutias; Hutias ; Cuban Hutias
 d Kuba-Baumratten; Hutias
 i Hutie

921 ***Capromys angelcabrerai***
  Rodentia - Capromyidae
 e Cabrera's Hutia
 i Hutia cubana

922 ***Capromys auritus***
  Rodentia - Capromyidae
 e Eared Hutia

 Capromys brownii syn.
 Geocapromnys brownii q.v.

923 ***Capromys garridoi***
  Rodentia - Capromyidae
 e Garrido's Hutia
 d Garridos Hutia

924 ***Capromys gundlachi***
  Rodentia - Capromyidae
 e Gundlach's Hutia

 Capromys ingrahami syn.
 Geocapromys ingrahami q.v.

925 ***Capromys melanurus***
  Rodentia - Capromyidae
 e Black-tailed Hutia; Bushy-tailed

  Hutia
 d Schwarzschwanzbaumratte;
  Hutiasata
 f Rat à queue noire

926 ***Capromys meridionalis***
  Rodentia - Capromyidae
 e Southern Hutia

927 ***Capromys nanus***
  Rodentia - Capromyidae
 e Tiny Hutia; Dwarf Hutia
 d Zwergbaumratte
 f Rat nain

928 ***Capromys pilorides***
  Rodentia - Capromyidae
 e Cuban Hutia; Desmarest's Hutia
 d Hutiaconga; Kuba-Baumratte
 f Rat poilu
 i Hutia di Desmarest; Hutia conga

929 ***Capromys prehensilis***
  Rodentia - Capromyidae
 e Prehensile-tailed Hutia
 d Hutiacarabali
 f Rat à queue préhensile
 i Hutia

930 ***Capromys sanfelipensis***
  Rodentia - Capromyidae
 e San Felipe Hutia
 i Hutia terrestre

 Capromys thoracatus syn.
 Geocapromys thoracatus q.v.

 Caracal syn. *Felis* q.v.

931 ***Cardiocranius***
  Rodentia - Dipodidae
 e Satunins Jerboas; Satunin's Five-toed
  Pygmy Jerboas; Satunin's Five-toed
  Dwarf Jerboas; Five-toed Pygmy
  Jerboas

d   Fünfzehige Zwergspringmäuse;
      Fünfzehen-Zwergspringmäuse
f   Gerboises naines à cinque doigts
i   Gerboa pigmei

932   Cardiocranius paradoxus
      Rodentia - Dipodidae
e   Satunins Jerboa; Satunin's Five-toed
      Pygmy Jerboa; Satunin's Five-toed
      Dwarf Jerboa; Five-toed Pygmy
      Jerboa
d   Fünfzehige Zwergspringmaus;
      Fünfzehen-Zwergspringmaus
f   Gerboise nain à cinque doigts
i   Gerboa pigmeo

933   Cardioderma
      Chiroptera - Megadermatidae
e   Heart-nosed False Vampires; African
      False Vampire Bats; Big-eared Bats;
      Heart-nosed Bats
d   Herznasenfledermäuse
f   Mégadermes de coeur; Mégadermes
      nez-en-coeur

934   Cardioderma cor
      Chiroptera - Megadermatidae
e   Heart-nosed False Vampire; African
      False Vampire Bat; Big-eared Bat;
      Heart-nosed Bat
d   Herznasenfledermaus
f   Mégaderme de coeur; Mégaderme
      nez-en-coeur

935   Carollia
      Chiroptera - Phyllostomidae
e   Short-tailed Bats; Leaf-nosed Bats;
      Short-tailed Leaf-nosed Bats;
      Neotropical Short-tailed Fruit Bats
d   Blattnasen-Federmäuse
f   Chauve-souris frugivores

936   Carollia brevicauda
      Chiroptera - Phyllostomidae
e   Silky Short-tailed Bat

d   Seidige Kurzschwanzblattnase

937   Carollia castanea
      Chiroptera - Phyllostomidae
e   Allen's Short-tailed Bat; Least Short-
      tailed Bat; Chestnut Short-tailed Bat;
      Allen's Short-tailed Leaf-nosed Bat

938   Carollia perspicillata
      Chiroptera - Phyllostomidae
e   Seba's Short-tailed Bat; Seba's Short-
      tailed Fruit Bat; Linneaus's Short-
      tailed Bat; Linneaus's Short-tailed
      Fruit Bat; Short-tailed Fruit Bat
d   Brillenblattnase
f   Vespertillion à nez plat; Chauve-
      souris fer-de-lance

939   Carollia sowelli
      Chiroptera - Phyllostomidae
e   Sowell's Short-tailed Bat

940   Carollia subrufa
      Chiroptera - Phyllostomidae
e   Hahn's Short-tailed Bat; Grey Short-
      tailed Bat

941   Carpomys
      Rodentia - Muridae
e   Luzon Rats; Luzon Tree Rats
d   Luzon-Ratten; Luzon-Baumratten
f   Rats des Philippines

942   Carpomys melanurus
      Rodentia - Muridae
e   Short-footed Luzon Rat; Short-footed
      Luzon Tree Rat

943   Carpomys phaeurus
      Rodentia - Muridae
e   White-bellied Luzon Tree Rat

944   Carterodon
      Rodentia - Echimyidae
e   Owl's Spiny Rats

945   *Carterodon sulcidens*
      Rodentia - Echimyidae
  e   Owl's Spiny Rat; Owl's Rat

946   *Casinycteris*
      Chiroptera - Pteropodidae
  e   Short-palated Fruit Bats; Short-palate
      Fruit Bats

947   *Casinycteris argynnis*
      Chiroptera - Pteropodidae
  e   Short-palated Fruit Bat; Short-palate
      Fruit Bat

948   *Castor*
      Rodentia - Castoridae
  e   Beavers
  d   Biber
  f   Castors
  i   Castori

949   *Castor canadensis*
      Rodentia - Castoridae
  e   American Beaver; Beaver (NA);
      North American Beaver
  d   Kanadischer Biber; Kanada-Biber
  f   Castor du Canada
  i   Castoro americano; Castoro del
      Canada; Castoro nord-americano;
      Castoro canadese

950   *Castor canadensis belugae*
      Rodentia - Castoridae
  e   Beluga River Beaver

951   *Castor canadensis caecator*
      Rodentia - Castoridae
  e   Newfoundland Beaver
  d   Neufundland Biber

952   *Castor canadensis canadensis*
      Rodentia - Castoridae
  e   Canadian Beaver; Canada Beaver
  d   Amerikanischer Biber

953   *Castor canadensis mexicanus*
      Rodentia - Castoridae
  e   Rio Grande Beaver
  d   Rio-Grande-Biber

954   *Castor canadensis michiganensis*
      Rodentia - Castoridae
  e   Michigan Beaver
  d   Waldbiber

955   *Castor fiber*
      Rodentia - Castoridae
  e   European Beaver; Asiatic Beaver;
      Eurasian Beaver
  d   Europäischer Biber; Europäischer
      Flussbiber; Biber; Eurasiatischer
      Biber; Eurasischer Biber
  f   Castor d'Europe; Castor commun;
      Castor; Castor d'Eurasie
  i   Castoro europeo; Castoro euro-
      asiatico

956   *Castor fiber albicus*
      Rodentia - Castoridae
  e   Eurasian Beaver; Elbe Beaver
  d   Elbe-Biber; Deutscher Elbebiber
  f   Castor de l'Elbe

957   *Castor fiber birulai*
      Rodentia - Castoridae
  e   Central Asiatic Beaver
  d   Mongolischer Biber

958   *Castor fiber fiber*
      Rodentia - Castoridae
  e   Scandinavian Beaver
  d   Schwedisch-Norwegischer Biber
  f   Castor de Scandinavie

959   *Castor fiber galliae*
      Rodentia - Castoridae
  e   Rhone Beaver
  d   Rhone-Biber
  f   Castor du Rhône; Castor rhodanien

960   *Castor fiber pohlei*
Rodentia - Castoridae
e   Siberian Beaver
d   Ural-Biber; Westsibirischer Biber

961   *Castor fiber subauratus*
Rodentia - Castoridae
e   Golden Beaver
d   Goldbiber

962   *Castor fiber tuvinicus*
Rodentia - Castoridae
e   Tuvinian Beaver
d   Tuvinischer Biber

963   *Castor fiber vistulanus*
Rodentia - Castoridae
e   Central European Beaver
d   Bjelorussischer Biber

964   **Castoridae**
Rodentia
e   Beavers
d   Biber ; Biberartige; Flussbiber ;
Schwimmnager ; Schwimmfüßer
f   Castors; Castoridés
i   Castoridi

965   *Catagonus*
Artiodactyla - Tayassuidae
e   Chaco Peccaries; Chacoan Peccaries
d   Chaco-Pekaris
f   Pécaris géants
i   Pecari gigante

966   *Catagonus wagneri*
Artiodactyla - Tayassuidae
e   Chaco Peccary; Chacoan Peccary;
Giant Peccary
d   Chaco-Pekari
f   Pécari géant
i   Pecari gigante; Pecari del Chaco

Catopuma syn. *Felis* q.v.

967   *Cavia*
Rodentia - Caviidae
e   Guinea-pigs; Cavies
d   Meerschweinchen
f   Cobayes; Cochons d'Inde
i   Porcellini d'India; Cavie

968   *Cavia aperea*
Rodentia - Caviidae
e   Aperea; Guianan Cavy; Brazilian
Guinea-pig; Aperea Wild Cavy
d   Aperea; Gemeines Meerschweinchen
f   Cobaye sauvage
i   Cavia selvatica

969   *Cavia fulgida*
Rodentia - Caviidae
e   Shiny Guinea-pig

970   *Cavia intermedia*
Rodentia - Caviidae
e   Intermediate Guinea-pig

971   *Cavia magna*
Rodentia - Caviidae
e   Greater Guinea-pig; Greater Cavy
d   Sumpfmeerschweinchen
f   Grand cobaye

972   *Cavia porcellus*
Rodentia - Caviidae
e   Guinea-pig; Domestic Guinea-pig
d   Meerschwein; Hausmeerschwein;
Zahmes Meerschwein; Cobaya;
Ferkelhase; Meerschweinchen
f   Cobaye; Cochon d'Inde
i   Cavia domestica

973   *Cavia tschudi*
Rodentia - Caviidae
e   Wild Cavy; Montane Guinea-pig
d   Tschudi-Meerschweinchen

974   **Caviidae**
Rodentia

e   Guinea-pigs; Cavies; Cavies and
    Guineapigs
d   Meerschweinchen
f   Caviidés
i   Cavidi

975   *Cebus*
Primates - Atelide
e   Capuchins; Capuchin Monkeys;
    Ring-tailed Monkeys; Sapajous
d   Kapuziner; Kapuzineraffen
f   Sajous; Sapajous; Singes capucins
i   Cebi

976   *Cebus albifrons*
Primates - Atelide
e   White-fronted Sapajou; White-
    fronted Capuchin; Brown Pale-
    fronted Capuchin; White-fronted
    Capuchin Monkey; Pale-fronted
    Capuchin
d   Weißstirnkapuziner
f   Sapajou à front blanc; Sajou à front
    blanc
i   Cebo dalla fronte bianca

977   *Cebus apella*
Primates - Atelide
e   Black-capped Capuchin; Brown
    Capuchin; Tufted Capuchin; Brown
    Capuchin Monkey; Hooded
    Capuchin; Tufted Capuchin Monkey;
    Brown Tufted Capuchin
d   Gehaupter Kapuziner; Apella;
    Faunaffe
f   Sapajou apelle; Sajou apelle; Sajou
    noir
i   Cebo dai cornetti; Cebo dai ciuffetti

978   *Cebus apella robustus*
Primates - Atelide
e   Robust Tufted Capuchin

979   *Cebus capucinus*
Primates - Atelide
e   White-faced Capuchin ; White-faced
    Capuchin Monkey; White-faced

    Sapajou; White-throated Capuchin;
    White-headed Capuchin
d   Weißschulteraffe; Kapuziner;
    Eigenticher Kapuziner;
    Weißschulterkapuziner
f   Sapajou capucin; Sajou capucin
i   Cebo cappuccino

980   *Cebus kaapori*
Primates - Atelide
e   Ka'apor Capuchin; Southern Bahian
    Masked Titi
i   Cebo dei Kaapor

981   *Cebus libidinosus*
Primates - Atelide
e   Black-striped Tufted Cpuchin
d   Schwarzstreifenkapuziner

982   *Cebus nigritus*
Primates - Atelide
e   Black Tufted Capuchin

***Cebus nigrivittatus* syn.** *C. olivaceus*
q.v.

983   *Cebus olivaceus*
Primates - Atelide
e   Weeper Capuchin; Weeping
    Capuchin; Wedge-capped Capuchin
    Monkey
d   Brauner Kapuziner
f   Singe musqué; Sapajou brun
i   Cebo olivaceo

984   *Cebus xanthosternus*
Primates - Atelide
e   Yellow-breasted Capuchin; Buffy-
    headed Tufted Capuchin; Golden-
    bellied Tufted Capuchin
d   Gelbbrustkapuziner;
    Goldbauchkapuziner
f   Sapajou à poitrine jaune

985   *Celaenomys*
Rodentia - Muridae

e   Luzon Shrew-like Rats; Luzon Grey
      Water Rats; Luzon Shrewrats; Blazed
      Luzon Shrewrats
d   Luzon-Spitzmausratten
f   Rats musaraignes

986   *Celaenomys silaceus*
      Rodentia - Muridae
e   Luzon Shrew-like Rat; Luzon Grey
      Water Rat; Luzon Shrewrat; Blazed
      Luzon Shrewrat
d   Luzon-Spitzmausratte
f   Rat musaraigne

      Centetes syn. *Tenrec* q.v.

987   *Centronycteris*
      Chiroptera - Emballonuridae
e   Shaggy Haired Bats; Thomas's Bats;
      Shaggy Bats

988   *Centronycteris maximiliani*
      Chiroptera - Emballonuridae
e   Maximilians Shaggy-haired Bat;
      Maximilian's Bat; Thomas's Bat;
      Shaggy Haired Bat; Shaggy Bat

989   *Centurio*
      Chiroptera - Phyllostomidae
e   Lattice-winged Bats; Central
      American Wrinkle-faced Bats;
      Greater Wrinkle-faced Bats;
      Wrinkle-faced Bats
d   Greisengesichte
f   Vespertillons ridés

990   *Centurio senex*
      Chiroptera - Phyllostomidae
e   Lattice-winged Bat; Central
      American Wrinkle-faced Bat; Greater
      Wrinkle-faced Bat; Wrinkle-faced
      Bat
d   Greisengesicht; Zenturionen-
      Fledermaus
f   Vespertillon ridé

991   *Cephalophus*

      Artiodactyla - Bovidae
e   Duikers; Forest Duikers; Maxwell's
      Duikers
d   Ducker ; Schopfducker
f   Céphalophes
i   Cefalofi

992   *Cephalophus adersi*
      Artiodactyla - Bovidae
e   Ader's Duiker; Zanzibar Duiker
d   Sansibar-Ducker
f   Céphalophe roux de Zanzibar

993   *Cephalophus callipygus*
      Artiodactyla - Bovidae
e   Peters's Duiker
d   Peters Ducker
f   Céphalophe de Peters
i   Cefalofo di Peters

994   *Cephalophus dorsalis*
      Artiodactyla - Bovidae
e   Bay Duiker
d   Schwarzrückenducker
f   Céphalophe à bande dorsale noire;
      Céphalophe bai; Céphalophe à dos
      noir; Céphalophe à bande noire
i   Cefalofo dalla schiena nera; Cefalofo
      baio; Cefalofo dorsale; Cefalofo a
      dorso nero

995   *Cephalophus harveyi*
      Artiodactyla - Bovidae
e   Harvey's Duiker; Harvey's Red
      Duiker
d   Harvey-Ducker
f   Céphalophe d'Harvey

996   *Cephalophus jentinki*
      Artiodactyla - Bovidae
e   Jentink's Duiker; Black-headed
      Duiker
d   Jentink-Ducker
f   Céphalophe de Jentink
i   Cefalofo di Jentink

997 *Cephalophus leucogaster*
Artiodactyla - Bovidae
e Gabon Duiker; White-bellied Duiker
d Gabun-Ducker; Weißbauchducker
f Céphalophe à ventre blanc
i Cefalofo ventre bianco

998 *Cephalophus maxwelli*
Artiodactyla - Bovidae
e Maxwell's Duiker
d Maxwell-Ducker
f Céphalophe de Maxwell

999 *Cephalophus monticola*
Artiodactyla - Bovidae
e Blue Duiker; Sundevall's Blue Duiker
d Blauducker; Rotfußducker; Blauböckchen
f Céphalophe bleu; Mboloko
i Cefalofo azzuro; Cefalofo di Maxwell

1000 *Cephalophus monticola simpsoni*
Artiodactyla - Bovidae
e Simpson's Duiker
d Simpson-Ducker

1001 *Cephalophus natalensis*
Artiodactyla - Bovidae
e Red Duiker; Red Forest Duiker; Natal Duiker
d Natal-Ducker; Rotducker
f Céphalophe rouge
i Cefalofo rosso del Natal; Cefalofo rosso

1002 *Cephalophus niger*
Artiodactyla - Bovidae
e Black Duiker
d Schwarzducker
f Céphalophe noir
i Cefalofo nero

1003 *Cephalophus nigrifrons*

Artiodactyla - Bovidae
e Black-fronted Duiker
d Schwarzstirnducker
f Céphalophe a front noir; Céphalophe rouge; Céphalophe à front roux
i Cefalofo dalla fronte nera

1004 *Cephalophus ogilbyi*
Artiodactyla - Bovidae
e Ogilby's Duiker; Fernando Po Duiker
d Fernando-Po-Ducker; Ogilby-Ducker
f Céphalophe d'Ogilby
i Cefalofo di Fernando Po; Cefalofo di Ogilby

1005 *Cephalophus ogilbyi brookei*
Artiodactyla - Bovidae
e Brooke's Duiker
f Céphalophe de Brooke

1006 *Cephalophus ogilbyi crusalbum*
Artiodactyla - Bovidae
e White-legged Duiker

1007 *Cephalophus rubidus*
Artiodactyla - Bovidae
e Ruwenzori Duiker; Ruwenzori Red Duiker
d Ruwenzori-Ducker
f Céphalophe du Ruwenzori

1008 *Cephalophus rufilatus*
Artiodactyla - Bovidae
e Red-flanked Duiker
d Rotflankenducker; Blaurückenducker
f Céphalophe aux flancs rouges; Céphalophe à flanc roux
i Cefalofo dai fianchi rosso

1009 *Cephalophus silvicultor*
Artiodactyla - Bovidae
e Yellow-backed Duiker; Light-backed Duiker; Giant Duiker
d Gelbrückenducker; Riesenducker
f Céphalophe géant; Céphalophe à dos

jaune
i   Cefalofo dei boschi; Cefalofo dalla
    schiena nera; Cefalofo dorso gialla

1010   *Cephalophus spadix*
Artiodactyla - Bovidae
e   Abbott's Duiker
d   Abbott-Ducker
f   Céphalophe d'Abbott
i   Cefalofo di Abbot

1011   *Cephalophus weynsi*
Artiodactyla - Bovidae
e   Weyn's Duiker
d   Weyns-Ducker
f   Céphalophe rouge

1012   *Cephalophus zebra*
Artiodactyla - Bovidae
e   Zebra Duiker; Zebra Antelope;
    Banded Duiker; Stripe-backed
    Duiker
d   Zebraducker; Streifenducker
f   Céphalophe rayé; Céphalophe zèbre
i   Cefalofo zebra

1013   *Cephalorhynchus*
Cete - Delphinidae
e   Commerson's Dolphins; Southern
    Dolphins; Subantarctic Dolphins;
    Piebald Dolphins
d   Schwarzweiße Delfine
f   Dauphins de Commerson
i   Cefalorinchi

1014   *Cephalorhynchus commersonii*
Cete - Delphinidae
e   Commerson's Dolphin; Piebald
    Dolphin; Piebald Porpoise; Black-
    and-white Dolphin
d   Commerson-Delfin; Jacobita; Tonina
f   Dauphin pie; Dauphin de
    Commerson; Jacobite
i   Cefalorinco di Commerson

1015   *Cephalorhynchus eutropia*

Cete - Delphinidae
e   White-bellied Dolphin; Black
    Dolphin; Chilean Dolphin
d   Flachkopfdelfin; Weißbauchdelfin;
    Chilenischer Delfin
f   Dauphin du Cap; Dauphin de Chili
i   Cefalorinco nero; Cefalorinco
    eutropia

1016   *Cephalorhynchus heavisidii*
Cete - Delphinidae
e   Heaviside's Dolphin; Benguela
    Dolphin
d   Heaviside-Delfin; Heavisides Delfin
f   Dauphin noir
i   Cefalorinco di Heaviside

1017   *Cephalorhynchus hectori*
Cete - Delphinidae
e   Hector's Dolphin; White-fronted
    Dolphin; White-headed Dolphin;
    New Zealand Dolphin (ANZ); Little
    Pied Dolphin(ANZ); White-front
    Dolphin (ANZ)
d   Hectors Delfin; Hector-Delfin
f   Dauphin d'Hector
i   Cefalorinco di Hector; Delfino di
    Ettore

1018   *Cephalorhynchus hectori hectori*
Cete - Delphinidae
e   South Island Hector's Dolphin (ANZ)

1019   *Cephalorhynchus hectori maui*
Cete - Delphinidae
e   Maui's Dolphin; North Island
    Hector's Dolphin (ANZ)

1020   *Ceratotherium*
Perissodactyla - Rhinocerotidae
e   White Rhinoceroses; Wide-mouthed
    Rhinoceroses; Square-lipped
    Rhinoceroses
d   Breitmaulnashörner
f   Rhinocéros blancs
i   Rinoceronti bianchi

1021 *Ceratotherium simum*
Perissodactyla - Rhinocerotidae
- *e* White Rhinoceros; Wide-mouthed Rhinoceros; Square-lipped Rhinoceros; Grass Rhinoceros
- *d* Breitmaulnashorn; Weißes Nashorn
- *f* Rhinocéros blanc; Rhinocéros de Burchell
- *i* Rinoceronte bianco; Rinoceronte camuso

1022 *Ceratotherium simum cottoni*
Perissodactyla - Rhinocerotidae
- *e* Northern White Rhinoceros
- *d* Nördliches Breitmaulnashorn
- *f* Rhinocéros blanc du Nord
- *i* Rinoceronte bianco del nord; Rinoceronte bianco settentrionale

1023 *Ceratotherium simum simum*
Perissodactyla - Rhinocerotidae
- *e* Southern White Rhinoceros
- *d* Südliches Breitmaulnashorn
- *f* Rhinocéros blanc du Sud
- *i* Rinoceronte bianco del sud; Rinoceronte bianco meridionale

1024 *Cercartetus*
Diprotodontia - Burramyidae
- *e* Dormouse Possums; Pygmy Possums; Pygmy Phalangers
- *d* Schlafbeutler ; Bilchbeutler
- *f* Phalangers loirs; Opossums pygmées
- *i* Possum ghiro pigmei; Opossum ghiro pigmei

1025 *Cercartetus caudatus*
Diprotodontia -
- *e* Long-tailed Pygmy Possum (ANZ); Long-tailed Pygmy Cuscus
- *d* Neuguinea-Bilchbeutler

1026 *Cercartetus concinnus*
Diprotodontia - Burramyidae
- *e* Southwestern Pygmy Phalanger; Southwestern Pygmy Possum; Western Pygmy Possum; Mundarda

(ANZ)
- *d* Dünnschwanz-Schlafbeutler

1027 *Cercartetus lepidus*
Diprotodontia - Burramyidae
- *e* Lesser Tasmanian Pygmy Phalanger; Little Tasmanian Pygmy Possum; Tasmanian Pygmy Possum; Little Pygmy Possum (ANZ)
- *d* Tasmanien-Bilchbeutler
- *i* Possum ghiro pigmeo; Opossum ghiro pigmeo

1028 *Cercartetus nanus*
Diprotodontia - Burramyidae
- *e* Dormouse Possum; Eastern Pygmy Possum (ANZ); Pygmy Phalanger; Dormouse Phalanger
- *d* Dickschwanz-Schlafbeutler

1029 *Cercocebus*
Primates - Cercopithecidae
- *e* Mangabeys; White Eyelid Monkeys; White-eyelid Mangabeys
- *d* Mangaben
- *f* Mangabeys; Cercocèbes; Singes mangabeys
- *i* Cercocebi

Cercocebus aethiops syn.
Chlorocebus aethiops q.v.

1030 *Cercocebus agilis*
Primates - Cercopithecidae
- *e* Agile Mangabey
- *d* Goldbauchmangabe
- *f* Cercocèbe agile; Cercocèbe à crête
- *i* Cercocebo agile

Cercocebus albigena syn.
Lophocebus albigena q.v.

Cercocebus aterrimus syn.
Lophocebus atterimus q.v.

1031 *Cercocebus atys*
Primates - Cercopithecidae
e Sooty Mangabey
d Rauchgraue Mangabe
f Singe vert mangabey; Cercocèbe enfumé
i Cercocebo moro

1032 *Cercocebus atys lunulatus*
Primates - Cercopithecidae
e White-crowned Mangabey
d Weißscheitelmangabe
f Cercocèbe à col blanc; Mangabey couronné; Cercocèbe couronné
i Cercocebo dal collare

1033 *Cercocebus chrysogaster*
Primates - Cercopithecidae
e Golden-bellied Mangabey
d Goldbauchmangabe
f Mangabey doré; Mangabey à ventre doré

1034 *Cercocebus galeritus*
Primates - Cercopithecidae
e Tana River Mangabey; Tana Mangabey
d Haubenmangabe; Kappenmangabe; Tana-Haubenmangabe; Tana Fluss-Mangabe
f Mangabey à ventre doré; Mangabey de la Tana
i Cercocebo dal berretto; Cercocebo del fiume Tana; Cercocebus crestato

1035 *Cercocebus sanjei*
Primates - Cercopithecidae
e Sanje Mangabey
d Sanje-Mangabe
f Mangabey sanje

1036 *Cercocebus torquatus*
Primates - Cercopithecidae
e Smoky Mangabey; Collared Mangabey; White-collared Mangabey

d Halsbandmangabe; Rotkopfmangabe
f Mangabey à collier blanc; Cercocèbe à collier blanc; Cercocèbe enfumé; Mangabey enfumé
i Cercocebo dal collare

Cercomys syn. *Thrichomys* q.v.

1037 **Cercopithecidae**
Primates
e Old World Monkeys; Old World Monkeys and Baboons; Guenon-like Monkeys
d Meerkatzen; Tieraffen; Meerkatzenartige
f Cercopithécidés
i Cercopicetidi; Scimmie del vecchio mondo; Scimmie cinocefale

1038 *Cercopithecus*
Primates - Cercopithecidae
e Guenons
d Meerkatzen
f Cercophitèques; Guenons
i Cercopitechi

1039 *Cercopithecus albogularis*
Primates - Cercopithecidae
e White-collared Guenon; White-throated Guenon; Samango Monkey; White-throated Monkey; Syke's Monkey
d Weißkehlmeerkatze; Sykes-Affe
f Sitka
i Cercopiteco a gola bianca

1040 *Cercopithecus ascanius*
Primates - Cercopithecidae
e Red-tailed Monkey; Black-cheeked White-nosed Monkey; Red-tail Monkey; Redtail
d Rotschwanzmeerkatze; Rotschwanz Affe
f Cercopithèque ascagne
i Cercopiteco nasobianco del Congo; Cercopiteco di Schmidt

1041 *Cercopithecus ascanius schmidti*
Primates - Cercopithecidae

e Schmidt's Guenon; Schmidt's
Monkey
d Schmidts Meerkatze
f Cercopithèque ascagne à nez blanc;
Cercopithèque ascagne blanc-nez de
Schmidt

1042 *Cercopithecus campbelli*
Primates - Cercopithecidae

e Campbell's Monkey; Campbell's
Guenon; Campbell's Mona Monkey;
Campbell's Mona
d Campbell-Meerkatze; Campbell-
Mona; Campbells Meerkatze
f Cercopithèque de Campbell
i Cercopiteco di Campbell

1043 *Cercopithecus cephus*
Primates - Cercopithecidae

e Moustached Monkey; True
Moustached Monkey; Moustached
Guenon
d Blaumaulmeerkatze; Blaumäulige
Meerkatze; Schnurrbartmeerkatze
f Moustac; Moustac de Buffon
i Cercopiteco Cefo; Cercopiteco dai
mustacchi; Cefo

1044 *Cercopithecus denti*
Primates - Cercopithecidae

e Dent's Mona Monkey; Dent's Guenon
d Dents Meerkatze
f Cercopithèque de Dent
i Cercopiteco coronato di Dent

1045 *Cercopithecus diana*
Primates - Cercopithecidae

e Diana Monkey; Roloway's Monkey
d Diana-Meeerkatze; Diana-Affe
f Cercopithèque diana
i Cercopiteco Diana; Scimmia Diana

1046 *Cercopithecus doggetti*
Primates - Cercopithecidae

e Silver Monkey
d Silberner Affe

1047 *Cercopithecus dryas*
Primates - Cercopithecidae

e Dryas Monkey; Dryas Guenon; Zaire
Diana Monkey; Salongo Monkey
d Dryas-Meerkatze; Zaire-Diana-
Meerkatze; Dryas-Affe; Salongo-
Affe
f Cercopithèque dryas
i Cercopiteco dryas; Cercopiteco
salonga

1048 *Cercopithecus erythrogaster*
Primates - Cercopithecidae

e Red-bellied Guenon; Red-bellied
Monkey; White-throated Guenon;
White-cheeked Guennon
d Rotbauchmeerkatze
f Hocheur à ventre rouge; Singe à
ventre rouge; Hocheur à ventre roux
i Cercopiteco dal ventre rosso;
Cercopiteco ventre rosso

1049 *Cercopithecus erythrotis*
Primates - Cercopithecidae

e Red-eared Monkey; Red-eared
Guenon; Russet-eared Guenon;
Russet-eared Monkey; Red-eared
Nose-spotted Monkey
d Rotnasenmeerkatze;
Rotohrmeerkatze
f Hocheur à nez rouge; Mustac aux
oreilles rouges; Moustac aux oreilles
rousses
i Cercopiteco dalle orecchie rosse;
Cercpiteco orecchie rosse

1050 *Cercopithecus hamlyni*
Primates - Cercopithecidae

e Hamlyn's Monkey; Owl-faced
Monkey; Hamlyn's Guenon; Owl-
faced Guenon; Hamlyn's Owl-faced
Monkey
d Eulenkopfmeerkatze; Hamlyn-
Meerkatze; Hamlyn-Affe

f   Cercopithèque à tête d'hibou
i   Cercopiteco di Hamlyn

1051   ***Cercopithecus kandti***
      Primates - Cercopithecidae
e   Golden Monkey
d   Goldener Affe
f   Singe doré

1052   ***Cercopithecus l'hoesti***
      Primates - Cercopithecidae
e   L'Hoest's Monkey; Mountain
      Monkey; L'Hoest's Guenon; L'Hoest
      Mountain Monkey; Hoest Mountain
      Monkey
d   L'Hoest-Meerkatze; L'Hoest's Affe
f   Cercopithèque de l'Hoest
i   Cercopiteco barbuto; Cercopiteco di
      L'Hoest

1053   ***Cercopithecus lowei***
      Primates - Cercopithecidae
e   Lowe's Mona; Lowe's Mona
      Monkey; Lowe's Monkey
d   Lowes Mona
f   Cercopithèque de Lowe

1054   ***Cercopithecus mitis***
      Primates - Cercopithecidae
e   Diademed Guenon; Blue Monkey;
      Mitis Monkey; Sykes's Diadem
      Monkey; Syke's Monkey
d   Diademmeerkatze; Blauer Affe;
      Echte Diademmeerkatze
f   Singe argenté; Cercopithèque à
      diadème
i   Cercopiteco dal diadema;
      Cercopiteco diadema; Cercopiteco a
      gola bianca

1055   ***Cercopithecus mona***
      Primates - Cercopithecidae
e   Mona Monkey
d   Mona-Meerkatze; Mona-Affe
f   Cercopithèqye mone; Mone
i   Cercopiteco mona

1056   ***Cercopithecus neglectus***
      Primates - Cercopithecidae
e   De Brazza's Monkey; Brazza's
      Monkey; Neglectus Monkey;
      Schlegel's Guenon; De Brazza's
      Guenon; Chestnut-browed Monkey
d   Brazza-Meerkatze
f   Cercopithèque de Brazza; Singe de
      Brazza
i   Cercopiteco di Brazza

1057   ***Cercopithecus nictitans***
      Primates - Cercopithecidae
e   Greater White-nosed Guenon;
      Greater White-nosed Monkey;
      White-nosed Guenon; Spot-nosed
      Monkey; Greater Spot-nosed
      Monkey; Greater Spot-nosed Guenon
d   Große Weißnasenmeerkatze
f   Cercopithèque hocheur; Hocheur;
      Hocheur blanc-nez; Pain à cacheter
i   Cercopiteco nasobianco maggiore

1058   ***Cercopithecus nictitans martini***
      Primates - Cercopithecidae
e   Putty-nosed Gueneon
f   Hocheur de Martin

1059   ***Cercopithecus nictitans stampfli***
      Primates - Cercopithecidae
e   Stampfl's Putty-nosed Guenon
f   Hocheur de Stampfl

      Cercopithecus nigroviridis syn.
      Allenopithecus nigroviridis q.v.

      Cercopithecus patas syn.
      Erythrocebus patas q.v.

1060   ***Cercopithecus petaurista***
      Primates - Cercopithecidae
e   Lesser White-nosed Guenon; Lesser
      White-nosed Monkey; Lesser Spot-
      nosed Monkey; Lesser Spot-nosed
      Guenon; Lesser Putty-nosed Monkey
d   Kleine Weißnasenmeerkatze

f   Hocheur blanc-nez; Hocheur blanc-nez de Bénin
i   Cercopiteco nasobianco minore

1061   *Cercopithecus pogonias*
Primates - Cercopithecidae
e   Crowned Guenon; Crested Mona Monkey; Crested Mona; Crowned Mona; Crowned Monkey; Bioko Crowned Guenon
d   Kronenmeerkatze
f   Cercopithèque pogonias
i   Cercopiteco coronato

1062   *Cercopithecus pogonias grayi*
Primates - Cercopithecidae
e   Gray's Guenon
d   Grays Kronenmeerkatze
f   Mone de Gray

1063   *Cercopithecus preussi*
Primates - Cercopithecidae
e   Preuss's Monkey; Preuss's Guenon
d   Preuss Meerkatze; Preuss Affe
f   Cercopithèque de Preuss
i   Cercopiteco di Preuss

1064   *Cercopithecus roloway*
Primates - Cercopithecidae
e   Roloway Monkey
d   Roloway-Meerkatze; Roloway-Affe
f   Cercopithèque de Roloway

Cercopithecus salongo syn. *C. dryas* q.v.

1065   *Cercopithecus sclateri*
Primates - Cercopithecidae
e   Sclater's Monkey
d   Vollbartmeerkatze
f   Cercopithèque de Sclater
i   Cercopiteco di Sclater

1066   *Cercopithecus solatus*
Primates - Cercopithecidae

e   Sun-tailed Monkey; Sun-tailed Guenon
d   Gabun-Meerkatze
f   Cercopithèque à queue dorée; Cercopithèque à queue soleil; Cercopithèque du Gabon
i   Cercopiteco dorato; Cercopiteco dalla coda dorata

Cercopithecus talapoin syn. *Miopithecus talapoin* q.v.

1067   *Cercopithecus wolfi*
Primates - Cercopithecidae
e   Wolf's Monkey; Wolf's Guenon; Wolf's Mona Monkey; Wolf's Mona
d   Wolf-Meerkatze; Wolfs Mona
f   Cercopithèque de Wolf
i   Cercopiteco coronato di Wolf

1068   *Cerdocyon*
Carnivora - Canidae
e   Crab-eating Foxes
d   Maikongs
f   Renards des savanes; Renards crabiers
i   Cerdocioni

1069   *Cerdocyon thous*
Carnivora - Canidae
e   Crab-eating Fox; Forest Fox; Common Zorro; Savannah Fox; Crab-eating Zorro
d   Maikong; Südamerikanischer Waldfuchs; Krabbenfuchs; Gemeiner Zorro
f   Renard crabier; Petit Loup; Renard des savanes
i   Cerdocione; Volpe dei boschi

1070   *Cervidae*
Artiodactyla
e   Deer ; Antlered Ruminants; Cervids; Deer, Elk. Moose and allies
d   Hirsche
f   Cerfs; Cervidés
i   Cervidi

1071   *Cervus*
          Artiodactyla - Cervidae
e   Red Deer ; Wapitis; Elks; Elk
d   Hirsche; Edelhirsche
f   Cerfs
i   Cervi

1072   *Cervus albirostris*
          Artiodactyla - Cervidae
e   Thorold's Deer; White-lipped Deer;
      Thorold's White-lipped Deer
d   Weißlippenhirsch
f   Cerf de Thorold; Cerf au museau
      blanc
i   Cervo a labbra bianche

1073   *Cervus alfredi*
          Artiodactyla - Cervidae
e   Visayan Spotted Deer; Prince Alfred
      Spotted Deer; Philippine Spotted
      Deer; Prince Alfred's Spotted Rusa
d   Prinz Alfreds Hirsch
f   Cerf de Prince Alfred
i   Cervo pomellato delle Filippine

      Cervus axis syn. *Axis axis* q.v.

      Cervus calamianensis syn. *Axis
      calamianensis* q.v.

      Cervus canadensis syn. *C. elaphus*
      q.v.

      Cervus dama syn. *Dama dama* q.v.

1074   *Cervus duvaucelii*
          Artiodactyla - Cervidae
e   Barasingha Deer; Swamp Deer;
      Barasingha; Duvaucel Deer
d   Barasingha; Zackenhirsch
f   Barasingha; Cerf de Duveaucel
i   Barasinga; Cervo di Duvaucel;
      Barasinga di Duvaucel

1075   *Cervus duvaucelii branderi*
          Artiodactyla - Cervidae

e   South Indian Swamp Deer
d   Mittelindischer Barasingha

1076   *Cervus duvaucelii duvaucelli*
          Artiodactyla - Cervidae
e   North Indian Swamp Deer
d   Nordindischer Barasingha

1077   *Cervus elaphus*
          Artiodactyla - Cervidae
e   Red Deer; European Red Deer;
      Common Deer
d   Hirsch; Edelhirsch; Rothirsch
f   Cerf; Cerf commun; Cerf élaphe;
      Cerf d'Europe; Cerf rouge; Gros cerf;
      Dix-cors; Élaphe d'Europe; Élaphe
      ordinaire; Élaphe vulgaire; Élaphe;
      Cerf noble
i   Cervo nobile; Cervo elafo; Cervo
      vomune

1078   *Cervus elaphus affinis*
          Artiodactyla - Cervidae
e   Bhutan Shou
d   Hanglu
f   Cerf musqué

1079   *Cervus elaphus alashanicus*
          Artiodactyla - Cervidae
e   Ala Shan Wapiti

1080   *Cervus elaphus atlanticus*
          Artiodactyla - Cervidae
e   Norwegian Red Deer
f   Cerf de l'Atlantique

1081   *Cervus elaphus bactrianus*
          Artiodactyla - Cervidae
e   Bactrian Deer; Bukharian Deer;
      Bactrian Red Deer
d   Buchara-Hirsch; Baktrischer
      Rothirsch
f   Cerf rouge du Turkestan
i   Cervo di Bukara; Cervo del
      Turkestan

1082 **Cervus elaphus barbarus**
Artiodactyla - Cervidae
e Barbary Stag; Barbary Deer; Barbary Red Deer
d Berber-Hirsch
f Cerf rouge
i Cervo dell'Atlante; Cervo berbero

1083 **Cervus elaphus canadensis**
Artiodactyla - Cervidae
e American Elk; Eastern Elk; White Deer; Wapiti (NA); Eastern Wapiti (NA)
d Ostwapiti; Östlicher Wapiti
f Cerf du Canada de l'Est; Wapiti de l'Est
i Cervo canadese

1084 **Cervus elaphus corsicanus**
Artiodactyla - Cervidae
e Sardinian Red Deer; Sardinian Deer; Corsican Red Deer
d Tyrrhenischer Rothirsch
f Cerf élaphe de Corse
i Cervo sardo; Cervo tirrenico

1085 **Cervus elaphus elaphus**
Artiodactyla - Cervidae
e Swedish Red Deer; West European Red Deer
d Schwedischer Rothirsch
f Cerf rouge européen
i Cervo rosso europeo

1086 **Cervus elaphus hanglu**
Artiodactyla - Cervidae
e Hanglu Deer; Kashmir Red Deer; Kashmir Deer; Kashmir Stag
d Kaschmir-Hirsch
f Cerf du Cachemire
i Hanglu; Cervo del Kashmir

1087 **Cervus elaphus hippelaphus**
Artiodactyla - Cervidae
e Middle-European Red Deer; Western Red Deer

d Mitteleuropäischer Rothirsch
f Cerf rouge des Carpathes; Cerf rouge alpin
i Cervo italiano; Cervo

1088 **Cervus elaphus hispanicus**
Artiodactyla - Cervidae
e Spanish Red Deer
d Iberia-Hirsch; Iberischer Hirsch
f Cerf d'Espagne

1089 **Cervus elaphus macneilli**
Artiodactyla - Cervidae
e MacNeill's Red Deer
d Szetschuan-Hirsch

1090 **Cervus elaphus maral**
Artiodactyla - Cervidae
e Maral Deer
d Kaukasus-Hirsch
f Cerf maral

1091 **Cervus elaphus merriami**
Artiodactyla - Cervidae
e Arizona Wapiti (NA); Arizona Elk; Merriam's Elk; Meriams Wapiti (NA); Arizona Red Deer; Merriams Red Deer

d Merriams Wapiti; Südwestlicher Wapiti

f Wapiti de l'Arizone; Cerf commun de l'Arizone; Cerf commun de Merriam; Cerf rouge de Merriam; Élaphe vulgaire de Merriam

1092 **Cervus elaphus montanus**
Artiodactyla - Cervidae
e Eastern Red Deer
d Karpaten-Hirsch
f Cerf de l'Est

1093 **Cervus elaphus nannodes**
Artiodactyla - Cervidae
e Tule Elk; Wapiti; Elk (NA)

d   Tule-Hirsch
f   Cerf de Tule

1094   ***Cervus elaphus nelsoni***
        Artiodactyla - Cervidae
e   Nelson's Elk
d   Felsengebirgs-Wapiti
f   Wapiti; Wapiti des Rocheuses
i   Wapiti; Alce delle Montagne
        Rocciose

1095   ***Cervus elaphus scoticus***
        Artiodactyla - Cervidae
e   Scottish Red Deer
d   Schottischer Hirsch
f   Cerf d'Écosse; Cerf écossais
i   Cervo scozzese

1096   ***Cervus elaphus sibiricus***
        Artiodactyla - Cervidae
e   Siberian Red Deer; Siberian Wapiti
d   Altai-Maral
f   Cerf de Sibérie

1097   ***Cervus elaphus wallichi***
        Artiodactyla - Cervidae
e   Wallich Deer; Schou's Deer
d   Tibetanischer Rothirsch; Shou
f   Cerf de Wallich

1098   ***Cervus elaphus xanthopygus***
        Artiodactyla - Cervidae
e   Manchurian Red Deer
d   Subra
f   Cerf chinois

1099   ***Cervus elaphus yarkandiensis***
        Artiodactyla - Cervidae
e   Yarkand Deer; Yarkand Stag;
        Yarkand Red Deer
d   Jarkand-Hirsch
f   Cerf yarkand

1100   ***Cervus eldii***

Artiodactyla - Cervidae
e   Eld's Deer; Brown-antlered Deer
d   Leierhirsch
f   Cerf d'Eld
i   Cervo di Eld; Tameng

1101   ***Cervus eldii eldii***
        Artiodactyla - Cervidae
e   Common Eld's Deer; Manipur
        Brown-antlered Deer; Sangai
d   Manipur-Leierhirsch
f   Cerf d'Eld commun

1102   ***Cervus eldii siamensis***
        Artiodactyla - Cervidae
e   Thailand Brown-antlered Deer
d   Siam-Leierhirsch
f   Cerf au cornes brunes de Thailande

1103   ***Cervus eldii thamin***
        Artiodactyla - Cervidae
e   Thamin Deer; Thamin
d   Birma-Leierhirsch
f   Cerf Thamin

Cervus kuhli syn. *Axis kuhli* q.v.

1104   ***Cervus mariannus***
        Artiodactyla - Cervidae
e   Philippine Sambur; Luzon Sambur;
        Philippine Deer; Philippine Brown
        Deer; Philippine Sambar; Luzon
        Sambar
d   Sambar; Philippinen-Sambar;
        Pferdehirsch; Philippinen-Hirsch;
        Luzon-Sambar
f   Sambar des Philippines
i   Cervo delle Filippine; Sambar delle
        Filippine

Cervus mesopotamica syn. *Dama dama* q.v.

1105   ***Cervus nippon***
        Artiodactyla - Cervidae
e   Japanese Deer; Sika Deer; Sika;

Japanese Sika; Japanese Sika Deer;
Formosan Sika; Chinese Sika Deer

d Sika; Sikahirsch; Japanischer Hirsch;
Japan-Sika

f Cerf sika; Cerf sika du Japon

i Cervo sika

1106 *Cervus nippon centralis*
Artiodactyla - Cervidae

e Hondo Sika Deer

f Sika d'Hondo

1107 *Cervus nippon dybowskii*
Artiodactyla - Cervidae

e Dybowski's Deer; Dybowsky's Sika
Deer

d Dybowski-Hirsch

f Cerf de Dybowski

1108 *Cervus nippon grassianus*
Artiodactyla - Cervidae

e Shansi Sika Deer; Shanxi Sika;
Shansi Sika

d Shanxi-Sikahirsch

f Sika de Shansi

1109 *Cervus nippon hortulorum*
Artiodactyla - Cervidae

e Pekin Sika Deer

f Sika de Pékin

1110 *Cervus nippon keramae*
Artiodactyla - Cervidae

e Keramama Sika Deer; Kerama Deer;
Black Sika Deer

d Kerama-Hirsch

f Sika de Kerama

1111 *Cervus nippon kopschi*
Artiodactyla - Cervidae

e South China Sika Deer; Kopsch's
Deer; South China Deer; Chinese
Sika Deer

d Südchinesischer Sikahirsch

f Sika du sud de la Chine; Sika de

Kopsch

1112 *Cervus nippon mageshimae*
Artiodactyla - Cervidae

e Megeshima Sika Deer

f Sika de Megeshima

1113 *Cervus nippon mandarinus*
Artiodactyla - Cervidae

e North China Sika Deer; Mandarin
Sika Deer; North China Sika;
Mandarin Sika

d Nordchinesischer Sikahirsch;
Nordchinesischer Sika; Mandarin-
Sikahirsch; Mandarin-Sika

f Sika mandarin; Cerf Sika mandarin

1114 *Cervus nippon mantchuricus*
Artiodactyla - Cervidae

e Manchurian Sika Deer

d Mandschurischer Sikahirsch

f Sika de Mandchourie

1115 *Cervus nippon pseudaxis*
Artiodactyla - Cervidae

e Tokhin Sika Deer; Vietnamese Sika
Deer

d Viatnamesischer Sikahirsch; Vietnam
Sikahirsch

f Sika du Tonkin; Cerf sika du
Vietnam

1116 *Cervus nippon taiouanus*
Artiodactyla - Cervidae

e Formosan Sika Deer

d Formosa-Sika; Taiwan-Sika

f Sika de Formose

1117 *Cervus nippon yakushimae*
Artiodactyla - Cervidae

e Yakushima Sika Deer; Maryland
Sika Deer

f Sika de Yakushima

1118 *Cervus nippon yesoensis*
Artiodactyla - Cervidae

e   Hokkaido Sika Deer
d   Hokkaido-Sikahirsch
f   Cerf sika d'Hokkaido

1119   *Cervus schomburgki*
       Artiodactyla - Cervidae
e   Schomburgk's Deer; Schomburgk's Barasingha; Schomburgk's Swamp Deer
d   Schomburgk-Hirsch; Schomburgk-Barasinga; Schomburg-Zackenhirsch
f   Barasingha de Schomburgk; Cerf de Schomburgk

1120   *Cervus timorensis*
       Artiodactyla - Cervidae
e   Sunda Sambur; Timor Deer; Rusa; Rusa Deer; Rus Deer; Sambur; Sambar; Sunda Sambar
d   Mähnenhirsch
f   Sambar de la Sonde

1121   *Cervus timorensis moluccensis*
       Artiodactyla - Cervidae
e   Moluccan Rusa
f   Cerf moluccan

1122   *Cervus timorensis russa*
       Artiodactyla - Cervidae
e   Javan Deer
d   Java-Mähnenhirsch
f   Cerf de Java; Cerf rusa

1123   *Cervus unicolor*
       Artiodactyla - Cervidae
e   Sambur; Sambar Deer; Sambur Deer; Indian Sambur; Indian Sambar
d   Indischer Sambar
f   Sambar
i   Sambar dell'India; Cervo sambar

1124   *Cervus unicolor basilanensis*
       Artiodactyla - Cervidae
e   Basilian Sambur; Basilian Sambar
f   Sambar de Basile

1125   *Cervus unicolor equinus*
       Artiodactyla - Cervidae
e   Malayan Sambur Deer; Malayan Sambar Deer
d   Sumatra-Pferdehirsch
f   Sambar de Malaisie

1126   *Cervus unicolor niger*
       Artiodactyla - Cervidae
e   Indian Sambur Deer; Indian Sambar Deer
d   Vorderindischer Sambar
f   Sambar d'Inde

1127   *Cervus unicolor nigricans*
       Artiodactyla - Cervidae
e   Blackish Sambur Deer
f   Sambar noir

1128   *Cervus unicolor swinhoeii*
       Artiodactyla - Cervidae
e   Swinhoe's Deer
f   Cerf de Swinhoe

1129   *Cervus unicolor unicolor*
       Artiodactyla - Cervidae
e   Ceylonese Sambur Deer; Ceylonese Sambar Deer
d   Ceylon-Pferdehirsch
f   Sambar de Ceylan

1130   *Chaerephon*
       Chiroptera - Molossidae
e   Lesser Free-tailed Bats
d   Freischwanzfledermäuse
f   Phyllanthes

1131   *Chaerephon aloysiisabaudia*
       Chiroptera - Molossidae
e   Duke of Abruzzi's Free-tailed Bat

1132   *Chaerephon ansorgei*
       Chiroptera - Molossidae
e   Ansorge's Free-tailed Bat

1133 **Chaerephon bemmeleni**
Chiroptera - Molossidae
e Gland-tailed Free-tailed Bat; Van Bemmelen's Free-tailed Bat

1134 **Chaerephon bivittata**
Chiroptera - Molossidae
e Spotted Free-tailed Bat

1135 **Chaerephon bregullae**
Chiroptera - Molossidae
e Fijian Mastiff Bat

1136 **Chaerephon chapini**
Chiroptera - Molossidae
e Chapin's Free-tailed Bat; Long-crested Free-tailed Bat
f Phyllanthe de Chapin

1137 **Chaerephon gallagheri**
Chiroptera - Molossidae
e Gallagher's Free-tailed Bat
d Gallaghers Freischwanzfledermaus

1138 **Chaerephon jobensis**
Chiroptera - Molossidae
e Northern Mastiff Bat; Northern Freetail Bat (ANZ)

1139 **Chaerephon johorensis**
Chiroptera - Molossidae
e Northern Free-tailed Bat; Dato Meldrum's Bat; Dato Mastif Bat

1140 **Chaerephon leucogaster**
Chiroptera - Molossidae
e Pale-bellied Free-tailed Bat

1141 **Chaerephon major**
Chiroptera - Molossidae
e Lappet-eared Free-tailed Bat; Large Lappet-eared Free-tailed Bat

1142 **Chaerephon nigeriae**
Chiroptera - Molossidae
e Nigerian Free-tailed Bat

1143 **Chaerephon plicata**
Chiroptera - Molossidae
e Wrinkle-lipped Free-tailed Bat; Wrinkled-lipped Bat; Wrinkle-lipped Bat
d Faltlippenfledermaus

1144 **Chaerephon plicata insularis**
Chiroptera - Molossidae
e Ceylon Wrinkle-lipped Bat
d Sri Lanka-Krauslippenfledermaus

1145 **Chaerephon pumila**
Chiroptera - Molossidae
e Little Free-tailed Bat; Cretzschmar's Little Free-tailed Bat; Free-tailed Bat

1146 **Chaerephon russata**
Chiroptera - Molossidae
e Russet Free-tailed Bat

1147 **Chaerephon solomonis**
Chiroptera - Molossidae
e Solomon's Free-tailed Bat

1148 **Chaerephon tomensis**
Chiroptera - Molossidae
e Sao Tomé Free-tailed Bat

1149 **Chaeropus**
Peramelemorphia - Peramelidae
e Pig-footed Bandicoots
d Schweinsfußnasenbeutler
f Bandicoots aux pieds de cochon
i Perameli; Bandicut a piedi di porco

1150 **Chaeropus ecaudatus**
Peramelemorphia - Peramelidae
e Pig-footed Bandicoot
d Schweinsfußnasenbeutler; Schweinsfuß; Schweinsfußbeuteldachs; Schweinsfußbandikut; Stutzbeutler

f   Bandicoot aux pieds de cochon;
     Bandicoot à pied de porc
i   Peramele; Bandicut a piedi di porco

Chaetocauda syn. *Dryomys* q.v.

1151   *Chaetodipus*
          Rodentia - Geomyidae
e   Pocket Mice; Coarse-haired Poxket
     Mice
d   Taschenmäuse
f   Souris à poche

1152   *Chaetodipus arenarius*
          Rodentia - Geomyidae
e   Little Desert Pocket Mouse; Sand
     Pocket Mouse

1153   *Chaetodipus artus*
          Rodentia - Geomyidae
e   Narrow-skulled Pocket Mouse

1154   *Chaetodipus baileyi*
          Rodentia - Geomyidae
e   Bailey's Pocket Mouse
d   Bailey-Taschenmaus
f   Souris à poche de Bailey

1155   *Chaetodipus californicus*
          Rodentia - Geomyidae
e   California Pocket Mouse
d   Kalifornische Taschenmaus
f   Souris à poche de Californie

1156   *Chaetodipus eremicus*
          Rodentia - Geomyidae
e   Chihuahan Desert Pocket Mouse

1157   *Chaetodipus fallax*
          Rodentia - Geomyidae
e   San Diego Pocket Mouse

1158   *Chaetodipus formosus*
          Rodentia - Geomyidae
e   Long-tailed Pocket Mouse

1159   *Chaetodipus goldmani*
          Rodentia - Geomyidae
e   Goldman's Pocket Mouse

1160   *Chaetodipus hispidus*
          Rodentia - Geomyidae
e   Hispid Pocket Mouse
d   Borstige Taschenmaus

1161   *Chaetodipus hispidus spilotus*
          Rodentia - Geomyidae
e   Kansas Pocket Mouse

1162   *Chaetodipus intermedius*
          Rodentia - Geomyidae
e   Rock Pocket Mouse

1163   *Chaetodipus lineatus*
          Rodentia - Geomyidae
e   Lined Pocket Mouse

1164   *Chaetodipus nelsoni*
          Rodentia - Geomyidae
e   Nelson's Pocket Mouse

1165   *Chaetodipus penicillatus*
          Rodentia - Geomyidae
e   Desert Pocket Mouse; Sonoran
     Desert Pocket Mouse

1166   *Chaetodipus pernix*
          Rodentia - Geomyidae
e   Sinaloan Pocket Mouse

1167   *Chaetodipus spinatus*
          Rodentia - Geomyidae
e   Spiny Pocket Mouse

1168   *Chaetomys*
          Rodentia - Erethizontidae
e   Thin-spined Porcupines; Bristle-
     spined Rats
d   Borstenbaumstachler
f   Porcs-épics épineux

1169 *Chaetomys subspinosus*
Rodentia - Erethizontidae
e Thin-spined Porcupine; Bristle-spined Rat
d Borstenbaumstachler
f Porc-épic épineux

1170 *Chaetophractus*
Ctngulata - Dasypodidae
e Hairy Armadillos; Quirquinchos
d Borstengürteltiere
f Tatous

1171 *Chaetophractus nationi*
Ctngulata - Dasypodidae
e Andean Hairy Armadillo; Hairy Armadillo; Quirquincho
d Braunborstengürteltier; Anden-Borstentier
f Tatou des Andes
i Armadillo peloso

1172 *Chaetophractus vellerosus*
Ctngulata - Dasypodidae
e Long-haired Armadillo; Screaming Armadillo; Screaming Hairy Armadillo; Little Hairy Armadillo
d Kleines Haariges Borstengürteltier; Schreiendes Haariges Gürteltier; Weißhaargürteltier
f Tatou velu des Andes
i Armadillo peloso piccolo

1173 *Chaetophractus villosus*
Ctngulata - Dasypodidae
e Large Hairy Armadillo; Peludo; Larger Hairy Armadillo; Big Hairy Armadillo
d Braunzottiges Gürteltier; Borstengürteltier; Braunzottiges Borstengürteltier; Braunborstengürteltier; Braunhaargürteltier
i Armadillo villoso; Armadillo peloso grande

Chalcomys syn. *Akodon* q.v.

1174 *Chalinolobus*
Chiroptera - Vespertilionidae
e Lobe-lipped Bats; Groove-lipped Bats; Wattled Bats
f Chalinolobes

1175 *Chalinolobus alboguttatus*
Chiroptera - Vespertilionidae
e Allen's Striped Bat

1176 *Chalinolobus argentatus*
Chiroptera - Vespertilionidae
e Silvered Bat; Common Glauconycterid

1177 *Chalinolobus beatrix*
Chiroptera - Vespertilionidae
e Beatrix's Bat

1178 *Chalinolobus curryi*
Chiroptera - Vespertilionidae
e Curry's Bat

1179 *Chalinolobus dwyeri*
Chiroptera - Vespertilionidae
e Complex Wattled Bat; Large-eared Mottled Bat; Large Pied Bat; Large-eared Pied Bat (ANZ)

1180 *Chalinolobus egeria*
Chiroptera - Vespertilionidae
e Bibundi Bat

1181 *Chalinolobus gleni*
Chiroptera - Vespertilionidae
e Glen's Wattled Bat

1182 *Chalinolobus gouldi*
Chiroptera - Vespertilionidae
e Gould's Bat; Gould's Wattled Bat (ANZ)

1183 *Chalinolobus kenyacola*
Chiroptera - Vespertilionidae
e Kenyan Wattled Bat

1184   *Chalinolobus morio*
Chiroptera - Vespertilionidae
e   Chocolate Bat; Chocolate Wattled
Bat (ANZ)

1185   *Chalinolobus neocaledonicus*
Chiroptera - Vespertilionidae
e   New Caledonian Wattled Bat
f   Chalinolobe néo-calédonien

1186   *Chalinolobus nigrogriseus*
Chiroptera - Vespertilionidae
e   Hoary Bat; Frosted Bat; Hoary
Wattled Bat (ANZ)

1187   *Chalinolobus picatus*
Chiroptera - Vespertilionidae
e   Pied Scotophilus; Pied Wattled Bat;
Little Pied Bat (ANZ)

1188   *Chalinolobus poensis*
Chiroptera - Vespertilionidae
e   Abo Bat

1189   *Chalinolobus superbus*
Chiroptera - Vespertilionidae
e   Pied Bat; African Pied Bat

1190   *Chalinolobus tuberculatus*
Chiroptera - Vespertilionidae
e   Long-tailed Bat; Long-tailed Wattled
Bat; New Zealand Long-tailed Bat;
Lobe-lipped Bat; New Zealand
Wattled Bat

1191   *Chalinolobus variegatus*
Chiroptera - Vespertilionidae
e   Butterfly Bat

1192   Cheirogaleidae
Primates
e   Lemurs; Mouse Lemurs; Dwarf
Lemurs; Dwarf and Mouse Lemurs
d   Zwerglemuren; Kleinlemuren
f   Cheirogaleidés

i   Lemuri pigmei; Chirogalei con le
orecchie pelose; Cheirogaleidi

1193   *Cheirogaleus*
Primates - Cheirogaleidae
e   Dwarf Lemurs; Fat-tailed Dwarf
Lemurs; Mouse Lemurs
d   Katzenmakis; Echte Katzenmakis
f   Chirogales; Lémurs nains

1194   *Cheirogaleus adipicaudatus*
Primates - Cheirogaleidae
e   Southern Fat-tailed Dwarf Lemur
d   Südlicher Katzenmaki

1195   *Cheirogaleus crossleyi*
Primates - Cheirogaleidae
e   Crossley's Mouse Lemur; Furry-
eared Dwarf Lemur
d   Haariger Zwergartiger Lemur

1196   *Cheirogaleus major*
Primates - Cheirogaleidae
e   Greater Dwarf Lemur
d   Großer Katzenmaki
f   Chirogale de Milius; Grand
Chirogale
i   Chirogaleo bruno

1197   *Cheirogaleus medius*
Primates - Cheirogaleidae
e   Fat-tailed Dwarf Lemur; Fat-tailed
Lemur; Western Fat-tailed Dwarf
Lemur
d   Mittlerer Katzenmaki;
Fettschwanzmaki; Fettschwanz-
Katzenmaki; Mittlerer
Fettschwanzmaki

f   Chirogale à queue grasse; Petit
Chirogale
i   Chirogaleo medio; Chirogaleo dalla
coda grossa

1198   *Cheirogaleus ravus*
Primates - Cheirogaleidae
e   Large Iron-grey Dwarf Lemur;

Greater Iron-grey Dwarf Lemur

d Großer Grauer Katzenmaki

1199 *Cheirogaleus sibreei*
Primates - Cheirogaleidae
e Sibree's Dwarf Lemur
d Sibrees Zwergartiger Lemur

Cheirogaleus trichotis syn.
Allocebus trichotis q.v.

1200 *Cheiromeles*
Chiroptera - Molossidae
e Naked Bats; Hairless Bats; Naked Bulldog Bats
d Halsbandfledermäuse
f Cheiromèles

1201 *Cheiromeles parvidens*
Chiroptera - Molossidae
e Lesser Hairless Bat

1202 *Cheiromeles torquatus*
Chiroptera - Molossidae
e Naked Bat; Hairless Bat; Philippine Naked Bat; Philippine Hairless Bat; Greater Naked Bat
d Halsbandfledermaus; Nacktfledermaus
f Cheiromèle nue

1203 *Chelemys*
Rodentia - Muridae
e Greater Long-clawed Mice
d Große Langkrallenmäuse
f Souris-taupes

1204 *Chelemys macronyx*
Rodentia - Muridae
e Andean Long-clawed Mouse; Field Mole Mouse

1205 *Chelemys megalonyx*
Rodentia - Muridae
e Large Long-clawed Mouse

Cheliones syn. *Meriones* q.v.

1206 *Chibchanomys*
Rodentia - Muridae
e Chibchan Water Mice
d Chibcha-Wassermäuse

1207 *Chibchanomys orcesi*
Rodentia - Muridae
e Orces's Chibchan Water Mouse

1208 *Chibchanomys trichotis*
Rodentia - Muridae
e Chibchan Water Mouse

1209 *Chilomys*
Rodentia - Muridae
e Colombian Forest Mice
d Kolumbianische Waldmäuse

1210 *Chilomys instans*
Rodentia - Muridae
e Colombian Forest Mouse
d Kolumbianische Waldmas

Chilonatalus syn. *Natalus* q.v.

Chilonycteris syn. *Pteronotus* q.v.

1211 *Chimarrogale*
Soricomorpha - Soricidae
e Asiatic Water Shrews; Asian Water Shrews
d Biberspitzmäuse
f Chimarogales

1212 *Chimarrogale hantu*
Soricomorpha - Soricidae
e Hantu Water Shrew; Malayan Water Shrew
d Hantu-Waldspitzmaus

1213 *Chimarrogale himalayica*
Soricomorpha - Soricidae

e   Himalajan Water Shrew
d   Himalaja-Wasserspitzmaus
f   Chimarogale de l'Himalaja

1214   *Chimarrogale phaeura*
Soricomorpha - Soricidae
e   Borneo Water Shrew; Sunda Water
    Shrew; Bornen Water Shrew
d   Borneo-Wasserspitzmaus

1215   *Chimarrogale platycephala*
Soricomorpha - Soricidae
e   Flat-headed Water Shrew; Japanese
    Water Shrew

1216   *Chimarrogale styani*
Soricomorpha - Soricidae
e   Styan's Water Shrew; Burmese Water
    Shrew; Montane Water Shrew

1217   *Chimarrogale sumatrana*
Soricomorpha - Soricidae
e   Sumatra Water Shrew

1218   *Chinchilla*
Rodentia - Chinchillidae
e   Chinchillas
d   Wollmäuse; Chinchillas; Eigentliche
    Chinchillas
f   Chinchillas
i   Cincillà

1219   *Chinchilla brevicaudata*
Rodentia - Chinchillidae
e   Short-tailed Chinchilla
d   Kurzschwanzchinchilla; Chinchilla;
    Große Chinchilla
f   Chinchilla à courte queue
i   Cincillà dalla coda corta

1220   *Chinchilla brevicaudata boliviana*
Rodentia - Chinchillidae
e   Lesser Short-tailed Chinchilla
d   Kleine Kurzschwanzchinchilla
i   Cincillà delle Ande

1221   *Chinchilla lanigera*
Rodentia - Chinchillidae
e   Chinchilla; Long-tailed Chinchilla
d   Wollmaus; Kleine Chinchilla;
    Langschawnzchinchilla
f   Chinchilla à longue queue
i   Cincillà; Cincillà dalla coda lunga

1222   Chinchillidae
Rodentia
e   Chinchillas; Viscachas; Viscachas
    and Chinchillas
d   Chinchillas; Eigentliche Chinchillas
f   Chinchillidés
i   Cincillidi

1223   *Chinchillula*
Rodentia - Muridae
e   Chinchilla Mice; Altiplano
    Chinchilla Mice
d   Altiplano-Chinchillamäuse
f   Rats chinchilla

1224   *Chinchillula sahamae*
Rodentia - Muridae
e   Chinchilla Mouse; Altiplano
    Chinchilla Mouse; Achula
d   Altiplano-Chinchillamaus
f   Rat chinchilla

1225   *Chionomys*
Rodentia - Muridae
e   Snow Voles
d   Schneemäuse
f   Campagnols
i   Arvicole delle nevi

1226   *Chionomys gud*
Rodentia - Muridae
e   Caucasian Snow Vole

1227   *Chionomys nivalis*
Rodentia - Muridae
e   European Snow Vole; Snow Mouse;
    Snow Vole

d   Alpenschneemaus; Alpenratte;
Alpenwühlmaus; Europäische
Schneemaus; Schneemaus

f   Campagnol des Alpes; Campagnol
des neiges; Campagnol à queue
blanche

i   Arvicola delle nevi; Campagnola
della neve

1228   *Chionomys roberti*
Rodentia - Muridae

e   Robert's Vole; Robert's Snow Vole

1229   *Chiroderma*
Chiroptera - Phyllostomidae

e   Big-eyed Bats; White-lined Bats;
Bats Without Nasal Bones

1230   *Chiroderma doriae*
Chiroptera - Phyllostomidae

e   Brazilian Big-eyed Bat

1231   *Chiroderma improvisum*
Chiroptera - Phyllostomidae

e   Guadeloupe White-lined Bat;
Guadeloupe Big-eyed Bat; Antillean
White-lined Bat

1232   *Chiroderma salvini*
Chiroptera - Phyllostomidae

e   Salvin's White-lined Bat; Salvin's
Big-eyed Bat

1233   *Chiroderma trinitatum*
Chiroptera - Phyllostomidae

e   Trinidad Big-eyed Bat; Trinidadian
White-lined Bat; Trinidad White-
lined Bat; Goodwin's White-lined
Bat; Little Big-eyed Bat

1234   *Chiroderma villosum*
Chiroptera - Phyllostomidae

e   Hairy White-lined Bat; Peters's
White-lined Bat; Greater White-lined
Bat; Shaggy White-lined Bat; Hairy
Big-eyed Bat; Big-eyed Bat

1235   *Chiromyscus*
Rodentia - Muridae

e   Asiatic Tree Mice; Fea's Tree Rats

d   Fea-Baumratten

1236   *Chiromyscus chiropus*
Rodentia - Muridae

e   Asiatic Tree Mouse; Fea's Tree Rat

d   Feas Baumratte; Fea-Baumratte

1237   *Chironax*
Chiroptera - Pteropodidae

e   Black-capped Fruit Bats

1238   *Chironax melanocephalus*
Chiroptera - Pteropodidae

e   Black-capped Fruit Bat

1239   *Chironectes*
Didelphimorphia - Didelphidae

e   Water Opossums; Yapoks; Yapocks

d   Schwimmbeutelratten;
Schwimmbeutler ; Wasseropossums

f   Yapoks; Yapocks

i   Chironetti

1240   *Chironectes minimus*
Didelphimorphia - Didelphidae

e   Water Opossum; Yapok; Yapock;
Aquatic Opossum

d   Schwimmbeutelratte;
Schwimmbeutler; Yapok;
Wasseropossum

f   Yapok; Yapock; Loutre marsupiale

i   Chironetto; Opossum dell'acqua

1241   *Chiropodomys*
Rodentia - Muridae

e   Pencil-tailed Tree Mice; Complex-
toothed Tree Mice

d   Pinselschwanzbaummäuse

f   Souris arboricoles; Souris d'arbre de
Malaysie; Souris de Malaisie

1242   *Chiropodomys calamianensis*
Rodentia - Muridae

e   Palawan Pencil-tailed Tree Mouse

1243   *Chiropodomys gliroides*
Rodentia - Muridae

e   Pencil-tailed Tree Mouse; Common
Pencil-tailed Tree Mouse; Javan
Pencil-tailed Mouse

d   Pinselschwanzbaummaus; Malaiische
Pinselschwanzbaummaus; Java-
Pinselschwanzbaummaus

f   Souris d'arbre; Souris arboricole

1244   *Chiropodomys karlkoopmani*
Rodentia - Muridae

e   Koopman's Pencil-tailed Tree Mouse;
Mentawai Pencil-tailed Tree Mouse;
Mentawai Pencilled-tailed Tree
Mouse

1245   *Chiropodomys major*
Rodentia - Muridae

e   Large Pencil-tailed Tree Mouse
d   Borneo-Pinselschwanzbaummaus

1246   *Chiropodomys muroides*
Rodentia - Muridae

e   Grey-bellied Pencil-tailed Tree
Mouse; Bornean Pencil-tailed Tree
Mouse

1247   *Chiropodomys pusillus*
Rodentia - Muridae

e   Small Pencil-tailed Tree Mouse;
Lesser Pencil-tailed Tree Mouse

1248   *Chiropotes*
Primates - Aatelidae

e   Bearded Sakis; Red-backed Sakis;
Bearded Saki Monkeys
d   Bartsakis
f   Sakis
i   Chiropote

1249   *Chiropotes albinasus*

Primates - Aatelidae

e   White-nosed Saki; White-nosed
Bearded Saki; White-nosed Bearded
Saki Monkey; Bearded Saki; Bearded
White-nosed Saki
d   Weißnasensaki; Zottelaffe
f   Saki à nez blanc
i   Saki dal naso bianco; Chiropote dal
naso bianco

1250   *Chiropotes satanas*
Primates - Aatelidae

e   Black-bearded Saki; Red-backed
Saki; Jacket Monkey; Black Saki;
Brown Bearded Saki; Bearded Saki;
Cuxius

d   Satansaffe
f   Satan; Saki Satan; Saki noir
i   Chiropote satanasso; Colobo satana

1251   *Chiruromys*
Rodentia - Muridae
e   New Guinea Tree Mice; Tree Mice
d   Baumratten
f   Rats à queue préhensile

1252   *Chiruromys forbesi*
Rodentia - Muridae

e   Greater Tree Mouse; Forbes's Tree
Mouse
d   Große Baumratte

Chiruromys kagi syn. *C. lamia* q.v.

1253   *Chiruromys lamia*
Rodentia - Muridae

e   Broad-skulled Tree Mouse; Broad-
headed Tree Mouse
d   Breitkopfbaumratte

Chiruromys shawmayeri syn. *C.
forbesi* q.v.

1254   *Chiruromys vates*
Rodentia - Muridae

e   Lesser Tree Mouse
d   Kleine Baumratte

1255   *Chlamyphorus*
Cingulata - Dasypodidae
e   Pichiciegos; Fairy Armadillos
d   Gürtelmulle
f   Chlamyphores
i   Clamifori; Pichicieghi

1256   *Chlamyphorus retusus*
Cingulata - Dasypodidae
e   Greater Pichiciego; Burmeister's
Armadillo; Chacoan Fairy Armadillo;
Greater Fairy Armadillo
d   Burmeister-Gürtelmull
f   Grand chlamyphore
i   Pichiciego maggiore

1257   *Chlamyphorus truncatus*
Cingulata - Dasypodidae
e   Lesser Pichiciego; Fairy Armadillo;
Lesser Fairy Armadillo; Pink Fairy
Armadillo; Pygmy Armadilo
d   Gürtelmull; Gürtelmaus; Kleiner
Gürtelmull; Schildwurf
f   Chlamyphore tronqué
i   Clamiforo troncato; Pichiciego
minore

1258   *Chlorocebus*
Primates - Cercopithecidae
e   Vervet Monkeys; African Green
Monkeys
d   Grüne Meerkatzen
f   Singes verts

1259   *Chlorocebus aethiops*
Primates - Cercopithecidae
e   Vervet Monkey; African Green
Monkey; Grivet; Grivet Monkey;
Vervet
d   Grüne Meerkatze; Graugrüne
Meerkatze
f   Grivet; Singe vert; Vervet; Singe
grivet

i   Cercopiteco grigioverde

1260   *Chlorocebus cynosurus*
Primates - Cercopithecidae
e   Malbrouk Monkey
f   Malbrouk

1261   *Chlorocebus djamdjamensis*
Primates - Cercopithecidae
e   Bale Monkey; Bale Mountains
Vervet
d   Ballen-Bergvervet
i   Cercopiteco delle montagne;
Cercopiteco delle montagne Bale;
Cercopiteco del Monte Bali

1262   *Chlorocebus pygerythrus*
Primates - Cercopithecidae
e   Vervet; Vervet Monkey; Savanna
Monkey; Savannah Monkey
f   Vervet bleu
i   Cercopiteco verde

1263   *Chlorocebus sabaeus*
Primates - Cercopithecidae
e   African Sabaeus Monkey; Green
Monkey
d   Gelbgrüne Meerkatze; Grüner Affe
f   Vervet vert
i   Cercopiteco gialloverde

1264   *Chlorocebus tantalus*
Primates - Cercopithecidae
e   Tantalus Monkey
d   Tantalus-Affe; Tantalus-Meerkatze
f   Vervet tantale
i   Cercopiteco tantalo

1265   *Chlorotalpa*
Chrysochloridea - Chrysochloridae
e   Golden Moles; African Golden
Moles

1266   *Chlorotalpa arendsi*
Chrysochloridea - Chrysochloridae
e   Arend's Golden Mole

1267   *Chlorotalpa duthiae*
Chrysochloridea - Chrysochloridae
e   Duthie's Golden Mole

1268   *Chlorotalpa leucorhina*
Chrysochloridea - Chrysochloridae
e   Congo Golden Mole

1269   *Chlorotalpa sclateri*
Chrysochloridea - Chrysochloridae
e   Sclater's Golden Mole

1270   *Chlorotalpa tytonis*
Chrysochloridea - Chrysochloridae
e   Somali Golden Mole

Chodsigoa syn. *Soriculus* q.v.

1271   *Choeroniscus*
Chiroptera - Phyllostomidae
e   Long-tailed Bats; Godman's Long-
        nosed Bats

1272   *Choeroniscus godmani*
Chiroptera - Phyllostomidae
e   Godman's Bat; Godman's Long-
        nosed Bat; Godman's Long-tailed
        Bat; Godman's Whiskered Long-
        nosed Bat

Choeroniscus inca syn. *C. minor* q.v.

**_Choeroniscus intermedius_ syn. *C.
minor* q.v.**

1273   *Choeroniscus minor*
Chiroptera - Phyllostomidae
e   Tailed Long-nosed Bat; Lesser Long-
        tailed Bat; Lesser Whiskered Long-
        nosed Bat; Tailed Bat; Trinidadian
        Long-nosed Bat; Trinidad Long-

nosed Bat; IntermediateLong-tailed
Bat

1274   *Choeroniscus periosus*
Chiroptera - Phyllostomidae
e   Greater Long-tailed Bat; Greater
        Whiskered Long-nosed Bat

1275   *Choeronycteris*
Chiroptera - Phyllostomidae
e   Hog-nosed Bats; Mexican Long-
        tongued Bats
d   Langnasenfledermäuse
f   Chauves-souris du Mexique

**_Choeronycteris harrisoni_ syn.
Musonycteris harrisoni q.v.**

1276   *Choeronycteris mexicana*
Chiroptera - Phyllostomidae
e   Hog-nosed Bat; Mexican Long-
        tongued Bat; Mexican Hog-nosed
        Bat
d   Langnasenfledermaus
f   Chauves-souris du Mexique

Choeropsis syn. *Hexaprotodon* q.v.

1277   *Choloepus*
Pilosa - Megalonychidae
e   Two-toed Sloths
d   Zweifingerfaultiere;
        Zweizehenfaultiere
f   Unaus
i   Colepi

1278   *Choloepus didactylus*
Pilosa - Megalonychidae
e   Linnaeus's Two-toed Sloth; Unau;
        Linné's Two-toed Sloth; Southern
        Two-toed Sloth
d   Unau; Zweifingerfaultier;
        Zweizehenfaultier; Südliches
        Zweifingerfaultier
f   Unau commun
i   Bradipo didattilo comune

1279   *Choloepus hoffmanni*
       Pilosa - Megalonychidae
  e Hoffmann's Two-toed Sloth;
       Hoffmann's Sloth
  d Hoffmans Zweifingerfaultier;
       Hoffmann-Zweifingerfaultier
  f Unau d'Hoffmann
  i Colepo di Hoffmann; Grande
       Bradipo di Hoffmann

1280   *Chroeomys*
       Rodentia - Muridae
  e Altiplano Mice
  d Altiplano-Mäuse

1281   *Chroeomys andinus*
       Rodentia - Muridae
  e Andean Altiplano Mouse

1282   *Chroeomys jelskii*
       Rodentia - Muridae
  e Jelski's Altiplano Mouse

1283   *Chrotogale*
       Carnivora - Viverridae
  e Owston's Civets; Owston's Banded
       Civets; Owston's Palm Civets
  d Fleckenroller
  f Civettes palmistes d'Owston
  i Civette palma

1284   *Chrotogale owstoni*
       Carnivora - Viverridae
  e Owston's Civet; Owston's Banded
       Civet; Owston's Palm Civet;
       Owston's Banded Palm Civet
  d Vietnamesischer Fleckenroller
  f Civette palmiste d'Owston
  i Civetta palma di Owston

1285   *Chrotomys*
       Rodentia - Muridae
  e Luzon Striped Rats; Luzon Back-
       striped Rats; Philippine Striped Rats
  d Streifenratte
  f Rats à bandes; Rats des Philippines

1286   *Chrotomys gonzalesi*
       Rodentia - Muridae
  e Isarog Striped Shrewrat; Mount
       Sarog Striped Rat
  d Isarog-Streifenratte

1287   *Chrotomys mindorensis*
       Rodentia - Muridae
  e Mindoro Striped Rat; Lowland
       Striped Shrewrat
  d Luzon-Flachlandstreifenratte

1288   *Chrotomys whiteheadi*
       Rodentia - Muridae
  e Luzon Striped Rat; Luzon Back-
       striped Rat; Luzon Montane Striped
       Shrewrat
  d Luzon-Streifenratte

1289   *Chrotopterus*
       Chiroptera - Phyllostomidae
  e Peters's Woolly False Vampire Bats;
       Woolly False Vampire Bats; Peters's
       False Vampires; Bit-eared Woolly
       Bars

1290   *Chrotopterus auritus*
       Chiroptera - Phyllostomidae
  e Peters's Woolly False V ampire Bat;
       Woolly False Vampire Bat; Peters's
       False Vampire; Big-eared Woolly
       Bat; Peters's False Vampire Bat

1291   **Chrysochloridae**
       Chrysochloridea
  e Golden Moles; Giant Golden Moles;
       Crisocloridi
  d Goldmulle; Goldmaulwürfe;
       Goldmullartige
  f Chrysochloridés; Taupes dorées
  i Talpe dorate

1292   *Chrysochloris*
       Chrysochloridea - Chrysochloridae
  e Golden Moles; Cape Golden Moles
  d Goldmulle; Kap-Goldmulle

f   Taupes dorées
i   Talpe dorate

1293   *Chrysochloris asiatica*
Chrysochloridea - Chrysochloridae
e   Cape Golden Mole; African Golden
Mole
d   Kapland-Goldmull; Kap-Goldmull
f   Taupe dorée du Cap
i   Talpa dorata del Capo

Chrysochloris fosteri syn. *C.
stuhlmanni* q.v.

1294   *Chrysochloris stuhlmanni*
Chrysochloridea - Chrysochloridae
e   Stuhlmann's Golden Mole
f   Taupe dorée de Stuhlmann
i   Talpa dorata di Stuhlmann

1295   *Chrysochloris visagiei*
Chrysochloridea - Chrysochloridae
e   Visagie's Golden Mole
f   Taupe dorée de Visagie
i   Talpa dorata di Visagie

1296   *Chrysocyon*
Carnivora - Canidae
e   Maned Wolves
d   Mähnenwölfe
f   Loups à crinière
i   Crisocioni

1297   *Chrysocyon brachyurus*
Carnivora - Canidae
e   Maned Wolf
d   Mähnenwolf
f   Loup à crinière
i   Crisocione; Lupo guara

1298   *Chrysospalax*
Chrysochloridea - Chrysochloridae
e   Giant Golden Moles; Large Golden
Moles
d   Riesengoldmulle; Riesenmulle

f   Grandes taupes dorées

1299   *Chrysospalax trevelyani*
Chrysochloridea - Chrysochloridae
e   Giant Golden Mole; Large Golden
Mole
d   Riesengoldmull
f   Grande taupe dorée

1300   *Chrysospalax villosus*
Chrysochloridea - Chrysochloridae
e   Sand Golden Mole; Rough-haired
Golden Mole

Cistugo syn. *Myotis* q.v.

Citellus syn. *Spermophilus* q.v.

1301   *Civettictis*
Carnivora - Viverridae
e   African Civets
d   Afrikanische Zibetkatzen
f   Civettes d'Afrique
i   Civette zibetto

1302   *Civettictis civetta*
Carnivora - Viverridae
e   African Civet; African Civet Cat
d   Afrika-Zibetkatze; Afrikanische
Zibetkatze

f   Civette d'Afrique; Civette africaine;
Civette à bandes
i   Civetta zibetto

Claviglis syn. *Graphiurus* q.v.

1303   *Clethrionomys*
Rodentia - Muridae
e   Bank Voles; Red-backed Voles; Red-
backed Mice
d   Rötelmäuse; Waldwühlmäuse
f   Campagnols
i   Arvicole

1304   ***Clethrionomys californicus***
Rodentia - Muridae
e   Western Red-backed Vole
d   Westliche Rötelmaus

1305   ***Clethrionomys californicus californicus***
Rodentia - Muridae
e   California Red-backed Mouse;
California Red-backed Vole
d   Kalifornische Rötelmaus

1306   ***Clethrionomys centralis***
Rodentia - Muridae
e   Tien Shan Red-backed Vole
d   T'ien-Shanische Rötelmaus

Clethrionomys frater syn. *C. centralis* q.v.

1307   ***Clethrionomys gapperi***
Rodentia - Muridae
e   Gapper's Red-backed Mouse;
Gapper' Red-backed Vole; Southern
Red-backed Vole; Red-backed Vole
(NA)
d   Gappers Rötelmaus
f   Campagnol à dos roux de Gapper

1308   ***Clethrionomys gapperi loringi***
Rodentia - Muridae
e   Loring's Red-backed Mouse
f   Campagnol à dos roux de Loring

1309   ***Clethrionomys glareolus***
Rodentia - Muridae
e   Bank Vole; Common Red-backed
Vole; Red-backed Vole
d   Rötelmaus; Waldwühlmaus; Ruttners
Rötelmaus; Nördliche
Flachlandrötelmaus; Harz-Maus;
Gewöhnliche Rötelmaus
f   Campagnol roussâtre; Campagnol du
Nord; Campagnol des bois;
Campagnol des grèves; Campagnol
des sables; Campagnol fauve;
Campagnol glaréole

i   Arvicola rossastra

1310   ***Clethrionomys glareolus glareolus***
Rodentia - Muridae
d   Mitteleuropäische Rötelmaus

1311   ***Clethrionomys glareolus istericus***
Rodentia - Muridae
d   Donau-Rötelmaus

1312   ***Clethrionomys glareolus ruttneri***
Rodentia - Muridae
d   Ruttner-Rötelmaus

1313   ***Clethrionomys glareolus skomarensis***
Rodentia - Muridae
e   Skomar Vole

Clethrionomys occidentalis syn. *C. californicus* q.v.

Clethrionomys rex syn. *C. rufocanus* q.v.

1314   ***Clethrionomys rufocanus***
Rodentia - Muridae
e   Korean Red-backed Mouse; Grey
Large-toothed Red-backed Vole;
Grey-sided Vole; Grey Red-backed
Vole
d   Grauseitenrötelmaus; Graurötelmaus;
Graue Rötelmaus; Östliche
Rötelmaus
f   Campagnol gris-roux de Sundevall;
Campagnol de Sundevall
i   Arvicola di Sundevall

1315   ***Clethrionomys rutilus***
Rodentia - Muridae
e   Ruddy Vole; Northern Red-backed
Vole; Northern Red-backed Mouse
d   Polar-Rötelmaus; Sibirische Polar-
Rötelmaus; Nordische Rötelmaus;
Sibirische Rötelmaus; Gelbbraune
Rötelmaus
f   Campagnol roux boréal; Campagnol

boréal; Campagnol rouge nordique
i   Arvicola boreale

1316   *Clethrionomys sikotanensis*
Rodentia - Muridae
e   Shikotan Vole

1317   *Cloeotis*
Chiroptera - Rhinolophidae
e   African Trident-nosed Bats;
Percival's Trident Bats
d   Kurzohrige Fledermäuse

1318   *Cloeotis percivali*
Chiroptera - Rhinolophidae
e   African Trident-nosed Bat; East
African Trident Bat; Percival's
Trident Bat; Short-eared Trident Bat

1319   *Clyomys*
Rodentia - Echimyidae
e   Spiny Rats
f   Rats à grosse tête

1320   *Clyomys bishopi*
Rodentia - Echimyidae
e   Bishop's Fossorial Spiny Rat

1321   *Clyomys laticeps*
Rodentia - Echimyidae
e   Broad-headed Spiny Rat
f   Rat à grosse tête

1322   *Coccymys*
Rodentia - Muridae
e   Brush Mice
f   Rats de Nouvelle Guinée

1323   *Coccymys albidens*
Rodentia - Muridae
e   White-toothed Brush Mouse; White-
toothed Mouse; White-toothed
Melomys
d   Melomysratte

1324   *Coccymys ruemmleri*
Rodentia - Muridae
e   Rümmler's Brush Mouse

Coelomys syn. *Mus* q.v.

1325   *Coelops*
Chiroptera - Rhinolophidae
e   Tailless Leaf-nosed Bats

1326   *Coelops frithi*
Chiroptera - Rhinolophidae
e   Tailless Leaf-nosed Bat; East Asian
Tailless Horseshoe Bat; East Asian
Tailless Leaf-nosed Bat

1327   *Coelops hirsutus*
Chiroptera - Rhinolophidae
e   Philippine Tailless Leaf-nosed Bat;
Philippine Tailless Roundleaf Bat

1328   *Coelops robinsoni*
Chiroptera - Rhinolophidae
e   Malayan Tailless Horseshoe Bat;
Malayan Tailless Leaf-nosed Bat

1329   *Coendou*
Rodentia - Erethizontidae
e   Tree Porcupines; Prehensile-tailed
Porcupines; Central and South
American American Porcupines
d   Greifstachler
f   Porcs-épics préhensiles; Porcs-épics
i   Coendu

1330   *Coendou bicolor*
Rodentia - Erethizontidae
e   Bicoloured Spined Porcupine;
Bicolour-spined Porcupine
d   Wolliger Greifstachler;
Wollgreifstachler

1331   *Coendou ichillus*
Rodentia - Erethizontidae
e   Long-tailed Hairy Dwarf Porcupine

1332   *Coendou insidiosus*
       Rodentia - Erethizontidae
   e   Bahia Hairy Dwarf Porcupine;
       Bristly Porcupine; Bristly Prehensile-
       tailed Porcupine; Woolly Prehensile-
       tailed Porcupine; Black-tailed
       Porcupine; Black-tailed Tree
       Porcupine; Pallid Hairy Dwarf
       Porcupine; Antillean Porcupine;
       West Indian Porcupine
   d   Antillen-Stachelschwein
   f   Porc-épic laineux; Porc-épic des
       Antilles

       Coendou koopmani syn. *C.*
       nycthemera q.v.

1333   *Coendou melanurus*
       Rodentia - Erethizontidae
   e   Black-tailed Hairy Dwarf Porcupine

1334   *Coendou mexicanus*
       Rodentia - Erethizontidae
   e   Mexican Hairy Dwarf Porcupine;
       Mexican Porcupine; Mexican
       Prehensile-tailed Porcupine; Coendu;
       Mexican Tree Porcupine
   d   Mexikanischer Baumstachler
   f   Porc-épic préhensile mexicain
   i   Coendu messicano

1336   *Coendou paragayensis*
       Rodentia - Erethizontidae
   e   Paraguay Hairy Dwarf Porcupine

1337   *Coendou prehensilis*
       Rodentia - Erethizontidae
   e   Brazilian Tree Porcupine; South
       American Porcupine; Brazilian
       Porcupine; Prehensile-tailed
       Porcupine
   d   Greifstachler; Cuandu
   f   Porc-épic préhensile
   i   Coendu dalla coda prensile

1338   *Coendou prehensilis longicaudis*
       Rodentia - Erethizontidae

   e   Trinidad Tree Porcupine

1340   *Coendou quichua*
       Rodentia - Erethizontidae
   e   Quichua Hairy Dwarf Porcupine

1341   *Coendou roosmalenorum*
       Rodentia - Erethizontidae
   e   Van Roosmalen's Dwarf Porcupine;
       Van Roosmalen's Hairy Dwarf
       Porcupine

1342   *Coendou rothschildi*
       Rodentia - Erethizontidae
   e   Rothschild's Porcupine

1343   *Coendou rufescens*
       Rodentia - Erethizontidae
   e   Upper Amazonian Porcupine; Stump-
       tailed Porcupine
   d   Bergstachler; Bergbaumstachler
   f   Porc-épic rougeâtre

1344   *Coendou spinosus*
       Rodentia - Erethizontidae
   e   Orange-spined Hairy Dwarf
       Porcupine; Atlantic Forest Hairy
       Dwarf Porcupine
   d   Südamerikanischer Greifstachler
   f   Porc-épic préhensile de Paraguay
   i   Coendu spinoso

1345   *Coendou vestitus*
       Rodentia - Erethizontidae
   e   Brown Hairy Dwarf Porcupine

       Coendou villosus syn. *C. spinosus*

1335   *Coendu nycthemera*
       Rodentia - Erethizontidae
   e   Koopman's Porcupine

       Coendu pallidus syn. *C. insidiosus*
       q.v.

1339 *Coendu pruinosus*
      Rodentia - Erethizontidae
e   Frosted Hairy Dwarf Porcupine

1346 *Coleura*
      Chiroptera - Emballonuridae
e   African Sheath-tailed Bats; Peters's
      Sheat-tailed Bats
f   Emballonures

1347 *Coleura afra*
      Chiroptera - Emballonuridae
e   Mozambique Sheath-tailed Bat; East
      African Sheath-tailed Bat; African
      Sheath-tailed Bat
f   Emballonure d'Afrique

1348 *Coleura seychellensis*
      Chiroptera - Emballonuridae
e   Seychelles Sheath-tailed Bat
i   Pipistrello delle Seychelles

      Colobotis syn. *Spermophilus* q.v.

1349 *Colobus*
      Primates - Cercopithecidae
e   Colobus Monkeys; Guerezas; Black-
      and-white Colobus Monkeys
d   Stummelaffen; Schwarzweiße
      Stummelaffen
f   Colobes; Colobes noir-et-blancs
i   Colobi

1350 *Colobus angolensis*
      Primates - Cercopithecidae
e   Angola Black-and-white Colobus;
      Angolan Colobus; Angola Colobus;
      Black-and-white Colobus; Angolan
      Colobus Monkey; Angola Pied
      Colobus
d   Angola-Stummelaffe
f   Colobe noir-et-blanc d'Angola;
      Colobe noir-et blanc angolais;
      Colobe d'Angola; Guéreza d'Angola
i   Colobo dell'Angola

1351 *Colobus guereza*
      Primates - Cercopithecidae
e   Eastern Black-and-white Colobus;
      Guereza; Mantled Guereza; Central
      African Colobus Monkey; Mantled
      Colobus; Guerezza Colobus;
      Magistrate Colobus
d   Guereza; Mantelaffe; Scheitelaffe
f   Colobe guéréza; Colobe aux épaules
      blanches; Guéreza du Kilimandjaro
i   Guereza; Colobo abissino; Guerezza
      dal mantello; Guereza bianco-e-nero;
      Scimmia guereza

1352 *Colobus guereza abyssinicus*
      Primates - Cercopithecidae
e   Abyssinian Black-and-white
      Colobus; Abyssinian Colobus
      Monkey; Northern Black-and-white
      Colobus
f   Colobe de l'Abyssinie

1353 *Colobus guereza kikuyuensis*
      Primates - Cercopithecidae
e   Northern Black Colobus

1354 *Colobus polykomos*
      Primates - Cercopithecidae
e   Southern Black-and-white Colobus;
      Western Black-and-white Colobus;
      Ursine Colobus; Western Pied
      Colobus; Colobus Monkey; King
      Colobus
d   Südliche Guerreza;
      Bärenstummelaffe;
      Weißbartstummelaffe; Südlicher
      Stummelaffe; Königscolobus
f   Colobe à longs poils; Colobe à
      camail; Colobe noir; Colobe blanc-
      et-noir d'Afrique occidentale; Colobe
      magistrat
i   Colobo orsino; Colobo bianco-nero;
      Scimmia catarrina; Scimmia bianca;
      Scimmia nera

1355 *Colobus polykomos ruwenzorii*
      Primates - Cercopithecidae
e   Ruwenzori Black-and-white Colobus

1356 *Colobus polykomos vellerosus*
Primates - Cercopithecidae
e White-thighed Colobus

1357 *Colobus satanas*
Primates - Cercopithecidae
e Black Colobus; Satanic Black
Colobus
d Schwarzer Guereza; Schwarzer
Stummelaffe; Schwarzer Colobus
f Guéreza noir; Colobe noir
i Colobo nero

1358 *Colobus vellerosus*
Primates - Cercopithecidae
e Geoffoy's Black-and-white Colobus;
Ursine Colobus; Geoffroy's Pied
Colobus
f Colobe magistrat
i Colobo velleroso

Colobus verus syn. *Proclobus verus*
q.v.

1359 *Colomys*
Rodentia - Muridae
e African Water Rats; African Tree
Rats
d Afrikanische Wasserratten
f Rats forestiers à ventre blanc

1360 *Colomys goslingi*
Rodentia - Muridae
e African Water Rat; Gosling's Swamp
Rat; Velvet Rat
d Afrikanische Waldbachmaus
f Rat forestier à ventre blanc

1361 *Condylura*
Erinaceomorpha - Talpidae
e Star-nosed Moles
d Sternmulle
f Condylures
i Talpe dal naso stellato

1362 *Condylura cristata*

Erinaceomorpha - Talpidae
e Star-nosed Mole
d Sternmull; Sternnasenmaulwurf
f Condylure étoilé
i Talpa dal naso stellato; Talpo dal
muso stellato

1363 *Conepatus*
Carnivora - Mustelidae
e Hog-nosed Skunks
d Ferkelskunks; Weißrüsselskunks;
Schweinsnasenskunks
f Moufettes à nez de cochon;
Mouffettes à nez de cochon (Qué);
Mofettes à nez de cochon
i Skunk dal naso di porco

Conepatus castaneus syn. *C.*
humboldtii q.v.

1364 *Conepatus chinga*
Carnivora - Mustelidae
e Argentine Skunk; Molina's Hog-
nosed Skunk; Chilean Skunk;
Andean Skunk; Andes Skunk; West
South American Hog-nosed Skunk
d Chile-Skunk; Anden-Skunk;
Chilenischer Skunk
f Moufette des Andes; Mouffette des
Andes (Qué); Moufette du Chili;
Mouffette du Chili (Qué); Mofette du
Chili
i Moffetta del Cile; Moffetta delle
Ande

1365 *Conepatus humboldtii*
Carnivora - Mustelidae
e Patagonian Skunk; Cordovan Skunk;
Humboldt's Hog-nosed Skunk;
Patagonian Hog-nosed Skunk
d Patagonischer Skunk
f Moufette de Patagonie; Mouffette de
Patagonie (Qué); Mofette de
Patagonie
i Skunk della Patagonia; Moffetta
della Patagonia; Zorrino della
Patagonia

1366 *Conepatus leuconotus*
Carnivora - Mustelidae
e Eastern Hog-nosed Skunk; White-
backed Hog-nosed Skunk; East North
American Hog-nosed Skunk
d Ferkelskunk; Östlicher Ferkelskunk
f Moufette à nez de cochon; Mouffette
à nez de cochon (Qué); Mofette à nez
de cochon
i Mofetta a naso di maiale

1367 *Conepatus mesoleucus*
Carnivora - Mustelidae
e Western Hog-nosed Skunk; Common
Hog-nosed Skunk; West North
American Hog-nosed Skunk
d Südlicher Ferkelskunk; Westlicher
Ferkelskunk
f Moufette à nez de porc épais;
Mouffette à nez de cochon (Qué);
Mofette à nez de cochon
i Zorrino comune

1368 *Conepatus mesoleucus telmalestes*
Carnivora - Mustelidae
e Big Thicket Hog-nosed Skunk

Conepatus rex syn. *C. chinga* q.v.

1369 *Conepatus semistriatus*
Carnivora - Mustelidae
e Amazonian Skunk; Striped Hog-
nosed Skunk; Central American Hog-
nosed Skunk
d Amazonas-Skunk
f Moufette d'Amazonie; Mouffette
d'Amazonie (Qué); Mofette
d'Amazonie
i Skunk della Patagonia

Congosorex syn. *Myosorex* q.v.

1370 *Conilurus*
Rodentia - Muridae
e Rabbit Rats (ANZ); Elsey's
Hapalotises

d Australische Kaninchenratten
f Rats lapins

1371 *Conilurus albipes*
Rodentia - Muridae
e White-footed Rabbit Rat (ANZ);
White-footed Tree Rat; White-footed
Hapalotis; Rabbit-eared Tree Rat;
Rabbit Rat (ANZ)
d Weißfußkaninchenratte
f Rat lapin aux pattes blanches

1372 *Conilurus penicillatus*
Rodentia - Muridae
e Brush-tailed Tree Rat (ANZ); Brush-
tailed Rabbit Rat (ANZ)
d Bürstenschwanzkaninchenratte

1373 *Connochaetes*
Artiodactyla - Bovidae
e Gnus; Wildebeests
d Gnus
f Gnous
i Gnu

1374 *Connochaetes gnou*
Artiodactyla - Bovidae
e White-tailed Gnu; Black Wildebeest
d Weißschwanzgnu
f Gnou à queue blanche
i Gnu dalla coda bianca

1375 *Connochaetes taurinus*
Artiodactyla - Bovidae
e Blue Wildebeest; White-bearded Gnu
d Streifengnu; Weißbartgnu; Blaues
Gnu
f Gnou à barbe blanche
i Gnu striato; Gnu; Gnu comune;
Wildebeest; Gnu azzuro

1376 *Connochaetes taurinus albojubatus*
Artiodactyla - Bovidae
e Eastern White-bearded Wildebeest

d   Östliches Weißbartgnu
i   Gnu dalla barba bianca

1377   *Connochaetes taurinus cooksoni*
         Artiodactyla - Bovidae
e   Cookson's Wildebeest
d   Cookson-Gnu

1378   *Connochaetes taurinus johnstoni*
         Artiodactyla - Bovidae
e   Blue Nyassa Wildebeest;
    Nysassaland Gnu
d   Weißbindengnu
f   Gnou de Nyassaland

1380   *Connochaetes taurinus taurinus*
         Artiodactyla - Bovidae
e   Southern White-bearded Wildebeest;
    Brindled Gnu; Brindled Wildebeest
d   Südliches Streifengnu; Südliches
    Blaues Gnu
f   Gnou bleu

1379   *Connochaetus taurinus mearnsi*
         Artiodactyla - Bovidae
e   Western White-bearded Wildebeest
d   Westliches Weißbartgnu; Westliches
    Streifengnu

1381   *Cormura*
         Chiroptera - Emballonuridae
e   Wagner's Sac-winged Bats; Chestnut
    Sac-winged Bats

1382   *Cormura brevirostris*
         Chiroptera - Emballonuridae
e   Wagner's Sac-winged Bat; Chestnut
    Sac-winged Bat

    Corvira syn. *Sturnira* q.v.

    Corynorhinus syn. *Plecotus* q.v.

1383   Craseonycteridae
         Chiroptera

e   Hog-nosed Bats; Butterfly Bats;
    Bumblebee Bats
d   Hummel-Fledermäuse
f   Chauves-souris bourdons
i   Pipistrelli-farfalla; Craseonicteridi

1384   *Craseonycteris*
         Chiroptera - Craseonycteridae
e   Hog-nosed Bats
d   Schweinsnasenfledermäuse
f   Chauves-souris bourdons
i   Pipistrelli-farfalla

1385   *Craseonycteris thonglongyai*
         Chiroptera - Craseonycteridae
e   Kitti's Hog-nosed Bat; Old World
    Hog-nosed Bat; Bumblebee Bat
d   Schweinsnasenfledermaus
f   Chauve-souris bourdon
i   Pipistrello-farfallo

1386   *Crateromys*
         Rodentia - Muridae
e   Bushy-tailed Rats; Bushy-tailed
    Cloud Rats
d   Borkenratten; Borkenkletterer
f   Rats des nuages

1387   *Crateromys australis*
         Rodentia - Muridae
e   Dinagat Bushy-tailed Cloud Rat;
    Dinagat Hairy-tailed Cloud Rat
d   Dinagat-Borkenratte

1388   *Crateromys heaneyi*
         Rodentia - Muridae
e   Panay Bush-tailed Cloud Rat
d   Panay-Borkenratte
f   Rat d'écorce de Panay; Rat d'écorce
    de Heaney

1389   *Crateromys paulus*
         Rodentia - Muridae
e   Ilin Bushy-tailed Cloud Rat; Ilin
    Island Bushy-tailed Cloud Rat; Ilin
    Island Cloud Rat; Ilin Cloud Rat; Ilin

Island Bushy-tailed Rat; Ilin Bushy-tailed Rat

d   Pauls Borkenratte; Ilin-Insel-Borkenratte; Ilin Borkenratte

f   Rat d'écorce de l'île Ilin; Rat d'écorce d'Ilin

1390   *Crateromys schadenbergi*
Rodentia - Muridae

e   Schadenberg's Great Rat; Luzon Bushy-tailed Cloud Rat

d   Schadenbergs Borkenratte; Buschschwanzborkenratte

Cratogeomys syn. *Pappogeomys* q.v.

1391   *Cremnomys*
Rodentia - Muridae

e   Indian Rats

d   Indische Felsenratten

f   Rats indiens

1392   *Cremnomys blanfordi*
Rodentia - Muridae

e   Blanford's Rat

d   Weißschwanzratte

1393   *Cremnomys cutchicus*
Rodentia - Muridae

e   Cutch Rat; Kutch Rat

1394   *Cremnomys elvira*
Rodentia - Muridae

e   Elvira Rat

1395   *Cricetomys*
Rodentia - Muridae

e   Giant Rats; Giant Pouched Rats; African Giant Pouched Rats

d   Riesenhamsterratten

f   Rats géants; Rats géants de Gambie; Cricétomes; Rats de Gambie

i   Ratti giganti africani

1396   *Cricetomys emini*
Rodentia - Muridae

e   Emin's Rat; Emin's Giant Rat; Giant Rat; Emin's Pouched Rat

d   Emin-Riesenhamsterratte; Ghana-Waldhamsterratte

f   Rat géant d'Emin; Cricétome de forêt

1397   *Cricetomys gambianus*
Rodentia - Muridae

e   Gambian Rat; Gambian Pouched Rat; African Giant Rat; Gambian Giant Pouched Rat; Giant Pouched Rat; Black Giant Pouched Rat; Gambian Giant Rat

d   Gambia-Riesenhamsterratte; Schwarze Riesenhamsterratte

f   Rat géant de Gambie; Rat de Gambie; Cricétome de savane

i   Ratto gigante della Gambia; Cricetomio del Gambia

1398   *Cricetulus*
Rodentia - Muridae

e   Eurasian Hamsters; Dwarf Hamsters; Rat-like Hamsters; Grey Long-tailed Hamsters

d   Graue Zwerghamster ; Zwerghamster

f   Hamsters nains gris

i   Criceti

1399   *Cricetulus alticola*
Rodentia - Muridae

e   Ladakhi Hamster; Short-tailed Tibetan Hamster; Ladakhi Short-tiled Hamster; Tibetan Dwarf Hamster

d   Tibetischer Zwerghamster; Indischer Zwerghamster; Ladakh-Zwerghamster

f   Hamster de Ladakh

i   Criceto di Ladak; Criceto nano del Tibet

1400   *Cricetulus barabensis*
Rodentia - Muridae

e   Striped Hamster; Striped Ratlike Hamster; Striped Dwarf Hamster

d   Barabischer Zwerghamster; Daurischer Zwerghamster; Chinesischer Streifenhamster; Gobi-

Zwerghamster; Transbaikal-
Zwerghamster

f Hamster de Chine

i Criceto nano striato; Criceto cinese
striato; Criceto del Gobi

Cricetulus curtatus syn.
Allocricetulus curtatus q.v.

Cricetulus eversmanni syn.
Allocricetulus eversmanni q.v.

Cricetulus griseus syn. *C.*
barabensis q.v.

1401 *Cricetulus kamensis*
Rodentia - Muridae

e Tibetan Hamster; Tibetan Rat-like
Hamster; Kam Dwarf Hamster

d Kam-Zwerghamster

f Hamster tibétain

i Criceto tibetano

1402 *Cricetulus longicaudatus*
Rodentia - Muridae

e Long-tailed Hamster; Lesser Long-
tailed Hamster; Long-tailed Dwarf
Hamster

d Langschwanz-Zwerghamster;
Langschwänziger Zwerghamster

f Hamster à longue queue

i Criceto dalla coda lunga

1403 *Cricetulus migratorius*
Rodentia - Muridae

e Migratory Hamster; Grey Hamster;
Grey Dwarf Hamster

d Jaik-Hamster; Grauer Zwerghamster;
Zwerghamster

f Hamster migrateur

i Criceto migratorio; Criceto migratore
grigio

Cricetulus obscurus syn. *C.*
barabensis q.v.

Cricetulus pseudogriseus syn. *C.*
barabensis q.v.

1404 *Cricetulus sokolovi*
Rodentia - Muridae

e Sokolov's Dwarf Hamster

d Sokolovs Zwerghamster

f Hamster nain de Sokolov

Cricetulus triton syn. *Tscherskia*
triton q.v.

1405 *Cricetus*
Rodentia - Muridae

e Common Hamsters; Black-bellied
Hamsters; East European Hamsters

d Feldhamster ; Hamster ; Gemeine
Hamster

f Hamsters; Hamsters communs;
Hamsters d'Europe

i Criceti comuni; Criceti giganti

1406 *Cricetus cricetus*
Rodentia - Muridae

e Common Hamster; Black-bellied
Hamster; East European Hamster;
European Hamster

d Feldhamster; Hamster; Gemeiner
Hamster; Großhamster; Osthamster;
Bunthamster; Schwarzbauchhamster;
Europäischer Feldhamster

f Cochon des blés; Hamster; Hamster
commun; Hamster d'Europe; Grand
Hamster

i Criceto comune; Criceto; Criceto
gigante

1407 *Crocidura*
Soricomorpha - Soricidae

e Musk White-toothed Shrews; White-
toothed Shrews

d Weißzahnspitzmäuse;
Feldspitzmäuse; Wimperspitzmäuse

f Crocidures; Musaraignes aux dents
blanches; Musettes

i Crocidure

Crocidura aequicauda syn. *C.*
attenuata q.v.

1408   *Crocidura aleksandrisi*
Soricomorpha - Soricidae
e   Alexandrian Shrew; Libyan Shrew

1409   *Crocidura allex*
Soricomorpha - Soricidae
e   Highland Shrew; Tom Thumb Shrew

1410   *Crocidura andamanensis*
Soricomorpha - Soricidae
e   Andaman Shrew; White-toothed
Andaman Shrew

1411   *Crocidura ansellorum*
Soricomorpha - Soricidae
e   Ansell's Shrew

1412   *Crocidura arabica*
Soricomorpha - Soricidae
e   Arabian Shrew

Crocidura arethusa syn. *C. fulvastra*
q.v.

1413   *Crocidura armenica*
Soricomorpha - Soricidae
e   Armenian Shrew

1414   *Crocidura attenuata*
Soricomorpha - Soricidae
e   Grey Shrew; Indochinese Shrew;
Asian Grey Shrew

1415   *Crocidura attila*
Soricomorpha - Soricidae
e   Hun Shrew

1416   *Crocidura baileyi*
Soricomorpha - Soricidae
e   Bailey's Shrew; Simien Shrew

Crocidura bartelsii syn. *C. monticola*
q.v.

1417   *Crocidura batesi*

Soricomorpha - Soricidae
e   Bates's Shrew; Bates's Forest Shrew
f   Crocidure de Bates

1418   *Crocidura beatus*
Soricomorpha - Soricidae
e   Mindanao Shrew

1419   *Crocidura beccarii*
Soricomorpha - Soricidae
e   Beccari's Shrew; Beccari Shrew

Crocidura bicolor syn. *C. fuscomurina* q.v.

Crocidura bolivari syn. *C. viaria* q.v.

1420   *Crocidura bottegi*
Soricomorpha - Soricidae
e   Bottego's Shrew; Bottego's Pygmy
Shrew

1421   *Crocidura bottegoides*
Soricomorpha - Soricidae
e   Bale Shrew; Tricoloured Shrew

1422   *Crocidura brunnea*
Soricomorpha - Soricidae
e   Thick-tailed Shrew

1423   *Crocidura buettikoferi*
Soricomorpha - Soricidae
e   Buettikofer's Shrew; Buettikoffer's
Forest Shrew

1424   *Crocidura caliginea*
Soricomorpha - Soricidae
e   African Foggy Shrew

1425   *Crocidura canariensis*
Soricomorpha - Soricidae
e   Canary Shrew
f   Crocidure des Canaries; Crocidure de
la Grande Canarie

1426   *Crocidura cinderella*
          Soricomorpha - Soricidae
     e   Cinderella Shrew
     d   Cinderella-Spitzmaus

1427   *Crocidura congobelgica*
          Soricomorpha - Soricidae
     e   Congo Shrew; Congo-White-toothed
          Shrew

1428   *Crocidura cossyrensis*
          Soricomorpha - Soricidae
     e   Pantellerian Shrew

1429   *Crocidura crenata*
          Soricomorpha - Soricidae
     e   Long-footed Shrew; Jumping Shrew

1430   *Crocidura crossei*
          Soricomorpha - Soricidae
     e   Crosse's Shrew; Crosse's Musk
          Shrew

1431   *Crocidura cyanea*
          Soricomorpha - Soricidae
     e   Reddish-grey Musk Shrew

1432   *Crocidura denti*
          Soricomorpha - Soricidae
     e   Dent's Shrew

1433   *Crocidura desperata*
          Soricomorpha - Soricidae
     e   Desperate Shrew

1434   *Crocidura dhofarensis*
          Soricomorpha - Soricidae
     e   Dhofarian Shrew; Dhofar Shrew

1435   *Crocidura dolichura*
          Soricomorpha - Soricidae
     e   Long-tailed Musk Shrew; African
          Long-tailed Shrew
     f   Crocidure à longue queue

1436   *Crocidura douceti*
          Soricomorpha - Soricidae
     e   Doucet's Musk Shrew

1437   *Crocidura dsinezumi*
          Soricomorpha - Soricidae
     e   Dsinezumi Shrew; Japanese White-
          toothed Shrew; Quelpart Shrew

1438   *Crocidura eisentrauti*
          Soricomorpha - Soricidae
     e   Eisentraut's Shrew

1439   *Crocidura elgonius*
          Soricomorpha - Soricidae
     e   Elgon Shrew; Mount Elgon Shrew

1440   *Crocidura elongata*
          Soricomorpha - Soricidae
     e   Elongated Shrew; Sulawesi Long-
          tailed Shrew

1441   *Crocidura erica*
          Soricomorpha - Soricidae
     e   Heather Shrew

1442   *Crocidura fischeri*
          Soricomorpha - Soricidae
     e   Fischer's Shrew

1443   *Crocidura flavescens*
          Soricomorpha - Soricidae
     e   Giant Musk Shrew; African Giant
          Shrew; Greater Musk Shrew; Greater
          Red Musk Shrew; South African
          Giant Shrew
     d   Riesenwimperspitzmaus
     f   Crocidure géante

1444   *Crocidura floweri*
          Soricomorpha - Soricidae
     e   Flower's Shrew
     f   Musaraigne de Flower

1445   *Crocidura foetida*
Soricomorpha - Soricidae
e   Lowland Bornean Shrew

1446   *Crocidura foxi*
Soricomorpha - Soricidae
e   Fox's Shrew

1447   *Crocidura fuliginosa*
Soricomorpha - Soricidae
e   Southeast Asian White-toothed
Shrew; Southeast Asian Shrew

1448   *Crocidura fuliginosa trichura*
Soricomorpha - Soricidae
e   Christmas Island Shrew

1449   *Crocidura fulvastra*
Soricomorpha - Soricidae
e   Savannah Shrew

1450   *Crocidura fumosa*
Soricomorpha - Soricidae
e   Smoky White-toothed Shrew

1451   *Crocidura fuscomurina*
Soricomorpha - Soricidae
e   Tiny Musk Shrew

1452   *Crocidura fuscomurina sansibarica*
Soricomorpha - Soricidae
e   Zanzibar Shrew

1453   *Crocidura glassi*
Soricomorpha - Soricidae
e   Glass's Shrew

1454   *Crocidura gmelini*
Soricomorpha - Soricidae
e   Gmelin's Shrew

1455   *Crocidura goliath*
Soricomorpha - Soricidae
e   Goliath Shrew; Giant Forest Shrew

d   Afrikanische Riesenspitzmaus
f   Musaraigne géante

1456   *Crocidura gracilipes*
Soricomorpha - Soricidae
e   Peters's Musk Shrew

1457   *Crocidura grandiceps*
Soricomorpha - Soricidae
e   Large-headed Shrew; Large-headed
Forest Shrew

1458   *Crocidura grandis*
Soricomorpha - Soricidae
e   Mount Malindang Shrew; Greater
Mindanao Shrew

1459   *Crocidura grassei*
Soricomorpha - Soricidae
e   Grasse's Shrew

1460   *Crocidura grayi*
Soricomorpha - Soricidae
e   Luzon Shrew

1461   *Crocidura greenwoodi*
Soricomorpha - Soricidae
e   Greenwood's Shrew

1462   *Crocidura gueldenstaedti*
Soricomorpha - Soricidae
e   Güldenstädt's Shrew

1463   *Crocidura harenna*
Soricomorpha - Soricidae
e   Harenna Shrew

1464   *Crocidura hildegardeae*
Soricomorpha - Soricidae
e   Hildegarde's Shrew

1465   *Crocidura hilliana*
Soricomorpha - Soricidae
e   Hill's Shrew

1466   *Crocidura hirta*
Soricomorpha - Soricidae
e   Zambezi Lesser Red Musk Shrew;
Lesser Red Musk Shrew

1467   *Crocidura hispida*
Soricomorpha - Soricidae
e   Andaman Spiny Shrew; Andaman
Islands Spiny Shrew

1468   *Crocidura horsfieldii*
Soricomorpha - Soricidae
e   Horsfield's Shrew
d   Horsfield-Spitzmaus

1469   *Crocidura hutanis*
Soricomorpha - Soricidae
e   Hutan Shrew

1470   *Crocidura jacksoni*
Soricomorpha - Soricidae
e   Jackson's Shrew

1471   *Crocidura jenkinsi*
Soricomorpha - Soricidae
e   Jenkins's Shrew

1472   *Crocidura kivuana*
Soricomorpha - Soricidae
e   Kivu Shrew

1473   *Crocidura lamottei*
Soricomorpha - Soricidae
e   Lamotte's Shrew

1474   *Crocidura lanosa*
Soricomorpha - Soricidae
e   Lemara Shrew; Long-haired Kivu
Shrew

1475   *Crocidura lasiura*
Soricomorpha - Soricidae
e   Ussuri White-toothed Shrew; Ussuri
Large White-toothed Shrew

1476   *Crocidura latona*
Soricomorpha - Soricidae
e   Latona Shrew

1477   *Crocidura lea*
Soricomorpha - Soricidae
e   Sulawesi Shrew; Lesser Black-footed
Shrew

1478   *Crocidura leucodon*
Soricomorpha - Soricidae
e   White-toothed Shrew; Bicoloured
White-toothed Shrew; Bicoloured
Shrew
d   Feldwimperspitzmaus;
Feldspitzmaus; Weißzähnige
Spitzmaus
f   Crocidure bicolore; Crocidure
blanche; Crocidure leucode;
Musaraigne bicolore; Musaraigne
leucode; Musaraigne des champs
i   Toporagno ventre bianco; Crocidura
ventre bianco; Crocidura dal ventre
bianco; Topino pancia bianca

1479   *Crocidura levicula*
Soricomorpha - Soricidae
e   Celebes Shrew; Sulawesi Tiny
Shrew; Brown Shrew

1480   *Crocidura littoralis*
Soricomorpha - Soricidae
e   Butiaba Naked-tailed Shrew

1481   *Crocidura longipes*
Soricomorpha - Soricidae
e   Savanna Swamp Shrew; Savannah
Swamp Shrew

1482   *Crocidura lucina*
Soricomorpha - Soricidae
e   Moorland Shrew; Lucy's Shrew
f   Musaraigne d'Éthiopie

1483   *Crocidura ludia*
       Soricomorpha - Soricidae
  e   Dramatic Shrew

1484   *Crocidura luna*
       Soricomorpha - Soricidae
  e   Grey-brown Musk Shrew; Greater
       Grey-brown Musk Shrew; Moonlight
       Shrew

1485   *Crocidura lusitania*
       Soricomorpha - Soricidae
  e   Mauritanian Shrew; Lusitanian
       Shrew
  f   Musaraigne Kulandar

1486   *Crocidura macarthuri*
       Soricomorpha - Soricidae
  e   MacArthur's Shrew

1487   *Crocidura macmillani*
       Soricomorpha - Soricidae
  e   MacMillan's Shrew

1488   *Crocidura macowi*
       Soricomorpha - Soricidae
  e   Macow's Shrew

1489   *Crocidura malayana*
       Soricomorpha - Soricidae
  e   Malayan Shrew; Malayan White-
       toothed Shrew

1490   *Crocidura manengubae*
       Soricomorpha - Soricidae
  e   Manenguba Shrew

1491   *Crocidura maquassiensis*
       Soricomorpha - Soricidae
  e   Maquassie Musk Shrew

1492   *Crocidura mariquensis*
       Soricomorpha - Soricidae
  e   Swamp Musk Shrew; Marico Shrew

1493   *Crocidura maurisca*
       Soricomorpha - Soricidae
  e   Dark Shrew

       Crocidura maxi syn. *C. monticola*
       q.v.

1494   *Crocidura mindorus*
       Soricomorpha - Soricidae
  e   Mindoro Shrew

       Crocidura minuta syn. *C. monticola*
       q.v.

1495   *Crocidura miya*
       Soricomorpha - Soricidae
  e   Ceylon Long-tailed Shrew; Sri
       Lankan Long-tailed Shrew
  d   Ceylon-Weißzahnspitzmaus; Sri
       Lanka-Weißzahnspitzmaus; Sri
       Lanka Langschwanzspitzmaus

1496   *Crocidura monax*
       Soricomorpha - Soricidae
  e   Rombo Shrew

1497   *Crocidura monticola*
       Soricomorpha - Soricidae
  e   Sunda Shrew; Max's Shrew; Minute
       Shrew

1498   *Crocidura montis*
       Soricomorpha - Soricidae
  e   Montane White-toothed Shrew

1499   *Crocidura muricauda*
       Soricomorpha - Soricidae
  e   Mouse-tailed Shrew; West African
       Long-tailed Shrew

1500   *Crocidura mutesae*
       Soricomorpha - Soricidae
  e   Uganda Large-toothed Shrew

1501   *Crocidura nana*
       Soricomorpha - Soricidae

e   Dwarf White-toothed Shrew

f   Crocidure naine

1502   *Crocidura nanilla*
Soricomorpha - Soricidae

e   Tiny White-toothed Shrew

1503   *Crocidura neglecta*
Soricomorpha - Soricidae

e   Neglected Shrew

1504   *Crocidura negrina*
Soricomorpha - Soricidae

e   Negros Shrew

1505   *Crocidura nicobarica*
Soricomorpha - Soricidae

e   Nicobar Shrew

1506   *Crocidura nigeriae*
Soricomorpha - Soricidae

e   Nigerian Shrew

1507   *Crocidura nigricans*
Soricomorpha - Soricidae

e   Black White-toothed Shrew

1508   *Crocidura nigripes*
Soricomorpha - Soricidae

e   Black-footed Shrew

1509   *Crocidura nigrofusca*
Soricomorpha - Soricidae

e   Tenebrous Shrew

1510   *Crocidura nimbae*
Soricomorpha - Soricidae

e   Nimba Shrew

1511   *Crocidura niobe*
Soricomorpha - Soricidae

e   Stony Shrew

1512   *Crocidura obscurior*

Soricomorpha - Soricidae

e   Obscure White-toothed Shrew

Crocidura occidentalis syn. *C. flavescens* q.v.

1513   *Crocidura olivieri*
Soricomorpha - Soricidae

e   Olivier's Shrew; Giant Musk Shrew

1514   *Crocidura orii*
Soricomorpha - Soricidae

e   Amami Shrew; Ryukyu White-toothed Shrew

1515   *Crocidura osorio*
Soricomorpha - Soricidae

e   Osorio Shrew

f   Crocidure de la Grande Canarie

1516   *Crocidura palawanenis*
Soricomorpha - Soricidae

e   Palawan Shrew

1517   *Crocidura paradoxura*
Soricomorpha - Soricidae

e   Paradox Shrew; Sumatran Long-tailed Shrew

1518   *Crocidura parvipes*
Soricomorpha - Soricidae

e   Small-footed Shrew; Short-legged Savanna Shrew; Short-legged Savannah Shrew

1519   *Crocidura pasha*
Soricomorpha - Soricidae

e   Pasha Shrew; Tiny Sahel Shrew

1520   *Crocidura pergrisea*
Soricomorpha - Soricidae

e   Pale Grey Shrew; Kashmir Rock Shrew

1521   *Crocidura phaeura*
          Soricomorpha - Soricidae
e   Guramba Shrew

1522   *Crocidura picea*
          Soricomorpha - Soricidae
e   Pitch Screw

1523   *Crocidura pitmani*
          Soricomorpha - Soricidae
e   Pitman's Shrew

1524   *Crocidura planiceps*
          Soricomorpha - Soricidae
e   Flat-headed Shrew

1525   *Crocidura poensis*
          Soricomorpha - Soricidae
e   Fraser's Musk Shrew

1526   *Crocidura polia*
          Soricomorpha - Soricidae
e   Fuscous Shrew

1527   *Crocidura pullata*
          Soricomorpha - Soricidae
e   Dusky White-toothed Shrew

1528   *Crocidura ramona*
          Soricomorpha - Soricidae
e   Negev Shrew

1529   *Crocidura religiosa*
          Soricomorpha - Soricidae
e   Egyptian Pygmy Shrew; Palestine
      Pygmy Shrew

1530   *Crocidura rhoditis*
          Soricomorpha - Soricidae
e   Temboan Shrew; Sulawesi White-
      headed Shrew

1531   *Crocidura roosevelti*
          Soricomorpha - Soricidae
e   Roosevelt's Shrew

1532   *Crocidura russula*
          Soricomorpha - Soricidae
e   White-toothed Shrew; Common
      White-toothed Shrew; Common
      European White-toothed Shrew;
      Greater White-toothed Shrew; Great
      White-toothed Shrew; European
      White-toothed Shrew
d   Hausspitzmaus;
      Hauswimperspitzmaus
f   Crocidure aranivore; Crocidure
      aranivore musette; Crocidure
      aranivore des sables; Leucode
      aranivore; Musaraigne musette;
      Crocidure musette; Musette des
      sables
i   Toporagno rossiccio; Toporagno
      maggiore; Crocidura rossiccia;
      Crocidura russula; Topino pettirosso

1533   *Crocidura selina*
          Soricomorpha - Soricidae
e   Moon Shrew

1534   *Crocidura serezkyensis*
          Soricomorpha - Soricidae
e   Serezkaya Shrew; Lesser Rock
      Shrew

          Crocidura sericea syn. *C. fulvastra*
          q.v.

1535   *Crocidura shantungensis*
          Soricomorpha - Soricidae
e   Shantung Shrew

1536   *Crocidura sibirica*
          Soricomorpha - Soricidae
e   Siberian Shrew; Siberian White-
      toothed Shrew

1537   *Crocidura sicula*
          Soricomorpha - Soricidae
e   Sicilian Shrew
f   Crocidure de Sicile

i Toporagno di Sicilia; Toporagno siciliano; Crocidura siciliana; Crocidura di Sicilia

1538 *Crocidura silacea*
Soricomorpha - Soricidae
e Lesser Grey-brown Musk Shrew

1539 *Crocidura smithi*
Soricomorpha - Soricidae
e Desert Musk Shrew
d Wüstenwimperspitzmaus
f Crocidure du désert

1540 *Crocidura somalica*
Soricomorpha - Soricidae
e Somali Shrew

1541 *Crocidura stenocephala*
Soricomorpha - Soricidae
e Narrow-headed Shrew; Kahuzi Swamp Shrew

Crocidura suahelae syn. *C. viaria* q.v.

1542 *Crocidura suaveolens*
Soricomorpha - Soricidae
e Lesser White-toothed Shrew
d Zwergwimperspitzmaus; Kleine Weißzahnspitzmaus; Mittelländische Spitzmaus; Gartenspitzmaus; Wimperspitzmaus
f Crocidure des jardins; Musaraigne des jardins
i Toporagno minore; Crocidura minore; Crocidura odorosa

1543 *Crocidura suaveolens cassiteridum*
Soricomorpha - Soricidae
e Scilly Shrew; Garden Shrew
f Musaraigne des jardins

1544 *Crocidura suaveolens coreae*
Soricomorpha - Soricidae
e Korean Shrew

1545 *Crocidura susiana*
Soricomorpha - Soricidae
e Iranian Shrew; Susiana Shrew

1546 *Crocidura tansaniana*
Soricomorpha - Soricidae
e Tanzanian Shrew

1547 *Crocidura tarella*
Soricomorpha - Soricidae
e Ugadan Shrew

1548 *Crocidura tarfayensis*
Soricomorpha - Soricidae
e Tarfaya Shrew; Sahara Shrew
f Musaraigne des Tarfaya

1549 *Crocidura telfordi*
Soricomorpha - Soricidae
e Telford's Shrew

1550 *Crocidura tenuis*
Soricomorpha - Soricidae
e Thin Shrew; Timor Shrew

Crocidura tephra syn. *C. viaria* q.v.

1551 *Crocidura thalia*
Soricomorpha - Soricidae
e Thalia Shrew
f Musaraigne d'Éthiopie

1552 *Crocidura theresae*
Soricomorpha - Soricidae
e Therese's Shrew

1553 *Crocidura thomensis*
Soricomorpha - Soricidae
e Sao Tomé Shrew; Sao Tomé Islanf Shrew

Crocidura trichura syn. *C. attenuata* q.v.

1554   *Crocidura turba*
         Soricomorpha - Soricidae
  e   Tumultuous Shrew

1555   *Crocidura ultima*
         Soricomorpha - Soricidae
  e   Ultimate Shrew

1556   *Crocidura usambarae*
         Soricomorpha - Soricidae
  e   Usambara Shrew
  f   Musaraigne de Tanzanie

1557   *Crocidura viaria*
         Soricomorpha - Soricidae
  e   Savanna Path Shrew; Savannah Path
         Shrew
  f   Crocidure de Bolivar

1558   *Crocidura voi*
         Soricomorpha - Soricidae
  e   Voi Shrew

1559   *Crocidura vorax*
         Soricomorpha - Soricidae
  e   Voracious Shrew

1560   *Crocidura vosmaeri*
         Soricomorpha - Soricidae
  e   Banka Shrew

1561   *Crocidura whitakeri*
         Soricomorpha - Soricidae
  e   Whittaker's Shrew; Lesser Maghreb
         Shrew
  f   Musaraigne de Whittaker

1562   *Crocidura wimmeri*
         Soricomorpha - Soricidae
  e   Wimmer's Shrew

1563   *Crocidura xantippe*
         Soricomorpha - Soricidae
  e   Vermiculate Shrew

1564   *Crocidura yankariensis*
         Soricomorpha - Soricidae
  e   Yankari Shrew

1565   *Crocidura zaphiri*
         Soricomorpha - Soricidae
  e   Zaphir's Shrew

1566   *Crocidura zarudnyi*
         Soricomorpha - Soricidae
  e   Zarudny's Shrew

1567   *Crocidura zimmeri*
         Soricomorpha - Soricidae
  e   Zimmer's Shrew; Upemba Shrew

1568   *Crocidura zimmermanni*
         Soricomorpha - Soricidae
  e   Zimmermann's Shrew;
         Zimmermann's White-toothed Shrew
  f   Crocidure de Zimmermann
  i   Crocidura di Creta; Toporagno
         cretese

1569   *Crocuta*
         Carnivora - Hyaenidae
  e   Spotted Hyenas; Laughing Hyenas
  d   Tüpfelhyänen
  f   Hyènes tachetées
  i   Iene macchiate

1570   *Crocuta crocuta*
         Carnivora - Hyaenidae
  e   Spotted Hyena; Laughing Hyena
  d   Tüpfelhyäne
  f   Hyène tachetée
  i   Iena macchiata

1571   *Crossarchus*
         Carnivora - Herpestidae
  e   Cusimanses; Mangues
  d   Kusimansen
  f   Cusimanses
  i   Crossarchi; Cusimanse

1572   *Crossarchus alexandri*
Carnivora - Herpestidae
e   Congo Cusimanse; Alexander's
Cusimanse
d   Kongo-Kusimanse
f   Mangue d'Alexandre
i   Crossarco di Alexander; Cusimanse
di Alexander

1573   *Crossarchus ansorgei*
Carnivora - Herpestidae
e   Angolan Cusimanse; Ansorge's
Cusimanse
d   Angola-Kusimanse
i   Crossarco di Ansorge; Cusimanse di
Ansorge

1574   *Crossarchus obscurus*
Carnivora - Herpestidae
e   Long-nosed Cusimanse; Cusimanse;
Dark Cusimanse; Dark Mongoose;
Common Cusimanse
d   Dunkelkusimanse
f   Mangouste brune; Mangue brune
i   Mangusta bruna; Cusimanso dal naso
lungo

Crossogale syn. *Chimarrogale* q.v.

1575   *Crossomys*
Rodentia - Muridae
e   Earless Water Rats
d   Schwimmratten; Moncktons
Schwimmratten
f   Rats de Nouvelle Guinée

1576   *Crossomys moncktoni*
Rodentia - Muridae
e   Earless Water Rat; Monckton's Water
Rat; Papuan Earless Water Rat
d   Moncktons Schwimmratte
f   Rat de Nouvelle Guinée

1577   *Crunomys*
Rodentia - Muridae
e   Philippine Swamp Rats; Flat-headed

Luzon Water Rats; Philippine
Shrewrats
d   Philippinische Sumpfratten
f   Rats des Philippines

1578   *Crunomys celebensis*
Rodentia - Muridae
e   Celebes Shrewrat; Celebes Swamp
Rat; Sulawesi Shrewmuse

1579   *Crunomys fallax*
Rodentia - Muridae
e   Northern Luzon Shrewrat; Northern
Luzon Shrewmouse; North Luxaon
Shrewmouse

1580   *Crunomys melanius*
Rodentia - Muridae
e   Mindanao Shrewrat; Southern
Shrewmouse; Southern Philippine
Shrewmouse; Leyte Shrewrat

Crunomys rabori syn. *C. melanius*
q.v.

1581   *Crunomys suncoides*
Rodentia - Muridae
e   Mount Kitanglad Shrewmouse

1582   *Cryptochloris*
Chrysochloridea - Chrysochloridae
e   Secretive Golden Moles
d   Winton-Goldmulle; De Winton-
Goldmulle
f   Taupes dorées de Winton

1583   *Cryptochloris wintoni*
Chrysochloridea - Chrysochloridae
e   De Winton's Golden Mole
d   Winton-Goldmull; De Winton-
Goldmull
f   Taupe dorée de Winton

1584   *Cryptochloris zyli*
Chrysochloridea - Chrysochloridae
e   Van Zyl's Golden Mole

d Van Zyl-Mull
f Taupe dorée de van Zyl

Cryptogale syn. **Geogale** q.v.

1585 **Cryptomys**
Rodentia - Bathyergidae
e Blesmols; Gray's Molerats; African
Molerats; Common Molerats
d Graumulle
f Rats-taupes

1586 **Cryptomys anselli**
Rodentia - Bathyergidae
e Ansell's Molerat
d Ansells Graumull

1587 **Cryptomys bocagei**
Rodentia - Bathyergidae
e Bocage's Molerat
d Bocages Graumull

1588 **Cryptomys damarensis**
Rodentia - Bathyergidae
e Damara Molerat; Damaraland
Molerat
d Damara-Graumull
f Rat taupe de Damaraland; Rat taupe
de Damara

1589 **Cryptomys foxi**
Rodentia - Bathyergidae
e Nigerian Molerat
d Nigerianischer Graumull

1590 **Cryptomys hottentotus**
Rodentia - Bathyergidae
e African Molerat; Blesmol
d Afrikanischer Grasmull
f Rat-taupe africain

1591 **Cryptomys hottentotus hottentotus**
Rodentia - Bathyergidae
e Common Molerat
d Hottentotten-Graumull

1592 **Cryptomys kafuensis**
Rodentia - Bathyergidae
e Kafue Molerat

1593 **Cryptomys mechowi**
Rodentia - Bathyergidae
e Mechow's Molerat; Zambian Giant
Molerat
d Riesengraumull

1594 **Cryptomys ochraceocinereus**
Rodentia - Bathyergidae
e Ochre Molerat; Central African
Molerat
d Zentralafrikanischer Graumull

1595 **Cryptomys zechi**
Rodentia - Bathyergidae
e Togo Molerat
d Zech-Graumull

1596 **Cryptoprocta**
Carnivora - Viverridae
e Fossas; Fossa Cats; Fosas; Foussas
d Fossas; Frettkatzen
f Cryptoproctes
i Fossa

1597 **Cryptoprocta ferox**
Carnivora - Viverridae
e Fossa; Reddish Fossa; Fosa; Foussa
d Fossa; Frettkatze
f Cryptoprocte; Cryptoprocte féroce;
Foussa
i Fossa

1598 **Cryptotis**
Soricomorpha - Soricidae
e Lesser American Short-tailed
Shrews; Little Shrews; Small-eared
Shrews
d Kleinohrspitzmäuse
f Petites musaraignes à queue courte

Cryptotis avia syn. *C. thomasi* q.v.

1599 **Cryptotis colombiana**
Soricomorpha - Soricidae
e Colombian Small-eared Shrew

1600 **Cryptotis endersi**
Soricomorpha - Soricidae
e Enders's Small-eared Shrew

1601 **Cryptotis goldmani**
Soricomorpha - Soricidae
e Goldman's Small-eared Shrew

1602 **Cryptotis goodwini**
Soricomorpha - Soricidae
e Goodwin's Small-eared Shrew;
Goodwin's Shrew

1603 **Cryptotis gracilis**
Soricomorpha - Soricidae
e Talamancan Small-eared Shrew

1604 **Cryptotis hondurensis**
Soricomorpha - Soricidae
e Honduran Small-eared Shrew

1605 **Cryptotis magna**
Soricomorpha - Soricidae
e Big Small-eared Shrew

1606 **Cryptotis mayensis**
Soricomorpha - Soricidae
e Maya Small-eared Shrew

1607 **Cryptotis mera**
Soricomorpha - Soricidae
e Darien Small-eared Shrew

1608 **Cryptotis meridensis**
Soricomorpha - Soricidae
e Merida Small-eared Shrew; Cloud
Forest Shrew

1609 **Cryptotis merriami**
Soricomorpha - Soricidae

e Merriam's Small-eared Shrew

1610 **Cryptotis mexicana**
Soricomorpha - Soricidae
e Mexican Small-eared Shrew

1611 **Cryptotis montivaga**
Soricomorpha - Soricidae
e Ecuadorean Small-eared Shrew; Grey
Andean Small-eared Shrew;
Ecuadorian Small-eared Shrew

1612 **Cryptotis nigrescens**
Soricomorpha - Soricidae
e Blackish Small-eared Shrew

1613 **Cryptotis parva**
Soricomorpha - Soricidae
e Least Shrew (NA); Lesser American
Short-tailed Shrew; Little Shrew;
Lesser Short-tailed Shrew (NA);
American Least Shrew; Small Brown
Shrew
d Nordamerikanische
Kleinohrspitzmaus
f Petite musaraigne; Petite musaraigne
à queue courte

1614 **Cryptotis peruviensis**
Soricomorpha - Soricidae
e Peruvian Small-eared Shrew

Cryptotis philllipsi syn. *C. mexicana*
q.v.

1615 **Cryptotis squamipes**
Soricomorpha - Soricidae
e Scaly-footed Small-eared Shrew;
Andean Long-clawed Shrew

1616 **Cryptotis tamensis**
Soricomorpha - Soricidae
e Paramo de Tama Small-eared Shrew

1617 **Cryptotis thomasi**
Soricomorpha - Soricidae

e   Thomas's Small-eared Shrew; Andean Small-eared Shrew; Colombian Small-eared Shrew

1618   *Ctenodactylidae*
       Rodentia -
e   Gundis; Gundis and Speke's Pectinators
d   Kammfinger ; Gundis
f   Goundis; Cténodactylidés
i   Gundi; Gondi; Ctenodactilidi

1619   *Ctenodactylus*
       Rodentia - Ctenodactylidae
e   Gundis; Common Gundis
d   Kammfinger ; Gundis
f   Goundis
i   Gondi; Gundi

1620   *Ctenodactylus gundi*
       Rodentia - Ctenodactylidae
e   Gundi; Common Gundi; North African Gundi
d   Kammfinger; Gundi; Gewöhnlicher Gundi
f   Goundi; Goundi de l'Atlas

1621   *Ctenodactylus vali*
       Rodentia - Ctenodactylidae
e   Val's Gundi; Desert Gundi
f   Goundi du désert

1622   *Ctenomys*
       Rodentia - Octodontidae
e   Tuco-tucos; Tucutucus
d   Kammratten; Tukotukos
f   Cténomys; Tuco-tucos

1623   *Ctenomys argentinus*
       Rodentia - Octodontidae
e   Argentine Tuco-tuco; Argentine Tucutucu

1624   *Ctenomys australis*
       Rodentia - Octodontidae

e   Southern Tuco-tuco; Southern Tucutucu

1625   *Ctenomys azarai*
       Rodentia - Octodontidae
e   Azara's Tuco-tuco; Azara's Tucutucu

1626   *Ctenomys boliviensis*
       Rodentia - Octodontidae
e   Bolivian Tuco-tuco; Bolivian Tucutucu

1627   *Ctenomys bonettoi*
       Rodentia - Octodontidae
e   Bonetto's Tuco-tuco; Bonetto's Tucutucu

1628   *Ctenomys brasiliensis*
       Rodentia - Octodontidae
e   Brazilian Tuco-tuco; Brazilian Tucutucu

1629   *Ctenomys colburni*
       Rodentia - Octodontidae
e   Colburn's Tuco-tuco; Colburn's Tucutucu

1630   *Ctenomys conoveri*
       Rodentia - Octodontidae
e   Conover's Tuco-tuco; Conover's Tucutucu

1631   *Ctenomys coyhaiquensis*
       Rodentia - Octodontidae
e   Coihaique Tuco-tuco; Coihaique Tucutucu

1632   *Ctenomys dorsalis*
       Rodentia - Octodontidae
e   Chacoan Tuco-tuco; Chacoan Tucutucu

1633   *Ctenomys emilianus*
       Rodentia - Octodontidae
e   Emily's Tuco-tuco; Emily's Tucutucu

1634 **Ctenomys frater**
Rodentia - Octodontidae
e Forest Tuco-tuco; Forest Tucutucu

1635 **Ctenomys fulvus**
Rodentia - Octodontidae
e Tawny Tuco-tuco; Tawny Tucutucu

1636 **Ctenomys haigi**
Rodentia - Octodontidae
e Haig's Tuco-tuco; Haig's Tucutucu
d Haigs Tukotuko

1637 **Ctenomys knighti**
Rodentia - Octodontidae
e Catamarca Tuco-tuco; Catamarca
Tucutucu; Knight's Tuco-tuco;
Knights Tucutucu
d Knight-Tukotuko
f Cténomys de Knight

1638 **Ctenomys latro**
Rodentia - Octodontidae
e Mottled Tuco-tuco; Mottled
Tucutucu

1639 **Ctenomys leucodon**
Rodentia - Octodontidae
e White-toothed Tuco-tuco; White-
toothed Tucutucu

1640 **Ctenomys lewisi**
Rodentia - Octodontidae
e Lewis's Tuco-tuco; Lewis's Tucutucu

1641 **Ctenomys magellanicus**
Rodentia - Octodontidae
e Magellanic Tuco-tuco; Magellanic
Tucutucu

1642 **Ctenomys maulinus**
Rodentia - Octodontidae
e Maule Tuco-tuco; Maule Tucutucu

1643 **Ctenomys mendocinus**

Rodentia - Octodontidae
e Mendoza Tuco-tuco; Mendoza
Tucutucu

1644 **Ctenomys minutus**
Rodentia - Octodontidae
e Tiny Tuco-tuco; Tiny Tucutucu

1645 **Ctenomys nattereri**
Rodentia - Octodontidae
e Natterer's Tuco-tuco; Natterer's
Tucutucu

1646 **Ctenomys occultus**
Rodentia - Octodontidae
e Furtive Tuco-tuco; Furtive Tucutucu

1647 **Ctenomys opimus**
Rodentia - Octodontidae
e Highland Tuco-tuco; Highland
Tucutucu

1648 **Ctenomys osvaldoreigi**
Rodentia - Octodontidae
e Osvaldo Reig's Tuco-tuco; Osvaldo
Reig's Tucutucu

1649 **Ctenomys pearsoni**
Rodentia - Octodontidae
e Pearson's Tuco-tuco; Pearson's
Tucutucu

1650 **Ctenomys perrensis**
Rodentia - Octodontidae
e Goya Tuco-tuco; Goya Tucutucu

1651 **Ctenomys peruanus**
Rodentia - Octodontidae
e Peruvian Tuco-tuco; Peruvian
Tucutucu

1652 **Ctenomys pontifex**
Rodentia - Octodontidae
e San Luis Tuco-tuco; San Luis
Tucutucu

1653 **Ctenomys porteousi**
Rodentia - Octodontidae
e Porteous's Tuco-tuco; Porteous's
Tucutucu

Ctenomys robustus syn. *C. fulvus*
q.v.

1654 **Ctenomys roigi**
Rodentia - Octodontidae
e Roig's Tuco-tuco; Roig's Tucutucu

1655 **Ctenomys saltarius**
Rodentia - Octodontidae
e Salta Tuco-tuco; Salta Tucutucu

1656 **Ctenomys sericeus**
Rodentia - Octodontidae
e Silky Tuco-tuco; Silky Tucutucu

1657 **Ctenomys sociabilis**
Rodentia - Octodontidae
e Social Tuco-tuco; Social Tucutucu

1658 **Ctenomys steinbachi**
Rodentia - Octodontidae
e Steinbach's Tuco-tuco; Steinbach's
Tucutucu
d Steinbachs Tukotuko

1659 **Ctenomys talarum**
Rodentia - Octodontidae
e Talas Tuco-tuco; Talas Tucutucu

1660 **Ctenomys torquatus**
Rodentia - Octodontidae
e Collared Tuco-tuco; Collared
Tucutucu

1661 **Ctenomys tuconax**
Rodentia - Octodontidae
e Robust Tuco-tuco; Robust Tucutucu

1662 **Ctenomys tucumanus**
Rodentia - Octodontidae

e Tucuman Tuco-tuco; Tucuman
Tucutucu

1663 **Ctenomys validus**
Rodentia - Octodontidae
e Strong Tuco-tuco; Strong Tucutucu

1664 **Cuniculus**
Rodentia - Agoutidae
e Pacas
d Pakas
f Pacas
i Paca

1665 **Cuniculus paca**
Rodentia - Agoutidae
e Paca; Aquatic Hare; Spotted Cavy;
Greater Paca; Agouti
d Paka
f Paca
i Grande paca; Paca

1666 **Cuniculus taczanowskii**
Rodentia - Agoutidae
e Mountain Paca
d Bergpaka
f Paca des montagnes
i Paca delle montagne

1667 **Cuon**
Carnivora - Canidae
e Dholes; Indian Dholes; Red Dogs
d Rotwölfe; Rothunde
f Cuons; Cuons d'Asie
i Cuon; Cyon

1668 **Cuon alpinus**
Carnivora - Canidae
e Asiatic Wild Dog; Indian Wild Dog;
Dhole; Indian Dhole; Red Dog;
Asian Wild Dog

d Rothund; Asiatischer Wildhund
f Cuon d'Asie; Chien sauvage d'Asie
i Cuon alpino; Cane selvatico asiatico;

Cane de di montagna; Dhole; Cane
selvatico indiano; Cuon Dhole

1669   *Cuon alpinus alpinus*
Carnivora - Canidae
e   Alpine Wolf; Asian Dhole
d   Alpenwolf

1670   *Cuon alpinus dukhumensis*
Carnivora - Canidae
e   Indian Red Wolf
d   Dekhan-Rothund

1671   *Cuon alpinus hesperius*
Carnivora - Canidae
e   Turkestani Red Wolf; West Asian
Dhole; Turkestani Alpine Wolf
d   Turkestanischer Alpenwolf

1672   *Cuon alpinus javanicus*
Carnivora - Canidae
e   Java Red Wolf
d   Java-Rothund

1673   *Cuon alpinus lepturus*
Carnivora - Canidae
e   Chinese Alpine Wolf; Chinese Dhole
d   Kiangsi-Alpenwolf

1674   *Cuscomys*
Rodentia - Abrocomidae
e   Cusco Rats

1675   *Cuscomys ashanika*
Rodentia - Abrocomidae
e   White-fronted Cusco Rat

1676   Cyclopedidae
Pilosa
e   Silky Anteaters
d   Zwergameisenbären
f   Fourmiliers didactyles
i   Formichieri didattili; Ciclopedidi

1677   *Cyclopes*
Pilosa - Cyclopedidae
e   Two-toed Anteaters; Pygmy
Anteaters; Silky Anteaters
d   Zwergameisenbären
f   Fourmiliers didactyles
i   Formichieri didattili

1678   *Cyclopes didactylus*
Pilosa - Cyclopedidae
e   Two-toed Anteater; Pygmy Anteater;
Silky Anteater
d   Zwergameisenbär
f   Fourmilier didactyle
i   Formichiere didattilo; Formichiere
nano

1679   *Cynictis*
Carnivora - Herpestidae
e   Yellow Mongooses
d   Fuchsmangusten
f   Mangoustes fauves
i   Manguste gialle

1680   *Cynictis penicillata*
Carnivora - Herpestidae
e   Yellow Mongoose; Bushy-tailed
Meerkat; Red Meerkat
d   Fuchsmanguste
f   Mangouste fauve
i   Mangusta fulva; Mangusta
penicillata; Mangusta gialla

1681   *Cynictis selousi*
Carnivora - Herpestidae
e   Grey Meerkat; Selous's Mongoose;
Selous's Meerkat
d   Trugmanguste
f   Mangouste de Selous
i   Mangusta di Selous

1682   *Cynocephalus*
Primates - Galeopithecidae
e   Flying Lemurs; Gliding Lemurs;
Colugos
d   Riesengleitflieger ; Gleitflieger

f Cynocéphales
i Lemuri volanti; Galeopithechi

1683 *Cynocephalus variegatus*
Primates - Galeopithecidae
e Malayan Flying Lemur; Malayan
Gliding Lemur; Flying Lemur;
Gliding Lemur; Sunda Flying Lemur;
Cobego; Colugo
d Temminck-Gleitflieger
i Lemure volante malese; Galeopiteco
della Malesia

1684 *Cynocephalus volans*
Primates - Galeopithecidae
e Philippine Gliding Lemur; Philippine
Flying Lemur
d Hundskopfgleiter; Philippinen-
Gleitflieger
i Galeopiteco delle Filippine

1685 *Cynogale*
Carnivora - Viverridae
e Otter Civets
d Hundsroller; Otterzivetten
f Civettes loutre de Sumatra
i Civette lontra

1686 *Cynogale bennettii*
Carnivora - Viverridae
e Otter Civet; WaterCivet
d Mampalon; Otterzivette
f Civette loutre de Sumatra
i Civetta lontra ; Mampolon

**   *Cynomops* syn. *Molossops* q.v.**

1687 *Cynomys*
Rodentia - Sciuridae
e Prairie Dogs; Prairie Marmots;
Barking Marmots
d Präriehunde
f Cynomys ; Chiens des prairies
i Cani della prateria

1688 *Cynomys gunnisoni*
Rodentia - Sciuridae
e Gunnison's Prairie Dog
d Gunnisons Präriehund
f Cynomys de Gunnison; Chien de
prairie de Gunnison; Chien de prairie
du Colorado

1689 *Cynomys leucurus*
Rodentia - Sciuridae
e White-tailed Prairie Dog
d Weißschwanzprairiehund
f Chien de prairie blanc; Chien de
prairie à queue blanche; Cynomys à
queue blanche
i Cane della prateria dalla coda bianca

1690 *Cynomys ludovicianus*
Rodentia - Sciuridae
e Plains Prairie Dog; Black-tailed
Prairie Dog
d Schwarzschwanzprairiehund
f Cynomys social; Chien de prairie à
queue noir; Cynomys à queue noire
i Cane della prateria dalla coda nera;
Cane della prateria della pianura

1691 *Cynomys ludovicianus arizonensis*
Rodentia - Sciuridae
e Arizona Prairie Dog

1692 *Cynomys mexicanus*
Rodentia - Sciuridae
e Mexican Prairie Dog
d Mexikanischer Präriehund
f Chien de prairie du Mexique; Chien
de prairie mexicain
i Cane di prateria del Messico

1693 *Cynomys parvidens*
Rodentia - Sciuridae
e Utah Prairie Dog
d Utah-Präriehund
f Chien de prairie de l'Utah

**   *Cynopithecus* syn. *Macaca* q.v.**

1694 **Cynopterus**
Chiroptera - Pteropodidae
e Short-nosed Fruit Bats; Dog-faced
Fruit Bats
d Kurznasenflughunde
f Cynoptères
i Cinotteri; Volpi volanti

1695 **Cynopterus brachyotis**
Chiroptera - Pteropodidae
e Malaysian Fruit Bat; Indian Short-
eared Fruit Bat; Lesser Dog-faced
Fruit Bat; Lesser Short-nosed Fruit
Bat; Dog-faced Fruit Bat

1696 **Cynopterus horsfieldi**
Chiroptera - Pteropodidae
e Short-nosed Fruit Bat; Larger Dog-
faced Fruit Bat; Horsfield's Fruit Bat;
Peg-toothed Short-nosed Fruit Bat;
Grey Horsfield's Fruit Bat;
Horsfield's Short-nosed Fruit Bat

1697 **Cynopterus luzoniensis**
Chiroptera - Pteropodidae
e Sulawesi Short-nosed Bat

1698 **Cynopterus minutus**
Chiroptera - Pteropodidae
e Small Short-nosed Fruit Bat

1699 **Cynopterus nusatenggara**
Chiroptera - Pteropodidae
e Nusa Tenggara Short-nosed Fruit Bat

1700 **Cynopterus sphinx**
Chiroptera - Pteropodidae
e Indian Fruit Bat; Short-nosed Fruit
Bat; Greater Short-nosed Fruit Bat
d Indischer Kurznasenflughund
f Cynoptère à nez court

1701 **Cynopterus sphinx ceylonensis**
Chiroptera - Pteropodidae
e Ceylon Short-nosed Fruit Bat
d Fruchtfressender Ceylon

Kurznasenflughund

1702 **Cynopterus sphinx sphinx**
Chiroptera - Pteropodidae
e Indian Short-nosed Fruit Bat
d Fruchtfressender Indischer
Kurznasenflughund

1703 **Cynopterus titthaecheilus**
Chiroptera - Pteropodidae
e Indonesian Short-nosed Fruit Bat;
Greater Dog-faced Fruit Bat

Cyromys syn. *Uromys* q.v.

1704 **Cystophora**
Carnivora - Phocidae
e Hooded Seals; Bladder-nosed Seals;
Crested Seals
d Blasenrobben; Klappenrobben;
Klappmützen; Mützenrobben
f Phoques à capuchon; Phoques
capuchonnés; Phoques à casque
i Cistofore

1705 **Cystophora cristata**
Carnivora - Phocidae
e Hooded Seal; Bladder-nosed Seal;
Crested Seal; Bladdernose
d Blasenrobbe; Klappenrobbe;
Klappmütze; Mützenrobbe
f Phoque à capuchon; Phoque
capuchonné; Phoque à casque;
Phoque de Gmelin; Phoque à crête
i Cistofora crestata; Foca dal capuccio

1706 **Cyttarops**
Chiroptera - Emballonuridae
e Short-eared Bats

1707 **Cyttarops alecto**
Chiroptera - Emballonuridae
e Short-eared Bat; Smoky Bat; Smoky
Sheath-tailed Bat

D

1708   Dacnomys
Rodentia - Muridae

e   Large-toothed Rats; Large-toothed
Giant Rats; Millard's Rats
d   Millard-Ratten

1709   Dacnomys millardi
Rodentia - Muridae

e   Large-toothed Rat; Large-toothed
Giant Rat; Millard's Rat
d   Millard-Ratte

1710   Dactylomys
Rodentia - Echimyidae

e   Coro-coros; Arboreal Rats;
Neotropical Bamboo Rats
d   Fingerratten

1711   Dactylomys boliviensis
Rodentia - Echimyidae

e   Bolivian Bamboo Rat

1712   Dactylomys dactylinus
Rodentia - Echimyidae

e   Amazon Bamboo Rat; Coro-coro
d   Eigentliche Fingerratte

1713   Dactylomys peruanus
Rodentia - Echimyidae

e   Peruvian Bamboo Rat

Dactylonax syn. *Dactylopsila* q.v.

1714   Dactylopsila
Diprotodontia - Petauridae

e   Striped Possums
d   Streifenbeutler ;
Streifenkletterbeutler ;

Streifenphalanger
f   Phalangers à pélage rayé; Possums
rayés; Trioks
i   Possum; Opossum

1715   Dactylopsila megalura
Diprotodontia - Petauridae

e   Large-tailed Possum; Great-tailed
Triok
d   Gleithörnchenbeutler
f   Triok à longue queue

1716   Dactylopsila palpator
Diprotodontia - Petauridae

e   Long-fingered Possum; Long-
fingered Striped Phalanger; Long-
fingered Triok
d   Kleiner Streifenbeutler;
Kleinstreifenbeutler
f   Triok aux longs doigts

1717   Dactylopsila tatei
Diprotodontia - Petauridae

e   Fergusson Island Possum; Fergusson
Striped Possum; Tate's Triok
d   Ferguson-Insel-Streifenbeutler
f   Possum rayé des îles Fergusson

1718   Dactylopsila trivirgata
Diprotodontia - Petauridae

e   Striped Phalanger; Striped Possum
(ANZ); Common Striped Possum;
Common Bushtail Possum

d   Großer Streifenbeutler;
Großstreifenbeutler
f   Phalanger au pélage rayé; Possum
rayé
i   Possum striato; Opossum striato

1719   Dama
Artiodactyla - Cervidae

e   Fallow Deer
d   Damhirsche
f   Daims
i   Daini

1720 **Dama dama**
Artiodactyla - Cervidae
e Fallow Deer
d Damhirsch
f Daim; Daim platycerque; Daim européen
i Daino

1721 **Dama dama dama**
Artiodactyla - Cervidae
e European Fallow Deer; Fallow Deer
d Europäischer Damhirsch
f Daim européen
i Daino europeo

1722 **Dama mesopotamica**
Artiodactyla - Cervidae
e Mesopotamian Fallow Deer; Persian Fallow Deer
d Mesopotamischer Damhirsch
f Daim persan; Daim de Mésopotamie
i Daino della Mesopotamia

1723 **Damaliscus**
Artiodactyla - Bovidae
e Topis; Blesboks; Sassabies
d Leierantilopen; Halbmondantilopen
f Damalisques
i Damalischi

1724 **Damaliscus hunteri**
Artiodactyla - Bovidae
e Hunter's Hartebeest; Hirola; Hunter's Antelope
d Hunters Leierantilope; Hunter-Antilope
f Hirola; Antilope de Hunter; Damalisque de Hunter
i Damalisco di Hunter; Damalisco Hirola

Damaliscus korrigum syn. *D. lunatus* q.v.

1725 **Damaliscus lunatus**
Artiodactyla - Bovidae

e Sassaby; South African Sassaby; Korrigum; Tiang; Topi; Tsessebe; Topi Antelope; Sassaby Antelope; Senegal Hartebeest
d Halbmondantilope; Leierantilope; Korrigum; Riesenleierantilope; Sassaby
f Damalisque; Topi; Korrigum; Sassaby; Tsessebe
i Damalisco; Damalisco comune; Tessebe

1726 **Damaliscus pygargus**
Artiodactyla - Bovidae
e Bontebok; South African Antelope
d Buntbock
f Damalisque à queue blanche
i Antilope pigarga

1727 **Damaliscus pygargus phillipsi**
Artiodactyla - Bovidae
e Blesbok
d Blessbock
f Blesbok
i Blesbok

1728 **Damaliscus pygargus pygargus**
Artiodactyla - Bovidae
e Bontebok
d Buntbock; Echter Buntbock
f Bontebok
i Bontebok; Antilope pigarga

Dasogale syn. *Setifer* q.v.

1729 **Dasycercus**
Dasyuromorphia - Dasyuridae
e Crested-tailed Marsupial Mice; Crest-tailed Marsupial Mice; Mulgaras (ANZ)
d Kammschwanzbeutelmäuse
f Souris marsupiales
i Mulgara

Dasycercus byrnei syn. *Dasyuroides byrnei* q.v.

1730   *Dasycercus cristicauda*
Dasyuromorphia - Dasyuridae
e   Crest-tailed Marsupial Mouse;
Mulgara (ANZ)
d   Südliche Kammschwanzbeutelmaus;
Süd-Kammschwanzbeutelmaus;
Mulgara
f   Mulgara
i   Mulgara

1731   *Dasycercus hilleri*
Dasyuromorphia - Dasyuridae
e   Ampurta
f   Amputa

Dasykaluta syn. *Antechinus* q.v.

1732   *Dasymys*
Rodentia - Muridae
e   Shaggy African Marsh Rats; Shaggy-
haired Rats; Marsh Rats; African
Marsh Rats; Water Rats; African
Water Rats; Shaggy Rats
d   Wollhaarratten
f   Rats des marécages

1733   *Dasymys foxi*
Rodentia - Muridae
e   Fox's Shaggy Rat

1734   *Dasymys incomtus*
Rodentia - Muridae
e   African Marsh Rat; Shaggy-haired
Rat; Marsh Rat; Shaggy Rat; Water
Rat; African Water Rat
d   Afrikanische Wollhaarratte
f   Rat hirsute

1735   *Dasymys montanus*
Rodentia - Muridae
e   Montane Shaggy Rat; Montane
Marsh Rat

1736   *Dasymys nudipes*
Rodentia - Muridae
e   Angolan Marsh Rat

1737   *Dasymys rufulus*
Rodentia - Muridae
e   West African Shaggy Rat

1738   Dasypodidae
Xenarthra
e   Armadillos
d   Gürteltiere
f   Dasypodidés; Tatous
i   Armadilli

1739   *Dasyprocta*
Rodentia - Agoutidae
e   Agoutis
d   Stummelschwanzagutis; Agutis;
Eigentliche Agutis
f   Agoutis
i   Aguti

Dasyprocta aguti syn. *D. leporina*
q.v.

Dasyprocta albida syn. *D. leporina*
q.v.

1740   *Dasyprocta azarai*
Rodentia - Agoutidae
e   Azara's Agouti
d   Azara-Aguti
f   Agouti d'Azara
i   Aguti di Azara

1741   *Dasyprocta coibae*
Rodentia - Agoutidae
e   Coiban Agouti; Coiba Island Agouti
d   Coiba-Aguti
f   Agouti de l'île de Coiba

1742   *Dasyprocta cristata*
Rodentia - Agoutidae
e   Crested Agouti
d   Haubenaguti

1743   *Dasyprocta fuliginosa*
Rodentia - Agoutidae

e   Smoky Agouti; Grey Agouti; Black Agouti

d   Mohrenaguti; Schwarzaguti

1744   ***Dasyprocta guamara***
Rodentia - Agoutidae

e   Orinoco Agouti

d   Orinoco-Aguti

1745   ***Dasyprocta kalinowskii***
Rodentia - Agoutidae

e   Kalinowski's Agouti

d   Kalinowski-Aguti

1746   ***Dasyprocta leporina***
Rodentia - Agoutidae

e   Orange-rumped Agouti; Brazilian Agouti; Red-rumped Agouti

d   Goldaguti; Eigentliches Aguti; Goldhase; Goldrückenaguti

f   Agouti doré

1747   ***Dasyprocta mexicana***
Rodentia - Agoutidae

e   Mexican Agouti; Mexican Black Agouti

d   Mexikanisches Aguti

1748   ***Dasyprocta prymnolopha***
Rodentia - Agoutidae

e   Black-rumped Agouti

d   Schwarzbauchaguti

1749   ***Dasyprocta punctata***
Rodentia - Agoutidae

e   Central American Agouti; Agouti; Common Agouti

d   Mittelamerikanisches Aguti

f   Agouti

i   Agouti punteggiato; Agouti

1750   ***Dasyprocta ruatanica***
Rodentia - Agoutidae

e   Roatan Island Agouti

d   Ruatan-Aguti

1751   ***Dasyprocta variegata***
Rodentia - Agoutidae

e   Brown Agouti

d   Braunes Aguti

Dasypterus syn. *Lasiurus* q.v.

1752   ***Dasypus***
Cingulata - Dasypodidae

e   Long-nosed Armadillos; Nine-banded Armadillos

d   Weichgürteltiere

f   Tatous

1753   ***Dasypus hybridus***
Cingulata - Dasypodidae

e   Mulita Armadillo; Southern Long-nosed Armadillo; Southern Lesser Long-nosed Armadillo

d   Kurzschwanzgürteltier; Südliches Siebenbindengürteltier

f   Tatou hybride

1754   ***Dasypus kappleri***
Cingulata - Dasypodidae

e   Kappler's Armadillo; Greater Long-nosed Armadillo; Great Long-nosed Armadillo

d   Kappler-Weichgürteltier

f   Tatou de Kappler

1755   ***Dasypus novemcinctus***
Cingulata - Dasypodidae

e   Nine-banded Armadillo; Texas Armadillo; Common Long-nosed Armadillo; Long-nosed Armadillo

d   Neunbindengürteltier; Tatu; Langschwänziges Gürteltier

f   Tatou à neuf bandes

i   Armadillo dalle nove fasce

1756   ***Dasypus novemcinctus mexicanus***
Cingulata - Dasypodidae

e   Mexican Nine-banded Armadillo

1757 *Dasypus pilosus*
Cingulata - Dasypodidae
e Hairy Armadillo; Hairy Long-nosed Armadillo
d Pelzgürteltier
f Tatou poilu
i Armadillo peloso dal naso lungo

1758 *Dasypus sabanicola*
Cingulata - Dasypodidae
e Northern Long-nosed Armadillo; Llanos Long-nosed Armadillo; Northern Lesser Long-nosed Armadillo
d Erwachtes Gürteltier; Nördliches Siebenbindengürteltier

1759 *Dasypus septemcinctus*
Cingulata - Dasypodidae
e Seven-banded Armadillo; Brazilian Long-nosed Armadillo; Lesser Long-nosed Armadillo; Brazilian Lesser Long-nosed Armadillo
d Siebenbindengürteltier
f Tatou à sept bandes
i Armadillo dalle sette fasce

1760 *Dasypus yepesi*
Cingulata - Dasypodidae
e Yepes Long-nosed Armadillo

1761 **Dasyuridae**
Dasyuromorphia
e Flesh-eating Marsupials; Australian Meat-and-Insect-eating Marsupials; Marsupial Mice, Marsupial Cats etc.; Dasyurids; Native Cats and Marsupial Mice; Pouched Mice and Rats
d Raubbeutler
f Dasyuridés
i Topi marsupiali ; Ratti marsupiali; Carnivori marsupiali; Dasiuridi

1762 *Dasyuroides*
Dasyuromorphia - Dasyuridae
e Kowaris

d Doppelkammbeutelmäuse
f Kowaris
i Topi marsupiali dalla coda crestata

1763 *Dasyuroides byrnei*
Dasyuromorphia - Dasyuridae
e Kowari (ANZ); Crest-tailed Marsupial Rat; Byrne's Pouched Mouse; Byrne's Marsupial Mouse
d Doppelkammbeutelmaus; Kowari
f Kowari; Rat marsupial à double crête
i Topo marsupiale dalla coda crestata; Kowari

1764 *Dasyurus*
Dasyuromorphia - Dasyuridae
e Eastern Native Cats; Native Cats; Marsupial Cats; Dasyures; Quolls
d Fleckenbeutelmarder ; Tüpfelbeutelmarder ; Beutelmarder ; Quolls
f Dasyures; Chats marsupials; Quolls
i Dasiuri

1765 *Dasyurus albopunctatus*
Dasyuromorphia - Dasyuridae
e New Guinea Marsupial Cat; New Guinea Quoll
d Neuguinea-Beutelmarder; Neuguinea-Quoll
f Satinello de Nouvelle Guinée; Dasyure de Nouvelle Guinée
i Quoll

1766 *Dasyurus geoffroii*
Dasyuromorphia - Dasyuridae
e Western Dasyure; Black-tailed Native Cat; Western Native Cat; Western Australian Native Cat; Chuditch; Western Quoll (ANZ); Geoffroy's Native Cat
d Schwarzschwanzfleckenbeutelmarder ; Schwarzschwanzbeutelmarder; Westquoll
f Dasyure; Quoll; Chat marsupial; Chuditch; Quoll occidental
i Chuditch

1767   *Dasyurus hallucatus*
Dasyuromorphia - Dasyuridae
e   Little Northern Dasyure; Satanellus;
Little Northern Native Cat; Northern
Quoll
d   Zwergfleckenbeutelmarder;
Zwergbeutelmarder; Nordquoll
f   Quoll nordique

1768   *Dasyurus maculatus*
Dasyuromorphia - Dasyuridae
e   Spotted Marten; Spotted Native Cat;
Large Spotted-tailed Native Cat;
Spotted-tailed Dasyure; Tiger Cat;
Spotted-tailed Quoll (ANZ); Large
Spotted Native Cat (ANZ); Tiger
Quoll
d   Fleckschwanzfleckenbeutelmarder;
Fleckschwanzbeutelmarder;
Riesenbeutelmarder; Beschmutztes
Quoll
f   Chat marsupial
i   Dasiuro gigante

1769   *Dasyurus spartacus*
Dasyuromorphia - Dasyuridae
e   Bronze Quoll
d   Bronzequoll
f   Quoll bronzé

1770   *Dasyurus viverrinus*
Dasyuromorphia - Dasyuridae
e   Eastern Dasyure; Eastern Australian
Native Cat; Eastern Quoll (ANZ);
Quoll (ANZ); Eastern Native Cat
(ANZ)
d   Tüpfelbeutelmarder; Ostquoll
f   Quoll oriental
i   Quoll orientale; Gatto nativo
maculato

1771   *Daubentonia*
Primates - Daubentoniidae
e   Aye-ayes
d   Fingertiere
f   Aiays
i   Ayè-ayè; Aiè aiè

1772   *Daubentonia madagascariensis*
Primates - Daubentoniidae
e   Aye-aye
d   Aaye-Aye; Fingertier
f   Aiay; Hay-Hay
i   Ayè-ayè; Aiè aiè; Scimmia picchio

1773   **Daubentoniidae**
Primates
e   Aye-ayes
d   Fingertiere; Aye-Ayes; Hay-Haye
f   Daubentoniidés
i   Ayè-ayè; Aiè aiè; Daubentonidi

1774   *Delanymys*
Rodentia - Muridae
e   Delany's Swamp Mice
d   Delanys Sumpfklettermäuse

1775   *Delanymys brooksi*
Rodentia - Muridae
e   Delany's Swamp Mouse; Delany's
Mouse
d   Delanys Sumpfklettermaus

1776   *Delomys*
Rodentia - Muridae
e   Atlantic Forest Rats
d   Atlantische Waldratten

1777   *Delomys dorsalis*
Rodentia - Muridae
e   Striped Atlantic Forest Rat

1778   *Delomys sublineatus*
Rodentia - Muridae
e   Pallid Atlantic Forest Rat

1779   *Delphinapterus*
Cete - Monodontidae
e   White Wales; Belugas
d   Weißwale
f   Delphinaptères; Dauphins blancs
i   Beluga

1780   *Delphinapterus leucas*
Cete - Monodontidae
e   White Whale; Beluga; Beluga
Whale; White Porpoise; Whitefish
d   Weißwal; Weißfisch; Beluga; Weißer
Delfin
f   Baleine blanche; Dauphin blanc;
Marsouin blanc; Bélouga; Béluga;
Delphinaptère blanc
i   Beluga; Balena bianca; Canarino di
mare

1781   **Delphinidae**
Cete
e   Dolphins; Marine Dolphins
d   Delfine; Delfinartige
f   Dauphins; Delphinidés
i   Delfini; Delfinidi

1782   *Delphinus*
Cete - Delphinidae
e   Dolphins; Common Dolphins
d   Delfine; Gemeine Delfine
f   Dauphins
i   Delfini

Delphinus bairdi syn. *D. delphis* q.v.

1783   *Delphinus capensis*
Cete - Delphinidae
e   Long-beaked Common Dolphin
d   Langschnäuziger Gemeiner Delfin
f   Dauphin commun longirostre
i   Delfino comune dal lungo rostro

1784   *Delphinus delphis*
Cete - Delphinidae
e   Common Dolphin; Common Ocean
Dolphin; Saddleback Porpoise;
White-bellied Porpoise; Short-beaked
Saddleback Dolphin; Saddleback
Dolphin; Short-beaked Common
Dolphin
d   Delfin; Gemeiner Delfin;
Gewöhnlicher Delfin
f   Dauphin commun; Dauphin des

anciens
i   Delfino comune

1785   *Delphinus tropicalis*
Cete - Delphinidae
e   Arabian Common Dolphin
d   Arabischer Gewöhnlicher Delfin;
Araber Gemeiner Delfin
f   Dauphin commun de l'Arabie

Deltamys syn. *Akodon* q.v.

1786   *Dendrogale*
Scandentia - Tupaiidae
e   Small Smooth-tailed Tree Shrews;
Smooth-tailed Tree Shrews
d   Bergtupajas
f   Tuapïas; Tupajas
i   Tupaie

1787   *Dendrogale melanura*
Scandentia - Tupaiidae
e   Southern Smooth-tailed Tree Shrew;
Borneo Smooth-tailed Tree Shrew;
Bornean Smooth-tailed Tree Shrew
d   Südliche Bergtupaja;
Schwarzschwanztupaia; Borneo-
Bergtupaja; Bergtupaja
f   Tupaïa du Sud; Tupaja du Sud

1788   *Dendrogale murina*
Scandentia - Tupaiidae
e   Small Smooth-tailed Tree Shrew;
Northern Smooth-tailed Tree Shrew;
Mainland Slender-tailed Tree Shrew
d   Nördliche Bergtupaja;
Mausspitzhörnchen
f   Tupaïa; Tupaja
i   Tupaia dei monti

1789   *Dendrohyrax*
Urenotheria - Procaviidae
e   Tree Dassies; Tree Hyraxes
d   Baumschliefer ; Waldschliefer
f   Damans des arbres
i   Provavie

1790	***Dendrohyrax arboreus***		Tree Kangaroo
	Urenotheria - Procaviidae	*d*	Doria-Baumkänguru
e	Tree Hyrax; Southern Tree Hyrax; Tree Dassie	*f*	Kangourou arboricole de Doria; Dendrolague unicolor
d	Steppenwaldbaumschliefer	*i*	Canguro arboricolo monocolore;
f	Daman des arbres; Daman arboricole		Canguro arboricola di Doria
i	Dendroirace arborea; Procavia arborea; Irace arboricolo		

1796 ***Dendrolagus goodfellowi***
Diprotodontia - Macropodidae
e Goodfellow's Tree Kangaroo
d Goodfellow-Baumkänguru
f Dendrolague de Goodfellow; Kangarou arboricole de Goodfellow
i Canguro arboricolo di Goodfellow

1791 ***Dendrohyrax dorsalis***
Urenotheria - Procaviidae
e Beecroft's Hyrax; Beecroft's Tree Hyrax; Western Tree Hyrax
d Baumschliefer
f Daman des arbres de Beecroft
i Procavia

1797 ***Dendrolagus goodfellowi pulcherrimus***
Diprotodontia - Macropodidae
e Golden-mantled Tree Kangaroo

1792 ***Dendrohyrax validus***
Urenotheria - Procaviidae
e Eastern Tree Hyrax; Kilimanjaro Tree Hyrax; Mountain Forest Tree Hyrax
d Bergwaldbaumschliefer
i Irace arboricolo orientale

1798 ***Dendrolagus inustus***
Diprotodontia - Macropodidae
e Grizzled Tree Kangaroo; Dusky Tree Kangaroo; Grizzled Grey Tree Kangaroo
d Graues Baumkänguru
f Kangarou arboricole gris; Dendrolague grisonnant
i Canguro arboricolo grigio

1793 ***Dendrolagus***
Diprotodontia - Macropodidae
e Tree Kangaroos
d Baumkängurus; Bärenkängurus
f Kangourous arboricoles; Dendrolagues
i Canguri arboricoli

1799 ***Dendrolagus lumholtzi***
Diprotodontia - Macropodidae
e Lumholtz's Tree Kangaroo; Boongary (ANZ)
d Lumholtz-Baumkänguru
f Kangourou arboricole de Lumholtz; Dendrolague de Lumholtz
i Canguro di Lumholtz

1794 ***Dendrolagus bennettianus***
Diprotodontia - Macropodidae
e Bennett's Tree Kangaroo
d Bennet-Baumkänguru
f Kangourou arboricole de Bennett; Dendrolague de Bennett

Dendrolagus deltae syn. *D. matschiei* q.v.

1800 ***Dendrolagus matschiei***
Diprotodontia - Macropodidae
e Matschie's Tree Kangaroo; Huon Tree Kangaroo; Ornate Tree Kangaroo

1795 ***Dendrolagus dorianus***
Diprotodontia - Macropodidae
e Doria's Tree Kangaroo; Unicoloured

d Matschie-Baumkänguru; Matschies Baumkänguru

f   Kangourou arboricole de Matschie
i   Canguro arboricolo di Matschie

1801   *Dendrolagus mbaiso*
Diprotodontia - Macropodidae
e   Dingiso; Dingiso Tree Kangaroo;
Black-and-white Whistling Tree
Kangaroo; Black-and-white Tree
Kangaroo; Bondegezou
d   Schwarzweißes Baumkänguru
f   Dingiso

1802   *Dendrolagus scottae*
Diprotodontia - Macropodidae
e   Tenkile Tree Kangaroo; Scott's Tree
Kangaroo
d   Schwarzes Baumkänguru
f   Dendrolague de Papousie-Nouvelle-
Guinée; Kangourou arboricole de
Papousie Nouvelle-Guinée

1803   *Dendrolagus spadix*
Diprotodontia - Macropodidae
e   Lowland Tree Kangaroo; Lowlands
Tree Kangaroo
d   Tiefland-Baumkänguru

1804   *Dendrolagus ursinus*
Diprotodontia - Macropodidae
e   Black Tree Kangaroo; Vogelkop
Tree Kangaroo; White-throated Tree
Kangaroo
d   Bären-Baumkänguru
f   Kangarou arboricole noir
i   Canguro arboricolo orsino nero;
Canguro arboricolo orsino; Canguro
arboricolo nero

1805   *Dendromus*
Rodentia - Muridae
e   Tree Mice; African Climbing Mice
d   Aalstrichklettermäuse; Klettermäuse
f   Souris des bananiers ; Souris
grimpantes

1806   *Dendromus insignis*

Rodentia - Muridae
e   Remarkable Climbing Mouse
d   Aalstrichklettermaus

1807   *Dendromus kahuziensis*
Rodentia - Muridae
e   Mount Kahuzi Climbing Mouse;
Kahuzi Climbing Mouse

1808   *Dendromus kivu*
Rodentia - Muridae
e   Kivu Climbing Mouse; Ruwenzori
Climbing Mouse

1809   *Dendromus lovati*
Rodentia - Muridae
e   Lovat's Climbing Mouse

1810   *Dendromus melanotis*
Rodentia - Muridae
e   Dark-eared Tree Mouse; Grey
Climbing Mouse; Grey Pygmy
Climbing Mouse; Dark-eared
Climbing Mouse

d   Afrikanische Klettermaus

1811   *Dendromus mesomelas*
Rodentia - Muridae
e   Mountain Tree Mouse; Brant's
Climbing Mouse

1812   *Dendromus messorius*
Rodentia - Muridae
e   Banana Climbing Mouse

1813   *Dendromus mystacalis*
Rodentia - Muridae
e   Lesser Climbing Mouse; Chestnut
Climbing Mouse

1814   *Dendromus nyikae*
Rodentia - Muridae
e   Nyika Climbing Mouse

1815 *Dendromus oreas*
Rodentia - Muridae
e Cameroon Climbing Mouse

1816 *Dendromus vernayi*
Rodentia - Muridae
e Vernay's Climbing Mouse

1817 *Dendroprionomys*
Rodentia - Muridae
e Velvet Climbing Mice
d Rousselot-Baummäuse

1818 *Dendroprionomys rousseloti*
Rodentia - Muridae
e Velvet Climbing Mouse
d Rousselot-Baummaus

1819 *Deomys*
Rodentia - Muridae
e Link Rats; Congo Forest Mice
d Insektenessende Waldmäuse;
Insektenfressende Waldmäuse;
Deomys Waldmäuse
f Rats à manteau roux

1820 *Deomys ferrugineus*
Rodentia - Muridae
e Congo Forest Mouse; Rusty Link Rat
d Insektenessende Waldmaus;
Insektenfressende Waldmaus;
Deomys-Waldmaus
f Rat à manteau roux

Depanycteris syn. *Diclidurus* q.v.

1821 *Dephomys*
Rodentia - Muridae
e Defua Rats
d Defua-Ratten

1822 *Dephomys defua*
Rodentia - Muridae
e Defua Rat
d Defua-Ratte

1823 *Dephomys eburnea*
Rodentia - Muridae
e Ivory Coast Rat

Dermanura syn. *Artibeus* q.v.

1824 *Desmana*
Erinaceomorpha - Talpidae
e Desmans; Russian Desmans
d Bisamrüssler ; Bisamspitzmäuse;
Bisamratten; Desmane
f Desmans
i Topi muschiati

1825 *Desmana moschata*
Erinaceomorpha - Talpidae
e Desman; Russian Desman
d Russischer Desman; Desman;
Südrussischer Bisamrüssler;
Wychochol; Bisamrüssler
f Desman de Moscovie; Musaraigne
musqé; Rat musqué
i Desman muschiato; Topo muschiato
della Russia; Desman

1826 *Desmodilliscus*
Rodentia - Muridae
e Dwarf Gerbils; Small Short-eared
Rats; Pouched Gerbils
d Brauers Rennmäuse
f Gerbilles de Brauer

1827 *Desmodilliscus braueri*
Rodentia - Muridae
e Brauer's Dwarf Gerbil; Dwarf Gerbil;
Pouched Gerbil; Brauer's Gerbil
d Kurzohrige Nacksohlenrennmaus;
Brauer-Rennmaus
f Gerbille de Brauer; Gerbille naine de
Buchanan

1828 *Desmodillus*
Rodentia - Muridae
e Cape Short-eared Gerbils; Cape
Short-tailed Gerbils; Short-eared
Rats; Namaqualand Gerbils

d   Kurzschwanzrennmäuse
f   Gerbilles à poche

1829   *Desmodillus auricularis*
Rodentia - Muridae
e   Cape Short-eared Gerbil; Cape Short-tailed Gerbil; Short-eared Rat; Short-tailed Gerbil
d   Südafrikanische Nacktsohlenrennmaus; Kurzschwanzrennmaus; Kurzschwanzgerbil; Kap-Kurzschwanzrennmaus
f   Gerbille à poche

1830   *Desmodus*
Chiroptera - Phyllostomidae
e   Vampire Bats; Common Vampire Bats
d   Gemeine Vampire; Echte Vampire
f   Vampires
i   Vampiri

1831   *Desmodus rotundus*
Chiroptera - Phyllostomidae
e   Vampire; True Vampire; Common Vampire Bat; South American Vampire Bat; Blood-sucking Bat; Vampire Bat
d   Gemeiner Vampir; Echter Vampir; Vampir; Großer Blutsauger; Gewöhnlicher Vampir
f   Vampire d'Azara
i   Vampiro; Pipistrello vampiro

1832   *Desmomys*
Rodentia - Muridae
e   Harrington's Rats
d   Harrington-Ratten

1833   *Desmomys harringtoni*
Rodentia - Muridae
e   Harrington's Rat
d   Harrington-Ratte

1834   *Diaemus*

Chiroptera - Phyllostomidae
e   White-winged Vampire Bats; Avian Vampire Bats
d   Weißschwingenvampire; Weißflügelvampire
f   Vampires aux ailes blanches

1835   *Diaemus youngi*
Chiroptera - Phyllostomidae
e   Spotted Vampire Bat; White-winged Vampire Bat; White-winged Vampire
d   Weißschwingenvampir; Weißflügelvampir
f   Vampire aux ailes blanches

1836   *Dicerorhinus*
Perissodactyla - Rhinocerotidae
e   Sumatran Rhinoceroses; Asiatic Two-horned Rhinoceroses
d   Sumatra-Nashörner; Halbpanzernashörner
f   Rhinocéros de Sumatra
i   Rinoceronte

1837   *Dicerorhinus sumatrensis*
Perissodactyla - Rhinocerotidae
e   Sumatran Rhinoceros; Asiatic Two-horned Rhinoceros; Sumatran Two-horned Rhinoceros
d   Sumatra-Nashorn
f   Rhinocéros de Sumatra
i   Rinoceronte di Sumatra

1838   *Dicerorhinus sumatrensis harrissoni*
Perissodactyla - Rhinocerotidae
e   Eastern Sumatran Rhinoceros
f   Rhinocéros de Sumatra oriental

1839   *Dicerorhinus sumatrensis lasiotis*
Perissodactyla - Rhinocerotidae
e   Burmese Sumatran Rhinoceros

1840   *Dicerorhinus sumatrensis sumatrensis*
Perissodactyla - Rhinocerotidae

e Western Sumatran Rhinoceros
f Rhinocéros de Sumatra occidental

1841 *Diceros*
Perissodactyla - Rhinocerotidae
e African Rhinoceroses; African Black Rhinoceroses
d Spitzmaulnashörner
f Rhinocéros noirs; Rhinocéros bicornes d'Afrique
i Rinoceronti neri

1842 *Diceros bicornis*
Perissodactyla - Rhinocerotidae
e African Rhinoceros; African Black Rhinoceros; Black Rhinoceros; Hood-lipped Rhinoceros; Hook-lipped Rhinoceros
d Spitzmaulnashorn; Schwarzes Nashorn
f Rhinocéros noir; Rhinocéros bicorne d'Afrique
i Rinoceronte nero

1843 *Diceros bicornis bicornis*
Perissodactyla - Rhinocerotidae
e Southwestern Black Rhinoceros
f Rhinocéros noir du Sud-Ouest

1844 *Diceros bicornis longipes*
Perissodactyla - Rhinocerotidae
e Western Black Rhinoceros; West African Black Rhinoceros
f Rhinocéros noir de l'Ouest

1845 *Diceros bicornis michaeli*
Perissodactyla - Rhinocerotidae
e East African Black Rhinoceros; Eastern Black Rhinoceros
f Rhinocéros noir de l'Est

1846 *Diceros bicornis minor*
Perissodactyla - Rhinocerotidae
e South Central African Black Rhinoceros
f Rhinocéros noir du Centre-Sud

Diceros simum syn. *Ceratotherium simum* q.v.

1847 *Diclidurus*
Chiroptera - Emballonuridae
e Ghost Bats; Sac-tailed Bats; White Bats
d Gespenstfledermäuse
f Diclidures

1848 *Diclidurus albus*
Chiroptera - Emballonuridae
e White Bat; Greater White Bat; Ghost Bat; Northern Ghost Bat

1849 *Diclidurus ingens*
Chiroptera - Emballonuridae
e Greater Ghost Bat

1850 *Diclidurus isabellus*
Chiroptera - Emballonuridae
e Isabelle's Ghost Bat

1851 *Diclidurus scutatus*
Chiroptera - Emballonuridae
e Lesser Ghost Bat

1852 *Dicotyles*
Artiodactyla - Tayassuidae
e Collared Peccaries
d Halsbandpekaris
f Pécaris à collier
i Pecari dal collare

1853 *Dicotyles tacaju*
Artiodactyla - Tayassuidae
e Collared Peccary; Javelina
d Halsbandpekari
f Pécari à collier
i Pecari dal collare; Javelina

1854 *Dicotyles tajacu nanus*
Artiodactyla - Tayassuidae
e Cozumel Collared Peccary

1855 *Dicrostonyx*
Rodentia - Muridae
e   Collared Lemmings; Arctic
Lemmings; Arctic Collared
Lemmings
d   Halsbandlemminge; Gabelkrall-
Lemminge
f   Lemmings arctiques
i   Lemming

Dicrostonyx exsul syn. *D.
groenlandicus* q.v.

1856 *Dicrostonyx groenlandicus*
Rodentia - Muridae
e   American Arctic Lemming; Northern
Collared Lemming; Nelson's
Collared Lemming; Ogilvie
Mountain Collared Lemming; Bering
Collared Lemming; Greenland
Collared Lemming; St. Lawrence
Island Collared Lemming; Unalaska
Collared Lemming; Collared White
Lemming; Arctic Lemming; Hoofed
Lemming
d   Grönländischer Halsbandlemming
f   Lemming arctique; Lemming à
collier; Lemming variable

1857 *Dicrostonyx hudsonius*
Rodentia - Muridae
e   Ungava Lemming; Labrador Collared
Lemming; Ungava Collared
Lemming; Hudson Bay Collared
Lemming
d   Hudson-Halsbandlemming
f   Lemming d'Ungava
i   Lemming dal collare del Labrador

Dicrostonyx kilangmiutak syn. *D.
groenlandicus* q.v.

Dicrostonyx nelsoni syn. *D.
groenlandicus* q.v.

Dicrostonyx nunatakensis syn. *D.
groenlandicus* q.v.

1858 *Dicrostonyx richardsoni*
Rodentia - Muridae
e   Richardson's Collared Lemming

Dicrostonyx rubricatus syn. *D.
groenlandicus* q.v.

1859 *Dicrostonyx torquatus*
Rodentia - Muridae
e   Arctic Lemming; Collared Lemming
d   Gehufte Eisfuchsmaus; Gabelkrall-
Lemming; Halsbandlemming;
Eigentlicher Halsbandlemming
f   Lemming arctique; Lemming à
collier

Dicrostonyx unalascensis syn. *D.
groenlandicus* q.v.

1860 *Dicrostonyx vinogradovi*
Rodentia - Muridae
e   Wrangel Collared Lemming;
Wrangel Lemming
d   Wrangel-Lemming

1861 **Didelphidae**
Didelphimorphia
e   Opossums; American Opossums
d   Beutelratten
f   Opossums d'Amérique; Sarigues
i   Didelfidi

1862 *Didelphis*
Didelphimorphia - Didelphidae
e   American Opossums; Large
American Opossums; Common
Opossums; Opossums
d   Opossums; Große Amerikanische
Opossums
f   Opossums; Sarigues
i   Opossum; Possum

1863 *Didelphis albiventris*
Didelphimorphia - Didelphidae
e   Azara's Opossum; White-eared
Opossum

d Weißohropossum

f Sarigue aux oreilles blanc

1864 *Didelphis aurita*
Didelphimorphia - Didelphidae

e Big-eared Opossum; Black-eared Opossum

d Großohropossum

f Sarigue du Sud-Est brézilien; Opossum commun du Brésil

Didelphis azarae syn. *D. albiventris* q.v.

1865 *Didelphis marsupialis*
Didelphimorphia - Didelphidae

e Common Opossum; Large American Opossum; Opossum (NA); North American Opossum; Southern Opossum; Manacou

d Südopossum; Südliches Opossum

f Opossum commun; Opossum australe

i Opossum d'America; Opossum comune

1866 *Didelphis virginiana*
Didelphimorphia - Didelphidae

e North American Opossum

d Nordamerikanisches Opossum

f Opossum de Virginie; Sarigue d'Amérique du Nord

i Opossum della Virginia

1867 *Didelphis virginiana virginiana*
Didelphimorphia - Didelphidae

e Virginia Opossum

d Virginisches Opossum; Virginia-Opossum

1868 *Dinaromys*
Rodentia - Muridae

e Martino's Snow Voles; Nehring's Snow Voles

d Bergmäuse

f Campagnols de Nehring; Campagnols de montagne

i Arvicole dei Balcani

1869 *Dinaromys bogdanovi*
Rodentia - Muridae

e Martino's Snow Vole; Nehring's Snow Vole; Balkan Snow Vole; Martino's Vole

d Bergmaus; Jugoslawische Schermaus; Nehrings Bergmaus

f Campagnol de montagne; Campagnol de Martino; Camopagnol de Nehring

i Arvicola dei Balcani

1870 Dinomyidae
Rodentia

e Pacaranas; False Pacas; Branick's Rats

d Pakaranartige; Falsche Pakaranas

f Pakaranas; Dinomyidés

i Paracane; Dinomidi

1871 *Dinomys*
Rodentia - Dinomyidae

e Pacaranas; False Pacas; Branick's Rats

d Falsche Pakas; Pakaranas

f Pakaranas

1872 *Dinomys branickii*
Rodentia - Dinomyidae

e Pacarana; False Paca; Branick's Rat

d Falscher Paka; Pakarana

f Pakarana

1873 *Diomys*
Rodentia - Muridae

e Manipur Mice; Crump's Mice

d Crump-Mäuse

1874 *Diomys crumpi*
Rodentia - Muridae

e Manipur Mouse; Crump's Mouse

d Crump-Maus

1875 *Diphylla*
Chiroptera - Phyllostomidae
e Hairy-legged Vampire Bats; Hairy-
legged Vampires
d Kammzahnflatterer ; Kleine
Blutsäuger
f Diphylles

1876 *Diphylla ecaudata*
Chiroptera - Phyllostomidae
e Hairy-legged Vampire Bat; Hairy-
legged Vampire
d Kammzahnflatterer; Kleiner
Blutsauger; Kammzahnvampir
f Diphylle sans queue

1877 *Diplogale*
Carnivora - Viverridae
e Hose's Civets
d Schlichtroller
f Civettes palmistes d'Hose

1878 *Diplogale hosei*
Carnivora - Viverridae
e Hose's Civet; Hose's Palm Civet
d Schlichtroller
f Civette palmiste d'Hose

1879 *Diplomesodon*
Soricomorpha - Soricidae
e Piebald Shrews; Sand Shrews;
Turkestan Desert Shrews
d Gescheckte Spitzmäuse; Scheckige
Spitzmäuse; Sandspitzmäuse;
Wüstenspitzmäuse

1880 *Diplomesodon pulchellum*
Soricomorpha - Soricidae
e Piebald Shrew; Sand Shrew;
Turkestan Desert Shrew
d Gescheckte Spitzmaus; Scheckige
Spitzmaus; Sandspitzmaus;
Gescheckte Wüstenspitzmaus

1881 *Diplomys*
Rodentia - Echimyidae

e Arboreal Soft-furred Spiny Rats
d Baumstachelratten

1882 *Diplomys caniceps*
Rodentia - Echimyidae
e Arboreal Soft-furred Spiny Rat
d Graukopfbaumstachelratte
f Rat épineux à tête grise

Diplomys darlingi syn. *D.
rufodorsalis* q.v.

1883 *Diplomys labilis*
Rodentia - Echimyidae
e Gliding Spiny Rat; Rufous Tree Rat
d Fliegende Panama-Baumstachelratte

1884 *Diplomys rufodorsalis*
Rodentia - Echimyidae
e Red-crested Spiny Rat; Red-crested
Tree Rat

1885 *Diplothrix*
Rodentia - Muridae
e Ryukyu Rats
d Riukiu-Ratten

1886 *Diplothrix legatus*
Rodentia - Muridae
e Ryukyu Rat
d Riukiu-Ratte

1887 *Dipodidae*
Rodentia
e Jerboas; Jerboas and Jumping Mice
d Springmäuse; Springnager;
Wüstenspringmäuse
f Gerboises; Dipodidés
i Dipodidi

Dipodillus syn. *Gerbillus* q.v.

1888 *Dipodomys*
Rodentia - Geomyidae
e Kangaroo Rats

d Taschenspringer ; Kängururatten
f Rats kangourous
i Ratti canguro; Topi canguro

1889 ***Dipodomys agilis***
Rodentia - Geomyidae
e Agile Kangaroo Rat
d Pazifik-Kängururatte

1890 ***Dipodomys agilis agilis***
Rodentia - Geomyidae
e Pacific Kangaroo Rat

1891 ***Dipodomys californicus***
Rodentia - Geomyidae
e Californian Kangaroo Rat
d Kalifornische Kängururatte

1892 ***Dipodomys compactus***
Rodentia - Geomyidae
e Gulf Coast Kangaroo Rat
d Golfküsten-Kängururatte

1893 ***Dipodomys compactus compactus***
Rodentia - Geomyidae
e Padre Island Kangaroo Rat

1894 ***Dipodomys deserti***
Rodentia - Geomyidae
e Desert Kangaroo Rat
d Wüstenkängururatte
f Rat kangourou du Pacifique

1895 ***Dipodomys elator***
Rodentia - Geomyidae
e Texas Kangaroo Rat
d Texas-Kängururatte
f Rat kangourou de Texas

Dipodomys elephantinus syn. *D. venustus elephantinus* q.v.

1896 ***Dipodomys gravipes***
Rodentia - Geomyidae

e San Quintin Kangaroo Rat
d San-Quintin-Kängururatte

1897 ***Dipodomys heermanni***
Rodentia - Geomyidae
e Heermann's Kangaroo Rat
d Heermanns Kägururatte

1898 ***Dipodomys heermanni berkeleyensis***
Rodentia - Geomyidae
e Berkeley Kangroo Rat

1899 ***Dipodomys heermanni dixoni***
Rodentia - Geomyidae
e Merced Kangaroo Rat

1900 ***Dipodomys heermanni morroensis***
Rodentia - Geomyidae
e Morro Bay Kangaroo Rat
i Topo canguro di Merriam Bay

1901 ***Dipodomys ingens***
Rodentia - Geomyidae
e Giant Kangaroo Rat
d Riesenkängururatte

1902 ***Dipodomys insularis***
Rodentia - Geomyidae
e San José Kangaroo Rat; San José Island Kangaroo Rat
d San-Jose-Kängururatte

1903 ***Dipodomys margaritae***
Rodentia - Geomyidae
e Margarita Island Kangaroo Rat; Santa Margarita Kangaroo Rat
d Margarita-Kängururatte

1904 ***Dipodomys merriami***
Rodentia - Geomyidae
e Merriams Kangaroo Rat
d Merriam-Kängururatte; Merriams Kängururatte
i Ratto canguro di Merriam

1905 *Dipodomys merriami collinus*
Rodentia - Geomyidae
e   Aguanga Kangaroo Rat

1906 *Dipodomys merriami parvus*
Rodentia - Geomyidae
e   San Bernardino Kangaroo Rat

1907 *Dipodomys microps*
Rodentia - Geomyidae
e   Chisel-toothed Kangaroo Rat; Great
Basin Kangaroo Rat
d   Meißelzahn-Kängururatte

1908 *Dipodomys nelsoni*
Rodentia - Geomyidae
e   Fresno Kangaroo Rat; Nelson's
Kangaroo Rat
d   Nelsons Kängururatte
f   Rat-kangourou de Fresno

1909 *Dipodomys nitratoides*
Rodentia - Geomyidae
e   San Joaquin Kangaroo Rat
d   Fresno-Kängururatte

1910 *Dipodomys nitratoides brevinasus*
Rodentia - Geomyidae
e   Short-nosed Kangaroo Rat

1911 *Dipodomys nitratoides nitratoides*
Rodentia - Geomyidae
e   Tipton Kangaroo Rat

1912 *Dipodomys ordii*
Rodentia - Geomyidae
e   Ord's Kangaroo Rat; Five-toed
Kangaroo Rat
d   Ord-Kängururatte
f   Rat kangourou d'Ord

Dipodomys ornatus syn. *D. phillipsii*
q.v.

1913 *Dipodomys panamintinus*

Rodentia - Geomyidae
e   Panamint Kangaroo Rat
d   Panamint-Kängururatte

Dipodomys paralius syn. *D. agilis*
q.v.

Dipodomys peninsularis syn. *D.
agilis* q.v.

1914 *Dipodomys phillipsii*
Rodentia - Geomyidae
e   Phillips's Kangaroo Rat
d   Phillips-Kängururatte

1915 *Dipodomys simulans*
Rodentia - Geomyidae
e   Dulzura Kangaroo Rat; San Diego
Kangaroo Rat

1916 *Dipodomys spectabilis*
Rodentia - Geomyidae
e   Banner-tailed Kangaroo Rat
d   Fahnenschwanzkängururatte

1917 *Dipodomys spectabilis baileyi*
Rodentia - Geomyidae
e   New Mexican Banner-tailed
Kangaroo Rat

1918 *Dipodomys stephensi*
Rodentia - Geomyidae
e   Stephens's Kangaroo Rat
d   Stephens-Kängururatte

1919 *Dipodomys venustus*
Rodentia - Geomyidae
e   Narrow-faced Kangaroo Rat
d   Santa Cruz-Kängururatte

1920 *Dipodomys venustus elephantinus*
Rodentia - Geomyidae
e   Big-eared Kangaroo Rat
d   Großohrkängururatte
f   Rat kangourou aux grands oreilles

1921 *Dipodomys venustus venustus*
Rodentia - Geomyidae
e Santa Cruz Kangaroo Rat

1922 *Dipus*
Rodentia - Dipodidae
e Hairy-footed Jerboas; Rough-legged Jerboas; Feather-footed Jerboas; Northern Three-toed Jerboas
d Dreizehige Springmäuse; Rauhfußspringmäuse; Rauhfüssige Wüstenspringmäuse
f Gerboises aux pattes rugeuses
i Gerboa dai piedi rogosi; Gerboe dai piedi rogosi

1923 *Dipus sagitta*
Rodentia - Dipodidae
e Hairy-footed Jerboa; Rough-legged Jerboa; Feather-footed Jerboa; Northern Three-toed Jerboa
d Rauhfüssige Wüstenspringmaus; Rauhfußspringmaus; Pfeilspringmaus; Rauhfüßige Pfeilspringmaus; Dreizehennordspringmaus
f Gerboise aux pattes rugeuses
i Gerboa dai piedi rogosi

1924 *Distoechurus*
Diprotodontia - Acrobatidae
e Pen-tailed Phalangers; Pen-tailed Possums; Feather-tailed Phalangers; Feather-tailed Posssums
d Federschwanzbeutler ; Pinselschwanzbeutler
i Possum dalla coda a penna; Opossum dalla coda a penna

1925 *Distoechurus pennatus*
Diprotodontia - Acrobatidae
e Pen-tailed Phalanger; Pen-tailed Possum; Feather-tailed Phalanger; Feather-tailed Posssum
d Federschwanzbeutler; Pinselschwanzbeutler
i Possum dalla coda a penna

1926 *Dobsonia*
Chiroptera - Pteropodidae
e Spinal-winged Fruit Bat; Bare-backed Fruit Bat; Naked-backed Fruit Bat
d Nacktrückenflughunde
f Chauve-souris à dos nu

1927 *Dobsonia anderseni*
Chiroptera - Pteropodidae
e Andersen's Bare-backed Fruit Bat

1928 *Dobsonia beauforti*
Chiroptera - Pteropodidae
e Beaufort's Naked-backed Fruit Bat; Beaufort's Bare-backed Fruit Bat

1929 *Dobsonia chapmani*
Chiroptera - Pteropodidae
e Negros Naked-backed Fruit Bat; Philippine Bare-backed Fruit Bat; Negros Bare-backed Fruit Bat

1930 *Dobsonia crenulata*
Chiroptera - Pteropodidae
e Halmahera Bare-backed Fruit Bat

1931 *Dobsonia emersa*
Chiroptera - Pteropodidae
e Biak Naked-backed Fruit Bat; Biak Bare-backed Fruit Bat

1932 *Dobsonia exoleta*
Chiroptera - Pteropodidae
e Celebes Naked-backed Bat; Sulawesi Naked-backed Bat; Sulawesi Naked-backed Fruit Bat; Celebes Bare-backed Fruit Bat

1933 *Dobsonia inermis*
Chiroptera - Pteropodidae
e Solomons Naked-backed Bat; Solomons Naked-backed Fruit Bat; Solomons Bare-backed Fruit Bat

1934 *Dobsonia magna*
Chiroptera - Pteropodidae
e Great Bare-backed Fruit Bat; Greater
Naked-backed Bat

1935 *Dobsonia minor*
Chiroptera - Pteropodidae
e Lesser Naked-backed Bat; Lesser
Naked-backed Fruit Bat; Lesser
Bare-backed Fruit Bat

1936 *Dobsonia moluccensis*
Chiroptera - Pteropodidae
e Bare-backed Fly-Fox; Molluccan
Naked-backed Fruit Bat; Mollucan
Bare-backed Fruit Bat; Bare-backed
Fruit Bat (ANZ)

1937 *Dobsonia pannietensis*
Chiroptera - Pteropodidae
e Panniet Naked-backed Fruit Bat; De
Vies's Bare-backed Fruit Bat

1938 *Dobsonia peronii*
Chiroptera - Pteropodidae
e Western Naked-backed Bat; Western
Naked-backed Fruit Bat; Western
Bare-backed Fruit Bat

1939 *Dobsonia praedatrix*
Chiroptera - Pteropodidae
e New Britain Naked-backed Bat; New
Britain Naked-backed Fruit Bat;
Bismarck Bare-backed Fruit Bat

1940 *Dobsonia viridis*
Chiroptera - Pteropodidae
e Greenish Naked-backed Bat;
Greenish Naked-backed Fruit Bat;
Greenish Bare-backed Fruit Bat

1941 *Dolichotis*
Rodentia - Caviidae
e Patagonian Hards; Patagonian
Cavies; Maras
d Maras; Pampashasen

f Maras; Lièvres des pampas; Lièvres
des pampas de Patagonie
i Marà

1942 *Dolichotis patagonum*
Rodentia - Caviidae
e Mara; Patagonian Mara; Patagonian
Cavy; Patagonian Hare
d Mara; Große Mara; Großer
Pampashase
f Mara; Lièvre des pampas; Lièvre des
pampas de Patagonie
i Marà; Lepre della Patagonia

1943 *Dolichotis salinicola*
Rodentia - Caviidae
e Salt Desert Cavy; Chacoan Mara;
Desert Cavy
d Kleiner Pampashase
i Coniglio delle saline

Dologale syn. *Helogale* q.v.

Dolomys syn. *Dinaromys* q.v.

1944 *Dorcatragus*
Artiodactyla - Bovidae
e Beiras
d Beiras
f Dorcotragues
i Beire

1945 *Dorcatragus megalotis*
Artiodactyla - Bovidae
e Beira; Beira Antelope
d Baeira
f Beira
i Beira

1946 *Dorcopsis*
Diprotodontia - Macropodidae
e Free Wallabies; New Guinea Forest
Wallabies; Forest Wallabies;
Dorcopsises
d Buschkängurus
f Wallabies de la Nouvelle Guinée;

Dorcopsis
i Wallaby delle foreste

1947 *Dorcopsis atrata*
Diprotodontia - Macropodidae
e Black Forest Wallaby; Black
Dorcopsis; Black Dorcopsis Wallaby
d Goodenough-Buschkänguru
f Dorcopsis noir des forèts; Wallaby
noir

1948 *Dorcopsis hageni*
Diprotodontia - Macropodidae
e Greater Forest Wallaby; Northern
New Guinea Wallaby; White-striped
Dorcopsis
d Hagen-Buschkänguru
f Wallaby du Nord de la Nouvelle
Guinée

1949 *Dorcopsis luctuosa*
Diprotodontia - Macropodidae
e Grey Dorcopsis
d Waldwallaby
f Dorcopsis gris

1950 *Dorcopsis muelleri*
Diprotodontia - Macropodidae
e Common Forest Wallaby; Brown
Dorcopsis; Brown Dorcopsis
Wallaby
d Streifenbuschkäguruh;
Neuguineisches Buschkänguru

Dorcopsis veterum syn. *D. muelleri*
q.v.

1951 *Dorcopsulus*
Diprotodontia - Macropodidae
e Forest Wallabies
d Buschkängurus
f Wallabies des montagnes
i Wallaby delle foreste

1952 *Dorcopsulus macleayi*
Diprotodontia - Macropodidae

e Papuan Forest Wallaby; New Guinea
Mountain Wallaby; New Guinea
Forest Mountain Wallaby; Macleay's
Dorcopsis
d Macleay-Bushkänguru
f Wallaby de Macleay
i Wallaby della foresta di Macleay

1953 *Dorcopsulus vanheurni*
Diprotodontia - Macropodidae
e Lesser Forest Wallaby; Common
Mountain Forest Wallaby
d Bergbuschkänguru; Kleines
Buschkänguru
i Piccolo Wallaby della foresta

1954 *Dremomys*
Rodentia - Sciuridae
e Long-nosed Squirrels; Red-cheeked
Squirrels
d Rotwangenhörnchen
f Écureuils terrestres

1955 *Dremomys everetti*
Rodentia - Sciuridae
e Bornean Mountain Ground Squirrel;
Bornean Long-nosed Squirrel
d Borneo-Berghörnchen
i Scoiattolo delle montagne del Borneo

1956 *Dremomys lokriah*
Rodentia - Sciuridae
e Orange-bellied Himalajan Squirrel;
Orange-bellied Squirrel; Himalajan
Ground Squirrel
d Orangebauch-Himalaja-Hörnchen

1957 *Dremomys pernyi*
Rodentia - Sciuridae
e Perny's Long-nosed Squirrel; Perny's
Ground Squirrel
d Perny-Langnasenhörnchen

1958 *Dremomys pyrrhomerus*
Rodentia – Sciuridae

e   Red-hipped Squirrel; Red-hipped
     Ground Squirrel

1959   *Dremomys rufigenis*
        Rodentia - Sciuridae
e   Southeastern Asiatic Squirrel; Asian
     Red-cheeked Squirrel; Red-cheeked
     Ground Squirrel
d   Echtes Rotwangenhörnchen

1960   *Dromiciops*
        Microbiotheria - Microbiotheriidae
e   Colocoloses; Mouse Opossums
d   Chiloe-Beutelratten
f   Monitos del montes
i   Moniti del Monte

1961   *Dromiciops gliroides*
        Microbiotheria - Microbiotheriidae
e   Colocolos; Monito del Monte; Mouse
     Opossum; Chiloe Opossum
d   Chiloe-Beutelratte; Monito del
     Monte
f   Monito del monte
i   Monito del Monte

1962   *Dryomys*
        Rodentia - Gliridae
e   Tree Dormice; Forest Dormice
d   Baumschläfer
f   Lérotins; Loirs des forêts
i   Ghiro delle foreste

1963   *Dryomys laniger*
        Rodentia - Gliridae
e   Woolly Dormouse
d   Türkischer Baumschläfer
f   Lérotin laineux

1964   *Dryomys niethammeri*
        Rodentia - Gliridae
e   Niethammer's Dormouse

1965   *Dryomys nitedula*
        Rodentia - Gliridae

e   Forest Dormouse; Tree Dormouse
d   Weinschläfer; Baumschläfer;
     Schlesischer Baumschläfer
f   Lérotin; Lérot dryas
i   Driomio

1966   *Dryomys sichuanensis*
        Rodentia - Gliridae
e   Chinese Dormouse
i   Driomio; Ghiro della Cina

1967   *Dugong*
        Uranotheria - Dugongidae
e   Dugongs; Sea Cows
d   Dugonge
f   Dugongs
i   Dugonghi

1968   *Dugong dugon*
        Uranotheria - Dugongidae
e   Dugong; Sea Cow
d   Dugong; Seeschwein; Pazifische
     Seekuh
f   Dugong; Dugon
i   Dugongo

1969   *Dugongidae*
        Uranotheria
e   Dugongs; Sea Cows; Dugongs and
     Seacows
d   Gabelschwanzseekühe; Dugonge
f   Dugongidés
i   Dugonghi; Dugonghidi

1970   *Dusicyon*
        Carnivora - Canidae
e   South American Foxes; Falkland
     Island Wolves
d   Falkland-Füchse
f   Loups des îles Falkland; Loups des
     Falklands
i   Volpi delle isole Falkland

1971   *Dusicyon australis*
        Carnivora - Canidae

 e Falkland Island Wolf; Antarctic
   Wolf; Falkland Island Fox; Falkland
   Wolf; Falkland Fox; Marine Wolf
 d Falkland-Fuchs; Falkland-Insel-
   Wolf; Falkland-Wolf; Falkland-Insel-
   Fuchs
 f Loup des îles Falkland; Loup des
   Falkland
 i Volpe delle isole Falkland

Dusicyon culpaeus syn. *Pseudalopex culpaeus* q.v.

Dusicyon fulvipes syn. *Pseudalopex griseus* q.v.

Dusicyon griseus syn. *Pseudalopex griseus* q.v.

Dusicyon gymnocercus syn. *Pseudalopex gymnocercus* q.v.

Dusicyon sechurae syn. *Pseudalopex sechurae* q.v.

Dusicyon thous syn. *Cerdocyon thous* q.v.

Dusicyon vetulus syn. *Lycalopex vetulus* q.v.

1972 **Dyacopterus**
   Chiroptera - Pteropodidae
 e Dyak Fruit Bats
 d Südostasiatische Fruchtfledermäuse

1973 **Dyacopterus spadiceus**
   Chiroptera - Pteropodidae
 e Dyak Fruit Bat
 d Südostasiatische Fruchtfledermaus

Dymecodon syn. *Urotrichus* q.v.

E

1974   Echimyidae
Rodentia
e   Spiny Rats; American Spiny Rats
d   Stachelratten
f   Échymyidés
i   Ratti spinosi; Echimidi

1975   Echimys
Rodentia - Echimyidae
e   Crested Spiny Rats; Arboreal Spiny
Rats; Spiny Tree Rats
d   Kammstachelratten
f   Rats épineux
i   Ratti spinosi

Echimys armatus syn. *Makalata armata* q.v.

1976   Echimys blainvillei
Rodentia - Echimyidae
e   Golden Atlantic Tree Rat

1977   Echimys braziliensis
Rodentia - Echimyidae
e   Red-nosed Tree Rat

1978   Echimys chrysurus
Rodentia - Echimyidae
e   White-crested Spiny Rat; White-faced Spiny Rat; White-faced Tree Rat

1979   Echimys dasythrix
Rodentia - Echimyidae
e   Drab Atlantic Tree Rat

1980   Echimys grandis
Rodentia - Echimyidae
e   Giant Tree Rat

1981   Echimys lamarum
Rodentia - Echimyidae
e   Pallid Atlantic Tree Rat

1982   Echimys macrurus
Rodentia - Echimyidae
e   Long-tailed Tree Rat

1983   Echimys nigrispinus
Rodentia - Echimyidae
e   Black-spined Atlantic Tree Rat

1984   Echimys pictus
Rodentia - Echimyidae
e   Painted Tree Rat

1985   Echimys rhipidurus
Rodentia - Echimyidae
e   Peruvian Tree Rat

1986   Echimys saturnus
Rodentia - Echimyidae
e   Dark Tree Rat

1987   Echimys semivillosus
Rodentia - Echimyidae
e   Speckled Tree Rat

1988   Echimys thomasi
Rodentia - Echimyidae
e   Giant Atlantic Tree Rat

1989   Echimys unicolor
Rodentia - Echimyidae
e   Unicoloured Tree Rat

Echinoprocta syn. *Coendu* q.v.

1990   Echinops
Soricomorpha - Tenrecidae
e   Lesser Hedgehog Tenrecs
d   Kleine Igeltanreks
f   Tenrecs-hérissons

1991 *Echinops telfairi*
Soricomorpha - Tenrecidae
e Small Madagascar Hedehog; Lesser
Hedgehog Tanrec; Small Madagascar
Hedgehog Tanrec
d Kleiner Igeltanrek
f Tenrec-hérisson

1992 *Echinosorex*
Erinaceomorpha - Erinaceidae
e Moonrats; Gymnures
d Haarigel ; Rattenigel
f Gymnures
i Ratti lunari

1993 *Echinosorex gymnura*
Erinaceomorpha - Erinaceidae
e Moonrat; Raffles's Gymnure;
Malayan Gymnure; Malayan Rat
Shrew
d Großer Haarigel; Großer Rattenigel
f Gymnure
i Ratto lunaro

1994 *Echiothrix*
Rodentia - Muridae
e Sulawesi Spiny Rats
d Sulawesi-Stachelratten

1995 *Echiothrix leucura*
Rodentia - Muridae
e Sulawesi Spiny Rat; Greater
Sulawesi Shrew Rat
d Sulawesi-Spitzmausratte; Sulawesi-
Spitzratte; Sulawesi-Stachelratte

1996 *Echymipera*
Peramelia - Peroryctidae
e Spiny Bandicoots; New Guinea
Spiny Bandicoot; Echymiperas;
Spiny-furred Bandicoots; Spiny New
Guinea Bandicoots
d Stachelnasenbeutler
f Echymiperas
i Bandicut spinosi

1997 *Echymipera clara*
Peramelia - Peroryctidae
e White-lipped Bandicoot; Clara's
Echymipera; Dymorphic Echymipera
d Japan-Stachelnasenbeutler; Clara-
Stachelnasenbeutler
f Echymipera de Clara

1998 *Echymipera davidi*
Peramelia - Peroryctidae
e Dacid's Echymipera
d Kiriwina-Stachelnasenbeutler
f Echymipera de David

1999 *Echymipera echinista*
Peramelia - Peroryctidae
e Menzie's Echymipera
d Menzies-Stachelnasenbeutler

2000 *Echymipera kalubu*
Peramelia - Peroryctidae
e Spiny Bandicoot; Kalubu
Echymipera; Spiny Echymipera;
Common Echymipera
d Flachstachelnasenbeutler
f Echymipera commun
i Bandicut spinoso della Nuova
Guinea

2001 *Echymipera rufescens*
Peramelia - Peroryctidae
e Rufescent Bandicoot; Long-nosed
Echymipera
d Dickkopfstachelnasenbeutler; Roter
Stachelnasenbeutler

2002 *Echymipera rufescens australis*
Peramelia - Peroryctidae
e Rufous Spiny Bandicoot

2003 *Ectophylla*
Chiroptera - Phyllostomidae
e White Bats
d Weiße Fledermäuse

2004   *Ectophylla alba*
Chiroptera - Phyllostomidae
e   White Bat; Honduran White Bat;
White Fruit Bat
d   Weiße Fledermaus

2005   *Eidolon*
Chiroptera - Pteropodidae
e   Straw-coloured Fruit Bats; Eidolon
Fruit Bats; Yellow-haired Fruit Bats
d   Palmenflughunde
f   Roussettes paillées

2006   *Eidolon dupreanum*
Chiroptera - Pteropodidae
e   Madagascar Fruit Bat; Madagascar
Straw-coloured Fruit Bat
f   Roussette malgache couleur paille

2007   *Eidolon helvum*
Chiroptera - Pteropodidae
e   Straw-coloured Fruit Bat; Yellow-
haired Fruit Bat; African Straw-
coloured Fruit Bat
d   Palmenflughund
f   Roussette paillée; Roussette paillé
africaine; Roussette jaune

2008   *Eira*
Carnivora - Mustelidae
e   Tayras
d   Tayras
f   Tayras
i   Tayre; Taira

2009   *Eira barbara*
Carnivora - Mustelidae
e   Tayra; Grey-headed Weasel; Grey-
headed Tayra
d   Tayra; Hyrare
f   Tayra
i   Tayra; Taira; Urone

2010   *Elaphodus*
Artiodactyla - Cervidae
e   Tufted Deer ; Tibetan Muntjacs

d   Schopfhirsche; Schopfmuntjaks
f   Cerfs à touffe
i   Elefadi

2011   *Elaphodus cephalophus*
Artiodactyla - Cervidae
e   Tufted Deer; Tibetan Muntjac
d   Schopfhirsch
f   Cerf à touffe
i   Elefado cefalofo

2012   *Elaphodus cephalophus*
cephalophus
Artiodactyla - Cervidae
e   Western Tufted Deer

2013   *Elaphurus*
Artiodactyla - Cervidae
e   Père David's Deer
d   David-Hirsche; Milu-Hirsche; Milus;
Davids Hirsche
f   Élaphures; Cerfs du Père David
i   Cervi di Padre David

2014   *Elaphurus davidianus*
Artiodactyla - Cervidae
e   Père David's Deer
d   Milu-Hirsch; David-Hirsch; Milu;
Davids Hirsch
f   Cerf du Père David
i   Cervo di Padre David

2015   **Elephantidae**
Proboboscidea
e   Elephants
d   Elefanten
f   Éléphantidés
i   Elefanti; Elefantidi

2016   *Elephantulus*
Macroscelidea - Macroscelididae
e   Long-eared Elephant Shrews; Rock
Elephant Shrews; Elephant Shrews;
Small Elephant Shrews ; Long-eared
Sengis

d   Elefantenspitzmäuse
f   Macroscélides
i   Toporagni elefanti

2017   *Elephantulus brachyrhynchus*
     Macroscelidea - Macroscelididae
e   Short-snouted Elephant Shrew
d   Kurznasenelefantenspitzmaus
i   Macroscelide dal muso corto;
     Toporagno dal muso corto

2018   *Elephantulus edwardii*
     Macroscelidea - Macroscelididae
e   Cape Sengi; Cape Elephant Shrew;
     Cape Rock Elephant Shrew
d   Kap-Rüsselspringer

2019   *Elephantulus fuscipes*
     Macroscelidea - Macroscelididae
e   Uganda Elephant Shrew; Dusky-
     footed Elephant Shrew; Dusky-
     footed Sengi

d   Schwarzfüßige Elefantenspitzmaus
f   Macroscélide à nez court

2020   *Elephantulus fuscus*
     Macroscelidea - Macroscelididae
e   Zambezi Elephant Shrew; Peters's
     Short-snouted Elephant Shrew;
     Dusky Elephant Shrew; Bushveld
     Elephant Shrew; Dusky Sengi

d   Dunkle Elefantenspitzmaus

2021   *Elephantulus intufi*
     Macroscelidea - Macroscelididae
e   Bushveld Elephant Shrew; Bushveld
     Sengi
d   Trockenland-Elefantenspitzmaus
i   Toporagno elefante delle regioni
     aride

2022   *Elephantulus myurus*
     Macroscelidea - Macroscelididae
e   Transvaal Elephant Shrew; Eastern

Rock Elephant Shrew; Eastern Rock
Sengi; Smith's Rock Elephant Shrew
d   Langschwanzrüsselspringer
f   Macroscélide des rochers

2023   *Elephantulus revoili*
     Macroscelidea - Macroscelididae
e   Somali Sengi; Somali Elephant
     Shrew
d   Somali Rüsselspringer

2024   *Elephantulus rozeti*
     Macroscelidea - Macroscelididae
e   North African Elephant Shrew; North
     African Sengi
d   Nordafrikanische Elefantenspitzmaus
f   Macroscélide d'Afrique du Nord;
     Macroscélide de Rozet

2025   *Elephantulus rufescens*
     Macroscelidea - Macroscelididae
e   Spectacled Elephant Shrew; East
     African Long-eared Elephant Shrew;
     Rufous Sengi; Rufous Elephant
     Shrew; Rufous Spectacled Elephant
     Shrew
d   Rote Elefantenspitzmaus
i   Macroscelide rossastro; Toporagno
     rossastro

2026   *Elephantulus rupestris*
     Macroscelidea - Macroscelididae
e   Rock Elephant Shrew; Western Rock
     Elephant Shrew; Smith's Elephant
     Shrew; Western Rock Sengi; Smith's
     Rock Elephant Shrew
d   Klippen-Elefantenspitzmaus
i   Toporagno elefanto delle rocce

2027   *Elephas*
     Uranotheria - Elephantidae
e   Asiatic Elephants
d   Asiatische Elefanten
f   Éléphants d'Asie
i   Elefanti asiatici

2028 *Elephas maximus*
Uranotheria - Elephantidae
e   Asiatic Elephant; Asian Elephant
d   Asiatischer Elefant
f   Éléphant d'Asie
i   Elefante asiatico; Elefante indiano

2029 *Elephas maximus bengalensis*
Uranotheria - Elephantidae
e   Indian Elephant
d   Indischer Elefant
f   Éléphant indien

2030 *Elephas maximus borneensis*
Uranotheria - Elephantidae
e   Borneo Elephant; Pygmy Elephant;
Borneo Pygmy Elephant
d   Borneo-Zwergelefant
f   Éléphant de Bornéo

2031 *Elephas maximus hirsutus*
Uranotheria - Elephantidae
e   Malayan Elephant
d   Malaya-Elefant
f   Éléphant Malaisie

2032 *Elephas maximus maximus*
Uranotheria - Elephantidae
e   Sri Lankan Elephant; Ceylon
Elephant
d   Ceylon-Elefant
f   Éléphant de Ceylan
i   Elefante di Ceylon

2033 *Elephas maximus sumatrensis*
Uranotheria - Elephantidae
e   Sumatran Elephant
d   Sumatra-Elefant
f   Éléphant de Sumatra

2034 *Elephas maximus vil-aliya*
Uranotheria - Elephantidae
e   Ceylon Marsh Elephant
d   Sri Lanka-Sumpfelefant

2035 *Eligmodontia*
Rodentia - Muridae
e   Gerbil Mice; Highland Desert Mice;
Highland Gerbil Mice
d   Wüstenmäuse;
Hochlandwüstenmäuse

Eligmodontia hypopageus syn. *E.
typus* q.v.

2036 *Eligmodontia moreni*
Rodentia - Muridae
e   Monte Gerbil Mouse

2037 *Eligmodontia morgani*
Rodentia - Muridae
e   Morgan's Gerbil Mouse

2038 *Eligmodontia puerulus*
Rodentia - Muridae
e   Andean Gerbil Mouse

2039 *Eligmodontia typus*
Rodentia - Muridae
e   Highland Gerbil Mouse; Highland
Desert Mouse
d   Bergwüstenmaus

2040 *Eliomys*
Rodentia - Gliridae
e   Garden Dormice; Golden Dormice
d   Gartenschläfer
f   Lérots

2041 *Eliomys melanurus*
Rodentia - Gliridae
e   Asian Garden Dormouse; Arabian
Garden Dormouse; Black-tailed
Dormouse; Israeli Garden Dormouse
d   Syrischer Gartenschläfer
f   Lérot d'Arabie
i   Ghiro del deserto

2042 *Eliomys quercinus*
Rodentia - Gliridae

e Garden Dormouse; Orchard
Dormouse
d Gartenschläfer; Eichelmaus
f Lérot; Lérot commun; Loir lérot
i Quercino; Topo quercino

2043 *Eliomys quercinus denticulatus*
Rodentia - Gliridae
e Desert Dormouse
d Wüstenschläfer

2044 *Eliurus*
Rodentia - Muridae
e Tufted-tail Rats; Tufftail Rats; Tuft-
tail Rats
d Bilchschwänze
f Rats arboricoles

2045 *Eliurus antsingy*
Rodentia - Muridae
e Antsingy Tufted-tail Rat

2046 *Eliurus ellermani*
Rodentia - Muridae
e Ellerman's Tufted-tail Rat

2047 *Eliurus majori*
Rodentia - Muridae
e Major's Tufted-tail Rat

2048 *Eliurus minor*
Rodentia - Muridae
e Lesser Tufted-tail Rat; Small Tufted-
tail Rat

2049 *Eliurus myoxinus*
Rodentia - Muridae
e Dormouse Tufted-tail Rat
d Großer Bilchschwanz

2050 *Eliurus pennicillatus*
Rodentia - Muridae
e White-tipped Tufted-tail Rat

2051 *Eliurus petteri*

Rodentia - Muridae
e Petter's Tufted-tail Rat

2052 *Eliurus tanala*
Rodentia - Muridae
e Tanala Tufted-tail Rat
d Tanala-Bilchschwanz

2053 *Elliurus webbi*
Rodentia - Muridae
e Webb's Tufted-tail Rat

2054 *Ellobius*
Rodentia - Muridae
e Mole Voles; Mole Lemmings
d Mull-Lemminge
f Rats-taupes
i Ellobi talpini

2055 *Ellobius alaicus*
Rodentia - Muridae
e Alai Mole Vole

2056 *Ellobius fuscocapillus*
Rodentia - Muridae
e Southern Mole Vole; Afghan Mole
Vole; Quetta Mole Vole; Afghan
Mole Lemming
d Südlicher Mull-Lemming
f Rat-taupe d'Afghanistan

2057 *Ellobius lutescens*
Rodentia - Muridae
e Transcaucasian Mole Vole
d Persischer Mull-Lemming;
Bergmull-Lemming; Südwestlicher
Mull-Lemming

2058 *Ellobius talpinus*
Rodentia - Muridae
e Mole Vole; Northern Mole Vole;
Mole Lemming; Mole-like Meadow
Mouse
d Nördlicher Mull-Lemming;
Gemeiner Mull-Lemming; Wurfmull

f   Rat-taupe; Ours terrestre
i   Ellobio talpino

2059   Ellobius tancrei
      Rodentia - Muridae
e   Zaisan Mole Vole

2060   Emballonura
      Chiroptera - Emballonuridae
e   Sac-winged Bats; Sheath Tailed Bats;
      Old World Sheath-tailed Bats
d   Glattnasenfreischwänze
f   Emballonures
i   Emballoniri

2061   Emballonura alecto
      Chiroptera - Emballonuridae
e   Philippine Sheath-tailed Bat; Small
      Asian Sheath-tailed Bat

2062   Emballonura atrata
      Chiroptera - Emballonuridae
e   Peters's Sheath-tailed Bat;
      Madagascar Sheath-tailed Bat

2063   Emballonura beccarii
      Chiroptera - Emballonuridae
e   Beccari's Sheath-tailed Bat

2064   Emballonura dianae
      Chiroptera - Emballonuridae
e   Rennell Island Sheath-tailed Bat;
      Large-eared Sheath-tailed Bat

2065   Emballonura furax
      Chiroptera - Emballonuridae
e   Greater Sheath-tailed Bat; New
      Guinea Sheath-tailed Bat

2066   Emballonura monticola
      Chiroptera - Emballonuridae
e   Lesser Sheath-tailed Bat

2067   Emballonura raffrayana
      Chiroptera - Emballonuridae

e   Raffray's Sheath-tailed Bat

Emballonura rivalis syn. *E. alecto*
q.v.

Emballonura rotensis syn. *E.*
semicaudata q.v.

2068   Emballonura semicaudata
      Chiroptera - Emballonuridae
e   Polynesian Sheath-tailed Bat

2069   Emballonura serii
      Chiroptera - Emballonuridae
e   Seri's Sheath-tailed Bat

2070   Emballonuridae
      Chiroptera
e   Free-tailed Bats; Reflex-winged Bats;
      Sac-winged Bats; Sharpnosed Bats;
      Sheath-tailed Bats; Ghost Bats;
      Pouched Bats
d   Glattnasenfreischwänze
f   Emballonuridés
i   Emballonuridi

Enchisthenes syn. *Artibeus* q.v.

2071   Enhydra
      Carnivora - Mustelidae
e   Sea Otters
d   Kamschatka-Biber ; Seeotter ;
      Meerotter
f   Enhydres
i   Lontre marine

2072   Enhydra lutris
      Carnivora - Mustelidae
e   Sea Otter
d   Kalan; Meerotter; Kamtschatka-
      Biber
f   Loutre de mer; Loutre marine
i   Lontra marina; Lontra del
      Camsciatca; Lontra di mare

2073 Enhydra lutris kenyoni
Carnivora - Mustelidae
e Northern Sea Otter; Alaskan Sea Otter

2074 Enhydra lutris lutris
Carnivora - Mustelidae
e Russian Sea Otter
d Kamschatka-Biber; Berings Seeotter
f Loutre de mer d'Alaska
i Lontra della Camsciacta

2075 Enhydra lutris nereis
Carnivora - Mustelidae
e Southern Sea Otter; California Sea Otter
d Kalifornischer Seeotter
f Loutre marine de Californie; Loutre de mer de Californie
i Lontra di mare meridionale

2076 Eoglaucomys
Rodentia - Sciuridae
e Kashmir Flying Squirrels
d Kaschmir-Gleithörnchen

2077 Eoglaucomys fimbriatus
Rodentia - Sciuridae
e Kashmir Flying Squirrel
d Kaschmir-Gleithörnchen

2078 Eolagurus
Rodentia - Muridae
e Yellow Steppe Lemmings
d Gelblemminge
f Lemmings jaunes des steppes
i Lemming della steppe

Eolagurus lutens syn. *E. przewalskii* q.v.
2079 Eolagurus luteus
Rodentia - Muridae
e Yellow Steppe Lemming
d Gelber Steppenlemming; Gelblemming

f Lagure jaune
i Lemming gialla della steppe

2080 Eolagurus przewalskii
Rodentia - Muridae
e Przewalski's Steppe Lemming
f Lemming brun
i Lemming della steppe di Przewalski

2081 Eonycteris
Chiroptera - Pteropodidae
e Dawn Bats; Dawn Fruit Bats; Nectar-eating Bats
f Éonyctères

2082 Eonycteris major
Chiroptera - Pteropodidae
e Greater Dawn Bat; Greater Nectar-eating Bat

2083 Eonycteris spelaea
Chiroptera - Pteropodidae
e Cave Fruit Bat; Dobson's Long-tongued Fruit Bat; Dobson's Long-tongued Dawn Bat; Dawn Bat; Lesser Dawn Bat; Cave-dwelling Nectar-eating Bat; Cave Dawn Bat

d Höhlen-Langzungenflughund
f Éonyctère des cavernes

2084 Eothenomys
Rodentia - Muridae
e Pratt's Voles; Chinese Voles; Père David's Voles; Oriental Voles; Dawn Meadow Mice; South Asian Voles

d Père Davids Wühlmäuse; Pater Davids Wühlmäuse; David-Wühlmäuse

2085 Eothenomys chinensis
Rodentia - Muridae
e Pratt's Vole

2086 *Eothenomys custos*
    Rodentia - Muridae
e   Southwest China Mole

2087 *Eothenomys eva*
    Rodentia - Muridae
e   Gansu Vole

2088 *Eothenomys inez*
    Rodentia - Muridae
e   Kolan Vole

2089 *Eothenomys melanogaster*
    Rodentia - Muridae
e   Père David's Vole
d   Schwarzbauchwühlnaus; David-
    Wühlmaus
f   Comagnol de Père David

2090 *Eothenomys olitor*
    Rodentia - Muridae
e   Chaotung Vole

2091 *Eothenomys proditor*
    Rodentia - Muridae
e   Yulung Shan Vole

2092 *Eothenomys regulus*
    Rodentia - Muridae
e   Royal Vole

2093 *Eothenomys shanseius*
    Rodentia - Muridae
e   Shansei Vole

2094 *Eozapus*
    Rodentia - Dipodidae
e   Chinese Jumping Mice; Szechwan
    Jumping Mice
d   Chinesische Hüpfmäuse

2095 *Eozapus setchuanus*
    Rodentia - Dipodidae
e   Szechwan Jumping Mouse; Chinese
    Jumping Mouse; Sichuan Jumping

    Mouse
d   Chinesische Hüpfmaus

Episoriculus syn. *Soriculus* q.v.

2096 *Epixerus*
    Rodentia - Sciuridae
e   Splendid-tailed Squirrels; African
    Palm Squirrels
d   Wilson-Riesenhörnchen ;
    Afrikanische Palmenhörnchen
f   Écureuils de Wilson
i   Scoiattoli gigante

2097 *Epixerus ebii*
    Rodentia - Sciuridae
e   Ebien Squirrel; Temminck's Giant
    Squirrel; Western Palm Squirrel;
    African Palm Squirrel; Ebien Palm
    Squirrel
d   Großes Rotschenkelhörnchen;
    Westafrikanisches Palmenhörnchen
f   Écureuil d'Ébien

2098 *Epixerus wilsoni*
    Rodentia - Sciuridae
e   Biafran Palm Squirrel; Gabon Palm
    Squirrel
d   Wilson-Riesenhörnchen; Wilsons
    Riesenhörnchen; Biafra
    Palmenhörnchen
f   Écureuil de Wilson

2099 *Epomophorus*
    Chiroptera - Pteropodidae
e   Epaulet Bats; Epauletted Fruit Bats;
    Epauletted Bats; Little Epauletted
    Fruit Bats
d   Epaulettenflughunde
f   Épomophores; Épomorphes
i   Epomofori

2100 *Epomophorus angolensis*
    Chiroptera - Pteropodidae
e   Angola Fruit Bat; Angola Epauletted
    Fruit Bat; Angolan Epauletted Fruit
    Bat

2101   *Epomophorus gambianus*
Chiroptera - Pteropodidae
e   Gambian Epauletted Fruit Bat
d   Gambia-Epaulettenflughund

2102   *Epomophorus grandis*
Chiroptera - Pteropodidae
e   Lesser Angolan Epauletted Fruit Bat

2103   *Epomophorus labiatus*
Chiroptera - Pteropodidae
e   Little Epauletted Fruit Bat;
Ethiopiean Epauletted Fruit Bat

2104   *Epomophorus minimus*
Chiroptera - Pteropodidae
e   East African Epauletted Fruit Bat

2105   *Epomophorus wahlbergi*
Chiroptera - Pteropodidae
e   Wahlberg's Epauletted Fruit Bat;
Wahlberg's Fruit Bat
d   Wahlberg-Epaulettenflughund
f   Épomophore de Wahlberg

2106   *Epomops*
Chiroptera - Pteropodidae
e   Epaulet Bats; Epaulet Fruit Bats;
Epauletted Fruit Bats; Gray's Epaulet
Bats; African Epauletted Bats
d   Epaulettenflughunde
f   Épomophores
i   Epomofori

2107   *Epomops buettikoferi*
Chiroptera - Pteropodidae
e   Büttikofer's Epauletted Fruit Bat;
Büttikofers Fruit Bat; Buetikofer's
Epauletted Bat
d   Büttikofers Epaulettenflughund;
Büttikofer-Epaulettenflughund
f   Épomophore de Büttikofer

2108   *Epomops dobsoni*
Chiroptera - Pteropodidae

e   Dobson's Fruit Bat; Donson's
Epauletted Fruit Bat

2109   *Epomops franqueti*
Chiroptera - Pteropodidae
e   Franquet's Fruit Bat; Franquet's
Singing Fruit Bat; Singing Fruit Bat;
Franquet's Epauletted Bat; Franquet's
Epauletted Fruit Bat

d   Franquets Epaulettenflughund;
Franquet-Epaulettenflughund

f   Épomophore de Franquet; Chien
volant aux épaulettes du Congo

i   Epomoforo di Franquet

2110   *Eptesicus*
Chiroptera - Vespertilionidae
e   Serotines; Brown Bats
d   Breitflügelfledermäuse
f   Sérotines

***Eptesicus andinus* syn. *E.
brasiliensis* q.v.**

2111   *Eptesicus bobrinskii*
Chiroptera - Vespertilionidae
e   Bobrinski's Serotine
d   Bobrinskis Fledermaus

2112   *Eptesicus bottae*
Chiroptera - Vespertilionidae
e   Botta's Serotine
d   Bottas Fledermaus
f   Sérotine de Botta

2113   *Eptesicus brasiliensis*
Chiroptera - Vespertilionidae
e   Brazilian Brown Bat; Andean Brown
Bat
2114   *Eptesicus darlingtoni*
Chiroptera - Vespertilionidae
e   Large Forest Bat

2115 **Eptesicus demissus**
Chiroptera - Vespertilionidae
e Surat Serotine; Surat Serotine Bat

2116 **Eptesicus diminutus**
Chiroptera - Vespertilionidae
e Diminuitive Serotine

2117 **Eptesicus floweri**
Chiroptera - Vespertilionidae
e Lesser Sudan Horn-skinned Bat;
Horn-skinned Bat

2118 **Eptesicus furinalis**
Chiroptera - Vespertilionidae
e Tropical Brown Bat; Argentine
Brown Bat

2119 **Eptesicus fuscus**
Chiroptera - Vespertilionidae
e Barn Bat; House Bat; Big Brown
Bat; Large Brown Bat
d Große Braune Fledermaus
f Sérotine de maison; Grande chauve-
souris brune

2120 **Eptesicus guadeloupensis**
Chiroptera - Vespertilionidae
e Guadeloupe Brown Bat; Guadeloupe
Big Brown Bat

2121 **Eptesicus hottentotus**
Chiroptera - Vespertilionidae
e Long-tailed House Bat

2122 **Eptesicus innoxius**
Chiroptera - Vespertilionidae
e Harmless Serotine

2123 **Eptesicus kobayashii**
Chiroptera - Vespertilionidae
e Kobayashi's Serotine

2124 **Eptesicus nasutus**
Chiroptera - Vespertilionidae

e Sind Serotine; Persian Serotine; Sind
Bat

2125 **Eptesicus nilssoni**
Chiroptera - Vespertilionidae
e Northern Bat; Nilsson's Serotine
d Nordfledermaus; Nordische
Fledermaus; Umberfledermaus;
Wanderfledermaus
f Sérotine boréale; Sérotine de
Nilsson; Vespérien boréal
i Serotino di Nilsson; Vespérugo
boréale

2126 **Eptesicus pachyotis**
Chiroptera - Vespertilionidae
e Thick-eared Bat

2127 **Eptesicus platyops**
Chiroptera - Vespertilionidae
e Lagos Serotine

Eptesicus sagitulla syn. *E.
darlingtoni* q.v.

2128 **Eptesicus serotinus**
Chiroptera - Vespertilionidae
e Common Serotine; Serotine;
Rattlemouse; House Bat; Serotine
Bat
d Spätfliegende Fledermaus;
Breitflügelfledermaus
f Sérotine; Sérotine commune; Grande
sérotine; Chauve-souris sérotine;
Vespérien sérotine; Sérotine
ordinaire; Vespertilion sérotine
i Serotino comune; Sertolino comune

2129 **Eptesicus tatei**
Chiroptera - Vespertilionidae
e Sombre Bat

Eptesicus walli syn. *E. nasutus* q.v.

2130 **Equidae**
Perissodactyla
e Horses; Horses Zebras and Asses;
Horses, Zebras, Asses and Burros

d   Pferde
f   Équidés
i   Cavalli; Asini selvatici; Equidi;
    Cavalli e simili

2131   *Equus*
Perissodactyla - Equidae
e   Horses
d   Pferde; Echte Pferde
f   Chevaux
i   Cavalli

2132   *Equus africanus*
Perissodactyla - Equidae
e   African Wild Ass; Wild Ass; African
    Ass; Nubian Wild Ass
d   Wildesel; Afrikanischer Wildesel;
    Nubischer Wildesel
f   Âne sauvage de l'Afrique
i   Asino selvatico africano

2133   *Equus asinus*
Perissodactyla - Equidae
e   Donkey; Burro (NA); Feral Burro
    (NA); Ass
d   Hausesel
f   Âne; Âne vrai; Onagre
i   Asino selvatico; Asino

2134   *Equus asinus atlanticus*
Perissodactyla - Equidae
e   Algerian Wild Ass
d   Nordafrikanischer Wildesel; Atlas-
    Wildesel
f   Âne sauvage d'Algérie

2135   *Equus asinus somaliensis*
Perissodactyla - Equidae
e   Somali Wild Ass
d   Somali-Wildesel
f   Âne de Somalie; Âne sauvage
    d'Afrique; Âne d'Abyssinie
i   Asino selvatico della Somalia

2136   *Equus burchelli*

Perissodactyla - Equidae
e   Burchell's Zebra; Common Zebra;
    Steppe Zebra

d   Burchell-Zebra; Steppenzebra;
    Pferdezebra
f   Zèbre de Burchell; Zèbre des steppes
i   Zebra comune; Zebra di Burchell;
    Zebra delle steppe

2137   *Equus caballus*
Perissodactyla - Equidae
e   Horse; Domestic Horse; Feral Horse
d   Wildpferd; Tarpan
f   Cheval Potok; Cheval sauvage
i   Cavallo domestico; Tarpan

2138   *Equus caballus gmelini*
Perissodactyla - Equidae
e   East Carpathian Horse
d   Steppentarpan; Südrussisches
    Steppenwildpferd
f   Tarpan; Tarpan européen; Tarpan des
    steppes
i   Tarpan

2139   *Equus ferus*
Perissodactyla - Equidae
e   Przewalski's Wild Horse;
    Przewalski's Horse; Asiatic Wild
    Horse; Mongolian Wild Horse

d   Przewalski-Pferd; Urwildpferd
f   Cheval de Przewalski; Cheval
    sauvage de Przewalski

i   Cavallo di Przewalski; Cavallo
    selvatico della Mongolia; Cavallo
    selvatico

2140   *Equus grevyi*
Perissodactyla - Equidae
e   Grévy's Zebra
d   Grevy-Zebra
f   Zèbre de Grévy
i   Zebra reale; Zebra di Grévy

2141   *Equus hemionus*
　　　　Perissodactyla - Equidae
e　Asiatic Wild Ass; Kulan
d　Asiatischer Wildesel; Asiatischer
　　　Halbesel; Halbesel
f　Hémione ; Hémippe; Âne sauvage de
　　　Mongolie
i　Asino selvatico asiatico; Emione;
　　　Onagro

2142   *Equus hemionus hemionus*
　　　　Perissodactyla - Equidae
e　Mongolian Wild Ass
d　Dschiggetai; Mongolischer Halbesel
f　Âne sauvage de Mongolie

2143   *Equus hemionus khur*
　　　　Perissodactyla - Equidae
e　Khur; Indian Wild Ass
d　Khur; Gorkhar
f　Khur

2144   *Equus hemionus kulan*
　　　　Perissodactyla - Equidae
e　Kulan
d　Kulan; Turkmenischer Halbesel
f　Kulan
i　Kulan

2145   *Equus hemionus luteus*
　　　　Perissodactyla - Equidae
e　Gobi Kulan
d　Gobi-Dschiggetai

2146   *Equus kiang*
　　　　Perissodactyla - Equidae
e　Kiang; Tibetan Wild Ass
d　Kiang; Tibetanischer Halbesel
f　Âne kiang; Kiang; Âne sauvage du
　　　Tibet
i　Kiang

2147   *Equus onager*
　　　　Perissodactyla - Equidae
e　Onager; Onager Ass; Persian Onager

d　Onager; Persischer Halbesel
f　Onagre
i　Onagro

2148   *Equus quagga*
　　　　Perissodactyla - Equidae
e　Quagga
d　Quagga
f　Quagga; Couagga
i　Zebra delle pianure; Quagga; Zebra
　　　delle steppe

2149   *Equus quagga antiquorum*
　　　　Perissodactyla - Equidae
e　Chapman's Zebra; Damaraland Zebra
d　Chapman-Zebra; Wahlberg-Zebra;
　　　Damara-Zebra
f　Zèbre de Chapman
i　Zebra di Chapman

2150   *Equus quagga boehmi*
　　　　Perissodactyla - Equidae
e　Grant's Zebra
d　Grant-Zebra; Böhm-Zebra
f　Zèbre de Grant
i　Zebra di Grant

2151   *Equus quagga crawshayi*
　　　　Perissodactyla - Equidae
e　Crawshay's Zebra
d　Crayshaw-Zebra
f　Zèbre de Crawshay
i　Zebra di Crawshay

2152   *Equus quagga selousi*
　　　　Perissodactyla - Equidae
e　Sealous's Zebra
d　Selous-Zebra
f　Zèbre de Selous
i　Zebra di Selous

2153   *Equus zebra*
　　　　Perissodactyla - Equidae
e　Mountain Zebra

d   Bergzebra
f   Zèbre vrai
i   Zebra delle montagne; Zebra di
    montagna

2154   *Equus zebra hartmanni*
    Perissodactyla - Equidae
e   Hartmann's Mountain Zebra;
    Hartmann's Zebra
d   Hartmann-Bergzebra
f   Zèbre de montagne de Hartmann;
    Zèbre de montagne
i   Zebra di Hartmann

2155   *Equus zebra zebra*
    Perissodactyla - Equidae
e   Cape Mountain Zebra
d   Kap-Bergzebra
f   Zèbre de montagne du Cap
i   Zebra di montagna del Capo; Zebra
    montana

2156   *Eremitalpa*
    Insectivora - Chrysochloridae
e   Grant's Moles; Grant's Golden Moles
d   Wüstengoldmulle
f   Taupes dorées de Grant
i   Talpe dorate di Grant

2157   *Eremitalpa granti*
    Insectivora - Chrysochloridae
e   Grant's Desert Golden Mole; Grant's
    Golden Mole
d   Wüstengoldmull; Grant-Goldmull
f   Taupe dorée de Grant
i   Talpa dorata di Grant

2158   *Eremitalpa granti namibensis*
    Insectivora - Chrysochloridae
e   Namib Golden Mole
d   Namib-Goldmull

2159   *Eremodipus*
    Rodentia - Dipodidae
e   Lichtenstein's Jerboas

d   Lichtensteins Wüstenspringmäuse
f   Gerboises de Lichtenstein
i   Gerboe di Lichtenstein; Gerboa di
    Lichtenstein

2160   *Eremodipus lichtensteini*
    Rodentia - Dipodidae
e   Lichtenstein's Jerboa
d   Lichtensteins Wüstenspringmaus
f   Gerboise de Lichtenstein
i   Gerboa di Lichtenstein

2161   *Erethizon*
    Rodentia - Erethizontidae
e   American Porcupines; North
    American Porcupines; Canadian
    Porcupines
d   Ursons; Baumstachelschweine;
    Nordamerikanische Baumstachler
f   Porcs-épics nord-américains
i   Porcospini

2162   *Erethizon dorsatum*
    Rodentia - Erethizontidae
e   American Porcupine; North
    American Porcupine; Canadian
    Porcupine; Common Porcupine
    (NA); Porcupine (NA)

d   Urson; Baumstachelschwein;
    Nordamerikanischer Baumstachler
f   Porc-épic nord-américain
i   Ursone; Porcospino del Canada;
    Porcospino arborea nordamericano

2163   *Erethizon dorsatum myops*
    Rodentia - Erethizontidae
e   Alaskan Porcupine

2164   Erethizontidae
    Rodentia
e   New World Porcupines
d   Baumstachler
f   Érizontidés
i   Istrici arborei; Eretizontidi

Ericulus syn. *Setifer* q.v.

2165   Erignathus
Carnivora - Phocidae
e   Leporine Seals; Bearded Seals; Great
Seals; Long-bodied Seals;
Squareflippers
d   Bartrobben
f   Érignathes; Phoque à barbe
i   Foche barbate

2166   Erignathus barbatus
Carnivora - Phocidae
e   Leporine Seal; Bearded Seal; Great
Seal; Long-bodied Seal;
Squareflipper; Ugruk; Ugrug
d   Bärtige Robbe; Bärtiger Seehund;
Bartrobbe; Blaurobbe
f   Phoque à barbe; Phoque barbu;
Grand phoque
i   Foca barbata

2167   Erignathus barbatus barbatus
Carnivora - Phocidae
e   Atlantic Bearded Seal
d   Atlantische Bartrobbe

2168   Erignathus barbatus nauticus
Carnivora - Phocidae
e   Pacific Bearded Seal
d   Pazifische Bartrobbe

2169   Erinaceidae
Insectivora -
e   Moonrats; Hedgehogs; Gymnures;
Moonrats and Hedgehogs
d   Igelartige; Igel ; Echte Igel;
Stacheligel
f   Érinacéidés; Hérissons
i   Erinaceidi

2170   Erinaceus
Erinaceomorpha - Erinaceidae
e   Hedgehogs; Common Hedgehogs;
Woodland Hedgehogs; Eursian
Hedgehogs

d   Kleinohrigel ; Igel
f   Hérissons communs
i   Ricci

2171   Erinaceus albiventris
Erinaceomorpha - Erinaceidae
e   Four-toed Hedgehog; African Pygmy
Hedgehog; White-bellied Hedgehog;
African Hedgehog
d   Weißbauchigel; Pruners Igel;
Steppenigel; Afrikanischer Zwergigel
f   Hérisson à ventre blanc
i   Riccio dal ventre bianco

2172   Erinaceus algirus
Erinaceomorpha - Erinaceidae
e   Algerian Hedgehog; North African
Hedgehog
d   Mittelmeer-Igel; Algerischer Igel;
Weißer Igel; Nordafrikanischer Igel
f   Hérisson d'Algérie
i   Riccio algerino

2173   Erinaceus algirus vagrans
Erinaceomorpha - Erinaceidae
e   Vagrant Hedgehog
d   Wanderigel

2174   Erinaceus amurensis
Erinaceomorpha - Erinaceidae
e   Manchurian Hedgehog; Amur
Hedgehog
d   Amur-Igel; Chinesischer Igel
f   Hérisson de l'Amur

2175   Erinaceus concolor
Erinaceomorpha - Erinaceidae
e   Eastern Hedgehog; East European
Hedgehog; Eastern European
Hedgehog; White-breasted Hedgehog

d   Ostigel; Osteuropäischer Igel;
Weißbrustigel
f   Hérisson oriental; Hérisson d'Europe
orientale; Hérisson de l'Europe de
l'Est
i   Riccio orientale

2176   *Erinaceus dauricus*
Erinaceomorpha - Erinaceidae
e   Daurian Hedgehog
d   Daurischer Igel
f   Hérisson daurien

2177   *Erinaceus europaeus*
Erinaceomorpha - Erinaceidae
e   Eurasian Hedgehog; European
Hedgehog; Northern Hedgehog;
Hedgehog

d   Igel; Eurasischer Igel; Gemeiner Igel
f   Hérisson; Hérisson commun;
Hérisson d'Europe; Hérisson
européen; Hérisson ordinaire
i   Riccio europeo occidentale; Riccio
europeo; Riccio

2178   *Erinaceus europaeus europaeus*
Erinaceomorpha - Erinaceidae
e   West European Hedgehog; Western
Hedgehog; Western European
Hedgehog
d   Braunbrustiger Igel; Westigel;
Westeuropäischer Igel;
Braunbrustigel
f   Hérisson de l'Europe de l'Ouest

2179   *Erinaceus europaeus hispanicus*
Erinaceomorpha - Erinaceidae
e   Spanish Hedgehog
d   Solanicher Kleinohrigel
f   Hérisson de Portugal

2180   *Erinaceus europaeus italicus*
Erinaceomorpha - Erinaceidae
e   Italian Hedgehog
d   Italienischer Igel
f   Hérisson sarde
i   Riccio occidentale

2181   *Erinaceus europaeus koreensis*
Erinaceomorpha - Erinaceidae
e   Korean Hedgehog
d   Korea-Igel

**2182   *Erinaceus europaeus
transcaucasicus***
Erinaceomorpha - Erinaceidae
e   Transcaucasian Hedgehog
d   Transkaukasischer Igel

2183   *Erinaceus frontalis*
Erinaceomorpha - Erinaceidae
e   South African Hedgehog; Southern
African Hedgehog
d   Kap-Igel
f   Hérisson d'Afrique du Sud
i   Riccio dell'Africa meridionale

2184   *Erinaceus hughi*
Erinaceomorpha - Erinaceidae
e   Hugh's Hedgehog
d   Hughs Igel

2185   *Erinaceus sclateri*
Erinaceomorpha - Erinaceidae
e   Somali Hedgehog
d   Sclaters Igel; Somalischer Igel

2186   *Eropeplus*
Rodentia - Muridae
e   Celebes Soft-furred Rats; Sulawesi
Soft-furred Rats
d   Sulawesi-Weichratten

2187   *Eropeplus canus*
Rodentia - Muridae
e   Celebes Soft-furred Rat; Celebes
Grey Rat; Sulawesi Soft-furred Rat;
Sulawesi Grey Rat
d   Graue Sulawesi-Ratte; Sulawesi-
Weichratte

2188   *Erophylla*
Chiroptera - Phyllostomidae
e   Brown Flower Bats

2189   *Erophylla sezekorni*
Chiroptera - Phyllostomidae
e   Buffy Flower Bat; Brown Flower Bat

2190   *Erythrocebus*
         Primates - Cercopithecidae
e   Patas Monkeys; Hussar Monkeys
d   Husarenaffen
f   Patas ; Singes rouges
i   Eritrocebi

2191   *Erythrocebus patas*
         Primates - Cercopithecidae
e   Patas Monkey; Hussar Monkey; Red Monkey; Military Monkey; Common Patas; Red Guenon
d   Husarenaffe
f   Singe rouge; Singe pleureur; Patas
i   Eritrocebo pata; Pata; Scimmia rossa; Eritrocebo

2192   *Erythrocebus patas patas*
         Primates - Cercopithecidae
e   Black-nosed Patas Monkey
d   Schwarznasenhusarenaffe

2193   *Erythrocebus patas pyrrhonotus*
         Primates - Cercopithecidae
e   White-nosed Patas Monkey; Nisnas Monkey
d   Weißnasenhusarenaffe

2194   *Eschrichtius*
         Cete - Eschrichtiidae
e   California Grey Whales; Pacific Grey Whales; Greyback Whales; Grey Whales
d   Grauwale
f   Baleines grises
i   Balene grigie

    Eschrichtius gibbosus syn. *E. robustus* q.v.

    Eschrichtius glaucus syn. *E. robustus* q.v.

2195   *Eschrichtius robustus*
         Cete - Eschrichtiidae
e   California Grey Whale; Pacific Grey

Whale; Greyback Whale; Grey Whale
d   Grauwal
f   Baleine grise; Baleine grise de Californie
i   Balena grigia

    Eubalaena syn. *Balaena* q.v.

2196   *Euchoreutes*
         Rodentia - Dipodidae
e   Long-eared Jerboas
d   Langohrspringer ; Riesenohrspringmäuse
f   Gerboises aux longues oreilles

2197   *Euchoreutes naso*
         Rodentia - Dipodidae
e   Long-eared Jerboa
d   Langohrspringer; Riesenohrspringmaus
f   Gerboise aux longues oreilles

2198   *Euderma*
         Chiroptera - Vespertilionidae
e   Spotted Bats; Pinto Bats
d   Gefleckte Fledermäuse
f   Oreillards maculés

2199   *Euderma maculatum*
         Chiroptera - Vespertilionidae
e   Spotted Bat; Pinto Bat
d   Gefleckte Fledermaus
f   Oreillard maculé

2200   *Eudiscopus*
         Chiroptera - Vespertilionidae
e   Disk-footed Bats

2201   *Eudiscopus denticulus*
         Chiroptera - Vespertilionidae
e   Disk-footed Bat

Eudromicia syn. *Cercartetus* q.v.

2202   **Eulemur**
Primates - Lemuridae
e   True Lemurs
d   Makis; Lemuren
f   Lémurs
i   Lemuri

2203   **Eulemur albifrons**
Primates - Lemuridae
e   White-fronted Lemur
d   Weißkonfrontiertes Lemur; Weißkopfmaki
f   Lémur à front blanc

2204   **Eulemur albocollaris**
Primates - Lemuridae
e   White-collared Lemur
f   Lémur à collier blanc; Lémur à barbe blanche

2205   **Eulemur cinereiceps**
Primates - Lemuridae
e   Grey-headed Lemur

2206   **Eulemur collaris**
Primates - Lemuridae
e   Red-collared Lemur
d   Schwarzkopfmaki
f   Lémur à collier; Lémur à fraise

2207   **Eulemur coronatus**
Primates - Lemuridae
e   Crowned Lemur; Mongoose Lemur
d   Kronenmaki; Gekröntes Lemur
f   Lémur couronné
i   Lemure coronato

2208   **Eulemur fulvus**
Primates - Lemuridae
e   Brown Lemur
d   Schwarzstirnmaki; Braunes Lemur; Brauner Maki
f   Lémur brun; Lémur fauve

i   Lemure bruno; Lemure fulvo; Lemure testa nera; Maki bruno

2209   **Eulemur fulvus mayottensis**
Primates - Lemuridae
e   Mayotte Lemur
d   Mayotte-Maki
f   Lémur brun de Mayotte; Lémur de Mayotte

2210   **Eulemur fulvus rufus**
Primates - Lemuridae
e   Red-fronted Brown Lemur
d   Rotstirnmaki
f   Lémur à front roux

2211   **Eulemur macaco**
Primates - Lemuridae
e   Black Lemur; Acoumba Lemur
d   Mohrenmaki; Akumba; Schwarzes Lemur
f   Lémur macaco
i   Lemure macaco; Maki macaco

2212   **Eulemur macaco flavifrons**
Primates - Lemuridae
e   Sclater's Lemur; Sclater's Black Lemur; Blue-eyed Lemur; Blue-eyed Black Lemur
d   Blauäugiger Maki; Sclaters Maki; Blauaugenlemur
f   Lémur aux yeux turquoises; Lémur noir de Sclater; Lémur flavifrons

2213   **Eulemur mongoz**
Primates - Lemuridae
e   Mongoose Lemur
d   Mongozmaki; Mongoz
f   Lémur mongos; Mongos; Lémur mongoz
i   Lemure mongoz; Machi mongoz; Mongoz

2214   **Eulemur rubriventer**
Primates - Lemuridae
e   Red-bellied Lemur

d   Rotbauchmaki; Rotbauchlemur
f   Lémur à ventre rouge
i   Lemure dal ventre rosso

2215   *Eulemur sanfordi*
      Primates - Lemuridae
e   Sanford's Lemur
d   Sanfords Lemur; Sanfords Maki
f   Lémur de Sandford

2216   *Eumetopias*
      Carnivora - Otariidae
e   Steller's Sea Lions; Northern Sea
      Lions
d   Seelöwen; Haarrobben; Stellersche
      Seelöwen
f   Otaries de Steller
i   Leoni marini di Steller

2217   *Eumetopias jubatus*
      Carnivora - Otariidae
e   Steller's Sea Lion; Northern Sea
      Lion; Sea King
d   Stellers Seelöwe; Nordischer
      Seelöwe; Stellerscher Seelöwe
f   Otarie de Steller
i   Leone marino di Steller; Otario di
      Steller

2218   *Eumops*
      Chiroptera - Molossidae
e   Mastiff Bats; Bonneted Bats
d   Bulldoggfledermäuse
f   Chauves-souris dogues

2219   *Eumops auripendulus*
      Chiroptera - Molossidae
e   Slouch-eared Bat; Black Bonneted
      Bat

2220   *Eumops bonariensis*
      Chiroptera - Molossidae
e   Peters's Mastiff Bat; Dwarf Bonneted
      Bat

2221   *Eumops dabbenei*

      Chiroptera - Molossidae
e   Big Bonneted Bat

2222   *Eumops glaucinus*
      Chiroptera - Molossidae
e   Wagner's Mastiff Bat; Chestnut
      Mastiff Bat; Wagner's Bonneted Bat

2223   *Eumops glaucinus floridanus*
      Chiroptera - Molossidae
e   Florida Mastiff Bat

2224   *Eumops hansae*
      Chiroptera - Molossidae
e   Sanborn's Mastiff Bat; Sanborn's
      Bonneted Bat

2225   *Eumops maurus*
      Chiroptera - Molossidae
e   Guianan Bonnetted Bat; Guianan
      Mastiff Bat; Guyanan Bonneted Bat

      Eumops nanus syn. *E. bonariensis*
      q.v.

2226   *Eumops patagonicus*
      Chiroptera - Molossidae
e   Argentine Dwarf Bonneted Bat

2227   *Eumops perotis*
      Chiroptera - Molossidae
e   Greater Mastiff Bat; Western
      Bonneted Bat
d   Bulldoggfledermaus

2228   *Eumops perotis californicus*
      Chiroptera - Molossidae
e   California Mastiff Bat; Western
      Mastiff Bat; Greater Western Mastiff
      Bat

      Eumops trumbulli syn. *E. perotis*
      q.v.

2229   *Eumops underwoodi*
      Chiroptera - Molossidae

e   Underwood's Mastiff Bat;
Underwood's Bonneted Bat

2230 Euneomys
Rodentia - Muridae
e   Chinchilla Mice
d   Chinchillamäuse

2231 Euneomys chinchilloides
Rodentia - Muridae
e   Patagonian Chinchilla Mouse
d   Patagonische Chinchillamaus

2232 Euneomys fossor
Rodentia - Muridae
e   Burrowing Chinchilla Mouse

2233 Euneomys mordax
Rodentia - Muridae
e   Biting Chinchilla Mouse

2234 Euneomys petersoni
Rodentia - Muridae
e   Peterson's Chinchilla Mouse

2235 Euoticus
Loridae - Loridae
e   Needle-clawed Bushbabies; Needle-clawed Galagos
d   Kielnagelgalagos
f   Galagos
i   Galagoni con unghie ad ago

2236 Euoticus elegantulus
Primates - Loridae
e   Western Needle-clawed Bushbaby; Needle-clawed Galago; Needle-clawed Bushbaby; Western Needle-clawed Galago; Western Needle-nailed Bushbaby; Elegant Needle-clawed Galago
d   Westlicher Kielnagelgalago
f   Galago élégant; Galago mignon
i   Galagone con unghie ad ago; Galagone dalle unghie aghiformi;

Galagone dalle umghie aguzze

Euoticus inustus syn. *E. pallidus* q.v.

2237 Euoticus pallidus
Primates - Loridae
e   Pallid Needle-clawed Galago; Pale Needle-nailed Bushbaby; Eastern Needle-clawed Bushbaby; Eastern Needle-nailed Bushbaby; Eastern Needle-clawed Galago
d   Östlicher Kielnagelglago; Südlicher Kielnagelgalago
f   Galago sombre; Galago du Congo
i   Galagone pallido

2238 Eupetaurus
Rodentia - Sciuridae
e   Woolly Flying Squirrels
d   Felsgleithörnchen
f   Écureuils volants cendrés
i   Scoiattoli volanti lanosi

2239 Eupetaurus cinereus
Rodentia - Sciuridae
e   Woolly Flying Squirrel
d   Felsgleithörnchen
f   Écureuil volant cendré
i   Scoiattolo volante lanoso

2240 Euphractus
Cingulata - Dasypodidae
e   Six-banded Armadillos; White-bristled Hairy Armadillos
d   Weißborstengürteltiere; Sechsbindengürteltiere; Borstengürteltiere
f   Tatous à six bandes
i   Armadilli dalle sei fasce

Euphractus nationi syn.
Chaetophractus nationi q.v.

2241 Euphractus sexcinctus
Xenarthra - Dasypodidae
e   Six-banded Armadillo; White-

bristled Hairy Armadillo; Yellow
Armadillo

d Weißborstengürteltier;
Sechsbindengürteltier

f Tatou à six bandes

i Armadillo dalle sei fasce; Armadillo
giallo

Euphractus vellerosus syn.
Chaetophractus vellerosus q.v.

Euphractus villosus syn.
Chaetophractus villosus q.v.

2242 *Eupleres*
       Carnivora - Viverridae

e Falanoucs; Small-toothed Mongooses

d Ameisenschleichkatzen; Falanuks

f Euplères

i Eupleridi

2243 *Eupleres goudoti*
       Carnivora - Viverridae

e Falanouc; Slender Falanouc; Small-
toothed Mongoose; Taller Falanouc

d Kleinfalanuk; Großfalanuk; Falanuk

f Euplère de Goudot

i Eupleride di Goudot

Eupleres major syn. *E. goudoti* q.v.

2244 *Euroscaptor*
       Erinaceomorpha - Talpidae

e Oriental Moles

f Taupes d'Asie

2245 *Euroscaptor grandis*
       Erinaceomorpha - Talpidae

e Greater Chinese Mole

2246 *Euroscaptor klossi*
       Erinaceomorpha - Talpidae

e Kloss's Mole; Thai Mole

2247 *Euroscaptor longirostris*

       Erinaceomorpha - Talpidae

e Long-nosed Mole

2248 *Euroscaptor micrura*
       Erinaceomorpha - Talpidae

e Himalajan Mole; Short-tailed Mole

d Ostmaulwurf

2249 *Euroscaptor mizura*
       Erinaceomorpha - Talpidae

e Japanese Mountain Mole

2250 *Euroscaptor parvidens*
       Erinaceomorpha - Talpidae

e Small-toothed Mole

2251 *Euryzygomatomys*
       Rodentia - Echimyidae

e Guiras; Suiras

d Suiras

f Rats épineux

2252 *Euryzygomatomys spinosus*
       Rodentia - Echimyidae

e Guira; Suira

d Suira

f Rat épineux

Eutamias syn. *Tamias* q.v.

Euxerus syn. *Xerus* q.v.

Evotomys syn. *Clethrionomys* q.v.

2253 *Exilisciurus*
       Rodentia - Sciuridae

e Bornean Pygmy Squirrels; Pygmy
Squirrels; Oriental Pygmy Squirrels

d Asiatische Zwerghörnchen

f Écureils pygmée

i Scoiattoli pigmei

2254 *Exilisciurus concinnus*
       Rodentia - Sciuridae

　　e　Philippine Pygmy Squirrel
　　d　Philippinen-Zwerghörnchen

2255　*Exilisciurus exilis*
　　　Rodentia - Sciuridae
　　e　Plain Pygmy Squirrel; Least Pygmy
　　　Squirrel
　　d　Zwerghörnchen
　　i　Scoiattolo nano di pianura

2256　*Exilisciurus whiteheadi*
　　　Rodentia - Sciuridae
　　e　Whitehead's Pygmy Squirrel;
　　　Whitehead's Dwarf Squirrel; Tufted
　　　Pygmy Squirrel
　　d　Whitehead-Zwerghörnchen;
　　　Quastenzwerghörnchen
　　f　Écureuil pygméede Whitehead

F

Falsistrellus syn. *Pipistrellus* q.v.

2257 *Felidae*
Carnivora
- e Cats
- d Katzen; Echte Katzen; Katzenartige Raubtiere
- f Chats; Chats vrais; Félidés
- i Leoni, Tigri, Pantere etc.; Felini; Felidi

2258 *Felis*
Carnivora - Felidae
- e Cats; Small Cats
- d Katzen
- f Chats
- i Gatti

2259 *Felis aurata*
Carnivora - Felidae
- e African Golden Cat; African Tiger Cat; Golden Cat
- d Afrikanische Goldkatze
- f Chat doré d'Afrique; Chat doré
- i Gatto dorato africano

2260 *Felis badia*
Carnivora - Felidae
- e Bay Cat; Bornean Red Cat; Bornean Bay Cat
- d Borneo-Goldkatze
- f Chat sauvage d'Asie; Chat bai de Bornéo; Chat doré de Borneo; Chat bai
- i Bay Cat del Borneo

2261 *Felis bengalensis*
Carnivora - Felidae
- e Leopard Cat; Bengal Cat
- d Bengal-Katze; Zwergkatze
- f Chat léopard; Chat du Bengale
- i Gatto leopardo; Gatto del Bengala

2262 *Felis bengalensis bengalensis*
Carnivora - Felidae
- e Bengal Leopard Cat
- d Hinterindische Bengal-Katze; Indische Bengalkatze
- f Chat léopard du Bengale; Chat de Chine
- i Gatto leopardo del Bengala

2263 *Felis bengalensis euptilura*
Carnivora - Felidae
- e Siberian Leopard Cat; Tshushima Cat
- f Chat léopard de Sibérie
- i Gatto leopardo della Siberia

2264 *Felis bengalensis iriomotensis*
Carnivora - Felidae
- e Iriomote Cat
- d Iriomote-Katze
- f Chat d'Irrimote
- i Gatto di Iriomonte

2265 *Felis bieti*
Carnivora - Felidae
- e Chinese Desert Cat; Chinese Leopard Cat; Chinese Mountain Cat
- d Graukatze; Gobi-Katze
- f Chat de Biet
- i Gatto del deserto della Cina

2266 *Felis braccata*
Carnivora - Felidae
- e Pantanal Cat
- f Chat Pantanal

2267 *Felis canadensis*
Carnivora - Felidae
- e Canadian Lynx; Canada Lynx; LionCougar
- d Kanadischer Luchs

f   Lynx du Canada; Loup-cervier (Qué)

i   Lince canadese

2268   *Felis caracal*
Carnivora - Felidae

e   Caracal; Caracal Lynx; Desert Lynx;
Persian Lynx; African Lynx

d   Wüstenluchs; Karakal

f   Caracal

i   Caracal; Lince africana; Lince del
deserto; Lince africana del deserto

2269   *Felis caracal michaelis*
Carnivora - Felidae

e   Turkmenian Caracal

d   Turkmenischer Karakal

2271   *Felis chaus*
Carnivora - Felidae

e   Jungle Cat; Chaus; Reed Cat; Swamp
Cat

d   Chaus; Dschungelkatze;
Katzenluchs; Luchskatze; Rohrkatze;
Sumpfluchs

f   Chat du jungle; Lynx des marais

i   Gatto della giungla; Gatto di palude

2272   *Felis chaus kelaarti*
Carnivora - Felidae

e   Ceylon Jungle Cat

d   Sri Lanka-Dschungelkatze

f   Chat du de Ceylan

2273   *Felis colocolo*
Carnivora - Felidae

e   Colocolo

d   Pampaskatze

f   Kudmu; Chat des Pampas

i   Gatto montano

2274   *Felis concolor*
Carnivora - Felidae

e   Puma; Cougar; Mountain Lion; Red
Tiger; Deer Tiger; Panther

d   Puma; Couguar; Kuguar; Berglöwe;
Silberlöwe

f   Puma; Leone argentato

i   Pantera; Cogouro; Leone di
montagna; Puma; Leone argentato

2275   *Felis concolor azteca*
Carnivora - Felidae

e   Mexican Puma

d   Mexikanischer Puma

f   Puma du Mexique

2276   *Felis concolor browni*
Carnivora - Felidae

e   Yuma Mountain Lion

d   Yuma-Puma

f   Puma de Yuma

2277   *Felis concolor cabrerae*
Carnivora - Felidae

e   Central South American Puma

2278   *Felis concolor californica*
Carnivora - Felidae

e   Californian Mountain Lion

d   Kalifornischer Puma

f   Puma de Californie

2279   *Felis concolor capricornensis*
Carnivora - Felidae

e   Eastern South American Puma

2280   *Felis concolor concolor*
Carnivora - Felidae

e   Northern South American Puma;
Brazilian Puma

d   Brasilianischer Puma

f   Puma du Brésil

2281   *Felis concolor coryi*
Carnivora - Felidae

e   Florida Puma; Florida Panther;
Florida Cougar; Florida Mountain
Lion; Florida Deer Tiger; Florida
Red Tiger

d   Florida-Puma; Florida-Kuguar;
Florida-Berglöwe; Florida-

Silberlöwe; Florida-Panther
f   Puma de la Floride; Cougouar de
Floride; Panthère de Floride
i   Puma della Florida; Pantera della
Florida

2282   *Felis concolor costaricensis*
Carnivora - Felidae
e   Central American Puma
d   Costa-Rica-Puma
f   Puma de l'Amérique centrale
i   Puma dell'America Centrale

2283   *Felis concolor cougar*
Carnivora - Felidae
e   Catamount; American Lion; Cougar;
Eastern Panther; Eastern Cougar;
North American Cougar
d   Ostamerikanischer Puma
f   Puma oriental; Cougouar de l'Est
i   Puma orientale

2284   *Felis concolor missoulensis*
Carnivora - Felidae
e   Missoula Cougar; Canadian Cougar
f   Puma de Missoun

2285   *Felis concolor oregonensis*
Carnivora - Felidae
e   Oregon Puma
f   Puma de l'Oregon

2286   *Felis concolor pearsoni*
Carnivora - Felidae
e   Patagonian Puma
f   Puma de Pearson

2287   *Felis concolor puma*
Carnivora - Felidae
e   Southern South American Puma
f   Puma du Chili
i   Puma; Coguaro; Leone di montagna

2288   *Felis concolor schorgeri*
Carnivora - Felidae

e   Wisconsin Puma
f   Puma de Wisconsin

2289   *Felis concolor stanleyana*
Carnivora - Felidae
e   Texas Puma
d   Texas-Puma

2290   *Felis geoffroyi*
Carnivora - Felidae
e   Geoffroy's Cat; Geoffroy's Ocelot
d   Kleinfleckkatze; Salzkatze;
Geoffroy-Katze
f   Chat de Geoffroy
i   Gatto di Geoffroy; Gatto di monte

2291   *Felis guigna*
Carnivora - Felidae
e   Kodkod; Guigna; Chilean Cat; Huina
d   Kodkod; Chilenische Waldkatze;
Nachtkatze
f   Kodkod
i   Kodkod

2292   *Felis jacobita*
Carnivora - Felidae
e   Andean Cat; Mountain Cat; Andean
Mountain Cat
d   Bergkatze; Anden-Katze
f   Chat des Andes
i   Gatto delle Ande

2293   *Felis lynx*
Carnivora - Felidae
e   Lynx; Northern Lynx; European
Lynx
d   Luchs; Gemeiner Luchs; Polarluchs;
Nordluchs; Eurasischer Luchs;
Europäischer Luchs
f   Lynx vulgaire; Lynx; Lynx boréal;
Lynx d'Europe
i   Lince; Lince euroasiatica; Lince
europea; Lince comune; Lupo
cerviero

2294   *Felis lynx isabellinus*
           Carnivora - Felidae
e   Turkestan Lynx
d   Turkestan-Luchs

2295   *Felis lynx stroganovi*
           Carnivora - Felidae
e   Eurasian Lynx

2296   *Felis lynx wrangeli*
           Carnivora - Felidae
e   Siberian Lynx
d   Sibirischer Luchs

2297   *Felis manul*
           Carnivora - Felidae
e   Manul; Pallas's Cat; Steppe Cat
d   Manul
f   Chat Manul
i   Manul; Gatto di Pallas; Gatto delle
           Steppe

2298   *Felis margarita*
           Carnivora - Felidae
e   Sand Cat; Dune Cat
d   Sandkatze; Barchan-Katze;
           Wüstenkatze
f   Chat des sables
i   Gatto delle sabbie

2299   *Felis margarita harrisoni*
           Carnivora - Felidae
e   Arabian Sand Cat
d   Arabische Sandkatze
i   Gatto delle sabbie arabo

2300   *Felis marmorata*
           Carnivora - Felidae
e   Marbled Cat
d   Marmelkatze; Marmorkatze
f   Chat marbré
i   Gatto marmorato; Gatto
           marmorizzato

2301   *Felis nigripes*

           Carnivora - Felidae
e   Black-footed Cat; Small Spotted Cat
d   Schwarzfußkatze
f   Chat aux pattes noires; Chat aux
           pieds noirs
i   Gatto dai piedi neri

2302   *Felis pajero*
           Carnivora - Felidae
e   Pampas Cat

2303   *Felis pardalis*
           Carnivora - Felidae
e   Painted Leopard; Ocelot
d   Ozelot; Pardelkatze
f   Ocelot
i   Ocelot; Gatopardo americano

2304   *Felis pardalis mearnsi*
           Carnivora - Felidae
e   Costa Rican Ocelot
d   Costa-Rica-Ozelot
f   Ocelot de Costa Rica
i   Ocelot di Costa Rica

2305   *Felis pardina*
           Carnivora - Felidae
e   Spanish Lynx; Iberian Lynx;
           Southern Lynx
d   Iberischer Luchs; Pardelluchs;
           Südluchs
f   Lynx pardelle; Lynx eurasien; Lynx
           d'Espagne; Loup-cervier

i   Lince pardina; Lince spagnola; Lince
           iberica

2306   *Felis planiceps*
           Carnivora - Felidae
e   Flat-headed Cat
d   Flachkopfkatze
f   Chat à tète plate
i   Gatto della testa piana; Gatto della
           testa piatta

2307   *Felis planiceps planiceps*
Carnivora - Felidae
e   Little Malayan Red Cat

2308   *Felis rubiginosa*
Carnivora - Felidae
e   Rusty-spotted Cat
d   Rostkatze
f   Chat rougeâtre; Chat léopard de
l'Inde
i   Gatto rugino; Gatto rugginoso

2309   *Felis rubiginosa phillipsi*
Carnivora - Felidae
e   Ceylon Rusty-spotted Cat

2310   *Felis rufa*
Carnivora - Felidae
e   Bobcat; Red Lynx; American Bobcat
d   Rotluchs
f   Lynx roux
i   Lince rossa; Bobcat; Gatto lince

2311   *Felis rufa baileyi*
Carnivora - Felidae
e   Southwestern Lynx

2312   *Felis rufa californicus*
Carnivora - Felidae
e   California Lynx

2313   *Felis rufa escuinapae*
Carnivora - Felidae
e   Mexican Lynx
i   Lince rossa del Messico

2314   *Felis rufa fasciatus*
Carnivora - Felidae
e   Oregon Bobcat; British Columbia
Lynx

2315   *Felis rufa floridanus*
Carnivora - Felidae
e   Florida Bobcat; Florida Lynx

2316   *Felis rufa gigas*
Carnivora - Felidae
e   Maine Lynx

2317   *Felis rufa pallescens*
Carnivora - Felidae
e   Rocky Mountains Lynx

2318   *Felis rufa peninsularis*
Carnivora - Felidae
e   Baja Lynx

2319   *Felis rufa rufa*
Carnivora - Felidae
e   Northeastern Lynx

2320   *Felis rufa superiorensis*
Carnivora - Felidae
e   Northwestern Lynx

2321   *Felis rufa texensis*
Carnivora - Felidae
e   Texas Lynx

2322   *Felis serval*
Carnivora - Felidae
e   Servaline Genet; Serval Cat; Serval
d   Serval; Buschkatze
f   Serval
i   Serval; Gatto serval

2323   *Felis serval constantina*
Carnivora - Felidae
e   Barbary Serval
d   Berber-Serval

2324   *Felis silvestris*
Carnivora - Felidae
e   Wild Cat; Feral Cat
d   Wilde Katze; Waldkatze;
Europäische Wildkatze; Wildkatze
f   Chat sauvage; Chat forestier
i   Gatto selvatico; Gatto selvatico
europeo

2325   *Felis silvestris cafra*
Carnivora - Felidae
e   Southern Africa Wild Cat
d   Südafrikanische Falbkatze

2326   *Felis silvestris caucasica*
Carnivora - Felidae
e   Caucasian Wild Cat

2327   *Felis silvestris caudata*
Carnivora - Felidae
e   Caspian Sea Wild Cat

2328   *Felis silvestris gordoni*
Carnivora - Felidae
e   Gordon's Cat; Gordon's Wild Cat; Oman Wild Cat
d   Oman-Falbkatze
f   Chat fauve d'Afrique; Chat de Gordon

2329   *Felis silvestris grampia*
Carnivora - Felidae
e   Scottish Wild Cat
i   Gatto selvatico scozzese

2330   *Felis silvestris libyca*
Carnivora - Felidae
e   African Wild Cat; Kaffir Cat
d   Nubische Falbkatze; Nubische Wildkatze
f   Chat forestier; Chat sauvage
i   Gatto fulvo; Gatto selvatico africano; Gatto fulvo d'Egitto; Gatto selvatico sardo

2331   *Felis silvestris silvestris*
Carnivora - Felidae
e   European Wild Cat; Forest Wild Cat
d   Mitteleuropäische Wildkatze; Waldwildkatze
f   Chat sauvage de l'Europe
i   Gatto selvatico europeo

2332   *Felis temminckii*
Carnivora - Felidae
e   Temminck's Cat; Temminck's Golden Cat; Asian Golden Cat; Asiatic Golden Cat
d   Asiatische Goldkatze
f   Chat doré; Chat doré d'Asie
i   Gatto dorato asiatico; Gatto di Temmink; Gatto dorato d'Asia

2333   *Felis temminckii tristis*
Carnivora - Felidae
e   Fontaine's Cat

2334   *Felis tigrina*
Carnivora - Felidae
e   Little Spotted Cat; Oncilla; Tiger Cat; American Tiger Cat; Little Tiger Cat
d   Tigerkatze; Ozelotkatze; Oncille
f   Chat-tigre tacheté
i   Gatto tigre; Oncilla

2335   *Felis viverrina*
Carnivora - Felidae
e   Fishing Cat; Yu Mao; Indian Fishing Cat
d   Fischkatze; Tüpfelkatze; Wagati-Katze
f   Chat viverrin; Chat pêcheur
i   Gatto pescatore

2336   *Felis wiedii*
Carnivora - Felidae
e   Tree Ocelot; Margay; Margay Cat
d   Langschwanzkatze; Bergozelot; Baumozelot; Margay
f   Margay
i   Margay; Gatto a coda lunga

2337   *Felis yaguarondi*
Carnivora - Felidae
e   Jaguarundi; Otter Cat; Weasel Cat; Spotless Cat; Jaguarondi
d   Jaguarundi; Wieselkatze
f   Jaguarundi

i Jaguarondi; Giaguarundi;
    Yagouarondi

2270 Feliscatus
    Carnivora - Felidae
e Domestic Cat
d Hauskatze
f Chat; Chat domestique
i Gatto domestico

2338 Felovia
    Rodentia - Ctenodactylidae
e Felou Gundis
d Senegal-Gundis
f Goundis de Félou

2339 Felovia vae
    Rodentia - Ctenodactylidae
e Felou Gundi
d Senegal-Gundi
f Goundi de Félou

Fennecus syn. *Vulpes* q.v.

2340 Feresa
    Cete - Delphinidae
e Blackfishes; Pygmy Killer Whales
d Zwerggrindwale
f Orques nains; Orques pygmées
i Ferese

2341 Feresa attenuata
    Cete - Delphinidae
e Slender Blackfish; Slender Pilot
    Whale; Pygmy Killer; Dwarf Killer
    Whale; Pygmy Killer Whale
d Zwerggrindwal; Kleinschwertwal
f Orque nain; Orque pygmée
i Orca pigmea; Feresa; Balena
    assassina

Feresa occulta syn. *F. attenuata* q.v.

2342 Feroculus
    Soricomorpha - Soricidae

e Kelaarts's Long-clawed Shrews;
    Keelart's Long-tailed Shrews
d Kelaarts Langkrallenspitzmäuse

2343 Feroculus feroculus
    Soricomorpha - Soricidae
e Kelaarts's Long-clawed Shrew;
    Keelart's Long-tailed Shrew
d Kelaarts Langkrallenspitzmaus

2344 Fossa
    Carnivora - Viverridae
e Fanalokas; Malagasy Civets
d Fanalokas
f Civettes fossanes
i Civette del Madagascar; Fanaloke

Fossa fossa syn. *F. fossana* q.v.

2345 Fossa fossana
    Carnivora - Viverridae
e Fanaloka; Malagasy Civet; Striped
    Civet; Madagascar Civet
d Fanaloka
f Civette fossane
i Civetta del Madagascar; Fanaloka

2346 Funambulus
    Rodentia - Sciuridae
e Palm Squirrels; Asiatic Striped Palm
    Squirrels; Indian Striped Squirrels
d Palmenhörnchen ; Gestreifte
    Palmenhörnchen
f Écureuils des palmes; Écureuils
    palmés

2347 Funambulus layardi
    Rodentia - Sciuridae
e Layard's Striped Squirrel; Layard's
    Palm Squirrel; Flame-striped Jungle
    Squirrel
d Layard-Palmenhörnchen; Gestreiftes
    Dschungelhörnchen

2348 Funambulus layardi signatus
    Rodentia - Sciuridae

e Western Flame-Striped Jungle Squirrel
d Westliches Gestreiftes Dschungelhörnchen

2349 ***Funambulus palmarum***
Rodentia - Sciuridae
e Palm Squirrel; Indian Palm Squirrel; Southern Palm Squirrel
d Indisches Palmenhörnchen
f Écureuil des palmes

2350 ***Funambulus palmarum brodiei***
Rodentia - Sciuridae
e Northern Ceylon Palm Squirrel
d Nördliches Sri Lanka-Palmenhörnchen

2351 ***Funambulus palmarum favonicus***
Rodentia - Sciuridae
e Western Ceylon Palm Squirrel
d Westliches Sri Lanka-Palmenhörnchen

2352 ***Funambulus palmarum kelaarti***
Rodentia - Sciuridae
e Eastern Ceylon Palm Squirrel
d Östliches Sri Lanka-Palmenhörnchen

2353 ***Funambulus palmarum olympius***
Rodentia - Sciuridae
e Highland Ceylon Palm Squirrel
d Sri Lanka-Hochlandpalmenhörnchen

2354 ***Funambulus pennanti***
Rodentia - Sciuridae
e Northern Palm Squirrel; Five-striped Palm Squirrel
d Nördliches Palmenhörnchen; Fünfstreifenpalmenhörnchen
f Écureuil aux rayures

2355 ***Funambulus sublineatus***
Rodentia - Sciuridae
e Dusky Striped Squirrel; Dusky Palm Squirrel

d Dunkles Palmenhörnchen

2356 ***Funambulus sublineatus obscurus***
Rodentia - Sciuridae
e Dusky-striped Jungle Squirrel; Ceylon Dusky-striped Jungle SquirrelJungle
d Gestreiftes Ceylon-Dschungelhörnchen

2357 ***Funambulus tristriatus***
Rodentia - Sciuridae
e Asiatic Striped Palm Squirrel; Jungle Striped Palm Squirrel; Jungle Palm Squirrel; Jungle Striped Squirrel; Western Ghats Squirrel
d Dschungelpalmenhörnchen

2358 ***Funisciurus***
Rodentia - Sciuridae
e African Striped Squirrels; Rope Squirrels
d Rotschenkelhörnchen
f Écureuils rayés d'Afrique; Funiscures
i Funiscuiri

2359 ***Funisciurus anerythrus***
Rodentia - Sciuridae
e Thomas's Tree Squirrel; Thomas's Rope Squirrel; Redless Squirrel
d Thomas-Rotschenkelhörnchen
f Funiscure à dos rayé

2360 ***Funisciurus bayonii***
Rodentia - Sciuridae
e Bocage's Tree Squirrel; Lunda Rope Squirrel
d Lunda-Rotschenkelhörnchen
f Funiscure de Bocage

2361 ***Funisciurus carruthersi***
Rodentia - Sciuridae
e Mountain Tree Squirrel; Carruther's Mountain Squirrel; Carruther's Mountain Tree Squirrel

d Carruthers-Rotschenkelhörnchen
f Écureil de Carruthers

2362 *Funisciurus congicus*
Rodentia - Sciuridae
e Kuhl's Tree Squirrel; Congo Rope
Squirrel; Congo Striped Tree Squirrel
d Kongo-Rotschenkelhörnchen
f Funiscure de Kuhl

2363 *Funisciurus isabella*
Rodentia - Sciuridae
e Gray's Four-striped Squirrel; Lady
Burton's Rope Squirrel; Lady
Burton's Squirrel
d Lady Burton-Rotschenkelhörnchen
f Écureuil à quatre raies; Écureuil
d'Isabella

2364 *Funisciurus lemniscatus*
Rodentia - Sciuridae
e Le Conte's Four-striped Squirrel;
Ribboned Rope Squirrel
d Westliches Rotschenkelhörnchen;
Gebändertes Rotschenkelhörnchen
f Écureuil à quatre raies de Le Conte

2365 *Funisciurus leucogenys*
Rodentia - Sciuridae
e Red-cheeked Squirrel; Orange-
headed Squirrel; Red-cheeked Rope
Squirrel
d Rotwangenhörnchen
f Funiscure à tète orange

2366 *Funisciurus pyrrhopus*
Rodentia - Sciuridae
e Cuvier's Fire-footed Squirrel;
Cuvier's Tree Squirrel; Fire-footed
Rope Squirrel; Red-footed Tree
Squirrel
d Feuerfußhörnchen; Cuvier-
Rotschenkelhörnchen
f Funiscure aux pattes rousses
i Funisciuro zamperosse

2367 *Funisciurus substriatus*

Rodentia - Sciuridae
e Kintampo Squirrel; De Winton's Tree
Squirrel; Kintampo Rope Squirrel;
Kintampo
d Kintampo-Rotschenkelhörnchen
f Funiscure de Kintampo

2368 Furipteridae
Chiroptera
e Smoky Thumbless Bats; Smoky
Bats; Thumbless Bats
d Stummeldaumen; Rauchfledermäuse;
Rauchfarbfledermäuse
f Furiptéridés
i Furipteridi

2369 *Furipterus*
Chiroptera - Furipteridae
e Little Smoky Bats; Smoky Bats;
Eastern South American Smoky
Bats; Panamanian Smoky Bats
d Stummeldaumen

2370 *Furipterus horrens*
Chiroptera - Furipteridae
e Cuvier's Smoky Bat; Cuvier's
Thumbless Bat; Cuvier's Pig Bat;
Eastern Smoky Bat; Thumbless Bat;
Smoky Bat
d Cuviers Stummeldaumen

G

2371   *Galago*
Primates - Loridae
e   Galagos; Bushbabies; Lesser
        Galagos; Lesser Bushbabies
d   Galagos; Ohrenmakis; Buschbabys
f   Galagos
i   Galagoni

2372   *Galago alleni*
Primates - Loridae
e   Allen's Bushbaby; Allen's Galago;
        Allen's Squirrel Galago
d   Buschwaldgalago; Allens Galago;
        Allens Bioko Bushbaby
f   Galago d'Allen
i   Galagone di Allen

2373   *Galago cameronensis*
Primates - Loridae
e   Cross River Squirrel Galago
d   Allens Kreuzfluss-Bushbaby

***Galago crassicaudatus* syn.**
Otolemur crassicaudata q.v.

2374   *Galago demidoff*
Primates - Loridae
e   Demidoff's Bushbaby; Demidoff's
        Dwarf Galago; Demidoff's Pymy
        Galago; Dwarf Galago; Pygmy
        Galago; Demidoff's Galago; Prince
        Demidoff's Galago
d   Zwerggalago; Prinz Demidoffs
        Bushbaby; Urwaldgalago
f   Galago de Demidoff
i   Galagone di Demidoff; Galagone
        nano

***Galago elegantulus* syn.** *Euoticus*
elegantulus q.v.

2375   *Galago gabonensis*
Primates - Loridae
e   Gabon Squirrel Galago
d   Allens Gabun-Bushbaby

2376   *Galago gallarum*
Primates - Loridae
e   Somali Galago; Eastern Lesser
        Bushbaby; Somali Bushbaby; Somali
        Lesser Galago
d   Somali-Galago; Somalischer
        Bushbaby
f   Galago de Somali
i   Galagone dei Galla

***Galago garnetti* syn.** *Otolemur*
garnetti q.v.

2377   *Galago granti*
Primates - Loridae
e   Grant's Dwarf Galago; Grant's
        Galago
d   Grants Bushbaby
i   Galagone di Grant

2378   *Galago matschiei*
Primates - Loridae
e   Spectacled Galago; Eastern Needle-
        nailed Bushbaby; Matschie's
        Bushbaby; Eastern Needle-clawed
        Galago
d   Östlicher Kielnagelgalago; Düsteres
        Bushbaby
f   Galago du Congo
i   Galagone di Matschie; Galagone
        dagli occhiali

2379   *Galago moholi*
Primates - Loridae
e   South African Galago; South African
        Lesser Bushbaby; Mohol; South
        African Lesser Galago; Lesser
        Bushbaby (CSA); Southern Lesser
        Bushbaby; Moholi Lesser Bushbaby

d Moholi; Moholi-Galago; Moholi-
Bushbaby
i Galagone moholi; Galagone minore

2380 *Galago nyasae*
Primates - Loridae
e Malawi Galago
d Malawi-Bushbaby

2381 *Galago orinus*
Primates - Loridae
e Amami Dwarf Galago; Taita
Mountain Galago; Uluguru Galago
d Ulluguru-Bushbaby
i Galagone dei Monti; Galagone dei
Monti Uluguru

2382 *Galago rondoensis*
Primates - Loridae
e Rondo Dwarf Galago; Rondo Galago
d Rondo-Bushbaby
i Galagone della foresta di Rondo

2383 *Galago senegalensis*
Primates - Loridae
e West African Lesser Bushbaby;
Lesser Galago; Senegal Galago;
Lesser Bushbaby; Senegal Bushbaby;
Northern Lesser Bushbaby; Northern
Lesser Galago
d Senegal-Galago; Senegal-Bushbaby;
Steppengalago
f Galago du Sénégal
i Galagone del Senegal; Galagone
minore; Galagone comune

2384 *Galago senegalensis braccatus*
Primates - Loridae
e Ugandan Bushbaby
d Gelbschenkelgalago

2385 *Galago thomasi*
Primates - Loridae
e Thomas's Dwarf Galago; Thomas's
Galago
d Thomas-Bushbaby

f Galago de Thomas
i Galagone di Thomas

2386 *Galago udzungwensis*
Primates - Loridae
e Matundu Dwarf Galago; Matundu
Galago
d Udzungwa-Bushbaby
i Galagone dei Monti Udzungwe

2387 *Galago zanzibaricus*
Primates - Loridae
e Zanzibar Bushbaby; Zanzibar Galago
d Sansibar-Galago; Sansibar-Bushbaby
f Galago du Zanzibar
i Galagone di Zanzibar

Galagoides syn. *Galago* q.v.

2388 *Galea*
Rodentia - Caviidae
e Yellow-toothed Cavies
d Wieselmeerschweinchen
f Cobayes aux dents jaunes

2389 *Galea flavidens*
Rodentia - Caviidae
e Brandt's Yellow-toothed Cavy

2390 *Galea monasteriensis*
Rodentia - Caviidae
e Muenster Guinea-pig
d Münstersches Meerschweinchen

2391 *Galea musteloides*
Rodentia - Caviidae
e Common Yellow-toothed Cavy
d Graues Wieselmeerschweinchen;
Großes Wieselmeerschweinchen

2392 *Galea spixii*
Rodentia - Caviidae
e Spix's Yellow-toothed Cavy;
Shipton's Mountain Cavy

Galea wellsi syn. *G. spixii* q.v.

2393 Galemys
Erinaceomorpha - Talpidae
e Pyrenean Desmans
d Pyrenäen-Bisamspitzmäuse
f Desmans des Pyrénées
i Desmani

2394 Galemys pyrenaicus
Erinaceomorpha - Talpidae
e Pyrenean Desman; Iberian Desman
d Pyrenäen-Bisamspitzmaus;
   Pyrenäen-Desman; Almizclero
f Desman des Pyrénées; Rat-trompette
i Desman dei Pirenei

2395 Galenomys
Rodentia - Muridae
e Garlepp's Mice
d Garlepp-Mäuse

2396 Galenomys garleppi
Rodentia - Muridae
e Garlepp's Mouse
d Garlepp-Maus

2397 Galeopithecidae
Orimates -
e Flying Lemurs
d Riesengleitflieger
f Lémurs volants
i Lemuri volante; Galeopitecidi

Galeopithecus syn. *Cynocephalus* q.v.

Galera syn. *Eira* q.v.

Galerella syn. *Herpestes* q.v.

Galeriscus syn. *Bdeogale* q.v.

2398 Galictis

Carnivora - Mustelidae
e Grisons
d Grisons
f Grisons
i Grigioni

2399 Galictis cuja
Carnivora - Mustelidae
e Lesser Grison; Little Grison
d Kleingrison
f Petit grison
i Grigione minore

2400 Galictis vittata
Carnivora - Mustelidae
e South American Grison; Allamand's
   Grison; Hurone Grison; Guiana
   Marten; Greater Grison; Grison
d Großgrison
f Grison d'Allamand; Grison
i Grigione maggiore; Grigione vittato

2401 Galidia
Carnivora - Herpestidae
e Ring-tailed Mongooses; Madagascar
   Ring-tailed Mongooses; Galidias
d Ringelmungos
f Galidies
i Galidi

2402 Galidia elegans
Carnivora - Herpestidae
e Ring-tailed Mongoose; Madagascar
   Ring-tailed Mongoose; Ring-tailed
   Galidia; Malagasy Ring-tailed
   Mongoose
d Ringelmungo; Ringelschwanzmungo
f Galidie à queue annelée
i Falide elegante

2403 Galidictis
Carnivora - Herpestidae
e Broad-striped Mongooses;
   Madagascar Broad-striped
   Mongooses; Striped Mongooses

d   Breitstreifenmungos
f   Galidies; Moungoustes à bandes

2404   *Galidictis faciata*
       Carnivora - Herpestidae
e   Broad-striped Galidia; Madagascar
       Broad-striped Mongoose; Broad-
       striped Mongoose
d   Breitstreifenmungo; Bändermungo
f   Galidie rayée

2405   *Galidictis grandidieri*
       Carnivora - Herpestidae
e   Giant-striped Mongoose;
       Grandidier's Mongoose

       Galidictis striata syn. *G. fasciata* q.v.

       Gatimyia syn. *Mus* q.v.

2406   *Gazella*
       Artiodactyla - Bovidae
e   Gazelles
d   Gazellen
f   Gazelles
i   Gazzelle

2407   *Gazella arabica*
       Artiodactyla - Bovidae
e   Arabian Gazelle
d   Arabische Gazelle
f   Gazelle d'Arabie

2408   *Gazella arabica bilkis*
       Artiodactyla - Bovidae
e   Queen of Sheba's Gazelle
d   Jemen-Gazelle
f   Gazelle de la reine de Saba

2409   *Gazella bennettii*
       Artiodactyla - Bovidae
e   Indian Gazelle; Jebeer Gazelle;
       Chinkara
d   Indische Gazelle
f   Gazelle indienne

2410   *Gazella cuvieri*
       Artiodactyla - Bovidae
e   Edmi Gazelle; Cuvier's Gazelle;
       Mountain Gazelle; Atlas Mountain
       Gazelle; Atlas Gazelle; Edmi
d   Cuvier-Gazelle
f   Edmi; Gazelle de Cuvier
i   Gazzella di Cuvier; Gazella Edmi

2411   *Gazella dama*
       Artiodactyla - Bovidae
e   Addra Gazelle; Dama Gazelle; Addra
d   Dama-Gazelle
f   Gazelle dama; Biche Robert
i   Gazzella dama

2412   *Gazella dama mhorr*
       Artiodactyla - Bovidae
e   Mhorr Gazelle
d   Mhorr-Gazelle
f   Gazelle dama du Sud marocain

2413   *Gazella dama ruficollis*
       Artiodactyla - Bovidae
e   Nubian Red-necked Gazelle
d   Rothalsgazelle

2414   *Gazella dorcas*
       Artiodactyla - Bovidae
e   Dorcas Gazelle
d   Dorkas-Gazelle
f   Dorcas; Gazelle dorcas
i   Gazzella dorcade; Gazella dorcas;
       Damalisco a fronte bianca; Gazella
       del deserto

2415   *Gazella dorcas isabella*
       Artiodactyla - Bovidae
e   Isabelline Gazelle
d   Isabella-Gazelle

2416   *Gazella dorcas litoralis*
       Artiodactyla - Bovidae
e   Eritrean Gazelle
d   Eritrea-Gazelle

2417   *Gazella dorcas pelzelni*
       Artiodactyla - Bovidae
e   Pelzeln Gazelle
d   Pelzeln-Gazelle
f   Gazelle de Pelzeln

2418   *Gazella gazella*
       Artiodactyla - Bovidae
e   Gazelle; Common Gazelle; Idmi;
      Mountain Gazelle
d   Echtgazelle; Edmi-Gazelle
f   Edmi; Gazelle de Montagne
i   Gazzella di montagna

2419   *Gazella granti*
       Artiodactyla - Bovidae
e   Grant's Gazelle
d   Grant-Gazelle
f   Gazelle de Grant
i   Gazzella di Grant

2420   *Gazella granti brighti*
       Artiodactyla - Bovidae
e   Bright's Gazelle

2421   *Gazella granti granti*
       Artiodactyla - Bovidae
e   Southern Grant's Gazelle
f   Gazelle de Grant du Sud

2422   *Gazella granti lacuum*
       Artiodactyla - Bovidae
e   Northern Grant's Gazelle
f   Gazelle de Grant du Nord

2423   *Gazella granti petersi*
       Artiodactyla - Bovidae
e   Peters's Gazelle; Tana
d   Peters-Gazelle

2424   *Gazella granti robertsi*
       Artiodactyla - Bovidae
e   Roberts's Gazelle
f   Gazelle de Roberts

2425   *Gazella gutturosa*
       Artiodactyla - Bovidae
e   Mongolian Gazelle; Zeren
d   Mongolei-Antilope; Nordchinesische
      Gazelle; Dseren; Mongolei-Gazelle;
      Mongolische Antilope; Mongolische
      Gazelle
f   Gazelle de Mongolie; Gazelle à
      queue blanche

2426   *Gazella leptoceros*
       Artiodactyla - Bovidae
e   Loder's Gazelle; Algerian Sand
      Gazelle; Sand Gazelle; Rhim; Rhim
      Gazelle
d   Dünengazelle
f   Gazelle aux cornes grêles; Gazelle
      leptocère; Gazelle rhim
i   Gazella bianca; Gazella Rhim

2427   *Gazella leptoceros leptoceros*
       Artiodactyla - Bovidae
e   Slender-horned Gazelle
f   Gazelle blanche

2428   *Gazella picticaudata*
       Artiodactyla - Bovidae
e   Tibetan Gazelle; Goa
d   Tibet-Gazelle
f   Gazelle du Tibet

2429   *Gazella przewalskii*
       Artiodactyla - Bovidae
e   Przewalski's Gazette
d   Przewalski-Gazelle
f   Gazelle Przewalski

2430   *Gazella rufifrons*
       Artiodactyla - Bovidae
e   Korin Gazelle; Korin; Red-fronted
      Gazelle
d   Rotstirngazelle
f   Gazelle à front roux
i   Gazella dalla fronte rossa

2431   *Gazella rufifrons tilonura*
     Artiodactyla - Bovidae
e   Heuglin Gazelle
d   Heuglin-Gazelle

2432   *Gazella rufina*
     Artiodactyla - Bovidae
e   Red Gazelle; Rufous Gazelle
d   Algerische Gazelle; Rotgazelle
f   Gazelle rouge

2433   *Gazella saudiya*
     Artiodactyla - Bovidae
e   Saudi Gazelle
d   Saudi-Gazelle
f   Gazelle saoudienne

2434   *Gazella soemmeringii*
     Artiodactyla - Bovidae
e   Sömmering's Gazelle
d   Soemmering-Gazelle
f   Gazelle de Sömmering
i   Gazella di Soemmering

2435   *Gazella soemmeringii berberana*
     Artiodactyla - Bovidae
e   Somali Sömmering's Gazelle

2436   *Gazella soemmeringii butteri*
     Artiodactyla - Bovidae
e   Borani Sömmering's Gazelle

2437   *Gazella soemmeringii soemmeringii*
     Artiodactyla - Bovidae
e   Sudan Sömmering's Gazelle

2438   *Gazella spekei*
     Artiodactyla - Bovidae
e   Speke's Gazelle; Dero; Speke Gazelle
d   Speke-Gazelle
f   Gazelle de Speke
i   Gazella di Speke

2439   *Gazella subgutturosa*
     Artiodactyla - Bovidae
e   Persian Gazelle; Goitered Gazelle; Black-tailed Gazelle
d   Kropfgazelle; Persische Gazelle
f   Gazelle à goitre

2440   *Gazella subgutturosa hilleriana*
     Artiodactyla - Bovidae
e   Mongolian Goitered Gazelle
d   Mongolische Kropfgazelle

2441   *Gazella subgutturosa marica*
     Artiodactyla - Bovidae
e   Arabian Sand Gazelle
d   Arabische Kropfgazelle
f   Gazelle des sables

2442   *Gazella subgutturosa subgutturosa*
     Artiodactyla - Bovidae
e   Persian Goitered Gazelle
d   Persische Kropfgazelle
f   Gazelle à goitre

2443   *Gazella thomsonii*
     Artiodactyla - Bovidae
e   Thomson's Gazelle
d   Thomson-Gazelle
f   Gazelle de Thomson
i   Gazella di Thomson

2444   *Gazella thomsonii albonotata*
     Artiodactyla - Bovidae
e   Mongalla Gazelle; Thommy

2445   *Genetta*
     Carnivora - Viverridae
e   Genets
d   Ginsterkatzen; Genetten
f   Genettes
i   Genette

2446   *Genetta abyssinica*
     Carnivora - Viverridae
e   Abyssinian Genet; Ethiopian Genet

d Sennar-Ginsterkatze; Äthiopische
Ginsterkatze
f Genette d'Éthiopie
i Genetta abissina

2447 *Genetta angolensis*
Carnivora - Viverridae
e Angola Genet; Angolan Genet;
Mozambique Genet; Miombo Genet
d Angola-Ginsterkatze
f Genette de Mozambique; Genette
d'Angola
i Genetta dell'Angola; Gatto genetta

2448 *Genetta bourtoni*
Carnivora - Viverridae
e Bourton's Genet
f Genette de Bourton

Genetta felina syn. *G. genetta* q.v.

2449 *Genetta genetta*
Carnivora - Viverridae
e Common Genet; European Genet;
Small-spotted Genet; Genet; Feline
Genet
d Europäische Genette;
Kleinfleckginstkatze;
Kleinfleckginsterkatze; Ginstkatze;
Ginsterkatze; Genette
f Genette d'Europe; Genette commune;
Genette vulgaire; Genette
i Genetta; Genetta comune

2450 *Genetta johnstoni*
Carnivora - Viverridae
e Johnston's Genet; Lehmann's Genet
d Johnston-Ginsterkatze
f Genette de Johnstone
i Genetta di Johnston

2451 *Genetta maculata*
Carnivora - Viverridae
e Rusty-spotted Genet
d Gefleckte Ginsterkatze

f Genette pardine
i Gatto genetta

Genetta mossambicus syn. *G. angolensis* q.v.

2452 *Genetta pardina*
Carnivora - Viverridae
e Forest Genet; Panther Genet; Pardine
Genet
d Pantherginsterkatze
f Genette des forêts
i Genetta pardina

2453 *Genetta poensis*
Carnivora - Viverridae
e King Genet

2454 *Genetta servalina*
Carnivora - Viverridae
e Servaline Genet
d Waldginsterkatze; Serval-
Ginsterkatze
f Genette servaline

2455 *Genetta thierryi*
Carnivora - Viverridae
e False Genet; Hausa Genet; Villier's
Genet
d Hausa-Ginsterkatze
i Genetta di Villiers; Genetta hausa

2456 *Genetta tigrina*
Carnivora - Viverridae
e Blotched Genet; Large-spotted
Genet; Cape Genet
d Großfleck-Ginsterkatze
f Genette tigrine
i Genetta tigrina

2457 *Genetta victoriae*
Carnivora - Viverridae
e Giant Genet; Giant Forest Genet;
Giant Servaline Genet
d Riesenginsterkatze

    f   Genette géante
    i   Genetta gigante

2458   *Geocapromys*
        Rodentia - Capromyidae
    e   Bahaman and Jamaican Hutias
    d   Stummelschwanz-Hutias; Eigentliche Ferkelratten; Ferkelratten
    f   Hutias
    i   Hutie

2459   *Geocapromys brownii*
        Rodentia - Capromyidae
    e   Brown's Hutia; Jamaican Hutia
    d   Jamaika-Ferkelratte
    f   Hutia de la Jamaïque; Hutia de Jamïque

2460   *Geocapromys ingrahami*
        Rodentia - Capromyidae
    e   Ingram's Hutia; Bahama Hutia; Bahama Island Hutia
    d   Bahama-Ferkelratte; Bahama-Insel-Ferkelratte; Bahama-Stummelschwanzhutia
    f   Hutia d'Ingraham; Rat d'Ingraham

2461   *Geocapromys thoracatus*
        Rodentia - Capromyidae
    e   Swan Island Hutia; Honduras Hutia; Little Swan Island Hutia
    d   Swan-Island-Hutia; Little-Swan-Island-Ferkelratte; Little-Swan-Island-Stummelschwanzferkelratte; Little-Swan-Island Hutia; Little-Swan-Island-Stummelschwanzhutia
    f   Rat grimpeur de Little Swan Island; Hutia de Little Swan Island; Rat de Little Swan Island

2462   *Geogale*
        Soricomorpha - Tenrecidae
    e   Burrowing Tenrecs; Burrowing Tanrecs
    d   Erdtanreks
    f   Géogales

2463   *Geogale aurita*
        Soricomorpha - Tenrecidae
    e   Burrowing Tenrecs; Burrowing Tanrec; Large-eared Tenrec; Large-eared Tanrec
    d   Erdtanrek
    f   Géogale

2464   **Geomyidae**
        Rodentia
    e   Pocket Gophers
    d   Taschenratten; Taschennager
    f   Gaufres à poche; Géomyidés
    i   Geomidi

2465   *Geomys*
        Rodentia - Geomyidae
    e   Eastern Pocket Gophers (NA); Eastern American Pocket Gophers
    d   Flachlandtaschenratten
    f   Gaufres à poche; Geomys
    i   Gopher

2466   *Geomys arenarius*
        Rodentia - Geomyidae
    e   Desert Pocket Gopher
    f   Geomys du désert

2467   *Geomys attwateri*
        Rodentia - Geomyidae
    e   Attwater's Pocket Gopher

2468   *Geomys breviceps*
        Rodentia - Geomyidae
    e   Baird's Pocket Gopher

2469   *Geomys bursarius*
        Rodentia - Geomyidae
    e   Plains Pocket Gopher; Dark Brown Prairie Gopher; Mississippi Valey Pocket Gopher
    d   Flachlandtaschenratte
    f   Gaufre à poche; Gaufre brun
    i   Gopher delle praterie

Geomys colonus syn. *G. pinetis* q.v.

Geomys cumberlandius syn. *G. pinetis* q.v.

Geomys fontanelus syn. *G. pinetis* q.v.

2470 *Geomys knoxjonesi*
Rodentia - Geomyidae
e   Jones's Pocket Gopher; Knox Jones's Pocket Gopher

2471 *Geomys personatus*
Rodentia - Geomyidae
e   Texas Pocket Gopher
d   Texas-Taschenratte

2472 *Geomys pinetis*
Rodentia - Geomyidae
e   Southeastern Pocket Gopher; Sherman's Pocket Gopher; Common Gopher
d   Südöstliche Taschenratte; Pinientaschenratte; Pinienflachlandtaschenratte; Südöstliche Pinientaschenratte
f   Gaufre à poche du Sud-est

2473 *Geomys pinetis goffi*
Rodentia - Geomyidae
e   Goff's Southeastern Pocket Gopher; Goff's Pocket Gopher
d   Goff-Pinientaschenratte; Goff-Pinienflachlandtaschenratte
f   Gaufre à poche de Goff

2474 *Geomys texensis*
Rodentia - Geomyidae
e   Central Texas Pocket Gopher

2475 *Geomys tropicalis*
Rodentia - Geomyidae
e   Tropical Pocket Gopher

2476 *Georychus*

Rodentia - Bathyergidae
e   Cape Blesmols; Cape Molerats
d   Kap-Blessmulle; Kap-Mullratten; Blessmulle
f   Rats-taupes du Cap

2477 *Georychus capensis*
Rodentia - Bathyergidae
e   Cape Molerat; Blesmol; Bles Mole
d   Kap-Blessmull; Blessmull; Kap-Mullratte
f   Rat-taupe du Cap

Geosciurus syn. *Xerus* q.v.

2478 *Geoxus*
Rodentia - Muridae
e   Long-clawed Molemice
d   Langkrallenmaulwurfsmäuse

2479 *Geoxus valdividianus*
Rodentia - Muridae
e   Long-clawed Molemouse
d   Langkrallenmaulwurfsmaus

Gerbillicus syn. *Tatera* q.v.

2480 *Gerbillurus*
Rodentia - Muridae
e   Southern Pygmy Gerbils; Hairy-footed Gerbils
d   Zwergrennmäuse; Namib-Rennmäuse
f   Gerbilles

2481 *Gerbillurus paeba*
Rodentia - Muridae
e   South African Pygmy Gerbil; Hairy-footed Gerbil; Sand Rat; Namib Paeba; Snowshoe Gerbil; Common Hairy-footed Gerbil
d   Südafrikanische Zwergrennmaus; Kalahari-Rennmaus
f   Gerbille aux pieds velus

2482   *Gerbillurus setzeri*
Rodentia - Muridae
e   Setzer's Hairy-footed Gerbil;
Southern Pygmy Gerbil
f   Gerbille de Setzer

2483   *Gerbillurus tytonis*
Rodentia - Muridae
e   Dune Hairy-footed Gerbil
f   Gerbille des dunes

2484   *Gerbillurus vallinus*
Rodentia - Muridae
e   Tassel-tailed Pygmy Gerbil; Brush-
tailed Hairy-footed Gerbil; Brush-
tailed Hairy Gerbil; Bushy-tailed
Hairy-footed Gerbil
d   Borstenschwanzrennmaus

2485   *Gerbillus*
Rodentia - Muridae
e   Typical Gerbils; Smaller Gerbils;
Pygmy Gerbils; Sand Rats; Northern
Pygmy Gerbils; Gerbils; Dipodils
d   Eigentliche Rennmäuse; Rennmäuse;
Rennratten; Zwergrennmäuse
f   Gerbilles; Dipodils
i   Gerbilli

2486   *Gerbillus acticola*
Rodentia - Muridae
e   Berbera Gerbil

2487   *Gerbillus agag*
Rodentia - Muridae
e   Agag Gerbil
f   Gerbille de l'Agag

2488   *Gerbillus allenbyi*
Rodentia - Muridae
e   Allenby's Gerbil
f   Gerbille d'Allenby

2489   *Gerbillus amoenus*
Rodentia - Muridae

e   Pleasant Gerbil; Charming Gerbil;
Charming Dipodil
d   Giza-Zwergrennmaus; Zwerg
Dippodil
f   Gerbille charmante

2490   *Gerbillus andersoni*
Rodentia - Muridae
e   Anderson's Gerbil; Bayoudi
d   Andersons Wüstenrennmaus;
Andersons Rennmaus
f   Gerbille d'Anderson

2491   *Gerbillus aquilus*
Rodentia - Muridae
e   Swarthy Gerbil
d   Dunkle Rennmaus

2492   *Gerbillus bilensis*
Rodentia - Muridae
e   Bilen Gerbil

2493   *Gerbillus bonhotei*
Rodentia - Muridae
e   Bonhote's Gerbil

2494   *Gerbillus bottai*
Rodentia - Muridae
e   Botta's Gerbil

2495   *Gerbillus brockmani*
Rodentia - Muridae
e   Brockman's Gerbil

2496   *Gerbillus burtoni*
Rodentia - Muridae
e   Large Egyptian Gerbil

2497   *Gerbillus campestris*
Rodentia - Muridae
e   Rock Gerbil; Large North African
Gerbil; North African Gerbil; Large
North African Dipodil
d   Feldrennmaus

f Gerbille des champs; Gerbille champêtre; Gerbille des rochers

2498 *Gerbillus cheesmani*
Rodentia - Muridae
e Cheesman's Gerbil
d Cheesmans Rennmaus
f Gerbille de Cheesman

2499 *Gerbillus cosensi*
Rodentia - Muridae
e Cosens's Gerbil

2500 *Gerbillus dalloni*
Rodentia - Muridae
e Dallon's Gerbil

2501 *Gerbillus dasyurus*
Rodentia - Muridae
e Wagner's Gerbil; Wagner's Dipodil; Rough-tailed Dipodil; Wadi Hof Gerbil
d Wagner-Rennmaus
f Gerbille de Wagner

2502 *Gerbillus diminutus*
Rodentia - Muridae
e Diminutive Gerbil

2503 *Gerbillus dongolanus*
Rodentia - Muridae
e Dongola Gerbil

2504 *Gerbillus dunni*
Rodentia - Muridae
e Somalia Gerbil; Dunn's Gerbil

2505 *Gerbillus famulus*
Rodentia - Muridae
e Black-tufted Gerbil

2506 *Gerbillus floweri*
Rodentia - Muridae
e Flower's Gerbil

2507 *Gerbillus garamantis*
Rodentia - Muridae
e Algerian Gerbil
d Nordafrikanische Rennmaus

2508 *Gerbillus gerbillus*
Rodentia - Muridae
e Lesser Egyptian Gerbil; Lesser Gerbil; Small Gerbil
d Kleine Ägyptische Rennmaus; Kleine Sandrennmaus
f Petite gerbille du sable; Petite gerbille; Petite gerbille d'Égypte; Petite gerbille du Sahara
i Gerbillo comune

2509 *Gerbillus gleadowi*
Rodentia - Muridae
e Indian Hairy-footed Gerbil
d Pakistan-Wüstenrennmaus; Indische Rauhfußrennmaus; Gleadows Rennmaus
f Gerbille velue de l'Inde

2510 *Gerbillus grobbeni*
Rodentia - Muridae
e Grobben's Gerbil

2511 *Gerbillus harwoodi*
Rodentia - Muridae
e Harwood's Gerbil

2512 *Gerbillus henleyi*
Rodentia - Muridae
e Pygmy Gerbil; Henley's Gerbil
f Gerbille de Henley

2513 *Gerbillus hesperinus*
Rodentia - Muridae
e Western Gerbil
f Gerbille hespérine

Gerbillus hirtipes syn. *G. gerbillus* q.v.

2514 *Gerbillus hoogstraali*
Rodentia - Muridae
e   Hoogstraal's Gerbil
f   Gerbille de Souss

2515 *Gerbillus jamesi*
Rodentia - Muridae
e   James's Gerbil

2516 *Gerbillus juliani*
Rodentia - Muridae
e   Julian's Gerbil

Gerbillus kaiseri syn. *G. simoni* q.v.

2517 *Gerbillus latastei*
Rodentia - Muridae
e   Small Egyptian Gerbil; Lataste's
     Gerbil; Hairy-footed Gerbil
d   Rauhfüssige Rennmaus
f   Gerbille de Lataste; Gerbille aux
     pattes poilues; Gerbille aux pieds
     velus

2518 *Gerbillus lowei*
Rodentia - Muridae
e   Lowe's Gerbil

2519 *Gerbillus mackilligini*
Rodentia - Muridae
e   Mackillingin's Gerbil; Mackillingin's
     Dipodil
f   Gerbille de Mackilligin

2520 *Gerbillus maghrebi*
Rodentia - Muridae
e   Greater Short-tailed Gerbil; Maghreb
     Gerbil
d   Große Kurzschwanzrennmaus
f   Grande gerbille à queue courte

2521 *Gerbillus mauritaniae*
Rodentia - Muridae
e   Mauritanian Gerbil

2522 *Gerbillus mesopotamiae*
Rodentia - Muridae
e   Mesopotamian Gerbil; Harrison's
     Gerbil
f   Gerbille de Harrison

2523 *Gerbillus muriculus*
Rodentia - Muridae
e   Barfur Gerbil

2524 *Gerbillus nancillus*
Rodentia - Muridae
e   Sudan Gerbil

2525 *Gerbillus nanus*
Rodentia - Muridae
e   Baluchistan Gerbil; Field Gerbil;
     Dwarf Naked-soled Gerbil;
     Baluchistan Dwarf Gerbil

d   Belutschistan-Rennmaus; Kleine
     Rennmaus
f   Gerbille naine; Gerbille du
     Baluchistan

2526 *Gerbillus nigeriae*
Rodentia - Muridae
e   Nigerian Gerbil
f   Gerbille de Nigérie

2527 *Gerbillus occiduus*
Rodentia - Muridae
e   Occidental Gerbil
f   Gerbille occidentale

Gerbillus paeba syn. *Gerbillurus
paeba* q.v.

2528 *Gerbillus percivali*
Rodentia - Muridae
e   Percival's Gerbil

2529 *Gerbillus perpallidus*
Rodentia - Muridae
e   Pale Gerbil; Pallid Gerbil

d Blasse Wüstenrennmaus; Ägyptische
Rennmaus
f Gerbille pâle

2530 *Gerbillus poecilops*
Rodentia - Muridae
e Large Aden Gerbil
f Grande gerbille de l'Aden

2531 *Gerbillus principulus*
Rodentia - Muridae
e Principal Gerbil

2532 *Gerbillus pulvinatus*
Rodentia - Muridae
e Cushioned Gerbil
f Gerbille d'Éthiopie

2533 *Gerbillus pusillus*
Rodentia - Muridae
e Least Gerbill
f Kigogo

2534 *Gerbillus pyramidum*
Rodentia - Muridae
e Greater Egyptian Gerbil; Greater
Gerbil; Pyramid Gerbil
d Pyramiden-Maus; Große Ägyptische
Rennmaus; Pyramiden-Rennmaus
f Grande gerbille

2535 *Gerbillus quadrimaculatus*
Rodentia - Muridae
e Four-spotted Gerbil

2536 *Gerbillus riggenbachi*
Rodentia - Muridae
e Riggenbach's Gerbil
f Gerbille de Riggenbach

2537 *Gerbillus rosalinda*
Rodentia - Muridae
e Rosalinda Gerbil
f Gerbille de Rosalinda

2538 *Gerbillus ruberrimus*
Rodentia - Muridae
e Little Red Gerbil

2539 *Gerbillus rupicola*
Rodentia - Muridae
e Rock-loving Gerbil

Gerbillus setzeri syn. *Gerbillurus
setzeri* q.v.

2540 *Gerbillus simoni*
Rodentia - Muridae
e Lesser Short-tailed Gerbil; Simon's
Dipodil; Zakaria's Gerbil
d Simon-Rennmaus
f Gerbille de Simon; Gerbille de
Kaiser

2541 *Gerbillus somalicus*
Rodentia - Muridae
e Somalian Gerbil; Somali Gerbil

2542 *Gerbillus stigmonyx*
Rodentia - Muridae
e Khartoum Gerbil

2543 *Gerbillus syrticus*
Rodentia - Muridae
e Sand Gerbil; Southern Pygmy Gerbil

2544 *Gerbillus tarabuli*
Rodentia - Muridae
e Tarabul's Gerbil

Gerbillus tytonis syn. *Gerbillurus
tytonis* q.v.

Gerbillus vallinus syn. *Gerbillurus
vallinus* q.v.

2545 *Gerbillus vivax*
Rodentia - Muridae
e Vivacious Gerbil

2546 ***Gerbillus watersi***
Rodentia - Muridae
e Waters's Gerbil

Gerbillus zakariai syn. *G. simoni*
q.v.

2547 ***Giraffa***
Artiodactyla - Giraffidae
e Giraffes
d Giraffen
f Girafes
i Giraffe

2548 ***Giraffa camelopardalis***
Artiodactyla - Giraffidae
e Giraffe
d Steppengiraffe; Langhalsgiraffe;
Giraffe
f Girafe
i Giraffa

2549 ***Giraffa camelopardalis angolensis***
Artiodactyla - Giraffidae
e Angolan Giraffe; Smoky Giraffe
d Angola-Giraffe
f Girafe d'Angola

2550 ***Giraffa camelopardalis antiquorum***
Artiodactyla - Giraffidae
e Kordofan Giraffe
d Kordofan-Giraffe
f Girafe de Kordofan

2551 ***Giraffa camelopardalis
camelopardalis***
Artiodactyla - Giraffidae
e Nubian Giraffe
d Nubische Giraffe
f Girafa de Nubie
i Giraffa del Nubia

2552 ***Giraffa camelopardalis giraffa***
Artiodactyla - Giraffidae
e Cape Giraffe; Southern Giraffe

d Kap-Giraffe
f Girafe du Cap

2553 ***Giraffa camelopardalis peralta***
Artiodactyla - Giraffidae
e Nigerian Giraffe; West African
Giraffe
d Tschad-Giraffe; Westafrikanische
Giraffe
f Girafe du Niger
i Giraffa del Ciad; Giraffa del Sahel

2554 ***Giraffa camelopardalis reticulata***
Artiodactyla - Giraffidae
e Reticulated Giraffe
d Netzgiraffe
f Girafe réticulée
i Giraffa samburu; Giraffa reticulata

2555 ***Giraffa camelopardalis rothschildii***
Artiodactyla - Giraffidae
e Rothschilds Giraffe; Baringo Giraffe
d Uganda-Giraffe
f Girafe de Rothschild; Girafe Baringo

2556 ***Giraffa camelopardalis thornicrofti***
Artiodactyla - Giraffidae
e Rhodesian Giraffe; Thornicroft's
Giraffe
f Girafe de Thornicroft; Girafe de la
Rhodésie

2557 ***Giraffa camelopardalis tippelskirchi***
Artiodactyla - Giraffidae
e Masai Giraffe
d Massai-Giraffe
f Girafe masai
i Giraffa masai

2558 ***Giraffidae***
Artiodactyla
e Giraffes; Giraffes and Okapis
d Giraffen
f Girafes; Girafidés
i Giraffidi

2559 *Glaucomys*
Rodentia - Sciuridae
e American Flying Squirrels; New-world Flying Squirrels
d Neuweltliche Gleithörnchen
f Polatouches
i Scoiattoli americani di volo

2560 *Glaucomys sabrinus*
Rodentia - Sciuridae
e Northern Flying Squirrel
d Nördliches Gleithörnchen
f Grand polatouche; Écureuil volant du Nord
i Glaucomio del nord

2561 *Glaucomys sabrinus coloratus*
Rodentia - Sciuridae
e Carolina Northern Flying Squirrel
d Carolina-Gleithörnchen

2562 *Glaucomys sabrinus fuscus*
Rodentia - Sciuridae
e Virginia Northern Flying Squirrel
d Virginia-Gleithörnchen

2563 *Glaucomys volans*
Rodentia - Sciuridae
e Eastern Flying Squirrel
d Assapan; Nordamerikanisches Zwerggleithörnchen; Südliches Gleithörnchen
f Écureuil volant du Sud; Petit polatouche
i Scoiatttolo volante

2564 *Glaucomys volans querceti*
Rodentia - Sciuridae
e Southern Flying Squirrel (NA)

Glauconycteris syn. *Chalinolobus* q.v.

2565 **Gliridae**
Rodentia

e Dormice
d Schläfer
f Gliridés
i Gliridi

2566 *Glironia*
Didelphimorphia - Didelphidae
e Brush-tailed Opossums; Bushy-tailed Opossums
d Buschschwanzbeutelratten

2567 *Glironia venusta*
Didelphimorphia - Didelphidae
e Brush-tailed Opossum; Bushy-tailed Opossum
d Buschschwanzbeutelratte; Peri-Buschschwanzbeutelratte

2568 *Glirurus*
Rodentia - Gliridae
e Japanese Dormice
d Japanische Schläfer
f Loirs du Japon; Muscardins du Japon

2569 *Glirurus japonicus*
Rodentia - Gliridae
e Japanese Dormouse
d Japanischer Schläfer
f Muscardin du Japon

2570 *Glis*
Rodentia - Gliridae
e Fat Dormice
d Siebenschläfer
f Loirs
i Ghiri

2571 *Glis glis*
Rodentia - Gliridae
e Fat Dormouse; Edible Dormouse; Common Dormouse; Squirrel-tailed Dormouse; Dormouse
d Große Haselmaus; Bilch; Siebenschläfer; Mitteleuropäischer Siebenschläfer

f   Loir; Loir commun; Loir gris; Loir
     ordinaire; Loir vulgaire
i   Ghiro

2572   *Glischropus*
      Chiroptera - Vespertilionidae
e   Thick-thumbed Bats

2573   *Glischropus javanicus*
      Chiroptera - Vespertilionidae
e   Javan Pipistrelle; Javan Thick-
     thumbed Bat; Javan Thick-thumbed
     Pipistrelle

2574   *Glischropus tylopus*
      Chiroptera - Vespertilionidae
e   Thick-thumbed Bat; Thick-thumbed
     Pipistrelle; Common Thick-thumbed
     Bat

2575   *Globicephala*
      Cete - Delphinidae
e   Pilot Whales
d   Grindwale; Rundkopfwale;
     Schwarzwale
f   Globicéphales
i   Globicefali

      Globicephala edwardi syn. *G. melas*
      q.v.

2576   *Globicephala macrorhynchus*
      Cete - Delphinidae
e   Short-finned Blackfish; Indian Pilot
     Whale; Short-finned Pilot Whale;
     Pothead; Pothead Whale; Tropical
     Pilot Whale; Shortfin Pilot Whale;
     Pilot Whale
d   Indischer Grindwal;
     Kurzflossengrindwal
f   Globicéphale d'Inde; Globicéphale
     tropical
i   Balaena pilota; Globicefalo di Gray;
     Globicefalo indiano; Balena pilota
     pinna corta; Balena pilota dalla pinna
     corta

2577   *Globicephala melas*
      Cete - Delphinidae
e   Blackfish; Atlantic Blackfish;
     Common Blackfish; Common Pilot
     Whale; Long-finned Pilot Whale;
     Northern Pilot Whale; Driving
     Whale; Calling Whale; Social Whale;
     Howling Whale
d   Schwarzwal; Grindwal;
     Gewöhnlicher Grindwal;
     Langflossengrindwal
f   Grinde; Globicéphale conducteur;
     Globicéphale noir; Globicéphale
     commun
i   Globicefalo dalle pinne lunghe;
     Globicefalo; Delfino pilota;
     Globicefalo nero

2578   *Glossophaga*
      Chiroptera - Phyllostomidae
e   Long-tongued Bats; Shrew-faced
     Bats
d   Langzungen; Langzüngler
f   Glossophages

      Glossophaga alticola syn. *G. leachi*
      q.v.

2579   *Glossophaga commissarisi*
      Chiroptera - Phyllostomidae
e   Commissari's Long-tongued Bat;
     Brown Long-tongued Bat
d   Blumenfledermaus

2580   *Glossophaga leachii*
      Chiroptera - Phyllostomidae
e   Davis's Long-tongued Bat; Gray's
     Long-tongued Bat; Western Long-
     tongued Bat
f   Glossophage de Leach

2581   *Glossophaga longirostris*
      Chiroptera - Phyllostomidae
e   Greater Long-tongued Bat; Miller's
     Long-tongued Bat; Trinidad Long-
     tongued Bat
f   Glossophage de Miller

Glossophaga morenoi syn. *G. leachii* q.v.

2582   Glossophaga soricina
Chiroptera - Phyllostomidae
e   Common Long-tongued Bat; Lesser Pallas's Longue-tongued Bat; Pallas's Long-tongued Bat; Pallas's Long-nosed Bat
d   Spitzmauslangzüngler
f   Glossophage de Pallas

Glyphonycteris syn. *Micronycteris* q.v.

2583   Glyphotes
Rodentia - Sciuridae
e   Bornean Squirrels; Thomas's Pygmy Squirrels; Sculptor Squirrels
d   Borneo-Zwerghörnchen

2584   Glyphotes simus
Rodentia - Sciuridae
e   Bornean Squirrel; Pygmy Ground Squirrel; Red-bellied Sculptor Squirrel; Sculptor Squirrel
d   Borneo-Zwerghörnchen

2585   Golunda
Rodentia - Muridae
e   Indian Bush Rats; Coffee Rats
d   Kaffeeratten

2586   Golunda ellioti
Rodentia - Muridae
e   Indian Bush Rat
d   Indische Buschratte

2587   Golunda ellioti newara
Rodentia - Muridae
e   Ceylon Highland Bushrat
d   Ceylon-Hochlandbuschratte

Gorgon syn. *Connochaetes* q.v.

2588   Gorilla
Primates - Hominidae
e   Gorillas
d   Gorillas
f   Gorilles
i   Gorilla

2589   Gorilla beringei
Primates - Hominidae
e   Eastern Gorilla; Mountain Gorilla
d   Berggorilla
f   Gorille de montagne; Gorille de l'Est
i   Gorilla di montagna

2590   Gorilla beringei graueri
Primates - Hominidae
e   Eastern Lowland Gorilla; Grauer's Gorilla
d   Östlicher Flachlandgorilla; Grauer Gorilla
f   Gorille des plaines occidentales
i   Gorilla di pianura occidentale

2591   Gorilla gorilla
Primates - Hominidae
e   Western Gorilla
d   Gorilla; Flachlandgorilla
f   Gorille
i   Gorilla

2592   Gorilla gorilla diehli
Primates - Hominidae
e   Cross River Gorilla
f   Gorille de la Cross; Gorille de Cross River; Gorille du rivière Cross

2593   Gorilla gorilla gorilla
Primates - Hominidae
e   Lowland Gorilla; Western Lowland Gorilla
d   Westlicher Flachlandgorilla
f   Gorille des plaines de l'Ouest; Gorille des plaines orientales
i   Gorilla di pianura; Gorilla orientale

2594   Gracilinanus
Didelphimorphia - Didelphidae
e   Gracile Mouse Opossums
d   Zwergbeutelratten; Zarte
Mäuseopossums

2595   Gracilinanus aceramarcae
Didelphimorphia - Didelphidae
e   Aceramarca Mouse Opossum;
Aceramarca Gracile Mouse Opossum
d   Aceramarca-Zwergbeutelratte; Zartes
Opossum; Acemeraca-Maus

2596   Gracilinanus agilis
Didelphimorphia - Didelphidae
e   Agile Gracile Mouse Opossum ;
Agile Mouse Opossum; South
American Mouse Opossum
d   Schlankzwergbeutelratte;
Bewegliches Zartes Mäuseopossum

2597   Gracilinanus dryas
Didelphimorphia - Didelphidae
e   Wood Sprite Gracile Mouse
Opossum
d   Höhenzwergbeutelratte

2598   Gracilinanus emiliae
Didelphimorphia - Didelphidae
e   Emilia's Gracile Mouse Opossum;
Long-tailed Gracile Mouse Opossum
d   Emilias Maus

2599   Gracilinanus ignitus
Didelphimorphia - Didelphidae
e   Red-bellied Gracile Mouse Opossum

2600   Gracilinanus marica
Didelphimorphia - Didelphidae
e   Northern Gracile Mouse Opossum
d   Marica-Zwergbeutelratte

2601   Gracilinanus microtarsus
Didelphimorphia - Didelphidae
e   Brazilian Gracile Mouse Opossum
d   Kleinfußzwergbeutelratte;

Brasilianisches Zartes Mausopossum

2602   Gracilinanus perijae
Didelphimorphia - Didelphidae
e   Sierra di Perijá Gracile Mouse
Opossum

Grammogale syn. *Mustela* q.v.

2603   Grammomys
Rodentia - Muridae
e   Thicket Rats; Forest Mice; Bush
Thicket Rats; African Thicket Rats
d   Afrikanische Buschklettermäuse;
Akazienmäuse
f   Muscardins; Souris arboricoles

2604   Grammomys aridulus
Rodentia - Muridae
e   Arid Thicket Rat

2605   Grammomys buntingi
Rodentia - Muridae
e   Bunting's Thicket Rat

2606   Grammomys caniceps
Rodentia - Muridae
e   Grey-headed Thicket Rat
d   Afrikanische Akazienmaus

2607   Grammomys cometes
Rodentia - Muridae
e   Mozambique Thicket Rat;
Mozambique Woodland Mouse

2608   Grammomys dolichurus
Rodentia - Muridae
e   Woodland Thicket Rat; Woodland
Mouse
d   Afrikanische Buschklettermaus

2609   Grammomys dryas
Rodentia - Muridae
e   Forest Thicket Rat

2610 **Grammomys gigas**
Rodentia - Muridae
e Giant Thicket Rat

2611 **Grammomys ibeanus**
Rodentia - Muridae
e Ruwenzori Thicket Rat

2612 **Grammomys macmillani**
Rodentia - Muridae
e Macmillan's Thicket Rat

2613 **Grammomys minnae**
Rodentia - Muridae
e Ethiopian Thicket Rat

2614 **Grammomys rutilans**
Rodentia - Muridae
e Shining Thicket Rat

Grampidelphis syn. *Grampus* q.v.

2615 **Grampus**
Cete - Delphinidae
e Risso's Dolphins; Grampuses
d Rundkopfdelfine
f Dauphins gris
i Grampi

2616 **Grampus griseus**
Cete - Delphinidae
e Risso's Dolphin; White-headed
Dolphin; Grey Dolphin; Grey
Grampus; Mottled Grampus; Risso's
White-headed Dolphin
d Rundkopfdelfin; Gestreifter Delfin;
Risso-Delfin; Gramper
f Dauphin gris; Dauphin gris de Risso;
Dauphin de Risso; Grampus
i Grampo griseo; Grampo; Grampo
grigio; Delfino di Risso

Grampus rectipinna syn. *G. griseus*
q.v.

2617 **Graomys**
Rodentia - Muridae
e Leaf-eared Mice; Chaco Mice
d Großohrmäuse; Blattohrmäuse

2618 **Graomys domorum**
Rodentia - Muridae
e Pale Leaf-eared Mouse

2619 **Graomys edithae**
Rodentia - Muridae
e Edith's Leaf-eared Mouse

2620 **Graomys griseoflavus**
Rodentia - Muridae
e Grey Leaf-eared Mouse
d Graugelbe Großohrmaus

2621 **Graomys olrogi**
Rodentia - Muridae
e Olrog's Chacao Mouse

2622 **Graomys pearsoni**
Rodentia - Muridae
e Pearson's Chaco Mouse

2623 **Graomys roigi**
Rodentia - Muridae
e Roig's Chaco Mouse

2624 **Graphiurus**
Rodentia - Gliridae
e African Dormice
d Afrikanische Bilche; Afrikanische
Schläfer ; Pinselschwanzbilche
f Loirs africains

2625 **Graphiurus christyi**
Rodentia - Gliridae
e Christy's Dormouse

2626 **Graphiurus crassicaudatus**
Rodentia - Gliridae
e Jentink's Dormouse

2627   *Graphiurus hueti*
Rodentia - Gliridae
e   Huet's Dormouse; Cameroon
Dormouse
d   Kamerun-Mausschläfer

2628   *Graphiurus kelleni*
Rodentia - Gliridae
e   Kellen's Dormouse

2629   *Graphiurus lorraineus*
Rodentia - Gliridae
e   Lorrain Dormouse

2630   *Graphiurus microtis*
Rodentia - Gliridae
e   Small-eared Dormouse

2631   *Graphiurus monardi*
Rodentia - Gliridae
e   Monard's Dormouse

2632   *Graphiurus murinus*
Rodentia - Gliridae
e   Woodland Dormouse; Common
African Dormouse
d   Pinselschwanzbilch; Afrikanischer
Mausschläfer
f   Lérot de savane
i   Minighiro africano; Minighiro dei
boschi

2633   *Graphiurus ocularis*
Rodentia - Gliridae
e   Spectacled Dormouse

2634   *Graphiurus olga*
Rodentia - Gliridae
e   Olga's Dormouse

2635   *Graphiurus parvus*
Rodentia - Gliridae
e   Savanna Dormouse; Lesser Savanna
Dormouse; African Pygmy
Dormouse; Savannah Dormouse;

Lesser Savannah Dormouse
d   Afrikanischer Zwergschläfer

2636   *Graphiurus platyops*
Rodentia - Gliridae
e   Rock Dormouse

2637   *Graphiurus rupicola*
Rodentia - Gliridae
e   Stone Dormouse

2638   *Graphiurus surdus*
Rodentia - Gliridae
e   Silent Dormouse

Grison syn. *Galictis* q.v.

Grisonella syn. *Galictis* q.v.

Guerlinguetus syn. *Sciurus* q.v.

2639   *Gulo*
Carnivora - Mustelidae
e   Wolverines; Gluttons
d   Vielfraße
f   Gloutons
i   Ghiottone

2640   *Gulo gulo*
Carnivora - Mustelidae
e   Wolverine; Glutton; Wolverene;
Carcajou
d   Vielfraß; Vielfraßmarder; Gemeiner
Vielfraß; Nordischer Vielfraß;
Bärenmarder; Järv
f   Glouton arctique; Glouton boréale;
Glouton
i   Ghiottone; Volverina

2641   *Gulo gulo gulo*
Carnivora - Mustelidae
e   Old World Wolverine
d   Europäischer Vielfraß
f   Glouton d'Europe

2642 **Gulo gulo luscus**
Carnivora - Mustelidae
e North American Wolverine; New
World Wolverine
d Nordamerikanischer Vielfraß
f Glouton d'Amérique du Nord
i Ghiottone; Volverina

2643 **Gulo gulo vancouverensis**
Carnivora - Mustelidae
e Vancouver Island Wolverine

Gunomys syn. *Bandicota* q.v.

2644 **Gymnobelideus**
Diprotodontia - Petauridae
e Leadbeater's Possums (ANZ;
Leadbeater's Opossums; Bass River
Possums
d Hörnchenbeutler ;
Hörnchenkletterbeutler
f Opposums de Leadbeater
i Possum di Leadbeater

2645 **Gymnobelideus leadbeateri**
Diprotodontia - Petauridae
e Leadbeater's Possum (ANZ);
Leadbeater's Fairy Possum;
Leadbeater's Phalanger
d Hörnchenbeutler;
Hörnchenkletterbeutler
f Opposum de Leadbeater
i Possum di Leadbeater

2646 **Gymnuromys**
Rodentia - Muridae
e Voalavoanalas
d Voalavoanalas
f Voalavoanalas américains

2647 **Gymnuromys roberti**
Rodentia - Muridae
e Voalavoanala
d Voalavoanala; Roberts
Nacktschwanzmaus
f Voalavoanala américain

Gymnurus syn. *Echinosorex* **q.b.**

Gyomys syn. *Pseudomys* q.v.

H

2648 Habromys
Rodentia - Muridae
e Crested-tailed Deer Mice

2649 Habromys chinanteco
Rodentia - Muridae
e Chinanteco Deer Mouse

Habromys ixtanli syn. **H. lepturus**
q.v.

2650 Habromys lepturus
Rodentia - Muridae
e Slender-tailed Deer Mouse

2651 Habromys lophurus
Rodentia - Muridae
e Crested-tailed Mouse; Crested-tailed
Deer Mouse
d Mähnenschwanzmaus

2652 Habromys simulatus
Rodentia - Muridae
e Jico Deer Mouse
f Souris mexicaine

2653 Hadromys
Rodentia - Muridae
e Manipur Bush Rats
d Manipur-Buschratten

2654 Hadromys humei
Rodentia - Muridae
e Manipur Bush Rat; Hume's Rat
d Manipur-Buschratte

Hadrosciurus syn. **Sciurus** q.v.

2655 Haeromys
Rodentia - Muridae
e Pygmy Tree Mice; Pygmy Tree Rats;
Ranee Mice
d Zwergbaummäuse

2656 Haeromys margarettae
Rodentia - Muridae
e Ranee Mouse; Greater Ranee Mouse
d Ranee-Baummaus

2657 Haeromys minahassae
Rodentia - Muridae
e Minahassa Ranee Mouse; Celebes
Pygmy Tree Mouse; Sulawesi Pygmy
Tree Mouse
d Sulawesi-Knirpsratte

2658 Haeromys pusillus
Rodentia - Muridae
e Lesser Ranee Mouse
d Kleine Raneemaus

2659 Halichoerus
Carnivora - Phocidae
e Grey Seals
d Kegelrobben
f Halichères; Phoques gris
i Foche grigie

2660 Halichoerus grypus
Carnivora - Phocidae
e Scilly Seal; Atlantic Seal; Horsehead
Seal; Grey Seal; Atlantic Grey Seal;
Hook-nosed Sea-pig
d Kegelrobbe; Dickschnauzige
Kegelrobbe; Krummnasige
Kegelrobbe; Langschauzige
Kegelrobbe; Rauhe Kegelrobbe;
Grauer Seehund; Utsel; Utzel; Urzel;
Urtzel
f Grand phoquie; Phoque gris;
Halichère gris
i Foca grigia

2661 **Halichoerus grypus balticus**
Carnivora - Phocidae
e Baltic Grey Seal
d Ostsee-Kegelrobbe

2662 **Hapalemur**
Primates - Lemuridae
e Gentle Lemurs; Broad-nosed
Lemurs; Bamboo Lemurs; Grey
Bamboo Lemurs; Lesser Bamboo
Lemurs
d Halbmakis
f Hapalémurs
i Apalemuri

2663 **Hapalemur alaotrensis**
Primates - Lemuridae
e Alaotran Gentle Lemur; Alaotran
Bamboo Lemur
d Alaotra Halbmaki
f Hapalémur gris d'Alaotra

2664 **Hapalemur aureus**
Primates - Lemuridae
e Golden Bamboo Lemur; Bandro
d Goldener Bambuslemur
f Hapalémur doré
i Apalemure dorato

2665 **Hapalemur griseus**
Primates - Lemuridae
e Grey Gentle Lemur; Grey Lemur;
Grey Bamboo Lemur
d Grauer Halbmaki; Kleiner Halbmaki;
Grauer Bambuslemur
f Hapalémur gris
i Apalemure grigio; Bocombal

2666 **Hapalemur occidentalis**
Primates - Lemuridae
e Sambirano Bamboo Lemur
d Sambirano-Halbmaki
f Hapalémur gris occidental

Hapalemur simus syn. *Prolemur
simus* q.v.

2667 **Hapalomys**
Rodentia - Muridae
e Marmoset Rats; Asiatic Climbing
Rats; Marmoset Mice; Asiatic
Climbing Mice
d Asiatische Klettermaus

2668 **Hapalomys delacouri**
Rodentia - Muridae
e Delacour's Marmoset Rat; Lesser
Marmoset Mouse

2669 **Hapalomys longicaudatus**
Rodentia - Muridae
e Marmoset Rat; Asiatic Climbing Rat;
Asiatic Climbing Mouse; Marmoset
Mouse
d Marmosettratte

2670 **Haplonycteris**
Chiroptera - Pteropodidae
e Philippine Pygmy Bats; Philippine
Pygmy Fruit Bats
d Sibuyanische Pygmäenflughunde

2671 **Haplonycteris fischeri**
Chiroptera - Pteropodidae
e Philippine Pygmy Bat; Philippine
Pygmy Fruit Bat

d Sibuyanische Pygmäenflughund

2672 **Harpiocephalus**
Chiroptera - Vespertilionidae
e Hairy-winged Bats; Tube-nosed Bats

2673 **Harpiocephalus harpia**
Chiroptera - Vespertilionidae
e Hairy-winged Bat; Tube-nosed Bat

Harpiola syn. *Murina* q.v.

2674 **Harpionycteris**
Chiroptera - Pteropodidae
e Harpy Fruit Bats

d	Spitzzahnflughunde
f	Harpionyctères

2675 *Harpionycteris celebensis*
Chiroptera - Pteropodidae
e Sulawesi Harpy Fruit Bat

2676 *Harpionycteris whiteheadi*
Chiroptera - Pteropodidae
e Whitehead's Harpy Fruit Bat; Harpy Fruit Bat
d Spitzzahnflughund
f Harpionyctère de Whitehead

2677 *Heimyscus*
Rodentia - Muridae
e African Smoky Mice
d Afrikanische Rauchmäuse

2678 *Heimyscus fumosus*
Rodentia - Muridae
e African Smoky Mouse
d Afrikanische Rauchmaus

2679 *Helarctos*
Carnivora - Ursidae
e Sun Bears; Malayan Sun Bears
d Malayen-Bären; Sonnenbären
f Ours malais ; Helarctos
i Orsi malesi

2680 *Helarctos malayanus*
Carnivora - Ursidae
e Malayan Bear; Malayan Sun Bear; Sun Bear
d Malayen-Bär; Sonnenbär
f Ours malais
i Orso malese; Biruang

Helictis syn. *Melogale* q.v.

2681 *Heliophobius*
Rodentia - Bathyergidae
e Sand Rats; Silvery Molerats

d Erdbohrer
f Rats-taupes

2682 *Heliophobius argenteocinereus*
Rodentia - Bathyergidae
e Sand Rat; Silvery Molerat; Silky Molerat; Blesmol
d Silbergrauer Erdbohrer
f Rat-taupe

2683 *Heliosciurus*
Rodentia - Sciuridae
e Sun Squirrels
d Graufußhörnchen ; Sonnenhörnchen
f Écureuils de Gambie; Écureuils d'Afrique noire
i Scoiattoli del sole

2684 *Heliosciurus gambianus*
Rodentia - Sciuridae
e Gambian Sun Squirrel; West African Sun Squirrel
d Graufußhörnchen
f Écureuil de Gambie; Hélioscure de Gambie
i Scoiattolo del sole di Gambia

2685 *Heliosciurus mutabilis*
Rodentia - Sciuridae
e Mutable Sun Squirrel
d Variables Sonnenhörnchen

2686 *Heliosciurus punctatus*
Rodentia - Sciuridae
e Small Sun Squirrel; Temminck's Spotted Squirrel
d Kleines Sonnenhörnchen

2687 *Heliosciurus rufobrachium*
Rodentia - Sciuridae
e Red-legged Sun Squirrel
d Rotfüßiges Sonnenhörnchen; Rotarmsonnenhörnchen
f Hélioscure aux pattes rousses

2688   *Heliosciurus ruwenzorii*
Rodentia - Sciuridae
e   Ruwenzori Sun Squirrel; Montane
Sun Squirrel
d   Ruwenzori-Sonnenhörnchen
f   Hélioscure du Ruwenzori

2689   *Heliosciurus undulatus*
Rodentia - Sciuridae
e   Zanj Sun Squirrel
d   Zanj-Sonnenhörnchen

2690   *Helogale*
Carnivora - Herpestidae
e   Dwarf Mongooses
d   Zwergmangusten
f   Mangoustes naines
i   Manguste nane

2691   *Helogale dybowskii*
Carnivora - Herpestidae
e   Pousargues Mongoose; Savannah
Mongoose
d   Listige Manguste
f   Mangouste de Dybowski
i   Mangusta nana di Dybowski

2692   *Helogale hirtula*
Carnivora - Herpestidae
e   Desert Dwarf Mongoose; Somali
Dwarf Mongoose; Dwarf Mongoose;
African Tropical Savannah
Mongoose; Desert Mongoose;
Eastern Dwarf Mongoose
d   Östliche Zwergmanguste
i   Mangusta nana somala

2693   *Helogale parvula*
Carnivora - Herpestidae
e   Dwarf Mongoose; Southern Dwarf
Mongoose; Desert Mongoose
d   Südliche Zwergichneumon;
Südlicher Zwergichneumon
f   Mangouste naine
i   Mangusta nana; Mangusta
meridionale

2694   *Hemibelideus*
Diprotodontia - Petauridae
e   Lemuroid Ringtail Possums
d   Lemurenringschwanzbeutler ;
Lemurenringbeutler

2695   *Hemibelideus lemuroides*
Diprotodontia - Petauridae
e   Lemuroid Ringtail Possum; Brush-
tipped Ringtail; Brushy-tailed
Ringtail; Brush-tipped Ring-tailed
Phalanger; Lemuroid Ringtail
d   Lemurenringschwanzbeutler;
Lemurenringbeutler

2696   *Hemicentetes*
Soricomorpha - Tenrecidae
e   Streaked Tenrecs; Streaked Tanrecs
d   Halbborstenigel ; Gelbstreifentanreks
f   Hemicentètes; Tenrecs zebrés
i   Tenrec

2697   *Hemicentetes nigriceps*
Soricomorpha - Tenrecidae
e   Highland Streaked Tenrec; Highland
Streaked Tanrec
d   Schwarzkopftanrek

2698   *Hemicentetes semispinosus*
Soricomorpha - Tenrecidae
e   Lowland Streaked Tenrec; Lowland
Streaked Tanrec; Streaked Tenrec;
Streaked Tanrec
d   Halbborstenigel; Gelbstreifentanrek;
Streifentanrek
f   Tenrec zebré des terres basses
i   Tenrec striato

2699   *Hemiechinus*
Erinceomorpha - Erinaceidae
e   Eared Hedgehogs; Long-eared
Hedgehogs; Long-eared Desert
Hedgehogs; Steppe Hedgehogs;
Desert Hedgehogs
d   Halbstachler ; Ohrenigel ;
Großohrigel ; Langohrigel

f   Hérissons oreillards
i   Ricci

2700   *Hemiechinus aethiopicus*
         Erinceomorpha - Erinaceidae
e   Desert Hedgehog; Ethiopian
      Hedgehog; African Desert
      Hedgehog; Ethiopian Long-eared
      Hedgehog
d   Äthiopischer Igel; Afrikanischer
      Wüstenigel; Äthiopischer
      Wüstenigel; Äthiopischer
      Langohrigel
f   Hérisson du désert d'Éthiope;
      Hérisson aux longues oreilles
      d'Éthiopie; Hérisson oreillard
      d'Éthiopie; Hérisson du désert
i   Riccio del deserto

2701   *Hemiechinus auritus*
         Erinceomorpha - Erinaceidae
e   Long-eared Hedgehog; Long-eared
      Desert Hedgehog
d   Langohrigel; Ohrenigel; Großohrigel;
      Europäischer Langohrigel
f   Hérisson oreillard
i   Riccio orecchiuto; Riccio dalle
      orecchie lunghe

2702   *Hemiechinus auritus aegypticus*
         Erinceomorpha - Erinaceidae
e   Egyptian Long-eared Hedgehog
d   Ägyptischer Langohrigel

2703   *Hemiechinus auritus dorotheae*
         Erinceomorpha - Erinaceidae
e   Cyprus Long-eared Hedgehog

2704   *Hemiechinus auritus persicus*
         Erinceomorpha - Erinaceidae
e   Iranian Long-eared Hedgehog
d   Persischer Langohrigel

2705   *Hemiechinus collaris*
         Erinceomorpha - Erinaceidae
e   Indian Long-eared Hedgehog

d   Indischer Langohrigel

Hemiechinus dauricus syn.
Erinaceus dauricus q.v.

Hemiechinus hughi syn. *Erinaceus
hughi* q.v.

2706   *Hemiechinus hypomelas*
         Erinceomorpha - Erinaceidae
e   Brandt's Hedgehog
d   Brandts Igel
f   Hérisson de Brandt

2707   *Hemiechinus micropus*
         Erinceomorpha - Erinaceidae
e   Indian Hedgehog; Pale Hedgehog
d   Indischer Igel; Indischer Wüstenigel

2708   *Hemiechinus nudiventris*
         Erinceomorpha - Erinaceidae
e   Bare-bellied Hedgehog; Southern
      Indian Hedgehog
d   Nacktbauchigel

Hemiechinus sylvaticus syn.
Erinaceus hughi q.v.

2709   *Hemigalus*
         Carnivora - Viverridae
e   Hardwick's Banded Palm Civets;
      Hardwick's Civets
d   Bänderroller
f   Civettes palmistes aux bandes
i   Civetta dalle palme fasciate

2710   *Hemigalus derbyanus*
         Carnivora - Viverridae
e   Banded Palm Civet; Civet
d   Bänderroller
f   Civette palmiste aux bandes de
      Derby; Marte des palmies; Martre
      des palmier
i   Civetta dalle palme fasciata

Hemigalus hosei syn. *Diplogale hosei* q.v.

2711 Hemitragus
Artiodactyla - Bovidae
e Tahrs
d Tahre
f Tahrs
i Tar

2712 Hemitragus hylocrius
Artiodactyla - Bovidae
e Nilgiri Tahr
d Nilgiri-Tahr
f Tahr de Nilgiri

2713 Hemitragus jayakari
Artiodactyla - Bovidae
e Arabian Tahr
d Arabischer Tahr
f Tahr d'Arabie

2714 Hemitragus jemlahicus
Artiodactyla - Bovidae
e Himalajan Tahr
d Himalaja-Tahr
f Tahr de l'Himalaja; Jharal
i Tar; Tar dell'Himalaja

Herpailurus syn. *Felis* q.v.

2715 Herpestes
Carnivora - Herpestidae
e Common Mongooses; Grey Mongooses; Typical Mongooses
d Echte Mungos; Mangusten
f Mangoustes
i Manguste

2716 Herpestes auropunctatus
Carnivora - Herpestidae
e Small Indian Mongoose; Gold-spotted Mongoose; Javan Mongoose; Gold-speckled Mongoose
d Goldstaubmanguste

f Petite Mangouste indienne

2717 Herpestes brachyurus
Carnivora - Herpestidae
e Short-tailed Mongoose
d Kurzschwanzmanguste
f Mangouste aux queue courte
i Mangusta a coda corta

2718 Herpestes brachyurus flavidens
Carnivora - Herpestidae
e Highland Ceylon Brown Mongoose
i Mangusta dalla coda corta

2719 Herpestes brachyurus fuscus
Carnivora - Herpestidae
e Indian Brown Mongoose; Gold-brown Mongoose
f Mangouste brune de l'Inde
i Mangusta a coda corta dell'India

2720 Herpestes brachyurus maccarthiae
Carnivora - Herpestidae
e Northern Ceylon Brown Mongoose

2721 Herpestes brachyurus rubidior
Carnivora - Herpestidae
e Western Ceylon Brown Mongoose

2722 Herpestes edwardsi
Carnivora - Herpestidae
e Indian Common Mongoose; Indian Grey Mongoose; Indian Mongoose; Grey Mongoose
d Indischer Mungo
f Mangouste d'Edwards; Mungo indien
i Mangusta grigia indiana; Mangusta grigia dell'India; Mangusta indiana grigia

2723 Herpestes edwardsi lanka
Carnivora - Herpestidae
e Common Ceylon Grey Mongoose
d Brauner Sri Lanka Hochland Mungo

2724   *Herpestes flavescens*
  Carnivora - Herpestidae
e Black Slender Mongoose
d Schwarze Schlankmanguste

2725   *Herpestes ichneumon*
  Carnivora - Herpestidae
e Ichneumon; Ichneumon Mongoose;
  African Mongoose; Egyptian
  Mongoose; Large Grey Mongoose
d Eigentlicher Ichneumon; Ichneumon;
  Europäische Manguste; Mungo
f Mangouste ichneumon; Mangouste;
  Rat de pharaon
i Mangusta egiziana; Icneumone;
  Mangusta icneumone

2726   *Herpestes ichneumon widdringtoni*
  Carnivora - Herpestidae
e Spanish Mongoose
d Spanischer Ichneumon
f Mangouste d'Espagne

2727   *Herpestes javanicus*
  Carnivora - Herpestidae
e Javan Mongoose; Small Asian
  Mongoose; Small Indian Mongoose
d Kleiner Mungo
f Mangouste de Java
i Mangusta di Giava

2728   *Herpestes naso*
  Carnivora - Herpestidae
e Greater Long-nosed Mongoose;
  Long-nosed Mongoose; Long-
  snouted Mongoose; Cameroon
  Mongoose
d Kongo-Manguste;
  Langnasenmanguste
f Mangouste des marais du Cameroun;
  Mangouste à long museau
i Mangusta dal muso lungo

2729   *Herpestes palustris*
  Carnivora - Herpestidae
e Bengal Mongoose; Spotted
  Mongoose

d Bengal-Manguste
f Mangouste indiene
i Piccola mangusta indiana

2730   *Herpestes pulverulentus*
  Carnivora - Herpestidae
e Cape Grey Mongoose; Small Grey
  Mongoose
d Graue Kap-Manguste; Kap-
  Manguste; Kap-Ichneumon;
  Kleininchneumon
f Mangouste grise du Cap
i Mangusta grigia del Capo

2731   *Herpestes sanguineus*
  Carnivora - Herpestidae
e Slender Mongoose; Common Slender
  Mongoose; Black-tipped Mongoose
d Rote Manguste; Schlankmungo;
  Schlankichneumon;
  Schlankmanguste

f Mangouste rouge aux queue noire;
  Mangouste rouge; Mangouste naine
i Mangusta rossastra

2732   *Herpestes semitorquatus*
  Carnivora - Herpestidae
e Collared Mongoose
d Halsbandmanguste
f Mangouste aux collier

2733   *Herpestes smithi*
  Carnivora - Herpestidae
e Ruddy Mongoose
d Indische Rotmanguste
f Mangouste vermeille
i Mangusta di Smith

2734   *Herpestes smithi zeylanicus*
  Carnivora - Herpestidae
e Ceylon Ruddy Mongoose

2735   *Herpestes swalius*
  Carnivora – Herpestidae

e   Namaqua Slender Mongoose
d   Namaqua-Manguste

2736   **Herpestes urva**
Carnivora - Herpestidae
e   Crab-eating Mongoose
d   Krabbenmanguste
f   Mangouste crabière

2737   **Herpestes vitticollis**
Carnivora - Herpestidae
e   Stripe-necked Mongoose
d   Halsstreifenmanguste
f   Mangouste aux cou rayé
i   Mangusta a collo striato

2738   **Herpestidae**
Carnivora
e   Mongooses
d   Mangusten
f   Mangoustes; Herpestidés
i   Manguste; Erpestidi

Herpetomys syn. *Microtus* q.v.

Hesperomys syn. *Calomys* q.v.

2739   **Hesperoptenus**
Chiroptera - Vespertilionidae
e   False Serotines

2740   **Hesperoptenus blanfordi**
Chiroptera - Vespertilionidae
e   Blanford's Bat; Lesser False Serotine;
Blanford's False Serotine

2741   **Hesperoptenus doriae**
Chiroptera - Vespertilionidae
e   Peters's False Serotine; False
Serotine Bat

2742   **Hesperoptenus gaskelli**
Chiroptera - Vespertilionidae
e   Gaskell's False Serotine; Sulawesi
False Serotine

2743   **Hesperoptenus tickelli**
Chiroptera - Vespertilionidae
e   Tickell's Bat; Tickell's False Serotine
d   Tickells Fledermaus

2744   **Hesperoptenus tomesi**
Chiroptera - Vespertilionidae
e   Large False Serotine; Large False
Serotine Bat

Hesperosciurus syn. *Sciurus* q.v.

2745   **Heterocephalus**
Rodentia - Bathyergidae
e   Naked Molerats; Naked Sand Rats
d   Nacktmulle
f   Rats nus des sables; Rats-taupes nus
i   Eterocefali

2746   **Heterocephalus glaber**
Rodentia - Bathyergidae
e   Naked Molerat; Naked Sand Rat
d   Nacktmull
f   Rat nu des sables; Rat-taupe nu
i   Eterocefalo glabro

Heterogeomys syn. *Orthogeomys*
q.v.

2747   **Heterohyrax**
Hyracoidea - Procaviidae
e   Rock Hyraxes; Grey Hyraxes;
Yellow-spotted Hyraxes; Bush
Hyraxes
d   Erdferkel ; Buschschliefer
f   Damans gris; Damans des steppes
i   Iraci delle steppe

2748   **Heterohyrax antineae**
Hyracoidea - Procaviidae
e   Hoggar Hyrax; Ahaggar Hyrax

2749   **Heterohyrax brucei**
Hyracoidea - Procaviidae
e   Bruce's Hyrax; Small-toothed Rock

Hyrax; Yellow-spotted Dassie;
Yellow-spotted Hyrax
- *d*   Steppenschliefer
- *f*   Daman des steppes
- *i*   Irace delle Steppe; Procavia delle
  steppe

2750   Heteromys
Rodentia - Geomyidae
- *e*   Spiny Pocket Mice; Forest Spiny
  Pocket Mice
- *d*   Taschenstachelmäuse;
  Waldstacheltaschenmäuse
- *f*   Souris épineuses aux poche

2751   Heteromys anomalus
Rodentia - Geomyidae
- *e*   Trinidad Spiny Pocket Mouse; Spiny
  Pocket Mouse
- *d*   Taschenstachelmaus
- *f*   Souris sentinelle

2752   Heteromys australis
Rodentia - Geomyidae
- *e*   Southern Spiny Pocket Mouse
- *d*   Südliche Taschenstachelmaus

2753   Heteromys desmarestianus
Rodentia - Geomyidae
- *e*   Desmarest's Spiny Pocket Mouse;
  Motzorongo Spiny Pocket Mouse;
  Santo Domingo Spiny Pocket Mouse;
  Long-tailed Spiny Pocket Mouse;
  Forest Spiny Pocket Mouse

2754   Heteromys gaumeri
Rodentia - Geomyidae
- *e*   Gaumer's Spiny Pocket Mouse

2755   Heteromys goldmani
Rodentia - Geomyidae
- *e*   Goldman's Spiny Pocket Mouse
- *d*   Goldman-Stacheltaschenmaus
- *f*   Souris aux poche de Goldman

Heteromys lepturus syn. *H.
desmarestianus* q.v.

Heteromys longicaudatus *H.
demarestianus* q.v.

2756   Heteromys nelsoni
Rodentia - Geomyidae
- *e*   Nelson's Spiny Pocket Mouse

Heteromys nigricaudatus syn. *H.
desmarestianus* q.v.

2757   Heteromys oresterus
Rodentia - Geomyidae
- *e*   Mountain Spiny Pocket Mouse

Heteromys temporalis syn. *H.
desmarestianus* q.v.

2758   Hexaporotodon
Artiodactyla - Hippopotamidae
- *e*   Pygmy Hippotamuses
- *d*   Zwergflusspferde
- *f*   Hippopotames nains
- *i*   Ippopotami pigmei

2759   Hexaprotodon liberiensis
Artiodactyla - Hippopotamidae
- *e*   Pygmy Hippotamus
- *d*   Zwergflusspferd
- *f*   Hippopotame nain
- *i*   Ippopotamo pigmeo; Ippopotamo
  nano

2760   Hexaprotodon liberiensis heslopio
Artiodactyla - Hippopotamidae
- *e*   Eastern Pygmy Hippopotamus; Niger
  Pygmy Hippopotamus
- *d*   Östliches Zwergflusspferd; Niger-
  Zwergflusspferd
- *f*   Hippopotame nain de l'Est

2761   Hippocamelus
Artiodactyla - Cervidae
- *e*   Guemals; Huemuls; Andean Deer

  d   Anden-Hirsche; Gabelhirsche
  f   Cerfs des Andes
  i   Huemul

2762   *Hippocamelus antisensis*
      Artiodactyla - Cervidae
  e   Peruvian Guemal; Peruvian Huemul;
      Andean Deer; Northern Guemal;
      Northern Huemul; North Andean
      Deer
  d   Nordanden-Hirsch; Guemal;
      Nördlicher Anden-Hirsch
  f   Cerf andin; Guémal; Taruca; Cerf
      gris; Cerf des Andes septentrional;
      Cerf andin septentrional; Guémal de
      Pérou
  i   Huemul; Antisensis; Huemul del
      nord; Cervo delle Ande

2763   *Hippocamelus bisulcus*
      Artiodactyla - Cervidae
  e   Chilean Guemal; Chilean Huemul;
      Southern Huemul; Southern Guemal
  d   Südanden-Hirsch; Huemul; Südlicher
      Anden-Hirsch
  f   Huemul; Guémal du Chili
  i   Huemul dal sud

2764   *Hippopotamidae*
      Artiodactyla
  e   Hippopotamuses
  d   Flusstiere; Flusspferde
  f   Hippopotamidés
  i   Ippopotami; Ippopotamidi

2765   *Hippopotamus*
      Artiodactyla - Hippopotamidae
  e   Hippopotamuses
  d   Großflusspferde; Nilpferde
  f   Hippopotames

2766   *Hippopotamus amphibius*
      Artiodactyla - Hippopotamidae
  e   Hippopotamus; African River Horse
  d   Großflusspferd; Nilpferd; Flusspferd
  f   Hippopotame; Hippopotame

      amphibie
  i   Ippopotamo; Ippopotamo anfibio

2767   *Hipposideros*
      Chiroptera - Rhinolophidae
  e   Leaf-nosed Bats; Old World Leaf-
      nosed Bats; Greater Horseshoe Bats;
      Roundleaf Bats
  d   Rundblattnasen;
      Rundblattnasenfedermäuse
  f   Phyllorines; Rhinolophes

2768   *Hipposideros abae*
      Chiroptera - Rhinolophidae
  e   Aba Leaf-nosed Bat; Aba Roundleaf
      Bat

2769   *Hipposideros armiger*
      Chiroptera - Rhinolophidae
  e   Himalajan Leaf-nosed Bat; Great
      Himalajan Leaf-nosed Bat; Great
      Roundleaf Horseshoe Bat; Great
      Roundleaf Bat
  d   Himalaja-Rundblattnase
  f   Phyllorhine de l'Himalaja

2770   *Hipposideros ater*
      Chiroptera - Rhinolophidae
  e   Dusky Leaf-nosed Bat; Dusky
      Roundleaf Bat; Dusky Leafnose Bat
      (ANZ)

2771   *Hipposideros ater aruensis*
      Chiroptera - Rhinolophidae
  e   Eastern Dusky Leaf-nosed Bat

2772   *Hipposideros ater gilberti*
      Chiroptera - Rhinolophidae
  e   Western Dusky Leaf-nosed Bat

2773   *Hipposideros beatus*
      Chiroptera - Rhinolophidae
  e   Benito Leaf-nosed Bat; Benito Dwarf
      Leaf-nosed Bat; Dwarf Leaf-nosed
      Bat; Benito Roundleaved Bat

2774   ***Hipposideros bicolor***
       Chiroptera - Rhinolophidae
   e   Bicoloured Leaf-nosed Bat; Two-
        coloured Leaf-nosed Bat; Bicoloured
        Roundleaf Bat

2775   ***Hipposideros breviceps***
       Chiroptera - Rhinolophidae
   e   Short-headed Roundleaf Bat; Pagai
        Roundleaf Bat

2776   ***Hipposideros caffer***
       Chiroptera - Rhinolophidae
   e   Sundevall's Leaf-nosed Bat;
        Sundevall's Roundleaf Bat
   d   Gewöhnliche Rundblattnase
   f   Phyllorine de Cafrérie; Rhinolophe
        de Cafrerie

2777   ***Hipposideros calcaratus***
       Chiroptera - Rhinolophidae
   e   Spurred Leaf-nosed Bat; Spurred
        Roundleaf Bat

2778   ***Hipposideros camerunensis***
       Chiroptera - Rhinolophidae
   e   Greater Cyclops Bat; Greater
        Roundleaf Bat

2779   ***Hipposideros cervinus***
       Chiroptera - Rhinolophidae
   e   Gould's Leaf-nosed Bat; Fawn
        Roundleaf Bat; Fawn-coloured
        Roundleaf Bat; Fawn Leaf-nosed Bat

2780   ***Hipposideros cineraceus***
       Chiroptera - Rhinolophidae
   e   Least Leaf-nosed Bat; Least Round-
        leaf Horseshoe Bat; Ashy Roundleaf
        Bat

2781   ***Hipposideros commersoni***
       Chiroptera - Rhinolophidae
   e   Commerson's Leaf-nosed Bat;
        Commerson's Giant Leaf-nosed Bat;
        Commerson's Roundleaf Bat
   d   Riesenrundblattnase

   f   Phyllorhine de Commerson

2782   ***Hipposideros coronatus***
       Chiroptera - Rhinolophidae
   e   Large Mindanao Roundleaf Bat

2783   ***Hipposideros corynophyllus***
       Chiroptera - Rhinolophidae
   e   Telefomin Roundleaf Bat

2784   ***Hipposideros coxi***
       Chiroptera - Rhinolophidae
   e   Cox's Leaf-nosed Bat; Cox's
        Roundleaf Bat

2785   ***Hipposideros crumeniferus***
       Chiroptera - Rhinolophidae
   e   Timor Leaf-nosed Bat; Timor
        Roundleaf Bat

       Hipposideros cupidus syn. *H.*
       calcaratus q.v.

2786   ***Hipposideros curtus***
       Chiroptera - Rhinolophidae
   e   Short-tailed Leaf-nosed Bat; Short-
        tailed Roundleaf Bat

2787   ***Hipposideros cyclops***
       Chiroptera - Rhinolophidae
   e   Cyclops Bat; Cyclops Leaf-nosed
        Bat; Cyclops Roundleaf Bat
   f   Phyllorhine laineuse

2788   ***Hipposideros demissus***
       Chiroptera - Rhinolophidae
   e   Makira Roundleaf Bat

2789   ***Hipposideros diadema***
       Chiroptera - Rhinolophidae
   e   Diadem Leaf-nosed Bat; Large
        Malatan Leaf-nosed Bat; Diadem
        Bat; Diadem Rounleaf Horseshoe
        Bat; Diadem Roundleaf Bat; Diadem
        Leafnose Bat (ANZ)
   d   Diadem-Rundblattnase

2790 ***Hipposideros diadema inornatus***
Chiroptera - Rhinolophidae
e   Arnhem Leaf-nosed Bat

2791 ***Hipposideros dinops***
Chiroptera - Rhinolophidae
e   Fierce Leaf-nosed Bat; Fierce
     Roundleaf Bat

2792 ***Hipposideros doriae***
Chiroptera - Rhinolophidae
e   Borneo Roundleaf Bat

2793 ***Hipposideros dyacorum***
Chiroptera - Rhinolophidae
e   Dayak Leaf-nosed Bat

2794 ***Hipposideros edwardshilli***
Chiroptera - Rhinolophidae
e   John Hill's Roundleaf Bat

2795 ***Hipposideros fuliginosus***
Chiroptera - Rhinolophidae
e   Sooty Leaf-nosed Bat; Sooty
     Roundleaf Bat

2796 ***Hipposideros fulvus***
Chiroptera - Rhinolophidae
e   Fulvous Leaf-nosed Bat; Fulvous
     Roundleaf Bat

2797 ***Hipposideros galeritus***
Chiroptera - Rhinolophidae
e   Cantor's Roundleaf Bat; Common
     Roundleaf Horseshoe Bat; Cantor's
     Leaf-nosed Bat

2798 ***Hipposideros galeritus brachyotis***
Chiroptera - Rhinolophidae
e   Dekhan Leaf-nosed Bat
d   Dekhan-Rundblattnasenfledermaus

2799 ***Hipposideros halophyllus***
Chiroptera - Rhinolophidae
e   Thailand Roundleaf Bat

2800 ***Hipposideros hypophyllus***
Chiroptera - Rhinolophidae
e   Kolar Roundleaf Bat

2801 ***Hipposideros inexpectatus***
Chiroptera - Rhinolophidae
e   Crested Leaf-nosed Bat; Crested
     Roundleaf Bat

2802 ***Hipposideros jonesi***
Chiroptera - Rhinolophidae
e   Jones's Leaf-nosed Bat; Jones's
     Roundleaf Bat

2803 ***Hipposideros lamottei***
Chiroptera - Rhinolophidae
e   Lamotte's Roundleaf Bat
f   Chauve-souris de Mont Nimba

2804 ***Hipposideros lankadiva***
Chiroptera - Rhinolophidae
e   Indian Roundleaf Bat

2805 ***Hipposideros lankadiva lankadiva***
Chiroptera - Rhinolophidae
e   Great Ceylon Leaf-nosed Bat
d   Große Ceylon-
     Rundblattnasenfledermaus

2806 ***Hipposideros larvatus***
Chiroptera - Rhinolophidae
e   Horsfield's Leaf-nosed Bat;
     Intermediate Roundleaf Bat;
     Horsfield's Roundleaf Bat;
     Intermediate Leaf-nosed Bat

2807 ***Hipposideros lekaguli***
Chiroptera - Rhinolophidae
e   Lekagul's Leaf-nosed Bat; Large
     Asian Roundleaf Bat

2808 ***Hipposideros lylei***
Chiroptera - Rhinolophidae
e   Shield-faced Leaf-nosed Bat; Shield-
     faced Roundleaf Bat

2809   *Hipposideros macrobullatus*
Chiroptera - Rhinolophidae
e   Big-eared Roundleaf Bat; Macros
Roundleaf Bat

2810   *Hipposideros madurae*
Chiroptera - Rhinolophidae
e   Maduran Roundleaf Bat

2811   *Hipposideros maggietaylorae*
Chiroptera - Rhinolophidae
e   Maggie's Leaf-nosed Bat; Maggie
Taylor's Roundleaf Bat; Maggie's
Roundleaf Bat

2812   *Hipposideros marisae*
Chiroptera - Rhinolophidae
e   Aellen's Leaf-nosed Bat; Aellen's
Roundleaf Bat

2813   *Hipposideros megalotis*
Chiroptera - Rhinolophidae
e   Big-eared Leaf-nosed Bat; Ethiopian
Large-eared Leaf-nosed Bat;
Ethiopian Large-eared Roundleaf
Bat; Large-eared Roundleaf Bat

2814   *Hipposideros muscinus*
Chiroptera - Rhinolophidae
e   Fly River Leaf-nosed Bat; Fly River
Roundleaf Bat

2815   *Hipposideros nequam*
Chiroptera - Rhinolophidae
e   Malayan Roundleaf Horshoe Bat;
Malayan Leaf-nosed Bat; Malay
Leaf-nosed Bat; Malayan Roundleaf
Bat

2816   *Hipposideros obscurus*
Chiroptera - Rhinolophidae
e   Philippine Forest Roundleaf Bat;
Philippine Roundleaf Bat

2817   *Hipposideros orbiculus*
Chiroptera - Rhinolophidae

e   Orb-faced Roundlef Bat

2818   *Hipposideros papua*
Chiroptera - Rhinolophidae
e   Geelvink Bay Leaf-nosed Bat; Biak
Roundleaf Bat; Geelvink Roundleaf
Bat

2819   *Hipposideros pomona*
Chiroptera - Rhinolophidae
e   Pomona Roundleaf Bat

2820   *Hipposideros pratti*
Chiroptera - Rhinolophidae
e   Pratt's Leaf-nosed Bat; Pratt's
Roundleaf Bat

2821   *Hipposideros pygmaeus*
Chiroptera - Rhinolophidae
e   Philippine Pygmy Roundleaf Bat

2822   *Hipposideros ridleyi*
Chiroptera - Rhinolophidae
e   Singapore Roundleaf Horseshoe Bat;
Ridley's Leaf-nosed Bat; Ridley's
Roundleaf Bat
d   Ridley-Rundblattnase

2823   *Hipposideros ruber*
Chiroptera - Rhinolophidae
e   Noack's Leaf-nosed Bat; Noack's
African Leaf-nosed Bat; Noack's
Roundleaf Bat
f   Phyllorhine rousse

2824   *Hipposideros sabanus*
Chiroptera - Rhinolophidae
e   Lawas Roundleaf Horsehoe Bat;
Sabah Leaf-nosed Bat; Least
Roundleaf Bat

2825   *Hipposideros schistaceus*
Chiroptera - Rhinolophidae
e   Split Roundleaf Bat

2826 **Hipposideros semoni**
Chiroptera - Rhinolophidae
e Semon's Leaf-nosed Bat; Wart-nosed Horseshoe Bat; Semon's Roundleaf Bat

2827 **Hipposideros speoris**
Chiroptera - Rhinolophidae
e Schneider's Leaf-nosed Bat; Schneider's Roundleaf Bat
d Schneiders Rundblattnasenfledermaus

2828 **Hipposideros stenotis**
Chiroptera - Rhinolophidae
e Lesser Wart-nosed Horseshoe Bat; Narrow-eared Leaf-nosed Bat; Narrow-eared Roundleaf Bat; Northern Leaf-nosed Bat (ANZ)

2829 **Hipposideros turpis**
Chiroptera - Rhinolophidae
e Lesser Leaf-nosed Bat; Lesser Roundleaf Bat

2830 **Hipposideros wollastoni**
Chiroptera - Rhinolophidae
e Wollaston's Leaf-nosed Bat; Wollaston's Roundleaf Bat

2831 **Hippotragus**
Artiodactyla - Bovidae
e Sable Antelopes; Roan and Sable Antelopes
d Rossantilopen; Pferdeböcke
f Hippotragues
i Antilopi

2832 **Hippotragus equinus**
Artiodactyla - Bovidae
e Roan Antelope
d Falbenantilope; Pferdeantilope
f Antilope rouanne; Hippotrague
i Antilope roana; Antilope equina

2833 **Hippotragus equinus bakeri**

Artiodactyla - Bovidae
e Sudan Roan Antelope

2834 **Hippotragus equinus cottoni**
Artiodactyla - Bovidae
e Angolan Roan Antelope

2835 **Hippotragus equinus koba**
Artiodactyla - Bovidae
e Western Roan Antelope
f Antilope cheval

2836 **Hippotragus equinus langheldi**
Artiodactyla - Bovidae
e East African Roan Antelope

2837 **Hippotragus niger**
Artiodactyla - Bovidae
e Sable Antelope; Black-Buck; Mbarapi Antelope
d Rappenantilope
f Antilope noire; Hippotrague noir
i Antilope nera

2838 **Hippotragus niger kirkii**
Artiodactyla - Bovidae
e Zambian Sable Antelope; Kirk's Sable Antelope

2839 **Hippotragus niger niger**
Artiodactyla - Bovidae
e Common Sable Antelope; Southern Sable Antelope
d Südafrikanische Rappenantilope

2840 **Hippotragus niger roosevelti**
Artiodactyla - Bovidae
e East African Sable Antelope; Eastern Sable Antelope; Roosevelt's Sable Antelope
d Ostafrikanische Rappenantilope

2841 **Hippotragus niger variani**
Artiodactyla - Bovidae
e Giant Sable Antelope; Angolan Sable

Antelope; Royal Sable Antelope

d   Riesenrappenantilope

f   Hippotrague noir géant

i   Antilope nera gigante; Antilope equina neroa

2842   *Histiotus*
Chiroptera - Vespertilionidae

e   Big-eared Brown Bats

2843   *Histiotus alienus*
Chiroptera - Vespertilionidae

e   Strange Big-eared Brown Bat

2844   *Histiotus humboldti*
Chiroptera - Vespertilionidae

e   Humboldt's Big-eared Bat

2845   *Histiotus macrotus*
Chiroptera - Vespertilionidae

e   Big-eared Brown Bat; Andean Big-eared Brown Bat

2846   *Histiotus montanus*
Chiroptera - Vespertilionidae

e   Small Big-eared Brown Bat

2847   *Histiotus velatus*
Chiroptera - Vespertilionidae

e   Tropical Big-eared Brown Bat

2848   *Histriophoca*
Carnivora - Phocidae

e   Ribbon Seals

d   Bandrobben

f   Phoques aux bandes

i   Foche dalle fasce

2849   *Histriophoca fasciata*
Carnivora - Phocidae

e   Ribbon Seal; Banded Seal

d   Bandrobbe

f   Phoque aux bandes; Phoque aux rubans

i   Foca dalle fasce

2850   *Hodomys*
Rodentia - Muridae

e   Allen's Woodrats

d   Allens Buschratten

2851   *Hodomys alleni*
Rodentia - Muridae

e   Allen's Woodrat

d   Allens Buschratte

2852   *Holochilus*
Rodentia - Muridae

e   Web-footed Rats; Marsh Rats; Web-footed Marsh Rats; South American Marsh Rats

d   Sumpfratten

2853   *Holochilus brasiliensis*
Rodentia - Muridae

e   Web-footed Marsh Rat

d   Brasilien-Sumpfratte

2854   *Holochilus chacarius*
Rodentia - Muridae

e   Chaco Marsh Rat

Holochilus magnus syn. *Lundomys molitor* q.v.

2855   *Holochilus sciureus*
Rodentia - Muridae

e   Marsh Rat; Common Marsh Rat

2856   *Hominidae*
Primates

e   Humans and Great Apes

d   Menschen

f   Hominidés

i   Ominidi

2857   *Homo*
Primates - Hominidae

e   Men; Humans

d   Menschen

f Hommes

i Uomi

2858 Homo sapiens
Primates - Hominidae

e Man; Human

d Mensch; Moderner Mensch

f Homme

i Uomo

2859 Hoplomys
Rodentia - Echimyidae

e Thick-spined Armoured Rats; Harsh-furred Spiny Rats; Armoured Rats; Porcupine Rats

d Lanzenratten

2860 Hoplomys gymnurus
Rodentia - Echimyidae

e Thick-spined Armoured Rat; Armoured Rat

d Lanzenratte

2861 Hyaena
Carnivora - Hyaenidae

e Strand Wolves; Striped Hyenas; Striped and Brown Hyenas

d Streifenhyänen

f Hyènes rayées

i Iene

2862 Hyaena hyaena
Carnivora - Hyaenidae

e Striped Hyena

d Gestreifte Hyäne; Streifenhyäne

f Hyène; Hyène rayée

i Iena striata

2863 Hyaenidae
Carnivora

e Hyaenas; Strand Wolves

d Hyänen

f Hyénidés; Hyènes

i Iene; Ienidi

2864 Hybomys
Rodentia - Muridae

e One-striped Mice; Back-striped Mice; Striped Mice

d Streifenwaldmäuse; Östliche Streifenmäuse

f Rats aux bande dorsale noire; Souris aux rayures noires

2865 Hybomys basilii
Rodentia - Muridae

e Father Basilio's Striped Mouse

2866 Hybomys eisentrauti
Rodentia - Muridae

e Eisentraut's Striped Mouse; Eisentraut's Hump-nosed Mouse

2867 Hybomys lunaris
Rodentia - Muridae

e Moon Striped Mouse; Ruwenzori Striped Mouse

d Östliche EinstreifenwaldmausHay

2868 Hybomys planifrons
Rodentia - Muridae

e Miller's Striped Mouse

2869 Hybomys trivirgatus
Rodentia - Muridae

e Temminck's Striped Mouse

d Dreistreifenwaldmaus

2870 Hybomys univittatus
Rodentia - Muridae

e Peters's Striped Mouse; Hump-nosed Mouse; One-striped Forest Mouse

d Westliche Einstreifenwaldmaus

2871 Hydrochoeridae
Rodentia

e Carpinchos; Capybaras; Water Hogs

d Riesennager ; Wasserschweine

f Hydrochéridés

i Capibare; Idrocheridi

2872 *Hydrochoerus*
      Rodentia - Hydrochoeridae
e   Carpinchos; Capybaras; Water Hogs
d   Wasserschweine
f   Carpinchos; Cabiais; Capybaras
i   Capibare

2873 *Hydrochoerus hydrochaeris*
      Rodentia - Hydrochoeridae
e   Water Pig; Capybara; Water Cavy;
      Giant Water Cavy
d   Wasserschwein; Capybara;
      Riesennager
f   Cabiai; Capybara
i   Capibara

2874 *Hydromys*
      Rodentia - Muridae
e   Australian Water Rats; Beaver Rats;
      Water Rats
d   Schwimmratten; Eigentliche
      Schwimmratten; Australische
      Schwimmratten
f   Rats d'eau d'Australie; Rats
      aquatiques d' Australie; Hydromines
i   Ratti d'acqua australiani

2875 *Hydromys chrysogaster*
      Rodentia - Muridae
e   Golden-bellied Beaver Rat; Eastern
      Water Rat (ANZ); Australian Water
      Rat; Beaver Rat; Golden-bellied
      Water Rat; Water Rat (ANZ); Golden
      Water Rat; Common Water Rat
      (ANZ); Australasian Water Rat
d   Australische Schwimmratte;
      Goldbauchschwimmratte
f   Rat aquatique d' Australie;
      Hydromine
i   Ratto d'acqua australiano

2876 *Hydromys habbema*
      Rodentia - Muridae
e   Mountain Water Rat
d   Gebirgsschwimmratte

2877 *Hydromys hussoni*

      Rodentia - Muridae
e   Western Water Rat
d   Westliche Schwimmratte

2878 *Hydromys neobritannicus*
      Rodentia - Muridae
e   New Britain Water Rat

2879 *Hydromys shawmayeri*
      Rodentia - Muridae
e   Shaw Mayer's Water Rat
d   Shaw Mayers Schwimmratte

2880 *Hydropotes*
      Artiodactyla - Cervidae
e   Chinese Water Deer
d   Wasserrehe
f   Hydropotes; Cerfs des marais
i   Idropoti

2881 *Hydropotes inermis*
      Artiodactyla - Cervidae
e   Chinese Water Deer
d   Wasserreh
f   Cerf des marais; Hydropote chinois
i   Idropote inerme

2882 *Hydropotes inermis argyropus*
      Artiodactyla - Cervidae
e   Korean Water Deer
d   Korea-Wasserreh; Koreanisches
      Wasserreh

2883 *Hydrurga*
      Carnivora - Phocidae
e   Leopard Seals
d   Seeleoparden
f   Phoques léopards
i   Foche leopardo

2884 *Hydrurga leptonyx*
      Carnivora - Phocidae
e   Leopard Seal
d   Seeleopard

f Phoque léopard
i Foca leopardo

2885 *Hyemoschus*
Artiodactyla - Tragulidae
e Water Chevrotains
d Eigentliche Hirschferkel;
Zwergmoschustiere;
Wassermoschustiere
f Chevrotains africains; Chevrotains aquatiques
i Iemosci

2886 *Hyemoschus aquaticus*
Artiodactyla - Tragulidae
e African Water Chevrotain; Water Chevrotain
d Afrikanisches Hirschferkel;
Wassermoschustier
f Chevrotain africain; Chevrotain aquatique; Tragulidé d'eau africain
i Iemosco acquatico

2887 *Hylobates*
Primates - Hominidae
e Gibbons
d Eigentliche Gibbons; Langmaraffen;
Gibbons
f Gibbons
i Gibboni

2888 *Hylobates agilis*
Primates - Hominidae
e Agile Gibbon; Dark-handed Gibbon;
Black-handed Gibbon
d Schlankgibbon; Schwarzhandgibbon;
Ungka
f Gibbon agile
i Gibbone agile

2889 *Hylobates agilis agilis*
Primates - Hominidae
e Mountain Agile Gibbon

2890 *Hylobates agilis unko*
Primates - Hominidae

e Lowland Agile Gibbon

2891 *Hylobates albibarbis*
Primates - Hominidae
e White-bearded Bornean Gibbon;
Bornean White-bearded Gibbon
d Weißbärtiger Gibbon; Weißbärtiger Borneo-Gibbon

2892 *Hylobates concolor*
Primates - Hominidae
e Black Gibbon; Crested Gibbon;
Black Crested Gibbon; Western
Black-crested Gibbon
d Westlicher Schwarzer Schopfgibbon;
Schopfgibbon; Schwarzer
Schopfgibbon; Weißwangengibbon
f Gibbon noir
i Gibbone dal ciuffo; Gibbone dai favoriti

2893 *Hylobates gabriellae*
Primates - Hominidae
e Buff-cheeked Gibbon; Pink-cheeked
Gibbon; Red-cheeked Gibbon;
Yellow-cheeked Crested Gibbon
d Gelbwangenschopfgibbon
i Gibbone dalle guance rosa

2894 *Hylobates hainanus*
Primates - Hominidae
e Hainan Gibbon; Hainan Black-
crested Gibbon; Eastern Black
Crested Gibbon
d Hainan-Gibbon; Östlicher Schwarzer
Schopfgibbon
f Gibbon de Hainan

2895 *Hylobates hoolock*
Primates - Hominidae
e Hoolock Gibbon; White-browed
Gibbon
d Hulock; Weißbrauengibbon
f Gibbon houlock; Houlock; Hoolock
i Hulock; Ulock

2896   *Hylobates klossii*
Primates - Hominidae
e   Kloss's Gibbon; Dwarf Gibbon;
Dwarf Siamang; Beeloh; Mentawai
Gibbon; Bilou
d   Kloss-Gibbon; Zwergsiamang; Biloh;
Mentawai-Gibbon; Biolou
f   Gibbon de Kloss; Siamang de Kloss
i   Gibbone delle Mentawai; Siamango
golnano

2897   *Hylobates lar*
Primates - Hominidae
e   White-handed Gibbon; Common
Gibbon; Lar Gibbon
d   Weißhandgibbon; Lar
f   Gibbon lar; Gibbon aux mains
blanches; Gibbon cendré
i   Gibbone dalle mani bianche; Lar;
Ilobate dalle mane bianche

2898   *Hylobates lar carpenteri*
Primates - Hominidae
e   Carpenter's Lar Gibbon
d   Carpenter-Weißhandgibbon

2899   *Hylobates lar entelloides*
Primates - Hominidae
e   Central Lar Gibbon

2900   *Hylobates lar vestitus*
Primates - Hominidae
e   Sumatran Lar Gibbon

2901   *Hylobates lar yunnanensis*
Primates - Hominidae
e   Yunnan Lar Gibbon

2902   *Hylobates leucogenys*
Primates - Hominidae
e   White-cheeked Gibbon; Northern
White-cheeked Gibbon
d   Weißwangenschopfgibbon
f   Gibbon aux joues blanches; Gibbon
aux joues pâles
i   Gibbone dalle guance bianche;

Gibbone dalle mani bianche

2903   *Hylobates moloch*
Primates - Hominidae
e   Silvery Gibbon; Javan Gibbon; Grey
Gibbon
d   Silbergibbon; Java-Gibbon; Silbriger
Gibbon
f   Gibbon cendré
i   Gibbone cenerino

2904   *Hylobates moloch moloch*
Primates - Hominidae
e   Western Silver Gibbon; Western
Javan Gibbon

2905   *Hylobates moloch pongoalsoni*
Primates - Hominidae
e   Eastern Silvery Gibbon; Central
Javan Gibbon

2906   *Hylobates muelleri*
Primates - Hominidae
e   Borneo Gibbon; Müller's Gibbon;
Bornean Gibbon; Müller's Bornean
Gibbon
d   Borneo-Gibbon; Grauer Gibbon;
Müllers Borneo-Gibbon
f   Gibbon de Müller
i   Gibbone di Müller; Gibbone del
Borneo

2907   *Hylobates muelleri abbotti*
Primates - Hominidae
e   Abbot's Gray Gibbon

2908   *Hylobates muelleri funereus*
Primates - Hominidae
e   Norther Grey Gibbon

2909   *Hylobates muelleri muelleri*
Primates - Hominidae
e   Müller's Grey Gibbon

2910   *Hylobates pileatus*
Primates - Hominidae

e Pileated Gibbon; Capped Gibbon
d Kappengibbon
f Gibbon aux bonnet
i Gibbone dal berretto

2911 *Hylobates siki*
Primates - Hominidae
e Southern White-cheeked Gibbon
d Südlicher Weißwangengibbon
f Gibbon aux favoris blanc du Sud

2912 *Hylobates* syn*dactylus*
Primates - Hominidae
e Siamang; Siamang Gibbon
d Siamang
f Siamang
i Siamango

2913 *Hylochoerus*
Artiodactyla - Suidae
e Giant Forest Pigs; Giant Forest Hogs
d Waldschweine; Riesenwaldschweine
f Hylochères
i Ilocheri

2914 *Hylochoerus meinertzhageni*
Artiodactyla - Suidae
e Giant Forest Pig; Giant Forest Hog; Cape Somali Hog
d Waldschwein; Riesenwaldschwein
f Hylochère
i Ilochero gigante

2915 *Hylochoerus meinertzhageni ivoriensis*
Artiodactyla - Suidae
e Western Forest Hog

2916 *Hylochoerus meinertzhageni rimator*
Artiodactyla - Suidae
e Central African Giant Forest Hog

2917 *Hylomys*
Erinceomorpha - Erinaceidae

e Lesser Gymnures; Asian Gymnures
d Kleine Rattenigel

2918 *Hylomys hainanensis*
Erinceomorpha - Erinaceidae
e Hainan Moonrat; Hainan Gymnure
d Hainan-Rattenigel

2919 *Hylomys sinensis*
Erinceomorpha - Erinaceidae
e Chinese Gymnure; Shrew Hedgehog; Shrew Gymnure
d Spitzmausigel
f Néotétracus

2920 *Hylomys suillus*
Erinceomorpha - Erinaceidae
e Lesser Moonrat; Lesser Gymnure; Short-tailed Gymnure; Pig-tailed Shrew
d Kleiner Rattenigel

2921 *Hylomyscus*
Rodentia - Muridae
e African Wood Mice
d Kleine Afrikanische Waldmäuse
f Souris des buissons

2922 *Hylomyscus aeta*
Rodentia - Muridae
e Beaded Wood Mouse
f Souris des buissons à dessous blanc

2923 *Hylomyscus alleni*
Rodentia - Muridae
e Allen's Wood Mouse

2924 *Hylomyscus baeri*
Rodentia - Muridae
e Baer's Wood Mouse

2925 *Hylomyscus carillus*
Rodentia - Muridae
e Angolan Wood Mouse

2926   ***Hylomyscus denniae***
            Rodentia - Muridae
e   Montane Wood Mouse; Climbing
      Wood Mouse

2927   ***Hylomyscus parvus***
            Rodentia - Muridae
e   Little Wood Mouse

2928   ***Hylomyscus stella***
            Rodentia - Muridae
e   Stella Wood Mouse
f   Souris des buissons à dessous gris

2929   ***Hylonycteris***
            Chiroptera - Phyllostomidae
e   Underwood's Long-nosed Bats;
      Costa Rican Long-nosed Bats;
      Underwood's Long-tongued Bats
d   Spitzmaus-Langzungenfledermäuse

2930   ***Hylonycteris underwoodi***
            Chiroptera - Phyllostomidae
e   Underwood's Long-nosed Bat; Costa
      Rican Long-nosed Bat; Underwood's
      Long-tongued Bat
d   Spitzmaus-Langzungenfledermaus

2931   ***Hylopetes***
            Rodentia - Sciuridae
e   Arrow-tailed Flying Squirrels;
      Pygmy Flying Squirrels; Indo-
      Malayan Flying Squirrels
d   Pfeilschwanzgleithörnchen
f   Écureils volants

2932   ***Hylopetes alboniger***
            Rodentia - Sciuridae
e   Particoloured Flying Squirrel
d   Schwarzweißes Gleithörnchen

2933   ***Hylopetes baberi***
            Rodentia - Sciuridae
e   Pygmy Kashmir Flying Squirrel;
      Smaller Kashmir Flying Squirrel;
      Kashmir Pygmy Flying Squirrel;

Afghan Flying Squirrel
d   Kaschmir-Zwerggleithörnchen

2934   ***Hylopetes bartelsi***
            Rodentia - Sciuridae
e   Bartels's Flying Squirrel
d   Bartels-Gleithörnchen

            Hylopetes fimbriatus syn.
            Eoglaucomys fimbriatus q.v.

2935   ***Hylopetes lepidus***
            Rodentia - Sciuridae
e   Indo-Malayan Flying Squirrel; Javan
      Lesser Flying Squirrel; Burmese
      Pygmy Flying Squirrel; Grey-
      cheeked Flying Squirrel
d   Javanisches Gleithörnchen;
      Grauwangengleithörnchen
f   Écureuil volant de Java

2936   ***Hylopetes nigripes***
            Rodentia - Sciuridae
e   Palawan Flying Squirrel
d   Palawan-Gleithörnchen

2937   ***Hylopetes phayrei***
            Rodentia - Sciuridae
e   Phayre's Flying Squirrel; Indochinese
      Flying Squirrel
d   Phayre-Gleithörnchen

2938   ***Hylopetes sipora***
            Rodentia - Sciuridae
e   Sipora Flying Squirrel
d   Sipura-Gleithörnchen

2939   ***Hylopetes spadiceus***
            Rodentia - Sciuridae
e   Red-cheeked Flying Squirrel
d   Rotwangengleithörnchen

2940   ***Hylopetes winstoni***
            Rodentia - Sciuridae
e   Sumatran Flying Squirrel; Winston

Flying Squirrel

d   Sumatra-Gleithörnchen

2941 *Hyomys*
Rodentia - Muridae
e   Rough-tailed Giant Rats
d   Hyomys-Ratten
f   Rats de Nouvelle Guinée

2942 *Hyomys dammermani*
Rodentia - Muridae
e   Western White-eared Giant Rat
d   Westliche Weißohrriesenratte

2943 *Hyomys goliath*
Rodentia - Muridae
e   Rough-tailed Giant Rat; Eastern White-eared Giant Rat; White-eared Giant Rat
d   Goliathratte

2944 *Hyosciurus*
Rodentia - Sciuridae
e   Long-snouted Squirrels; Celebes Long-nosed Squirrels; Sulawesi Long-nosed Squirrels
d   Ferkelhörnchen
f   Écureuils de Heinrich

2945 *Hyosciurus heinrichi*
Rodentia - Sciuridae
e   Long-snouted Squirrel; Celebes Long-nosed Squirrel; Sulawesi Long-nosed Squirrel; Montane Long-nosed Squirrel; Southern Long-nosed Squirrel
d   Bergferkelhörnchen
f   Écureuil de Heinrich

2946 *Hyosciurus ileile*
Rodentia - Sciuridae
e   Lowland Long-nosed Squirrels; Northern Long-nosed Squirrel
d   Flachlandferkelhörnchen

2947 *Hyperacrius*

Rodentia - Muridae
e   Kashmir Voles; Punjab Voles
d   Kaschmir-Wühlmäuse

2948 *Hyperacrius fertilis*
Rodentia - Muridae
e   True's Vole; Burrowing Vole
d   True-Wühlmaus

2949 *Hyperacrius wynnei*
Rodentia - Muridae
e   Murree Vole

2950 *Hyperoodon*
Cete - Hyperoodontidae
e   Bottle-nosed Whales; Bottlenose Whales
d   Entenwale; Döglinge
f   Baleines à bec; Hypérodons
i   Iperodonti

2951 *Hyperoodon ampullatus*
Cete - Hyperoodontidae
e   Common Bottle-nosed Whale; North Atlantic Bottle-nosed Whale; Northern Bottle-nosed Whale; Bottlenose; Northern Bottlenose Whale; Botttle-nose Dolphin; Northern Bottlenose
d   Dögling; Nordischer Entenwal; Nordischer Dögling; Butzkopf; Entenwal; Nördlicher Entenwal
f   Hypérodon du Nord; Hypérodon boréal; Hyperodon
i   Iperodonte dal rostro; Iperodonte boreale; Iperodonte del nord; Iperodonte comune

2952 *Hyperoodon planifrons*
Cete - Hyperoodontidae
e   Flat-headed Bottle-nosed Whale; Flower's Bottle-nosed Whale; Southern Bottle-nosed Whale; Antarctic Bottlenose; Flat-fronted Bottle-nose; Flat-faced Bottle-nose; Pacific Beaked Whale; Southern Bottlenose Whale

d   Südlicher Entenwal
f   Hypérodon du Sud; Hypérodon
      australe
i   Iperodonte australe

2953   Hyperoodontidae
      Cete
e   Beaked Whales
d   Schnabelwale
f   Balaines à bec
i   Iperodontidi

2954   *Hypogeomys*
      Rodentia - Muridae
e   Votsotsas; Malagasy Giant Rats
d   Votsotsas
f   Vositses; Rats sauteur géants de
      Madagascar
i   Ratti giganti del Madagascar

2955   *Hypogeomys antimena*
      Rodentia - Muridae
e   Votsotsa; Malagasy Giant Rat
d   Votsotsa
f   Vositse; Rat sauteur géant; Rat
      sauteur de Madagascar; Rat sauteur
      malgache; Rat malgache géant
i   Votsotsa; Ratto gigante del
      Madagascar

      Hyposugo syn. *Pipistrellus* q.v.

2956   *Hypsignathus*
      Chiroptera - Pteropodidae
e   Hammer-headed Bats; Hammer-
      headed Fruit Bats; Big-lipped Bats
d   Hammerkopfflughunde;
      Hammerköpfe
f   Hypsignathes

2957   *Hypsignathus monstrosus*
      Chiroptera - Pteropodidae
e   Hammer-headed Fruit Bat;
      Hammerhead Bat
d   Hammerkopfflughund; Hammerkopf
f   Hypsignathe monstreux; Chien

volant à tête en marteau

Hypsimys syn. *Akodon* q.v.

2958   *Hypsiprymnodon*
      Diprotodontia - Macropodidae
e   Musky Rat Kangaroos (ANZ); Musk
      Rat Kangaroos
d   Moschusrattenkängurus;
      Greiffußhüpfer
f   Rats musqués kangourou
i   Ratti canguro muschiati; Topi
      canguro muschiati

2959   *Hypsiprymnodon moschatus*
      Diprotodontia - Macropodidae
e   Musky Rat Kangaroo (ANZ); Musk
      Rat Kangaroo
d   Moschusrattenkänguru;
      Greiffußhüpfer
f   Rat musqué kangourou; Rat-
      kangourou musqué
i   Ratto canguro muschiato; Topo
      canguro muschiato

2960   Hystricidae
      Rodentia
e   Porcupines; Old World Porcupines
d   Altwelt-Stachelschweine;
      Stachelschweine; Erdstachelschweine
f   Hystricidés; Porcs-épics
i   Istrici; Istricidi

2961   *Hystrix*
      Rodentia - Hystricidae
e   Large Porcupines; Crested
      Porcupines; Old World Porcupine;
      Short-tailed Porcupines
d   Stachelschweine; Eigentliche
      Stachelschweine
f   Porcs-épics
i   Istrici

2962   *Hystrix africaeaustralis*
      Rodentia - Hystricidae
e   Cape Porcupine; South African
      Porcupine; South African Crested

Porcupine
d Südafrikanisches Stachelschwein
f Porc-épic d'Afrique du Sud; Porc-épic d'Afrique; Porc-épic du Cap
i Istrice sudafricano; Istrice del Sudafrica

2963 Hystrix brachyura
Rodentia - Hystricidae

e Malayan Porcupine; East Asian Porcupine; Crestless Porcupine; Crestless Himalajan Porcupine
d Kurzschwanzstachelschwein; Nepal-Stachelschwein
f Porc-épic chinois; Porc-Épic himalayan

2964 Hystrix crassispinis
Rodentia - Hystricidae

e Thick-spined Porcupine
d Borneo-Stachelschwein
f Porc-épic malais; Porc-épic de Malaisie; Porc-épic de Bornéo

2965 Hystrix cristata
Rodentia - Hystricidae

e Common Porcupine; Crested Porcupine; North African Crested Porcupine; African Crested Porcupine
d Gewöhnliches Stachelschwein; Europäisches Stachelschwein; Nordafrikanisches Stachelschwein; Gemeines Stachelschwein; Westafrikanisches Stachelschwein; Stachelschwein
f Porc-épic commun; Porc-épic d'Europe; Porc-épic à crête; Porc-épic
i Porcospino; Istrice; Istrice europeo

Hystrix hodgsoni syn. *H. brachyura* q.v.

2966 Hystrix indica
Rodentia - Hystricidae

e Indian Crested Porcupine; India Porcupine

d Indisches Stachelschwein
f Porc-épic indien

2967 Hystrix javanica
Rodentia - Hystricidae

e Sunda Porcupine
d Sunda-Stachelschwein; Java-Stachelschwein

2968 Hystrix pumila
Rodentia - Hystricidae

e Philippine Porcupine
d Zwergstachelschwein
f Porc-épic d'Indonésie; Porc-Épic des Philippines

2969 Hystrix sumatrae
Rodentia - Hystricidae

e Sumatran Porcupine
d Sumatra-Stachelschwein
f Porc-épic de Sumatra

I

2970   *Ia*
Chiroptera - Vespertilionidae
e   Great Pipistrelles; Great Evening
Bats

2971   *Ia io*
Chiroptera - Vespertilionidae
e   Great Pipistrelle; Great Evening Bat

2972   *Ichneumia*
Carnivora - Herpestidae
e   White-tailed Mongooses
d   Weißschwanzichneumons
f   Mangoustes à queue blanche
i   Manguste dalla coda bianca

2973   *Ichneumia albicauda*
Carnivora - Herpestidae
e   White-tailed Mongoose
d   Weißschwanzichneumon;
Weißschwanzmanguste
f   Mangouste à queue blanche
i   Mangusta coda bianca; Mangusta
dalla coda bianca

2974   *Ichthyomys*
Rodentia - Muridae
e   Fish-eating Rats; Aquatic Rats; Crab-
eating Rats
d   Fischratten
i   Ratti

2975   *Ichthyomys hydrobates*
Rodentia - Muridae
e   Fish-eating Rat; Crab-eating Rat;
Common Crab-eatig Rat
d   Fischratte

2976   *Ichthyomys pittieri*
Rodentia - Muridae
e   Pittier's Crab-eating Rat

2977   *Ichthyomys stolzmanni*
Rodentia - Muridae
e   Stolzmann's Crab-eating Rat

2978   *Ichthyomys tweedii*
Rodentia - Muridae
e   Tweedy's Crab-eating Rat; Northern
Crab-eating Rat

Ictidomys syn. *Spermophilus* q.v.

2979   *Ictonyx*
Carnivora - Mustelidae
e   African Polecats; Striped Polecats
d   Zorillas
f   Zorilles; Zorilles communs; Zorillas;
Zorillas communs

Ictonyx libyca syn. *Poecilictis libyca*
q.v.

2980   *Ictonyx striatus*
Carnivora - Mustelidae
e   African Polecat; African Striped
Polecat; Zoril; Zorille; Zorillo;
Zorilla; Striped Polecat; Mariput;
Cape Polecat
d   Zorilla; Bandiltis; Kap-Iltis; Kap-
Skunk
f   Zorille commun; Zorille; Zorilla;
Zorilla commun
i   Zorilla; Zorilla comune; Zorilla del
Capo

2981   *Idionycteris*
Chiroptera - Vespertilionidae
e   Allen's Big-eared Bats

2982   *Idionycteris phyllotis*
Chiroptera - Vespertilionidae
e   Allen's Big-eared Bat; Allen's Long-
eared Bat; Lappet-browed Bat;
Allen's Lappet-browed Bat

2983 ***Idiurus***
Rodentia - Anomaluridae
e Pygmy Flying Squirrels; Small
African Flying Squirrels; Flying
Mouse Squirrels; Pygmy Scaly-tailed
Flying Squirrels
d Gleitbilche
f Anomalures nains
i Glini volanti

2984 ***Idiurus macrotis***
Rodentia - Anomaluridae
e Long-eared Small Flying Squirrel;
Large-eared Small Flying Squirrel;
Long-eared Scaly-tailed Flying
Squirrel; Long-eared Flying Squirrel;
Small Flying Squirrel

d Großohrgleitbilch
f Anomalure nain aux longues oreilles;
Anomalure nain
i Glino volante dalle grandi orecchie

2985 ***Idiurus zenkeri***
Rodentia - Anomaluridae
e Zenker's Flying Squirrel; Small
Flying Squirrel; Pygmy Scaly-tailed
Flying Squirrel
d Zenker-Gleitbilch
f Anomalure nain de Zenker; Écureuil
volant de Zenker

2986 ***Indopacetus***
Cete - Hyperoodontidae
e Longman's Beaked Whales
d Longman-Schnabelwale
f Mésoplodons de Longman
i Mesoplodonti di Longman

2987 ***Indopacetus pacificus***
Cete - Hyperoodontidae
e Longman's Beaked Whale; Indo-
Pacific Beaked Whale
d Pazifischer Schnabelwal; Longman-
Schnabelwal

f Mésoplodon de Longman

i Mesoplodonte di Longman

2988 ***Indri***
Primates - Indriide
e Indri Lemurs; Indris; Endrinas;
Babakotos
d Indris
f Indris; Babakato
i Indri

2989 ***Indri indri***
Primates - Indriide
e Indri; Babakoto
d Indri
f Indri
i Indri; Babakota

2990 **Indriidae**
Primates
e Avahis; Leaping Lemurs; Woolly
Lemurs; Indris and Sifakas
d Indriartige; Indris
f Indriidés; Indridés
i Lemuri saltatori; Indri; Indridi

2991 ***Inia***
Cete - Iniidae
e Amazon Dolphins; Amazonian
Dolphins; Pink River Dolphins
d Inias; Amazonas-Delfine
f Inies
i Inie

2992 ***Inia geoffrensis***
Cete - Iniidae
e Boto; Amazon Dolphin; Amazonian
Dolphin; Amazon Porpoise; Pink
River Dolphin; Boutu; Inia; Amazon
River Dolphin
d Amazonas-Delfin; Süßwasserwal;
Flussdelfin; Orinoko-Delfin; Boto;
Inia
f Inie de Geoffroy; Dauphin de
l'Amazone; Boutou
i Inia; Boto

2993   ***Inia geoffrensis boliviensis***
           Cete - Iniidae
   e   Rio Madeira River Dolphin

2994   ***Inia geoffrensis humboldtiana***
           Cete - Iniidae
   e   Orinoco River Dolphin

2995   **Iniidae**
           Cete
   e   Amazon Dolphins; Amazon River
           Dolphins
   d   Inias
   f   Iniidés
   i   Inidi

2996   ***Iomys***
           Rodentia - Sciuridae
   e   Horsfield's Flying Squirrels
   d   Horsfield-Gleithörnchen
   i   Scoiattoli volanti di Horsfield

2997   ***Iomys horsfieldi***
           Rodentia - Sciuridae
   e   Horsfield's Flying Squirrel; Javanese
           Flying Squirrel; Javan Flying
           Squirrel
   d   Horsfield-Gleithörnchen
   i   Scoiattolo volante di Horsfield

2998   ***Iomys sipora***
           Rodentia - Sciuridae
   e   Sipora Flying Squirrel; Mentawai
           Flying Squirrel
   d   Mentawai-Gleithörnchen

2999   ***Irenomys***
           Rodentia - Muridae
   e   Chilean Rats; Chilean Climbing Mice
   d   Chile-Ratten
   f   Rats chiliens

3000   ***Irenomys tarsalis***
           Rodentia - Muridae
   e   Chilean Rat; Chilean Climbing

           Mouse
   d   Chile-Ratte
   f   Rat chilien

3001   ***Isolobodon***
           Rodentia - Capromyidae
   e   Laminar-toothed Hutias
   f   Rats grimpeurs de Porto Rico

3002   ***Isolobodon portoricensis***
           Rodentia - Capromyidae
   e   Puerto Rico Hutia; Puerto Rican
           Isolobodon; Allen's Hutia; Allen's
           Isolobodon; Hispaniolan Isolobodon;
           Haitian Isolobodon; Puerto Rican
           Hutia
   d   Puerto-Rico-Ferkelratte; Puerto-
           Rico-Hutia; Allen-Ferkelratte; Allen-
           Hutia
   f   Rat grimpeur de Porto Rico; Hutia de
           Porto Rico; Rat grimpeur d'Allen;
           Hutia d'Allen

3003   ***Isoodon***
           Peramelemorphia - Peramelidae
   e   Short-nosed Bandicoots
   d   Kurznasenbeutler

3004   ***Isoodon auratus***
           Peramelemorphia - Peramelidae
   e   Golden Bandicoot
   d   Goldener Kurznasenbeutler
   f   Bandicoot doré
   i   Bandicut dorato

3005   ***Isoodon auratus auratus***
           Peramelemorphia - Peramelidae
   e   Mainland Golden Bandicoot

3006   ***Isoodon auratus barrowensis***
           Peramelemorphia - Peramelidae
   e   Barrow Island Golden Bandicoot

3007   ***Isoodon macrourus***
           Peramelemorphia - Peramelidae

e Brindled Northern Bandicoot;
Brindled North Australian Bandicoot;
Brindled Short-nosed Northern
Bandicoot; Brindled Short-nosed
North Australian Bandicoot;
Northern Brown Bandicoot (ANZ);
Brindled Bandicoot

d Großer Kurznasenbeutler;
Großkurznasenbeutler; Brauner
Nord-Kurznasenbeutler

f Bandicoot brun du Nord

3008 ***Isoodon obesulus***
Peramelemorphia - Peramelidae

e Short-nosed Bandicoot; Southern
Brown Bandicoot (ANZ); Quenda

d Kleiner Kurznasenbeutler;
Kleinkurznasenbeutler; Brauner Süd-
Kurznasenbeutler

f Bandicoot brun du Sud

i Bandicut marrone del Sud

3009 ***Isothrix***
Rodentia - Echimyidae

e Toros; Arboreal Rats; Brush-tailed
Rats

f Rats arboricoles

3010 ***Isothrix bistriatus***
Rodentia - Echimyidae

e Yellow-crowned Brush-tailed Rat

3011 ***Isothrix pagurus***
Rodentia - Echimyidae

e Plains Brush-tailed Rat

3012 ***Isothrix sinnamarensis***
Rodentia - Echimyidae

e Sinnamary Brush-tailed Rat

3013 ***Isthmomys***
Rodentia - Muridae

e Isthmus Rats

3014 ***Isthmomys flavidus***
Rodentia - Muridae

e Yellow Deer Mouse; Yellow Isthmus
Rat

3015 ***Isthmomys pirrensis***
Rodentia - Muridae

e Mount Pirri Deer Mouse; Mount Pirri
Isthmus Rat

J

3016 **Jaculus**
Rodentia - Dipodidae
e Jerboas; Desert Jerboas; Hairy-footed Jerboas
d Wüstenspringmäuse
f Gerboises du désert
i Topi saltatori

3017 **Jaculus blanfordi**
Rodentia - Dipodidae
e Blanford's Jerboa; Greater Three-toed Jerboa; Persian Brush-footed Jerboa
d Blanfords Wüstenspringmaus

Jaculus deserti syn. *J. jaculus* q.v.

3018 **Jaculus jaculus**
Rodentia - Dipodidae
e Lesser Egyptian Jerboa; Lesser Jerboa
d Kleine Wüstenspringmaus; Kleine Ägyptische Wüstenspringmaus
f Gerboise des steppes; Petite gerboise d'Égypte; Gerboise d'Égypte; Gerboise du désert
i Gerboa del deserto

3019 **Jaculus jaculus vocator**
Rodentia - Dipodidae
e Lesser Arabian Jerboa; Arabian Lesser Jerboa
d Kleine Arabische Wüstenspringmaus

3020 **Jaculus orientalis**
Rodentia - Dipodidae
e Greater Egyptian Jerboa; Desert Jerboa
d Große Wüstenspringmaus; Größere Ägyptische Wüstenspringmaus

f Grande gerboise
i Topo delle piramidi maggiore

3021 **Jaculus turcmenicus**
Rodentia - Dipodidae
e Turkmen Jerboa
d Turkmen-Wüstenspringmaus

Jaguarius syn. *Panthera* q.v.

Jentinkia syn. *Bassariscus* q.v.

3022 **Juliomys**
Rodentia - Muridae
e Julio Mice

3023 **Juliomys rimofrons**
Rodentia - Muridae
e Long-haired Julio Mouse

3024 **Juscelinomys**
Rodentia - Muridae
e Juscelin's Mice; Burrowing Mice
d Maulwurfsmäuse

3025 **Juscelinomys candango**
Rodentia - Muridae
e Candango Mouse; Brazilian Burrowing Mouse

3026 **Juscelinomys guaporensis**
Rodentia - Muridae
e Rio Guapore Burrowing Mouse

3027 **Juscelinomys huanchacae**
Rodentia - Muridae
e Huanchaca Burrowing Mouse

K

3028 Kadarsanomys
Rodentia - Muridae
e Sody'a Tree Rats
d Sody-Baumratten

3029 Kadarsanomys sodyi
Rodentia - Muridae
e Sody's Tree Rat
d Sody-Baumratte

3030 Kannabateomys
Rodentia - Echimyidae
e Sody's Tree Rats; Atlantic Bamboo Rats
d Bambusfingerratten
f Rats du bambou

3031 Kannabateomys amblyonyx
Rodentia - Echimyidae
e Sody's Tree Rat; Atlantic Bamboo Rat; Southern Bamboo Rat
d Bambusfingerratte
f Rat du bambou

Karstomys syn. *Apodemus* q.v.

3032 Kerivoula
Chiroptera - Vespertilionidae
e Painted Bats; Woolly Bats; Trumpet-eared Bats

3033 Kerivoula aerosa
Chiroptera - Vespertilionidae
e Dubious Trumpet-eared Bat

3034 Kerivoula africana
Chiroptera - Vespertilionidae
e Tanzanian Woolly Bat

d Zwergflatterer; Langschwanzfledermaus

3035 Kerivoula agnella
Chiroptera - Vespertilionidae
e Louisiade Trumpet-eared Bat; St. Aignan's Trumpet-eared Bat

3036 Kerivoula argentata
Chiroptera - Vespertilionidae
e Damara Woolly Bat

3037 Kerivoula atrox
Chiroptera - Vespertilionidae
e Groove-toothed Bat

3038 Kerivoula cuprosa
Chiroptera - Vespertilionidae
e Copper Woolly Bat

3039 Kerivoula eriophora
Chiroptera - Vespertilionidae
e Ethiopian Woolly Bat

3040 Kerivoula flora
Chiroptera - Vespertilionidae
e Flores Woolly Bat

3041 Kerivoula hardwickei
Chiroptera - Vespertilionidae
e Hardwicke's Bat; Forest Bat; Hardwicke's Forest Bat

3042 Kerivoula hardwickei malpasi
Chiroptera - Vespertilionidae
e Malpas's Bat
d Malpas-Fledermaus

3043 Kerivoula intermedia
Chiroptera - Vespertilionidae
e Small Woolly Bat

3044 Kerivoula jagorii
Chiroptera - Vespertilionidae
e Peters's Trumpet-eared Bat

3045 *Kerivoula lanosa*
    Chiroptera - Vespertilionidae
 e Lesser Woolly Bat

3046 *Kerivoula minuta*
    Chiroptera - Vespertilionidae
 e Least Forest Bat; Least Woolly Bat

3047 *Kerivoula muscina*
    Chiroptera - Vespertilionidae
 e Fly River Trumpet-eared Bat

3048 *Kerivoula myrella*
    Chiroptera - Vespertilionidae
 e Bismarck Trumpet-eared Bat;
    Bismarck's Trumpet-eared Bat

3049 *Kerivoula papillosa*
    Chiroptera - Vespertilionidae
 e Papillose Bat; Papillose Woolly Bat

3050 *Kerivoula papuensis*
    Chiroptera - Vespertilionidae
 e Golden-tipped Bat (ANZ); Papuan
    Trumpet-eared Bat

3051 *Kerivoula pellucida*
    Chiroptera - Vespertilionidae
 e Clear-winged Bat; Clear-winged
    Woolly Bat

3052 *Kerivoula phalaena*
    Chiroptera - Vespertilionidae
 e Spurrell's Woolly Bat

3053 *Kerivoula picta*
    Chiroptera - Vespertilionidae
 e Painted Bat; Asian Painted Bat
 d Schmetterlingsfledermaus; Bunte
    Fledermaus

3054 *Kerivoula smithi*
    Chiroptera - Vespertilionidae
 e Smith's Woolly Bat

3055 *Kerivoula whiteheadi*
    Chiroptera - Vespertilionidae
 e Whitehead's Woolly Bat

3056 *Kerodon*
    Rodentia - Caviidae
 e Mocos; Rock Cavies
 d Bergmeerschweinchen ; Mokos;
    Felsenmeerschweinchen
 f Cobayes des rochers
 i Cavie delle rocce; Mocos

3057 *Kerodon acrobata*
    Rodentia - Caviidae
 e Acrobat Cavy

3058 *Kerodon rupestris*
    Rodentia - Caviidae
 e Moco; Rock Cavy
 d Bergmeerschweinchen; Moko;
    Felsenmeerschweinchen
 f Cobaye des rochers
 i Cavia delle rocce

Kivumys syn. *Lophuromys* q.v.

3059 *Kobus*
    Artiodactyla - Bovidae
 e Waterbucks; Kobs
 d Wasserböcke
 f Kobs; Cobs

3060 *Kobus ellipsiprymnus*
    Artiodactyla - Bovidae
 e Waterbuck; Common Waterbuck;
    Ringed Waterbuck
 d Wasserbock; Hirschantilope;
    Gemeiner Wasserbock
 f Kob de Buffon; Cob à croissant
 i Cobo d'acqua; Antilope d'acqua;
    Cobo

3061 *Kobus ellipsiprymnus defassa*
    Artiodactyla - Bovidae
 e Defassa Waterbuck; Northern

Waterbuck; East African Defassa
Waterbuck

  d   Defassa-Wasserbock

  f   Cob defassa; Kob defassa

  i   Cobo defassa

3062 *Kobus ellipsiprymnus*
      ellipsiprymnus
      Artiodactyla - Bovidae

  e   Ellipsen Waterbuck; Southern
Common Waterbuck

  d   Ellipsen-Wasserbock

3064 *Kobus ellipsiprymnus penricei*
      Artiodactyla - Bovidae

  e   Angolan Defassa Waterbuck

3063 *Kobus ellipsiprymnus ugandae*
      Artiodactyla - Bovidae

  e   Ugandan Defassa Waterbuck

3065 *Kobus kob*
      Artiodactyla - Bovidae

  e   Kob; Buffon's Kob

  d   Moorantilope; Kob

  f   Kob de Buffon; Cob de Buffon

  i   Antilope Kob; Kob

3066 *Kobus kob kob*
      Artiodactyla - Bovidae

  e   Western Kob

  d   Senegal-Kobantilope

3067 *Kobus kob leucotis*
      Artiodactyla - Bovidae

  e   White-eared Kob

  d   Weißohrkob

  f   Cob aux oreilles blanches; Kob aux
oreilles blanches

3068 *Kobus kob thomasi*
      Artiodactyla - Bovidae

  e   Uganda Kob

  d   Uganda-Kob

  f   Cob de Thomas; Kob de Thomas

  i   Antilope Kob di Thomas; Antilope
dell'Unganda

3069 *Kobus leche*
      Artiodactyla - Bovidae

  e   Lechwe

  d   Litchi-Moorantilope; Litschi-
Wasserbock; Letschwe; Litschi;
Litschi-Antilope

  f   Lechwe; Cob lechwe; Kob lechwe

  i   Cobo lichi; Antilope lichi; Antilope
lechwe; Lichi; Antilope di palude;
Antilope cobo

3070 *Kobus leche kafuensis*
      Artiodactyla - Bovidae

  e   Kafue Lechwe; Kafue Flats Lechwe

  d   Kafue-Litschi; Kafue-Litschi-
Wasserbock; Kafue-Letschwe-
Wasserbock

3071 *Kobus leche leche*
      Artiodactyla - Bovidae

  e   Red Kobus

  d   Litschi; Roter Letschwe-Wasserbock;
Roter Litschi; Rote Litschi-Antilope;
Roter Letschwe

  f   Cob rouge; Kob rouge

3072 *Kobus leche robertsi*
      Artiodactyla - Bovidae

  e   Roberts's Lechwe Antelope;
Roberts's Lechwe; Roberts's Lechwe
Waterbuck; Kavambwa Lechwe
Waterbuck; Kavambwa Lechwe;
Luena Lechwe Waterbuck; Luena
Lechwe

  d   Roberts Letschwe-Wasserbock;
Roberts Litschi-Wasserbock;
Roberts-Litschi; Roberts-
Wasserbock; Luena-Litschi; Luena-
Wasserbock; Luena-Antilope;
Kavambwa-Moorantilope; Roberts-
Letschwe Wasserbock

  f   Lechwe de Roberts; Cob lechwe de
Roberts; Kob lechwe de Kavambwa;
Lechwe de Luena; Kob lechwe de
Luena

3073 **_Kobus leche smithemani_**
Artiodactyla - Bovidae
e Black Lechwe
d Schwarzer Litchi; Schwarzer
Letschwe-Wasserbock; Schwarzer
Litschi-Wasserbock

3074 **_Kobus megaceros_**
Artiodactyla - Bovidae
e Nile Lechwe; Mrs. Gray's Lechwe;
Mrs. Gray's Kob
d Weißnackenmoorantilope; Frau
Grays Wasserbock
f Kob de Mrs. Gray; Cob de Mrs. Gray
i Lichi del Nilo; Lichi di Mrs. Gray

3075 **_Kobus vardonii_**
Artiodactyla - Bovidae
e Puku
d Puku
f Cob de Vardon; Kob de Vardon
i Puku

3076 **_Kogia_**
Cete - Kogiidae
e Lesser Sperm Whales; Pygmy Sperm
Whales; Short-headed Sperm
Whales; Small Sperm Whales
d Zwergpottwale
f Cachalots pygmées
i Cogie

3077 **_Kogia breviceps_**
Cete - Kogiidae
e Pygmy Sperm Whale; Lesser Sperm
Whale; Short-headed Sperm Whale;
Lesser Cachalot; Dwarf Sperm
Whale
d Zwergpottwal
f Kogia; Cachalot pygmée
i Cogia di De Blainville

3078 **_Kogia sima_**
Cete - Kogiidae
e Owen's Pygmy Sperm Whale
d Kleinpottwal; Kleiner Pottwal

f Cachalot nain
i Cogia di Owen

3079 **Kogiidae**
Cete
e Sperm Whales
d Pottwale
f Kogidés
i Cogie; Kogidi

3080 **_Komodomys_**
Rodentia - Muridae
e Komodo Rats
d Komodo-Ratten

3081 **_Komodomys rintjanus_**
Rodentia - Muridae
e Komodo Rat
d Komodo-Ratte

Koopmania syn. *Artibeus* q.v.

3082 **_Kunsia_**
Rodentia - Muridae
e South American Giant Rats
d Südamerikanische Riesenratten

3083 **_Kunsia fronto_**
Rodentia - Muridae
e Fossorial Giant Rat

3084 **_Kunsia tomentosus_**
Rodentia - Muridae
e Woolly Giant Rat

L

3085 *Laephotis*
Chiroptera - Vespertilionidae
e African Long-eared Bats

3086 *Laephotis angolensis*
Chiroptera - Vespertilionidae
e Angolan Long-eared Bat; Angola
Long-eared Bat

3087 *Laephotis botswanae*
Chiroptera - Vespertilionidae
e Botswanan Long-eared Bat

3088 *Laephotis namibensis*
Chiroptera - Vespertilionidae
e Namib Long-eared Bat

3089 *Laephotis wintoni*
Chiroptera - Vespertilionidae
e De Winton's Long-eared Bat

3090 *Lagenodelphis*
Cete - Delphinidae
e Bornean Dolphins
d Borneo-Delfine
f Dauphins de Fraser
i Lagenodelfini

3091 *Lagenodelphis hosei*
Cete - Delphinidae
e Bornean Dolphin; Fraser's Dolphin;
Sarawak Dolphin; Short-snouted
Whitebelly; Fraser's Porpoise;
Shortsnout Dolphin
d Borneo-Delfin; Frasers Delfin
f Dauphin de Fraser
i Lagenodelfino

3092 *Lagenorhynchus*
Cete - Delphinidae
e Short-beaked Dolphins; Ploughshare-
headed Dolphins; White-sided and
White-beaked Dolphins
d Weißschnauzendelfine;
Kurzschnauzendelfine
f Lagénorhynques
i Lagenorinci

3093 *Lagenorhynchus acutus*
Cete - Delphinidae
e White-sided Dolphin; Atlantic
White-sided Dolphin; Atlantic
White-sided Porpoise; Jumpere;
White-sided Bottlenose
d Weißseitendelfin; Nordischer
Kurzschnauzendelfin; Nordischer
Kurzschnauzenspringer; Nordischer
Delfin
f Dauphin aux flancs blancs;
Lagénorhynque à bec pointu
i Lagenorinco acuto

3094 *Lagenorhynchus albirostris*
Cete - Delphinidae
e White-beaked Dolphin; White-
beaked Porpoise; Whitebeak
Dolphin; White-headed Dolphin
d Weißschnauzendelfin;
Langfinnendelfin;
Weißschnauzenspringer;
Weißschnauziger Springer
f Dauphin à bec blanc; Lagénorhynque
à rostre blanc; Dauphin à bec blanc
de l'Atlantique; Lagénorhynqye à bec
blanc; Lagénorhynqye à bec blanc de
l'Atlantique
i Lagenorinco dal becco bianco;
Lagenorinco rostrobianco

3095 *Lagenorhynchus australis*
Cete - Delphinidae
e Peale's Dolphin; Peale's Porpoise;
Black-chinned Dolphin; Blackchin
Dolphin
d Peales Delfin; Peale-Delfin; Peales
Tümmler

f   Dauphin à menton noir; Dauphin de
     Peale
i   Lagenorinco di Peale; Lagenorinco
     australe

3096   *Lagenorhynchus cruiciger*
     Cete - Delphinidae
e   Cruciger White-sided Dolphin;
     Southern White-sided Dolphin;
     Hourglass Dolphin
d   Kreuzbanddelfin; Stundenglas-Delfin
f   Dauphin sablier; Dauphin crucigère
i   Lagenorinco dalla croce

3097   *Lagenorhynchus obliquidens*
     Cete - Delphinidae
e   Gray's White-sided Dolphin; North
     Pacific White-sided Dolphin; Pacific
     White-sided Dolphin; Hookfin
     Dolphin; Striped Dolphin; Pacific
     Striped Dolphin; White-striped
     Dolpin; Pacific White-striped
     Dolphin; Striped Porpoise; Pacific
     Striped Porpoise; White-sided
     Dolphin
d   Weißstreifendelfin; Gestreifter
     Pazifik-Delfin
f   Lagénorhynque de Gill; Dauphin aux
     flancs blancs du Pacifique; Dauphin
     aux flancs blancs
i   Lagenorinco dai denti obliqui

3098   *Lagenorhynchus obscurus*
     Cete - Delphinidae
e   Southern Striped Porpoise; Dusky
     Dolphin
d   Dunkeldelfin; Schwarzdelfin
f   Dauphin obscur; Dauphin sombre
i   Lagenorinco scuro

3099   *Lagenorhynchus obscurus fitzroyi*
     Cete - Delphinidae
e   South American Dusky Dolpin

3100   *Lagenorhynchus obscurus obscurus*
     Cete - Delphinidae
e   South African Dusky Dolphin

3101   *Lagidium*
     Rodentia - Chinchillidae
e   Mountain Viscachas; Lagidiums;
     Mountain Chinchillas; Peruvian
     Hares
d   Hasenmäuse; Viscachas
f   Lagostomes des montagnes;
     Viscaches des montagnes
i   Viscachas

3102   *Lagidium peruanum*
     Rodentia - Chinchillidae
e   Northern Viscacha; Peruvian
     Mountain Viscacha
d   Peruanische Hasenmaus; Peru-
     Hasenmaus

3103   *Lagidium viscacia*
     Rodentia - Chinchillidae
e   Mountain Viscacha; Mountain
     Chinchilla; Southern Viscacha;
     Southern Mountain Viscacha
d   Cuvier-Hasenmaus
f   Lagostome des montagnes
i   Viscaccia di Cuvier; Viscacha;
     Vizcacha; Viscaccia di montange

3104   *Lagidium wolffsohni*
     Rodentia - Chinchillidae
e   Wolffsohn's Viscacha
d   Südliche Hasenmaus

3105   *Lagorchestes*
     Diprotodontia - Macropodidae
e   Hare Wallabies
d   Hasenkängurus; Hasenspringer
f   Lièvres- wallabies; Wallabies-lièvres
i   Canguri lepre

3106   *Lagorchestes asomatus*
     Diprotodontia - Macropodidae
e   Central Hare Wallaby; Least Hare
     Wallaby
d   Östliches Hasenkänguru
f   Wallaby-lièvre du Centre; Lièvre-
     wallaby du Centre

3107 *Lagorchestes conspicillatus*
Diprotodontia - Macropodidae
e Spectacled Hare Wallaby (ANZ);
Spectacled Kangaroo
d Brillenhasenkänguru
f Wallaby-lièvre à lunettes; Lièvre-
wallaby à lunettes
i Canguro lepre dagli occhiali

**3108 *Lagorchestes conspicillatus
leichardti***
Diprotodontia - Macropodidae
e Leichardt's Spectacled Hare Wallaby
d Leichardts Brillenkänguru

3109 *Lagorchestes hirsutus*
Diprotodontia - Macropodidae
e Western Hare Wallaby (ANZ);
Wurrup; Rufous Hare Wallaby
d Zottelhasenkänguru; Hasenspringer
f Wallaby-lièvre de l'ouest; Lièvre-
wallaby de l'Ouest
i Canguro lepre occidentale

3110 *Lagorchestes hirsutus bernieri*
Diprotodontia - Macropodidae
e Bernier Island Hare Wallaby
d Bernier-Hasenkänguru

3111 *Lagorchestes hirsutus dorreae*
Diprotodontia - Macropodidae
e Dorre Island Hare Wallaby
d Dorré-Hasenkänguru
f Lièvre-Wallaby de l'île de Dorré

3112 *Lagorchestes leporides*
Diprotodontia - Macropodidae
e Brown Hare Wallaby; Eastern Hare
Wallaby; Eastern Hare Wallaby
(ANZ)
d Langohrhasenkänguru
f Wallaby-lièvre de l'Est; Lièvre-
wallaby de l'Est; Lièvre-wallaby aux
longues oreilles; Wallaby-lièvre aux
longues oreilles

3113 *Lagostomus*
Rodentia - Chinchillidae
e Plains Viscachas
d Hasenmäuse; Viscachas
f Lagostomes des Pampas; Chinchillas
i Viscacce delle Pampas

3114 *Lagostomus maximus*
Rodentia - Chinchillidae
e Plains Viscacha
d Hasenmaus; Viscacha;
Flachlandviscacha
f Lagostome des Pampas
i Viscaccia delle Pampas; Viscaccia di
pianura; Viscacchia

3115 *Lagostrophus*
Diprotodontia - Macropodidae
e Banded Hare Wallabies (ANZ);
Striped Hare Wallabies
d Bänderkängurus; Gestreifte
Hasenkängurus; Gebänderte
Hasenkängurus
f Wallabies rayés
i Canguri striati

3116 *Lagostrophus fasciatus*
Diprotodontia - Macropodidae
e Banded Hare Wallaby (ANZ);
Striped Hare Wallaby; Munning
d Bänderkänguru; Gebändertes
Hasenkänguru
f Wallaby rayé; Wallaby-lièvre rayé
i Canguro striato

3117 *Lagothrix*
Primates - Atelidae
e Woolly Monkeys
d Wollaffen
f Singes laineux; Lagotriches
i Scimmie lanose

3118 *Lagothrix cana*
Primates - Atelidae
e Grey Woolly Monkey

d   Grauer Wollaffe
i   Scimmia lanosa grigia

Lagothrix flavicauda syn. *Oreonax flavicauda* q.v.

3119   Lagothrix lagothricha
Primates - Atelidae
e   Humboldt's Common Woolly
Monkey ; Common Woolly Monkey;
Humboldt's Woolly Monkey;
Barrigudo; Caparro; Brown Woolly
Monkey
d   Großer Wollaffe; Wollaffe;
Eigentlicher Wollaffe
f   Lagotriche de Humboldt; Lagotriche
commun
i   Scimmia lanosa; Lagotrice; Lagotice
di Humboldt

3120   Lagothrix lugens
Primates - Atelidae
e   Colombian Woolly Monkey
d   Columbischer Bergwolf

3121   Lagothrix poeppigii
Primates - Atelidae
e   Silvery Woolly Monkey
d   Brauner Wollaffe

3122   Lagurus
Rodentia - Muridae
e   Steppe Lemmings
d   Steppenleminge; Graulemminge
f   Lemmings des steppes
i   Lemming delle steppe

3123   Lagurus lagurus
Rodentia - Muridae
e   Steppe Lemming; Common Steppe
Lemming
d   Steppenlemming; Graulemming;
Rauchschwänzige Maus;
Schwertelmaus; Grauer
Steppenlemming
f   Lemming des steppes

i   Lemming delle steppe

3124   Lama
Artiodactyla - Camelidae
e   Guanacos; Llamas
d   Kleinkamele; Lamas; Llamas
f   Lamas
i   Lama

3125   Lama glama
Artiodactyla - Camelidae
e   Llama
d   Lama
f   Lama
i   Lama

3126   Lama guanicoe
Artiodactyla - Camelidae
e   Guanaco
d   Guanako; Huanoko
f   Guanaco
i   Guanaco

Lama pacos syn. *Vicugna pacos* q.v.

Lama vicugna syn. *Vicugna vicugna* q.v.

3127   Lamottemys
Rodentia - Muridae
e   Mount Oku Rats; Mount Oku Mice
d   Oku-Ratten
f   Rats du Mont Oku

3128   Lamottemys okuensis
Rodentia - Muridae
e   Mount Oku Rat; Mount Oku Mouse
d   Oku-Ratte
f   Rat du Mont Oku

Lampronycteris syn. *Micronycteris* q.v.

Laomys syn. *Zyzomys* q.v.

3129 *Lariscus*
Rodentia - Sciuridae
e Malayan Black-striped Squirrels;
Striped Ground Squirrels; Three-
striped Palm Squirrels
d Schwarzstreifenhörnchen
f Écureuils terrestres

3130 *Lariscus hosei*
Rodentia - Sciuridae
e Four-striped Ground Squirrel
d Vierstreifenerdhörnchen
f Écureuil de Hose; Écureil terrestre
aux quatre bandes; Écureuil terrestre
à quatre raies; Écureuil de terre à
quatre raies

3131 *Lariscus insignis*
Rodentia - Sciuridae
e Malayan Black-striped Squirrel;
Three-striped Ground Squirrel
d Dreistreifenhörnchen
f Écureuil terrestre à trois bandes;
Écureil terrestre à trois bandes;
Écureuil terrestre à trois bandes de
Sarawak

3132 *Lariscus niobe*
Rodentia - Sciuridae
e Niobe Ground Squirrel
d Niobe-Schwarzstreifenhörnchen

3133 *Lariscus obscurus*
Rodentia - Sciuridae
e Mentawai Three-striped Squirrel
d Mentawai-Dreistreifenhörnchen

3134 *Lasionycteris*
Chiroptera - Vespertilionidae
e Silver-haired Bats
d Silberhaarfledermäuse
f Chauves-souris argentées

3135 *Lasionycteris noctivagans*
Chiroptera - Vespertilionidae
e Silver-haired Bat

d Silberhaarfledermaus
f Chauve-souris argentée

Lasiopodomys syn. *Microus* q.v.

3136 *Lasiorhinus*
Diprotodontia - Vombatidae
e Soft-furred Wombats; Hairy-nosed
Wombats; Long-eared Wombats
d Haarnasenwombats
f Wombats aux narines poilues
i Lasiorini

Lasiorhinus barnardi syn. *L. krefti*
q.v.

Lasiorhinus gillespiei syn. *L. krefti*
q.v.

3137 *Lasiorhinus krefftii*
Diprotodontia - Vombatidae
e Queensland Hairy-nosed Wombat;
Northern Hairy-nosed Wombat;
Common Northern Soft-furred
Wombat; Denliquin Soft-furred
Wombat; Moonie River Soft-furred
Wombat
d Nördlicher Haarnasenwombat;
Moonie-Wombat; Deniliquin-
Haarnasenwombat; Deniliquin-
Breitstirnwombat; Haarnasenwombat
f Wombat à nez poilu; Wombat aux
narines polues du Nord; Wombat aux
narines polues du Queensland;
Wombat de Deniliquin
i Vombato dal naso peloso; Vombato
dal naso peloso del Queensland

3138 *Lasiorhinus latifrons*
Diprotodontia - Vombatidae
e Soft-furred Wombat; Hairy-nosed
Wombat; Long-eared Wombat;
Southern Hairy-nosed Wombat
d Südlicher Haarnasenwombat;
Breitstirnwombat
f Wombat aux narines poilues
i Lasiorino

3139   *Lasiurus*
  Chiroptera - Vespertilionidae
  e   Red Bats; Hoary Bats; Yellow Bats;
        Hoary-and-red Bats; Hairy-tailed
        Bats
  d   Rote Fledermäuse

3140   *Lasiurus atratus*
  Chiroptera - Vespertilionidae
  e   Mourning Bat

3141   *Lasiurus blossevilli*
  Chiroptera - Vespertilionidae
  e   Western Red Bat; Desert Red Bat

3142   *Lasiurus borealis*
  Chiroptera - Vespertilionidae
  e   Red Bat; Eastern Red Bat
  d   Rote Fledermaus; New York-
        Fledermaus
  f   Chauve-souris rousse; Chauve-souris
        boreale; Chauve-souris à queue
        chevelue

  Lasiurus brachyotis syn. *L. borealis*
  q.v.

3143   *Lasiurus castaneus*
  Chiroptera - Vespertilionidae
  e   Tacarcuna Bat

3144   *Lasiurus cinereus*
  Chiroptera - Vespertilionidae
  e   Hoary Bat
  d   Silberfledermaus; Weißgraue
        Fledermaus; Schimmelfledermaus;
        Eisgraue Fledermaus

  f   Chauve-souris cendrée
  i   Pipistrello cenerino

3145   *Lasiurus cinereus semotus*
  Chiroptera - Vespertilionidae
  e   Hawaian Hoary Bat

Lasiurus degelidus syn. *L. borealis*
q.v.

3146   *Lasiurus ebenus*
  Chiroptera – Vespertilionidae
  e   Ebony Bat

3147   *Lasiurus ega*
  Chiroptera – Vespertilionidae
  e   Southern Yellow Bat

3148   *Lasiurus egregious*
  Chiroptera – Vespertilionidae
  e   Big Red Bat

3149   *Lasiurus intermedius*
  Chiroptera – Vespertilionidae
  e   Northern Yellow Bat
  f   Chauve souris jaune du Nord

3150   *Lasiurus intermedius floridanus*
  Chiroptera – Vespertilionidae
  e   Yellow Bat
  f   Chauve-souris jaune de la Floride

Lasiurus minor syn. *L. borealis* q.v.

Lasiurus pfeifferi syn. *L. borealis*
q.v.

3151   *Lasiurus seminolus*
  Chiroptera - Vespertilionidae
  e   Seminole Mahogany Bat; Seminole
        Bat

3152   *Lasiurus xanthinus*
  Chiroptera - Vespertilionidae
  e   Western Yellow Bat

3153   *Latidens*
  Chiroptera - Pteropodidae
  e   Salim Ali's Fruit Bats

3154   *Latidens salimalii*
           Chiroptera - Pteropodidae
e   Salim Ali's Fruit Bat

3155   *Lavia*
           Chiroptera - Megadermatidae
e   African Yellow-winged Bats;
           Yellow-winged Bats
d   Gelbflügelige Großblattnasen
f   Mégadermes aux ailes orangées
i   Falsi vampiri dalle ali gialli

3156   *Lavia frons*
           Chiroptera - Megadermatidae
e   African Yellow-winged Bat; Yellow-
           winged Bat
d   Gelbflügelige Großblattnase
f   Mégaderme aux ailes orangées
i   Falso vampiro dalle ali gialli

           Leggada syn. *Mus* q.v.

           Leggadilla syn. *Mus* q.v.

3157   *Leggadina*
           Rodentia - Muridae
e   Australian Native Mice
d   Australische Kleinmäuse
f   Souris d'Australie; Souris des forêts

3158   *Leggadina forresti*
           Rodentia - Muridae
e   Forrest's Mouse (ANZ); Forrest's
           Territory Mouse; Forrest Mouse
d   Forrest-Kleinmaus

3159   *Leggadina lakedownensis*
           Rodentia - Muridae
e   Lakeland Downs Mouse; Lakeland
           Downs Short-tailed Mouse
d   Lakeland-Downs-Kleinmaus

3160   *Leimacomys*
           Rodentia - Muridae

e   Groove-toothed Forest Mice;
           Büttner's Togo Mice
d   Büttners Waldmäuse

3161   *Leimacomys buettneri*
           Rodentia - Muridae
e   Groove-toothed Forest Mouse;
           Büttner's Togo Mouse; Büttner's
           Mouse; Togo Mouse
d   Büttners Waldmaus; Büttner-
           Baummaus
f   Souris des bois de Buettner

3162   *Lemniscomys*
           Rodentia - Muridae
e   Striped Grass Mice; Spotted Grass
           Mice; Striped Mice; Single-striped
           Field Mice; Zebra Rats; Striped
           Grass Rats

d   Streifengrasmäuse; Grasmäuse
f   Rats rayés d'Afrique; Rats des
           prairies à bande simple
i   Topi striati africani

3163   *Lemniscomys barbarus*
           Rodentia - Muridae
e   Barbary Striped Grass Mouse;
           African Striped Grass Mouse;
           Chipmunk Mouse; Striped Field
           Mouse

d   Afrikanische Streifengrasmaus;
           Mehrstreifengrasmaus
f   Rat rayé de Barbarie; Souris rayée de
           Barbarie

3164   *Lemniscomys barbarus barbarus*
           Rodentia - Muridae
e   Algerian Mouse
d   Algerische Vielstreifengrasmaus

3165   *Lemniscomys barbarus zebra*
           Rodentia - Muridae
e   Zebra Mouse; Zebra Rat
d   Streifengrasmaus

3166 *Lemniscomys bellieri*
          Rodentia - Muridae
e   Bellier's Striped Grass Mouse

3167 *Lemniscomys griselda*
          Rodentia - Muridae
e   Griselda's Striped Grass Mouse
d   Einstreifengrasmaus

3168 *Lemniscomys hoogstraali*
          Rodentia - Muridae
e   Hoogstraal's Striped Grass Mouse

3169 *Lemniscomys linulus*
          Rodentia - Muridae
e   Senegal One-striped Grass Mouse

3170 *Lemniscomys macculus*
          Rodentia - Muridae
e   Buffoon Striped Grass Mouse

3171 *Lemniscomys mittendorfi*
          Rodentia - Muridae
e   Mittendorf's Striped Grass Mouse

3172 *Lemniscomys rosalia*
          Rodentia - Muridae
e   Single-striped Grass Mouse; Single-
          striped Grass Rat

3173 *Lemniscomys roseveari*
          Rodentia - Muridae
e   Rosevear's Striped Grass Mouse
f   Rat rayé de Zambie

3174 *Lemniscomys striatus*
          Rodentia - Muridae
e   Typical Striped Grass Mouse;
          Spotted Grass Mouse
d   Tüpfelgrasmaus
i   Topo striato africano

3175 *Lemniscus*
          Rodentia - Muridae

e   Sagebrush Voles
f   Campagnols des sauges

3176 *Lemniscus curtatus*
          Rodentia - Muridae
e   Sagebrush Vole
f   Campagnol des sauges

3177 *Lemnus*
          Rodentia - Muridae
e   Lemmings; True Lemmings; Brown
          Lemmings
d   Lemminge; Echte Lemminge
f   Lemmings
i   Lemming

3178 *Lemnus amurensis*
          Rodentia - Muridae
e   Amur Lemming
d   Amur-Lemming

          Lemnus chrysogaster syn. *Lemnus
          sibiricus* q.v.

3179 *Lemnus lemnus*
          Rodentia - Muridae
e   Norway Lemming; Norwegian
          Lemming
d   Gemeiner Lemming; Gewöhnlicher
          Lemming; Europäischer Lemming;
          Lemming; Berglemming;
          Norwegischer Berglemming
f   Lemming des toundras; Lemming
          des toundras de Norvège; Lemming
          des toundras du Nord; Lemming
i   Lemming comune

          Lemnus schisticolor syn. *Myopus
          schisticolor* q.v.

3180 *Lemnus sibiricus*
          Rodentia - Muridae
e   Brown Lemming; Point Barrow
          Brown Lemming; Siberian Brown
          Lemming

d Sibirischer Lemming
i Lemming della Siberia

3181 *Lemur*
Primates - Lemuridae
e Lemurs; Ring-tailed Lemurs
d Echte Makis; Kattas
f Lémurs
i Lemuri

3182 *Lemur catta*
Primates - Lemuridae
e Ring-tailed Lemur
d Katta; Katzenlemur;
Ringelschwanzmaki
f Lémur catta; Maki catta
i Lemure catta; Catta

Lemur coronatus syn. *Eulemur coronatus* q.v.

Lemur fulvus syn. *Eulemur fulvus* q.v.

Lemur macaco syn. *Eulemur macaco* q.v.

Lemur mongoz syn. *Eulemur mongoz* q.v.

Lemur rubriventer syn. *Eulemur rubriventer* q.v.

Lemur variegatus syn. *Varecia variegata* q.v.

3183 Lemuridae
Primates
e Lemurs; Large Lemurs
d Makis; Lemuren
f Lémuridés
i Lemuri; Scimmie e Proscimmie;
Lemuridi

3184 *Lenomys*
Rodentia - Muridae

e Trefoil-toothed Giant Rats; Sulawesi
Giant Rats; Celebes Giant Rats;
Celebes Rats
d Kleezahnriesenratten
f Rats des Célèbes

Lenomys longicaudus syn. *L. meyeri* q.v.

3185 *Lenomys meyeri*
Rodentia - Muridae
e Trefoil-toothed Giant Rat; Sulawesi
Giant Rat
d Sulawesi-Riesenratte

3186 *Lenothrix*
Rodentia - Muridae
e Grey Tree Rats
d Graue Baumratten

3187 *Lenothrix canus*
Rodentia - Muridae
e Grey Tree Rat
d Graue Baumratte

3188 *Lenoxus*
Rodentia - Muridae
e Andean Rats; Peruvian Rats
d Anden-Ratten

3189 *Lenoxus apicalis*
Rodentia - Muridae
e Andean Rat; Peruvian Rat
d Peru-Ratte

Leo syn. *Panthera* q.v.

3190 *Leontopithecus*
Primates - Calltrichidae
e Lion Tamarins; Maned Tamarins;
Lion-headed Marmosets; Golden
Marmosets; Golden Lion Marmosets;
Golden Lion Tamarins; Lion
Marmosets
d Löwenäffchen

f   Singes-lions

i   Scimmie leonine; Leontocebi

3191   ***Leontopithecus caissara***
      Primates - Calltrichidae

e   Black-faced Lion Tamarin; Black-
     headed Lion Tamarin; Superagui
     Lion Tamarin

d   Schwarzkopflöwenäffchen

f   Tamarin-lion de Caixa; Tamarin-lion
     à face noire

i   Leontocebo dalla testa nera

3192   ***Leontopithecus chrysomelas***
      Primates - Calltrichidae

e   Golden-headed Lion Tamarin ;
     Golden-headed Tamarin

d   Goldkopflöwenäffchen

f   Tamarin-lion à tête dorée; Singe-lion
     à tête dorée

i   Leontoceba dalla testa dorata

3193   ***Leontopithecus chrysopygus***
      Primates - Calltrichidae

e   Golden-rumped Lion Tamarin; Black
     Lion Tamarin

d   Rotsteißlöwenäffchen;
     Goldsteißlöwenäffchen

f   Tamarin-lion à croupe dorée; Singe-
     lion à queue jane

i   Leontocebo dalla groppa rossa

3194   ***Leontopithecus rosalia***
      Primates - Calltrichidae

e   Golden Lion Tamarin; Lion Tamarin;
     Lion-headed Marmoset

d   Goldgelbes Löwenäffchen; Großes
     Löwenäffchen

f   Petit singe-lion; Tamarin-lion doré

i   Leontocebo rosalia; Scimmia leonina

      Leopardus syn. *Felis* q.v.

3195   ***Leopoldamys***
      Rodentia - Muridae

e   Long-tailed Giant Rats

d   Asiatische Langschwanzriesenratten;
     Langschwanzriesenratten

3196   ***Leopoldamys edwardsi***
      Rodentia - Muridae

e   Edwards's Rat; Edwards's Long-
     tailed Giant Rat; Milne-Edwards
     Long-tailed Giant Rat

3197   ***Leopoldamys neilli***
      Rodentia - Muridae

e   Neill's Rat; Neill's Long-tailed Giant
     Rat

3198   ***Leopoldamys sabanus***
      Rodentia - Muridae

e   Long-tailed Giant Rat; Noisy Rat;
     Common Long-tailed Giant Rat

3199   ***Leopoldamys siporanus***
      Rodentia - Muridae

e   Mentawai Long-tailed Giant Rat;
     Mentawai Long-tailed Rat

3200   ***Lepilemur***
      Primates - Lemuridae

e   Sportive Lemurs; Weasel Lemurs

d   Wieselmakis

f   Lépilémurs

i   Lepilemuri

3201   ***Lepilemur dorsalis***
      Primates - Lemuridae

e   Grey-backed Sportive Lemur

d   Graurückenmaki

f   Lépilémur à dos gris

i   Lepilemure dalla schiena grigia

3202   ***Lepilemur edwardsi***
      Primates - Lemuridae

e   Milne-Edwards's Sportive Lemur;
     Edwards's Sportive Lemur

d   Milne-Edwards-Wieselmaki;
     Edwards Lemur

f   Lépilémur de Milne-Edwards

i   Lepilemure di Milne-Edwards

3203   ***Lepilemur leucopus***
Primates - Lemuridae
e   White-footed Sportive Lemur
d   Weißfußwieselmaki
f   Lépilémur aux pieds blancs
i   Lepilemure dai piedi bianchi

3204   ***Lepilemur microdon***
Primates - Lemuridae
e   Small-toothed Sportive Lemur;
Light-necked Sportive Lemur
d   Kleinzahnwieselmaki
f   Lépilémur aux petits dents
i   Lepilemure dal collo chiaro

3205   ***Lepilemur mustelinus***
Primates - Lemuridae
e   Greater Sportive Lemur; Weasel
Lemur; Sportive Lemur; Weasel
Sportive Lemur
d   Großer Wieselmaki
f   Lépilmur mustélin; Lépilémur à
queue rouge
i   Lepilemure mustelino; Lepilemure
sportivo; Lemure donnola

3206   ***Lepilemur ruficaudatus***
Primates - Lemuridae
e   Red-tailed Sportive Lemur
d   Kleiner Wieselmaki
f   Lépilémur à queue rousse; Lépilémur
à queue rouge
i   Lepilemure dalla coda rossa

3207   ***Lepilemur septentrionalis***
Primates - Lemuridae
e   Northern Sportive Lemur
d   Nördlicher Wieselmaki
f   Lépilémur septentrional
i   Lepilemure settentrionale

3208   **Leporidae**
Lagomorpha
e   Hares; Hares and Rabbits; European
Hares
d   Hasenartige; Hasen; Echte Hasen;

Hasen und Kaninchen
f   Léporidés
i   Leporidi

3209   ***Leporillus***
Rodentia - Muridae
e   Sticknest Rats (ANZ); Australian
Sticknest Rats; House-building
Jerboa Rats
d   Australische Häschenratten;
Häschenratten; Zweignestratten
f   Léporilles; Rats nicheurs; Rats
australiens
i   Leporilli

3210   ***Leporillus apicalis***
Rodentia - Muridae
e   White-tailed Sticknest Rat (ANZ);
White-tipped Hapalotis; White-
tipped Sticknest Rat; Lesser Sticknest
Rat; Lesser Australian Sticknest Rat;
White-tailed Hapalotis; Tillikin;
White-tailed Sticknest Rat
d   Kleine Zweignestratte; Kleine
Häschenratte
f   Petit rat lièvre; Petite léporille

3211   ***Leporillus conditor***
Rodentia - Muridae
e   Greater Sticknest Rat; Sticknest Rat
d   Langohrhäschenratte; Große
Zweignestratte
f   Rat architect
i   Leporillo costruttore

Leptailurus syn. *Felis* q.v.

Leptogale syn. *Microgale* q.v.

3212   ***Leptomys***
Rodentia - Muridae
e   New Guinea Water Rats
d   Neuguinea-Wasserratten
f   Rats aquatiques de Nouvelle Guinée;
Rats de Nouvelle Guinée

3213   *Leptomys elegans*
Rodentia - Muridae
e   New Guinea Water Rat; Long-footed
Hydromyine; Long-footed Water
Rat; Large Leptomys
d   Langfußwasserratte

3214   *Leptomys ernstmayri*
Rodentia - Muridae
e   Ernst Mayr's Water Rat; Ernst Mayr's
Leptomys

3215   *Leptomys signatus*
Rodentia - Muridae
e   Fly River Water Rat; Fly River
Leptomys

3216   *Leptonychotes*
Carnivora - Phocidae
e   Weddell Seals
d   Weddell-Robben
f   Phoques de Weddell
i   Foche di Weddell

3217   *Leptonychotes weddelli*
Carnivora - Phocidae
e   Weddell Seal; Wedell's Seal
d   Wedell-Robbe
f   Phoque de Weddell
i   Foca di Weddelll

3218   *Leptonycteris*
Chiroptera - Phyllostomidae
e   Saussure's Long-nosed Bats; Long-
nosed Bats
f   Petites chauve-souris à long nez

3219   *Leptonycteris curasoae*
Chiroptera - Phyllostomidae
e   Southern Long-nosed Bat; Lesser
Long-nosed Bat; Curacao Long-
nosed Bat
f   Chauve-souris à long nez du Sud

3220   *Leptonycteris nivalis*

Chiroptera - Phyllostomidae
e   Saussure's Long-nosed Bat; Mexican
Long-nosed Bat; Big Long-nosed Bat
f   Grande chauve-souris à long nez de
Mexique

3221   *Leptonycteris yerbabuanae*
Chiroptera - Phyllostomidae
e   North American Long-nosed Bat
f   Chauve-souris à long nez du Nord

3222   *Lepus*
Lagomorpha - Leporidae
e   Hares; True Hares; Hares and
Jackrabbits
d   Echte Hasen; Eigentliche Hasen;
Hasen
f   Lièvres; Lièvres typiques
i   Lepri

3223   *Lepus alleni*
Lagomorpha - Leporidae
e   Antelope Jackrabbit
d   Antilopenhase
f   Lièvre antilope

3224   *Lepus americanus*
Lagomorpha - Leporidae
e   Snowshoe Hare; Varying Hare
d   Schneeschuhhase
f   Lièvre américain; Lièvre d'Amérique;
Lièvre de l'Amérique du Nord;
Lièvre variable
i   Lepre variabile del Nordamerica;
Lepre dalle zampe bianche

3225   *Lepus americanus macfarlani*
Lagomorpha - Leporidae
e   Alaskan Snowshoe Hare
i   Lepre polare

3226   *Lepus arcticus*
Lagomorpha - Leporidae
e   Arctic Hare
d   Arktisher Hase; Polarhase

f   Lièvre arctique
i   Lepre artica; Lepre del Canada

3227   *Lepus brachyurus*
Lagomorpha - Leporidae
e   Japanese Hare
d   Japanischer Hase; Kurzschwanzhase
f   Lièvre du Japon

3228   *Lepus californicus*
Lagomorpha - Leporidae
e   Black-tailed Jackrabbit; Black
Jackrabbit; Jackass Rabit
d   Kalifornischer Eselhase;
Schwarzschwanz Hase
f   Lièvre de Californie
i   Jack Rabbit dalla coda nera

3229   *Lepus callotis*
Lagomorpha - Leporidae
e   White-sided Jackrabbit
d   Mexiko-Hase

3230   *Lepus capensis*
Lagomorpha - Leporidae
e   Cape Hare
d   Kap-Hase; Marokko-Hase; Angola-
Hase
f   Lièvre du Cap; Lièvre brun
i   Lepre del Capo; Lepre africana

3231   *Lepus capensis mediterraneus*
Lagomorpha - Leporidae
e   Mediterranean Hare
d   Mittelmeer Hase
f   Lièvre de la Mediterranée
i   Lepre sarda; Lepre comune; Lepre
italica

3232   *Lepus castroviejoi*
Lagomorpha - Leporidae
e   Broom Hare
d   Besenhase
f   Lièvre de Castroviejo
i   Lepre di Castroviejo

3233   *Lepus comus*
Lagomorpha - Leporidae
e   Yunnan Hare; Woolly Hare
d   Tibetanischer Wollhase; Yunnan
Hase

3234   *Lepus coreanus*
Lagomorpha - Leporidae
e   Korean Hare
d   Korea-Hase; Koreanischer Hase

3235   *Lepus corsicanicus*
Lagomorpha - Leporidae
e   Corsican Hare
d   Korsika-Hase
f   Lièvre de Corse
i   Lepre appenninica; Lepre italica;
Lepre italo-corsa

**_Lepus crawshayi_ syn. *L. saxatilis*
q.v.**

3236   *Lepus europaeus*
Lagomorpha - Leporidae
e   European Hare; Brown Hare;
European Brown Hare
d   Europäischer Feldhase; Gemeiner
Feldhase; Westlicher Feldhase;
Gemeiner Hase; Feldhase
f   Lièvre d'Europe; Lièvre d'Europe
occidentale; Lièvre européen; Lièvre
commun; Lièvre des bois; Lièvre
brun; Lièvre ordinaire; Lièvre
vulgaire; Lièvre
i   Lepre europea; Lepre comune
europea; Lepre; Lepre comune

3237   *Lepus fagani*
Lagomorpha - Leporidae
e   Ethiopian Hare; Abyssinian Hare
d   Äthiopischer Hase; Abessinischer
Hase
f   Lièvre éthiopien
i   Lepre abissina

3238   *Lepus flavigularis*
Lagomorpha - Leporidae

e   Tehuantepec Jackrabbit
d   Tehuantepec-Eselhase; Tuantepec-Hase

3239   *Lepus granatensis*
Lagomorpha - Leporidae
e   Granada Hare; Iberian Hare
d   Iberischer Hase; Granada-Hase
f   Lièvre Ibérique
i   Lepre iberica

3240   *Lepus hainanus*
Lagomorpha - Leporidae
e   Hainan Hare
d   Hainan-Hase
i   Lepre Hainan

3241   *Lepus insularis*
Lagomorpha - Leporidae
e   Black Jackrabbit
d   Schwarzer Eselhase; Schwarzer Hase
i   Lepre negra

3242   *Lepus mandshuricus*
Lagomorpha - Leporidae
e   Manchurian Hare
d   Mandschurischer Hase

Lepus mexicanus syn. *L. callotis* q.v.

Lepus monticularis syn. *Bunolagus monticularis* q.v.

3243   *Lepus nigricollis*
Lagomorpha - Leporidae
e   Indian Hare; Black-naped Hare
d   Schwarznackenhase; Indischer Hase
f   Lièvre à collier noir

3244   *Lepus nigricollis singhala*
Lagomorpha - Leporidae
e   Ceylon Black-naped Hare
d   Schwarzgefleckter Sri Lanka-Hase

3245   *Lepus oiostolus*
Lagomorpha - Leporidae
e   Woolly Hare
d   Wolliger Hase

3246   *Lepus othus*
Lagomorpha - Leporidae
e   Alaskan Hare
d   Tundrahase; Alaska-Hase
f   Lièvre polaire; Lièvre d'Alaska

3247   *Lepus othus othus*
Lagomorpha - Leporidae
e   Tundra Hare

3248   *Lepus peguensis*
Lagomorpha - Leporidae
e   Burmese Hare; Siamese Hare
d   Pegu-Hase; Siam-Hase; Birmesischer Hase; Birmanischer Hase
i   Lepre della Birmania

3249   *Lepus saxatilis*
Lagomorpha - Leporidae
e   Scrub Hare; Southern Bush Hare; Whyte's Hare
d   Afrikanischer Berghase; Buschhase; Strauchhase; Berghase; Whyte-Hase; Savannenhase; Malawi-Hase
f   Lièvre des rochers; Lièvre de Whyte; Lièvre des buissons
i   Lepre sudafricana; Lepre di Whyte; Lepre di Crayshaw

Lepus siamensis syn. *L. peguensis* q.v.

3250   *Lepus sinensis*
Lagomorpha - Leporidae
e   Chinese Hare; East Chinese Hare
d   Chinesischer Hase
f   Lièvre chinois

3251   *Lepus starki*
Lagomorpha – Leporidae

    e   Ethiopian Highland Hare
    d   Äthiopischer Hochlandhase

3252   ***Lepus timidus***
        Lagomorpha - Leporidae
    e   Mountain Hare; Arctic Hare;
        Variable Hare; Alpine Hare; Polar
        Hare; Hill Hare; Varying Hare
    d   Eurasischer Schneehase; Nordischer
        Schneehase; Veränderlicher Hase;
        Schneehase; Gebirgshase
    f   Lièvre changeant; Lièvre timide;
        Lièvre variable; Lièvre montagnard;
        Lièvre des montagnes
    i   Lepre variabile; Lepre bianca; Lepre
        alpina; Lepre blu

3253   ***Lepus timidus hibernicus***
        Lagomorpha - Leporidae
    e   Irish Mountain Hare; Northern
        Ireland Irish Hare
    f   Lièvre d'Irlande

3254   ***Lepus timidus scoticus***
        Lagomorpha - Leporidae
    e   Scottish Mountain Hare
    f   Lièvre d'Écosse

3255   ***Lepus timidus timidus***
        Lagomorpha - Leporidae
    e   Scandinavian Hare
    f   Lièvre variable de Scandinavie

3256   ***Lepus timidus varronis***
        Lagomorpha - Leporidae
    e   Blue Hare
    f   Lièvre variable des Alpes; Blanchon
    i   Lepre variabile

3257   ***Lepus tolai***
        Lagomorpha - Leporidae
    e   Tolai Hare
    d   Tolai-Hase
    f   Lièvre Tolai

3258   ***Lepus townsendii***
        Lagomorpha - Leporidae
    e   White-tailed Jackrabbit
    d   Prairiehase; Weißschwanzeselhase;
        Weißschwanzhase
    f   Lièvre de Townsend
    i   Coniglio saltanto dalla coda bianca

3259   ***Lepus townsendii townsendii***
        Lagomorpha - Leporidae
    e   Western White-tailed Jackrabbit;
        Utah White-tailed Jackrabbit

3260   ***Lepus victoriae***
        Lagomorpha - Leporidae
    e   African Savanna Hare
    d   Mosambique-Hase

        Lepus whytei syn. *L. saxatilis* q.v.

3261   ***Lepus yarkandensis***
        Lagomorpha - Leporidae
    e   Yarkand Hare
    d   Jarkand-Hase; Yarkand-Hase

3262   ***Lestodelphys***
        Didelphimorphia - Didelphidae
    e   Patagonian Opossums
    d   Patagonien-Zwergbeutelratten;
        Patagonische Beutelratten

3263   ***Lestodelphys halli***
        Didelphimorphia - Didelphidae
    e   Patagonian Opossum
    d   Patagonien-Zwergbeutelratte;
        Patagonische Beutelratte

3264   ***Lestoros***
        Paucituberculata - Caenolestidae
    e   Peruvian Rat Opossums; Incan Shew
        Opossums
    d   Peru-Opossummäuse; Peruanische
        Mausopossums; Spitzmausopossums
    f   Opossums musaraignes

3265 *Lestoros inca*
Paucituberculata - Caenolestidae
e   Peruvian Rat Opossum; Incan Shew
Opossum; Shrew Possum
d   Peru-Opossummaus; Peruanische
Mausopossum; Inka-
Spitzmausopossum
f   Opossum musaraigne inca

Leuconoe syn. *Myotis* q.v.

3266 *Liberiictis*
Carnivora - Herpestidae
e   Liberian Mongooses
d   Liberia-Mangusten
i   Manguste della Liberia

3267 *Liberiictis kuhni*
Carnivora - Herpestidae
e   Kuhn's Mongoose; Liberian
Mongoose; Kuhn's Kusimanse
d   Liberia-Manguste
i   Mangusta della Liberia; Mangusta di
Kuhn

Lichanotus syn. *Avahi* q.v.

3268 *Lichonycteris*
Chiroptera - Phyllostomidae
e   Brown-nosed Bats; Dark Long-
tongued Bats

3269 *Lichonycteris obscura*
Chiroptera - Phyllostomidae
e   Brown Long-nosed Bat; Dusky
Long-tongued Bat; Dark Long-
tongued Bat

3270 *Limnogale*
Soricomorpha - Tenrecidae
e   Water Shrews; Web-footed Tanrecs;
Marsh Tenrecs; Aquatic Tenrecs
d   Wassertanreks; Wassertenreks
f   Limnogales

3271 *Limnogale mergulus*
Soricomorpha - Tenrecidae
e   Water Shrew; Web-footed Tanrec;
Marsh Tenrec; Aquatic Tenrec;
Water Tenrec
d   Wassertanrek; Wassertenrek
f   Limnogale

3272 *Limnomys*
Rodentia - Muridae
e   Mindano Mountain Rats
d   Mindanao-Ratten

Limnomys mearnsi syn. *L. sibuanus*
q.v.

3273 *Limnomys sibuanus*
Rodentia - Muridae
e   Mindano Mountain Rat
d   Mindanao-Ratte

3274 *Liomys*
Rodentia - Geomyidae
e   Spiny Pocket Mice
d   Stacheltaschenmäuse
f   Souris épineuses à poche

3275 *Liomys adspersus*
Rodentia - Geomyidae
e   Panamanian Spiny Pocket Mouse

Liomys annectens syn. *L. pictus* q.v.

Liomys crispus syn. *L. salvini* q.v.

3276 *Liomys irroratus*
Rodentia - Geomyidae
e   Mexican Spiny Pocket Mouse;
Mexican Pocket Mouse
d   Mexikanische Taschenmaus
f   Souris épineuse à poche du Mexique;
Souris aux abajoues épineuses;
Souris aux abajoues épineuses du
Mexique

3277 **Liomys pictus**
Rodentia - Geomyidae
e Painted Spiny Pocket Mouse
d Gemalte Stacheltaschenmaus

3278 **Liomys salvini**
Rodentia - Geomyidae
e Salvin's Spiny Pocket Mouse

3279 **Liomys spectabilis**
Rodentia - Geomyidae
e Jaliscan Spiny Pocket Mouse

3280 **Lionycteris**
Chiroptera - Phyllostomidae
e Little Long-tongued Bats; Chestnut Long-tongued Bats

3281 **Lionycteris spurrelli**
Chiroptera - Phyllostomidae
e Little Long-tongued Bat; Chestnut Long-tongued Bat

Liponycteris syn. *Taphozous* q.v.

3282 **Lipotes**
Cete - Lipotidae
e Chinese Dolphins; White Flag Dolphins; Yangtze River Dolphins
d Chinesische Flussdelfine
f Dauphins d'eau douce de Chine
i Delfini lacustri cinesi; Lipoti

3283 **Lipotes vexillifer**
Cete - Lipotidae
e Chinese Lake Dolphin; Chinese River Dolphin; White Fin Dolphin; White Flag Dolphin; Yangtze River Dolphin; Beiji; Baijitun; Baiji
d Chinesischer Flussdelfin; Baiji; Jangtse-Delfin; Yangtsi-Delfin
f Dauphin d'eau douce de Chine; Dauphin de Chine; Baiji; Dauphin du Yangtsé; Dauphin lacustre de Chine
i Lipote; Delfino dalla pinna bianca; Delfino lacustre cinese

3284 **Lipotidae**
Cete
e Chinese Dolphins; Yangtze River Dolphins
d Chinesische Delfine
f Lipotidés
i Delfini lacustri cinesi; Lipotidi

3285 **Lissodelphis**
Cete - Delphinidae
e Right Whale Dolphins
d Glattdelfine
f Lissodelphes
i Lissodelfini

3286 **Lissodelphis borealis**
Cete - Delphinidae
e Northern Right Whale Dolphin; Pacific Right Whale Dolphin; Northern Right Whale Porpoise; Pacific Right Whale Porpoise
d Nördlicher Glattdelfin
f Dauphin du Nord; Lissodelphe boréale; Dauphin aptère boréal
i Lissodelfino boreale

3287 **Lissodelphis peronii**
Cete - Delphinidae
e Péron's Dolphin; Southern Right Whale Dolphin; White-bellied Right Whale Dolphin
d Südlicher Glattdelfin
f Dauphin de Péron; Lissodelphe australe; Dauphin-baleine austral
i Lissodelfino australe

Lissonycteris syn. *Rousettus* q.v.

3288 **Litocranius**
Artiodactyla - Bovidae
e Gerenuks; Giraffe Gazelles; Waller's Gazelles
d Giraffengazellen; Gerenuks
f Gazelles-girafes; Guérénouks
i Gerenuk

3289 *Litocranius walleri*
Artiodactyla - Bovidae
e Gerenuk; Waller's Gazelle; Giraffe Gazelle
d Giraffengazelle; Gerenuk
f Gazelle-girafe; Guérénouk
i Antilope gerenuk; Antilope giraffa; Gerenuk

3290 *Litocranius walleri sclateri*
Artiodactyla - Bovidae
e Northern Gerenuk; Northern Waller's Gazelle

3291 *Litocranius walleri walleri*
Artiodactyla - Bovidae
e Southern Gerenuk; Southern Waller's Gazelle

3292 *Lobodon*
Carnivora - Phocidae
e Crabeater Seals; Crab-eating Seals
d Krabenesser ; Krabbenfresser
f Phoques crabiers
i Foche carcinofaghe

3293 *Lobodon carcinophagus*
Carnivora - Phocidae
e Crabeater Seal; Crab-eating Seal
d Krabenesser; Krabenfresser
f Phoque crabier
i Foca carcinofaga; Foca cancrivora

3294 *Lonchophylla*
Chiroptera - Phyllostomidae
e Long-tongued Bats; Nectar Bats

3295 *Lonchophylla bokermanni*
Chiroptera - Phyllostomidae
e Bokermann's Nectar Bat

3296 *Lonchophylla dekeyseri*
Chiroptera - Phyllostomidae
e Dekeyser's Nectar Bat

3297 *Lonchophylla handleyi*
Chiroptera - Phyllostomidae
e Handley's Nectar Bat

3298 *Lonchophylla hesperia*
Chiroptera - Phyllostomidae
e Western Nectar Bat

3299 *Lonchophylla mordax*
Chiroptera - Phyllostomidae
e Brazilian Long-tongued Bat; Goldman's Long-tongued Bat; Goldman's Nectar Bat

3300 *Lonchophylla robusta*
Chiroptera - Phyllostomidae
e Panama Long-tongued Bat; Rusty Long-tongued Bat; Orange Nectar Bat

3301 *Lonchophylla thomasi*
Chiroptera - Phyllostomidae
e Thomas's Long-tongued Bat; Thomas's Nectar Bat

3302 *Lonchorhina*
Chiroptera - Phyllostomidae
e Sword-nosed Bats; Tomes's Long-eared Bats
d Schwertnasen
f Vespertillions de Tomes

3303 *Lonchorhina aurita*
Chiroptera - Phyllostomidae
e Sword-nosed Bat; Tomes's Long-eared Bat; Tome's Sword-nosed Bat; Common Sword-nosed Bat
d Schwertnase
f Vespertillion de Tomes

3304 *Lonchorhina fernandezi*
Chiroptera - Phyllostomidae
e Fernandez's Sword-nosed Bat

3305 **Lonchorhina inusitata**
Chiroptera - Phyllostomidae
e Strange Sword-nosed Bat

3306 **Lonchorhina marinkellei**
Chiroptera - Phyllostomidae
e Marinkelle's Sword-nosed Bat

3307 **Lonchorhina orinocensis**
Chiroptera - Phyllostomidae
e Orinoco Sword-nosed Bat

3308 **Lonchothrix**
Rodentia - Echimyidae
e Tuft-tailed Spiny Spiny Tree Rats

3309 **Lonchothrix emiliae**
Rodentia - Echimyidae
e Tuft-tailed Spiny Spiny Tree Rat

Lontra syn. *Lutra* q.v.

3310 **Lophiomys**
Rodentia - Muridae
e Maned Rats; Crested Rats; Maned Hamsters; Crested Hamsters
d Mähnenratten
f Hamsters d'Imhaus
i Ratti dalla criniera

3311 **Lophiomys imhausi**
Rodentia - Muridae
e Maned Rat; Crested Rat; Maned Hamster; Crested Hamster
d Mähnenratte
f Hamster d'Imhaus
i Ratto dalla criniera

3312 **Lophocebus**
Primates - Cercopithecidae
e Grey-cheeked Mangabeys; Crested Mangabeys
d Mantelmangeben
f Cercocèbes aux joues grises; Mangabeys aux joues blanches; Mangabeys aux joues grises
i Cercocebi

3313 **Lophocebus albigena**
Primates - Cercopithecidae
e Grey-cheeked Mangabey; White-cheeked Mangabey
d Mantelmangabe; Grauwangenmangabe
f Cercocèbe aux joues grises; Mangabey aux joues blanches; Mangabey aux joues grises
i Cercocebo dal mantello; Cercocebo crestato

3314 **Lophocebus albigena johnstoni**
Primates - Cercopithecidae
e Johnston's Mangabey
f Cercocèbe aux joues grises de Johnston

3315 **Lophocebus aterrimus**
Primates - Cercopithecidae
e Black Mangabey; Black Crested Mangabey
d Schopfmangabe
f Cercocèbe noir; Mangabey noir; Cercocèbe à collier
i Cercocebo dal ciuffo; Cercocebo nero

3316 **Lophocebus opdenboschi**
Primates - Cercopithecidae
e Opdenbosch's Mangabey
d Opdenboschs Mangabe

3317 **Lophuromys**
Rodentia - Muridae
e Brush-furred Rats; Red-bellied Mice; Brush-furred Mice
d Afrikanische Bürstenhaarmäuse
f Rats hérissés; Rats à pélage en brosse

3318 **Lophuromys angolensis**
Rodentia - Muridae
e Angolan Brush-furred Rat; Angolan Brush-furred Mouse

3319   *Lophuromys aquilus*
Rodentia - Muridae
e   Blackish Brush-furred Rat; Blackish
Brush-furred Mouse

3320   *Lophuromys brevicaudus*
Rodentia - Muridae
e   Short-tailed Brush-furred Rat; Short-
tailed Brush-furred Mouse

3321   *Lophuromys brunneus*
Rodentia - Muridae
e   Brown Brush-furred Rat; Brown
Brush-furred Mouse

3322   *Lophuromys chrysopus*
Rodentia - Muridae
e   Golden-footed Brush-furred Rat;
Golden-footed Brush-furred Mouse

3323   *Lophuromys cinereus*
Rodentia - Muridae
e   Grey Brush-furred Rat; Grey Brush-
furred Mouse

3324   *Lophuromys dieterleni*
Rodentia - Muridae
e   Dieterlen's Brush-furred Rat

3325   *Lophuromys dudui*
Rodentia - Muridae
e   Dudu's Brush-furred Rat; Dudu's
Brush-furred Mouse

3326   *Lophuromys flavopunctatus*
Rodentia - Muridae
e   Yellow-spotted Brush-furred Rat;
Yellow-spotted Brush-furred Mouse
d   Ostafrikanische Bürstenhaarmaus

3327   *Lophuromys huttereri*
Rodentia - Muridae
e   Hutterer's Brush-furred Rat

3328   *Lophuromys luteogaster*
Rodentia - Muridae

e   Yellow-bellied Brush-furred Rat;
Yellow Brush-furred Mouse; Yellow-
bellied Brush-furred Mouse

3329   *Lophuromys medicaudatus*
Rodentia - Muridae
e   Medium-tailed Brush-furred Rat;
Medium-tailed Brush-furred Mouse

3330   *Lophuromys melanonyx*
Rodentia - Muridae
e   Black-clawed Brush-furred Rat;
Black-clawed Mouse

3331   *Lophuromys nudicaudus*
Rodentia - Muridae
e   Fire-bellied Brush-furred Rat; Fire-
bellied Brush-tailed Mouse

3332   *Lophuromys rahmi*
Rodentia - Muridae
e   Rahm's Brush-furred Rat; Rahm's
Brush-tailed Mouse

3333   *Lophuromys roseveari*
Rodentia - Muridae
e   Rosevear's Brush-furred Rat;
Rosevear's Brush-furred Mouse

3334   *Lophuromys sikapusi*
Rodentia - Muridae
e   Rusty-bellied Brush-furred Rat;
Rusty-bellied Rat; Rusty-bellied
Brush-furred Mouse
d   Westafrikanische Bürstenhaarmaus
f   Rat hérissé à dessus brun-sépia

3335   *Lophuromys verhageni*
Rodentia - Muridae
e   Verhagen's Brush-furred Rat;
Verhagen's Brush-furred Mouse

3336   *Lophuromys woosnami*
Rodentia - Muridae
e   Woosnam's Brush-furred Rat;
Woosnam's Brush-furred Mouse

d Woosnams Bürstenhaarmaus
f Rat à pélage en brosse de Woosnam

3337 *Lophuromys zena*
Rodentia - Muridae
e Zena Brush-furred Rat; Zena Brush-furred Mouse

3338 *Lorentzimys*
Rodentia - Muridae
e New Guinea Kangaroo Mice; New Guinea Jumping Mice
d Neuguinea-Springmäuse
f Rats de Nouvelle Guinée

3339 *Lorentzimys nouhuysi*
Rodentia - Muridae
e New Guinea Kangaroo Mouse; New Guinea Jumping Mouse; Long-footed Tree Mouse
d Neuguinea-Springmaus

3340 Loridae
Primates
e Lorises; Bushbabies; Slender-faced Strepsirhines; Angwantibos, Lorises and Pottos; Galagos and Lorises
d Loris
f Lorisidés
i Lorisidi; Lorisini

3341 *Loris*
Primates - Loridae
e Lorises; Slender Lorises; Ceylon Sloths
d Lori
f Loris
i Lori

3342 *Loris lydekkerianus*
Primates - Loridae
e Slow Loris; Grey Slender Loris
i Lori gracile

3343 *Loris lydekkerianus nycticeboides*

Primates - Loridae
e Horton Plains Slender Loris; Ceylon Mountain Slender Loris

3344 *Loris tardigradus*
Primates - Loridae
e Red Slender Loris; Slender Loris
d Schlanklori
f Loris grêle
i Lori gracile minore; Lori gracile

3345 *Loxodonta*
Uranotheria - Elephantidae
e African Elephants
d Afrikanische Elefanten
f Éléphants d'Afrique
i Elefanti africani

3346 *Loxodonta africana*
Uranotheria - Elephantidae
e African Elephant
d Kap-Elefant
f Éléphant d'Afrique
i Elefante africano; Elefante di boscaglia

3347 *Loxodonta africana africana*
Uranotheria - Elephantidae
e South African Bush Elephant; African Bush Elephant; African Savanna Elephant; African Savannah Elephant
d Steppenelefant
f Éléphant de la savane; Éléphant de buisson; Éléphant de savane
i Elefante di savana

3348 *Loxodonta africana pumilio*
Uranotheria - Elephantidae
e Pygmy African Elephant
f Éléphant nain; Éléphant pygmée

3349 *Loxodonta cyclotis*
Uranotheria - Elephantidae
e African Forest Elephant; Forest

Elephant
d Waldelefant
f Éléphant de forêt
i Elefante di foresta; Elefante africano di foresta

3350 *Loxodontomys*
Rodentia - Muridae
e Big-eared Mice

3351 *Loxodontomys micropus*
Rodentia - Muridae
e Southern Big-eared Mouse

3352 *Lundomys*
Rodentia - Muridae
e Greater Marsh Rats

3353 *Lundomys molitor*
Rodentia - Muridae
e Greater Marsh Rat

3354 *Lutra*
Carnivora - Mustelidae
e Otters; River Otters; Typical Otters
d Fischotter ; Otter
f Loutres
i Lontre

Lutra annectens syn. *L. longicaudis* q.v.

3355 *Lutra canadensis*
Carnivora - Mustelidae
e Canadian River Otter; North American River Otter; Northern River Otter; River Otter (NA); Virginia Otter; Florida Otter
d Nordamerikanischer Fischotter; Kanadischer Fischotter; Florida-Otter; Virginia-Otter; Kanada-Otter; Amerikanischer Fischotter
f Loutre du Canada
i Lontra canadese; Lontra di fiume del Nord America; Lontra di fiume nordamericana

3356 *Lutra canadensis lataxina*
Carnivora - Mustelidae
e Southeastern River Otter

3357 *Lutra canadensis pacifica*
Carnivora - Mustelidae
e Southwestern River Otter; River Otter

3358 *Lutra canadensis sonora*
Carnivora - Mustelidae
e Sonoran River Otter; Southwestern River Otter

3359 *Lutra felina*
Carnivora - Mustelidae
e Chungungo; Marine Otter; South American Otter; Southern Sea Otter; Sea Cat; Chunhungo
d Küstenotter; Chingungo; Chilenischer Otter; Meerotter
f Loutre de mer; Loutre de mer méridionale
i Lontra marina; Lontra felina; Lontra del Cile

3360 *Lutra longicaudis*
Carnivora - Mustelidae
e Neotropical River Otter; Neotropical Otter; Southern River Otter; South American River Otter; South American Otter; Washback; La Plata Otter; Chilean Otter; Long-tailed Otter
d Südamerikanischer Otter; Südamerikanischer Fischotter; La Plata-Otter; Mittelamerikanischer Otter; Mittelamerikanischer Fischotter; Costa-Rica-Otter
f Loutre à longue queue; Loutre de Costa Rica; Loutre de l'Amérique du Sud
i Lontra a coda lunga del Centro e sud America; Lontra dell'America centrale; Lontra di Costa Rica; Lontra del rio de La Plata; Lontra di Costa Rica; Lontra di fiume del Sudamerica

3361 **Lutra longicaudis enudris**
Carnivora - Mustelidae
e Brazilian River Otter
d Brasilianischer Flussotter
f Loutre du Brésil
i Lontra del Brasile

3362 **Lutra longicaudis incarum**
Carnivora - Mustelidae
e Peruvian Otter
d Peru-Otter
f Loutre du Pérou
i Lontra del Peru

3363 **Lutra lutra**
Carnivora - Mustelidae
e Otter; Common Otter; Eurasian
Otter; European Otter; Old World
Otter; River Otter; Land Otter;
European River Otter; Eurasian River
Otter
d Gemeiner Fischotter; Fischotter;
Flussotter; Gemeiner Otter;
Gemeiner Flussotter; Eurasischer
Fischotter; Bagdad-Otter;
Europäischer Fischotter; Otter
f Loutre commune; Loutre d'Europe;
Loutre de rivière; Loutre vulgaire;
Loutre ordinaire; Loutre
i Lontra; Lontra comune; Lontra di
fiume; Lontra europea; Lontra
europea e asiatica; Lutra

3364 **Lutra lutra ceylonica**
Carnivora - Mustelidae
e Ceylon Otter

3365 **Lutra maculicollis**
Carnivora - Mustelidae
e Spotted-necked Otter; Speckle-
throated Otter; Spot-necked Otter;
Rhodesia Otter; African Spotted-neck
River Otter
d Fleckenhalsotter; Kongo-Otter;
Rhodesien-Otter
f Loutre à cou tacheté
i Lontra dal collo macchiato; Lontra di

fiume del Congo e Rodesia

Lutra perspicillata syn. *Lutrogale
perspicillata* q.v.

Lutra platensis syn. *Lontra
longicaudis* q.v.

3366 **Lutra provocax**
Carnivora - Mustelidae
e Southern River Otter; Washback;
Huillin; Chilean River Otter
d Südlicher Flussotter
f Loutre du Chili; Huillin
i Lontra di fiume meridionale; Lontra
del Cile; Lontra di fiume del Cile e
Argentina

3367 **Lutra sumatrana**
Carnivora - Mustelidae
e Hairy-nosed Otter; Sumatran Otter
d Haarnasenotter
f Loutre de Sumatra
i Lontra dal naso peloso

Lutreola syn. *Mustela* q.v.

3368 **Lutreolina**
Didelphimorphia - Didelphidae
e Thick-tailed Opossums; Little Water
Opossums; Lutrine Opossums
d Dickschwanzbeutelratten

3369 **Lutreolina crassicaudata**
Didelphimorphia - Didelphidae
e Thick-tailed Opossum; Little Water
Opossum; Lutrine Opossum; Red
Opossum
d Dickschwanzbeutelratte

3370 **Lutrogale**
Carnivora - Mustelidae
e Smooth-coated Otters
d Indische Fischotter
f Loutres d'Asie
i Lontre asiatiche

3371   *Lutrogale perspicillata*
Carnivora - Mustelidae
e   Smooth-coated Otter; Smooth Indian
Otter; Smooth-coated Indian Otter;
Indian Smooth-coated Otter; Asian
Smooth-coated Otter
d   Indischer Fischotter; Weichfellotter
f   Loutre d'Asie
i   Lontra asiatica

3372   *Lycalopex*
Carnivora - Canidae
e   Hoary Foxes
d   Brasilianischer Kampfüchse

3373   *Lycalopex vetulus*
Carnivora - Canidae
e   Hoary Fox; Small-toothed Dog;
Hoary Zorro
d   Brasilianischer Kampfuchs

3374   *Lycaon*
Carnivora - Canidae
e   African Hunting Dogs; Cape Hunting
Dogs; African Wild Dogs
d   Hyänenhunde; Afrikanische
Wildhunde
f   Chiens chasseurs; Chiens sauvages
africains
i   Licaoni

3375   *Lycaon pictus*
Carnivora - Canidae
e   African Hunting Dog; Cape Hunting
Dog; Hunting Dog; African Wild
Dog; Wild Dog; Painted Dog
d   Afrikanischer Wildhund;
Steppenhund; Hyänenhund; Gemalter
Hund; Jagdhyäne
f   Chien chasseur; Chien sauvage
africain; Loup peint; Cynihène;
Chien sauvage d'Afrique; Chien
viverrin
i   Licaone; Cane iena; Cane selvatico
africano

3376   *Lyncodon*

Carnivora - Mustelidae
e   Patagonian Weasels
d   Zwerggrisons
i   Lincodonti

3377   *Lyncodon patagonicus*
Carnivora - Mustelidae
e   Patagonian Weasel
d   Zwerggrison; Patagonisches Wiesel
i   Donnola della Patagonia; Lincodonte

Lynx syn. *Felis* q.v.

Lyroderma syn. *Megaderma* q.v.

M

3378 *Macaca*
Primates - Cercopithecidae
e Macaques
d Makaken
f Macaques
i Macachi

3379 *Macaca arctoides*
Primates - Cercopithecidae
e Stump-tailed Macaque; Bear Macaque; Red-faced Stump-tailed Macaque
d Bärenmakak; Stumpfschwanzmakak
f Macaque à face rouge; Macaque brun
i Macaco orsino

3380 *Macaca assamensis*
Primates - Cercopithecidae
e Assam Macaque; Assamese Macaque; Himalajan Macaque
d Bergrhesus; Assam-Rhesus; Assam-Makak
f Macaque d'Assam
i Macaco dell'Assam; Macaco dell'Himalaja

Macaca brunnescens syn. *M. ochreata* q.v.

3381 *Macaca cyclopis*
Primates - Cercopithecidae
e Formosa Macaque; Taiwan Macaque; Taiwanese Macaque; Formosan Rock Macaquw
d Formosa-Makak; Rundgesichtmakak; Formosa Rhesus
f Macaque de Formose
i Macaco di Taiwan; Macaco di Formosa

3382 *Macaca fascicularis*
Primates - Cercopithecidae
e Crab-eating Macaque; Long-tailed Macaque; Kera; Cynomolgus Macaque; Java Monkey; Java Macaque; Cynomolgus Monkey
d Javaner Affe; Langschwanzmakak
f Macaque crabier; Macaque de l'Indonésie; Macaque de Buffon
i Macaco cinomolgo; Cinomolgo; Macaco di Giava

3383 *Macaca fascicularis atriceps*
Primates - Cercopithecidae
e Dark-crowned Long-tailed Macaque

3384 *Macaca fascicularis aurea*
Primates - Cercopithecidae
e Burmese Long-tailed Macaque

3385 *Macaca fascicularis condorensis*
Primates - Cercopithecidae
e Con Song Long-tailed Macaque

3386 *Macaca fascicularis fusca*
Primates - Cercopithecidae
e Simeulue Monkey

3387 *Macaca fascicularis karimondjawae*
Primates - Cercopithecidae
e Kemujan Long-tailed Macaque

3388 *Macaca fascicularis lasiae*
Primates - Cercopithecidae
e Lasia Long-tailed Macaque

3389 *Macaca fascicularis philippinensis*
Primates - Cercopithecidae
e Philippine Macaque

3390 *Macaca fuscata*
Primates - Cercopithecidae
e Japanese Macaque; Japanese Snow Macaque; Snow Monkey
d Rotgesichtsmakak; Japan-Makak

f   Macaque japonais
i   Macaco del Giappone

3391   ***Macaca fuscata yakui***
      Primates - Cercopithecidae
e   Yaku-Shima Japanese Macaque
f   Macaque de Yakushima

3392   ***Macaca hecki***
      Primates - Cercopithecidae
e   Heck's Macaque
f   Macaque à crète

      Macaca irus syn. *M. fascicularis* q.v.

3393   ***Macaca leonina***
      Primates - Cercopithecidae
e   Northern Pig-tailed Macaque
f   Macaque à queue de cochon du Nord

3394   ***Macaca maura***
      Primates - Cercopithecidae
e   Moor Macaque
d   Mohrenmakak
f   Macaque des Célèbes
i   Macaco nero

3395   ***Macaca mulatta***
      Primates - Cercopithecidae
e   Rhesus Macaque; Rhesus Monkey
d   Rhesusaffe; Rhesusmakak
f   Rhésus; Macaque rhésus
i   Reso; Macaco reso

3396   ***Macaca nemestrina***
      Primates - Cercopithecidae
e   Pig-tailed Macaque; Pigtail Macaque; Sunda Pig-tailed Macaque; Beruk
d   Schweinsaffe; Schweinsschwanzaffe
f   Singe à queue de cochon; Macaque à queue de cochon des îles de la Sonde; Nemestrinus
i   Macaco nemestrina; Macaco dalla coda di porco

3397   ***Macaca nigra***
      Primates - Cercopithecidae
e   Celebes Macaque; Crested Macaque; Sulawesi Crested Black Macaque; Celebes Ape; Celebes Crested Macaque; Sulawesi Crested Macaque; Black Ape; Celebes Black Macaque
d   Schopfmakak; Schopfpavian; Schopfaffe
f   Cynopithèque nègre; Macaque nègre
i   Cinopiteco; Cinopiteco nero; Macaco dei Celebes; Gorilla nero dei Celebes

3398   ***Macaca nigrescens***
      Primates - Cercopithecidae
e   Gorontalo Macaque
f   Macaque de Temminck

3399   ***Macaca ochreata***
      Primates - Cercopithecidae
e   Booted Macaque; Sulawesi Booted Macaque; Muna-Butung Macaque; Ochre Macaque; Celebes Ashy-black Monkey
d   Graumakak; Grauarmmakak
f   Macaque de Muna; Macaque des Célèbes aux bras gris
i   Macaco a braccie grigie

3400   ***Macaca pagensis***
      Primates - Cercopithecidae
e   Mentawai Macaque; Bokkoi; Pagai Island Macaque
d   Mentawai-Makak
i   Macaco delle Mentawai

3401   ***Macaca radiata***
      Primates - Cercopithecidae
e   Bonnet Monkey; Bonnet Macaque; Madras Macaque
d   Indischer Hutaffe
f   Bonnet chinois; Macaque bonnet chinois; Macaque commun
i   Macaco dal berretto indiano; Macaco dal berretto

3402   ***Macaca siberu***
Primates - Cercopithecidae
e   Siberut Macaque

3403   ***Macaca silenus***
Primates - Cercopithecidae
e   Lion-tailed Macaque; Wanderoo; Liontail Macaque
d   Wanderu; Bartaffe; Wanderaffe
f   Ouanderou; Macaque ouanderoo
i   Uanderù; Scimmia dalla barba bianca; Sileno

3404   ***Macaca sinica***
Primates - Cercopithecidae
e   Toque Macaque; Toque Monkey
d   Ceylon-Hutaffe
f   Macaque couroné; Macaque à toque
i   Macaco dal berretto di Sri Lanka; Macaco di Ceylon

Macaca speciosa syn. *M. arctoides* q.v.

3405   ***Macaca sylvanus***
Primates - Cercopithecidae
e   Barbary Ape; Barbary Macaque; Rock Ape; Magot
d   Berber-Affe; Magot; Magog; Gibraltar-Affe
f   Macaque de Barbarie; Magot; Magot de Gibraltar
i   Bertuccia

3406   ***Macaca thibetana***
Primates - Cercopithecidae
e   Père David's Macaque; Tibetan Stump-tailed Macaque; Tibetan Macaque; Milne-Edward's Macaque; Short-tailed Tibetan Macaque
d   Tibet-Makak; Tibetanischer Bärenmakak
f   Macaque du Tibet
i   Macaco tibetano; Macaca di Padre David

3407   ***Macaca tonkeana***

Primates - Cercopithecidae
e   Tonkean Macaque; Tonkean Black Ape
d   Tonkin-Makak; Tonkean-Makak
f   Macaque de Tonkean
i   Macaco di Tonkean

3408   ***Macroderma***
Chiroptera - Megadermatidae
e   Australian Giant False Vampire Bats; Giant False Vampires; Australian False Vampire Bats; Ghost Bat; Australian False Vampires
d   Australische Gespenstfledermäuse
f   Macrodermes; Macrodermes d'Australie
i   Falsi vampiri australiani

3409   ***Macroderma gigas***
Chiroptera - Megadermatidae
e   Australian Giant False Vampire Bat; Ghost Bat (ANZ); Australian Ghost Bat; False Vampire; Australian False Vampire; Australian False Vampire Bat
d   Australische Gespenstfledermaus; Australische Großblattnase
f   Macroderme d'Australie
i   Falso vampiro australiano; Pipistrello fantasma

3410   ***Macrogalidia***
Carnivora - Viverridae
e   Celebes Palm Civets; Brown Palm Civets; Sulawesi Palm Civets
d   Celebes-Roller ; Sulawesi-Roller
f   Civettes palmistes brunes des Célèbes
i   Civette delle palme di Celebes

3411   ***Macrogalidia musschenbroeki***
Carnivora - Viverridae
e   Celebes Palm Civet; Brown Palm Civet; Sulawesi Palm Civet
d   Celebes-Roller; Sulawesi-Roller
f   Civette palmiste brune des Célèbes
i   Civetta delle palme di Celebes

Macrogeomys syn. *Orthogeomys* q.v.

3412   *Macroglossus*
Chiroptera - Pteropodidae
e   Long-tongued Fruit Bats
d   Langzungenflughunde; Afrikanischer Langzungenflughund
f   Macroglosses

Macroglossus lagochilus syn. *M. minimus* q.v.

3413   *Macroglossus minimus*
Chiroptera - Pteropodidae
e   Small Long-tongued Fruit Bat; Asiatic Long-tongued Fruit Bat; Common Long-tongued Fruit Bat; Long-tongued Nectar Bat; Lesser Long-tongued Fruit Bat; Nectarivorous Fruit Bat; Lesser Long-tongued Nectar Bat; Northern Blossom Bat
d   Zwerglangzungenflughund
f   Macroglosse minime

3414   *Macroglossus sobrinus*
Chiroptera - Pteropodidae
e   Greater Long-tongued Fruit Bat; Hill Long-tongued Fruit Bat; Greater Long-tongued Nectar Bat

3415   *Macrophyllum*
Chiroptera - Phyllostomidae
e   Long-legged Bats
d   Langbeinfledermäuse
f   Macrophylles

3416   *Macrophyllum macrophyllum*
Chiroptera - Phyllostomidae
e   Long-legged Bat; Wied's Long-legged Bat
d   Langbeinfledermaus
f   Macrophylle

3417   Macropodidae
Diprotodontia

e   Wallabies and Kangaroos
d   Kängurus; Springbeutler
f   Kangourous
i   Canguri; Canguri e Ouallabie; Macropodidi

3418   *Macropus*
Diprotodontia - Macropodidae
e   Red and Grey Kangaroos; Great Grey Kangaroos; Foresters
d   Riesenkängurus; Großkängurus; Graue Riesenkängurus
f   Grands kangourous; Grands wallabies
i   Canguri

3419   *Macropus agilis*
Diprotodontia - Macropodidae
e   Sandy Wallaby; Agile Wallaby (ANZ); River Wallaby; Sand Wallaby
d   Flinkes Känguru; Sandwallaby; Flinkkänguru
f   Wallaby agile

3420   *Macropus antilopinus*
Diprotodontia - Macropodidae
e   Antelope Kangaroo; Antilopine Wallaroo (ANZ)
d   Antilopenkänguru
f   Wallaby antilope
i   Canguro gigante antilopina

3421   *Macropus bernardus*
Diprotodontia - Macropodidae
e   Bernard's Wallaroo; Black Wallaroo (ANZ)
d   Schwarzes Bergkänguru
f   Wallaby noir

3422   *Macropus dorsalis*
Diprotodontia - Macropodidae
e   Black-striped Wallaby
d   Rückstreifenkänguru; Aalstrichwallaby; Rückenstreifenwallaby
f   Wallaby à raie noire

3423 *Macropus eugenii*
Diprotodontia - Macropodidae
e Darby's Wallaby; Tammar Wallaby
(ANZ); Tammar; Dama Pademelon
d Eugene-Filander; Tammar; Derby-
Wallaby; Tammar-Wallaby; Derby-
Känguru
f Wallaby de l'île Eugène; Wallaby
thétis; Wallaby de l'île Kangourou
i Wallaby dell'isola Eugeni

3424 *Macropus fuliginosus*
Diprotodontia - Macropodidae
e Western Grey Kangaroo
d Westliches Graues Riesenkänguru
f Kangourou gris; Kangourou gris de
l'Ouest
i Canguro grigio occidentale; Canguro
gigante grigio occidentale

3425 *Macropus giganteus*
Diprotodontia - Macropodidae
e Great Grey Kangaroo; Scrub
Kangaroo; Forester
d Graues Riesenkänguru; Graues
Großkänguru; Östliches Graues
Riesenkänguru
f Kangourou géant
i Canguro gigante; Canguro gigante
grigio orientale; Canguro grigio
orientale; Canguro gigante orientale

3426 *Macropus giganteus tasmanicus*
Diprotodontia - Macropodidae
e Tasmanian Grey Kangaroo; Eastern
Grey Kangaroo (ANZ)
d Tasmanisches Riesenkänguru
f Kangourou gris de l'Est

3427 *Macropus greyi*
Diprotodontia - Macropodidae
e Toolache; Toolache Wallaby; Grey's
Wallaby
d Östliches Irma-Wallaby; Toolache-
Wallaby; Grey-Wallaby; Toolache-
Känguru; Grey-Känguru
f Wallaby de Grey

3428 *Macropus irma*
Diprotodontia - Macropodidae
e Black-gloved Wallaby; Western
Brush Wallaby
d Irma-Wallaby; Westliches Irma-
Wallaby
f Wallaby d'Irma

3429 *Macropus parma*
Diprotodontia - Macropodidae
e Parma Wallaby; White-fronted
Wallaby
d Parma-Filander; Parma-Känguru;
Parma-Wallaby
f Wallaby de Parma
i Wallaby Parma

3430 *Macropus parryi*
Diprotodontia - Macropodidae
e Whiptail; Prettyface (ANZ); Pretty-
faced Wallaby; Pretty Face Wallaby;
Whiptail Wallaby (ANZ)
d Hübschgesichtkänguru;
Hübschgesichtwallaby
f Wallaby de Parry

3431 *Macropus robustus*
Diprotodontia - Macropodidae
e Wallaroo (ANZ); Euro; Scrubber;
Hill Kangaroo; Will Wallaroo;
Commonon Wallaroo (ANZ)
d Bergkänguru; Berggroßkänguru
f Wallarou; Kangourou euro; Wallarou
euro
i Wallaroo; Canguro euro; Canguro
delle colline

3432 *Macropus robustus cervinus*
Diprotodontia - Macropodidae
e Deer Wallaroo
d Hirschkänguru

3433 *Macropus robustus robustus*
Diprotodontia - Macropodidae
e New South Wales Wallaroo
d Südostliches Bergkänguru

3434 *Macropus robustus woodwardi*
Diprotodontia - Macropodidae
e Red-necked Wallaroo

3435 *Macropus rufogriseus*
Diprotodontia - Macropodidae
e Brush Kangaroo; Brush Wallaby;
Red Wallaby
d Rotnackenkänguru; Rothalskänguru
f Wallaby a cou roux
i Canguro rosso; Wallaby della foresta

3436 *Macropus rufogriseus bennetti*
Diprotodontia - Macropodidae
e Bennett's Wallaby
d Bennet-Wallaby; Bennett-Käguruh
f Kangourou de Bennet; Wallaby de
Bennet
i Wallaby di Bennet

3437 *Macropus rufogriseus frutica*
Diprotodontia - Macropodidae
e Eastern Brush Wallaby
d Tasmanisches Bennet-Känguru

3438 *Macroscelides*
Macroscelidea - Macroscelididae
e Short-eared Elephant Shrews
d Elefantenspitzmäuse;
Kurtzohrrüsselspringer ;
Rüsselspringer
f Macroscélides
i Macroscelidi

3439 *Macroscelides proboscideus*
Macroscelidea - Macroscelididae
e Short-eared Elephant Shrew; Round-
eared Elephant Shrew; Round-eared
Sengi
d Kurzohrrüsselspringer;
Elefantenspitzmaus
f Macroscélide
i Macroscelide dalla proboscide

3440 **Macroscelididae**

Macroscelidea
e Elephant Shrews
d Rüsselspringer
f Macroscélidés
i Toporagni elefante; Macroscelidi

3441 *Macrotarsomys*
Rodentia - Muridae
e Big-footed Mice; Forest Mice
d Inselmäuse; Inselratten

3442 *Macrotarsomys bastardi*
Rodentia - Muridae
e Bastard Big-footed Mouse

3443 *Macrotarsomys ingens*
Rodentia - Muridae
e Greater Big-footed Mouse

3444 *Macrotis*
Peramalemorpha - Peramelidae
e Rabbit Bandicoots; Rabbit-eared
Bandicoots; Bilbies (ANZ)
d Kaninchennasenbeutler
f Bandicoots-lapins; Bilbies

3445 *Macrotis lagotis*
Peramalemorpha - Peramelidae
e Rabbit Bandicoot; Rabbit-eared
Bandicoot; Common Rabbit
Bandicoot; Common Rabbit-eared
Bandicoot; Bilby (ANZ); Greater
Bilby; Large-eared Pelagalea; Pinkie;
Dalgyte; Bilby
d Großer Kaninchennasenbeutler
f Bandicoot-lapin; Bilby
i Bandicut-coniglio; Bilby

3446 *Macrotis lagotis lagotis*
Peramalemorpha - Peramelidae
e Western Bilby
i Paramele dalle orecchie; Bilby
maggiore

3447 **Macrotis leucura**
Peramelemorpha - Peramelidae
e Lesser Rabbit Bandicoot; Lesser
Rabbit-eared Bandicoot; White-tailed
Rabbit Bandicoot; Lesser Bilby
(ANZ); Yallara; White-tailed Rabbit-
eared Bandicoot; White-tailed Bilby

d Kleiner Kaninchennasenbeutler
f Bandicoot-lapin mineur; Petit
bandicoot lapin; Petit péramèle lapin
i Bandicut-congiglio dalla coda bianca

3448 **Macrotus**
Chiroptera - Phyllostomidae
e Big-eared Bats; American Leaf-
nosed Bat; Leaf-nosed Bats
d Großohrmäuse

3449 **Macrotus californicus**
Chiroptera - Phyllostomidae
e California Leaf-nosed Bat

3450 **Macrotus waterhouseii**
Chiroptera - Phyllostomidae
e Waterhouse's Leaf-nosed Bat

3451 **Macruromys**
Rodentia - Muridae
e New Guinea Jumping Rats; New
Guinea Rats; Small-toothed Rats
d Neuguinea-Ratten;
Langschwanzratten

3452 **Macruromys elegans**
Rodentia - Muridae
e Western Small-toothed Rat; Lesser
Small-toothed Rat
d Westliche Indonesien-Ratte

3453 **Macruromys major**
Rodentia - Muridae
e Eastern Small-toothed Rat; Greater
Small-toothed Rat
d Östliche Indonesien-Ratte

3454 **Madoqua**
Artiodactyla - Bovidae
e Dikdiks
d Windspielantilopen; Dikdiks;
Tapirböckchen
f Dik-diks
i Dik-dik

3455 **Madoqua guentheri**
Artiodactyla - Bovidae
e Günther's Dikdik
d Günther-Dikdik; Rüsseldikdik
i Dik-dik di Günther

3456 **Madoqua kirki**
Artiodactyla - Bovidae
e Kirk's Dikdik
d Kirk-Dikdik
f Dik-dik de Kirk
i Dik-dik di Kirk

3457 **Madoqua kirki damarensis**
Artiodactyla - Bovidae
e Damara Dikdik
d Damara-Dikdik

3458 **Madoqua piacentinii**
Artiodactyla - Bovidae
e Silver Dikdik; Piacentini's Dikdik
d Silberdikdik

3459 **Madoqua saltiana**
Artiodactyla - Bovidae
e Salt's Dikdik
d Salts Dikdik; Eritrea-Dikdik;
Rotbauchdikdik; Kleindikdik
f Dik-dik de Salt
i Dik-dik di Salt; Dik-dik d'Eritrea

3460 **Madoqua saltiana erlangeri**
Artiodactyla - Bovidae
e Erlanger Dikdik
d Erlanger-Dikdik

3461   *Madoqua saltiana hararensis*
           Artiodactyla - Bovidae
   e   Harar Dikdik

3462   *Madoqua saltiana lawrencei*
           Artiodactyla - Bovidae
   e   Lawrence Dikdik

3463   *Madoqua saltiana phillipsi*
           Artiodactyla - Bovidae
   e   Phillip's Dikdik
   i   Dik-dik di Phillips

3464   *Madoqua saltiana swaynei*
           Artiodactyla - Bovidae
   e   Swayne's Dikdik
   d   Swaynes Dikdik
   f   Dik-dik de Swayne
   i   Dik-dik di Swayne

3465   *Makalata*
           Rodentia - Echimyidae
   e   Armoured Spiny Rats

3466   *Makalata armata*
           Rodentia - Echimyidae
   e   Armoured Spiny Rat; Brazilian Spiny
           Tree Rat
   d   Sania

           Makalata didelphoides syn. *M.*
           armata q.v.

3467   *Makalata occasius*
           Rodentia - Echimyidae
   e   Bare-tailed Tree Rat

3468   *Malacomys*
           Rodentia - Muridae
   e   African Swamp Rats; Big-eared Rats;
           Long-legged Marsh Rats; Long-
           footed Rats; Big-eared Swamp Rats;
           Milne-Edwards's Swamp Rats
   d   Afrikanische Langfußwaldratten
   f   Rats des marais africains

3469   *Malacomys cansdalei*
           Rodentia - Muridae
   e   Cansdale's Swamp Rat

3470   *Malacomys edwardsi*
           Rodentia - Muridae
   e   Edwards's Swamp Rat; Milne-
           Edwards's Swamp Rat

3471   *Malacomys longipes*
           Rodentia - Muridae
   e   Big-eared Swamp Rat

3472   *Malacomys lukolelae*
           Rodentia - Muridae
   e   Lukolela Swamp Rat

3473   *Malacomys verschureni*
           Rodentia - Muridae
   e   Verschuren's Swamp Rat

3474   *Malacothrix*
           Rodentia - Muridae
   e   Gerbil Mice; Long-eared Mice;
           Large-eared Mice
   d   Langohrmäuse; Großohrmäuse
   f   Souris-gerbilles

3475   *Malacothrix typica*
           Rodentia - Muridae
   e   Gerbil Mouse; Long-eared Mouse;
           Mouse Gerbil; Large-eared Mouse
   d   Langohrmaus; Großohrmaus
   f   Souris-gerbille

3476   *Mallomys*
           Rodentia - Muridae
   e   Giant Tree Rats; Woolly Rats
   d   Mallomys-Ratten
   f   Rats de Nouvelle Guinée

3477   *Mallomys aroensis*
           Rodentia - Muridae
   e   De Vies's Woolly Rat

3478 *Mallomys gunung*
Rodentia - Muridae
e Alpine Woolly Rat

3479 *Mallomys istapantap*
Rodentia - Muridae
e Subalpine Woolly Rat

3480 *Mallomys rothschildi*
Rodentia - Muridae
e Rothschild's Woolly Rat; Giant Tree Rat; Smooth-tailed Giant Rat; Black-eared Giant Rat; New Guinea Complex-toothed Rat; Giant New Guinea Rat
d Schwarzohrriesenratte

3481 *Mandrillus*
Primates - Cercopithecidae
e Forest Baboons; Mandrills; Drills
d Backenfurchenpaviane
f Mandrills
i Mandrilli

3482 *Mandrillus leucophaeus*
Primates - Cercopithecidae
e Drill
d Drill
f Drill
i Drillo

3483 *Mandrillus sphinx*
Primates - Cercopithecidae
e Mandrill
d Mandrill
f Mandrill
i Mandrillo

3484 **Manidae**
Cimolesta
e Scaly Anteaters; Pangolins
d Schuppentiere; Tannenzapfenschupper
f Manidés; Pangolins
i Pangolini; Manidi

3485 *Manis*
Cimolesta - Manidae
e Scaly Anteaters; Pangolins; Asian Pangolis
d Schuppentiere
f Pangolins
i Pangolini

3486 *Manis crassicaudata*
Cimolesta - Manidae
e Indian Pangolin; Scaly Anteater
d Ceylanisches Schuppentier; Vorderindisches Schuppentier; Indisches Schuppentier; Indisches Gürteltier; Vorderindisch-Ceylanisches Schupptentier
f Pangolin indien
i Pangolino indiano

Manis gigantea syn. *Smutsia gigantea* q.v.

3487 *Manis javanica*
Cimolesta - Manidae
e Malayan Pangolin; Sunda Pangolin
d Hinterindisches Schuppentier; Javanisches Schuppentier; Javanisch-Malaiisches Schuppentier; Malaiisches Schuppentier
f Pangolin malais; Pangolin javanais
i Pangolino malese

3488 *Manis pentadactyla*
Cimolesta - Manidae
e Chinese Pangolin
d Chinesisches Schuppentier; Chinesisches Ohrenschuppentier
f Pangolin de Chine; Pangolin indien à queue courte
i Pangolino cinese

Manis temminckii syn. *Smutsia temmincki* q.v.

Manis tetradactyla syn. *Uromanis tetradactyla* q.v.

Manis tricuspis syn. *Phataginus tricuspis* q.v.

3489   *Margaretamys*
Rodentia - Muridae
e   Margareta Rats
d   Margareta-Ratten

3490   *Margaretamys beccarii*
Rodentia - Muridae
e   Beccari's Margareta Rat

3491   *Margaretamys elegans*
Rodentia - Muridae
e   Elegant Margareta Rat

3492   *Margaretamys parvus*
Rodentia - Muridae
e   Little Margaret Rat

3493   *Marmosa*
Didelphimorphia - Didelphidae
e   Murine Opossums; Mouse Opossums
d   Zwergbeutelratten; Mausopossums
f   Souris-opossums; Sarigues
i   Opossum topo

Marmosa aceramarcae syn. *Gracilinanus acermarcae* q.v.

Marmosa agilis syn. *Gracilinanus agilis* q.v.

Marmosa alstoni syn. *Micoureus alstoni* q.v.

3494   *Marmosa andersoni*
Didelphimorphia - Didelphidae
e   Anderson's Mouse Opossum
d   Andersons Mausopossum

3495   *Marmosa bishopi*
Didelphimorphia - Didelphidae
e   Bishop's Slender Mouse Opossum

3496   *Marmosa canescens*
Didelphimorphia - Didelphidae
e   Greyish Mouse Opossum
d   Gaumer-Zwergbeutelratte;
         Grauliches Mausopossum

Marmosa carri syn. *M. fuscata* q.v.

3497   *Marmosa cracens*
Didelphimorphia - Didelphidae
e   Slim-faced Slender Mouse Opossum

3498   *Marmosa dorothea*
Didelphimorphia - Didelphidae
e   Dorothy's Slender Mouse Opossum
d   Dorothys Schlankes Mausopossum

Marmosa dryas syn. *Gracilinanus dryas* q.v.

3499   *Marmosa fuscata*
Didelphimorphia - Didelphidae
e   Grey-bellied Slender Mouse
         Opossum
d   Graubauchmausopossum

3500   *Marmosa handleyi*
Didelphimorphia - Didelphidae
e   Handley's Slender Mouse Opossum
d   Handleys Schlankes Mausopossum
f   Sarigue de Colombie
i   Opossum della Colombia; Possum
         della Colombia

3501   *Marmosa impavida*
Didelphimorphia - Didelphidae
e   Andean Slender Mouse Opossum;
         Pale Mouse Opossum; Highland
         Mouse Opossum
d   Anden Schlankes Mausopossum

3502   *Marmosa incana*
Didelphimorphia - Didelphidae
e   Grey Slender Mouse Opossum
d   Graues Schlankes Mausopossum

3503   ***Marmosa invicta***
Didelphimorphia - Didelphidae
e   Slaty Slender Mouse Opossum; Slaty
Mouse Opossum

3504   ***Marmosa juninensis***
Didelphimorphia - Didelphidae
e   Junin Slender Mouse Opossum

3505   ***Marmosa lepida***
Didelphimorphia - Didelphidae
e   Little Rufous Mouse Opossum

Marmosa marica syn. *Gracilinanus
marica* q.v.

3506   ***Marmosa mexicana***
Didelphimorphia - Didelphidae
e   Mexican Mouse Opossum
d   Schwarzringzwergbeutelratte;
Mexikanisches Mausopossum
f   Opossum mexicain

Marmosa microtarsus syn.
Gracilinanus microtarsus q.v.

Marmosa mitis syn. *Marmosa
robinsoni* q.v.

3507   ***Marmosa murina***
Didelphimorphia - Didelphidae
e   Murine Mouse Opossum; Common
Mouse Opossum
d   Mauszwergbeutelratte; Äneas-
Beutelratte; Gemeines Mausopossum
i   Opossum topo; Marmosa murina

3508   ***Marmosa neblina***
Didelphimorphia - Didelphidae
e   Cerro Neblino Slender Mouse
Opossum

3509   ***Marmosa noctivaga***
Didelphimorphia - Didelphidae
e   White-bellied Sleder Mouse
Opossum

d   Nacht-Zwergbeutelratte; Weißbauch
Schlankes Mausopossum

3510   ***Marmosa parvidens***
Didelphimorphia - Didelphidae
e   Delicate Slender Mouse Opossum
d   Empfindliches Schlankes
Mausopossum

3511   ***Marmosa paulensis***
Didelphimorphia - Didelphidae
e   Sao Paulo Slender Mouse Opossum

3512   ***Marmosa pinheiroi***
Didelphimorphia - Didelphidae
e   Pinheiro's Slender Mouse Opossum

3513   ***Marmosa robinsoni***
Didelphimorphia - Didelphidae
e   Robinson's Mouse Opossum; Pale-
bellied Mouse Opossum
d   Südamerikanische Zwergbeutelratte;
Blassbauchzwergbeutelratte;
Robinson-Zwergbeutelratte

3514   ***Marmosa rubra***
Didelphimorphia - Didelphidae
e   Red Mouse Opossum
d   Rote Zwergbeutelratte; Rotes
Mausopossum

3515   ***Marmosa tyleriana***
Didelphimorphia - Didelphidae
e   Tyler's Mouse Opossum
d   Tyleria-Zwergbeutelratte; Tylers
Opossum-Maus; Tylers
Mausopossum

Marmosa velutina syn. *Thylamys
velutinus* q.v.

3516   ***Marmosa xerophila***
Didelphimorphia - Didelphidae
e   Dryland Mouse Opossum
d   Patagonisches Opossum

Marmosops syn. *Marmosa* q.v.

3517   *Marmota*
Rodentia - Sciuridae
e   Marmots; Woodchucks
d   Murmeltiere
f   Marmottes
i   Marmotte

3518   *Marmota baibacina*
Rodentia - Sciuridae
e   Grey Marmot
d   Altaisches Murmeltier; Graues
      Murmeltier
i   Marmotta altaica grigia

3519   *Marmota bobak*
Rodentia - Sciuridae
e   Bobac; Bobac Marmot; Bobak;
      Bobak Marmot
d   Bobak; Steppenmurmeltier;
      Russisches Murmeltier
f   Marmotte bobac; Marmotte des
      steppes
i   Marmotta bobak; Marmotta asiatica
      bobak; Marmotta delle steppe;
      Marmotta di Siberia; Marmotta
      asiatica

3520   *Marmota broweri*
Rodentia - Sciuridae
e   Alaska Marmot
d   Alaska-Murmeltier
f   Marmotte d'Alaska

3521   *Marmota caligata*
Rodentia - Sciuridae
e   Hoary Marmot
d   Eisgraues Murmeltier
f   Marmotte grise; Marmotte des
      Rocheuses
i   Marmotta grigia

3522   *Marmota caligata caligata*
Rodentia - Sciuridae
e   Northern Hoary Marmot

3523   *Marmota camtschatica*
Rodentia - Sciuridae
e   Black-capped Marmot
d   Kamtschatka-Murmeltier;
      Kappenmurmeltier;
      Schwarzhutmurmeltier
f   Marmotte du Kamschatka

3524   *Marmota caudata*
Rodentia - Sciuridae
e   Long-tailed Marmot; Kashmir
      Marmot
d   Langschwänzige Murmeltier;
      Langschwanzmurmeltier
f   Marmotte à longue queue
i   Marmotta dalla coda lunga

3525   *Marmota flaviventris*
Rodentia - Sciuridae
e   Yellow-bellied Marmot; Yellow-
      footed Marmot; Rockchuck;
      Mountain Marmot; Yellowbelly;
      Yellow Groundhog; Yellow Whistler
d   Gelbbäuchiges Murmeltier;
      Gelbbauchmurmeltier
f   Marmotte à ventre jaune; Marmotte à
      ventre fauve
i   Marmotta dal ventre giallo

3526   *Marmota himalayana*
Rodentia - Sciuridae
e   Himalajan Marmot; Tibetan Snow
      Pig
d   Tibetisches Murmeltier; Himalaja-
      Murmeltier
f   Marmotte des Himalajas; Marmotte
      de l'Himalaja
i   Marmotta dell'Himalaja

3527   *Marmota marmota*
Rodentia - Sciuridae
e   Alpine Marmot
d   Murmeltier; Alpenmurmeltier;
      Mankei; Murmele; Marmotte
f   Marmotte vulgaire; Marmotte des
      Alpes; Marmotte
i   Marmotta; Marmotta alpina

3528 **Marmota menzbieri**
Rodentia - Sciuridae
e Menzbier's Marmot
d Menzbiers Murmeltier
f Marmotte de Menzbier

3529 **Marmota monax**
Rodentia - Sciuridae
e Woodchuck; Groundhog (NA);
Whistlepig
d Waldmurmeltier
f Monax; Marmotte commune;
Siffleux
i Marmotta monax; Cane di terra

3530 **Marmota monax ochracea**
Rodentia - Sciuridae
e Yukon Woodchuck

3531 **Marmota monax rufescens**
Rodentia - Sciuridae
e Rufescent Woodchuck

3532 **Marmota olympus**
Rodentia - Sciuridae
e Olympic Marmot; Olympic Mountain
Marmot
d Olympisches Murmeltier
f Marmotte Olympic

3533 **Marmota sibirica**
Rodentia - Sciuridae
e Siberian Marmot; Tarbagan Marmot
d Sibirisches Murmeltier; Tarbagan
f Marmotte de Mongolie; Marmotte de
Sibérie

3534 **Marmota vancouverensis**
Rodentia - Sciuridae
e Vancouver Marmot; Vancouver
Island Marmot
d Vancouver-Murmeltier
f Marmotte de l'île Vancouver

3535 **Martes**

Carnivora - Mustelidae
e Martens; Martens and Fishers; Sables
d Marder ; Eigentliche Marder ;
Edelmarder
f Martes; Martres

3536 **Martes americana**
Carnivora - Mustelidae
e American Marten; Canadian Sable
d Fichtenmarder; Amerikanischer
Zobel
f Marte américaine; Marte d'Amérique;
Martre d'Amérique; Martre
américaine
i Martora canadese; Zibellino canadese

3537 **Martes americana actuosa**
Carnivora - Mustelidae
e Alaska Marten

3538 **Martes americana humboldtensis**
Carnivora - Mustelidae
e Humboldt Marten; Marten

3539 **Martes americana nesophila**
Carnivora - Mustelidae
e Pine Marten (NA); American Pine
Marten

3540 **Martes flavigula**
Carnivora - Mustelidae
e Yellow-throated Marten; Indian
Marten; Asiatic Yellow-throated
Marten; South Indian Marten
d Charsa-Marder; Buntmarder
f Marte à gorge jaune; Marte de l'Inde;
Martre de l'Inde; Martre à gorge
jaune; Marte asiatique à gorge jaune;
Martre asiatique à gorge jaune; Marte
à gorge jaune de l'Inde du Sud
i Martora di Charsa; Martora dalla
gola gialla

3541 **Martes flavigula aterrima**
Carnivora - Mustelidae

e   Northern Yellow-throated Marten

d   Nördlicher Buntmarder

3542   *Martes flavigula chrysospila*
      Carnivora - Mustelidae

e   Taiwanese Marten

3543   *Martes foina*
      Carnivora - Mustelidae

e   Beech Marten; Rock Marten; Stone Marten; House Marten

d   Hausmarder; Steinmarder; Europäischer Steinmarder

f   Fouine; Marte fouine; Martre fouine; Fouine d'Europe

i   Faina; Martora di Francia

3544   *Martes foina intermedia*
      Carnivora - Mustelidae

e   Stone Marten

f   Marte fouine; Martre fouine

3545   *Martes gwatkinsii*
      Carnivora - Mustelidae

e   Nilgiri Marten

d   Indischer Charsa; Charsa

f   Marte de l'Inde du Sud; Martre de l'Inde du Sud

3546   *Martes martes*
      Carnivora - Mustelidae

e   Common Marten; European Pine Marten; Sweet Marten; Marten; Pine Marten

d   Westeuropäischer Baummarder; Edelmarder; Baummarder

f   Marte; Marte ordinaire; Marte commune; Marte vulgaire; Marte des pins; Martre; Martre ordinaire; Martre commune; Martre vulgaire; Martre des pins

i   Martora dei boschi; Martora di Prussia; Martora

3547   *Martes melampus*
      Carnivora - Mustelidae

e   Japanese Marten

d   Japanischer Marder

f   Marte du Japon; Martre du Japon

i   Martora giapponese

3548   *Martes pennanti*
      Carnivora - Mustelidae

e   Fisher; Pekan; Virginian Polecat

d   Fischer; Pekan; Virginia-Iltis; Fischermarder

f   Marte de Pennant; Pékan

i   Martora di Pennant; Puzzola della Virginia; Pekan

3549   *Martes pennanti pacifica*
      Carnivora - Mustelidae

e   Pacific Fisher

3550   *Martes zibellina*
      Carnivora - Mustelidae

e   Sable

d   Zobel; Zobelwiesel; Sobol

f   Zibeline; Marte zibeline

i   Zibellino; Zibellino della Siberia

3551   *Massoutiera*
      Rodentia - Ctenodactylidae

e   Mzab Gundis; Lataste's Gundis

d   Langhaargundis

f   Goundis du Mzab

3552   *Massoutiera mzabi*
      Rodentia - Ctenodactylidae

e   Mzab Gundi; Lataste's Gundi

d   Langhaargundi; Sahara-Gundi

f   Goundi du Sahara; Goundi du Mzab

3553   *Mastomys*
      Rodentia - Muridae

e   Multimammate Mice; Multimammate Rats

d   Vielzitzenmäuse; Vielzitzenratten

f   Rats aux mammelles multiples

i   Ratti africani

3554 *Mastomys angolensis*
Rodentia - Muridae
e Angolan Multimammate Mouse

3555 *Mastomys awashensis*
Rodentia - Muridae
e Awash Multimammate Mouse

3556 *Mastomys coucha*
Rodentia - Muridae
e Southern Multimammate Mouse;
Coucha Multimammate Rat
d Coucha-Vielzitzenmaus

3557 *Mastomys erythroleucus*
Rodentia - Muridae
e Guinea Multimammate Mouse

3558 *Mastomys hildebrandtii*
Rodentia - Muridae
e Hildebrandt's Multimammate Mouse

3559 *Mastomys natalensis*
Rodentia - Muridae
e Natal Multimammate Mouse
d Natal-Vielzitzenmaus

3560 *Mastomys pernanus*
Rodentia - Muridae
e Dwarf Multimammate Mouse

3561 *Mastomys shortridgei*
Rodentia - Muridae
e Shortridge's Multimammate Mouse

3562 *Mastomys verheyeni*
Rodentia - Muridae
e Verheyen's Multimammate Mouse

3563 *Maxomys*
Rodentia - Muridae
e Oriental Spiny Rats
d Südostasiatische Stachelratten

3564 *Maxomys alticola*
Rodentia - Muridae
e Mountain Spiny Rat

3565 *Maxomys baeodon*
Rodentia - Muridae
e Small Spiny Rat; Bornean Small
Spiny Rat

3566 *Maxomys bartelsii*
Rodentia - Muridae
e Bartel's Spiny Rat

3567 *Maxomys dollmani*
Rodentia - Muridae
e Dollmann's Spiny Rat

3568 *Maxomys hellwaldii*
Rodentia - Muridae
e Hellwald's Spiny Rat

3569 *Maxomys hylomyoides*
Rodentia - Muridae
e Sumatran Spiny Rat

3570 *Maxomys inas*
Rodentia - Muridae
e Malayan Mountain Spiny Rat

3571 *Maxomys inflatus*
Rodentia - Muridae
e Fat-nosed Spiny Rat

3572 *Maxomys moi*
Rodentia - Muridae
e Mo's Spiny Rat

3573 *Maxomys musschenbroeki*
Rodentia - Muridae
e Musschenbroek's Spiny Rat;
Musschenbroek's Rat

3574 *Maxomys ochraceiventer*
Rodentia - Muridae

e	Chestnut-bellied Spiny Rat
d	Indonesische Stachelratte

3575 **Maxomys pagensis**
Rodentia - Muridae
e Pagai Spiny Rat

3576 **Maxomys panglima**
Rodentia - Muridae
e Palawan Spiny Rat

3577 **Maxomys rajah**
Rodentia - Muridae
e Rajah Spiny Rat; Brown Spiny Rat;
Rajah Rat
d Rajah-Stachelratte

3578 **Maxomys surifer**
Rodentia - Muridae
e Red Spiny Rat; Yellow Rajah Rat
d Rote Stachelratte

3579 **Maxomys wattsi**
Rodentia - Muridae
e Watts's Spiny Rat

3580 **Maxomys whiteheadi**
Rodentia - Muridae
e Whitehead's Spiny Rat; Whitehead's
Rat
d Whiteheads Stachelratte

Mayalurus syn. *Felis* q.v.

3581 **Mayermys**
Rodentia - Muridae
e Shaw Mayer's Mice; One-toothed
Shrewmice
d Shaw-Mayer-Mäuse
f Rats de Nouvelle Guinée

3582 **Mayermys ellermani**
Rodentia - Muridae
e Shaw Mayer's Mouse; One-toothed
Shrewmouse

d Ellermans Maus

3583 **Mazama**
Artiodactyla - Cervidae
e Brocket Deer ; Brockets
d Mazamas; Spießhirsche; Mazama-
Hirsche
i Mazama

3584 **Mazama americana**
Artiodactyla - Cervidae
e Red Brocket; Red Brocket Deer;
Brocket Deer
d Großmazama; Rotmazama
f Daguet rouge
i Renna rossa

3585 **Mazama americana cerasina**
Artiodactyla - Cervidae
e Large Red Brocket; Large Brocket
Deer
d Guatemalesischer Roter Spießhirsch
f Daguet rouge de Guatemala
i Mazama grande

3586 **Mazama americana temama**
Artiodactyla - Cervidae
e Mexican Red Brocket Deer
d Mexikanischer Großmazama

3587 **Mazama bricenii**
Artiodactyla - Cervidae
e Merioa Brocket
d Merioa-Mazama
f Mazame nain gris

3588 **Mazama chunyi**
Artiodactyla - Cervidae
e Dwarf Brocket
d Zwergmazama; Kleinstmazama;
Kleiner Graumazama
f Daguet nain
i Piccolo Mazama

3589 *Mazama gouazoupira*
Artiodactyla - Cervidae
e Grey Brocket; Brown Brocket; Grey-brown Brocket Deer; Grey Brocket Deer; Brown Brocket Deer
d Graumazama; Braunmazama
f Daguet gris; Daguet brun

3590 *Mazama nana*
Artiodactyla - Cervidae
e Pygmy Brocket; Pygmy Brocket Deer; Bororo
d Kleinmazama
f Bororo

3591 *Mazama pandora*
Artiodactyla - Cervidae
e Yucatan Brown Brocket Deer; Yucatan Brown Brocket
f Daguet brun du Yucatan

3592 *Mazama rufina*
Artiodactyla - Cervidae
e Little Red Brocket
d Kleiner Rotmazama
f Petit Daguet rouge

3593 *Megadendromus*
Rodentia - Muridae
e Nikolaus's Mice
d Riesenbaummäuse

3594 *Megadendromus nikolausi*
Rodentia - Muridae
e Nikolaus's Mouse
d Riesenbaummaus

3595 *Megaderma*
Chiroptera - Megadermatidae
e False Vampire Bats; False Vampires; Asian False Vampire Bats
d Eigentliche Großblattnasen; Ziernasen
f Mégadermes

3596 *Megaderma lyra*
Chiroptera - Megadermatidae
e Greater False Vampire; Greater False Vampire Bat
d Lyra-Fledermaus
f Mégaderme lyre

3597 *Megaderma lyra lyra*
Chiroptera - Megadermatidae
e Indian False Vampire; Indian False Vampire Bat
d Indischer Falscher Vampir

3598 *Megaderma spasma*
Chiroptera - Megadermatidae
e Lesser False Vampire; Malayan False Vampire; Lesser False Vampire Bat; Malayan False Vampire Bat; Asian False Vampire Bat
d Malaiischer Falscher Vampir
f Mégaderme spasma

3599 *Megaderma spasma ceylonense*
Chiroptera - Megadermatidae
e Ceylon False Vampire Bat
d Sri Lanka Falscher Vampir

3600 *Megadermatidae*
Chiroptera
e Large-winged Bats; False Vampire Bats; False Vampires; Yellow-winged Bats
d Klaffmäuler; Großblattnasen; Falsche Vampire
f Megadermatidés
i Megadermatidi

3601 *Megadontomys*
Rodentia - Muridae
e Thomas's Deer Mice; Giant Deer Mice
d Thomas-Hirschmäuse

3602 *Megadontomys cryophilus*
Rodentia - Muridae
e Oaxaca Giant Deer Mouse

3603   *Megadontomys nelsoni*
           Rodentia - Muridae
     e   Nelson's Giant Deer Mouse; Nelson's
           Deer Mouse

3604   *Megadontomys thomasi*
           Rodentia - Muridae
     e   Thomas's Deer Mouse; Thomas's
           Giant Deer Mouse
     d   Thomas-Hirschmaus

3605   *Megaerops*
           Chiroptera - Pteropodidae
     e   Tailless Short-nosed Fruit Bats;
           Tailless Fruit Bats

3606   *Megaerops ecaudatus*
           Chiroptera - Pteropodidae
     e   Temminck's Tailless Fruit Bat

3607   *Megaerops kusnotoi*
           Chiroptera - Pteropodidae
     e   Javan Tailless Fruit Bat

3608   *Megaerops niphanae*
           Chiroptera - Pteropodidae
     e   Ratanaworabhan's Fruit Bat;
           Northern Tailless Fruit Bat

3609   *Megaerops wetmorei*
           Chiroptera - Pteropodidae
     e   White-collared Fruit Bat

3610   *Megaleia*
           Diprotodontia - Macropodidae
     e   Red Kangaroos
     d   Rote Riesenkängurus
     f   Kangourous roux

3611   *Megaleia rufa*
           Diprotodontia - Macropodidae
     e   Red Kangaroo; Plains Kangaroo;
           Blue Flier
     d   Rotes Riesenkänguru;
           Rotgroßkänguru

     f   Kangourou roux; Grand Kangourou
           rouge

3612   *Megaloglossus*
           Chiroptera - Pteropodidae
     e   African Long-tongued Fruit Bats;
           Woermann's Bats
     d   Afrikanische Langzungenflughunde
     f   Mégaglosses

3613   *Megaloglossus woermanni*
           Chiroptera - Pteropodidae
     e   African Long-tongued Fruit Bat;
           Woermann's Long-tongued Fruit Bat;
           Woermann's Bat; Nectar Bat
     d   Afrikanische Langzungenflughund
     f   Mégaglosse de Woermann

3614   *Megalomys*
           Rodentia - Muridae
     e   Swamp Rice Rats; West Indian Giant
           Rice Rats; Giant Rice Rats
     d   Westindische Riesenreisratten;
           Antillen-Reisratten
     f   Raz du riz des Antilles

3615   *Megalomys demarestii*
           Rodentia - Muridae
     e   Antillean Muskrat; Antillean Giant
           Rice Rat; Martinique Musk Rat;
           Martinique Island Giant Rice Rat;
           Martinique Island Swamp Rice Rat;
           Martinique Swamp Rice Rat;
           Martinique Rice Rat
     d   Martinique Riesenreisratte;
           Martinique-Reisratte
     f   Raz du riz géant de la Martinique;
           Rat du riz de la Martinique; Raz du
           riz géant des Antilles

3616   *Megalomys luciae*
           Rodentia - Muridae
     e   Santa Lucia Giant Rice Rat; St. Lucia
           Musk Rat; St. Lucia Giant Rice Rat;
           St. Lucia Island Muskrat; St. Lucia
           Muskrat; St. Lucia Island Giant Rice
           Rat; St. Lucia Giant Rice Rat; St.
           Lucia Island Swamp Rice Rat; St.

Lucia Swamp Rice Rat; St. Lucia
Island Rice Rat; St. Lucia Rice Rat
d St. Lucia Riesenreisratte; St. Lucia-
Reisratte
f Rat du riz géant de l'île Sainte Lucie;
Rat du riz géant de Sainte-Lucie

3617 *Megalonychidae*
Pilosa -
e Two-toed Sloths
d Zweizehenfaultiere
f Mégalonychidés
i Megalonichidi

3618 *Megaptera*
Cete - Balaenopteridae
e Hunch-backed Whales; Hump
Whales; Hump-backed Whales;
Humpbacks
d Buckelwale; Langflossenwale
f Mégaptères; Baleines à bosse
i Megattere

3619 *Megaptera novaeangliae*
Cete - Balaenopteridae
e Hunch-backed Whale; Hump Whale;
Hump-backed Whale; Humpback;
Humpback Whale
d Buckelwal; Langflossenwal;
Knurrwal; Langflossiger Finnfisch
f Mégaptère; Baleine à bosse; Jubarte;
Baleine longimane; Baleine tampon;
Rorqual; Rorqual à bosse
i Megattera; Balenottera longimana

Megascapheus syn. *Thomomys* q.v.

3620 *Megasorex*
Soricomorpha - Soricidae
e Giant Mexican Shrews; Mexican
Shrews
d Große Wüstenspitzmäuse
f Grandes musaraignes du désert

3621 *Megasorex gigas*
Soricomorpha - Soricidae

e Giant Mexican Shrew; Merriam's
Desert Shrew; Mexican Shrew
d Große Wüstenspitzmaus;
Mexikanische Wüstenspitzmaus
f Grande musaraigne du désert

3622 *Melanomys*
Rodentia - Muridae
e Dark Rice Rats
d Dunkle Reisratten

3623 *Melanomys caligonosus*
Rodentia - Muridae
e Dusky Rice Rat

3624 *Melanomys robustulus*
Rodentia - Muridae
e Robust Dark Rice Rat

3625 *Melanomys zunigae*
Rodentia - Muridae
e Zuniga's Dark Rice Rat

3626 *Melasmothrix*
Rodentia - Muridae
e Celebes Long-clawed Mice; Lesser
Shrewrats; Sulawesan Shrewrats
d Kleine Sulawesi-Spitzmausratten
f Rats des Célèbes

3627 *Melasmothrix naso*
Rodentia - Muridae
e Celebes Long-clawed Mouse; Lesser
Shrewrat; Sulawesan Shrewrat
d Kleine Sulawesi-Spitzmausratte;
Kleine Celebes-Spitzmausratte
f Rat des Célèbes

3628 *Meles*
Carnivora - Mustelidae
e Eurasian Badgers
d Dachse; Echte Dachse
f Blaireaux
i Tassi eurasiatici

3629   *Meles meles*
Carnivora - Mustelidae
e   Eurasian Badger; Old World Badger; Badger; Common Badger; European Badger

d   Dachsbär; Europäischer Dachs; Gemeiner Dachs; Dachs; Eurasischer Dachs; Grimbart; Gräving
f   Blaireau commun; Blaireau d'Eurasie; Blaireau d'Europe; Blaireau; Blaireau ordinaire; Blaireau vulgaire; Blaireau européen

i   Tasso; Tascio; Tasso europeo; Tasso eurasiatico; Melogna

3630   *Meles meles albogularis*
Carnivora - Mustelidae
e   Tibetan Badger

3631   *Meles meles anaguma*
Carnivora - Mustelidae
e   Japanese Badger
d   Japanischer Dachs
f   Blaireau japonais; Tanuki
i   Tasso giapponese

3632   *Meles meles arenarius*
Carnivora - Mustelidae
e   Kazakstan Badger

3633   *Meles meles danicus*
Carnivora - Mustelidae
e   Danish Badger

3635   *Meles meles leptorhynchus*
Carnivora - Mustelidae
e   Russian Badger

3634   *Meles meles leucurus*
Carnivora - Mustelidae
e   Chinese Badger

3636   *Meles meles marianensis*
Carnivora - Mustelidae

e   Spanish Badger

3637   *Meles meles rhodius*
Carnivora - Mustelidae
e   Rhodes Badger

3638   *Mellivora*
Carnivora - Mustelidae
e   Honey Badgers; Ratels
d   Honigdachse
f   Ratels
i   Tassi del miele

3639   *Mellivora capensis*
Carnivora - Mustelidae
e   Honey Badger; Ratel
d   Honigdachs; Ratel
f   Ratel; Ratel du Cap
i   Tasso del miele; Tasso mellivora; Ratele; Mellivora

3640   *Melogale*
Carnivora - Mustelidae
e   Ferret-Badgers; Pahmi
d   Sonnendachse
f   Blaireaux-furets; Hélicres
i   Tassi del sole; Tassi-topo; Tassi furetto

3641   *Melogale everetti*
Carnivora - Mustelidae
e   Everett's Ferret-Badger
d   Everett-Sonnendachs

3642   *Melogale moschata*
Carnivora - Mustelidae
e   Chinese Ferret-Badger; Oriental Ferret-Badger; East Asia Ferret-Badger
d   China-Sonnendachs; Chinesischer Sonnendachs; Chinesischer Dachs
f   Blaireau de Chine; Blaireau-furet chinois
i   Tasso cinese delle mele; Tasso del sole; Tasso-topo

3643   **Melogale moschata sorella**
        Carnivora - Mustelidae
e   Southeast China Ferret-Badger

3644   **Melogale orientalis**
        Carnivora - Mustelidae
e   Javan Ferret-Badger
d   Java-Sonnendachs; Javanischer
        Sonnendachs

3645   **Melogale personata**
        Carnivora - Mustelidae
e   Burmese Ferret-Badger; Brown
        Ferret-Badger; Bornean Ferret-
        Badger; Southeast Asia Ferret-
        Badger; Large-toothed Ferret-
        Badger; Indian Ferret-Badger
d   Birma-Sonnendachs; Birmanischer
        Sonnendachs
f   Blaireau de Birmanie; Blaireau-furet
        de Java
i   Tasso furetto della Birmanie

3646   **Melomys**
        Rodentia - Muridae
e   Banana Rats; Mosaic-tailed Rats;
        Naked-tailed Rats; Melomys Rats;
        Melomyses
d   Melomysratten;
        Mosaikschwanzmäuse
f   Souris à queue en mosaique ; Rats
        des bananes

3647   **Melomys aerosus**
        Rodentia - Muridae
e   Dusky Mosaic-tailed Rat; Dusky
        Melomys; Seram Naked Tail Rat
d   Dunkle Melomysratte

3648   **Melomys bannisteri**
        Rodentia - Muridae
e   Bannister's Melomys

3649   **Melomys bougainville**
        Rodentia - Muridae
e   Bougainville Mosaic-tailed Rat;
        Bougainville Melomys

Melomys bruijnii syn.
Pogonomelomys bruijnii q.v.

3650   **Melomys burtoni**
        Rodentia - Muridae
e   Little Melomys; Burton's Melomys
d   Kleine Melomysratte;
        Graslandmelomysratte

3651   **Melomys capensis**
        Rodentia - Muridae
e   Cape York Melomys; Cape York
        Mosaic-tailed Rat

3652   **Melomys caurinus**
        Rodentia - Muridae
e   Short-tailed Talaud Melomys

3653   **Melomys cervinipes**
        Rodentia - Muridae
e   Buff-footed Rat; Fawn-footed
        Melomys; Fawn-footed Mosaic-
        tailed Rat
d   Mosaikschwanzratte

3654   **Melomys cooperae**
        Rodentia - Muridae
e   Cooper's Melomys

3655   **Melomys fellowsi**
        Rodentia - Muridae
e   Red-bellied Melomys; Red-bellied
        Mosaic-tailed Rat

3656   **Melomys fraterculus**
        Rodentia - Muridae
e   Manusela Mosaic-tailed Rat;
        Manusela Melomys

3657   **Melomys fulgens**
        Rodentia - Muridae
e   Orange Melomys

3658   **Melomys gracilis**
        Rodentia – Muridae

e   Slender Mosaic-tailed Rat; Long-
      tailed Melomys

3659   Melomys gressitti
      Rodentia - Muridae
e   Gressit's Melomys

3660   Melomys howi
      Rodentia - Muridae
e   How's Melomys

3661   Melomys lanosus
      Rodentia - Muridae
e   Thornton Park Melomys; Large-
      scaled Mosaic-tailed Rat; Large-
      scaled Melomys

3662   Melomys leucogaster
      Rodentia - Muridae
e   White-bellied Melomys; White-
      bellied Mosaic-tailed Rat

3663   Melomys levipes
      Rodentia - Muridae
e   Long-nosed Melomys; Long-nosed
      Mosaic-tailed Rat
d   Langnasenmelomysratte

      Melomys littoralis syn. *M. burtoni*
      q.v.

3664   Melomys lorentzii
      Rodentia - Muridae
e   Long-footed Melomys; Lorentz's
      Mosaic-tailed Rat; Lorentz's
      Melomys
d   Lorentz-Melomysratte;
      Langfußmelomysratte

3665   Melomys lutillus
      Rodentia - Muridae
e   Grassland Mosaic-tailed Rat;
      Grassland Melomys (ANZ)

3666   Melomys matambuai
      Rodentia - Muridae

e   Manus Melomys

      Melomys mayeri syn.
      Pogonomelomys mayeri q.v.

3667   Melomys mollis
      Rodentia - Muridae
e   Thomas's Mosaic-tailed Rat;
      Thomas's Melomys

3668   Melomys moncktoni
      Rodentia - Muridae
e   Southern Melomys; Monckton's
      Mosaic-tailed Rat; Monckton's
      Melomys
d   Südliche Melomysratte

3669   Melomys obiensis
      Rodentia - Muridae
e   Obi Mosaic-tailed Rat; Obi Naked-
      tail Rat; Obi Melomys
d   Obi-Melomysratte

3670   Melomys platyops
      Rodentia - Muridae
e   Lowland Melomys; Lowland
      Mosaic-tailed Rat
d   Tiefland-Melomysratte

      Melomys ponceleti syn. *Solomys*
      ponceleti q.v.

3671   Melomys rattoides
      Rodentia - Muridae
e   Large Mosaic-tailed Rat; Large
      Melomys
d   Große Mosaikschwanzratte

3672   Melomys rubex
      Rodentia - Muridae
e   Highland Melomys; Mountain
      Mosaic-tailed Rat; Mountain
      Melomys

3673   Melomys rubicola
      Rodentia - Muridae

e   Bramble Cay Melomys; Bramble
     Cay Mosaic-tailed Rat

3674 **Melomys rufescens**
      Rodentia - Muridae
e   Rufescent Melomys; Black-tailed
      Mosaic-tailed Rat; Black-tailed
      Melomys
d   Schwarzschwanzmelomysratte

      Melomys salebrosus syn. *Solomys
      salebrosus* q.v.

      Melomys sapientis syn. *Solomys
      sapientis* q.v.

      Melomys sevia syn. *Pogonomelomys
      sevia* q.v.

3675 **Melomys spechti**
      Rodentia - Muridae
e   Specht's Mosaic-tailed Rat; Specht's
      Melomys

3676 **Melonycteris**
      Chiroptera - Pteropodidae
e   Black-bellied Fruit Bats

      Melonycteris aurantius syn. *M.
      woodfordi aurantius* q.v.

3677 **Melonycteris melanops**
      Chiroptera - Pteropodidae
e   Black-bellied Fruit Bat; Bismarck
      Blossom Bat

3678 **Melonycteris woodfordi**
      Chiroptera - Pteropodidae
e   Solomon Islands Long-tongued Fruit
      Bat; Woodford's Fruit Bat; Solomon
      Long-tongued Fruit Bat; Woodford's
      Blossom Bat

3679 **Melonycteris woodfordi aurantius**
      Chiroptera - Pteropodidae
e   Orange Fruit Bat

3680 **Melursus**
      Carnivora - Ursidae
e   Indian Bears; Sloth Bears
d   Lippenbären
f   Ours de l'Inde; Ours aux longues
      lèvres
i   Orsi labiati

3681 **Melursus ursinus**
      Carnivora - Ursidae
e   Indian Bear; Sloth Bear
d   Lippenbär
f   Ours de l'Inde; Ours aux longues
      lèvres
i   Orso labiato

3682 **Melursus ursinus inornatus**
      Carnivora - Ursidae
e   Sri Lankan Sloth Bear; Ceylon Sloth
      Bear

3683 **Melursus ursinus ursinus**
      Carnivora - Ursidae
e   Indian Sloth Bear

3684 **Menetes**
      Rodentia - Sciuridae
e   Berdmore's Squirrels; Berdmore's
      Palm Squirrels; Indochinese Ground
      Squirrels
d   Berdmore-Palmenhörnchen
f   Écureuils terrestres
i   Scoiattoli di Berdmore

3685 **Menetes berdmorei**
      Rodentia - Sciuridae
e   Berdmore's Squirrel; Berdmore's
      Palm Squirrel; Indochinese Ground
      Squirrel
d   Berdmore-Palmenhörnchen ;
      Berdmore-Erdstreifenhörnchen
f   Écureuil des palmiers
i   Scoiattolo di Berdmore

3686 **Mephitis**
      Carnivora - Mustelidae

e   Striped Skunks; Striped and Hooded
     Skunks
d   Streifenskunks; Kanada-Skunks
f   Skunks; Moufettes; Mouffettes
     (Qué); Mofettes
i   Moffette

3687   *Mephitis macroura*
     Carnivora - Mustelidae
e   Hooded Skunk
d   Langschwanzskunk; Haubenskunk
f   Moufette de Mexico; Mouffette de
     Mexico (Qué); Mofettte de Mexico
i   Moffetta dalla lunga coda; Skunk dal
     cappuccio

3688   *Mephitis mephitis*
     Carnivora - Mustelidae
e   Striped Skunk; Skunk
d   Streifenskunk
f   Skunk commun; Moufette rayée;
     Moufette; Mouffette (Qué);
     Mouffette rayée (Qué); Mofette rayée
i   Moffetta comune; Skunk striato

3689   *Mephitis mephitis hudsonica*
     Carnivora - Mustelidae
e   Northern Plains Striped Skunk

3690   *Meriones*
     Rodentia - Muridae
e   Jirds; Sandrats; Gerbils
d   Sandmäuse; Wüstenmäuse;
     Rennmäuse; Rennratten
f   Mériones
i   Merioni

3691   *Meriones arimalius*
     Rodentia - Muridae
e   Arabian Jird

Meriones calurus syn. *Sekeetamys
calurus* q.v.

Meriones caudatus syn. *M. libycus*
q.v.

3692   *Meriones chengi*
     Rodentia - Muridae
e   Cheng's Jird

3693   *Meriones crassus*
     Rodentia - Muridae
e   Sundevall's Jird; Swinhoe's Jird;
     Silky Jird; Gentle Jird; Jerusalem Jird
d   Sundevall-Rennmaus
f   Mérione du désert; Mérione de
     Sundevall; Gerbille de Jérusalem

3694   *Meriones crassus perpallidus*
     Rodentia - Muridae
e   Pallid Sundevall's Jird
d   Ägyptische Sundevall-Rennmaus

3695   *Meriones dahli*
     Rodentia - Muridae
e   Dahl's Jird

3696   *Meriones hurrianae*
     Rodentia - Muridae
e   Indian Desert Gerbil; Indian Desert
     Jird
d   Indische Wüstenmaus
f   Rat des déserts d'Inde

3697   *Meriones libycus*
     Rodentia - Muridae
e   Libyan Jird
d   Rotschwänzige Rennmaus; Libysche
     Rennmaus; Libysche Rennratte
f   Mérione libyen; Mérione de Libye

3698   *Meriones meridianus*
     Rodentia - Muridae
e   Mid-day Gerbil; Southern Jird; Little
     Chinese Jird; Mid-day Jird; Chinese
     Gerbil
d   Mittagsrennmaus; Mittagsrennratte
f   Mérione du Sud

3699   *Meriones meridianus nogaiorum*
     Rodentia - Muridae

e   Nogai Mid-day Gerbil
d   Kalmückische Mittagsrennmaus

3700   *Meriones meridianus penicilliger*
Rodentia - Muridae
e   Kara-Kum Mid-day Gerbil
d   Türkmenische Mittagsrennmaus

3701   *Meriones meridianus urianchaicus*
Rodentia - Muridae
e   Mongolia Mid-day Gerbil
d   Mongolische Mittagsrennmaus

3702   *Meriones persicus*
Rodentia - Muridae
e   Persian Jird
d   Persische Wüstenmaus; Persische
Rennmaus; Persische Rennratte
f   Mérione de Perse

Meriones przewalskii syn.
Brachiones przewalskii q.v.

3703   *Meriones rex*
Rodentia - Muridae
e   King Jird
d   Königswüstenmaus
f   Mérione royal

3704   *Meriones sacramenti*
Rodentia - Muridae
e   Buxton's Jird; Negev Jird; Negev
Gerbil
d   Negev-Rennmaus
f   Mérione du Negev

3705   *Meriones shawi*
Rodentia - Muridae
e   Shaw's Jird; Morocco Jird
d   Shaw-Wüstenmaus; Shaws
Rennmaus
f   Mérione de Shaw; Gerbille de Shaw

3706   *Meriones shawi grandis*
Rodentia - Muridae

d   Shaws Große Wüstenmaus

3707   *Meriones shawi shawi*
Rodentia - Muridae
e   Egyptian Shaw's Jird
d   Shaws Kleine Rennmaus aus
Ägypten

3708   *Meriones tamariscinus*
Rodentia - Muridae
e   Tamarisk Gerbil; Tamarisk Jird;
Grebenchikova Gerbil
d   Tamariskenrennmaus;
Ringelrennmaus; Schenkelmaus;
Wüstenmaus

3709   *Meriones tristrami*
Rodentia - Muridae
e   Tristram's Jird; Asia Minor Gerbil
d   Tristram-Rennratte; Tristram-
Wüstenmaus; Tristrams Rennmaus
f   Mérione de Tristram
i   Merione di Tristram

3710   *Meriones unguiculatus*
Rodentia - Muridae
e   Clawed Jird; Mongolian Gerbil;
Mongolian Jird
d   Mongolische Rennmaus;
Krallenrennmaus; Mongolische
Rennratte; Mongolen-Rennmaus;
Mongolische Wüstenrennmaus
f   Mérione de Mongolie; Gerbille de
Mongolie
i   Gerbillo della Mongolia

3711   *Meriones vinogradovi*
Rodentia - Muridae
e   Vinogradov's Jird; Vinogradov's
Gerbil
d   Winagradows Rennmaus;
Vinagradovs Rennmaus
f   Mérione de Vinogradov

3712   *Meriones zarudnyi*
Rodentia - Muridae
e   Zarudny's Jird; Zarudny's Gerbil

d Afghanische Rennmaus; Sarudny-
Rennmaus

Mesechinus syn. *Erinaceus* q.v.

3713 **Mesembriomys**
Rodentia - Muridae
e Rabbit Rats; Rabbit-eared Rats;
Mantbuls; Tree Rats
d Australische Baumratten

3714 **Mesembriomys gouldi**
Rodentia - Muridae
e Long-haired Hapalotis; Long-haired
Rabbit Rat; Shaggy Rabbit Rat;
Black-footed Tree Rat
d Schwarzfußbaumratte

3715 **Mesembriomys macrurus**
Rodentia - Muridae
e Golden-backed Rabbit Rat; Western
Rabbit Rat; Golden-backed Tree Rat
(ANZ)

3716 **Mesocricetus**
Rodentia - Muridae
e Golden Hamsters; Goldhamsters
d Mittelhamster
f Hamsters dorés
i Criceti dorati

3717 **Mesocricetus auratus**
Rodentia - Muridae
e Golden Hamster
d Goldhamster; Gelbhamster
f Hamster doré
i Criceto dorato

3718 **Mesocricetus auratus auratus**
Rodentia - Muridae
e Syrian Golden Hamster; Golden
Syrian Hamster; Syrian Hamster
d Syrischer Goldhamster
i Criceto siriano

3719 **Mesocricetus brandti**
Rodentia - Muridae
e Brandt's Hamster; Turkish Hamster
d Brandts Hamster; Türkischer
Goldhamster
f Hamster de Turquie; Hamster
géorgien
i Criceto turco; Criceto di Brandt;
Criceto dell'Azerbaijan

3720 **Mesocricetus newtoni**
Rodentia - Muridae
e Romanian Hamster
d Dobrucha-Hamster; Rumänischer
Goldhamster
f Hamster de Newton; Hamster de
Roumanie; Hamster de la Dobroudja;
Hamster roumain; Hamster doré de
Roumanie
i Criceto di Romania; Criceto rumeno

3721 **Mesocricetus raddei**
Rodentia - Muridae
e Ciscaucasian Hamster
d Raddes Hamster; Dagestanischer
Hamster; Schwarzbrusthamster
f Hamster du Daghestan
i Criceto del Caucaso; Criceto
georgiano; Criceto ciscaucasico

3722 **Mesomys**
Rodentia - Echimyidae
e Hedgehog Rats; Spiny Tree Rats
f Rat arboricole

Mesomys didelphoides syn.
Makalata armata q.v.

3723 **Mesomys hispidus**
Rodentia - Echimyidae
e Spiny Tree Rat; Common Spiny Tree
Rat

3724 **Mesomys leniceps**
Rodentia - Echimyidae
e Woolly-headed Spiny Tree Rat

3725   **Mesomys obscurus**
Rodentia - Echimyidae
e   Dusky Spiny Tree Rat

3726   **Mesomys occultus**
Rodentia - Echimyidae
e   Furtive Spiny Tree Rat

3727   **Mesomys stimulax**
Rodentia - Echimyidae
e   Surinam Spiny Tree Rat

3728   **Mesophylla**
Chiroptera - Phyllostomidae
e   McConnell's Bats

3729   **Mesophylla macconnelli**
Chiroptera - Phyllostomidae
e   McConnell's Bat

3730   **Mesoplodon**
Cete - Hyperoodontidae
e   Beaked Whales; Cowfishes; Sword-toothed Dolphins
d   Zweizahnwale; Spitzschnauzendelfine; Sowerbys Wale
f   Mésoplodons; Baleines de Sowerby; Baleines à bec de Sowerby
i   Mesoplodonte

3731   **Mesoplodon bidens**
Cete - Hyperoodontidae
e   Sowerby's Whale; Sowerby's Beaked Whale; North Sea Beaked Whale
d   Sowerbys Wal; Sowerby-Zweizahnwal; Sowerby-Schnabelwal; Nordsee-Schnabelwal; Nordsee-Flosser; Flosser
f   Mésopolodon de Sowerby; Baleine de Sowerby; Baleine à bec de Sowerby
i   Mesoplodonte bidente; Mesoplodonte di Sowerby

3732   **Mesoplodon bowdoini**

Cete - Hyperoodontidae
e   Bowdoin's Beaked Whale; Andrew's Beaked Whale; Splaytooth Beaked Whale; Deep-crested Whale
d   Andrew-Schnabelwal; Bowdoin-Schnabelwal
f   Mésoplodon de Andrew
i   Mesoplodonte di Andrew; Mesoplodonte di Bowdoin

3733   **Mesoplodon carlhubbsi**
Cete - Hyperoodontidae
e   Hubbs's Beaked Whale; Hubbs's Swordtooth; Arch-beaked Whale
d   Hubbs-Schnabelwal
f   Mésoplodon de Hubbs
i   Mesoplodonte di Hubbs

3734   **Mesoplodon densirostris**
Cete - Hyperoodontidae
e   Blainville's Beaked Whale; Atlantic Beaked Whale; Tropical Beaked Whale; Blainville's Whale; Blainville's Dense-beaked Whale; De Blainville's Beaked Whale; Dense-beaked Whale
d   Blainville-Zweizahnwal; Blainville-Schnabelwal
f   Baleine à bec de Blainville; Mésoplodon de Blainville
i   Mesoplodonte densirostro; Mesoplodonte di De Blainville; Mesoplodonte di Blainville

3735   **Mesoplodon europaeus**
Cete - Hyperoodontidae
e   Antillean Beaked Whale; European Beaked Whale; Gervais's Beaked Whale; Gulf Stream Beaked Whale; Gervais's Whale; Cowfish; Antillean Swordtooth
d   Gervais-Zweizahnwal; Europäischer Schnabelwal; Gervais-Schnabelwal
f   Mésoplodon de Gervais; Baleine à bec de Gervais
i   Mesoplodonte europeo; Mesoplodonte di Gervais

Mesoplodon gervaisi syn. *M. europaeus* q.v.

3736   Mesoplodon ginkgodens
Cete - Hyperoodontidae
e   Japanese Beaked Whale; Japanese Ginkgo-toothed Beaked Whale; Ginkgo-toothed Whale; Ginkgo-toothed Beaked Whale
d   Japanischer Schnabelwal; Gingko-Schnabelwal
f   Mésoplodon ginko; Mésoplodon du Japon; Mésoplodon de Nishiwaki
i   Mesoplodonte dai denti a gingko; Mesoplodonte di Nishiwaki

3737   Mesoplodon grayi
Cete - Hyperoodontidae
e   Gray's Whale; Gray's Beaked Whale; New Zealand Scamperdown Whale; Scamperdown Whale; Southern Beaked Whale; Scamperdown Beaked Whale
d   Gray-Zweizahnwal; Camperdown-Wal; Gray-Schnabelwal; Südlicher Schnabelwal; Scamperdown-Wal
f   Baleine à bec de Gray; Mésoplodon de Gray
i   Mesoplodonte di Gray

3738   Mesoplodon hectori
Cete - Hyperoodontidae
e   Hector's Beaked Whale; Skew-beaked Whale; Deep-socketed Whale; New Zealand Beaked Whale
d   Hector-Schnabelwal
f   Mésoplodon d'Hector
i   Mesoplodonte di Hector

3739   Mesoplodon layardii
Cete - Hyperoodontidae
e   Layard's Whale; Layard's Beaked Whale; Strap-toothed Whale; Strap-toothed Beaked Whale
d   Layard-Wal; Layard-Zweizahnwal
f   Mésoplodon de Layard; Baleine à bec de Layard
i   Mesoplodonte di Layard

3740   Mesoplodon mirus
Cete - Hyperoodontidae
e   True's Beaked Whale; Wonderful Beaked Whale
d   True-Zweizahnwal; True-Wal
f   Mésoplodon de True; Baleine à bec de True
i   Mesoplodonte mirabile; Mesoplodonte di True

3741   Mesoplodon perrini
Cete - Hyperoodontidae
e   Perrin's Beaked Whale
d   Perrin-Schnabelwal
f   Mésoplodon de Perrin
i   Mesoplodonte di Perrin

3742   Mesoplodon peruvianus
Cete - Hyperoodontidae
e   Pygmy Beaked Whale; Lesser Beaked Whale
d   Zwergschnabelwal; Humboldt-Schnabelwal; Peruanischer Schnabelwal
f   Mésoplodon de Pérou; Mésoplodon nain; Mésoplodon pygmée
i   Mesoplodonte pigmeo

3743   Mesoplodon stejnegeri
Cete - Hyperoodontidae
e   Bering Sea Beaked Whale; Stejneger's Beaked Whale; Sabre-toothed Beaked Whale
d   Stejnegers Zweizahnwal; Stejneger-Schnabelwal; Beringmeer-Schnabelwal; Säbelzahnwal; Stejnegeger-Schnabelwal
f   Mésoplodon de Stejneger
i   Mesoplodonte dei Stejneger

3744   Mesoplodon traversii
Cete - Hyperoodontidae
e   Spade-toothed Whale; Bahamonde's Beaked Whale
d   Travers-Schnabelwal; Bahamonde-Schnabelwal

f Mésoplodon de Bahamond

i Mesoplodonte di Travers

Mesospalax syn. *Nannospalax* q.v.

Metachirops syn. *Philander* q.v.

3745 Metachirus
Didelphimorphia - Didelphidae
e Rat-tailed Opossums; Brown Four-eyed Opossums; Brown-masked Opossums; Long Bare-tailed Opossums
d Nacktschwanzbeutelratten
f Opossums à queue de rat
i Metachiri

3746 Metachirus nudicaudatus
Didelphimorphia - Didelphidae
e Rat-tailed Opossum; Brown Four-eyed Opossum; Brown-masked Opossum; Long Bare-tailed Opossum
d Nacktschwanzbeutelratte; Braunes Opossum
f Opossum à queue de rat

Metavampyressa syn. *Vampyressa* q.v.

3747 Micoureus
Didelphimorphia - Didelphidae
e Woolly Mouse Opossums
d Wollige Zwergbeutelratten

3748 Micoureus alstoni
Didelphimorphia - Didelphidae
e Alston's Woolly Mouse Opossum; Alston's Opossum; Ashy Opossum; Alston's Mouse Opossum
d Alston-Zwergbeutelratte; Aschgraue Zwergbeutelratte; Wolliges Opossum

3749 Micoureus constantiae
Didelphimorphia - Didelphidae
e Pale-bellied Woolly Mouse Opossum

3750 Micoureus demerarae
Didelphimorphia - Didelphidae
e Long-furred Woolly Mouse Opossum; Ashy Mouse Opossum

3751 Micoureus phaea
Didelphimorphia - Didelphidae
e Little Woolly Mouse Opossum

3752 Micoureus regina
Didelphimorphia - Didelphidae
e Short-furred Woolly Mouse Opossum

3753 Microakodontomys
Rodentia - Muridae
e Intermediate Lesser Grass Mice

3754 Microakodontomys transitorius
Rodentia - Muridae
e Intermediate Lesser Grass Mouse

3755 Microbiotheriidae
Microbiotheria
e Monitos del Monte
d Bergopossums
f Microbiotheridés
i Microbioteridi

3756 Microcavia
Rodentia - Caviidae
e Mountain Cavies
d Zwergmeerschweinchen
f Cobayes; Cobayes de montagne

3757 Microcavia australis
Rodentia - Caviidae
e Southern Mountain Cavy
d Südliches Zwergmeerschweinchen

3758 Microcavia niata
Rodentia - Caviidae
e Andean Mountain Cavy

3759   ***Microcavia shiptoni***
          Rodentia - Caviidae
e   Shipton's Mountain Cavy

3760   ***Microcebus***
          Primates - Cheirogaleidae
e   Mouse Lemurs; Mouse Dwarf
          Lemurs
d   Zwergmakis; Mausmakis
f   Microcèbes
i   Microcebi

3761   ***Microcebus berthae***
          Primates - Cheirogaleidae
e   Berthe's Mouse Lemur
d   Berthes Maki

          Microcebus coquereli syn. *Mirza
          coquereli* q.v.

3762   ***Microcebus murinus***
          Primates - Cheirogaleidae
e   Miller's Mouse Lemur; Lesser Mouse
          Lemur; Grey Mouse Lemur
d   Mausmaki; Eigentlicher Mausmaki;
          Grauer Mausmaki
f   Microcèbe mignon; Chirogale
          mignon
i   Microcebo murino

3763   ***Microcebus myoxinus***
          Primates - Cheirogaleidae
e   Pygmy Mouse Lemur
d   Zwergmausmaki
f   Microcèbe pygmée
i   Microcebo pigmeo

3764   ***Microcebus ravelobensis***
          Primates - Cheirogaleidae
e   Golden-brown Mouse Lemur;
          Golden Mouse Lemur
d   Goldbrauner Zwergmaki
f   Microcèbe brun-doré
i   Microcebo bruno-dorato

3765   ***Microcebus rufus***
          Primates - Cheirogaleidae
e   Russet Mouse Lemur; Brown Mouse
          Lemur; Brown Lesser Mouse Lemur;
          Rufous Mouse Lemur; Eastern
          Rufous Mouse Lemur; Red Mouse
          Lemur
d   Roter Mausmaki
f   Microcèbe roux
i   Microcebo rosso

3766   ***Microcebus sambiranensis***
          Primates - Cheirogaleidae
e   Sambirano Mouse Lemur
d   Sambirano-Mausmaki

3767   ***Microcebus tavarata***
          Primates - Cheirogaleidae
e   Northern Rufous Mouse Lemur
d   Nördlicher Mausmaki

3768   ***Microdillus***
          Rodentia - Muridae
e   Somali Pygmy Gerbils; Somalia
          Pygmy Gerbils
d   Somali-Zwergrennmäuse
f   Gerbilles pygmées

3769   ***Microdillus peeli***
          Rodentia - Muridae
e   Somali Pygmy Gerbil; Somalia
          Pygmy Gerbil
d   Somali-Zwergrennmaus
f   Gerbille pygmée

3770   ***Microdipodops***
          Rodentia - Geomyidae
e   Kangaroo Mice; Pygmy Kangaroo
          Rats
d   Kängurumäuse
f   Souris kangourous

3771   ***Microdipodops megacephalus***
          Rodentia – Geomyidae

e   Dark Kangaroo Mouse
d   Dunkle Kängurumaus

3772   ***Microdipodops pallidus***
Rodentia - Geomyidae
e   Pale Kangaroo Mouse
d   Blasse Kängurumaus
f   Souris kangourou pâle

3773   ***Microgale***
Soricomorpha - Tenrecidae
e   Tenrecs; Tanrecs; 0; Long-tailed
Tenrecs; Shrewlike Tanrecs;
Shrewlike Tenrecs; Shrew Tenrecs;
Shrew Tanrecs
d   Kleintanreks; Langschwanztanreks;
Spitzmaustanreks
f   Microgales

3774   ***Microgale brevicaudata***
Soricomorpha - Tenrecidae
e   Short-tailed Shrew Tenrec; Short-
tailed Shrew Tanrec

3775   ***Microgale cowani***
Soricomorpha - Tenrecidae
e   Cowan's Shrew Tenrec; Cowan's
Shrew Tanrec
d   Kleintankrek
f   Musaraigne de Cowan

3776   ***Microgale dobsoni***
Soricomorpha - Tenrecidae
e   Dobson's Shrew Tenrec; Dobson's
Shrew Tanrec
f   Musaraigne de Dobson

3777   ***Microgale drouhardi***
Soricomorpha - Tenrecidae
e   Striped Shrew Tenrec; Striped Shrew
Tanrec

3778   ***Microgale dryas***
Soricomorpha - Tenrecidae
e   Tree Shrew Tenrec; Tree Shrew
Tanrec; Woodnymph Tanrec;

Woodnymph Tenrec

3779   ***Microgale fotsifotsy***
Soricomorpha - Tenrecidae
e   Pale-footed Shrew Tenrec; Pale-
footed Shrew Tanrec
f   Musaraigne à visage pâle

3780   ***Microgale gracilis***
Soricomorpha - Tenrecidae
e   Gracile Shrew Tenrec; Gracile Shrew
Tanrec; Long-nosed Shrew Tanrec;
Long-nosed Shrew Tenrec

3781   ***Microgale longicaudata***
Soricomorpha - Tenrecidae
e   Long-tailed Tenrec; Long-tailed
Tanrec; Lesser Long-tailed Shrew
Tanrec; Lesser Long-tailed Shrew
Tenrec
d   Langschwanztanrek
f   Microgale à longue queue

3782   ***Microgale monticola***
Soricomorpha - Tenrecidae
e   Montane Shrew Tenrec; Montane
Shrew Tanrec

3783   ***Microgale nasoloi***
Soricomorpha - Tenrecidae
e   Nasolo's Shrew Tenrec; Nasolo's
Shrew Tanrec

3784   ***Microgale parvula***
Soricomorpha - Tenrecidae
e   Little Long-tailed Tenrec; Little
Long-tailed Tanrec; Pygmy Shrew
Tanrec; Pygmy Shrew Tenrec; Dark
Shrew Tanrec; Dark Shrew Tenrec;
Dusky Shrew Tenrec; Dusky Shrew
Tanrec
d   Kleiner Langschwanztanrek
f   Musaraigne pygmée

3785   ***Microgale principula***
Soricomorpha - Tenrecidae

e   Greater Long-tailed Shrew Tenrec;
Greater Long-tailed Shrew Tanrec

Microgale pulla syn. ***M. parvula*** q.v.

3786   ***Microgale pusilla***
Soricomorpha - Tenrecidae
e   Least Shrew Tenrec; Least Shrew
Tanrec

3787   ***Microgale soricoides***
Soricomorpha - Tenrecidae
e   Soricine Shrew Tenrec; Soricine
Shrew Tanrec

3788   ***Microgale taiva***
Soricomorpha - Tenrecidae
e   Taiva Shrew Tenrec; Taiva Shrew
Tanrec

3789   ***Microgale talazaci***
Soricomorpha - Tenrecidae
e   Talazac's Shrew Tenrec; Talazac's
Shrew Tanrec
f   Musaraigne de Talazac

3790   ***Microgale thomasi***
Soricomorpha - Tenrecidae
e   Thomas's Shrew Tenrec; Thomas's
Shrew Tanrec
d   Thomas-Langschwanztanrek
f   Musaraigne de Thomas

3791   ***Microhydromys***
Rodentia - Muridae
e   Smaller Water Rats; Lesser Water
Rats; Lesser Shrewmice
d   Furchenzahn-Spitzmausratten

3792   ***Microhydromys musseri***
Rodentia - Muridae
e   Musser's Shrewmouse

3793   ***Microhydromys richardsoni***
Rodentia - Muridae
e   Smaller Water Rat; Lesser Water

Rat; Groove-toothed Shrewmouse
d   Furchenzahn-Spitzmausratte

Microlagus syn. ***Sylvilagus*** q.v.

3794   ***Micromys***
Rodentia - Muridae
e   Harvest Mice; Old World Harvest
Mice; Eurasian Harvest Mice
d   Zwergmäuse
f   Souris des moissons; Rats des
moissons
i   Topolini delle risaie

Micromys danubianus syn. *M.
minutus* q.v.

3795   ***Micromys minutus***
Rodentia - Muridae
e   Harvest Mouse; Old World Harvest
Mouse; Eurasian Harvest Mouse
d   Europäische Zwergmaus;
Zwergmaus; Hafermaus; Eurasische
Zwergmaus; Eurasiatische
Zwergmaus
f   Souris des moissons; Rat des
moissons; Souris naine; Rat nain
i   Topolino delle risaie

Micronomus syn. *Mormopterus* q.v.

3796   ***Micronycteris***
Chiroptera - Phyllostomidae
e   Little Big-eared Bats; Small Big-
eared Bats; Big-eared Bats; Large-
eared Bats
d   Großohrfledermäuse

3797   ***Micronycteris behni***
Chiroptera - Phyllostomidae
e   Behn's Big-eared Bat

3798   ***Micronycteris brachyotis***
Chiroptera - Phyllostomidae
e   Yellow-throated Bat; Dobson's Big-
eared Bat; Yellow-throated Big-eared
Bat; Orange-throated Big-eared Bat

3799   *Micronycteris brosseti*
Chiroptera - Phyllostomidae
e   Brosset's Big-eared Bat

3800   *Micronycteris daviesi*
Chiroptera - Phyllostomidae
e   Davies's Big-eared Bat; Davies's
Long-eared Bat; Bartica Bat

3801   *Micronycteris hirsuta*
Chiroptera - Phyllostomidae
e   Hairy Big-eared Bat; Peters's Big-
eared Bat; Hairy Little Big-eared Bat

3802   *Micronycteris homezi*
Chiroptera - Phyllostomidae
e   Homez's Big-eared Bat

3803   *Micronycteris matses*
Chiroptera - Phyllostomidae
e   Matses's Big-eared Bat

3804   *Micronycteris megalotis*
Chiroptera - Phyllostomidae
e   Brazilian Big-eared Bat; Little Big-
eared Bat
d   Großohrige Fledermaus;
Großohrfledermaus

3805   *Micronycteris microtis*
Chiroptera - Phyllostomidae
e   Common Big-eared Bat

3806   *Micronycteris minuta*
Chiroptera - Phyllostomidae
e   White-bellied Big-eared Bat;
Gervais's Big-eared Bat; White-
bellied Big-eared Bat; Tiny Big-
eared Bat

3807   *Micronycteris nicefori*
Chiroptera - Phyllostomidae
e   Niceforo Forest Bat; Niceforo's Big-
eared Bat; Niceforo's Bat

3808   *Micronycteris pusilla*

Chiroptera - Phyllostomidae
e   Least Big-eared Bat

3809   *Micronycteris schmidtorum*
Chiroptera - Phyllostomidae
e   Schmidt's Big-eared Bat; Schmidt's
Little Big-eared Bat
d   Schmidts Großohrfledermaus

3810   *Micronycteris sylvestris*
Chiroptera - Phyllostomidae
e   Tricoloured Forest Bat; Tricolor
Forest Bat; Tricoloured Bat; Costa
Rican Big-eared Bat; Large-eared
Forest Bat; Tricoloured Big-eared
Bat

3811   *Microperorcytes*
Peramelemorphia - Peroryctidae
e   New Guinea Mouse Bandicoots;
Mouse Bandicoots
d   Mausnasenbeutler; Neuguinea-
Nasenbeutler

3812   *Microperorcytes murina*
Peramelemorphia - Peroryctidae
e   New Guinea Mouse Bandicoot;
Mouse Bandicoot
d   Eigentlicher Mausnasenbeutler

3813   *Microperoryctes longicaudata*
Peramelemorphia - Peroryctidae
e   Striped Bandicoot; Long-tailed New
Guinea Bandicoot
d   Langschwänziger Neuguinea-
Nasenbeutler; Langschwanz-
Neuguinea-Nasenbeutler;
Langschwanznasenbeutler

3814   *Microperoryctes papuensis*
Peramelemorphia - Peroryctidae
e   Papuan Bandicoot; Papua Bandicoot
d   Mura-Neuguinea-Nasenbeutler;
Mura-Nasenbeutler

3815   *Micropotamogale*
Insectivora - Tenrecidae

e   Small African Water Shrews; Dwarf
      African Water Shrews; Small Otter
      Shrews; Dwarf Otter Shrews; Otter
      Shrews
d   Zwergotterspitzmäuse; Kleine
      Otterspitzmäuse
f   Micropotamogales

3816   *Micropotamogale lamottei*
      Insectivora - Tenrecidae
e   Lesser Watershrew; Nimba
      Watershrew; Nimba Otter Shrew
d   Zwergotterspitzmaus; Kleine
      Otterspitzmaus
f   Micropotamogale de Lamotte

3817   *Micropotamogale ruwenzorii*
      Insectivora - Tenrecidae
e   Ruwenzori Otter Shrew; Small
      African Otter Shrew; Dwarf Otter
      Shrew; Ruwenzori Potamogale;
      Ruwenzori Least Otter Shrew
d   Ruwenzori-Zwergotterspitzmaus;
      Ruwenzori-Otterspitzmaus
f   Micropotamogale du Mont
      Ruwenzori; Micropotamogale du
      Ruwenzori

Micropteron syn. *Mesoplodon* q.v.

3818   *Micropteropus*
      Chiroptera - Pteropodidae
e   Dwarf Epaulet Bats; Dwarf Epaulet
      Fruit Bats; Lesser Epaulet Fruit Bat;
      Dwarf Epauletted Bat; Dwarf
      Epauletted Fruit Bats; Lesser
      Epauleted Fruit Bats
d   Zwergepaulettenflughunde
f   Microptères

3819   *Micropteropus intermedius*
      Chiroptera - Pteropodidae
e   Hayman's Epauletted Fruit Bat;
      Hayman's Dwarf Epauletted Fruit Bat

3820   *Micropteropus pusillus*
      Chiroptera - Pteropodidae

e   Dwarf Epauletted Fruit Bat; Lesser
      Epauleted Fruit Bat; Peters's Dwarf
      Epauletted Fruit Bat
d   Zwergepaulettenflughund
f   Petite Mictoptère

3821   *Microryzomys*
      Rodentia - Muridae
e   Small Rice Rats
d   Kleine Reisratten; Kleinsreisratten

3822   *Microryzomys altissimus*
      Rodentia - Muridae
e   Highland Small Rice Rat

3823   *Microryzomys minutus*
      Rodentia - Muridae
e   Forest Small Rice Rat

3824   *Microsciurus*
      Rodentia - Sciuridae
e   Neotropical Dwarf Squirrels; Dwarf
      Tree Squirrels; American Pygmy
      Squirrels; Dwarf Squirrels
d   Neuweltliche Zwerghörnchen
f   Microsciures

3825   *Microsciurus alfari*
      Rodentia - Sciuridae
e   Dwarf Tree Squirrel; Alfaro's Pygmy
      Squirrel; Central American Dwarf
      Squirrel
d   Zentralamerikanisches
      Zwerghörnchen
f   Microsciure nain

Microsciurus boquetensis syn. *M.
mimulus* q.v.

3826   *Microsciurus flaviventer*
      Rodentia - Sciuridae
e   Amazon Dwarf Squirrel
d   Amazonisches Zwerghörnchen

3827   *Microsciurus mimulus*
      Rodentia - Sciuridae

e   Boquete Pygmy Squirrel; Cloud-
forest Pygmy Squirrel; Western
Dwarf Squirrel; Western Pygmy
Squirrel

d   Westliches Zwerghörnchen

3828   *Microsciurus santanderensis*
Rodentia - Sciuridae

e   Santander Dwarf Squirrel;
Colombian Dwarf Squirrel

d   Santander-Zwerghörnchen

3829   *Microsorex*
Soricomorpha - Soricidae

e   American Pygmy Shrews

d   Zwergspitzmäuse

f   Musaraignes pygmées

3830   *Microsorex hoyi*
Soricomorpha - Soricidae

e   Pygmy Shrew (NA); American
Pygmy Shrew; Thompson's Pygmy
Shrew

d   Amerikanische Zwergspitzmaus

f   Musaraigne pygmée; Musaraigne
pygmee américaine; Musaraigne
pygmée d'Amérique

Microspalax syn. *Nannospalax* q.v.

3831   *Microtus*
Rodentia - Muridae

e   Meadow Voles; Common Field
Voles; Voles; Meadow Mice

d   Feldmäuse; Wühlmäuse

f   Campagnol

i   Arvicole

3832   *Microtus abbreviatus*
Rodentia - Muridae

e   Insular Vole

Microtus afghanus syn.
Blanfordimys afghanus q.v.

3833   *Microtus agrestis*

Rodentia - Arvicola

e   Common Fieldmouse; Short-tailed
Vole; Short-tailed Field Mouse;
Short-tailed Field Vole; Common
Field Vole; European Field Vole;
Northern Grass Mouse; Field Vole

d   Erdmaus; Ackermaus; Wühlmaus;
Gemeine Feldmaus

f   Campagnol agreste

i   Arvicola agreste

3834   *Microtus arvalis*
Rodentia - Muridae

e   Common Vole

d   Feldmaus; Europäische Feldmaus

f   Campagnol fauve; Campagnol
ordinaire; Campagnol commun;
Campagnol vulgaire; Campagnol des
champs

i   Arvicola campestre; Topo
campagnolo comune

3835   *Microtus arvalis orcadensis*
Rodentia - Muridae

e   Orkney Vole

d   Orkney-Feldmaus

3836   *Microtus arvalis sarniensis*
Rodentia - Muridae

e   Guernsey Vole

3837   *Microtus bavaricus*
Rodentia - Muridae

e   Bavarian Pine Vole

d   Bayerische Feldmaus

f   Campagnol souterrain de Bavarie

i   Arvicola della Baviera; Topo
bavarese dalle orecchie piccole

Microtus bedfordi syn. *Proedromys
bedfordi* q.v.

3838   *Microtus brandtii*
Rodentia - Muridae

e   Brandt's Vole

d   Brandts Steppenwühlmaus; Brandt-

Steppenwühlmaus;
Steppenwühlmaus
f   Campagnol de Brandt

3839   *Microtus breweri*
Rodentia - Muridae
e   Beach Vole
d   Strandwühlmaus; Küstenwühlmaus

3840   *Microtus cabrerae*
Rodentia - Muridae
e   Cabrera's Vole
d   Cabrera-Maus
f   Campagnol de Cabrera; Campagnol de la Méditerranée; Campagnol méditerranéen
i   Arvicola di Cabrera

3841   *Microtus californicus*
Rodentia - Muridae
e   California Vole
d   Kalifornische Wühlmaus

3842   *Microtus californicus aestuarinus*
Rodentia - Muridae
e   California Meadow Vole

3843   *Microtus californicus mohavensis*
Rodentia - Muridae
e   Mohave River Mole

3844   *Microtus californicus sanpabloensis*
Rodentia - Muridae
e   San Pablo Vole

3845   *Microtus californicus scirpensis*
Rodentia - Muridae
e   Armagosa Vole

3846   *Microtus californicus stephensi*
Rodentia - Muridae
e   Stephens's California Vole

3847   *Microtus californicus vallicola*
Rodentia - Muridae

e   Owen's Valley Rat; Owen's Valley Vole

3848   *Microtus canicaudus*
Rodentia - Muridae
e   Grey-tailed Vole

3849   *Microtus chrotorrhinus*
Rodentia - Muridae
e   Rock Vole
d   Gelbnasige Wühlmaus
f   Campagnol des rochers

Microtus clarkei syn. *Volemys clarkei* q.v.

3850   *Microtus daghestanicus*
Rodentia - Muridae
e   Daghestan Pine Vole

3851   *Microtus dogramacii*
Rodentia - Muridae
e   Dogramai's Vole

3852   *Microtus duodecimcostatus*
Rodentia - Muridae
e   Mediterranean Pine Vole
d   Mittelmeer-Kleinwühlmaus
f   Campagnol provençal
i   Arvicola iberica

Microtus epiroticus syn. *M. rossiameridionalis* q.v.

3853   *Microtus evoronensis*
Rodentia - Muridae
e   Evorsk Vole
f   Campagnol de Sibérie orientale

3854   *Microtus felteni*
Rodentia - Muridae
e   Felten's Vole; Balkan Pine Vole
f   Campagnol de Felten

3855   *Microtus fortis*
        Rodentia - Muridae
e   Reed Vole
d   Fernöstliche Feldmaus;
        Schilfwühlmaus

        Microtus fulviventer syn. *M. mexicanus* q.v.

3856   *Microtus fuscus*
        Rodentia - Muridae
e   Plateau Vole

3857   *Microtus gerbei*
        Rodentia - Muridae
e   Gerbe's Vole; Pyrenean Pine Vole
f   Campagnol des Pyrénées

3858   *Microtus gregalis*
        Rodentia - Muridae
e   Narrow-skulled Vole; Narrow-headed Vole
d   Langschädlige Feldmaus; Zwiebelmaus; Schmalköpfige Feldmaus; Schmalschädelwühlmaus
f   Campagnol des hauteurs

3859   *Microtus guatemalensis*
        Rodentia - Muridae
e   Guatemalan Vole

        Microtus gud syn. *Chionomys gud* q.v.

3860   *Microtus guentheri*
        Rodentia - Muridae
e   Günther's Vole; Levant Vole
d   Mittelmeer-Feldmaus; Balkan-Wühlmaus
f   Campagnol du Levant; Campagnol levantin; Campagnol de Günther
i   Arvicola di Günther

3861   *Microtus hyperboreus*
        Rodentia - Muridae
e   North Siberian Vole

3862   *Microtus irani*
        Rodentia - Muridae
e   Persian Vole

3863   *Microtus irene*
        Rodentia - Muridae
e   Chinese Scrub Vole

3864   *Microtus juldaschi*
        Rodentia - Muridae
e   Juniper Vole

3865   *Microtus kermanensis*
        Rodentia - Muridae
e   Baluchistan Vole

3866   *Microtus kirgisorum*
        Rodentia - Muridae
e   Tien Shan Mole

3867   *Microtus leucurus*
        Rodentia - Muridae
e   Blyth's Vole

3868   *Microtus limnophilus*
        Rodentia - Muridae
e   Lacustrine Vole

3869   *Microtus longicaudus*
        Rodentia - Muridae
e   Long-tailed Vole
d   Langschwänzige Wühlmaus
f   Campagnol longicaude

3870   *Microtus lusitanicus*
        Rodentia - Muridae
e   Lusitanian Pine Vole
d   Iberien-Wühlmaus
f   Campagnol basque
i   Arvicola di Portogallo

3871   *Microtus majori*
        Rodentia - Muridae
e   Major's Pine Vole

3872   Microtus mandarinus
Rodentia - Muridae
e   Mandarin Vole
d   Winogradows Wühlmaus;
       Vinogradovs Wühlmaus

3873   Microtus maximowiczii
Rodentia - Muridae
e   Maximovicz's Vole

3874   Microtus mexicanus
Rodentia - Muridae
e   Mexican Vole
d   Mexikanische Wühlmaus
f   Campagnol du Mexique

3875   Microtus middendorffii
Rodentia - Muridae
e   Middendorf's Vole
d   Middendorfs Feldmaus

3876   Microtus miurus
Rodentia - Muridae
e   Singing Vole
d   Alaska-Wühlmaus
f   Campagnol chanteur

3877   Microtus mogollonesis
Rodentia - Muridae
e   Mogolon Vole

3878   Microtus montanus
Rodentia - Muridae
e   Montane Vole
d   Rocky-Mountains-Wühlmaus
f   Campagnol montagnard

3879   Microtus montebelli
Rodentia - Muridae
e   Korean Meadow Vole; Japanese
       Grass Vole
d   Japanische Feldmaus

3880   Microtus mujanensis

Rodentia - Muridae
e   Muisk Vole

3881   Microtus multiplex
Rodentia - Muridae
e   Alpine Pine Vole; Fatio's Pine Vole
d   Fatio-Kleinwühlmaus;
       Alpenkleinwühlmaus
f   Campagnol de Fatio
i   Arvicola di Fatio

3882   Microtus nasarovi
Rodentia - Muridae
e   Nasarov's Vole; Nasrov's Pine Vole

Microtus nivalis syn. *Chionomys
nivalis* q.v.

3883   Microtus oaxacensis
Rodentia - Muridae
e   Tarabundi Vole

3884   Microtus obscurus
Rodentia - Muridae
e   Altai Vole

3885   Microtus ochrogaster
Rodentia - Muridae
e   Prairie Vole
d   Prairiewühlmaus
i   Arvicola della Prateria; Topolino di
       campagna; Topo campagnolo

3886   Microtus ochrogaster haydenii
Rodentia - Muridae
e   Western Upland Mouse

3887   Microtus ochrogaster ludovicianus
Rodentia - Muridae
e   Louisiana Vole; Lousiana Yellow-
       bellied Meadow Vole; Louisiana
       Yellow-bellied Vole; Luisiana
       Yellow-bellied Mouse; Louisiana
       Prairie Vole
d   Louisiana-Gelbbauchfeldmaus;
       Louisiana Prairie Wühlmaus

f   Campagnole à ventre jaune de Louisiane; Campagnol des prairies de Louisiane

3888   *Microtus oeconomus*
Rodentia - Muridae

e   Tundra Vole; Root Vole
d   Sibirische Magazinmaus; Nordische Wühlmaus; Nordische Wühlratte; Wurzelmaus; Ökonomische Maus
f   Campagnol économe; Campagnol des prés; Campagnol nordique
i   Arvicola del Nord; Arvicola nordica

3889   *Microtus oeconomus mehelyi*
Rodentia - Muridae

e   Pannonian Root Vole
d   Slowakische Wühlmaus

3890   *Microtus oregoni*
Rodentia - Muridae

e   Creeping Vole
d   Oregon-Wühlmaus
f   Campagnol de l'Orégon

3891   *Microtus pennsylvanicus*
Rodentia - Muridae

e   Meadow Mouse; Eastern Meadow Mouse; Eastern Meadow Vole; Meadow Vole
d   Wiesenfeldmaus; Wiesenmaus; Wiesenwühlmaus
f   Campagnol de Pennsylvanie
i   Arvicola di città

3892   *Microtus pennsylvanicus drummondi*
Rodentia - Muridae

e   Drummon Meadow Mouse

3893   *Microtus pennsylvanicus nesophilus*
Rodentia - Muridae

e   Gull Island Vole; Gull Island Meadow Vole; Gull Vole; Gull Islands Meadow Mouse; Gull Meadow Mouse

d   Gull-Insel-Wiesenmaus; Gull-Insel-Wiesenwühlmaus; Gull-Wiesenmaus; Gull-Wiesenwühlmaus
f   Campagnol des îles Gull; Campagnol des Gull

3894   *Microtus pinetorum*
Rodentia - Muridae

e   American Pine Vole; Woodland Vole; Pine Vole (NA)
d   Kiefernwühlmaus
f   Campagnol du pin

3895   *Microtus quasiater*
Rodentia - Muridae

e   Jalapan Pine Vole

3896   *Microtus richardsoni*
Rodentia - Muridae

e   Water Vole; American Water Vole; Richardon's Vole
d   Richardson-Wühlmaus; Richardsons Wühlmaus
f   Campagnol de Richardson

Microtus roberti syn. *Chionomys roberti* q.v.

3897   *Microtus rossiameridionalis*
Rodentia - Muridae

e   Southern Vole; Sibling Vole
d   Südfeldmaus
f   Campagnol d'Ondrias
i   Arvicola di Ognev

3898   *Microtus sachalinensis*
Rodentia - Muridae

e   Sakhalin Vole
d   Sachalinische Feldmaus

3899   *Microtus savii*
Rodentia - Muridae

e   Savi's Pine Vole
d   Savi-Kleinwühlmaus; Italienische Kleinwühlmaus; Untergrundmaus
f   Campagnol de Savi

i   Campagnolo di Savi; Arvicola di
     Savi; Arvicola sotteranea di Savi

3900   *Microtus schelkovnikovi*
         Rodentia - Muridae
e   Shelkovnikov's Pine Vole

3901   *Microtus sikimensis*
         Rodentia - Muridae
e   Sikkim Vole

3902   *Microtus socialis*
         Rodentia - Muridae
e   Social Vole
d   Gesellige Feldmaus; Kurzschwänzige
     Feldmaus; Tulpenmaus; Levante-
     Wühlmaus

     Microtus subarvalis syn. *M.*
     rossiameridionalis q.v.

3903   *Microtus subterraneus*
         Rodentia - Muridae
e   European Pine Vole; Common Pine
     Vole; Pine Vole
d   Kleine Wühlmaus; Kleinäugige
     Wühlmaus; Kurzohrmaus;
     Kleinwühlmaus
f   Campagnol souterrain
i   Arvicola sotteranea

3904   *Microtus tatricus*
         Rodentia - Muridae
e   Tatra Pine Vole
d   Tatra-Kleinwühlmaus
f   Campagnol des Tatras
i   Arvicola dei Tatra

3905   *Microtus thomasi*
         Rodentia - Muridae
e   Thomas's Pine Vole
d   Balkan-Kurzohrmaus
f   Campagnol de Thomas
i   Arvicola di Thomas

3906   *Microtus townsendii*
         Rodentia - Muridae
e   Townsend's Vole
d   Townsend-Wühlmaus
f   Campagnol de Townsend

3907   *Microtus transcaspicus*
         Rodentia - Muridae
e   Transcaspian Vole

3908   *Microtus umbrosus*
         Rodentia - Muridae
e   Zempoaltepec Vole

     Microtus ungurensis syn. *M.*
     maximowiczii q.v.

3909   *Microtus xanthognathus*
         Rodentia - Muridae
e   Chestnut-cheeked Vole; Yellow-
     cheeked Vole; Taiga Vole
d   Gelbwangenwühlmaus
f   Campagnol aux joues jaunes

     Microxus syn. *Akodon* q.v.

     Milithronycteris syn. *Hesperoptenus*
     q.v.

3910   *Millardia*
         Rodentia - Muridae
e   Indian Soft-furred Rats; Soft-furred
     Rats; Soft-furred Field Rats; Asian
     Soft-furred Rats
d   Millard-Ratten; Millard-Mäuse;
     Asiatische Weichratten

3912   *Millardia gleadowi*
         Rodentia - Muridae
e   Sand-coloured Rat; Sand-coloured
     Soft-furred Rat
d   Pakistanische Metadratte

3913   *Millardia kathleenae*
         Rodentia - Muridae
e   Miss Ryley's Soft-furred Rat

3911 *Millardia kondana*
Rodentia - Muridae
e Kondana Soft-furred Rat

3914 *Millardia meltada*
Rodentia - Muridae
e Soft-furred Field Rat; Common Soft-furred Rat; Soft-furred Rat; Indian Soft-furred Rat
d Indische Weichfellmaus
f Rat d'herbe

3915 *Mimetillus*
Chiroptera - Vespertilionidae
e Moloney's Flat-headed Bats

3916 *Mimetillus moloneyi*
Chiroptera - Vespertilionidae
e Moloney's Flat-headed Bat

3917 *Mimon*
Chiroptera - Phyllostomidae
e Gray's Spear-nosed Bats; Little Spear-nosed Bats; Hairy-nosed Bats
d Haarnasenfledermäuse

3918 *Mimon bennetti*
Chiroptera - Phyllostomidae
e Bennet's Spear-nosed Bat; Cozumel Spear-nosed Bat

3919 *Mimon cozumelae*
Chiroptera - Phyllostomidae
e Cozumel Golden Bat

3920 *Mimon crenulatum*
Chiroptera - Phyllostomidae
e Hairy Spear-nosed Bat; Geoffroy's Hairy Bat; Striped Spear-nosed Bat; Striped Hairy-nosed Bat; Spear-nosed Bat
d Gestreifte Haarnasenfledermaus

Mimon koepkeae syn. *M. crenulatum* q.v.

Mindanaomys syn. *Batomys* q.v.

3921 *Miniopterus*
Chiroptera - Vespertilionidae
e Long-winged Bats; Long-fingered Bats; Bent-winged Bats; Clinging Bats
d Langflügelfledermäuse
f Minioptères
i Miniotteri

3922 *Miniopterus australis*
Chiroptera - Vespertilionidae
e Bent-winged Bat; Little Long-fingered Bat; Little Bentwing Bat (ANZ); Little Bent-winged Bat
f Minioptère australasien

Miniopterus celebensis syn. *M. tristis* q.v.

3923 *Miniopterus fraterculus*
Chiroptera - Vespertilionidae
e Lesser Long-fingered Bat

3924 *Miniopterus fuscus*
Chiroptera - Vespertilionidae
e Southeast Asian Long-fingered Bat; Ryukyu Bent-winged Bat

3925 *Miniopterus inflatus*
Chiroptera - Vespertilionidae
e Greater-Long-fingered Bat; High-crowned Bat

Miniopterus insularis syn. *M. tristis* q.v.

3926 *Miniopterus macrocneme*
Chiroptera - Vespertilionidae
e Small Melanesian Bent-winged Bat

Miniopterus macrodens syn. *M. magnater* q.v.

3927   *Miniopterus magnater*
Chiroptera - Vespertilionidae
e   Western Long-fingered Bat; Western
       Bent-winged Bat

3928   *Miniopterus majori*
Chiroptera - Vespertilionidae
e   Major's Long-fingered Bat

3929   *Miniopterus manavi*
Chiroptera - Vespertilionidae
e   Malagasy Least Long-fingered Bat

3930   *Miniopterus medius*
Chiroptera - Vespertilionidae
e   Medium Bent-winged Bat

Miniopterus melanesiensis syn. *M.
tristis* q.v.

3931   *Miniopterus minor*
Chiroptera - Vespertilionidae
e   Least Long-fingered Bat

Miniopterus oceanensis syn. *M.
schreibersi* q.v.

3932   *Miniopterus propritristis*
Chiroptera - Vespertilionidae
e   Large Melanesian Bent-winged Bat

3933   *Miniopterus pusillus*
Chiroptera - Vespertilionidae
e   Small Bent-winged Bat

3934   *Miniopterus robustior*
Chiroptera - Vespertilionidae
e   Loyalty Bent-winged Bat
f   Mnioptère des îles Loyautés;
       Mnioptère robuste

3935   *Miniopterus schreibersi*
Chiroptera - Vespertilionidae
e   Long-winged Bat; Shreibers's Bat;
       Common Bentwing Bat (ANZ);

Schreibers's Long-fingered Bat;
Clinging Bat
d   Langflügelige Fledermaus;
       Langflügel; Gemeine
       Langflügelfledermaus;
       Langflügelfledermaus
f   Mnioptère de Schreibers; Mnioptère
       aux longues ailes; Mnioptère
i   Miniottero di Schreibers; Miniottero

Miniopterus solomonensis syn. *M.
australis* q.v.

3936   *Miniopterus tristis*
Chiroptera - Vespertilionidae
e   Great Bent-winged Bat

Miniopterus yaveyamae syn. *M.
fuscus* q.v.

3937   *Miopithecus*
Primates - Cercopithecidae
e   Talapoins; Talapoin Monkeys
d   Zwergmeerkatzen
f   Talapoins
i   Cercopithechi nani

3938   *Miopithecus ogouensis*
Primates - Cercopithecidae
e   Northern Talapoin ; Gabon Talapoin
d   Nördliche Zwergmeerkatze
f   Talapoin du Nord
i   Cercopiteco nano del Nord

3939   *Miopithecus talapoin*
Primates - Cercopithecidae
e   Talapoin; Southern Talapoin;
       Angolan Talapoin; Dwarf Monkey;
       Talapoin Monkey; Dwarf Guenon;
       Southern Talapoin Monkey

d   Zwergmeerkatze
f   Talapoin
i   Cercopiteco nano del sud;
       Cercopiteco nano; Talapoin

3940 *Mirounga*
Carnivora - Phocidae
e Elephant Seals
d See-Elefanten; Elefantenrobben;
   Meerwölfe; Morungas
f Éléphants de mer; Éléphants marins
i Elefanti marini

3941 *Mirounga angustirostris*
Carnivora - Phocidae
e Northern Elephant Seal; Californian
   Elephant Seal

d Nördlicher See-Elefant
f Éléphant-de-mer du Nord; Éléphant
   marin du Nord; Éléphant-de-mer
   boréal
i Elefante marino del Nord; Elefante
   marino settentrionale

3942 *Mirounga leonina*
Carnivora - Phocidae
e South Atlantic Elephant Seal;
   Southern Elephant Seal; Crabeter
   Seal
d Südlicher See-Elefant
f Éléphant-de-mer du Sud; Éléphant
   marin du Sud; Éléphant-de-mer
   austral
i Elefante marino del sud; Elefanto
   marino meridionale

3943 *Mirza*
Primates - Cheirogalidae
e Coquerels Mouse Lemurs
d Coquerels Zwergmakis
f Microcèbes de Coquerel
i Microcebi di Cocquerel

3944 *Mirza coquereli*
Primates - Cheirogalidae
e Coquerel's Mouse Lemur; Coquerel's
   Dwarf Lemur; Greater Mouse Lemur
d Coquerels Zwergmaki; Rattenmaki
f Microcèbe de Coquerel; Mirza de
   Coquerel

i Microcebo di Coquerel

Mogera syn. *Talpa* q.v.

3945 Molossidae
Chiroptera
e Free-tailed Mastiff Bats; Free-tailed
   Bats
d Bulldoggfledermäuse;
   Faltlippenfledermäuse
f Molosses; Molossidés; Chauves-
   souris à queue libre
i Molossidi

3946 *Molossops*
Chiroptera - Molossidae
e Dog-faced Bats; Malagas Free-tailed
   Bats

3947 *Molossops abrasus*
Chiroptera - Molossidae
e Cinnamon Dog-faced Bat

3948 *Molossops aequatorianus*
Chiroptera - Molossidae
e Equatorial Dog-faced Bat

Molossops brachymeles syn. *M.*
abrasus q.v.

3949 *Molossops greenhalli*
Chiroptera - Molossidae
e Trinidadian Free-tailed Bat; Trinidad
   Free-tailed Bat; Trinidadian Dog-
   faced Bat; Trinidad Dog-faced Bat;
   Greenhall's Dog-faced Bat

3950 *Molossops matogrossensis*
Chiroptera - Molossidae
e Mato Grosso Dog-faced Bat; South
   American Flat-headed Bat

3951 *Molossops mexicanus*
Chiroptera - Molossidae
e Mexican Dog-faced Bat

3952   *Molossops neglectus*
    Chiroptera - Molossidae
 e   Rufous Dog-faced Bat

3953   *Molossops paranus*
    Chiroptera - Molossidae
 e   Brown-bellied Dog-faced Bat

3954   *Molossops planirostris*
    Chiroptera - Molossidae
 e   White-bellied Mastiff Bat; Southern
    Dog-faced Bat

3955   *Molossops temminckii*
    Chiroptera - Molossidae
 e   Dwarf Dog-faced Bat

3956   *Molossus*
    Chiroptera - Molossidae
 e   Velvety Free-tailed Bats; Mastiff
    Bats
 d   Samtfledermäuse
 f   Molosses
 i   Pipistrelli coda di topo

3957   *Molossus ater*
    Chiroptera - Molossidae
 e   Red Mastiff Bat; Black Mastiff Bat
 i   Pipistrello coda di topo grande

3958   *Molossus barnesi*
    Chiroptera - Molossidae
 e   Barnes's Mastiff Bat

    Molossus bondae syn. *B. currentium*
    q.v.

    Molossus coibensis syn. *M. molossus*
    q.v.

3959   *Molossus currentium*
    Chiroptera - Molossidae
 e   Bonda Mastiff Bat
 d   Bonda-Samtfledermaus

Molossus espiritosantensis syn. *M.*
molossus q.v.

Molossus macdougalli syn. *M.*
pretiosus q.v.

3960   *Molossus molossus*
    Chiroptera - Molossidae
 e   Giant Velvety Free-tailed Bat; Kerr's
    Mastiff Bat; Pallas's Mastiff Bat;
    Velvety Free-tailed Bat; Rat Bat
 d   Große Samtfledermaus
 f   Molosse géant; Molosse commun

3961   *Molossus pretiosus*
    Chiroptera - Molossidae
 e   Miller's Mastiff Bat

3962   *Molossus sinaloae*
    Chiroptera - Molossidae
 e   Sinaloan Mastiff Bat; Allen's Mastiff
    Bat

3963   *Monachus*
    Carnivora - Phocidae
 e   Monk Seals
 d   Mönchsrobben
 f   Phoques-moines
 i   Foche monache

3964   *Monachus monachus*
    Carnivora - Phocidae
 e   Mediterranean Seal; Mediterranean
    Monk Seal; Monk Seal; Pied Seal
 d   Mönchsrobbe; Mittelmeer-
    Mönchsrobbe; Seemönch;
    Mittelmeer-Robbe; Seejungfrau
 f   Phoque-moine; Phoque à ventre
    blanc; Phoque-moine de la
    Méditerranée
 i   Foca monaca; Foca mediterranea;
    Foca monaco del mediterraneo;
    Vitello marino; Bue marino; Foca dal
    ventre bianco

3965   *Monachus schauinslandi*
    Carnivora - Phocidae

e   Laysan Monk Seal; Hawaiian Monk
    Seal
d   Laysan-Mönchsrobbe; Hawaii-
    Mönchsrobbe
f   Phoque-moine de Hawai
i   Foca monaca di Laysan

3966 *Monachus tropicalis*
    Carnivora - Phocidae
e   Caribbean Monk Seal; West Indian
    Monk Seal; West Indian Seal;
    Jamaican Seal
d   Karibische Mönchsrobbe;
    Westindische Mönchsrobbe
f   Phoque des Indes occidentales;
    Phoque-moine des Caraïbes; Phoque
    des Caraïbes; Phoque des Indes
i   Foca monaca dei Caraibi

3967 *Monodelphis*
    Marsupialia - Didelphidae
e   Short-tailed Opossums; Short Bare-
    tailed Opossums
d   Spitzmausbeutelratten;
    Kurzschwanzopossums
i   Opossum dalla coda corta; Possum;
    Opossum

3968 *Monodelphis adusta*
    Marsupialia - Didelphidae
e   Cloudy Short-tailed Opossum;
    Osgoods Short-tailed Opossum;
    Sepia Short-tailed Opossum

d   Kolumbien-Spitzmausbeutelratte;
    Sepia Kurschwanzopossum

3969 *Monodelphis americana*
    Marsupialia - Didelphidae
e   Three-striped Short-tailed Opossum
d   Dreistreifenspitzmausbeutelratte

3970 *Monodelphis brevicaudata*
    Marsupialia - Didelphidae
e   Seba's Short-tailed Oposssum; Red-
    legged Short-tailed Opossum; Red-
    sided Short-tailed Opossum

d   Kurzschwanzspitzmausbeutelratte
i   Opossum dalla coda corta; Possum
    dalla coda corta

3971 *Monodelphis dimidiata*
    Marsupialia - Didelphidae
e   Eastern Short-tailed Opossum;
    Southern Short-tailed Opossum
d   Grauspitzmausbeutelratte; Südliches
    Kurzschwanzopossum

3972 *Monodelphis domestica*
    Marsupialia - Didelphidae
e   Grey Short-tailed Opossum
d   Hausspitzmausbeutelratte;
    Kurzschwanz-Zwergopossum
f   Opossum gris

3973 *Monodelphis emiliae*
    Marsupialia - Didelphidae
e   Emilia's Short-tailed Opossum
d   Emilias Kurzschwanzopossum

3974 *Monodelphis iheringi*
    Marsupialia - Didelphidae
e   Ihering's Short-tailed Opossum
d   Iherings Kurzschwanzopossum

3975 *Monodelphis kunsi*
    Marsupialia - Didelphidae
e   Pygmy Short-tailed Opossum
d   Kuns-Spitzmausbeutelratte

3976 *Monodelphis maraxina*
    Marsupialia - Didelphidae
e   Marajo Short-tailed Opossum
d   Marajo-Spitzmausbeutelratte;
    Marajo-Kurzschwanzopossum

3977 *Monodelphis orinoci*
    Marsupialia - Didelphidae
e   Orinoco Short-tailed Opossum

3978 *Monodelphis osgoodi*
    Marsupialia - Didelphidae

   e   Osgood's Short-tailed Opossum

   d   Osgoods Kurzschwanzopossum

3979 *Monodelphis rubida*
       Marsupialia - Didelphidae
   e   Chestnut-striped Short-tailed
       Opossum

3980 *Monodelphis scalops*
       Marsupialia - Didelphidae
   e   Red-headed Short-tailed Opossum;
       Long-nosed Short-tailed Opossum
   d   Rotkopfspitzmausbeutelratte

3981 *Monodelphis sorex*
       Marsupialia - Didelphidae
   e   Soricine Short-tailed Opossum;
       Shrewish Short-tailed Opossum
   d   Zwergspitzmausbeutelratte

3982 *Monodelphis theresa*
       Marsupialia - Didelphidae
   e   Theresa's Short-tailed Opossum
   d   Theresas Kurzschwanz Opossum

       Monodelphis touan syn. *M.*
       brevicaudata q.v.

3983 *Monodelphis unistriata*
       Marsupialia - Didelphidae
   e   One-striped Short-tailed Opossum
   d   Streifenspitzmausbeutelratte;
       Einstreifenspitzmausbeutelratte;
       Einstreifenkurzschwanzopossum

       Monodia syn. *Gerbillus* q.v.

3984 *Monodon*
       Cete - Monodontidae
   e   Narwhals; Narwhales; Narwals
   d   Narwale; Einhornwale
   f   Unicornes; Licornes de mer
   i   Narvali

3985 *Monodon monoceros*

       Cete - Monodontidae
   e   Narwhal; Narwhale; Narwal; Unicorn
       Whale; Sea Unicorn
   d   Narwal; Einhornwal
   f   Unicorne; Licorne de mer; Narval
   i   Narvalo

3986 **Monodontidae**
       Cete
   e   White Whales; White Whales and
       Narwhals; Belugas and Narwhals;
       Single-toothed Whales
   d   Gründelwale; Weißwale
   f   Monodontidés
   i   Narvali; Monodontidi

3987 *Monophyllus*
       Chiroptera - Phyllostomidae
   e   Long-tongued Bats; Single Leaf Bats

3988 *Monophyllus plethodon*
       Chiroptera - Phyllostomidae
   e   Barbados Long-tongued Bat; Insular
       Long-tongued Bat; Lesser Antillean
       Long-tongued Bat
   f   Fer-de-lance de Barbade

3989 *Monophyllus redmani*
       Chiroptera - Phyllostomidae
   e   Jamaican Long-tongued Bat; Leach's
       Long-tongued Bat; Leach's Single-
       Leaf Bat; Greater Antillean Long-
       tongued Bat

3990 *Monticolomys*
       Rodentia - Muridae
   e   Malagasy Mountain Mice

3991 *Monticolomys koopmani*
       Rodentia - Muridae
   e   Malagasy Mountain Mouse

       Montisciurus syn. *Paraxerus* q.v.

3992 *Mops*
Chiroptera - Molossidae
e Greater Free-tailed Bats
d Freischwanzfledermäuse
f Molosses

3993 *Mops brachypterus*
Chiroptera - Molossidae
e Sierra Leone Free-tailed Bat

3994 *Mops condylurus*
Chiroptera - Molossidae
e Angolan Free-tailed Bat

3995 *Mops congicus*
Chiroptera - Molossidae
e Medje Free-tailed Bat

3996 *Mops demonstrator*
Chiroptera - Molossidae
e Mongalla Free-tailed Bat

3997 *Mops leucostigma*
Chiroptera - Molossidae
e Pale-marked Free-tailed Bat

3998 *Mops midas*
Chiroptera - Molossidae
e Midas Free-tailed Bat
f Molosse de Midas

3999 *Mops mops*
Chiroptera - Molossidae
e Malayan Free-tailed Bat

4000 *Mops nanulus*
Chiroptera - Molossidae
e Dwarf Free-tailed Bat

4001 *Mops niangarae*
Chiroptera - Molossidae
e Niangara Free-tailed Bat

4002 *Mops niveiventer*
Chiroptera - Molossidae
e White-bellied Free-tailed Bat

4003 *Mops petersoni*
Chiroptera - Molossidae
e Peterson's Free-tailed Bat

4004 *Mops sarasinorum*
Chiroptera - Molossidae
e Sulawesi Free-tailed Bat

4005 *Mops spurrelli*
Chiroptera - Molossidae
e Spurrell's Free-tailed Bat

4006 *Mops thersites*
Chiroptera - Molossidae
e Railer Bat

4007 *Mops trevori*
Chiroptera - Molossidae
e Trevor's Free-tailed Bat

4008 **Mormoopidae**
Chiroptera
e Leaf-chinned Bats; Naked-backed Bats; Moustached Bats; Ghost-faced Bats
d Kinnblattfledermäuse
f Mormoopidés
i Mormopidi

4009 *Mormoops*
Chiroptera - Mormoopidae
e Leaf-chinned Bats; Ghost-faced Bats
d Eigentliche Kinnblattfledermäuse

4010 *Mormoops blainvillei*
Chiroptera - Mormoopidae
e Blainville's Leaf-chinned Bat; Antillean Ghost-faced Bat

Mormoops cuvieri syn. *M. blainvillei* q.v.

4011 *Mormoops megalophylla*
Chiroptera - Mormoopidae
e Peters's Leaf-chinned Bat; Peters's
Ghost-faced Bat; Ghost-faced Bat;
Old Man Bat

4012 *Mormopterus*
Chiroptera - Molossidae
e Little Mastiff Bats
f Molosses

4013 *Mormopterus acetabulosus*
Chiroptera - Molossidae
e Natal Wrinkle-lipped Bat; Natal
Free-tailed Bat
f Petit Molosse

4014 *Mormopterus beccarii*
Chiroptera - Molossidae
e Beccari's Mastiff Bat; Beccari's
Freetail Bat (ANZ); Beccari's Free-
tailed Bat

4015 *Mormopterus doriae*
Chiroptera - Molossidae
e Sumatran Mastiff Bat

4016 *Mormopterus jugularis*
Chiroptera - Molossidae
e Peters's Wrinkle-lipped Bat

4017 *Mormopterus kalinowskii*
Chiroptera - Molossidae
e Kalinowski's Mastiff Bat

4018 *Mormopterus loriae*
Chiroptera - Molossidae
e Little Northern Free-tailed Bat

4019 *Mormopterus minutus*
Chiroptera - Molossidae
e Little Goblin Bat
f Chauve-souris de la Jata

4020 *Mormopterus norfolkensis*

Chiroptera - Molossidae
e Norfolk Island Scurrying Bat;
Eastern Little Mastiff Bat; Norfolk
Island Bat; Eastern Freetail Bat
(ANZ); East-coast Free-tailed Bat
d Winzige Fledermaus

4021 *Mormopterus petrophilus*
Chiroptera - Molossidae
e Roberts's Flat-headed Bat

4022 *Mormopterus phrudus*
Chiroptera - Molossidae
e Incan Little Mastiff Bat

4023 *Mormopterus planiceps*
Chiroptera - Molossidae
e Little Flat Bat; Southern Free-tailed
Bat; Southern Freetail Bat (ANZ)

4024 *Mormopterus setiger*
Chiroptera - Molossidae
e Peters's Flat-headed Bat

4025 *Moschidae*
Artiodactyla
e Musk Deer
d Moschustiere
f Moschidés
i Cervi muschiati; Moschidi

Moschiola syn. *Tragulus* q.v.

4026 *Moschus*
Artiodactyla - Moschidae
e Musk Deer
d Moschushirsche; Moschustiere
f Porte-muscs
i Cervi muschiati

4027 *Moschus berezovskii*
Artiodactyla - Moschidae
e Chinese Forest Musk Deer
d Chinesischer Moschushirsch

4028 **Moschus chrysogaster**
Artiodactyla - Moschidae
e Forest Musk Deer; Alpine Musk
Deer; Himalajan Musk Deer
d Himalaja-Moschushirsch
i Mosco alpino

4029 **Moschus fuscus**
Artiodactyla - Moschidae
e Dusky Musk Deer
d Schwarzer Moschushirsch

4030 **Moschus moschiferus**
Artiodactyla - Moschidae
e Musk Deer; Siberian Musk Deer
d Sibirischer Moschushirsch
f Porte-musc; Musc; Chevrotain porte-
musc
i Mosco della Siberia

4031 **Moschus moschiferus sachalinensis**
Artiodactyla - Moschidae
e Sakhalin Musk Deer

Moschus sibiricus syn. *M.*
moschiferus q.v.

Moschus sifanicus syn. *Moschus*
chrysogaster q.v.

4032 **Mosia**
Chiroptera - Eballonuridae
e Dark Sheath-tailed Bats

4033 **Mosia nigrescens**
Chiroptera - Eballonuridae
e Dark Sheath-tailed Bat

4034 **Mungos**
Carnivora - Herpestidae
e Banded Mongooses; Striped
Mongooses; African Mungos
d Zebramangusten; Mungos
f Mangues
i Manguste

4035 **Mungos gambianus**
Carnivora - Herpestidae
e Gambian Mongoose
d Gambia-Manguste; Gambia-Mungo
f Mangue de Gambie
i Mangusta della Gambia

4036 **Mungos mungo**
Carnivora - Herpestidae
e Banded Mongoose; Striped
Mongoose; Band Mongoose
d Zebramanguste
f Mangue rayé; Mangouste rayée
i Mungo; Mangusta fasciata; Mangusta
striata

4037 **Mungotictis**
Carnivora - Herpestidae
e Narrow-striped Mongooses;
Madagascar Narrow-striped
Mongooses
d Schmalstreifenmungos
f Mangoustes

4038 **Mungotictis decemlineata**
Carnivora - Herpestidae
e Narrow-striped Mongoose;
Madagascar Narrow-striped
Mongoose; Ten-striped Mongoose
d Schmalstreifenmungo
f Mangouste à dix raies; Mangouste
malgache; Galidie mongoz

Mungotictis lineata syn. *M.*
decemlineata q.v.

Mungotictis substriatus syn. *M.*
decemlineata q.v.

4039 **Muntiacus**
Artiodactyla - Cervidae
e Barking Deer ; Rib-faced Deer ;
Muntjacs; Muntjaks
d Muntjaks
f Muntjacs; Muntjaks
i Muntjak

4040   *Muntiacus atherodes*
        Artiodactyla - Cervidae
e   Bornean Yellow Muntjac; Bornean
      Yellow Muntjak
d   Borneo-Muntjak
i   Muntjak giallo

4041   *Muntiacus crinifrons*
        Artiodactyla - Cervidae
e   Black Muntjac; Hairy-fronted
      Muntjac; Black Muntjak; Hairy-
      fronted Muntjak
d   Schwarzer Muntjak
f   Muntjac noir; Muntjak noir
i   Muntjak nero

4042   *Muntiacus feae*
        Artiodactyla - Cervidae
e   Fea's Muntjac; Fea's Muntjak
d   Fea-Muntjak; Tenasserim-Muntjak
f   Muntjac de Bornéo; Muntjak de
      Bornéo
i   Muntjak di Fea

4044   *Muntiacus muntjak*
        Artiodactyla - Cervidae
e   Barking Deer; Rib-faced Deer;
      Muntjac; Muntjak
d   Muntjak; Indischer Muntjak
f   Muntjac; Muntjak; Cerf aboyeur
i   Muntjak comune

4045   *Muntiacus muntjak muntjak*
        Artiodactyla - Cervidae
e   Javan Muntjak; Javan Muntjac
d   Java-Muntjak
f   Muntjac de Java; Muntjak de Java

4046   *Muntiacus muntjak vaginalis*
        Artiodactyla - Cervidae
e   Indian Muntjak; Indian Muntjac
d   Nordindischer Muntjak
f   Muntjac d'Inde; Muntjak d'Inde

Muntiacus pleiharicus syn. *M.
atherodes* q.v.

4047   *Muntiacus putaoensis*
        Artiodactyla - Cervidae
e   Leaf Muntjak; Leaf Muntjac
d   Putao-Muntjak
f   Muntjac de Putao; Muntjak de Putao

4048   *Muntiacus reevesi*
        Artiodactyla - Cervidae
e   Chinese Muntjac; Reeves's Muntjac;
      Chinese Muntjak; Reeves's Muntjak
d   Chinesischer Muntjak;
      Zwergmuntjak
f   Muntjac de Chine; Muntjak de
      Chine; Muntiac de Reeves; Muntjak
      de Reeves
i   Muntjak cinese; Muntjak della Cina

4049   *Muntiacus trungsonensis*
        Artiodactyla - Cervidae
e   Trung Son Muntjac; Trung Son
      Muntjak
d   Trung-Son-Muntjak

4050   *Muntiacus vuqyangensis*
        Artiodactyla - Cervidae
e   Giant Muntjak; Giant Muntjac
d   Riesenmutjak
i   Muntjak gigante

4043   *Muntjakus gongshanensi*
        Artiodactyla - Cervidae
e   Gong Shan Muntjak; Gong Shan
      Muntjac
d   Gongshan-Muntjak
i   Muntjak di Gongshan

4051   *Murexia*
        Dasyurimorphia - Dasyuridae
e   New Guinea Marsupial Rats; Long-
      tailed Marsupial Mice; New Guinea
      Marsupial Mice; Long-tailed
      Dasyures
d   Neuguinea-Beutelmäuse

4052 **Murexia longicaudata**
Dasyurimorphia - Dasyuridae
e Short-haired Marsupial Mouse;
Long-tailed New Guinea Marsupial
Mouse; Short-furred Dasyure
d Langschwanz-Neuguinea-
Beutelmaus

4053 **Murexia rothschildi**
Dasyurimorphia - Dasyuridae
e Broad-striped Marsupial Mouse;
Rothschild's New Guinea Marsupial
Mouse; Broad-striped Dasyure
d Rothschild-Neuguinea-Beutelmaus

4054 **Muriculus**
Rodentia - Muridae
e Striped-back Mice
d Äthiopische Streifenmäuse

4055 **Muriculus imberbis**
Rodentia - Muridae
e Stripe-backed Mouse
d Rückstreifenmaus

4056 **Muridae**
Rodentia -
e Mice; Rats, Mice, Voles,
Gerbils,Hamsters and Lemmings;
Rats and Mice
d Mäuse; Eigentliche Mäuse;
Langschwanzmäuse; Mäuseartige
f Muridés
i Topi; Ratti; Muridi

4057 **Murina**
Chiroptera - Vespertilionidae
e Tube-nosed Bats; Insectivorous Bats;
Tube-nosed Insectivorous Bats
d Röhrennasen;
Schlauchnasenfledermäuse

4058 **Murina aenea**
Chiroptera - Vespertilionidae
e Bronzed Bat; Bronze Tube-nosed Bat

4059 **Murina aurata**
Chiroptera - Vespertilionidae
e Little Tube-nosed Bat
d Kleine Röhrennase; Ussurische
Röhrennase

4060 **Murina cyclotis**
Chiroptera - Vespertilionidae
e Round-eared Tube-nosed Bat

4061 **Murina cyclotis eileenae**
Chiroptera - Vespertilionidae
e Ceylon Tube-nosed Bat
d Sri Lanka-Röhrennase; Sri Lanka-
Schlauchnasenfledermaus

4062 **Murina florium**
Chiroptera - Vespertilionidae
e Flores Tube-nosed Bat

4063 **Murina fusca**
Chiroptera - Vespertilionidae
e Dusky Tube-nosed Bat

4064 **Murina grisea**
Chiroptera - Vespertilionidae
e Peters's Tube-nosed Bat

4065 **Murina huttoni**
Chiroptera - Vespertilionidae
e Hutton's Tube-nosed Bat

4066 **Murina leucogaster**
Chiroptera - Vespertilionidae
e Great Tube-nosed Bat; Greater Tube-
nosed Bat; Siberian Tube-nosed Bat
d Große Sibirische Röhrennase;
Sibirische Röhrennase; Große
Röhrennase

4067 **Murina puta**
Chiroptera - Vespertilionidae
e Taiwan Tube-nosed Bat

4068   *Murina rozendaali*
            Chiroptera - Vespertilionidae
    e   Gilded Tube-nosed Bat

4069   *Murina ryukyuana*
            Chiroptera - Vespertilionidae
    e   Ryukyu Tube-nosed Bat

4070   *Murina silvatica*
            Chiroptera - Vespertilionidae
    e   Forest Tube-nosed Bat

4071   *Murina suilla*
            Chiroptera - Vespertilionidae
    e   Brown Tube-nosed Bat

4072   *Murina tenebrosa*
            Chiroptera - Vespertilionidae
    e   Gloomy Tube-nosed Bat

4073   *Murina tubinaris*
            Chiroptera - Vespertilionidae
    e   Scully's Tube-nosed Bat

4074   *Murina ussuriensis*
            Chiroptera - Vespertilionidae
    e   Ussuri Tube-nosed Bat

4075   *Mus*
            Rodentia - Muridae
    e   Mice; House Mice; Old World Mice
    d   Mäuse
    f   Souris ; Souris de l'ancien monde

            Mus abbotti syn. *M. macedonicus*
            q.v.

4076   *Mus baoulei*
            Rodentia - Muridae
    e   Baoule's Mouse

4077   *Mus booduga*
            Rodentia - Muridae
    e   Little Indian Field Mouse

    d   Indische Maus

4078   *Mus bufo*
            Rodentia - Muridae
    e   Toad Mouse
    f   Souris d'Afrique

4079   *Mus callewaerti*
            Rodentia - Muridae
    e   Callewaert's Mouse

4080   *Mus caroli*
            Rodentia - Muridae
    e   Ryukyu Mouse
    d   Reisfeldmaus

4081   *Mus cervicolor*
            Rodentia - Muridae
    e   Fawn-coloured Mouse
    d   Beige Hausmaus; Falbmaus

4082   *Mus cervicolor fulviventris*
            Rodentia - Muridae
    e   Ceylon Field Mouse
    d   Sri Lanka-Feldmaus

4083   *Mus cookii*
            Rodentia - Muridae
    e   Cook's Mouse
    d   Cook-Maus

4084   *Mus crociduroides*
            Rodentia - Muridae
    e   Sumatran Shrew-like Mouse;
            Sumatran Shrew Mouse
    d   Sumatra-Hausmaus
    f   Souris de Sumatra

4085   *Mus domesticus*
            Rodentia - Muridae
    e   House Mouse; Western European
            House Mouse; Western House
            Mouse
    d   Hausmaus; Westliche Hausmaus
    f   Souris; Souris domestique; Souris

commune; Souris vulgaire

i   Topolino domestico; Topolino delle case; Topo domestico

4086   ***Mus domesticus gentilulus***
       Rodentia - Muridae
e   Jordan Mouse
d   Jordanische Hausmaus

4087   ***Mus famulus***
       Rodentia - Muridae
e   Servant Mouse

4088   ***Mus fernandoni***
       Rodentia - Muridae
e   Ceylon Spiny Mouse; Sri Lankan Spiny Mouse
d   Sri Lanka Stachelmaus

       Mus fulvidentris syn. *M. booduga* q.v.

4089   ***Mus goundae***
       Rodentia - Muridae
e   Gounda Mouse

4090   ***Mus haussa***
       Rodentia - Muridae
e   Hausa Mouse

       Mus hortulanus syn. *M. spicilegus* q.v.

       Mus imberbis syn. *Muriculus imberbis* q.v.

4091   ***Mus indutus***
       Rodentia - Muridae
e   Desert Pygmy Mouse
d   Wüstenzwergmaus

4092   ***Mus kasaicus***
       Rodentia - Muridae
e   Kasai Mouse

       Mus lepidoides syn. *M. booduga* q.v.

4093   ***Mus macedonicus***
       Rodentia - Muridae
e   Macedoninan Mouse
d   Makedonische Hausmaus
f   Souris de Macédoine
i   Topolino orientale a coda corta

4094   ***Mus mahomet***
       Rodentia - Muridae
e   Mahomet Mouse

4095   ***Mus mattheyi***
       Rodentia - Muridae
e   Matthey's Mouse

4096   ***Mus mayori***
       Rodentia - Muridae
e   Mayor's Mouse
d   Mayor-Maus

4097   ***Mus mayori mayori***
       Rodentia - Muridae
e   Ceylon Spiny Rat
d   Sri Lanka-Stachelratte

4098   ***Mus mayori pococki***
       Rodentia - Muridae
e   Bicoloured Spiny Rat
d   Zweifarbige Stachelratte

4099   ***Mus minutoides***
       Rodentia - Muridae
e   Pygmy Mouse; African Pygmy Mouse
d   Afrikanische Knirpsmaus; Afrikanische Zwergmaus
f   Souris pygmée africain; Souris naine d'Afrique
i   Topo pigmeo

       Mus molossinus syn. *M. musculus* q.v.

4100 **_Mus musculoides_**
Rodentia - Muridae
e   Temminck's Mouse
d   Temminck-Zwergmaus

4101 **_Mus musculus_**
Rodentia - Muridae
e   East European House Mouse; Eastern
House Mouse
d   Gartenmaus; Gelbbauchmaus;
Tabakmaus; Östliche Hausmaus
f   Rat souris; Souris grise; Souris
domestique; Souris
i   Topolino delle case; Topo domestico;
Topo comune; Topolino dei granai;
Topo; Topo casalino; Topo di casa

4102 **_Mus musculus bactrianus_**
Rodentia - Muridae
e   Southwestern Asian House Mouse
d   Baktrische Hausmaus
f   Souris sauvage d'Afghanistan; Souris
orientale

4103 **_Mus musculus castaneus_**
Rodentia - Muridae
e   Southeastern Asian House Mouse;
Indian House Mouse
d   Indische Hausmaus

4104 **_Mus musculus muralis_**
Rodentia - Muridae
e   St. Kilda House Mouse; St. Kilda
Island House Mouse; St. Kilda Island
Mouse; St. Kilda Mouse
d   St. Kilda Hausmaus
f   Souris domestique de l'île St. Kilda;
Souris commune de St. Kilda; Souris
vulgaire de St. Kilda; Rat souris de
St. Kilda; Souris de St. Kilda

4105 **_Mus musculus musculus_**
Rodentia - Muridae
e   Eastern European House Mouse
d   Nördliche Hausmaus

f   Souris domestique de l'Europe de
l'Est

4106 **_Mus musculus orientalis_**
Rodentia - Muridae
e   Egyptian House Mouse
d   Ägyptische Hausmaus
f   Souris domestique d'Égypte

4107 **_Mus neavei_**
Rodentia - Muridae
e   Neave's Mouse

4108 **_Mus orangiae_**
Rodentia - Muridae
e   Orange Mouse

4109 **_Mus oubanguii_**
Rodentia - Muridae
e   Oubangui Mouse

4110 **_Mus pahari_**
Rodentia - Muridae
e   Gairdner's Shrewmouse; Sikkim
Mouse

4111 **_Mus phillipsi_**
Rodentia - Muridae
e   Phillips's Mouse

4112 **_Mus platythrix_**
Rodentia - Muridae
e   Flat-haired Mouse; Indian Brown
Spiny Mouse
f   Souris aux cheveux plats

4113 **_Mus saxicola_**
Rodentia - Muridae
e   Rock-loving Mouse

4114 **_Mus setulosus_**
Rodentia - Muridae
e   Peters's Mouse

4115 Mus setzeri
Rodentia - Muridae
e Setzer's Pygmy Mouse

4116 Mus shortridgei
Rodentia - Muridae
e Shortridge's Mouse

4117 Mus sorella
Rodentia - Muridae
e Thomas's Pygmy Mouse

4118 Mus spicilegus
Rodentia - Muridae
e Mound-building Mouse
d Nördliche Ährenmaus; Ährenmaus
f Souris des steppes
i Topolino delle steppe

Mus spretoides syn. *M. macedonicus* q.v.

4119 Mus spretus
Rodentia - Muridae
e Algerian Mouse; North African House Mouse
d Algerische Hausmaus
f Souris d'Afrique du Nord
i Topo selvatico algerino; Topolino algerino

4120 Mus tenellus
Rodentia - Muridae
e Tender Mouse

4121 Mus terricolor
Rodentia - Muridae
e Earth-coloured Mouse

4122 Mus triton
Rodentia - Muridae
e Grey-bellied Pygmy Mouse
d Tritonsmaus

4123 Mus vulcani
Rodentia - Muridae
e Volcano Mouse; Javan Shrew Mouse
d Java-Hausmaus

4124 Muscardinus
Rodentia - Gliridae
e Hazelmice; Hazel Dormice
d Haselmäuse; Bilche
f Muscardins
i Moscardini; Nocciolini

4125 Muscardinus avellanarius
Rodentia - Gliridae
e Hazel Dormouse; Dormouse
d Haselschläfer; Kleine Haselmaus; Nordiche Haselmaus; Bilch; Kleine Schlafmaus; Haselmaus
f Muscardin; Loir muscardin; Muscardins des noisetiers
i Moscardino; Nocciolino

4126 Musonycteris
Chiroptera - Phyllostomidae
e Banana Bats; Colima Long-nosed Bats
d Bananenfledermäuse
f Chauves-souris des bananes

4127 Musonycteris harrisoni
Chiroptera - Phyllostomidae
e Banana Bat; Trumpet-npsed Bat
d Bananenfledermaus
f Chauves-souri des bananes

4128 Mustela
Carnivora - Mustelidae
e Weasels
d Erdmarder ; Stinkmarder
f Belettes

4129 Mustela africana
Carnivora - Mustelidae
e Amazon Weasel; Tropical Weasel
d Tropisches Wiesel
f Belette tropicale

4130 *Mustela altaica*
Carnivora - Mustelidae
e   Alpine Weasel; Mountain Weasel;
Pale Weasel; Altai Weasel
d   Alpenwiesel; Altai-Wiesel
f   Belette des Alpes; Belette de
montagne
i   Donnola degli Altai

4131 *Mustela erminea*
Carnivora - Mustelidae
e   Common Stoat; European Stoat;
Stoat; Ermine; Greater Weasel;
Short-tailed Weasel; Weasel
d   Großes Wiesel; Hermelin; Wiesel;
Großwiesel; Kurzschwanzwiesel
f   Hermine; Herminette; Belette
i   Ermellino

4132 *Mustela erminea muricus*
Carnivora - Mustelidae
e   Short-tailed Weasel; Ermine (NA)

4133 *Mustela eversmanni*
Carnivora - Mustelidae
e   Eversmann's Polecat; Russian
Polecat; Steppe Polecat; Turkestan
Polecat; Asiatic Polecat; Russian
Fitch
d   Steppeniltis; Weißer Iltis; Sibirischer
Iltis
f   Putois des steppes
i   Puzzola delle steppe; Puzzola bianca;
Puzzola chiara

4134 *Mustela felipei*
Carnivora - Mustelidae
e   Colombian Weasel
d   Kolumbianisches Wiesel
f   Belette colombienne

4135 *Mustela formosana*
Carnivora - Mustelidae
e   Taiwan Mountain Weasel
d   Taiwanisches Hochgebirgs-
Zwergwiesel

4136 *Mustela frenata*
Carnivora - Mustelidae
e   Long-tailed Weasel; American
Ermine; White Wesel
d   Langschwanzwiesel
f   Belette à longue queue
i   Donnola dalla lunga coda

4137 *Mustela furo*
Carnivora - Mustelidae
e   Domestic Ferret; Ferret; European
Ferret; Polecat-Ferret
d   Frettchen
f   Furet
i   Furetto

4138 *Mustela kathiah*
Carnivora - Mustelidae
e   Yellow-bellied Weasel; Pine Weasel
d   Gelbbauchwiesel
f   Belette à ventre jaune
i   Donnola dal ventre giallo

4139 *Mustela lutreola*
Carnivora - Mustelidae
e   European Mink; Old World Mink;
Marsh Otter
d   Sumpfotter; Europäischer Nerz;
Nordeuropäischer Nerz; Nörz;
Kleiner Fischotter; Krebsotter
f   Vison; Vison d'Europe; Putois vison;
Petite loutre
i   Visone europeo

4140 *Mustela lutreolina*
Carnivora - Mustelidae
e   Indonesian Mountain Weasel
d   Indonesisches Bergwiesel

Mustela minuta syn. *M. nivalis* q.v.

4141 *Mustela nigripes*
Carnivora - Mustelidae
e   Black-footed Ferret
d   Schwarzfußiltis

f Putois d'Amérique; Putois
d'Amérique aux pieds noirs; Putois
aux pieds noir
i Puzzola dai piedi neri; Furetto dai
piedi neri

4142 *Mustela nivalis*
Carnivora - Mustelidae
e Weasel; European Common Weasel;
Least Weasel; Lesser Weasel
d Kleines Wiesel; Mauswiesel;
Mitteleuropäisches Mauswiesel;
Hermännchen
f Belette; Belette commune; Belette
ordinaire; Belette d'Europe; Belette
franche; Belette vulgaire
i Donnola; Mustelo delle nevi

4143 *Mustela nivalis minuta*
Carnivora - Mustelidae
e Pygmy Weasel
d Zwergwiesel
i Donnola pigmea

4144 *Mustela nivalis rixosa*
Carnivora - Mustelidae
e Lesser Weasel
d Kleinstwiesel
f Belette pygmée

4145 *Mustela nudipes*
Carnivora - Mustelidae
e Bare-footed Weasel; Malaysian
Weasel; Barefoot Weasel; Malayan
Weasel
d Nacktfußwiesel
f Belette malaise

4146 *Mustela putorius*
Carnivora - Mustelidae
e Fitch; Polecat; European Polecat;
Western Polecat; Black Fitch;
German Fitch
d Stinkmarder; Iltis; Europäischer
Waldiltis; Waldiltis; Ratz; Schwarzer
Iltis; Europäischer Iltis
f Belette putois; Putois; Putois

d'Europe; Putois commun; Putois
ordinaire; Putois vulgaire
i Puzzola; Puzzola europea;
Canepùzzo; Canepuzzigli;
Canepuzziglio; Puzzola scura;
Puzzola nera

Mustela rixosa syn. *M. nivalis rixosa*
q.v.

4147 *Mustela sibirica*
Carnivora - Mustelidae
e Siberian Weasel
d Sibirischer Nerz; Erdmarder;
Feuermarder; Feuerwiesel;
Sibirisches Feuerwiesel; Sibirisches
Wiesel; Kolonok
f Belette de Sibérie; Vison de Sibérie;
Putois sibérien
i Visone della Siberia; Donnola
siberiana; Kolinsky

4148 *Mustela sibirica itatsi*
Carnivora - Mustelidae
e Japanese Weasel

4149 *Mustela sibirica manchurica*
Carnivora - Mustelidae
e Manchurian Weasel

4150 *Mustela sibirica sibirica*
Carnivora - Mustelidae
e Kolinsky
d Kolinsky
i Kolinsky siberiano

4151 *Mustela strigidorsa*
Carnivora - Mustelidae
e Back-striped Weasel
d Rückenstreifenwiesel

4152 *Mustela vison*
Carnivora - Mustelidae
e American Mink; Eastern Mink; New-
world Mink; North American Mink;
Mink (NA)

d Amerikanischer Nerz;
    Nordamerikanischer Nerz; Mink;
    Kanadischer Marder; Vison
f Vison; Vison d'Amérique; Vison
    américain; Marte du Canada; Putois
    vison; Martre du Canada
i Visone americano

4153 *Mustela vison letifera*
    Carnivora - Mustelidae
e Minnesota Mink

4154 Mustelidae
    Carnivora
e Mustelids; Weasels, Badgers, Skunks
    and Otters; Weasels and allies;
    Otters, Weasels and Badgers
d Marderartige Raubtiere; Marder ;
    Marderartige
f Belettes; Mustélidés
i Mustelidi; Lontre

    Mycteromys syn. *Mus* q.v.

4155 *Mydaus*
    Carnivora - Mustelidae
e Stink Badgers
d Stinkdachse; Malaische Stinkdachse;
    Philippinen-Stinkdachse
f Blaireaux
i Tassi

4156 *Mydaus javanensis*
    Carnivora - Mustelidae
e Sunda Stink Badger; Indonesian
    Stink Badger; Malayan Stink Badger;
    Teledu
d Stinkdachs; Java-Stinkdachs;
    Malaiischer Stinkdachs; Teledu;
    Sunda-Stinkdachs
f Télagon; Blaireau de Java
i Midao di Giava; Tasso fetido;
    Teledù; Tasso odoroso; Teledù
    indonesiano

4157 *Mydaus marchei*
    Carnivora - Mustelidae

e Palawan Stink Badger
d Philippinen-Stinkdachs; Palawan-
    Stinkdachs
f Blaireau de Palawan
i Tasso odoroso di Palawan

4158 *Mylomys*
    Rodentia - Muridae
e African Groove-toothed Rats;
    African Three-toed Grass Rats
d Afrikanische Furchenzahnratten

4159 *Mylomys dybowskii*
    Rodentia - Muridae
e African Groove-toothed Rat
d Dybowski-Ratte

    Mynomes syn. *Microtus* q.v.

4160 *Myocastor*
    Rodentia - Echimyidae
e Nutrias; Swamp Beavers; Coypus
d Nutrias; Biberratten; Sumpfbiber ;
    Schweifbiber
f Ragondins
i Castore delle paludi

4161 *Myocastor coypus*
    Rodentia - Echimyidae
e Nutria; Swamp Beaver; Coypu
d Nutria; Biberratte; Sumpfbiber;
    Schweifbiber
f Ragondin; Loutre d'Amérique;
    Myocaster
i Nutria; Castorino; Castoro delle
    paludi

4162 *Myoictis*
    Dasyuromorphia - Dasyuridae
e Pouched Mice; New Guinea Striped
    Dasyures; Three-striped Dasyures
d Streifenbeutelmarder

4163 *Myoictis melas*
    Dasyuromorphia - Dasyuridae
e Striped Native Cat; Pouched Mouse;

Three-striped Marsupial Mouse;
Three-striped Dasyure
d Streifenbeutelmarder

4164 *Myomimus*
Rodentia - Gliridae
e Mouse-like Dormice; Asiatic
Dormice; Mouse-tailed Dormice
d Asiatische Mausschläfer ;
Mausschläfer
f Myomimes; Muscardins
i Ghiri

4165 *Myomimus personatus*
Rodentia - Gliridae
e Mouse-tailed Dormouse; Ognev's
Dormouse; Asiatic Dormouse;
Mouse-like Dormouse
d Dünnschwanzmausschläfer;
Turkmenischer Mausschläfer;
Mausbilch
f Myomime à queue fine; Loir d'Ognev
i Ghiro di Ognev

4166 *Myomimus roachi*
Rodentia - Gliridae
e Roach's Mouse-tailed Dormouse
f Loir de Roach
i Ghiro asiatico

4167 *Myomimus setzeri*
Rodentia - Gliridae
e Setzer's Mouse-tailed Dormouse
i Ghiro di Setzer

4168 *Myomys*
Rodentia - Muridae
e African Mice
d Afrikanische Myomysmäuse
f Rats fouisseurs

4169 *Myomys albipes*
Rodentia - Muridae
e Ethiopian White-footed Mouse

4170 *Myomys daltoni*
Rodentia - Muridae
e Dalton's Mouse
d Daltons Maus

4171 *Myomys derooi*
Rodentia - Muridae
e Deroo's Mouse

4172 *Myomys fumatus*
Rodentia - Muridae
e African Rock Mouse

4173 *Myomys ruppi*
Rodentia - Muridae
e Rupp's Mouse

4174 *Myomys verreauxii*
Rodentia - Muridae
e Verreaux's Mouse; African Soft-
furred Rat

4175 *Myomys yemeni*
Rodentia - Muridae
e Yemeni Mouse

Myomyscus syn. *Praomys* q.v.

4176 *Myonycteris*
Chiroptera - Pteropodidae
e Collared Fruit Bats; Little Collared
Fruit Bats
d Kragenflughunde
f Myonyctères

4177 *Myonycteris brachycephala*
Chiroptera - Pteropodidae
e Sao Tomé Collared Fruit Bat

4178 *Myonycteris relicta*
Chiroptera - Pteropodidae
e East African Little Collared Fruit Bat
f Roussette du Kenya

4179   ***Myonycteris torquata***
Chiroptera - Pteropodidae
e   Little Collared Fruit Bat; Collared
Fruit Bat
d   Schmaler Kragenflughund
f   Myonyctère à collier

4180   ***Myoprocta***
Rodentia - Agoutidae
e   Tailed Agoutis; Acouchis; Acuchis
d   Geschwänzte Agutis; Acouchis
f   Acouchis; Acouchis de Buffon

4181   ***Myoprocta acouchy***
Rodentia - Agoutidae
e   Acouchi; Acuchi; Red Acouchi
d   Geschwänztes Aguti; Acouchi
f   Acouchi; Acouchi de Buffon

4182   ***Myoprocta exilis***
Rodentia - Agoutidae
e   Green Acouchi
d   Grünes Aguti

Myoprocta pratti syn. *M. acouchy*
q.v.

4183   ***Myopterus***
Chiroptera - Molossidae
e   African Free-tailed Bats

4184   ***Myopterus daubentonii***
Chiroptera - Molossidae
e   Daubenton's Free-tailed Bat

4185   ***Myopterus whitleyi***
Chiroptera - Molossidae
e   Bini Free-tailed Bat

4186   ***Myopus***
Rodentia - Muridae
e   Red-backed Lemmings; Wood
Lemmings; Grey Lemmings
d   Waldlemminge
f   Lemmings des forêts

i   Lemming delle foreste

4187   ***Myopus schisticolor***
Rodentia - Muridae
e   Red-backed Lemming; Wood
Lemming; Grey Lemming
d   Waldlemming
f   Lemming des forêts
i   Lemming delle foreste

4188   ***Myosciurus***
Rodentia - Sciuridae
e   African Pygmy Squirrels
d   Afrikanische Zwerghörnchen
f   Écureuils nains du Gabon
i   Scoiattoli pigmei

4189   ***Myosciurus pumilio***
Rodentia - Sciuridae
e   African Pygmy Squirrel; Pygmy
Squirrel
d   Afrikanisches Zwerghörnchen
f   Écureuil nain du Gabon
i   Scoiattolo pigmeo

4190   ***Myosorex***
Soricomorpha - Soricidae
e   Mouse Shrews; Forest Shrews
d   Spitzmäuse

4191   ***Myosorex babaulti***
Soricomorpha - Soricidae
e   Babault's Mouse Shrew; Livu Forest
Shrew

4192   ***Myosorex blarina***
Soricomorpha - Soricidae
e   Montane Mount Shrew; Ruwenzori
Forest Shrew; Ruwenzori Mouse

4193   ***Myosorex cafer***
Soricomorpha - Soricidae
e   Dark-footed Forest Shrew
d   Dunkelfüssige Waldspitzmaus

4194 **Myosorex eisentrauti**
Soricomorpha - Soricidae
e Eisentraut's Mouse Shrew;
Eisentraut's Forest Shrew

4195 **Myosorex geata**
Soricomorpha - Soricidae
e Geata Mouse Shrew; Uluguru Forest
Shrew

4196 **Myosorex kihaulei**
Soricomorpha - Soricidae
e Udzungwa Mouse Shrew

4197 **Myosorex longicaudatus**
Soricomorpha - Soricidae
e Long-tailed Forest Shrew

4198 **Myosorex norae**
Soricomorpha - Soricidae
e Aberdare Shrew; Aberdare Mole
Shrew
d Maulwurfsspitzmaus

4199 **Myosorex okuensis**
Soricomorpha - Soricidae
e Oku Mouse Shrew; Cameroon Forest
Shrew

4200 **Myosorex polli**
Soricomorpha - Soricidae
e Poll's Shrew

4201 **Myosorex polulus**
Soricomorpha - Soricidae
e Mount Kenya Shrew; Kenya
Mountain Shrew; Mount Kenya Mole
Shrew

4202 **Myosorex rumpii**
Soricomorpha - Soricidae
e Rumpi Mouse Shrew; Rumpi Forest
Shrew

4203 **Myosorex schalleri**
Soricomorpha - Soricidae

e Schaller's Mouse Shrew; Schaller's
Forest Shrew

4204 **Myosorex sclateri**
Soricomorpha - Soricidae
e Sclater's Tiny Mouse Shrew; Sclater's
Forest Shrew

4205 **Myosorex tenuis**
Soricomorpha - Soricidae
e Thin Mouse Shrew; Zuurbron Forest
Shrew

4206 **Myosorex varius**
Soricomorpha - Soricidae
e Forest Shrew; Common Forest Shrew
d Waldspitzmaus

4207 **Myospalax**
Rodentia - Muridae
e Zokors; Rodent Moles; Molerats;
Molemice; Asiatic Molerats
d Mullmäuse; Blindmulle
f Zokkors; Hamsters-taupes

4208 **Myospalax aspalax**
Rodentia - Muridae
e Siberian Zokor; Altai Zokor; North
Alatai Zokor; False Zokor

d Daurischer Blindmull
f Zokkor de Transbaikalie

Myospalax baileyi syn. *M. fontanieri*
q.v.

Myospalax cansus syn. *M. fontanieri*
q.v.

4209 **Myospalax epsilanus**
Rodentia - Muridae
e Manchurian Zokor
d Mandschurischer Blindmull

4210 **Myospalax fontanierii**
Rodentia - Muridae

e   Common Chinese Zokor; Chinese Zokor

d   Chinesischer Blindmull

4211   *Myospalax myospalax*
Rodentia - Muridae

e   Siberian Zokor

d   Altaische Mullmaus; Blindmull; Zokor; Sibirischer Blindmull

f   Rat-taupe; Zokkor sibérien; Hamster taupe sibérien

4212   *Myospalax psilurus*
Rodentia - Muridae

e   Transbaikal Zokor

d   Transbaikal-Blindmull

f   Hamster-taupe chinois

4213   *Myospalax rothschildi*
Rodentia - Muridae

e   Rothschild's Zokor

d   Rothschild-Blindmull

4214   *Myospalax smithii*
Rodentia - Muridae

e   Smith's Zokor

d   Smith-Blindmull

4215   *Myotis*
Chiroptera - Vespertilionidae

e   Common Bats; Mouse-eared Bats; Little Brown Bats; Myotises

d   Mausohrfledermäuse; Mausohren; Nachtfledermäuse

f   Murins

i   Vespertili

4216   *Myotis abei*
Chiroptera - Vespertilionidae

e   Sakhalin Myotis

4217   *Myotis adversus*
Chiroptera - Vespertilionidae

e   Great-footed Bat; Large-footed Bat; Large-footed Myotis (ANZ)

d   Kleine Braungraue Fledermaus

4218   *Myotis aelleni*
Chiroptera - Vespertilionidae

e   Southern Myotis

4219   *Myotis albescens*
Chiroptera - Vespertilionidae

e   Paraguay Myotis; Silver-haired Myotis; Silver-tipped Myotis

4220   *Myotis alcathoe*
Chiroptera - Vespertilionidae

e   Whiskered Bat; Alcatho's Myotis

d   Nymphenfledermaus

f   Murin des Balkans; Murin d'Alcathoe

4221   *Myotis altarium*
Chiroptera - Vespertilionidae

e   Szechwan Myotis; Sichuan Myotis

4222   *Myotis annamiticus*
Chiroptera - Vespertilionidae

e   Annamite Myotis

4223   *Myotis annectans*
Chiroptera - Vespertilionidae

e   Hairy-faced Bat

   Myotis araxenus syn. *M. schaubi* q.v.

   Myotis argentatus syn. *M. albescens* q.v.

4224   *Myotis atacamensis*
Chiroptera - Vespertilionidae

e   Atacama Myotis

4225   *Myotis ater*
Chiroptera - Vespertilionidae

e   Moluccan Whiskered Bat; Small Black Myotis

4226 *Myotis auriculus*
Chiroptera - Vespertilionidae
e Mexican Long-eared Myotis;
Southwestern Myotis
f Vespertillon du Sud-Ouest

4227 *Myotis australis*
Chiroptera - Vespertilionidae
e Small-footed Myotis; Australian
Myotis

4228 *Myotis austroriparius*
Chiroptera - Vespertilionidae
e Southeastern Myotis; Mississippi
Myotis; Southeastern Bat

4229 *Myotis bechsteini*
Chiroptera - Vespertilionidae
e Bechstein's Bat
d Bechsteinsche Fledermaus;
Bechstein-Fledermaus
f Vespertilion de Bechstein; Murin de
Bechstein
i Vespertilio di Bechstein

4230 *Myotis blythii*
Chiroptera - Vespertilionidae
e Lesser Mouse-eared Bat
d Spitzohrige Fledermaus; Kleines
Mausohr; Kleinmausohr
f Petit Murin
i Vespertilio minore; Vespertilio di
Blyth

4231 *Myotis blythii oxygnathus*
Chiroptera - Vespertilionidae
e Monticelli Bat
i Vespertilio di Monticelli

4232 *Myotis bocagei*
Chiroptera - Vespertilionidae
e Rufous Mouse-eared Bat; Bocage's
Banana Bat
f Murin

4233 *Myotis bombinus*

Chiroptera - Vespertilionidae
e Far Eastern Myotis
d Östliche Fransenfledermaus

4234 *Myotis brandtii*
Chiroptera - Vespertilionidae
e Brandt's Bat
d Große Bartfledermaus; Brandt-
Fledermaus
f Murin de Brandt; Vespertilion de
Brandt
i Vespertilio di Brandt

4235 *Myotis californicus*
Chiroptera - Vespertilionidae
e California Bat; California Myotis
d Kalifornisches Mausohr
f Chauve-souris de la Californie

4236 *Myotis capaccinii*
Chiroptera - Vespertilionidae
e Long-fingered Bat; Capaccini's Bat;
Long-fingered Myotis
d Langfußfledermaus; Langfüßige
Fledermaus
f Vespertilion de Cappaccini; Murin de
Cappaccini; Chauve-souris aux longs
doigts
i Vespertilio di Capaccini

4237 *Myotis chiloensis*
Chiroptera - Vespertilionidae
e Chilean Myotis

4238 *Myotis chinensis*
Chiroptera - Vespertilionidae
e Large Myotis

4239 *Myotis ciliolabrum*
Chiroptera - Vespertilionidae
e Western Small-footed Bat; Least
Brown Bat; Small-footed Myotis;
Western Small-footed Myotis; Small-
footed Bat
f Verspertillon pygmée de l'Ouest

4240 *Myotis cobanensis*
Chiroptera - Vespertilionidae
e Guatemalan Myotis

4241 *Myotis csorbai*
Chiroptera - Vespertilionidae
e Csorba's Myotis

4242 *Myotis dasycneme*
Chiroptera - Vespertilionidae
e Pond Bat; Rough-legged Water Bat
d Teichfledermaus
f Chauve-souris dasycnème; Chauve-souris velue; Vespertilion des marais; Vespertilion dasycnème; Murin des marais; Vespertillion brun

i Vespertilio dasicneme

4243 *Myotis daubentoni*
Chiroptera - Vespertilionidae
e Daubenton's Bat; Water Bat
d Wasserfledermaus; Fränkische Wasserfledermaus; Rotkurzohr
f Vespertilion de Daubenton; Murin de Daubenton
i Vespertilio del Daubenton

4244 *Myotis dominicensis*
Chiroptera - Vespertilionidae
e Dominican Myotis
f Myotis de Dominique

4245 *Myotis elegans*
Chiroptera - Vespertilionidae
e Elegant Myotis
d Elegantes Mausohr

4246 *Myotis emarginatus*
Chiroptera - Vespertilionidae
e Geoffroy's Bat; Notch-eared Bat
d Wimperfledermaus; Ausgerandete Fledermaus; Gewimperte Fledermaus

f Vespertilion cilié; Vespertilion échancré; Vespertilion aux oreilles

échancrées; Murin aux oreilles échancrées
i Vespertilio smarginato

4247 *Myotis evotis*
Chiroptera - Vespertilionidae
e Long-eared Myotis; Long-eared Bat (NA)
f Vespertilion aux longues oreilles

4248 *Myotis findleyi*
Chiroptera - Vespertilionidae
e Findley's Myotis

4249 *Myotis formosus*
Chiroptera - Vespertilionidae
e Hodgson's Bat; Hodgson's Myotis
f Vespertillon d'Hodgson

4250 *Myotis fortidens*
Chiroptera - Vespertilionidae
e Cinnamon Myotis; Cinnamon Bat

4251 *Myotis frater*
Chiroptera - Vespertilionidae
e Fraternal Myotis
d Langschwänziges Mausohr

4252 *Myotis gomantongensis*
Chiroptera - Vespertilionidae
e Gomantong Myotis

4253 *Myotis goudoti*
Chiroptera - Vespertilionidae
e Malagasy Mouse-eared Bat
f Chauve-souris malgache aux oreilles de souris

4254 *Myotis grisescens*
Chiroptera - Vespertilionidae
e Grey Myotis; Howell's Bat; Grey Bat
d Graue Mausohrfledermaus
i Vespertilio grigio

4255 **Myotis hajastanicus**
Chiroptera - Vespertilionidae
e Armenian Myotis

4256 **Myotis hasseltii**
Chiroptera - Vespertilionidae
e Lesser Large-footed Bat; Hasselt's
Myotis; Van Hasselt's Bat
d Van Hasselts Fledermaus

4257 **Myotis horsfieldii**
Chiroptera - Vespertilionidae
e Horsfield's Bat; Horsfield's Myotis

4258 **Myotis hosonoi**
Chiroptera - Vespertilionidae
e Hosono's Myotis

4259 **Myotis ikonnikovi**
Chiroptera - Vespertilionidae
e Ikonnikov's Bat; Ikonnikov's Myotis
d Ikonnikows Mausohr;
Kurzohrfledermaus
f Vespertilion d'Ikonnikow; Murin
d'Ikonnikov

4260 **Myotis insularum**
Chiroptera - Vespertilionidae
e Insular Myotis

4261 **Myotis keaysi**
Chiroptera - Vespertilionidae
e Hairy-legged Myotis

4262 **Myotis keenei**
Chiroptera - Vespertilionidae
e Keen's Acadian Bat; Keen's Bat;
Keen's Myotis; Northern Myotis
d Keens Mausohr
f Vespertilion de Keen; Chauve-souris
de Keen

Myotis larensis syn. *M. nesopulos*
q.v.

4263 **Myotis leibii**
Chiroptera - Vespertilionidae
e Eastern Small-footed Myotis; Small-
footed Myotis
f Chauve-souris pygmée

4264 **Myotis lesueuri**
Chiroptera - Vespertilionidae
e Lesueur's Wing-gland Bat; Lesueur's
Hairy Bat

4265 **Myotis levis**
Chiroptera - Vespertilionidae
e Yellowish Myotia

4266 **Myotis longipes**
Chiroptera - Vespertilionidae
e Kashmir Cave Bat; Kashmir Cave
Myotis

4267 **Myotis lucifugus**
Chiroptera - Vespertilionidae
e Little Brown Bat (NA); Little Brown
Myotis; American Mouse-eared Bat;
Arizona Myotis
d Kleine Braune Fledermaus
f Petite chauve-souris brune;
Vespertillon brun
i Piccolo pipistrello marrone del
Nordamerica; Vespertilio marrone
del Nordamerica

4268 **Myotis lucifugus carissima**
Chiroptera - Vespertilionidae
e Yellowstone Bat

4269 **Myotis lucifugus occultus**
Chiroptera - Vespertilionidae
e Occult Little Brown Bat

4270 **Myotis macrodactylus**
Chiroptera - Vespertilionidae
e Big-footed Myotis

4271   *Myotis macrotarsus*
       Chiroptera - Vespertilionidae
  e Pallid Large-footed Myotis

4272   *Myotis martiniquensis*
       Chiroptera - Vespertilionidae
  e Schwartz's Myotis
  f Murin martiniquais

4273   *Myotis milleri*
       Chiroptera - Vespertilionidae
  e Miller's Myotis
  f Vespertillon de Miller

4274   *Myotis montivagus*
       Chiroptera - Vespertilionidae
  e Burmese Whiskered Bat; Large
       Brown Myotis

4275   *Myotis morrisi*
       Chiroptera - Vespertilionidae
  e Morris's Bat

4276   *Myotis muricola*
       Chiroptera - Vespertilionidae
  e Whiskered Myotis; Nepalese
       Whiskered Bat; Wall-roosting Myotis

4277   *Myotis myotis*
       Chiroptera - Vespertilionidae
  e Brown Bat; Common Bat; Little
       Brown Bat; European Little Brown
       Bat; Mouse-eared Bat; Greater
       Mouse-eared Bat; Large Mouse-
       eared Bat
  d Mäuseohr; Rattenartige Fledermaus;
       Große Fledermaus; Großmausohr;
       Großmausohrfledermaus; Mausohr;
       Großes Mausohr;
       Mausohrfledermaus;
       Riesenfledermaus
  f Grande chauve-souris murine;
       Chauve-souris murine; Murin;
       Vespertilion murin; Grand murin
  i Vespertilio maggiore; Vespertilio
       comune

4278   *Myotis mystacinus*
       Chiroptera - Vespertilionidae
  e Whiskercd Bat; European Whiskered
       Bat
  d Bartige Fledermaus; Schnauzbartige
       Fledermaus; Kleine Bartfledermaus;
       Bartfledermaus
  f Chauve-souris mustac; Vespertilion à
       moustache; Vespertilion mustac;
       Murin à moustaches
  i Vespertilio mustacchino

4279   *Myotis natteri*
       Chiroptera - Vespertilionidae
  e Natterer's Bat; Red-grey Bat;
       Reddish-grey Bat
  d Fransenfledermaus; Gefranste
       Fledermaus; Natterers Fledermaus;
       Europäische Fransenfledermaus
  f Vespertilion de Natterer; Murin de
       Natterer
  i Vespertilio di Natterer

4280   *Myotis nesopolus*
       Chiroptera - Vespertilionidae
  e Curacao Myotis

4281   *Myotis nigricans*
       Chiroptera - Vespertilionidae
  e Black Myotis; Little Black Myotis

4282   *Myotis nipalensis*
       Chiroptera - Vespertilionidae
  e Asian Whiskered Bat

4283   *Myotis oreias*
       Chiroptera - Vespertilionidae
  e Singapore Whiskered Bat

4284   *Myotis oxyotus*
       Chiroptera - Vespertilionidae
  e Montane Myotis

4285   *Myotis ozensis*
       Chiroptera - Vespertilionidae
  e Honshu Myotis

4286 **Myotis peninsularis**
Chiroptera - Vespertilionidae
e Peninsular Myotis

4287 **Myotis pequinius**
Chiroptera - Vespertilionidae
e Peking Myotis

4288 **Myotis planiceps**
Chiroptera - Vespertilionidae
e Flat-headed Myotis

4289 **Myotis pruinosus**
Chiroptera - Vespertilionidae
e Frosted Myotis

4290 **Myotis ricketti**
Chiroptera - Vespertilionidae
e Rickett's Big-footed Bat; Rickett's
Big-footed Myotis

4291 **Myotis ridleyi**
Chiroptera - Vespertilionidae
e Ridley's Bat; Malay Pipistrelle;
Ridley's Myotis

4292 **Myotis riparius**
Chiroptera - Vespertilionidae
e Riparian Myotis

4293 **Myotis rosseti**
Chiroptera - Vespertilionidae
e Thick-thumbed Myotis

4294 **Myotis ruber**
Chiroptera - Vespertilionidae
e Red Myotis

4295 **Myotis schaubi**
Chiroptera - Vespertilionidae
e Schaub's Myotis

4296 **Myotis scotti**
Chiroptera - Vespertilionidae

e Scott's Mouse-eared Bat

4297 **Myotis seabrai**
Chiroptera - Vespertilionidae
e Angola Wing-gland Bat; Angolan
Hairy Bat

4298 **Myotis septentrionalis**
Chiroptera - Vespertilionidae
e Northern Myotis
f Chauve-souris nordique; Vespertillon
nordique

4299 **Myotis sicarius**
Chiroptera - Vespertilionidae
e Mandelli's Mouse-eared Bat

4300 **Myotis siligorensis**
Chiroptera - Vespertilionidae
e Himalajan Whiskered Bat; Small-
toothed Myotis

4301 **Myotis simus**
Chiroptera - Vespertilionidae
e Velvety Myotis

4302 **Myotis sodalis**
Chiroptera - Vespertilionidae
e Indiana Myotis; Indiana Bat; Social
Bat; Social Myotis

4303 **Myotis stalkeri**
Chiroptera - Vespertilionidae
e Kai Myotis

4304 **Myotis thysanodes**
Chiroptera - Vespertilionidae
e Fringed Bat; Fringed Myotis
f Chauve-souris à queue frangée

4305 **Myotis tricolor**
Chiroptera - Vespertilionidae
e Cape Hairy Bat; Little Brown Bat
(CSA)

4306   *Myotis velifer*
       Chiroptera - Vespertilionidae
  e Mexican Brown Bat; Cave Myotis;
       Cave Bat
  d Höhlenfledermaus; Texanische
       Höhlenfledermaus

4307   *Myotis vivesi*
       Chiroptera - Vespertilionidae
  e Fish-eating Bat; Fish-eating Myotis

4308   *Myotis volans*
       Chiroptera - Vespertilionidae
  e Long-legged Bat (NA); Long-legged
       Myotis
  f Chauve-souris aux longues pattes

4309   *Myotis welwitschi*
       Chiroptera - Vespertilionidae
  e Welwitsch's Bat; Welwitsch's Hairy
       Bat
  d Welwitsch-Fledermaus
  f Vespertilion de Welwitsch; Myotis
       de Welwitsch

4310   *Myotis yanbarensis*
       Chiroptera - Vespertilionidae
  e Okinawa Myotis

4311   *Myotis yesoensis*
       Chiroptera - Vespertilionidae
  e Yoshiyuki's Myotis

4312   *Myotis yumanensis*
       Chiroptera - Vespertilionidae
  e Yuma Bat; Yuma Myotis
  d Yuma-Myotis
  f Chauve-souris de Yuma

       Myotomis syn. *Otomys* q.v.

       Myoxus syn. *Glis* q.v.

4313   *Myrmecobius*
       Dasyuromorphia - Dasyuridae

  e Numbats (ANZ); Marsupial
       Anteaters; Banded Anteaters
  d Ameisenbeutler ; Numbat
  f Myrmecobies; Numbats; Fourmiliers
       marsupials
  i Numbat

4314   *Myrmecobius fasciatus*
       Dasyuromorphia - Dasyuridae
  e Numbat (ANZ); Marsupial Anteater;
       Banded Anteater
  d Ameisenbeutler
  f Fourmilier marsupial rayé;
       Myrmecobie à bandes
  i Numbat; Formichiere fasciato

4315   *Myrmecobius fasciatus fasciatus*
       Dasyuromorphia - Dasyuridae
  e Banded West Australia Numbat
  f Numbat de l'Ouest australien

4316   *Myrmecobius fasciatus rufus*
       Dasyuromorphia - Dasyuridae
  e Rusty South Australia Numbat
  f Numbat rouille

4317   *Myrmecophaga*
       Pilosa - Myrmecophagidae
  e Great Anteaters; Giant Anteaters
  d Große Ameisenbären;
       Ameisenfresser
  f Grands fourmilliers
  i Formichieri giganti

4318   *Myrmecophaga tridactyla*
       Pilosa - Myrmecophagidae
  e Great Anteater; Giant Anteater
  d Großer Ameisenbär; Yurumi;
       Schnabeligel
  f Grand fourmillier; Tamanoir
  i Formichiere gigante

4319   **Myrmecophagidae**
       Xenarthra
  e Anteaters; American Anteaters

d Ameisenbären; Ameisenfresser
f Fourmiliers
i Formichieri americani;
Mirmecofagidi

Mysateles syn. *Capromys* q.v.

4320 Mystacina
Chiroptera - Mystacinidae
e New Zealand Short-tailed Bats
d Neuseeland-Fledermäuse
f Mystacines
i Mistacinidi

4321 Mystacina robusta
Chiroptera - Mystacinidae
e New Zealand Greater Short-tailed
Bat; Greater New Zealand Short-
tailed Bat; New Zealand Robust
Short-tailed Bat
d Große Neuseeland-Fledermaus;
Kräftige Neuseeland-Fledermaus
f Grande chauve-souris de Nouvelle-
Zélande; Grande mystacine;
Mystacine robuste; Mystacine
robuste de Nouvelle-Zélande

4322 Mystacina tuberculata
Chiroptera - Mystacinidae
e New Zealand Lesser Short-tailed Bat;
Lesser New Zealand Short-tailed Bat
d Neuseeland-Fledermaus;
Kurzschwanzfledermaus
f Mystacine tuberculée

4323 Mystacinidae
Chiroptera
e New Zealand Short-tailed Bats
d Neuseeland-Fledermäuse
f Mystacinidés

4324 Mystromys
Rodentia - Muridae
e White-tailed Mice
d Weißchwänzige Hamster ;
Afrikanische Hamster

f Hamsters à queue blanche
i Criceti del Sudafrica

4325 Mystromys albicaudatus
Rodentia - Muridae
e White-tailed Mouse; White-tailed Rat
d Weißschwänziger Hamster; Afrika-
Hamster; Afrikanischer Hamster
f Hamster à queue blanche
i Criceto del Sudafrica

4326 Myzopoda
Chiroptera - Myzopodidae
e Sucker-footed Bats; Old World
Sucker-footed Bats; Golden Bats
d Madagassische
Haftscheibenfledermäuse
f Vespertilions dorés

4327 Myzopoda aurita
Chiroptera - Myzopodidae
e Sucker-footed Bat; Old World
Sucker-footed Bat; Golden Bat;
Madagascar Sucker-footed Bat
d Madagassische
Haftscheibenfledermaus
f Vespertilion doré

4328 Myzopodidae
Chiroptera
e Sucker-footed Bats; Old World
Sucker-footed Bats; Golden Bats
d Madagassische
Haftscheibenfledermäuse
f Myzopodidés
i Mizopodidi

N

4329 *Naemorhedus*
Artiodactyla - Bovidae
e   Gorals; Himalajan Chamois ; Serows
d   Waldziegenantelopen;
      Ziegenantilopen; Gorale
f   Bouquetins du Népal
i   Goral

4330 *Naemorhedus baileyi*
Artiodactyla - Bovidae
e   Red Goral
d   Roter Goral
f   Goral roux

4331 *Naemorhedus caudatus*
Artiodactyla - Bovidae
e   Chinese Goral
d   Nordchinesischer Goral;
      Langschwanzgoral; Langschwänzige
      Ziegenantilope
f   Bouquetin du Népal à longue queue

4332 *Naemorhedus goral*
Artiodactyla - Bovidae
e   Goral; Common Goral; Manchurian
      Goral; Grey Goral
d   Goral
f   Goral
i   Goral

4333 *Naemorhedus goral goral*
Artiodactyla - Bovidae
e   Grey Goral
d   Kaschmir-Goral
f   Goral gris

4334 *Nandinia*

Carnivora - Nandiniidae
e   African Palm Civets
d   Pardelroller
f   Nandinies
i   Nandinie

4335 *Nandinia binotata*
Carnivora - Nandiniidae
e   African Palm Civet; Two-spotted
      Palm Civet; Tree Civet; Palm Civet
d   Pardelroller
f   Nandinie à deux taches; Nandinie
i   Nandinia; Civetta delle Palme

4336 **Nandiniidae**
Carnivora
e   African Palm Civets
d   Palmenroller
f   Nandinies
i   Nandinidi

Nannomys syn. *Mus* q.v.

4337 *Nannosciurus*
Rodentia - Sciuridae
e   Oriental Pygmy Squirrels; Dwarf
      Squirrels; Black-eared Squirrels
d   Asiatische Zwerghörnchen ; Braune
      Zwerghörnchen
f   Écureuils pygmées
i   Scoiattoli pigmei

4338 *Nannosciurus melanotis*
Rodentia - Sciuridae
e   Brown Dwarf Squirrel; Black-eared
      Pygmy Squirrel; Black-eared Squirrel
d   Braunes Zwerghörnchen
f   Écureuil pygmée brun
i   Sciattolo pigmeo

4339 *Nannospalax*
Rodentia - Muridae
e   Lesser Blind Molerats
d   West-Blindmäuse

f Souris-taupes; Spalax ; Rats-taupes
i Spalaci

4340 *Nannospalax ehrenbergi*
Rodentia - Muridae
e Ehrenberg's Molerat; Palestine Molerat
d Ehrenberg-Blindmaus
f Spalax d'Ehrenberg; Spalax d'Égypte

4341 *Nannospalax leucodon*
Rodentia - Muridae
e Lesser Molerat; Hungarian Molerat
d West-Blindmull; Bergblindmaus; Pontische Blindmaus; Kleine Blindmaus; Gemeiner Blindmull; Stumpfschnauzenmull; West-Blindmaus
f Spalax occidental; Spalax de Hongrie
i Spalace occidentale

4342 *Nannospalax nehringi*
Rodentia - Muridae
e Nehring's Blind Molerat
f Spalax de Nehring
i Spalace ungherese

4343 *Nanonycteris*
Chiroptera - Pteropodidae
e Dwarf Fruit Bats; Veldkamp's Bats

4344 *Nanonycteris veldkampi*
Chiroptera - Pteropodidae
e Veldkamp's Bat; Veldkamps Dwarf Fruit Bat; Little Flying Cow

4345 *Napaeozapus*
Rodentia - Dipodidae
e Woodland Jumping Mice
d Waldhüpfmäuse
f Zapodes des bois; Souris sauteuses des bois

4346 *Napaeozapus insignis*
Rodentia - Dipodidae

e Woodland Jumping Mouse
d Waldhüpfmaus
f Zapode des bois; Souris sauteuse des bois

4347 *Nasalis*
Primates - Cercopithecidae
e Proboscis Monkeys; Long-nosed Monkeys; Snub-nosed Monkeys
d Nasenaffen
f Nasiques
i Rinopitechi

4348 *Nasalis concolor*
Primates - Cercopithecidae
e Pig-tailed Leaf Monkey; Pig-tailed Snub-nosed Monkey; Pig-tailed Langur; Mentawai Islands Snub-nosed Monkey; Simakobu; Simakobu Monkey; Mentawai Islands Snub-nosed Langur; Snub-nosed Langur
d Pageh-Stumpfnasenaffe; Kurzschwanzstumpfnase; Pageh-Stumpfnase
f Entelle de Pagai; Nasique des îles Pagai
i Rinopiteco di Pagai

4349 *Nasalis larvatus*
Primates - Cercopithecidae
e Proboscis Monkey; Long-nosed Monkey
d Nasenaffe
f Nasique; Long-nez
i Nasica; Nasica del Borneo

***Nasilio* syn.** *Elephantulus* q.v.

***Nasillus* syn.** *Uropsilus* q.v.

4350 *Nasua*
Carnivora - Procyonidae
e Coatis; Coatimundis
d Nasenbären; Coatis
f Coatis
i Nasue; Coati

4351   *Nasua narica*
Carnivora - Procyonidae
e   White-nosed Coati
d   Weißrüsselbär; Nasenbär; Coati;
Weißrüsselnasenbär
f   Coati brun; Coati à nez blanc
i   Nasua del naso bianco; Coati a naso
bianco

4352   *Nasua narica nelsoni*
Carnivora - Procyonidae
e   Nelson's Coati
d   Nelson-Nasenbär

4353   *Nasua nasua*
Carnivora - Procyonidae
e   Coati; South American Coati;
Coatimundi; Common Coati;
Ringtailed Coati
d   Südamerikanischer Nasenbär;
Gewöhnlicher Nasenbär
f   Nasua; Coati a queue annelée
i   Coatimundi

4354   *Nasua nasua solitaria*
Carnivora - Procyonidae
e   Coatimundi
d   Honduras-Nasenbär
f   Coati de l'île Cozumel
i   Nasua rosso; Coati rosso

4355   *Nasuella*
Carnivora - Procyonidae
e   Little Coatimundis
d   Gebirgsnasenbären; Bergnasenbären
f   Coatis des montagnes

4356   *Nasuella olivacea*
Carnivora - Procyonidae
e   Little Coatimundi; Mountain Coati;
Dwarf Coati
d   Gebirgsnasenbär; Bergnasenbär
f   Coati de montagne

4357   **Natalidae**

Chiroptera
e   Natalid Bats; Funnel-eared Bats;
Long-legged Bats
d   Trichterohren
f   Natalidés
i   Natalidi

4358   *Natalus*
Chiroptera - Natalidae
e   Funnel-eared Bats
d   Trichterohren ;
Trichterohrenfledermäuse
f   Natalides

Natalus brevimanus syn. *N.
micropus* q.v.

Natalus espiritosantensis syn. *N.
stramineus* q.v.

4359   *Natalus lepidus*
Chiroptera - Natalidae
e   Gervais's Long-legged Bat; Gervais's
Funnel-eared Bat
f   Chauve-souris papillon
i   Pipistrello farfalla

Natalus macer syn. *N. micropus* q.v.

Natalus major syn. *N. stramineus*
q.v.

4360   *Natalus micropus*
Chiroptera - Natalidae
e   Jamaican Long-legged Bat; Cuban
Funnel-eared Bat

4361   *Natalus stramineus*
Chiroptera - Natalidae
e   Mexican Funnel-eared Bat
d   Mexikanische
Trichterohrenfledermaus;
Mexikanisches Trichterohr

f   Vespertilion couleur de paille;
Chauve-souris couleur de paille;
Chauve-souris couleur paille du

Mexique; Natalide paillée
i Natalus stramineo

4362 *Natalus tumidifrons*
Chiroptera - Natalidae
e Bahaman Funnel-eared Bat

4363 *Natalus tumidirostris*
Chiroptera - Natalidae
e Trinidadian Funnel-eared Bat;
Trinidad Funnel-eared Bat; Trinidad
Long-legged Bat; Trinidadian Long-
legged Bat

4364 *Neacomys*
Rodentia - Muridae
e Spiny Rice Mice; Spiny Mice; Bristly
Mice
d Südamerikanische Borstenratten

4365 *Neacomys duboti*
Rodentia - Muridae
e Dubost's Bristly Mouse

4366 *Neacomys guianae*
Rodentia - Muridae
e Guiana Bristly Mouse

4367 *Neacomys minutus*
Rodentia - Muridae
e Small Bristly Mouse

4368 *Neacomys musseri*
Rodentia - Muridae
e Musser's Bristly Mouse

4369 *Neacomys paracou*
Rodentia - Muridae
e Paracou Bristly Mouse

4370 *Neacomys pictus*
Rodentia - Muridae
e Painted Bristly Mouse

4371 *Neacomys spinosus*

Rodentia - Muridae
e Bristly Mouse; Spiny Mouse;
Common Bristly Mouse
d Borstenreisratte

4372 *Neacomys tenuipes*
Rodentia - Muridae
e Narrow-footed Bristly Mouse

4373 *Nectogale*
Soricomorpha - Soricidae
e Web-footed Water Shrews; Tibetan
Water Shrews; Szechuan Water
Shrews; Elegant Water Shrews
d Gebirgsbachspitzmäuse
f Nectogales
i Nectogale

4374 *Nectogale elegans*
Soricomorpha - Soricidae
e Web-footed Water Shrew; Tibetan
Water Shrew; Szechuan Water
Shrew; Elegant Tibetan Water
Shrew; Elegant Water Shrew
d Gebirgsbachspitzmaus
f Nectogale élégant
i Nectogale del Tibet

4375 *Nectomys*
Rodentia - Muridae
e Neotropical Water Rats; Water Rats
d Südamerikanische Wasserratten;
Neotropische Wasserratten

4376 *Nectomys melanius*
Rodentia - Muridae
e Small-footed Water Rat

4377 *Nectomys palmipes*
Rodentia - Muridae
e Trinidad Water Rat

Nectomys parvipes syn. *N. melanius*
q.v.

4378   *Nectomys squamipes*
Rodentia - Muridae
e   Guiana Water Rat; South American
Water Rat; Neotropical Water Rat
d   Südamerikanische Wasserratte

4379   *Nelsonia*
Rodentia - Muridae
e   Nelson's Woodrats; Nelson's
Diminutive Woodrats; Diminutive
Woodrats
d   Zwergwaldratten

4380   *Nelsonia goldmani*
Rodentia - Muridae
e   Nelson and Goldman's Woodrat

4381   *Nelsonia neotomodon*
Rodentia - Muridae
e   Nelson's Woodrat; Nelson's
Diminutive Woodrat; Diminutive
Woodrat
d   Zwergwaldratte

Neodon syn. *Microtus* q.v.

4382   *Neofelis*
Carnivora - Felidae
e   Clouded Leopards
d   Nebelparder
f   Panthères nébuleuses
i   Leopardi nebulosi

4383   *Neofelis nebulosa*
Carnivora - Felidae
e   Clouded Leopard
d   Nebelparder
f   Panthère longibande; Panthère
nébuleuse
i   Leopardo nebuloso; Pantera
nebuloso; Leopardo delle nevi

4384   *Neofelis nebulosa brachyurus*
Carnivora - Felidae
e   Formosan Clouded Leopard;
Taiwanese Clouded Leopard; Taiwan
Clouded Leopard
d   Taiwanischer Nebelparder
f   Panthère nébuleuse de Taiwan

4385   *Neofelis nebulosa macrosceloides*
Carnivora - Felidae
e   Nepal Clouded Leopard
f   Panthère nébuleuse du Népal

4386   *Neofiber*
Rodentia - Muridae
e   Round-tailed Muskrats; Florida
Water Rats
d   Florida-Wasserratten
f   Rats d'eau de Floride

4387   *Neofiber alleni*
Rodentia - Muridae
e   Round-tailed Muskrat; Florida Water
Rat; Roundtail Muskrat
d   Florida-Wasserratte
f   Rat d'eau de Floride

4388   *Neohydromys*
Rodentia - Muridae
e   Short-tailed Shrewmice; Mottle-
tailed Shrewmice
d   Fleckschwanzschwimmratten

4389   *Neohydromys fuscus*
Rodentia - Muridae
e   Short-tailed Shrewmouse; Mottle-
tailed Shrewmouse
d   Fleckschwanzschwimmratte

Neohylomys syn. *Hylomys* q.v.

4390   *Neomys*
Soricomorpha - Soricidae
e   Brown-toothed Water Shrews; Water
Shrews; Eurasian Water Shrews; Old
World Water Shrews
d   Wasserspitzmäuse
f   Crossopes de Miller; Musaraignes de
Miller
i   Toporagni aquatici

4391 *Neomys anomolus*
Soricomorpha - Soricidae
e Mediterranean Water Shrew; Miller's Water Shrew; Southern Water Shrew
d Alpenwasserspitzmaus
f Crossope de Miller; Musaraigne de Miller
i Toporagno aquatico di Miller

4392 *Neomys fodiens*
Soricomorpha - Soricidae
e British Water Shrew; Eurasian Water Shrew; European Water Shrew; Old World Water Shrew; White-breasted Water Shrew; Water Shrew; Northern Water Shrew
d Wasserspitzmaus; Große Wasserspitzmaus
f Crossope aquatique; Crossope; Musaraigne aquatique; Musaraigne ciliée; Musaraigne d'eau; Musaraigne porte-rame
i Toporagno aquatico; Toporagno d'acqua; Toporagno d'acqua europeo

4393 *Neomys schelkovnikovi*
Soricomorpha - Soricidae
e Transcaucasian Water Shrew

Neonycteris syn. *Micronycteris* q.v.

4394 *Neophascogale*
Dasyuromorphia - Dasyuridae
e Long-clawed Marsupial Mice; Speckled Dasyures
d Lorentz-Pinselschwanzbeutelmäuse; Spitzhörnchenbeutler
i Topi marsupiale dai lunghi artigli

4395 *Neophascogale lorentzi*
Dasyuromorphia - Dasyuridae
e Long-clawed Marsupial Mouse; Speckled Dasyure
d Lorentz-Pinselschwanzbeutelmaus; Neuguinea-Spitzhörnchenbeutler

i Topo marsupiale dai lunghi artigli

4396 *Neophoca*
Carnivora - Otariidae
e Australian Sea Lions ; Tasmanian Sea Lions
d Australische Seelöwen
f Lions de mer d'Australie
i Leone marini australiani

4397 *Neophoca cinerea*
Carnivora - Otariidae
e Australian Sea Lion; White-capped Sea Lion; Grey Sea Lion
d Australischer Seelöwe
f Lion de mer d'Australie
i Leone marino australiano

4398 *Neophocaena*
Cete - Phocoenidae
e Black Finless Porpoises; Southeast Asian Porpoises
d Indische Braunfische; Glattschweinswale
f Marsouins aptères
i Neofocene

4399 *Neophocaena phocaenoides*
Cete - Phocoenidae
e Finless Porpoise; Asiatic Black Finless Porpoise; Indian Black Finless Porpoise; Finless Indian Porpoise; Black Finless Porpoise
d Indischer Schweinswal; Neutümmler; Finnenloser Schweinswal; Glattschweinswal
f Marsouin de l'Inde; Marsouin de Cuvier; Marsouin aptère; Marsouin noir; Baleine à bosse
i Neofocena

4400 *Neophocaena phocaenoides asiaorientalis*
Cete - Phocoenidae
e Yangtse Finless Porpoise
d Jiangtzhu-Schweinswal

4401 *Neophocaena phocaenoides phocaenoides*
Cete - Phocoenidae
e   Indo-Pacific Finless Porpoise

4402 *Neophocaena phocaenoides sunameri*
Cete` - Phocoenidae
e   Chinese Finless Porpoise
d   Japaniches Glattschweinswal

Neoplatymops syn. *Molossops* q.v.

4403 *Neopteryx*
Chiroptera - Pteropodidae
e   Small-toothed Fruit Bats

4404 *Neopteryx frosti*
Chiroptera - Pteropodidae
e   Small-toothed Fruit Bat

Neoromicis syn. *Pipistrellus* q.v.

Neosciurus syn. *Sciurus* q.v.

Neotamias syn. *Tamias* q.v.

Neotetracus syn. *Hylomys* q.v.

4405 *Neotoma*
Rodentia - Muridae
e   Brushrats; Woodrats; Mountain Rats; Cave Rats; Pack Rats; Trade Rats
d   Buschratten
f   Rats des bois; Néotomes
i   Ratti dei boschi

4406 *Neotoma albigula*
Rodentia - Muridae
e   White-throated Woodrat
d   Wüstenratte; Wüstenbuschratte
f   Rat des steppes

Neotoma alleni syn. *Hodomys alleni* q.v.

4407 *Neotoma angustapalata*
Rodentia - Muridae
e   Tamaulipan Woodrat

4408 *Neotoma anthonyi*
Rodentia - Muridae
e   Anthony's Woodrat; Todos Santos Packrat

4409 *Neotoma bryanti*
Rodentia - Muridae
e   Bryant's Woodrat

4410 *Neotoma bunkeri*
Rodentia - Muridae
e   Bunker's Woodrat

4411 *Neotoma chrysomelas*
Rodentia - Muridae
e   Nicaraguan Woodrat

4412 *Neotoma cinerea*
Rodentia - Muridae
e   Bushtail Woodrat; Bushy-tailed Woodrat; Packrat
d   Buschschwanzratte
f   Rat à queue touffue; Néotome à queue touffue

4413 *Neotoma cinerea rupicola*
Rodentia - Muridae
e   Pale Bushy-tailed Woodrat

4414 *Neotoma devia*
Rodentia - Muridae
e   Arizona Woodrat

4415 *Neotoma floridana*
Rodentia - Muridae
e   Eastern Pack Rat; Florida Pack Rat; Trade Rat; Eastern Woodrat
d   Florida-Buschratte
f   Néotome des Appalaches
i   Ratto dei boschi della Florida

4416 *Neotoma fuscipes*
Rodentia - Muridae
e Dusky-footed Woodrat

4417 *Neotoma fuscipes annectans*
Rodentia - Muridae
e San Francisco Dusky-footed
Woodrat; Dusky-footed Woodrat

4418 *Neotoma fuscipes luciana*
Rodentia - Muridae
e Monterey Dusky-footed Woodrat

4419 *Neotoma fuscipes riparia*
Rodentia - Muridae
e San Joaquin Valley Dusky-footed
Woodrat; Riparian Dusky-footed
Woodrat; Riparian Woodrat

4420 *Neotoma goldmani*
Rodentia - Muridae
e Goldman's Woodrat

Neotoma latifrons syn. *N. albigula*

4421 *Neotoma lepida*
Rodentia - Muridae
e Desert Woodrat
f Néotome du désert
i Ratto dei boschi

4422 *Neotoma lepida intermedia*
Rodentia - Muridae
e San Diego Desert Woodrat

4423 *Neotoma magister*
Rodentia - Muridae
e Alleghenny Woodrat

4424 *Neotoma martinensis*
Rodentia - Muridae
e San Martin Island Woodrat

4425 *Neotoma mexicana*
Rodentia - Muridae
e Mexican Woodrat

4426 *Neotoma micropus*
Rodentia - Muridae
e Southern Plains Woodrat

Neotoma montezumae syn. *N. albigula* q.v.

4427 *Neotoma nelsoni*
Rodentia - Muridae
e Nelson's Woodrat

4428 *Neotoma palatina*
Rodentia - Muridae
e Bolaos Woodrat

4429 *Neotoma phenax*
Rodentia - Muridae
e Sonoran Woodrat

4430 *Neotoma stephensi*
Rodentia - Muridae
e Stephens's Woodrat

4431 *Neotomodon*
Rodentia - Muridae
e Volcano Mice; Mexican Volcano
Mice
d Mexikanische Vulkanmäuse

4432 *Neotomodon alstoni*
Rodentia - Muridae
e Volcano Mouse; Mexican Volcano
Mouse
d Vulkanmaus

4433 *Neotomys*
Rodentia - Muridae
e Marsh Rats; Andean Swamp Rats;
Red-nosed Mice
d Anden-Sumpfratten

4434 *Neotomys ebriosus*
Rodentia - Muridae

e Marsh Rat; Andean Swamp Rat;
     Red-nosed Mouse
d Anden-Sumpfratte

4435 *Neotragus*
          Artiodactyla - Bovidae
e Pygmy Antelopes; Dwarf Antelopes;
     Royal Antelopes; Zanzibar Antelopes
d Kleinsteinböckchen ; Sansibar-
     Antilopen; Moschusböckchen
f Antilopes pygmées
i Neotragi

4436 *Neotragus batesi*
          Artiodactyla - Bovidae
e Bates's Dwarf Antelope; Dwarf
     Antelope; Bates's Pygmy Antelope
d Bates-Böckchen
f Antilope de Bates
i Neotrago di Bates

4437 *Neotragus moschatus*
          Artiodactyla - Bovidae
e Suni; Suni Antelope
d Moschusböckchen; Suni-Böckchen
f Suni; Antilope musquée
i Neotrago muschiato; Neotrago Suni

4438 *Neotragus moschatus*
          livingstonianus
          Artiodactyla - Bovidae
e Livingstone's Suni

4439 *Neotragus moschatus moschatus*
          Artiodactyla - Bovidae
e Zanzibar Suni
f Suni de Zanzibar

4440 *Neotragus moschatus zuluensis*
          Artiodactyla - Bovidae
e Zulu Suni

          Nesogale syn. *Microgale* q.v.

4441 *Nesokia*

          Rodentia - Muridae
e Pest Rats; Short-tailed Molerats;
     Short-tailed Bandicoot Rats; Asiatic
     Bandicoot Rats
d Kurzschwanzmaulwurfsratten;
     Pestratten; Nesokia-Ratten
f Rats bandicoots

4442 *Nesokia bunnii*
          Rodentia - Muridae
e Bunn's Short-tailed Bandicoot Rat

4443 *Nesokia indica*
          Rodentia - Muridae
e Pest Rat; Short-tailed Molerat; Short-
     tailed Bandicoot Rat; Asiatic
     Bandicoot Rat
d Kurzschwanzmaulwurfsratte;
     Pestratte; Indische Pestratte
f Rat à queue courte

4444 *Nesolagus*
          Lagomorpha - Leporidae
e Sumatran Hares; Short-eared
     Rabbits; Sumatran Rabbits
d Sumatra-Kaninchen
f Lapins de Sumatra

4445 *Nesolagus netscheri*
          Lagomorpha - Leporidae
e Sumatran Hare; Short-eared Rabbit;
     Sumatran Rabbit; Sumatra Short-
     eared Rabbit
d Sumatra-Kaninchen;
     Kurzohrkaninchen
f Lapin de Sumatra

4446 *Nesolagus timminsi*
          Lagomorpha - Leporidae
e Annamite Striped Rabbit
d Streifenkaninchen

4447 *Nesomys*
          Rodentia - Muridae
e Island Mice

d   Inselratten; Rote Waldratten
f   Rats rouge

4448   *Nesomys rufus*
Rodentia - Muridae
e   Island Mouse; Red Forest Rat
d   Inselratte; Lamberton-Inselratte
f   Rat rouge de la forêt orientale

Nesonycteris syn. *Melonycteris* q.v.

Nesoromys syn. *Rattus* q.v.

Nesoryctes syn. *Oryzorictes* q.v.

4449   *Nesoryzomys*
Rodentia - Muridae
e   Galapagos Mice
d   Galapagos-Mäuse
f   Souris des îles Galapagos

4450   *Nesoryzomys darwini*
Rodentia - Muridae
e   Darwin's Galapagos Mouse; Darwin's
Rice Rat
d   Darwin-Reisratte
f   Rat du riz de Darwin

4451   *Nesoryzomys fernandinae*
Rodentia - Muridae
e   Fernandina Galapagos Mouse;
Fernandina Rice Rat
f   Rat de l'île Fernandina

4452   *Nesoryzomys indefessus*
Rodentia - Muridae
e   Indefatigable Galapagos Mouse;
Indefatigable Rice Mouse; Santa
Cruz Rice Rat; Santa Cruz Island
Rice Rat; Indefatigable Island Rice
Rat
d   Santa-Cruz-Insel-Reisratte; Santa-
Cruz-Reisratte
f   Rat du riz de l'île Santa Cruz

4453   *Nesoryzomys swarthi*
Rodentia - Muridae
e   Santiago Galapagos Mouse; James
Rice Rat; Santiago Island Rice Rat;
Santiago Rice Rat; San Salvador
Island Rice Rat; San Salvador Rice
Rat; James Island Rice Rat
d   Santiago-Insel-Reisratte; Santiago-
Reisratte

4454   *Nesoscaptor*
Erinaceomorpha - Talpidae
e   Ryukyu Moles
d   Riukiu-Maulwürfe; Senkaku-
Maulwürfe

4455   *Nesoscaptor uchidai*
Erinaceomorpha - Talpidae
e   Ryukyu Mole; Senkaku Mole
d   Riukiu-Maulwurf; Senkaku-
Maulwurf

Nesotragus syn. *Neotragus* q.v.

4456   *Neurotrichus*
Erinaceomorpha - Talpidae
e   American Shrew Moles; Shrew
Moles
d   Amerikanische Spitzmausmaulwürfe;
Amerikanische Spitzmulle
f   Taupes naines
i   Talpe toporagno americane

4457   *Neurotrichus gibbsii*
Erinaceomorpha - Talpidae
e   Shrew Mole; American Shrew Mole
d   Amerikanischer Spitzmausmaulwurf;
Amerikanischer Spitzmull
f   Taupe naine de Gibbs; Taupe naine;
Taupe de Gibbs
i   Talpa toporagno americana

4458   *Neusticomys*
Rodentia - Muridae
e   Fish-eating Mice; Aquatic Mice;
Fish-eating Rats
d   Fischmäuse; Fischratten

4459   *Neusticomys monticolus*
          Rodentia - Muridae
   e   Montane Fish-eating Rat; Fish-eating
          Mouse
   d   Bergfischratte

4460   *Neusticomys mussoi*
          Rodentia - Muridae
   e   Musso's Fish-eating Rat
   d   Mussos Fischratte

4461   *Neusticomys oyapocki*
          Rodentia - Muridae
   e   Oyapock's Fish-eating Rat
   d   Oyapocks Fischratte

4462   *Neusticomys peruviensis*
          Rodentia - Muridae
   e   Peruvian Fish-eating Rat
   d   Peruanische Fischratte

4463   *Neusticomys venezuelae*
          Rodentia - Muridae
   e   Venezuelan Fish-eating Rat
   d   Venezuela-Fischratte

4464   *Nilopegamys*
          Rodentia - Muridae
   e   Ethiopian Water Mice
   d   Äthiopische Wassermäuse
   f   Souris d'eau éthiopienne
   i   Topi d'acqua etiopi

4465   *Nilopegamys plumbeus*
          Rodentia - Muridae
   e   Ethiopian Water Mouse
   d   Äthiopische Wassermaus
   f   Souris d'eau éthiopienne
   i   Topo d'acqua etiope

4466   *Ningaui*
          Dasyuromorphia - Dasyuridae
   e   Ningauis
   d   Ningauis

   f   Ningauis
   i   Ningaui

4467   *Ningaui ridei*
          Dasyuromorphia - Dasyuridae
   e   Wongai Ningaui
   d   Wongai-Ningaui
   i   Ningaui dell'entroterra; Ningaui
          Wonga

4468   *Ningaui timealeyi*
          Dasyuromorphia - Dasyuridae
   e   Pilbara Ningaui
   d   Pilbara-Ningaui

4469   *Ningaui yvonneae*
          Dasyuromorphia - Dasyuridae
   e   Southern Ningaui
   d   Südlicher Ningaui; Südningaui
   f   Ningaui du Sud

4470   *Niviventer*
          Rodentia - Muridae
   e   White-bellied Rats
   d   Asiatische Weißbauchratten;
          Weißbauchratten

4471   *Niviventer andersoni*
          Rodentia - Muridae
   e   Anderson's White-bellied Rat

4472   *Niviventer brahma*
          Rodentia - Muridae
   e   Brahma White-bellied Rat

4473   *Niviventer confucianus*
          Rodentia - Muridae
   e   Chinese White-bellied Rat
   d   Chinesische Weißbauchratte

4474   *Niviventer coxingi*
          Rodentia - Muridae
   e   Coxing's White-bellied Rat

4475 *Niviventer cremoriventer*
  Rodentia - Muridae
  e Dark-tailed Tree Rat; Pencil-tailed
    Rat
  d Dunkelschwänzige Weißbauchratte

4476 *Niviventer culturatus*
  Rodentia - Muridae
  e Oldfield White-bellied Rat

4477 *Niviventer eha*
  Rodentia - Muridae
  e Smoke-bellied Rat

4478 *Niviventer excelsior*
  Rodentia - Muridae
  e Large White-bellied Rat

4479 *Niviventer fulvescens*
  Rodentia - Muridae
  e Chestnut White-bellied Rat; Chestnut
    Rat; Chestnut Spiny Rat

4480 *Niviventer hinpoon*
  Rodentia - Muridae
  e Limestone Rat

4481 *Niviventer langbianis*
  Rodentia - Muridae
  e Lang Bian White-bellied Rat

4482 *Niviventer lepturus*
  Rodentia - Muridae
  e Narrow-tailed White-bellied Rat

4483 *Niviventer niviventer*
  Rodentia - Muridae
  e Hodgson's White-bellied Rat
  d Himalaja-Weißbauchratte

4484 *Niviventer rapit*
  Rodentia - Muridae
  e Long-tailed Mountain Rat

4485 *Niviventer tenaster*
  Rodentia - Muridae
  e Tenasserim White-bellied Rat

4486 *Noctilio*
  Chiroptera - Noctilionidae
  e Hare-lipped Bats; Bulldog Bats;
    Fishing Bats
  d Hasenmäuler
  f Noctilions
  i Pipistrelli pescatori

4487 *Noctilio albiventris*
  Chiroptera - Noctilionidae
  e Southern Bulldog Bat; Lesser
    Bulldog Bat; Lesser Fishing Bat
  d Kleines Hasenmaul; Kleine
    Hasenmaulfledermaus
  f Noctilion du Sud
  i Pipistrello pescatore minore

4488 *Noctilio leporinus*
  Chiroptera - Noctilionidae
  e Bulldog Bat; Mexican Bulldog Bat;
    Fisherman Bat; Greater Bulldog Bat;
    Big Fishing Bat; Greater Fishing Bat;
    Fishing Bat
  d Großes Hasenmaul; Amerikanische
    Hasenmaulfledermaus; Große
    Hasenmaulfledermaus;
    Fischerfledermaus
  f Noctilion pécheur
  i Pipistrello pescatore maggiore

4489 **Noctilionidae**
  Chiroptera
  e Bulldog Bats; Fisherman Bats; Hare-
    lipped Bats; Mastiff Bats; Fishing
    Bats
  d Hasenmäuler;
    Hasenmaulfledermäuse;
    Fischerfledermäuse
  f Noctilionidés
  i Pipistrelli pescatori; Noctilionidi

Nomascus syn. *Hylobates* q.v.

Notamacropus syn. *Macropus* q.v.

4490   Notiomys
Rodentia - Muridae
e   Long-clawed Mice; Long-Clawed
South American Mice; Molemice;
Edwards's Long-clawed Mice
d   Langkrallenmäuse

4491   Notiomys edwardsii
Rodentia - Muridae
e   Edwards's Long-clawed Mouse

Notiomys macronyx syn. *Chelemys
macronyx* q.v.

Notiomys megalonyx syn. *Chelemys
megalonyx* q.v.

Notiomys valdivianus syn. *Geoxus
valdivianus q.v*

4492   Notiosorex
Soricomorpha - Soricidae
e   Grey Shrews; Desert Shrews
d   Wüstenspitzmäuse
f   Musaraignes du désert

4493   Notiosorex cockrumi
Soricomorpha - Soricidae
e   Cockrum's Desert Shrew
i   Topo del deserto

4494   Notiosorex crawfordi
Soricomorpha - Soricidae
e   Desert Grey Shrew; Desert Shrew;
Crawford's Desert Shrew
d   Graue Wüstenspitzmaus
f   Musaraigne du désert

4495   Notiosorex evotis
Soricomorpha - Soricidae
e   Large-eared Desert Shrew;
Crawford's Desert Shrew

Notiosorex gigas syn. *Megasorex*

gigas q.v.

4496   Notiosorex villai
Soricomorpha - Soricidae
e   Villa's Grey Shrew

Notocitellus syn. *Spermophilus* q.v.

4497   Notomys
Rodentia - Muridae
e   Jerboa Mice; Kangaroo Mice;
Australian Kangaroo Mice;
Australian Hopping Mice
d   Australische Hüpfmäuse;
Kängurumäuse; Hüpfmäuse
f   Souris sauteuses d'Australie; Souris
fauves; Souris sauteuses
i   Topi saltatori

4498   Notomys alexis
Rodentia - Muridae
e   Hopping Mouse; Spinifex Hopping
Mouse (ANZ)
i   Topo saltatore dello spinifex;
Dargawarra

4499   Notomys amplus
Rodentia - Muridae
e   Short-tailed Hopping Mouse; Short-
tailed Australian Hopping Mouse;
Large Desert Hopping Mouse; Large
Desert Jerboa Mouse; Australian
Jerboa Rat; Short-tailed Australian
Jerboa Rat; Short-tailed Jerboa Rat;
Brazenor's Hopping Mouse; Short-
tailed Australian Kangaroo Mouse;
Short-tailed Kangaroo Mouse
d   Große Australische Hüpfmaus;
Große Hüpfmaus; Große
Australische Kängurumaus; Große
Kängurumaus; Kurzschwänzige
Australische Hüpfmaus;
Kurzschwanzhüpfmaus;
Kurzschwanzkängurumaus
f   Souris kangourou d'Australie à queue
courte; Souris sauteuse d'Australie à
queue courte; Souris sauteuse
d'Australie du grand déset

4500   *Notomys aquilo*
          Rodentia - Muridae
  e   Cape York Hopping Mouse;
        Northern Hopping Mouse
  d   Nordhüpfmaus

          Notomys carpentarius syn. *N. aquilo*
          q.v.

4501   *Notomys cervinus*
          Rodentia - Muridae
  e   Fawn-coloured Hapalotis; Neck-
        pouched Hopping Mouse; Fawn-
        coloured Hopping Mouse; Fawn
        Hopping Mouse (ANZ)
  d   Rehbraune Australische Hüpfmaus;
        Kitz-Hüpfmaus

4502   *Notomys fuscus*
          Rodentia - Muridae
  e   Dusky Hopping Mouse

4503   *Notomys longicaudatus*
          Rodentia - Muridae
  e   Long-tailed Hapalotis; Long-tailed
        Hopping Mouse; Long-tailed
        Australian Hopping Mouse; Long-
        tailed Australian Jerboa Mouse;
        Long-tailed Jerboa Rat; Long-tailed
        Australian Jerboa Rat; Long-tailed
        Australian Kangaroo Mouse; Long-
        tailed Kangaroo Mouse
  d   Langschwanzhüpfmaus;
        Langschwänzige Australische
        Hüpfmaus; Langschwänzige
        Australische Kängurumaus;
        Langschwanzkängurumaus
  f   Souris sauteuse d'Australie à longue
        queue; Souris kangourou d'Australie
        à longue queue

4504   *Notomys macrotis*
          Rodentia - Muridae
  e   Big-eared Hopping Mouse; Big-eared
        Australian Hopping Mouse; Big-
        eared Australian Jerboa Mouse; Big-
        eared Jerboa Mouse; Big-eared
        Australian Jerboa Rat; Big-eared

          Jerboa Rat; Big-eared Australian
          Kangaroo Mouse; Big-eared
          Kangaroo Mouse
  d   Großohrhüpfmaus; Großohrige
        Australische Hüpfmaus; Großohrige
        Australische Kängurumaus;
        Großohrkängurumaus
  f   Souris sauteuse d'Australie aux
        grandes oreilles; Souris kangourou
        d'Australie aux grandes oreilles;
        Souris sauteuse d'Australie oreillard;
        Souris kangourou d'Australie
        oreillard

          Notomys megalotis syn. *N. macrotis*
          q.v.

4505   *Notomys mitchelli*
          Rodentia - Muridae
  e   Mitchell's Hapalotis; Mitchell's
        Hopping Mouse (ANZ)
  d   Mitchells Hüpfmaus

4506   *Notomys mordax*
          Rodentia - Muridae
  e   Darling Downs Hopping Mouse;
        Darling Downs Jerboa Mouse;
        Darling Downs Jerboa Rat; Darling
        Downs Kangaroo Mouse; Darling
        Downs Australian Hopping Mouse;
        Darling Downs Australian Australian
        Jerboa Mouse; Darling Downs
        Australian Jerboa Rat; Darling
        Downs Australian Kangaroo Mouse
  d   Darling-Downs-Hüpfmaus; Darling-
        Downs-Kängurumaus
  f   Souris sauteuse d'Australie de
        Darling Downs; Souris kangourou
        d'australie des Darling Downs

4507   *Notopteris*
          Chiroptera - Pteropodidae
  e   Blossom Bats
  d   Langschwanzflughunde
  f   Notoptères

4508   *Notopteris macdonaldi*
          Chiroptera - Pteropodidae

e	Fijian Blossom Bat
d	Langschwanzflughund

4509   ***Notopteris neocaledonica***
Chiroptera - Pteropodidae
e   New Caledonian Blossom Bat
f   Notoptère de Nouvelle Calédonie

4510   ***Notoryctes***
Notoryctemorphia - Notoryctidae
e   Marsupial Moles; Pouched Mole
d   Beutelmulle; Beutelmaulwürfe
f   Notoryctes ; Taupes marsupiaux;
Taupes marsupiales
i   Talpe marsupiali

4511   ***Notoryctes caurinus***
Notoryctemorphia - Notoryctidae
e   Northwestern Marsupial Mole;
Marsupial Mole; Northern Marsupial
Mole
d   Kleiner Beutelmull
f   Petite taupe marsupiale

4512   ***Notoryctes typhlops***
Notoryctemorphia - Notoryctidae
e   Southern Marsupial Mole; Greater
Marsupial Mole
d   Großer Beutelmull
f   Grande taupe marsupial

4513   ***Notoryctidae***
Notoryctemorphhia -
e   Marsupial Moles; Pouched Moles
d   Beutelmulle; Beutelmaulwürfe
f   Notoryctidés; Taupes marsupiaux;
Taupes marsupiales
i   Talpe marsupiali; Notorictidi

4514   ***Nyctalus***
Chiroptera - Vespertilionidae
e   Noctule Bats; Noctules
d   Abendsegler ; Abendfledermäuse;
Waldfledermäuse
f   Noctules; Nyctales

i   Nottole

4515   ***Nyctalus aviator***
Chiroptera - Vespertilionidae
e   Birdlike Noctule

4516   ***Nyctalus azoreum***
Chiroptera - Vespertilionidae
e   Azores Noctule
d   Azoren-Abendsegler
f   Noctule des Açores

4517   ***Nyctalus lasiopterus***
Chiroptera - Vespertilionidae
e   Giant Noctule; Fatio's Bat; Greater
Noctule
d   Riesenabendsegler; Großabendsegler
f   Noctule géante; Noctule de Fatio;
Grande Noctule
i   Nottola gigante

4518   ***Nyctalus leisleri***
Chiroptera - Vespertilionidae
e   Lesser Noctule; Leisler's Bat; Hairy-
armed Bat; Hairy-sided Bat; Leisler's
Noctule; Hairy-winged Bat
d   Kleinabendsegler; Rauharmige
Fledermaus; Rauharmfledermaus;
Kleiner Abendsegler
f   Noctule de Leisler; Nyctale de
Leisler; Vespérien de Leisler;
Vespérugo de Leisler
i   Nottola di Leisler

4519   ***Nyctalus montanus***
Chiroptera - Vespertilionidae
e   Mountain Noctule

4520   ***Nyctalus noctula***
Chiroptera - Vespertilionidae
e   Common Noctule; Noctule; Great
Bat; Noctule Bat
d   Großer Abendsegler; Eurasischer
Abendsegler; Frühfliegende
Fledermaus; Waldfledermaus;
Speckmaus; Abendsegler

f Noctule commune; Chauve-souris noctule; Vespérien noctule; Vespérugo noctule; Vespertilion noctule
i Nottola comune; Nottola

4521 *Nyctereutes*
Carnivora - Canidae
e Raccoon Dogs; Raccoon-like Dogs
d Marderhunde
f Chiens viverrins
i Cani procioni

4522 *Nyctereutes procyonoides*
Carnivora - Canidae
e Raccoon Dog; Raccoon-like Dog; Asian Racoon Dog
d Marderhund; Waschbärhund; Obstfuchs; Seefuchs
f Chien viverrin
i Cane procione; Tanuki; Procione; Marmotta cinese; Tasso giapponese; Procione lavatore

4523 *Nyctereutes procyonoides viverrinus*
Carnivora - Canidae
e Japanese Racoon Dog
i Cane viverrino

4524 Nycteridae
Chiroptera
e Hispid Bats; Slit-faced Bats; Hollow-faced Bats
d Schlitznasen; Stirngrubenfledermäuse; Hohlnasen
f Nycteridés
i Nitteridi

4525 *Nycteris*
Chiroptera - Nycteridae
e Slit-faced Bats; Hollow-faced Bats
d Schlitznasen
f Nyctères
i Nitteridi

4526 *Nycteris arge*

Chiroptera - Nycteridae
e Bates's Slit-faced Bat
d Bates-Schlitznase
f Chauve-souris à queue bifurquée

4527 *Nycteris gambiensis*
Chiroptera - Nycteridae
e Gambian Slit-faced Bat
d Gambia-Schlitznase

4528 *Nycteris grandis*
Chiroptera - Nycteridae
e Great Slit-faced Bat; Large Slit-faced Bat
d Große Schlitznase; Riesenhohlnase; Schlitznase
f Grande Nyctère

4529 *Nycteris hispida*
Chiroptera - Nycteridae
e Hairy Slit-faced Bat; Hispid Sliy-faced Bat
d Rauhhaarschlitznase; Behaarte Schlitznase
f Nyctère hérissée

4530 *Nycteris intermedia*
Chiroptera - Nycteridae
e Intermediate Slit-faced Bat
d Mittlere Schlitznase

4531 *Nycteris javanica*
Chiroptera - Nycteridae
e Javan Slit-faced Bat; Javanese Slit-faced Bat; Malaysian Hollow-faced Bat
d Java-Hohlnase
f Nyctère de Java

4532 *Nycteris macrotis*
Chiroptera - Nycteridae
e Dobson's Slit-faced Bat; Large-eared Slit-faced Bat; West African Slit-faced Bat; Greater Slit-faced Bat
d Großohrschlitznase

4533 *Nycteris madagascariensis*
         Chiroptera - Nycteridae
   e   Madagascar Slit-faced Bat

4534 *Nycteris major*
         Chiroptera - Nycteridae
   e   Ja Slit-faced Bat
   d   Ja-Schlitznase

         Nycteris minor syn. *Lasiurus borealis* q.v.

4535 *Nycteris nana*
         Chiroptera - Nycteridae
   e   Dwarf Slit-faced Bat
   d   Zwergschlitznase

         Nycteris pfeifferi syn. *Lasiurus borealis* q.v.

         Nycteris seminolus syn. *Lasiurus seminolus* q.v.

4536 *Nycteris thebaica*
         Chiroptera - Nycteridae
   e   Egyptian Slit-faced Bat; Geoffroy's Slit-faced Bat; Egyptian Hollow-faced Bat
   d   Geoffroy-Schlitznase; Großohrhohlnase; Ägyptische Schlitznasenfledermaus; Ägyptische Schlitznase
   f   Nyctère de la Thébaïde; Nyctère de Geoffroy
   i   Nitteride di Tebe

4537 *Nycteris tragata*
         Chiroptera - Nycteridae
   e   Malayan Slit-faced Bat
   d   Malaiische Schlitznase

4538 *Nycteris woodi*
         Chiroptera - Nycteridae
   e   Wood's Slit-faced Bat
   d   Wood-Schlitznase

4539 *Nycticebus*
         Primates - Loridae
   e   Slow Lemurs; Slow Lorises
   d   Plumploris
   f   Nycticèbes
   i   Nitticebi

4540 *Nycticebus bengalensis*
         Primates - Loridae
   e   Bengal Slow Loris
   d   Bengalischer Plumplori
   f   Loris lent du Bengale
   i   Nitticebo dell Indocina

4541 *Nycticebus coucang*
         Primates - Loridae
   e   Sunda Slow Loris; Slow Loris
   d   Sunda-Plumplori; Eigentlicher Plumplori; Großer Plumplori; Plumplori
   f   Nycticèbe parresseux; Nycticèbe
   i   Nitticebo della Sonda; Nitticebo coucang; Lori lento; Nitticebo

4542 *Nycticebus pygmaeus*
         Primates - Loridae
   e   Lesser Slow Loris; Pygmy Slow Loris; Pygmy Loris; Lesser Loris
   d   Kleiner Lori; Zwergplumplori
   f   Nycticèbe pygmé; Loris paresseux pygmée
   i   Nitticebo pigmeo

         Nycticeinops syn. *Nycticeius* q.v.

4543 *Nycticeius*
         Chiroptera - Vespertilionidae
   e   Evening Bats; Twilight Bats; Broad-nosed Bats
   d   Abendsegler
   f   Chauves-souris vespérales

4544 *Nycticeius balsoni*
         Chiroptera - Vespertilionidae
   e   Inland Broad-nosed Bat

Nycticeius cubanus syn. *N. humeralis* q.v.

4545 *Nycticeius greyi*
Chiroptera - Vespertilionidae
e   Little Broad-nosed Bat (ANZ);
Grey's Bat

4546 *Nycticeius humeralis*
Chiroptera - Vespertilionidae
e   Black-shouldered Bat; Twilight Bat;
Evening Bat

4547 *Nycticeius orion*
Chiroptera - Vespertilionidae
e   Eastern Broad-nosed Bat

4548 *Nycticeius rueppelli*
Chiroptera - Vespertilionidae
e   Rüppell's Broad-nosed Bat; Greater
Broad-nosed Bat (ANZ)

4549 *Nycticeius sanborni*
Chiroptera - Vespertilionidae
e   Northern Broad-nosed Bat; Little
Northern Broad-nosed Bat

4550 *Nycticeius schlieffeni*
Chiroptera - Vespertilionidae
e   Schlieffen's Bat; Schlieffen's
Twilight Bat

Nyctiellus syn. *Natalus* q.v.

4551 *Nyctimene*
Chiroptera - Pteropodidae
e   Tube-nosed Bats; Tube-nosed Fruit
Bats
d   Röhrennasenflughunde
f   Nyctimènes
i   Nittimeni

4552 *Nyctimene aello*
Chiroptera - Pteropodidae
e   Broad-striped Tube-nosed Bat;
Broad-striped Tube-nosed Fruit Bat;

Greater Tube-nosed Fruit Bat; Dark
Tube-nosed Fruit Bat

4553 *Nyctimene albiventer*
Chiroptera - Pteropodidae
e   Common Tube-nosed Bat; Tube-
nosed Bat; Common Tube-nosed
Fruit Bat; Demonic Tube-nosed Fruit
Bat; Umboi Tube-nosed Fruit Bat;
Bismarck Tube Bat

4554 *Nyctimene bougainville*
Chiroptera - Pteropodidae
e   Solomons Tube-nosed Fruit Bat

4555 *Nyctimene bougainville malaitensis*
Chiroptera - Pteropodidae
e   Malaita Island Tube-nosed Bat;
Malaita Tube-nosed Bat; Malaita
Tube-nosed Fruit Bat

Nyctimene celaeno syn. *N. aello* q.v.

4556 *Nyctimene cephalotes*
Chiroptera - Pteropodidae
e   Large-headed Tube-nosed Fruit Bat;
Pallas's Tube-nosed Bat; Pallas's
Tube-nosed Fruit Bat
d   Großkopf Röhrennasenflughund
f   Nyctimène à grosse tête

4557 *Nyctimene certans*
Chiroptera - Pteropodidae
e   Mountain Tube-nosed Fruit Bat

4558 *Nyctimene cyclotis*
Chiroptera - Pteropodidae
e   Round-eared Tube-nosed Bat;
Round-eared Tube-nosed Fruit Bat

4559 *Nyctimene draconilla*
Chiroptera - Pteropodidae
e   Dragon Tube-nosed Fruit Bat

4560 *Nyctimene major*
Chiroptera - Pteropodidae

e   Giant Tube-nosed Fruit Bat; Greatest
     Tube-nosed Bat; Island Tube-nosed
     Fruit Bat
d   Großer Röhrennasenflughund
f   Nyctimène géante

Nyctimene malaitensis syn. *N.*
bougainville malaitensis q.v.

Nyctimene masalai syn. *N. albventer*
q.v.

4561   *Nyctimene minutus*
     Chiroptera - Pteropodidae
e   Lesser Tube-nosed Bat; Lesser Tube-
     nosed Fruit Bat

4562   *Nyctimene rabori*
     Chiroptera - Pteropodidae
e   Philippine Tube-nosed Fruit Bat;
     Rabor's Tube-nosed Bat
d   Philippinischer Röhrennasenflughund

4563   *Nyctimene robinsoni*
     Chiroptera - Pteropodidae
e   Queensland Tube-nosed Bat;
     Queensland Tube-nosed Fruit Bat;
     Eastern Tube-nosed Bat (ANZ)
d   Robinson-Röhrennasenflughund
f   Nyctimène de Robinson

4564   *Nyctimene santacrucis*
     Chiroptera - Pteropodidae
e   Nendo Tube-nosed Fruit Bat; Santa
     Cruz Bat
f   Nyctimène de Santa Cruz;
     Nyctimène du Nendo

Nyctimene vizcaccia syn. *N.*
albiventer q.v.

4565   *Nyctinomops*
     Chiroptera - Molossidae
e   New World Free-tailed Bats

4566   *Nyctinomops aurispinosus*
     Chiroptera - Molossidae

e   Peale's Free-tailed Bat

Nyctinomops europs syn. *N.*
laticaudatus q.v.

4567   *Nyctinomops femorasaccus*
     Chiroptera - Molossidae
e   Pocketed Free-tailed Bat

Nyctinomops gracilis syn. *N.*
laticaudatus q.v.

4568   *Nyctinomops laticaudatus*
     Chiroptera - Molossidae
e   Broad-eared Bat; Espirito Santo
     Free-tailed Bat

4569   *Nyctinomops macrotis*
     Chiroptera - Molossidae
e   Big Free-tailed Bat

Nyctinomops similis syn. *N.*
aurispinosus q.v.

Nyctinomops yucatanicus syn. *N.*
laticaudatus q.v.

4570   *Nyctomys*
     Rodentia - Muridae
e   Sumichrast's Vesper Rats; Vesper
     Rats
d   Sumichrast-Abendratten
i   Ratti vespertini di Sumichrast

4571   *Nyctomys sumichrasti*
     Rodentia - Muridae
e   Sumichrast's Vesper Rat; Vesper Rat;
     Vesper Mouse
d   Sumichrast-Abendratte
i   Ratto vespertino di Sumichrast

4572   *Nyctophilus*
     Chiroptera - Vespertilionidae
e   New Guinea Big-eared Bats;
     Australian Big-eared Bats; Long-
     eared Bats; Nyctophyluses

4573 *Nyctophilus arnhemensis*
Chiroptera - Vespertilionidae
e Arnhem Long-eared Bat; Arnhem
Land Long-eared Bat; Northern
Long-eared Bat; Arnhem
Nyctophilus

4574 *Nyctophilus bifax*
Chiroptera - Vespertilionidae
e Northern Nyctophilus

4575 *Nyctophilus geoffroyi*
Chiroptera - Vespertilionidae
e Geoffroy's Nyctophilus; Lesser
Long-eared Bat (ANZ); Lesser
Nyctophilus

4576 *Nyctophilus gouldi*
Chiroptera - Vespertilionidae
e Gould's Long-eared Bat; Gould's
Nyctophilus

4577 *Nyctophilus heran*
Chiroptera - Vespertilionidae
e Sunda Long-eared Bat; Sunda
Nyctophilus

Nyctophilus lophorhina syn. *N. microtis* q.v.

4578 *Nyctophilus microdon*
Chiroptera - Vespertilionidae
e Small-toothed Long-eared Bat;
Lamington Free-eared Bat; Small-
toothed Nyctophilus

4579 *Nyctophilus microtis*
Chiroptera - Vespertilionidae
e Papuan Long-eared Bat; New Guinea
Long-eared Bat; Small-eared
Nyctophilus

4580 *Nyctophilus nebulosus*
Chiroptera - Vespertilionidae
e New Caledonia Long-eared Bat; New
Caledonia Nyctophilus
f Nyctophile nébuleux; Nyctophile

néo-calédonienne

4581 *Nyctophilus timoriensis*
Chiroptera - Vespertilionidae
e Australian Big-eared Bat; Greater
Long-eared Bat (ANZ); Eastern
Long-eared Bat; Greater Nyctophilus

4582 *Nyctophilus walkeri*
Chiroptera - Vespertilionidae
e Little Territory Long-eared Bat;
Pygmy Long-eared Bat; Pygmy
Nyctophilus

O

4583   *Ochotona*
Lagomorpha - Ochotonidae
e   Rock Conies; Rock Rabbits; Pikas;
Mouse Hares; Whistling Hares
d   Pfeifhasen; Pikas
f   Pikas
i   Pika; Lepri fischianti

4584   *Ochotona alpina*
Lagomorpha - Ochotonidae
e   Altai Pika; Alpine Pika
d   Altaischer Pfeifhase; Alpenpfeifhase;
Altai-Pfeifhase; Altai-Pika
f   Pika de l'Altaï
i   Pika alpina

4585   *Ochotona cansus*
Lagomorpha - Ochotonidae
e   Gansu Pika; Kansu Pika; Kansu
Piping Hare; Kansu Mouse Hare;
Kansu Cony; Grey Pika; Grey Piping
Hare; Grey Cony; Grey Mouse Hare

d   Gansu-Pika; Kansu-Pfeifhase;
Kansu-Pika
f   Pika du Kansu; Lièvre criard de
Kansu; Lièvre criard de Kan-Sou;
Pika du Kan-Sou; Pika de Gansu

4586   *Ochotona collaris*
Lagomorpha - Ochotonidae
e   Alaskan Pika; Collared Pika
d   Alaska-Pfeifhase; Alaska-Pfeifer;
Halsbandpika
f   Pika à collier

4587   *Ochotona curzoniae*
Lagomorpha - Ochotonidae
e   Black-lipped Pika; Plateau Pika

d   Schwarzlippenpika
i   Pika dalle labbra nere; Pika degli
altipiani

4588   *Ochotona daurica*
Lagomorpha - Ochotonidae
e   Daurian Pika
d   Daurischer Pfeifhase
f   Pika de Daourie

4589   *Ochotona erythrotis*
Lagomorpha - Ochotonidae
e   Chinese Red Pika
d   Chinesischer Roter Pika

4590   *Ochotona forresti*
Lagomorpha - Ochotonidae
e   Forrest's Pika

4591   *Ochotona gaoligongensis*
Lagomorpha - Ochotonidae
e   Gaoligong Pika
d   Gaoligong-Pika
f   Pika de Gaoligong

4592   *Ochotona gloveri*
Lagomorpha - Ochotonidae
e   Glover's Pika

4593   *Ochotona himalayana*
Lagomorpha - Ochotonidae
e   Himalajan Pika
d   Himalaja-Pika

4594   *Ochotona hyperborea*
Lagomorpha - Ochotonidae
e   Northern Pika
d   Nördlicher Pfeifhase; Nordischer
Pfeifhase; Japanischer Pfeifhase;
Nordpika
f   Pika du Nord

4595   *Ochotona iliensis*
Lagomorpha – Ochotonidae

e   Ili Pika
d   Ili-Pika

4596 *Ochotona koslowi*
    Lagomorpha - Ochotonidae
e   Kozlov's Pika
d   Koslows Pika
f   Pika de Kozlov

4597 *Ochotona ladacensis*
    Lagomorpha - Ochotonidae
e   Ladak Pika
d   Ladakh-Pika

4598 *Ochotona macrotis*
    Lagomorpha - Ochotonidae
e   Large-eared Pika
d   Großohriger Pika

4599 *Ochotona muliensis*
    Lagomorpha - Ochotonidae
e   Muli Pika
d   Muli-Pika
f   Pika de Muli

    Ochotona nepalensis syn. *O. roylei*
    q.v.

4600 *Ochotona nubrica*
    Lagomorpha - Ochotonidae
e   Nubra Pika
d   Nubra-Pika

4601 *Ochotona pallasi*
    Lagomorpha - Ochotonidae
e   Pallas's Pika
d   Mongolischer Pfeifhase
f   Pika de Pallas

4602 *Ochotona princeps*
    Lagomorpha - Ochotonidae
e   Pika (NA); American Pika; Mouse
    Hare; Rocky Mountains Pika;
    Common Pika; American Piping
    Hare; American Cony; American

    Mouse Hare
d   Amerikanischer Pfeifhase;
    Nordamerikanischer Pika;
    Amerikanischer Pika
f   Pika d'Amérique; Lièvre criard
    d'Amérique; Pika américain
i   Pika nordamericano

4603 *Ochotona princeps muiri*
    Lagomorpha - Ochotonidae
e   Yosemite Cony; Yosemite Pika

4604 *Ochotona princeps obscura*
    Lagomorpha - Ochotonidae
e   Big Horn Mountain Pika

4605 *Ochotona princeps wasatchensis*
    Lagomorpha - Ochotonidae
e   Wasatch Pika

4606 *Ochotona pusilla*
    Lagomorpha - Ochotonidae
e   Steppe Pika; Dwarf Pika
d   Steppenpfeifhase; Zwerg Pika;
    Russischer Zwergpfeifhase;
    Zwergpfeifhase; Asiatischer Pika;
    Zwerghase; Steppenpika
f   Pika des steppes; Ochotona; Pika
    asiatique; Lagomys asiatique; Lièvre
    nain; Petit lièvre

4607 *Ochotona roylei*
    Lagomorpha - Ochotonidae
e   Indian Pika; Royle's Pika
d   Großohriger Pfeifhase; Himalaja-
    Pfeifhase; Royles Pika; Großohrpika
f   Pika de Royle

4608 *Ochotona rufescens*
    Lagomorpha - Ochotonidae
e   Afghan Pika
d   Rötlicher Pfeifhase; Afghanischer
    Pika
f   Pika afghan

4609   *Ochotona rutila*
Lagomorpha - Ochotonidae
e   Red Pika; Turkestan Red Pika
d   Roter Pfeifhase
f   Pika roux

4610   *Ochotona thibetana*
Lagomorpha - Ochotonidae
e   Moupin Pika
d   Tibetischer Pika; Moupin-Pika
i   Pika del Tibet

4611   *Ochotona thomasi*
Lagomorpha - Ochotonidae
e   Thomas's Pika
d   Thomas-Pika

4612   *Ochotonidae*
Lagomorpha -
e   Rock Conies; Rock Rabbits;
Whistling Hares; Piping Hares; Little
Chief Hares; Pikas; Rat Hares;
Mouse Hares
d   Pfeifhasen; Baumläufer
f   Ochotonidés; Pikas
i   Ocotonidi

4613   *Ochrotomys*
Rodentia - Muridae
e   Golden Mice
d   Goldmäuse

4614   *Ochrotomys nuttalli*
Rodentia - Muridae
e   Golden Mouse
d   Goldmaus

4615   *Ochrotomys nuttalli aureolus*
Rodentia - Muridae
e   Common Golden Mouse

4616   *Ochrotomys nuttalli nuttalli*
Rodentia - Muridae
e   Lewis's Golden Mouse

4617   *Octodon*
Rodentia - Octodontidae
e   South American Bush Rats; Degus
d   Strauchratten
f   Octodons; Octodontes; Dégus; Rats
en queue de trompète
i   Degu

4618   *Octodon bridgesi*
Rodentia - Octodontidae
e   Bridge's Degu
d   Walddegu

4619   *Octodon degus*
Rodentia - Octodontidae
e   Degu; Common Degu
d   Degu; Strauchratte
f   Dégu du Chili
i   Degu

4620   *Octodon lunatus*
Rodentia - Octodontidae
e   Moon-toothed Degu
d   Küstendegu

4621   *Octodon pacificus*
Rodentia - Octodontidae
e   Isla Mocha Degu
d   Pazifik-Degu

4622   *Octodontidae*
Rodentia
e   Octodonts; Octodont Rodents; Hedge
Rats; Degus, etc.; Degus and Tuco-
tucos; Tucutucuses
d   Trugratten; Degus
f   Octodontidés
i   Octodontidi

4623   *Octodontomys*
Rodentia - Octodontidae
e   Boris; Mountain Degus
d   Pinselschwanzratten
f   Boris; Octodontes

4624 **Octodontomys gliroides**
Rodentia - Octodontidae
e Bori; Mountain Degu
d Bori; Pinselschwanzratte
f Bori

4625 **Octomys**
Rodentia - Octodontidae
e Viscacha Rats
d Viscacharatten
f Rats minimes; Octodontes

4626 **Octomys mimax**
Rodentia - Octodontidae
e Viscacha Rat; Mountain Viscacha Rat
d Viscacharatte
f Rat minime

4627 **Odobenus**
Carnivora - Phocidae
e Walruses
d Walrosse
f Morses
i Trichechi

4628 **Odobenus rosmarus**
Carnivora - Phocidae
e Walrus; Sea Horse
d Gemeines Walross; Walross; Polarmeer-Walross
f Morse; Cheval-marin; Vache-marin
i Tricheco

4629 **Odobenus rosmarus divergens**
Carnivora - Phocidae
e Pacific Walrus; Illiger Pacific Walrus
d Pazifisches Walross
f Morse du Pacifique

4630 **Odobenus rosmarus laptevi**
Carnivora - Phocidae
e Laptev Sea Walrus
d Leptewsee-Walross

4631 **Odobenus rosmarus rosmarus**
Carnivora - Phocidae
e Atlantic Walrus
d Atlantisches Walross
f Morse de l'Atlantique

4632 **Odocoileus**
Artiodactyla - Cervidae
e American Deer ; White-tailed Deer and Mule Deer
d Amerika-Hirsche
f Cerfs d'Amérique
i Cervi americani

Odocoileus bezoarticus syn.
Ozotocereus bezoarticus q.v.

Odocoileus dichotomus syn.
Blastocerus dichotomus q.v.

4633 **Odocoileus hemionus**
Artiodactyla - Cervidae
e Mule Deer
d Großohrhirsch; Maultierhirsch; Schwarzwedelhirsch
f Cerf mulet; Cerf hemione; Cerf à queue noire
i Cervo Mulo

4634 **Odocoileus hemionus coileus**
Artiodactyla - Cervidae
e California Mule Deer
d Kalifornischer Maultierhirsch
f Cerf mulet de Californie

4635 **Odocoileus hemionus columbianus**
Artiodactyla - Cervidae
e Black-tailed Deer
d Kolumbia-Maultierhirsch
f Cerf à queue noire de Colombie
i Cervo dalla coda nera

4636 **Odocoileus hemionus hemionus**
Artiodactyla – Cervidae

e   Rocky Mountains Mule Deer
d   Felsengebirge-Maultierhirsch

4637   *Odocoileus hemionus sitkensis*
Artiodactyla - Cervidae
e   Sitka Deer
d   Sitka-Rotwild; Sitka-Maultierhirsch
f   Cerf Sitka; Cerf de Sitka

4638   *Odocoileus virginianus*
Artiodactyla - Cervidae
e   White-tailed Deer
d   Weißwedelhirsch
f   Cariacou; Cerf de la Virginie;
     Chevreuil de Virginie
i   Cervo della Virginia

4639   *Odocoileus virginianus clavium*
Artiodactyla - Cervidae
e   Key Deer
d   Keys Weßwedelhirsch
f   Cerf des Cayes de Floride

4640   *Odocoileus virginianus leucurus*
Artiodactyla - Cervidae
e   Columbian White-tailed Deer;
     White-tailed Deer
d   Columbia-Weißwedelhirsch
f   Cerf à queue blanche de Colombie

4641   *Odocoileus virginianus mayensis*
Artiodactyla - Cervidae
e   Guatemala White-tailed Deer
d   Mittelamerikanischer
     Weißwedelhirsch
f   Cerf de Virginie du Guatemala
i   Cervo a coda bianca del Guatemala

4642   *Odocoileus virginianus virginianus*
Artiodactyla - Cervidae
e   Virginia Deer; Virginia White-tailed
     Deer
d   Virginia-Hirsch; Virginianischer
     Weißwedelhirsch

f   Cerf à queue blanche de Virginie

4643   *Oecomys*
Rodentia - Muridae
e   Arboreal Rice Rats
d   Zweifarbreisratten

4644   *Oecomys auyantepui*
Rodentia - Muridae
e   North Amazonian Arboreal Rice Rat

4645   *Oecomys bicolor*
Rodentia - Muridae
e   Bicoloured Arboreal Rice Rat

4646   *Oecomys cleberi*
Rodentia - Muridae
e   Cleber's Arboreal Rice Rat

4647   *Oecomys concolor*
Rodentia - Muridae
e   Unicoloured Arboreal Rice Rat

4648   *Oecomys flavicans*
Rodentia - Muridae
e   Yellow Arboreal Rice Rat

4649   *Oecomys marmorae*
Rodentia - Muridae
e   Marmore Arboreal Rice Rat

4650   *Oecomys paricola*
Rodentia - Muridae
e   Brazilian Arboreal Rice Rat; South
     Amazonian Arboreal Rice Rat

4651   *Oecomys phaeotis*
Rodentia - Muridae
e   Dusky Arboreal Rice Rat

4652   *Oecomys rex*
Rodentia - Muridae
e   King Arboreal Rice Rat

4653 **Oecomys roberti**
Rodentia - Muridae
e Robert's Arboreal Rice Rat

4654 **Oecomys rutilus**
Rodentia - Muridae
e Red Arboreal Rice Rat

4655 **Oecomys speciosus**
Rodentia - Muridae
e Venezuelan Arboreal Rice Rat

4656 **Oecomys superans**
Rodentia - Muridae
e Foothill Arboreal Rice Rat

4657 **Oecomys trinitatis**
Rodentia - Muridae
e Trinidad Rice Rat; Big Arboreal Rice
Rat

4658 **Oenomys**
Rodentia - Muridae
e Rufous-nosed Rats
d Afrikanische Rotnasenratten;
Rotnasenratten
f Rats à museau roux

4659 **Oenomys hypoxanthus**
Rodentia - Muridae
e Rufous-nosed Rat
d Afrikanische Rotnasenratte;
Rotnasenratte
f Rat à museau roux

4660 **Oenomys ornatus**
Rodentia - Muridae
e Ghana Rufous-nosed Rat

4661 **Okapia**
Artiodactyla - Giraffidae
e Okapis
d Okapis
f Okapis
i Okapi

4662 **Okapia johnstoni**
Artiodactyla - Giraffidae
e Okapi
d Okapi; Waldgiraffe
f Okapi; Okapia
i Okapi

4663 **Olallamys**
Rodentia - Echimyidae
e Olalla Rats

4664 **Olallamys albicauda**
Rodentia - Echimyidae
e White-tailed Olalla Rat

4665 **Olallamys edax**
Rodentia - Echimyidae
e Greedy Olalla Rat

4666 **Oligoryzomys**
Rodentia - Muridae
e Pygmy Rice Rats
d Zwerggreisratten

4667 **Oligoryzomys andinus**
Rodentia - Muridae
e Andean Pygmy Rice Rat

4668 **Oligoryzomys arenalis**
Rodentia - Muridae
e Sandy Pygmy Rice Rat

4669 **Oligoryzomys chacoensis**
Rodentia - Muridae
e Chacoan Pygmy Rice Rat

4670 **Oligoryzomys delticola**
Rodentia - Muridae
e Delta Pygmy Rice Rat

4671 **Oligoryzomys destructor**
Rodentia - Muridae
e Destructive Pygmy Rice Rat

4672   *Oligoryzomys eliurus*
Rodentia - Muridae
e   Brazilian Pygmy Rice Rat

4673   *Oligoryzomys flavescens*
Rodentia - Muridae
e   Yellow Pygmy Rice Rat

4674   *Oligoryzomys fulvescens*
Rodentia - Muridae
e   Fulvous Pygmy Rice Rat; Northern
Pygmy Rice Rat

4675   *Oligoryzomys griseolus*
Rodentia - Muridae
e   Greyish Pygmy Rice Rat

4676   *Oligoryzomys longicaudatus*
Rodentia - Muridae
e   Long-tailed Pygmy Rice Rat

4677   *Oligoryzomys magellanicus*
Rodentia - Muridae
e   Magellanic Pygmy Rice Rat

4678   *Oligoryzomys microtis*
Rodentia - Muridae
e   Small-eared Pygmy Rice Rat

4679   *Oligoryzomys nigripes*
Rodentia - Muridae
e   Black-footed Pygmy Rice Rat

4680   *Oligoryzomys stramineus*
Rodentia - Muridae
e   Straw-coloured Pygmy Rice Rat

4681   *Oligoryzomys vegetus*
Rodentia - Muridae
e   Sprightly Pygmy Rice Rat

4682   *Oligoryzomys victus*
Rodentia - Muridae
e   St. Vincent Pygmy Rice Rat

4683   *Olisthomys*
Rodentia - Sciuridae
e   Morris's Flying Squirrels

4684   *Olisthomys morrisi*
Rodentia - Sciuridae
e   Morris's Flying Squirrel

4685   *Ommatophoca*
Carnivora - Phocidae
e   Ross's Seals; Ross Seals
d   Ross-Robben
f   Phoques de Ross
i   Foche di Ross

4686   *Ommatophoca rossi*
Carnivora - Phocidae
e   Ross's Seal; Antarctic Seal; Big-eyed
Seal; Ross Seal
d   Ross-Robbe; Rossmeer-Robbe
f   Phoque de Ross
i   Foca di Ross

Oncifelis syn. *Felis* q.v.

4687   *Ondatra*
Rodentia - Muridae
e   Muskrats
d   Bisamratten
f   Ondatras; Rats musqués
i   Ondatre

4688   *Ondatra zibethicus*
Rodentia - Muridae
e   Muskrat; Musquash
d   Bisamratte; Bisambiber;
Wohlriechende Wasserratte;
Zibetratte
f   Ondatra; Rat musqué
i   Ondatra ; Topo muschiato

4689   *Ondatra zibethicus ripensis*
Rodentia - Muridae
e   Pecos River Muskrat

Onotragus syn. *Kobus* q.v.

4690 Onychogalea
Diprotodontia - Macropodidae
e Nail-tailed Wallabies; Bridled Nail-tailed Wallabies; Silky Wallabies
d Kurznagelkängurus; Nagelkängurus
f Onychogales; Wallabys à queue cornue

4691 Onychogalea fraenata
Diprotodontia - Macropodidae
e Nail-tailed Wallaby; Bridled Nail-tailed Wallaby; Bridled Wallaby; Bridled Nailtail Wallaby (ANZ)
d Kurznagelkänguru; Zügelkänguru
f Wallaby bridé; Wallaby bridé à queue cornue
i Uallabia delle briglie

4692 Onychogalea lunata
Diprotodontia - Macropodidae
e Crescent Wallaby (ANZ); Crescent-marked Wallaby; Crescent Nail-tailed Wallaby; Lunated Nail-tailed Wallaby; Wurrung; Crescent Nailtail Wallaby
d Mondnagelkänguru; Mondnagelschwanz-Känguru
f Wallaby à queue cordée; Wallaby à queue ongle-de-lune; Onychogale à queue ongle-de-lune; Wallaby à queue cornue
i Uallabia dall'unghia lunata

4693 Onychogalea unguifera
Diprotodontia - Macropodidae
e Northern Nail-tailed Wallaby (ANZ); Sandy Nail-tailed Wallaby
d Flachnagelkänguru
f Wallaby de Fawn à queue cornue

4694 Onychomys
Rodentia - Muridae
e Grasshopper Mice; Scorpion Mice
d Grashüpfermäuse
f Onychomys ; Souris sauterelles

4695 Onychomys arenicola
Rodentia - Muridae
e Mearns's Grasshopper Mouse

4696 Onychomys leucogaster
Rodentia - Muridae
e Northern Grasshopper Mouse
d Nördliche Grashüpfermaus
f Onychomys du Nord; Souris sauterelle

4697 Onychomys torridus
Rodentia - Muridae
e Southern Grasshopper Mouse; Southern Scorpion Mouse
d Südliche Grashüpfermaus
f Onychomys du Sud

4698 Onychomys torridus tularensis
Rodentia - Muridae
e Tulare Grasshopper Mouse

4699 Orcaella
Cete - Delphinidae
e Irrawaddy Dolphins; Irrawaddy River Dolphins
d Irrawaddi-Delfine
f Orcelles
i Orcelle

4700 Orcaella brevirostris
Cete - Delphinidae
e Irrawaddy Dolphin; Irrawaddy River Dolphin; Snubfin Dolphin
d Irrawaddi-Delfin
f Orcelle de l'Irrawaddy; Orcelle; Orcelle de l'Irraouaddi; Orcelle de l'Iraouadi
i Orcella

4701 Orcinus
Cete - Delphinidae
e Killer Whales
d Schwertwale

f   Épaulards; Orques; Orques épaulards

i   Orche

4702   Orcinus orca
Cete - Delphinidae

e   Orca; Common Killer Whale; Killer Whale; Great Killer Whale

d   Schwertwal; Mörderdelfin; Mörderwal; Butzkopf; Orca; Großer Schwertwal

f   Épaulard; Orque épaulard; Gladiateur

i   Orca; Balena assassina; Delfino gladiatore

Oreailurus syn. *Felis* q.v.

4703   Oreamnos
Artiodactyla - Bovidae

e   Mountain Goats (NA); North American Mountain Goats

d   Schneeziegen

f   Chèvres des montagnes

i   Capre di montagna

4704   Oreamnos americanus
Artiodactyla - Bovidae

e   Mountain Goat (NA); North American Mountain Goat; Mountains Goat

d   Schneeziege; Richtige Schneegemse

f   Chèvre des montagnes; Chèvre des Montagnes Rocheuses

i   Capra di montagna; Capra delle nevi

4705   Oreonax
Primates - Atelidae

e   Yellow-tailed Woolly Monkeys

d   Gelbschwanzwollaffen

f   Singes laineux à queue jaune

i   Scimmie lanose dalla coda gialla

4706   Oreonax flavicauda
Primates - Atelidae

e   Hendee's Woolly Monkey; Peruvian Mountains Woolly Monkey; Yellow-tailed Woolly Monkey

d   Gelbschwanzwollaffe

f   Singe laineux à queue jaune; Lagotriche à queue jaune

i   Scimmia lanosa dalla coda gialla; Scimmia lanosa di Hener

4707   Oreotragus
Artiodactyla - Bovidae

e   Klipspringers; Klipboks

d   Klippspringer

f   Oréotragues

i   Saltarupe

4708   Oreotragus oreotragus
Artiodactyla - Bovidae

e   Klipspringer; Klipbok

d   Klippspringer

f   Oréotrague

i   Saltarupe; Sassa

Orientallac syn. *Allactaga* q.v.

4709   Ornithorhynchidae
Monotremata -

e   Platypuses; Duck-billed Platypuses

d   Schnabeltiere; Schnabelsäuger

f   Ornythorhynques; Ornithorhynchidés

i   Ornitorinchidi

4710   Ornithorhynchus
Platypoda - Ornithorhynchidae

e   Duckbills; Duck-billed Platypuses; Platypuses

d   Schnabeltiere

f   Ornythorhynques

i   Ornitorinchi

4711   Ornithorhynchus anatinus
Platypoda - Ornithorhynchidae

e   Duckbill; Duckmole; Platypus; Duck-billed Platypus

d   Schnabeltier; Platypus

f   Ornythorhynque

i   Ornitorinco

4712   **Orthogeomys**
        Rodentia - Geomyidae
e   Giant Pocket Gophers; Tuzas
d   Riesentaschenratten
f   Geomys

4713   **Orthogeomys cavator**
        Rodentia - Geomyidae
e   Chiriqui Pocket Gopher

4714   **Orthogeomys cherriei**
        Rodentia - Geomyidae
e   Cherrie's Pocket Gopher

4715   **Orthogeomys cuniculus**
        Rodentia - Geomyidae
e   Oaxacan Pocket Gopher

4716   **Orthogeomys dariensis**
        Rodentia - Geomyidae
e   Darien Pocket Gopher

4717   **Orthogeomys grandis**
        Rodentia - Geomyidae
e   Tuza; Giant Pocket Gopher; Large
        Pocket Gopher
d   Hamsterratte; Riesentaschenratte

4718   **Orthogeomys heterodus**
        Rodentia - Geomyidae
e   Variable Pocket Gopher

4719   **Orthogeomys hispidus**
        Rodentia - Geomyidae
e   Hispid Pocket Gopher

4720   **Orthogeomys lanius**
        Rodentia - Geomyidae
e   Big Pocket Gopher

4721   **Orthogeomys matagalpae**
        Rodentia - Geomyidae
e   Nicaraguan Pocket Gopher

4722   **Orthogeomys thaeleri**
        Rodentia - Geomyidae
e   Thaeler's Pocket Gopher

4723   **Orthogeomys underwoodi**
        Rodentia - Geomyidae
e   Underwood's Pocket Gopher

4724   **Orycteropodidae**
        Tubulidentata
e   Aardvarks; Earth Hogs; Earth Bears
d   Erdferkel
f   Oryctéropidés
i   Oritteropidi

4725   **Orycteropus**
        Tubulidentata - Orycteropodidae
e   Aardvarks; Earth Hogs; Earth Bears;
        Ant Bears
d   Erdferkel
f   Oryctéropes; Fourmilliers
i   Oritteropi

4726   **Orycteropus afer**
        Tubulidentata - Orycteropodidae
e   Aardvark; Earth Hog; Earth Bear;
        Ant Bear
d   Erdferkel
f   Oryctérope; Fourmillier
i   Oritteropo; Orso delle formiche;
        Maiale delle formiche

4727   **Oryctolagus**
        Lagomorpha - Leporidae
e   Rabbits; Old World Rabbits
d   Altweltliche Wildkaninchen ;
        Wildkaninchen
f   Lapins typiques
i   Conigli selvatici

4728   **Oryctolagus cuniculus**
        Lagomorpha - Leporidae
e   European Rabbit; Old World Rabbit;
        European Grey Rabbit; Common
        Rabbit; Rabbit

d Europäisches Wildkaninchen;
   Mitteleuropäisches Wildkaninchen;
   Europäisches Kaninchen; Kanickel;
   Wildkaninchen; Kaninchen

f Lapin; Lapin commun; Lapin
   ordinaire; Lapin vulgaire; Lapin des
   bois; Lièvre lapin; Lièvre de garenne;
   Lapin de garenne

i Coniglio selvatico; Coniglio
   domestico; Conniglio

4729 *Oryctolagus cuniculus huxleyi*
       Lagomorpha - Leporidae

e Wild Rabbit

d Porto-Santo-Kaninchen

i Coniglio selvatico

4730 *Oryx*
       Artiodactyla - Bovidae

e Oryxes; Gemsboks; Gemsbuks

d Spießböcke; Oryx-Antilopen

f Oryx

i Orici

4731 *Oryx dammah*
       Artiodactyla - Bovidae

e Scimitar Oryx; Scimitar-horned
   Oryx; Saharan Oryx

d Säbelantilope

f Oryx algazelle; Oryx blanc; Oryx de
   Libye

i Orice dalle corna a sciabola; Orice
   dalle corna a scimitarra; Orice
   algazel

4732 *Oryx gazella*
       Artiodactyla - Bovidae

e Oryx; Gemsbok

d Oryx; Spießbock; Säbelantilope

f Oryx; Oryx Gemsbock

i Orice gazella; Antilope camoscio;
   Orice

4733 *Oryx gazella beisa*
       Artiodactyla - Bovidae

e Beisa; Beisa Oryx

d Ostafrikanischer Spießbock; Eritrea
   Spießbock; Ostafrikanischer Oryx;
   Beisa Antilope; Nordafrikanische
   Beisa-Antilope

f Oryx beisa; Oryx d'Afrique de l'Est

i Orice beisa

4734 *Oryx gazella blainei*
       Artiodactyla - Bovidae

e Angolan Gemsbok

4735 *Oryx gazella callotis*
       Artiodactyla - Bovidae

e Fringe-eared Oryx

d Büschelohrspießbock

f Oryx aux oreilles frangées

4736 *Oryx gazella gazella*
       Artiodactyla - Bovidae

e Gemsbok

d Südafrikanischer Spießbock

f Gemsbok

i Orice del Kalahari

4737 *Oryx leucoryx*
       Artiodactyla - Bovidae

e Arabian Oryx

d Weiße Oryx

f Oryx d'Arabie

i Orice bianco d'Arabia; Orice
   d'Arabia; Orice araba; Orice bianco

Oryx tao syn. *O. dammah* q.v.

4738 *Oryzomys*
       Rodentia - Muridae

e Rice Rats

d Reisratten

f Rats du riz

i Ratti del riso

4739 *Oryzomys albigularis*
       Rodentia - Muridae

e Tomes's Rice Rat; Mount Pirri Rice
   Rat

4740   ***Oryzomys alfaroi***
        Rodentia - Muridae
    e Alfaro's Rice Rat

4741   ***Oryzomys auriventer***
        Rodentia - Muridae
    e Ecuadorean Rice Rat

4742   ***Oryzomys balneator***
        Rodentia - Muridae
    e Peruvian Rice Rat

4743   ***Oryzomys bolivaris***
        Rodentia - Muridae
    e Bolivar Rice Rat; Long-Whiskered
        Rice Rat

4744   ***Oryzomys buccinatus***
        Rodentia - Muridae
    e Paraguayan Rice Rat

        Oryzomys capito syn. *O.*
        megacephalus q.v.

4745   ***Oryzomys caracolus***
        Rodentia - Muridae
    e Caracol Rice Rat

4746   ***Oryzomys chapmani***
        Rodentia - Muridae
    e Chapman's Rice Rat

4747   ***Oryzomys couesi***
        Rodentia - Muridae
    e Coués's Rice Rat
    d Coués-Reisratte
    f Rat du riz de Coués

4748   ***Oryzomys devius***
        Rodentia - Muridae
    e Boquete Rice Rat; Montane Rice Rat

4749   ***Oryzomys dimidiatus***
        Rodentia - Muridae

    e Thomas's Rice Rat; Thomas's Water
        Rat; Nicaraguan Rice Rat

4750   ***Oryzomys emmonsae***
        Rodentia - Muridae
    e Emmons's Rice Rat

4751   ***Oryzomys galapagoensis***
        Rodentia - Muridae
    e Galapagos Rice Rat; Galapagos
        Islands Rice Rat
    d Galapagos-Inseln Reisratte;
        Galapagos-Reisratte

    f Rat du riz des îles Galapagos; Rat du
        riz des Galapagos

        Oryzomys gatunensis syn. *O. couesi*
        q.v.

4752   ***Oryzomys gorgasi***
        Rodentia - Muridae
    e Gorgas's Rice Rat

4753   ***Oryzomys hammondi***
        Rodentia - Muridae
    e Hammond's Rice Rat

4754   ***Oryzomys intectus***
        Rodentia - Muridae
    e Colombian Rice Rat

4755   ***Oryzomys intermedius***
        Rodentia - Muridae
    e Intermediate Rice Rat

4756   ***Oryzomys keaysi***
        Rodentia - Muridae
    e Keay's Rice Rat

4757   ***Oryzomys kelloggi***
        Rodentia - Muridae
    e Kellogg's Rice Rat

4758   *Oryzomys lamia*
            Rodentia - Muridae
e   Monster Rice Rat

            Oryzomys legatus syn. *O. russatus*
            q.v.

4759   *Oryzomys levipes*
            Rodentia - Muridae
e   Light-footed Rice Rat

4760   *Oryzomys macconnelli*
            Rodentia - Muridae
e   MacConnell's Rice Rat

4761   *Oryzomys megacephalus*
            Rodentia - Muridae
e   Large-headed Rice Rat; Terrestial
            Rice Rat

4762   *Oryzomys melanotis*
            Rodentia - Muridae
e   Black-eared Rice Rat

4763   *Oryzomys nelsoni*
            Rodentia - Muridae
e   Nelson's Rice Rat; Maria Madre
            Island Rice Rat; Maria Madre Rice
            Rat
d   Maria-Madre-Insel-Reisratte; Maria-
            Madre-Reisratte; Nelson-Reisratte
f   Rat du riz de Nelson; Rat du riz de
            l'île Maria Madre; Rat du riz de
            Maria Madre

4764   *Oryzomys nitidus*
            Rodentia - Muridae
e   Elegant Rice Rat

4765   *Oryzomys oniscus*
            Rodentia - Muridae
e   Sowbug Rice Rat

4766   *Oryzomys palustris*
            Rodentia - Muridae

e   Marsh Rice Rat
d   Sumpfreisratte; Amerikanische
            Wasserratte
f   Rat du riz

4767   *Oryzomys palustris natator*
            Rodentia - Muridae
e   Silver Rice Rat

4768   *Oryzomys palustris sanibeli*
            Rodentia - Muridae
e   Sanibel Island Rice Rat

            Oryzomys pirrensis syn. *O.
            albigularis* q.v.

4769   *Oryzomys polius*
            Rodentia - Muridae
e   Grey Rice Rat

            Oryzomys pyrrhorhinos syn.
            Wiedomys pyrrhorhinos q.v.

4770   *Oryzomys ratticeps*
            Rodentia - Muridae
e   Rat-headed Rice Rat

4771   *Oryzomys rhabdops*
            Rodentia - Muridae
e   Striped Rice Rat

            Oryzomys rivularis syn. *O. couesi*
            q.v.

4772   *Oryzomys rostratus*
            Rodentia - Muridae
e   Long-nosed Rice Rat; Rusty Rice Rat

4773   *Oryzomys russatus*
            Rodentia - Muridae
e   Big-headed Rice Rat

            Oryzomys russulus syn. *O. alfaroi*
            q.v.

4774 **Oryzomys saturatior**
Rodentia - Muridae
e Cloud Forest Rice Rat

4775 **Oryzomys seuanezi**
Rodentia - Muridae
e Seuánez's Rice Rat

Oryzomys simplex syn.
Pseudoryzomys simplex q.v.

4776 **Oryzomys subflavus**
Rodentia - Muridae
e Terraced Rice Rat

4777 **Oryzomys talamancae**
Rodentia - Muridae
e Talamancan Rice Rat

4778 **Oryzomys tatei**
Rodentia - Muridae
e Tate's Rice Rat

4779 **Oryzomys xantheolus**
Rodentia - Muridae
e Yellowish Rice Rat

4780 **Oryzomys yunganus**
Rodentia - Muridae
e Yungas Rice Rat

4781 **Oryzorictes**
Soricomorpha - Tenrecidae
e Rice Tenrecs; Mole Tanrecs
d Reistanreks; Reiswühler
f Tenrecs

4782 **Oryzorictes hova**
Soricomorpha - Tenrecidae
e Hova Rice Tenrec; Mole-like Rice Tanrec; Rice Tenrec; Mole Tenrec
d Maulwurfartiger Reistanrek

Oryzorictes talpoides syn. *O. hova* q.v.

4783 **Oryzorictes tetradactylus**
Soricomorpha - Tenrecidae
e Four-toed Rice Tenrec
d Vierfingertanrek

4784 **Osbornictis**
Carnivora - Viverridae
e Congo Water Civets; Water Civets; Aquatic Genets
d Wasserschleichkatzen; Wasserzivetten
f Genettes aquatiques
i Genette acquatiche

4785 **Osbornictis piscivora**
Carnivora - Viverridae
e Congo Water Civet; Water Civet; Aquatic Genet; Fishing Genet
d Wasserschleichkatze; Wasserzivette
f Genette aquatique
i Genetta acquatica; Civetta acquatica

4786 **Osgoodomys**
Rodentia - Muridae
e Michoacan Deer Mice
d Michoacan-Hirschmäuse

4787 **Osgoodomys banderanus**
Rodentia - Muridae
e Michoacan Deer Mouse
d Michoacan-Hirschmaus

Osphranter syn. *Macropus* q.v.

4788 **Otaria**
Carnivora - Otariidae
e South American Sea Lions; Southern Sea Lions; Patagonian Sea Lions
d Mähnenrobben
f Otaries à crinière; Otaries à crinière d'Amérique du Sud
i Leoni marini sudamericani; Foche dalle orecchie

Otaria byronia syn. *O. flavescens* q.v.

4789   *Otaria flavescens*
Carnivora - Otariidae
- *e*   South American Sea Lion; Southern Sea Lion; Patagonian Sea Lion
- *d*   Mähnenrobbe; Peruanischer Seelöwe; Südamerikanischer Seelöwe
- *f*   Otarie à crinière; Otarie à crinière d'Amérique du Sud
- *i*   Otaria della criniera; Leone marino meridionale; Leone marino sudamericano; Foca delle rocce

4790   Otariidae
Carnivora
- *e*   Eared Seals; Sea Lions and Fur Seals; Sea Lions; Eared Seals and Sealions
- *d*   Ohrenrobben
- *f*   Otariidés
- *i*   Otaridi; Arctocefali

Othriomys syn. *Microtus* q.v.

Otocolobus syn. *Felis* q.v.

4791   *Otocyon*
Carnivora - Canidae
- *e*   Big-eared Foxes; Bat-eared Foxes
- *d*   Löffelhunde; Löffelfüchse
- *f*   Chiens oreillards; Otocyons; Renards aux oreilles de chauve-souris
- *i*   Oreciioni

4792   *Otocyon megalotis*
Carnivora - Canidae
- *e*   African Big-eared Fox; Big-eared Fox; Big-eared Dog; Bat-eared Fox; Delandi's Fox; Cape Fox; Motlosi
- *d*   Löffelhund; Löffelfuchs
- *f*   Chien oreillard; Otocyon; Renard aux oreilles de chauve-souris
- *i*   Otocione

4793   *Otocyon megalotis megalotis*
Carnivora – Canidae

- *e*   Northern Bat-eared Fox

4794   *Otocyon megalotis virgatus*
Carnivora - Canidae
- *e*   Southern Bat-eared Fox

4795   *Otolemur*
Primates - Loridae
- *e*   Greater Galagos; Greater Bushbabies; Thick-tailed Bushbabies
- *d*   Riesengalagos
- *f*   Galagos à queue touffue; Galagos à queue épaisse
- *i*   Galagoni giganti

4796   *Otolemur crassicaudatus*
Primates - Loridae
- *e*   Greater Galago; Greater Bushbaby; Thick-tailed Bushbaby; Thick-tailed Galago; Fat-tailed Bushbaby; Small-eared Bushbaby
- *d*   Riesengalago; Buschwaldgalago
- *f*   Galago à queue touffue; Galago à queue épaisse; Galago à grosse queue
- *i*   Galagone gigante; Galagone; Galago dalla coda grassa; Otolemure; Galagone maggiore

4797   *Otolemur crassicaudatus argentatus*
Primates - Loridae
- *e*   Black Galago; Silver Bushbaby

4798   *Otolemur crassicaudatus crassicaudatus*
Primates - Loridae
- *e*   Brown Galago; Brown Greater Galago

4799   *Otolemur garnetti*
Primates - Loridae
- *e*   Small-eared Galago; Garnett's Galago; Garnett's Greater Bushbaby; Garnett's Greater Galago; Northern Greater Galago; Garnett's Bushbaby
- *d*   Großer Galago; Garnetts Galago
- *f*   Grand galago
- *i*   Galagone di Garnett

4800 **Otomops**
Chiroptera - Molossidae
e Otomops Bats; Big-eared Free-tailed Bats
d Großohrfledermäuse

4801 **Otomops formosus**
Chiroptera - Molossidae
e Javan Mastiff Bat

4802 **Otomops johnstonei**
Chiroptera - Molossidae
e Aior Mastiff Bat

4803 **Otomops martiensseni**
Chiroptera - Molossidae
e Big-eared Free-tailed Bat; Martienssen's Free-tailed Bat; Giant Mastiff Bat; Large-eared Free-tailed Bat
d Martienssen-Großohrfledermaus

4804 **Otomops papuensis**
Chiroptera - Molossidae
e Big-eared Mastiff Bat; Papuan Mastiff Bat

4805 **Otomops secundus**
Chiroptera - Molossidae
e Mantled Mastiff Bat

4806 **Otomops wroughtoni**
Chiroptera - Molossidae
e Wroughton's Free-tailed Bat

4807 **Otomys**
Rodentia - Muridae
e Groove-toothed Rats; Swamp Rats; Vlei Rats; African Swamp Rats
d Ohrenratten; Lamellenzahnratten
f Rats des marais à queue courte; Rats des marais; Rats des marécages

4808 **Otomys anchietae**
Rodentia - Muridae
e Angolan Vlei Rat

4809 **Otomys angoniensis**
Rodentia - Muridae
e Angoni Vlei Rat

4810 **Otomys denti**
Rodentia - Muridae
e Dent's Vlei Rat

4811 **Otomys irroratus**
Rodentia - Muridae
e Vlei Rat; Common Vlei Rat; Saunders's Vlei Rat
d Afrikanische Lamellenzahnratte; Afrikanische Ohrenratte

4812 **Otomys laminatus**
Rodentia - Muridae
e Laminate Vlei Rat

4813 **Otomys maximus**
Rodentia - Muridae
e Large Vlei Rat

4814 **Otomys occidentalis**
Rodentia - Muridae
e Western Vlei Rat

Otomys saundersiae syn. *O. irroratus* q.v.

4815 **Otomys sloggetti**
Rodentia - Muridae
e Sloggett's Vlei Rat

4816 **Otomys tropicalis**
Rodentia - Muridae
e Tropical Vlei Rat

4817 **Otomys typus**
Rodentia - Muridae
e Typical Vlei Rat

4818   *Otomys uninsulcatus*
           Rodentia - Muridae
e   Bush Vlei Rat; African Swamp Rat

4819   *Otonycteris*
           Chiroptera - Vespertilionidae
e   Desert Long-eared Bats
d   Hemprichs Schlitznasenfledermäuse
f   Oreillards de Hemprich

4820   *Otonycteris hemprichi*
           Chiroptera - Vespertilionidae
e   Hemprich's Long-eared Bat;
      Hemprich's Arrow-eared Bat;
      Hemprich's Big-eared Bat; Desert
      Long-eared Bat
d   Hemprichs Schlitznasenfledermaus
f   Oreillard de Hemprich

4821   *Otonyctomys*
           Rodentia - Muridae
e   Yucatan Vesper Rats; Hatt's Vesper
      Rats
d   Yucatan-Vesperratten

4822   *Otonyctomys hatti*
           Rodentia - Muridae
e   Yucatan Vesper Rat; Hatt's Vesper
      Rat
d   Yucatan-Vesperratte; Yucatan-
      Abendratte

4823   *Otopteropus*
           Chiroptera - Pteropodidae
e   Luzon Fruit Bats

4824   *Otopteropus cartilagonodus*
           Chiroptera - Pteropodidae
e   Luzon Fruit Bat

Otosciurus syn. *Sciurus* q.v.

Otospermophilus syn. *Spermophilus*
q.v.

4825   *Ototylomys*
           Rodentia - Muridae
e   Big-eared Climbing Rats; Tree Rats;
      Big-eared Climbing Mice
d   Großohrkletterratten
f   Rats grimpants

4826   *Ototylomys phyllotis*
           Rodentia - Muridae
e   Big-eared Climbing Rat; Tree Rat;
      Big-eared Climbing Mouse
d   Großohrkletterratte
f   Rat grimpant

4827   *Ourebia*
           Artiodactyla - Bovidae
e   Oribis; Ourebies
d   Oribis; Bleichböckchen
f   Ourébis
i   Oribi

4830   *Ourebia ourebi kenyae*
           Artiodactyla - Bovidae
e   Kenya Oribi Antelope
d   Kenia-Oribi

4828   *Ourebia ourebia*
           Artiodactyla - Bovidae
e   Oribi; Gambian Oribi
d   Oribi; Bleichböckchen
f   Ourébi
i   Oribi

4829   *Ourebia ourebia haggardi*
           Artiodactyla - Bovidae
e   Haggard Oribi
d   Haggard-Oribi

4831   *Ovibos*
           Artiodactyla - Bovidae
e   Muskoxen; Musk Sheep
d   Moschusochsen
f   Boeufs musqués
i   Buoi muschiati

4832 *Ovibos moschatus*
Artiodactyla - Bovidae
e Muskox; Musk Sheep
d Bisamochs; Moschusochse; Östlicher Moschusochse; Schafochse
f Boeuf musqué
i Bue muschiato

4833 *Ovibos moschatus moschatus*
Artiodactyla - Bovidae
e Alaska Muskox
d Alaska-Moschusochse

4834 *Ovibos moschatus wardi*
Artiodactyla - Bovidae
e Greenland Muskox
d Grönland-Moschusochse

4835 *Ovis*
Artiodactyla - Bovidae
e Sheep ; Moutain Sheep
d Schafe; Wildschafe
f Moutons
i Pecore

4836 *Ovis ammon*
Artiodactyla - Bovidae
e Argali; Argali Sheep; Wild Sheep
d Archar; Argali; Europäischer Mufflon; Wildschaf; Riesenwildschaf
f Mouflon
i Muflone

4837 *Ovis ammon ammon*
Artiodactyla - Bovidae
e Altai Sheep
d Altai-Wildschaf
i Muflone asiatico; Argali

4838 *Ovis ammon bocharensis*
Artiodactyla - Bovidae
e Urical Mouflon; Bokhara Sheep
d Bokhara-Wildschaf

4839 *Ovis ammon cycloceros*
Artiodactyla - Bovidae
e Ustuyrt Mountain Sheep
d Kreishornschaf

4840 *Ovis ammon gmelini*
Artiodactyla - Bovidae
e Armenian Sheep
d Armenisches Wildschaf

4841 *Ovis ammon hodgsoni*
Artiodactyla - Bovidae
e Himalajan Mouflon; Great Tibetan Sheep
d Himalaja-Schaf
f Mouflon d'Asie; Mouflon des montagnes
i Muflone dell'Himalaja

4842 *Ovis ammon nigrimontana*
Artiodactyla - Bovidae
e Kara Tau; Kara Tau Argali
d Kara-Tau-Wildschaf
f Mouflon de Karatau

4843 *Ovis ammon polii*
Artiodactyla - Bovidae
e Marco Polo Sheep; Great Pamir Sheep
d Pamir-Wildschaf
f Argali de Marco Polo
i Pecora di Marco Polo; Argali de Pamir

4844 *Ovis aries*
Artiodactyla - Bovidae
e Sheep; Domestic Sheep
d Hausschaf
f Mouflon européen
i Pecora domestica

4845 *Ovis aries ophion*
Artiodactyla - Bovidae
e Red Sheep; Red Mouflon

f   Mouflon de Chypre
i   Muflone di Cipro

4846   *Ovis aries platyura*
Artiodactyla - Bovidae
e   Persian Lamb
d   Breitschwanz
f   Mouflon de Perse
i   Agnello di Persia; Karakul; Karakul
afghano

4847   *Ovis aries steatopyga*
Artiodactyla - Bovidae
e   Calgan Lamb
i   Kalgan

4848   *Ovis canadensis*
Artiodactyla - Bovidae
e   American Bighorn; Bighorn Sheep
(NA); Mountain Sheep; Mexican
Bighorn Sheep; Desert Bighorn
d   Dickhornschaf; Felsengebirgs-Schaf
f   Mouflon d'Amérique; Mouflon du
Canada; Mouflon du Mexique
i   Pecora delle Montagne Rocciose;
Bighorn

4849   *Ovis canadensis auduboni*
Artiodactyla - Bovidae
e   Badlands Bighorn Sheep (NA);
Audubon's Bighorn Sheep; American
Bighorn Sheep; Badlands American
Bighorn; Audubon's Mountain Sheep
d   Audubons Dickhornschaf; Badlands
Schneeschaf; Audubon-
Dickhornschaf; Audubon-
Schneeschaf
f   Mouton Bighorn d'Audubon;
Mouflon Bighorn d'Audubon;
Mouflon du Canada d'Audubon;
Mouton du Canada d'Audubon

4850   *Ovis canadensis californiana*
Artiodactyla - Bovidae
e   Rimrock Bighorn Sheep (NA);
California Bighorn Sheep; Sierra
Nevada Bighorn Sheep

4851   *Ovis canadensis canadensis*
Artiodactyla - Bovidae
e   Rocky Mountains Sheep; Rocky
Mountains Bighorn Sheep
d   Felsengebirge-Dickhornschaf
f   Mouflon des Montagnes Rocheuses

4852   *Ovis canadensis cremnobates*
Artiodactyla - Bovidae
e   Peninsular Bighorn Sheep
d   Halbinsel-Dickhornschaf

4853   *Ovis dalli*
Artiodactyla - Bovidae
e   Dall Sheep; White Sheep; Dall's
Sheep
d   Weißes Dickhornschaf; Dall-Schaf;
Alaska-Schneeschaf
f   Mouflon de Dall
i   Pecora di Dall

4854   *Ovis nivicola*
Artiodactyla - Bovidae
e   Siberian Bighorn; Snow Sheep
d   Kamschatka-Dickhornschaf;
Kamschatka-Schneeschaf;
Schneeschaf; Sibirisches
Schneeschaf
f   Mouflon des neiges
i   Pecora delle nevi

4855   *Ovis orientalis*
Artiodactyla - Bovidae
e   Mouflon
d   Orientalisches Wildschaf; Mufflon;
Muffelwild
f   Mouflon méditerranéen; Mouflon de
Corse; Mouflon de Sardaigne;
Mouflon d'Europe
i   Muflone; Argali caucaso; Argali del
Caucaso

4856   *Ovis orientalis ophion*
Artiodactyla - Bovidae
e   Cyprus Mouflon; Cyprian Mouflon;
Cyprian Wild Sheep

d   Zypern-Mufflon; Zyprisches
Mufflon; Zypern-Wildschaf
f   Mouflon de Chypre
i   Muffione di Sardegna; Muflone

4857   *Ovis vignei*
Artiodactyla - Bovidae
e   Urial; Asiatic Wild Sheep; Shapo
d   Asiatisches Steppenschaf; Urial;
Urial-Schaf
f   Urial
i   Pecora della steppa

4858   *Ovis vignei vignei*
Artiodactyla - Bovidae
e   Kashmir Mouflon; Ladakh Urial
d   Ladakh-Schaf
f   Mouflon de Kashmir; Urial de Ladak
i   Muflone del Kashmir

4859   *Oxymycterus*
Rodentia - Muridae
e   Burrowing Mice; South American
Burrowing Mice; Hocicudos
d   Südamerikanische Grabmäuse;
Langnasenmäuse

4860   *Oxymycterus akodontius*
Rodentia - Muridae
e   Argentine Hocicudo

4861   *Oxymycterus amazonicus*
Rodentia - Muridae
e   Amazon Hocicudo

4862   *Oxymycterus angularis*
Rodentia - Muridae
e   Angular Hocicudo

4863   *Oxymycterus caparaoe*
Rodentia - Muridae
e   Caparao Hocicudo

4864   *Oxymycterus delator*
Rodentia - Muridae

e   Spy Hocicudo

4865   *Oxymycterus hiska*
Rodentia - Muridae
e   Small Hocicudo

4866   *Oxymycterus hispidus*
Rodentia - Muridae
e   Hispid Hocicudo

4867   *Oxymycterus hucucha*
Rodentia - Muridae
e   Quechuan Hocicudo

Oxymycterus iheringi syn.
Brucepattersonius iheringi q.v.

4868   *Oxymycterus inca*
Rodentia - Muridae
e   Incan Hocicudo
d   Inka-Grabmaus

4869   *Oxymycterus nasutus*
Rodentia - Muridae
e   Long-nosed Hocicudo

4870   *Oxymycterus paramensis*
Rodentia - Muridae
e   Paramo Hocicudo

4871   *Oxymycterus roberti*
Rodentia - Muridae
e   Robert's Hocicudo

4872   *Oxymycterus rufus*
Rodentia - Muridae
e   Red Hocicudo
f   Souris à museau rouge

Oxymycterus rutilans syn.
Oxymycterus rufus q.v.

4873   *Ozotoceros*
Artiodactyla - Cervidae

    e   Pampas Deer
    d   Pampashirsche; Kamphirsche
    f   Cerfs des Pampas
    i   Cervi delle Pampas

4874   ***Ozotoceros bezoarticus***
       Artiodactyla - Cervidae
    e   Pampas Deer; Venado
    d   Pampashirsch; Kamphirsch
    f   Cerf des Pampas
    i   Cervo delle Pampas

P

4875 Pachycrocuta
Carnivora - Hyaenidae
e Brown Hyaenas
d Schabrackenhyänen
f Hyènes brunes
i Iene brune

4876 Pachycrocuta brunnea
Carnivora - Hyaenidae
e Brown Hyena; Beach Wolf; Strand Wolf
d Schabrackenhyäne; Braune Hyäne
f Hyène brune
i Iena bruna

Pachyura syn. *Suncus* q.v.

4877 Pachyuromys
Rodentia - Muridae
e Fat-tailed Gerbils; Fat-tailed Mice; Duprasis
d Dickschwanzmäuse
f Souris à grosse queue; Gerbilles à grosse queue
i Topi dalla coda grassa; Gerbilli della coda grassa

4878 Pachyuromys duprasi
Rodentia - Muridae
e Fat-tailed Gerbil; Fat-tailed Mouse; Duprasi; Hammer-tailed Gerbil
d Dickschwanzmaus; Fettschwanzrennmaus
f Souris à grosse queue; Souris à queue en massue
i Topo dalla coda grassa; Gerbillo della coda grassa; Duprasi

4879 Pachyuromys duprasi natronensis
Rodentia - Muridae
e Egyptian Fat-tailed Gerbil; Egyptian Duprasi

4880 Pagophilus
Carnivora - Phocidae
e Harp Seals
d Grönland-Robben
f Phoques du Groenland
i Foche della Groenlandia

4881 Pagophilus groenlandica
Carnivora - Phocidae
e Saddleback; Harp Seal; Saddle-backed Seal; Greenland Seal; Fiddle-backed Seal
d Grönland-Robbe; Sattelrobbe; Grönländischer Seehund; Mondfleckiger Seehund
f Phoque du Groenland
i Foca della Groenlandia

4882 Paguma
Carnivora - Viverridae
e Himalayan Palm Civets; Masked Palm Civets
d Larvenroller
f Civettes palmistes à masque
i Civette delle palme mascherata

4883 Paguma larvata
Carnivora - Viverridae
e Himalayan Palm Civet; Masked Palm Civet; Gem-faced Civet
d Larvenroller
f Civette palmiste à masque
i Civetta delle palme mascherata; Zibetto mascherato

4884 Paguma larvata taivana
Carnivora - Viverridae
e Formosan Gem-faced Cat

4885 Paguma larvata tytlerii
Carnivora - Viverridae
e Andaman Masked Palm Civet

4886 *Paguma larvata wroughtoni*
Carnivora - Viverridae
e Kashmir Masked Palm Civet
d Kaschmir-Larvenroller

4887 *Palawanomys*
Rodentia - Muridae
e Palawan Soft-furred Mountain Rats
d Palawan-Ratten

4888 *Palawanomys furvus*
Rodentia - Muridae
e Palawan Soft-furred Mountain Rat;
     Palawan Rat
d Palawan-Ratte

     Pallasiinus syn. *Microtus* q.v.

     Pallasiomys syn. *Meriones* q.v.

4889 *Pan*
Primates - Hominidae
e Chimpanzees
d Schimpansen
f Chimpanzés
i Scimpanzi

4890 *Pan paniscus*
Primates - Hominidae
e Pygmy Chimpanzee; Bonobo
d Zwergschimpanse; Bonobo
f Chimpanzé nain; Bonobo;
     Chimpanzé pygmée
i Bonobo; Scimpanzé pigmeo

4891 *Pan troglodytes*
Primates - Hominidae
e Chimpanzee
d Schimpanse
f Chimpanzé
i Scimpanzé

4892 *Pan troglodytes schweinfurthi*
Primates - Hominidae
e Eastern Chimpanzee; Schweinfurth's

Chimpsanzee
d Ostafrikanischer Schimpanse
f Chimpanzé aux poils longs
i Scimpanzé orientale

4893 *Pan troglodytes troglodytes*
Primates - Hominidae
e Central Common Chimpanzee;
     Common Chimpanzee; Tschego;
     Koolakomba; Black-skin
     Chimpanzee
d Zentralafrikanischer Schimpanse
f Chimpanzé tschego
i Scimpanzé

4894 *Pan troglodytes vellerosus*
Primates - Hominidae
e Nigerian Chimpanzee
d Ostnigeria-Westkamerun-
     Schimpanse
f Chimpanzé de Nigéria

4895 *Pan troglodytes verus*
Primates - Hominidae
e Western Common Chimpanzee
d Westafrikanischer Schimpanse
f Chimpanzé verus

4896 *Panthera*
Carnivora - Felidae
e Big Cats
d Panther; Eigentliche Großkatzen
f Panthères
i Grandi felini

4897 *Panthera leo*
Carnivora - Felidae
e Lion
d Löwe
f Lion; Lion d'Afrique
i Leone

4898 *Panthera leo azandica*
Carnivora - Felidae
e Congo Lion

d Kongo-Löwe
f Lion du Nord-Est du Congo
i Leone del Congo

4899 *Panthera leo bleyenberghi*
Carnivora - Felidae
e Angola Lion; Katanga Lion
d Angola-Löwe; Katanga-Löwe
f Lion de Katanga
i Leone del Katanga

4900 *Panthera leo goojiratensis*
Carnivora - Felidae
e Indian Lion
d Indischer Löwe
f Lion d'Inde
i Leone indiano

4901 *Panthera leo krugeri*
Carnivora - Felidae
e Transvaal Lion; South African Lion
d Transvaal-Löwe
f Lion du Transvaal; Lion blanc
i Leone del transvaal

4902 *Panthera leo leo*
Carnivora - Felidae
e Barbary Lion; African Lion
d Berber-Löwe; Atlas-Löwe
f Lion de l'Atlas; Lion de Barbarie;
Lion d'Afrique du Nord
i Leone di Barberia; Leone dell'Atlante

4903 *Panthera leo massaicus*
Carnivora - Felidae
e Massai Lion
d Massai-Löwe
f Lion des Massaïs
i Leone dei Masai

4904 *Panthera leo melanochaita*
Carnivora - Felidae
e Cape Lion; Black-maned Lion

d Kap-Löwe
f Lion du Cap
i Leone del Capo

4905 *Panthera leo nubica*
Carnivora - Felidae
e Black-maned Nubian Lion; East
African Lion
d Ostafrikanischer Löwe
f Lion de l'Afrique de l'Est
i Leone nubico

4906 *Panthera leo persica*
Carnivora - Felidae
e Asian Lion; Asiatic Lion
d Persischer Löwe; Asiatischer Löwe
f Lion d'Asie; Lion de Perse
i Leone asiatico

4907 *Panthera leo senegalensis*
Carnivora - Felidae
e Senegalese Lion; Senegal Lion
d Senegal-Löwe
f Lion du Sénégal
i Leone del Senegal

4908 *Panthera leo somaliensis*
Carnivora - Felidae
e Somali Lion
d Somalischer Löwe
f Lion de Somalie

***Panthera nebulosa* syn.** *Neofelis
nebulosa* q.v.

4909 *Panthera onca*
Carnivora - Felidae
e Jaguar
d Jaguar
f Jaguar
i Giaguaro

4910 *Panthera onca arizonensis*
Carnivora – Felidae

e   Arizona Jaguar; Jaguar (NA)
d   Arizona-Jaguar

4911   *Panthera onca centralis*
       Carnivora - Felidae
e   Panama Jaguar
d   Panama-Jaguar

4912   *Panthera onca goldmani*
       Carnivora - Felidae
e   Yucatan Jaguar
d   Yukatan-Jaguar

4913   *Panthera onca hernandesi*
       Carnivora - Felidae
e   Mexican Jaguar

4914   *Panthera onca onca*
       Carnivora - Felidae
e   Amazon Jaguar; Amazon Leopard
d   Amazonas-Jaguar

4915   *Panthera onca palustris*
       Carnivora - Felidae
e   Paraguay Jaguar
d   Parana-Jaguar

4916   *Panthera onca peruviana*
       Carnivora - Felidae
e   Peruvian Jaguar
d   Peru-Jaguar

4917   *Panthera pardus*
       Carnivora - Felidae
e   Panther; Leopard
d   Leopard
f   Léopard; Panthère
i   Leopardo; Pantera

4918   *Panthera pardus adersi*
       Carnivora - Felidae
e   Zanzibar Leopard; Zanzibar Island
       Leopard; Zanzibar Island Panther;
       Zanzibar

d   Sansibar-Leopard
f   Léopard de l'île Zanzibar; Léopard de
       Zanzibar; Panthère de l'île Zanzibar;
       Panthère de Zanzibar
i   Leopardo di Zanzibar

4919   *Panthera pardus adusta*
       Carnivora - Felidae
e   Ethiopian Leopard

4920   *Panthera pardus antinorii*
       Carnivora - Felidae
e   Eritrean Leopard
d   Eritrea-Leopard

4921   *Panthera pardus chui*
       Carnivora - Felidae
e   Uganda Leopard
d   Uganda-Leopard

4922   *Panthera pardus ciscaucasica*
       Carnivora - Felidae
e   Caucasus Leopard
d   Kaukasus-Leopard
i   Leopardo caucasico; Leopardo del
       Caucaso

4923   *Panthera pardus dathei*
       Carnivora - Felidae
e   Iranian Leopard
d   Mittelpersischer Leopard

4924   *Panthera pardus delacouri*
       Carnivora - Felidae
e   Indochina Leopard
d   Hinterindischer Leopard
f   Panthère d'Indochine

4925   *Panthera pardus fusca*
       Carnivora - Felidae
e   Indian Leopard
d   Indischer Leopard
f   Panthère indienne

4926 *Panthera pardus ituriensis*
Carnivora - Felidae
e Congo Leopard
d Kongo-Leopard

4927 *Panthera pardus japonensis*
Carnivora - Felidae
e Northern Chinese Leopard; Chinese Leopard; North China Leopard
d Chinesischer Leopard
f Panthère de Chine; Panthère de Chine du Nord
i Leopardo della Cina

4928 *Panthera pardus jarvisi*
Carnivora - Felidae
e Sinai Leopard
d Sinai-Leopard

4929 *Panthera pardus lankae*
Carnivora - Felidae
e Sri Lankan Leopard; Ceylon Leopard
d Ceylon-Leopard

4930 *Panthera pardus leopardus*
Carnivora - Felidae
e Senegal Leopard
d Westafrikanischer Waldleopard; Westafrikanischer Leopard

4931 *Panthera pardus melanotica*
Carnivora - Felidae
e South African Leopard
d Südafrikanischer Leopard

4932 *Panthera pardus melas*
Carnivora - Felidae
e Java Leopard
d Java-Leopard
f Panthère de Java; Panthère noire
i Pantera nera

4933 *Panthera pardus millardi*
Carnivora - Felidae
e Kashmir Leopard

d Kaschmir-Leopard

4934 *Panthera pardus nannopardus*
Carnivora - Felidae
e Somali Leopard

4935 *Panthera pardus nimr*
Carnivora - Felidae
e Arabian Leopard; South Arabaian Leopard
i Leopardo arabo

4936 *Panthera pardus orientalis*
Carnivora - Felidae
e Amur Leopard
d Amur-Leopard
f Panthère de l'Amour; Léopard de l'Amour
i Leopardo dell'Amur

4937 *Panthera pardus panthera*
Carnivora - Felidae
e Barbary Leopard; Berber Leopard

4938 *Panthera pardus pardus*
Carnivora - Felidae
e Sudan Leopard
d Nordafrikanischer Leopard
f Panthère mouchetée d'Afrique

4939 *Panthera pardus pernigra*
Carnivora - Felidae
e Tibetan Leopard
d Nepal-Leopard

4940 *Panthera pardus reichenowi*
Carnivora - Felidae
e Cameroon Leopard
d Kamerun-Leopard

4941 *Panthera pardus saxicolor*
Carnivora - Felidae
e North Persian Leopard; Persian Leopard
d Nordpersischer Leopard

f	Panthère de Perse
i	Leopardo persiano

4942 *Panthera pardus shortridgei*
Carnivora - Felidae
e Central African Leopard
d Zentralafrikanischer Leopard

4943 *Panthera pardus sindica*
Carnivora - Felidae
e Baluchistan Leopard
d Belutschistan-Leopard

4944 *Panthera pardus suahelica*
Carnivora - Felidae
e Eastern African Leopard
d Ostafrikanischer Leopard

4945 *Panthera pardus tulliana*
Carnivora - Felidae
e Anatolian Leopard
d Kleinasiatischer Leopard

4946 *Panthera tigris*
Carnivora - Felidae
e Tiger
d Tiger
f Tigre
i Tigre

4947 *Panthera tigris altaica*
Carnivora - Felidae
e Amur Tiger; Siberian Tiger;
Manchurian Tiger; North-east China
Tiger; Ussuri Tiger
d Sibirischer Tiger
f Tigre de Sibérie; Tigre de l'Amour
i Tigre siberiana

4948 *Panthera tigris amoyensis*
Carnivora - Felidae
e South China Tiger; Amoy Tiger;
Chinese Lion
d Chinesischer Tiger
f Tigre de Chine

i Tigre cinese

4949 *Panthera tigris balica*
Carnivora - Felidae
e Bali Tiger; Balinese Tiger
d Bali-Tiger
f Tigre de Bali
i Tigre di Bali

4950 *Panthera tigris corbetti*
Carnivora - Felidae
e Indochina Tiger
d Indochinesischer Tiger; Indochina-
Tiger
f Tigre d'Indochine
i Tigre indo-cinese

4951 *Panthera tigris sondaica*
Carnivora - Felidae
e Java Tiger; Javan Tiger
d Java-Tiger; Sunda-Tiger
f Tigre de Java
i Tigre di Giava

4952 *Panthera tigris sumatrae*
Carnivora - Felidae
e Sumatran Tiger
d Sumatra-Tiger
f Tigre de Sumatra
i Tigre di Sumatra

4953 *Panthera tigris tigris*
Carnivora - Felidae
e Bengal Tiger; Indian Tiger
d Königstiger; Bengal-Tiger; Indischer
Tiger
f Tigre du Bengale; Tigre royal
i Tigre del Bengala

4954 *Panthera tigris virgata*
Carnivora - Felidae
e Caspian Tiger; Turan Tiger;
Hyrcanian Tiger; Persian Tiger
d Kaspischer Tiger; Kaspi-Tiger;
Turan-Tiger; Persischer Tiger

f Tigre de la Caspienne; Tigre de Perse

i Tigre del Caspio

4955 *Panthera uncia*
Carnivora - Felidae

e Snow Leopard; Ounce

d Irbis; Schneeleopard

f Léopard des neiges; Panthére des neiges; Once

i Leopardo delle nevi; Irbis; Pantera delle nevi

4956 *Pantholops*
Artiodactyla - Bovidae

e Chirus; Tibetan Antelopes; Orongos

d Tibet-Antilopen; Tschirus; Orongos; Chirus

f Antilopes du Tibet

i Antilopi di Hodgson

4957 *Pantholops hodgsonii*
Artiodactyla - Bovidae

e Chiru; Tibetan Antelope

d Tibet-Antilope; Tchiru; Chiru; Orongo

f Antilope du Tibet

i Antilope di Hodgson; Chiru; Antilope tibetana; Pantalopo di Hodgson

4958 *Papagomys*
Rodentia - Muridae

e Flores Island Giant Tree Rats; Flores Giant Rats

d Flores-Riesenratten

4959 *Papagomys armandvillei*
Rodentia - Muridae

e Flores Island Giant Tree Rat; Flores Complex-toothed Rat; Flores Giant Rat; Flores Giant Tree Rat

d Flores-Riesenratte

4960 *Papio*
Primates - Cercopithecidae

e Baboons; Savana Baboons; Savannah

Baboon

d Paviane

f Papions; Babouins

i Babbuini

4961 *Papio anubis*
Primates - Cercopithecidae

e Olive Baboon

d Grüner Pavian; Anubis-Pavian

f Babouin doguera; Babouin vert-olive; Papion anubis

i Babbuino verde; Babbuino olivastro; Anubi

4962 *Papio cynocephalus*
Primates - Cercopithecidae

e Yellow Baboon

d Gelber Pavian; Steppenpavian; Gelber Babuin; Babuin-Pavian

f Babouin jaune; Babouin cynocéphale

i Babbuino giallo; Babbuino

4963 *Papio cynocephalus ruhei*
Primates - Cercopithecidae

e Webb's Baboon

d Webb-Pavian

4964 *Papio hamadryas*
Primates - Cercopithecidae

e Hamadryas Baboon; Hamadryas Sacred Baboon; Sacred Baboon

d Mantelpavian; Bergaffe

f Tartarin; Hamadryas

i Amadriade

4965 *Papio papio*
Primates - Cercopithecidae

e Guinea Baboon; Western Baboon

d Guinea-Pavian; Sphinx-Pavian; Roter Pavian

f Babouin de Guinée

i Babuino della Guinea

4966 *Papio ursinus*
Primates - Cercopithecidae

e   Chacma Baboon
d   Bärenpavian; Tschakma
f   Babouin chacma; Chacma
i   Babuino nero; Chacma

4967   *Pappogeomys*
Rodentia - Geomyidae
e   Buller and Alcorn's Pocket Gophers;
Mexican Pocket Gophers
d   Tuzas; Gebirgstaschenratten; Gelbe
Taschenratten; Zimtfarbene
Taschenratten
f   Geomys

4968   *Pappogeomys alcorni*
Rodentia - Geomyidae
e   Alcorn's Pocket Gopher

4969   *Pappogeomys bulleri*
Rodentia - Geomyidae
e   Buller's Pocket Gopher

4970   *Pappogeomys castanops*
Rodentia - Geomyidae
e   Yellow Pocket Gopher; Yellow-faced
Pocket Gopher
d   Mexikanische Taschenratte
f   Rat à poche mexicain

4971   *Pappogeomys fumosus*
Rodentia - Geomyidae
e   Smoky Pocket Gopher

4972   *Pappogeomys gymnurus*
Rodentia - Geomyidae
e   Llano Pocket Gopher

4973   *Pappogeomys merriami*
Rodentia - Geomyidae
e   Merriam's Pocket Gopher

4974   *Pappogeomys neglectus*
Rodentia - Geomyidae
e   Querétaro Pocket Gopher

4975   *Pappogeomys tylorhinus*
Rodentia - Geomyidae
e   Naked-nosed Pocket Gopher

4976   *Pappogeomys zinseri*
Rodentia - Geomyidae
e   Zinser's Pocket Gopher

4977   *Paracoelops*
Chiroptera - Rhinolophidae
e   Vietnam Leaf-nosed Bats
d   Blattnasen

4978   *Paracoelops megalotis*
Chiroptera - Rhinolophidae
e   Vietnam Leaf-nosed Bat

4979   *Paracrocidura*
Soricomorpha - Soricidae
e   African Shrews; Large-headed
Shrews; Montane Shrews
d   Kongo-Wimperspitzmäuse

4980   *Paracrocidura graueri*
Soricomorpha - Soricidae
e   Grauer's Shrew; Itombwe Large-
headed Shrew; Grauer's Montane
Shrew

4981   *Paracrocidura maxima*
Soricomorpha - Soricidae
e   Greater Shrew; East African
Montane Shrew

4982   *Paracrocidura shoutedeni*
Soricomorpha - Soricidae
e   Schouteden's Shrew; Lesser Large-
headed Shrew
d   Kongo-Wimperspitzmaus

4983   *Paradipus*
Rodentia - Dipodidae
e   Comb-toed Jerboas
d   Kammzehenspringmäuse;

Kammzehige Springmäuse
f Gerboises aux doigts pectinés

4984 Paradipus ctenodactylus
Rodentia - Dipodidae
e Comb-toed Jerboa
d Kammzehenspringmaus;
Kammzehige Springmaus
f Gerboise aux doigts pectinés

4985 Paradoxurus
Carnivora - Viverridae
e Toddy Cats; Asian Palm Civets;
Small-toothed Civets; Musangs
d Palmenroller ; Rollmarder ; Musangs
f Civettes palmistes
i Paradossure

4986 Paradoxurus hermaphroditus
Carnivora - Viverridae
e Toddy Cat; Asian Palm Civet;
Common Palm Civet; Indian Palm
Civet; Common Indian Palm Cat
d Palmenroller; Fleckenmusang
f Civette palmiste hermaphrodite
i Civetta delle palme; Musang

4987 Paradoxurus hermaphroditus
lignicolor
Carnivora - Viverridae
e Mentawai Palm Civet
d Mentawai-Palmenroller

4988 Paradoxurus jerdoni
Carnivora - Viverridae
e Jerdon's Palm Civet
d Jerdon-Musang
f Civette palmiste de Jerdon

Paradoxurus philippinensis syn. *P.*
hermaphroditus q.v.

4989 Paradoxurus zeylonensis
Carnivora - Viverridae
e Golden Palm Civet

d Goldmusang
f Civette palmiste dorée
i Civetta dorata della palma

Paraechinus syn. *Hemiechinus* q.v.

4990 Parahydromys
Rodentia - Muridae
e Coarse-haired Water Rats
d Gebirgsschwimmratten

4991 Parahydromys asper
Rodentia - Muridae
e Coarse-haired Hydromyine; Coarse-
haired Water Rat; Waterside Rat
d Bergwasserratte; Gebirgswasserratte

Paralactaga syn. *Allactaga* q.v.

4992 Paraleptomys
Rodentia - Muridae
e New Guinea False Water Rats;
Montane Water Rats; Hydromines
d Neuguinea-Bergwasserratten
f Rats de Nouvelle Guinée

4993 Paraleptomys rufilatus
Rodentia - Muridae
e Red-sided Hydromyine; Montane
Water Rat; Northern Hydromine

4994 Paraleptomys wilhelmina
Rodentia - Muridae
e Short-footed Hydromyine; Short-
haired Hydromyine; Short-haired
Water Rat
d Kurzhaarwasserratte

Paramanis syn. *Manis* q.v.

Parameriones syn. *Meriones* q.v.

Paramicrogale syn. *Microgale* q.v.

Parantechinus syn. *Antechinus* q.v.

4995   *Paranyctimene*
           Chiroptera - Pteropodidae
e   Lesser Tube-nosed Fruit Bats;
           Unstriped Tube-nose Bats

4996   *Paranyctimene raptor*
           Chiroptera - Pteropodidae
e   Unstriped Tube-nosed Bat; Long-
           tailed Fruit Bat

4997   *Paranyctimene tenax*
           Chiroptera - Pteropodidae
e   Greater Unstriped Tube-nosed Bat

           Paranyctis syn. *Cynictis* q.v.

           Paraonyx syn. *Aonyx* q.v.

4998   *Parascalops*
           Erinaceomorpha - Talpidae
e   Hairy-tailed Moles; Brewer's Moles
d   Haarschwanzmaulwürfe;
           Bürstenmaulwürfe
f   Taupes à queue chevelue

4999   *Parascalops breweri*
           Erinaceomorpha - Talpidae
e   Hairy-tailed Mole; Brewer's Hairy-
           tailed Mole; Brewer's Mole
d   Haarschwanzmaulwurf;
           Bürstenmaulwurf
f   Taupe à queue chevelue

           Parascaptor syn. *Talpa* q.v.

           Paratatera syn. *Tatera* q.v.

5000   *Paraxerus*
           Rodentia - Sciuridae
e   Bush Squirrels; African Bush
           Squirrels; East African Bush
           Squirrels
d   Afrikanische Buschhörnchen
f   Écureils de brousse
i   Scoiattoli africani

5001   *Paraxerus alexandri*
           Rodentia - Sciuridae
e   Alexander's Bush Squirrel
d   Alexander-Hörnchen
f   Écureil d'Alexandre

5002   *Paraxerus boehmi*
           Rodentia - Sciuridae
e   Böhm's Bush Squirrel; Böhm's
           African Bush Squirrel
d   Boehm-Hörnchen
f   Écureuil de Böhm

5003   *Paraxerus cepapi*
           Rodentia - Sciuridae
e   Smith's Bush Squirrel; South African
           Tree Squirrel; Yellow-footed
           Squirrel; Yellow-footed Bush
           Squirrel
d   Smith-Buschhörnchen
f   Écureil de Smith; Écureuil de brousse
           aux pieds jaunes; Écureuil aux pieds
           jaunes
i   Scoiattolo africano di Smith

5004   *Paraxerus cooperi*
           Rodentia - Sciuridae
e   Cooper's Green Squirrel; Cooper's
           Mountain Squirrel
d   Cooper-Hörnchen; Cooper-
           Buschhörnchen

5005   *Paraxerus flavovittis*
           Rodentia - Sciuridae
e   Striped Bush Squirrel
d   Gestreiftes Buschhörnchen

5006   *Paraxerus lucifer*
           Rodentia - Sciuridae
e   Black-and-red Bush Squirrel;
           Tanganyka Mountain Squirrel;
           Black-and-red Squirrel
d   Schwarzrotes Buschhörnchen

5007   *Paraxerus ochraceus*
           Rodentia - Sciuridae

   e   Huet's Bush Squirrel; Ochre Bush
       Squirrel
   d   Ockerbuschhörnchen

5008   *Paraxerus palliatus*
      Rodentia - Sciuridae
   e   Mantled African Bush Squirrel; Red
       Bush Squirrel

5009   *Paraxerus palliatus frerei*
      Rodentia - Sciuridae
   e   Red-bellied Coast Squirrel; Chindi

5010   *Paraxerus palliatus tongensis*
      Rodentia - Sciuridae
   e   Tonga Red Squirrel
   d   Rotes Tonga-Eichhörnchen

5011   *Paraxerus poensis*
      Rodentia - Sciuridae
   e   Green Squirrel; Small Green
       Squirrel; Fernando Po Squirrel
   d   Grünhörnchen; Grünes
       Buschhörnchen
   f   Écureuil à manteau; Écureil de
       Fernando Po

5012   *Paraxerus vexillarius*
      Rodentia - Sciuridae
   e   Swynnerton's Bush Squirrel
   d   Swynnerton-Hörnchen

5013   *Paraxerus vincenti*
      Rodentia - Sciuridae
   e   Vincent's Bush Squirrel
   d   Vincent-Hörnchen

      Pardictis syn. *Prionodon* q.v.

      Pardofelis syn. *Felis* q.v.

5014   *Parotomys*
      Rodentia - Muridae
   e   Karroo Rats; Whistling Rats
   d   Pfeifratten

   f   Rats du Karoo

5015   *Parotomys brantsii*
      Rodentia - Muridae
   e   Brandt's Whistling Rat
   d   Karru-Ratte; Brants Pfeifratte
   f   Rat siffleur de Brandt

5016   *Parotomys littledalei*
      Rodentia - Muridae
   e   Littledale's Karoo Rat; Littledale's
       Whistling Rat; Karroo Rat

5017   *Paruromys*
      Rodentia - Muridae
   e   Sulawesi Giant Rats
   d   Sulawesi-Riesenratten

5018   *Paruromys dominator*
      Rodentia - Muridae
   e   Sulawesi Giant Rat

5019   *Paruromys ursinus*
      Rodentia - Muridae
   e   Sulawesi Bear Rat

5020   *Paulamys*
      Rodentia - Muridae
   e   Flores Long-nosed Rats
   d   Flores-Langnasenratten

5021   *Paulamys naso*
      Rodentia - Muridae
   e   Flores Long-nosed Rat; Flores Island
       Mouse
   d   Flores-Langnasenratte

5022   *Pearsonomys*
      Rodentia - Muridae
   e   Pearson's Long-clawed Mice

5023   *Pearsonomys annectens*
      Rodentia - Muridae
   e   Pearson's Long-clawed Mouse

Pecari syn. *Dicotyles* q.v.

5024   Pectinator
Rodentia - Ctenodactylidae
e   Speke's Pectinators; East African
Gundis; Bushy-tailed Rats
d   Speke-Kammfinger
f   Pectinators de Speke
i   Pettinatori

5025   Pectinator spekei
Rodentia - Ctenodactylidae
e   Speke's Pectinator; East African
Gundi; Bushy-tailed Rat; Bushy-
tailed Gundi
d   Speke-Kammfinger;
Buschschwanzgundi
f   Pectinator de Speke
i   Pettinatore di Speke

5026   Pedetes
Rodentia - Pedetidae
e   African Jumping Hares; Cape
Jumping Hares; Springhares;
Springhaases
d   Springhasen
f   Lièvres sauteurs
i   Lepri saltatrici del Capo

5027   Pedetes capensis
Rodentia - Pedetidae
e   African Jumping Hare; Cape
Jumping Hare; Springhare;
Springhaas (CSA); East Cape
Jumping Hare; Jumping Hare
d   Südafrikanischer Springhase;
Springhase
f   Lièvre sauteur d'Afrique du Sud;
Lièvre sauteur
i   Lepre saltatrice del Capo; Springhaas

5028   Pedetidae
Rodentia
e   African Jumping Hares; Cape
Jumping Hares; Spring Hares;
Springhaases
d   Springhasen

f   Lièvres sauteurs
i   Pedetidi

Pediolagus syn. *Dolichotis* q.v.

Pedomys syn. *Microtus* q.v.

5029   Pelea
Artiodactyla - Bovidae
e   Rheboks; Rhebucks; Ribboks
d   Rehböcke; Rehantilopen;
Rehböckchen
f   Pélées
i   Pelea

5030   Pelea capreolus
Artiodactyla - Bovidae
e   Rhebok; Grey Rhebok; Vaal Rhebok;
Rhebuck; Grey Rhebuck; Vaal
Rhebuck; Ribbok; Grey Ribbok;
Vaal Ribbok; Common Rhebok
d   Rehantilope; Rehbock
f   Pélée; Antilope-chevreuil
i   Pelea; Antilope capriolo; Pelea di
montagna

5031   Pelomys
Rodentia - Muridae
e   Creek Rats; Groove-toothed Swamp
Rats
d   Pelomys-Grasratten

5032   Pelomys campanae
Rodentia - Muridae
e   Bell Groove-toothed Swamp Rat
d   Angola-Pelomys-Grasratte

Pelomys dybowskyii syn. *Mylomys dybowskyii* q.v.

5033   Pelomys fallax
Rodentia - Muridae
e   Creek Rat; Creek Groove-toothed
Swamp Rat
d   Ostafrikanische Pelomys-Grasratte

Pelomys harringtoni syn. *Desmomys harringtoni* q.v.

5034 Pelomys hopkinsi
Rodentia - Muridae
e   Hopkins's Groove-toothed Swamp Rat; Papyrus Rat

5035 Pelomys isseli
Rodentia - Muridae
e   Issel's Groove-toothed Swamp Rat

5036 Pelomys minor
Rodentia - Muridae
e   Least Groove-toothed Swamp Rat
d   Kleine Pelomys-Grasratte

5037 Pentalagus
Lagomorpha - Leporidae
e   Ryukyu Rabbits; Amami Rabbits
d   Riukiu-Kaninchen
f   Lapins de Ryukyu
i   Conigli di Ryuky

5038 Pentalagus furnessi
Lagomorpha - Leporidae
e   Ryukyu Rabbit; Liukiu Rabbit; Amami Rabbit
d   Riukiu-Kaninchen
f   Lapin de Ryukyu
i   Coniglio di Amami; Coniglio di Ryukyu

5039 Penthetor
Chiroptera - Pteropodidae
e   Lucas's Short-faced Fruit Bats; Lucas's Short-nosed Fruit Bats; Dusky Fruit Bats

5040 Penthetor lucasi
Chiroptera - Pteropodidae
e   Lucas's Short-faced Fruit Bat; Lucas's Short-nosed Fruit Bat; Dusky Fruit Bat

5041 Peponocephala

Cete - Delphinidae
e   Broad-beaked Dolphins
d   Breitschnabeldelfine
f   Péponocéphales
i   Pepenocefali

5042 Peponocephala electra
Cete - Delphinidae
e   Melon-headed Whale; Blackfish Dolphin; Broad-beaked Dolphin; Hawaiian Blackfish; Little Blackfish; Many-toothed Blackfish; Melon-headed Blackfish; Melon-headed Dolphin
d   Breitschnabeldelfin; Melonenkopfdelfin
f   Dauphin d'Électre; Péponocéphale
i   Pepenocefalo

5043 Peradorcas
Diprotodontia - Macropodidae
e   Pygmy Rock Wallabies
d   Zwergsteinkängurus
f   Petites pétrogales

5044 Peradorcas concinna
Diprotodontia - Macropodidae
e   Pygmy Rock Wallaby; Nabarlek (ANZ); Little Rock Wallaby
d   Zwergsteinkänguru
f   Petite Pétrogale; Petit Wallaby des rochers

5045 Perameles
Peramelemorphia - Peramelidae
e   Long-nosed Bandicoots; Common and Long-nosed Bandicoots; Gunn's Bandicoots; Slender Bandicoots
d   Langnasenbeutler ; Beuteldachse
f   Bandicoots; Pérambles
i   Perameli nasuti

5046 Perameles bougainvillei
Peramelemorphia - Peramelidae
e   Little Mari Bandicoot; Barred Bandicoot; Little Barred Bandicoot;

Mari; Little Mari; Long-nosed
Bandicoot; Western Barred
Bandicoot

d   Rauhaar-Langnasenbeutler;
Bänderlangnasenbeutler;
Westaustralischer
Streifenbeuteldachs; Bougainville-
Langnasenbeutler

f   Bandicoot de Bougainville;
Bandicoot à long nez de Bougainville

i   Peramele nasuto di Bougainville;
Topo gigante striato occidentale

5047   Perameles bougainvillei fasciata
Peramelemorphia

e   Eastern Barred Bandicoot; Western
Striped Bandicoot; South-east
Australian Bougainville's Long-
nosed Bandicoot

d   Östlicher Bougainville-
Langnasenbeutler; Östlicher
Bougainville-Langnasenbeutler;
Östlicher Rauhaarlangnasenbeutler;
Östlicher Rauhaarbandikut

f   Bandicoot à long nez de Bougainville
d'Australie Sud-est; Péramèle à long
nez de Bougainville d'Australie Sud-
est

5048   Perameles eremiana
Peramelemorphia -

e   Desert Bandicoot; Orange-backed
Bandicoot; Desert Long-nosed
Bandicoot

d   Wüstenlangnasenbeutler;
Wüstenlangnasenbeuteldachs;
Wüstenbandikut;
Wüstennasenbeutler;
Wüstennasenbeuteldachs

f   Bandicoot à long nez du désert;
Péramèle du désert

Perameles fasciata syn. *P.*
bougainvillei fasciata q.v.

5049   Perameles gunni
Peramelemorphia - Peramelidae

e   Eastern Barred Bandicoot (ANZ);

Tasmanian Barred Bandicoot; Striped
Bandicoot; Gunn's Bandicoot

d   Tasmanischer Langnasenbeutler;
Tasmanien-Langnasenbeutler

i   Peramele nasuto di Tasmania

5050   Perameles nasuta
Peramelemorphia - Peramelidae

e   Long-nosed Bandicoot (ANZ); Long-
nosed Perameles

d   Groß-Langnasenbeutler; Großer
Langnasenbeutler; Nasenbeuteldachs

f   Bandicoot à nez long

i   Peramele nasuto grande

5051   Peramelidae
Peramelia

e   Australian Bandicoots and Bilbies;
Bandicoots; Short-nosed Bandicoots

d   Nasenbeutler ; Beuteldachse;
Bandikuts

f   Bandicoots; Pérámélidés

i   Peramelidi; Bandicut

Perimyotis syn. *Pipistrellus* q.v.

5052   Perodicticus
Primates - Loridae

e   Bosman's Pottos

d   Pottos

f   Pottos de Calabar

i   Potti

5053   Perodicticus potto
Primates - Loridae

e   Bosman's Potto; Potto

d   Potto

f   Potto; Potto de Bosman

i   Potto

5054   Perognathus
Rodentia - Geomyidae

e   Pocket Mice; Silky Pocket Mice

d   Eigentliche Taschenmäuse;
Eigentliche Taschenratten

f   Souris à poche
i   Perognati

5055   *Perognathus alticola*
Rodentia - Geomyidae
e   Mount Pinos Pocket Mouse; White-eared Pocket Mouse

5056   *Perognathus alticola inexpectatus*
Rodentia - Geomyidae
e   Tehachapi Pocket Mouse

5057   *Perognathus amplus*
Rodentia - Geomyidae
e   Arizona Pocket Mouse

5058   *Perognathus fasciatus*
Rodentia - Geomyidae
e   Olive-backed Pocket Mouse; Maximilian Pocket Mouse
d   Wyoming-Taschenmaus
f   Souris aux abajoues des plaines

5059   *Perognathus flavescens*
Rodentia - Geomyidae
e   Plains Pocket Mouse
d   Flachlandtaschenmaus

5060   *Perognathus flavescens perniger*
Rodentia - Geomyidae
e   Dusky Pocket Mouse

5061   *Perognathus flavus*
Rodentia - Geomyidae
e   Silky Pocket Mouse
d   Seidentaschenmaus; Gelbe Kängurumaus
f   Souris à poche soyeuse

5062   *Perognathus inornatus*
Rodentia - Geomyidae
e   San Joaquin Pocket Mouse

5063   *Perognathus inornatus psammophilus*

Rodentia - Geomyidae
e   Salinas Pocket Mouse

5064   *Perognathus longimembris*
Rodentia - Geomyidae
e   Little Pocket Mouse

5065   *Perognathus longimembris bangsi*
Rodentia - Geomyidae
e   Palm Springs Pocket Mouse

5066   *Perognathus longimembris brevinasus*
Rodentia - Geomyidae
e   Los Angeles Pocket Mouse

5067   *Perognathus longimembris internationalis*
Rodentia - Geomyidae
e   Jacumba Pocket Mouse

5068   *Perognathus longimembris pacificus*
Rodentia - Geomyidae
e   Pacific Pocket Mouse

5069   *Perognathus merriami*
Rodentia - Geomyidae
e   Merriam's Pocket Mouse

5070   *Perognathus parvus*
Rodentia - Geomyidae
e   Great Basin Pocket Mouse; Yellow-eared Pocket Mouse
d   Kleine Taschenmaus
f   Souris aux abajoues des pinèdes

Perognathus xanthonotus syn. *P. parvus* q.v.

5071   *Peromyscus*
Rodentia - Muridae
e   White-footed Mice; Deer Mice
d   Weißfußmäuse; Hirschmäuse
f   Souris aux pattes blanches;

Péromysques; Souris de chasse

i   Peromischi

Peromyscus allophylus syn. *P. gymnotis* q.v.

Peromyscus altilaneus syn. *P. guatemalensis* q.v.

5072   ***Peromyscus attwateri***
        Rodentia - Muridae

e   Texas Mouse; Texas Deer Mouse

d   Texas-Maus

5073   ***Peromyscus aztecus***
        Rodentia - Muridae

e   Aztec Mouse; Aztec Deer Mouse

d   Aztek-Maus

Peromyscus banderanus syn. *Osgoodomys banderanus* q.v.

5074   ***Peromyscus boylii***
        Rodentia - Muridae

e   Brush Mouse; Brush Deer Mouse

5075   ***Peromyscus boylii stephani***
        Rodentia - Muridae

e   San Esteban Island Mouse; San Esteban Island Deer Mouse

d   San-Esteban-Maus

5076   ***Peromyscus bullatus***
        Rodentia - Muridae

e   Perote Mouse; Pereote Deer Mouse

5077   ***Peromyscus californicus***
        Rodentia - Muridae

e   California Mouse; California Deer Mouse

d   Kalifornische Maus

f   Souris de Californie

5078   ***Peromyscus californicus benitoensis***
        Rodentia - Muridae

e   Parasitic Mouse

Peromyscus caniceps syn. *P. eremicus farterculus* q.v.

Peromyscus chinanteco syn. *Habromys chinanteco* q.v.

5079   ***Peromyscus crinitus***
        Rodentia - Muridae

e   Canyon Mouse; Canyon Deer Mouse

d   Canyon-Maus

Peromyscus dickeyi syn. *P. merriami dickeyi* q.v.

5080   ***Peromyscus difficilis***
        Rodentia - Muridae

e   Zacatecan Deer Mouse

d   Felsenmaus

5081   ***Peromyscus eremicus***
        Rodentia - Muridae

e   Cactus Mouse; Cactus Deer Mouse

d   Kaktusmaus

5082   ***Peromyscus eremicus eremicus***
        Rodentia - Muridae

e   San Lorenzo Deer Mouse

5083   ***Peromyscus eremicus fraterculus***
        Rodentia - Muridae

e   Burt's Deer Mouse

d   Burts Hirschmaus

5084   ***Peromyscus eva***
        Rodentia - Muridae

e   Eva's Desert Mouse; Eva's Deer Mouse

d   Evas Wüstenmaus

Peromyscus flavidus syn. *Isthmomys flavidus* q.v.

Peromyscus floridanus syn.
Podomys floridanus q.v.

5085   Peromyscus furvus
Rodentia - Muridae
e   Blackish Deer Mouse
d   Schwarze Hirschmaus

5086   Peromyscus gossypinus
Rodentia - Muridae
e   Cotton Mouse; Cotton Deer Mouse
d   Baumwollmaus
f   Souris de coton

5087   Peromyscus gossypinus allapaticola
Rodentia - Muridae
e   Key Largo Cotton Mouse; Keys
Cotton Mouse; Key Largo Mouse

5088   Peromyscus gossypinus restrictus
Rodentia - Muridae
e   Chadwick Beach Cotton Mouse;
Chadwick BeachCotton Deer Mouse;
Chadwick Beach Cotton White-
footed Mouse; Chadwick Beach Deer
Mouse; Chadwick Beach White-
footed Mouse; Chadwick Beach
Mouse
d   Chadwick-Beach-Baumwollmaus;
Chadwick-Beach-
Baumwollhirschmaus; Chadwick-
Beach-Baumwollweißfußmaus;
Chadwick-Beach-Hirschmaus;
Chadwick-Beach-Weißfußmaus;
Chadwick-Beach-Maus
f   Souris de coton de Chadwick Beach;
Souris de Chadwick Beach; Souris
aux pattes blanches du coton de
Chadwick Beach

5089   Peromyscus grandis
Rodentia - Muridae
e   Big Deer Mouse; Giant Deer Mouse
d   Große Hirschmaus

5090   Peromyscus gratus
Rodentia - Muridae

e   Osgood's Mouse; Saxicolous Deer
Mouse
d   Osgoods Maus

5091   Peromyscus guardia
Rodentia - Muridae
e   Angel Island Mouse; Angel Island
Deer Mouse
d   Angel-Island-Maus

5092   Peromyscus guatemalensis
Rodentia - Muridae
e   Guatemalan Deer Mouse
d   Guatemala-Hirschmaus

5093   Peromyscus gymnotis
Rodentia - Muridae
e   Naked-ear Deer Mouse; Naked-eared
Deer Mouse
d   Nacktohrige Hirschmaus

5094   Peromyscus hooperi
Rodentia - Muridae
e   Hooper's Mouse
d   Hoopers Maus

Peromyscus interparietalis syn. *P.*
eremicus eremicus q.v.

5095   Peromyscus keeni
Rodentia - Muridae
e   Northwestern Deer Mouse; Sitka
Mouse; Columbian Mouse
d   Columbische Maus; Keens Maus;
Sitka-Maus
f   Souris de Sitka

Peromyscus latirostris syn. *P. furvus*
q.v.

Peromyscus lepturus syn. *Habromys*
lepturus q.v.

5096   Peromyscus leucopus
Rodentia - Muridae
e   White-footed Mouse; White-footed

Deer Mouse
d   Weißfußmaus
f   Souris aux pattes blanches
i   Peromisco dai piedi bianchi

5097   *Peromyscus leucopus ammodytes*
Rodentia - Muridae
e   Muskeget Island Beach Mouse

5098   *Peromyscus leucopus aridulus*
Rodentia - Muridae
e   Badlands White-footed Mouse

5099   *Peromyscus leucopus noveboracensis*
Rodentia - Muridae
e   Northern White-footed Mouse

5100   *Peromyscus levipes*
Rodentia - Muridae
e   Nimble-footed Mouse; Southern Brush Mouse; Southern Deer Mouse
d   Flinkfüßige Maus

Peromyscus lophurus syn.
Hambromys lophurus q.v.

5101   *Peromyscus madrensis*
Rodentia - Muridae
e   Tres Marias Island Mouse; Tres Marias Deer Mouse
d   Tres-Marias-Island-Maus

5102   *Peromyscus maniculatus*
Rodentia - Muridae
e   Deer Mouse; Prairie Deer Mouse; North American Deer Mouse
d   Hirschmaus
f   Souris du soir; Souris sylvestre

5103   *Peromyscus maniculatus anacapae*
Rodentia - Muridae
e   Anacapa Island Deer Mouse

5104   *Peromyscus maniculatus*

clementinus
Rodentia - Muridae
e   San Clemente Deer Mouse

5105   *Peromyscus maniculatus osgoodi*
Rodentia - Muridae
e   Osgood White-footed Deer Mouse

5106   *Peromyscus mayensis*
Rodentia - Muridae
e   Maya Mouse; Maya Deer Mouse
d   Maya-Maus

5107   *Peromyscus megalops*
Rodentia - Muridae
e   Brown Deer Mouse
d   Braune Hirschmaus

5108   *Peromyscus mekisturus*
Rodentia - Muridae
e   Puebla Deer Mouse
d   Puebla-Hirschmaus

5109   *Peromyscus melanocarpus*
Rodentia - Muridae
e   Zempoaltepec Mouse; Zempoaltepec Deer Mouse
d   Zempoaltec-Hirschmaus

5110   *Peromyscus melanophrys*
Rodentia - Muridae
e   Plateau Mouse; Plateau Deer Mouse
d   Hochebenenmaus

5111   *Peromyscus melanotis*
Rodentia - Muridae
e   Black-eared Mouse; Black-eared Deer Mouse
d   Schwarzohrige Maus

5112   *Peromyscus melanurus*
Rodentia - Muridae
e   Black-tailed Mouse; Black-tailed Deer Mouse
d   Schwarzschwänzige Maus

5113 *Peromyscus merriami*
Rodentia - Muridae
e Merriam's Mouse; Mesquite Mouse;
Merriams Deer Mouse
d Merriams Maus

5114 *Peromyscus merriami dickeyi*
Rodentia - Muridae
e Dickey's Deer Mouse
d Dickeys Hirschmaus

5115 *Peromyscus mexicanus*
Rodentia - Muridae
e Mexican Deer Mouse; Naked-footed
Deer Mouse
d Mexikanische Hirschmaus

5116 *Peromyscus nasutus*
Rodentia - Muridae
e Northern Rock Mouse; Northern
Rock Deer Mouse
d Nördliche Felsenmaus

Peromyscus nudipes syn. *P. mexicanus* q.v.

Peromyscus oaxacensis syn. *P. aztecus* q.v.

5117 *Peromyscus ochraventer*
Rodentia - Muridae
e El Carrizo Deer Mouse
d El-Carrizo-Hirschmaus

Peromyscus oreas syn. *P. keeni* q.v.

5118 *Peromyscus pectoralis*
Rodentia - Muridae
e White-ankled Mouse; White-ankled
Deer Mouse
d Weißfesselmaus

5119 *Peromyscus pembertoni*
Rodentia - Muridae
e Pemberton's Deer Mouse;
Pemberton's Beach Mouse;

Pemberton's White-footed Mouse;
Pemberton's Mouse
d Pembertons Hirschmaus; Pembertons
Weißfußmaus; Pemberton-Maus
f Souris aux pattes blanches de
Pemberton; Souris de Pemberton

5120 *Peromyscus perfulvus*
Rodentia - Muridae
e Marsh Mouse; Marsh Deer Mouse
d Sumpfmaus

Peromyscus pirrensis syn.
Isthmomys pirrensis q.v.

5121 *Peromyscus polionotus*
Rodentia - Muridae
e Oldfield Mouse; Oldfield Deer
Mouse; Beach Mouse
d Küstenmaus

5122 *Peromyscus polionotus allophrys*
Rodentia - Muridae
e Choctawhatchee Beach Mouse
d Choctawhatchee-Küstenmaus

5123 *Peromyscus polionotus ammobates*
Rodentia - Muridae
e Alabama Beach Mouse
d Alabama-Küstenmaus

5124 *Peromyscus polionotus decoloratus*
Rodentia - Muridae
e Pale Beach Mouse ; Pallid Beach
Mouse; Pallid Beach Deer Mouse;
Pallid Beach White-footed Mouse;
Pallid Brown-backed Deer Mouse;
Pallid Brown-backed White-footed
Mouse; Pallid Brown-backed Mouse;
Pallid Oldfield Mouse

d Blasse Küstenmaus; Helle
Strandmaus; Helle Strandhirschmaus;
Helle Strandweißfußmaus; Helle
Braunrückenstrandmaus; Helle
Braunrückenmaus
f Souris de plage pâle; Souris aux

pattes blanches de plage pâle; Souris
à dos brun pâle

**5125 *Peromyscus polionotus
leucocephalus***
Rodentia - Muridae
e Santa Rosa Beach Mouse
d Santa-Rosa-Küstenmaus

5126 *Peromyscus polionotus niveiventris*
Rodentia - Muridae
e South-eastern Beach Mouse
d Südöstliche Küstenmaus

5127 *Peromyscus polionotus peninsularis*
Rodentia - Muridae
e St. Andrews Beach Mouse
d St. Andrews-Küstenmaus

5128 *Peromyscus polionotus phasma*
Rodentia - Muridae
e Anastasia Island Beach Mouse
d Anastasia-Island-Küstenmaus

5129 *Peromyscus polionotus polionotus*
Rodentia - Muridae
e Beach Mouse

5130 *Peromyscus polionotus trissyllepsis*
Rodentia - Muridae
e Perdido Key Beach Mouse
d Perdido-Key-Küstenmaus

5131 *Peromyscus polius*
Rodentia - Muridae
e Chihuahuan Mouse; Chihuahuan
Deer Mouse
d Chihuahua-Maus

5132 *Peromyscus pseudocrinitus*
Rodentia - Muridae
e False Canyon Mouse; False Canyon
Deer Mouse
d Falsche Canyon-Maus

5133 *Peromyscus sejugis*
Rodentia - Muridae
e Sant Cruz Mouse; Santa Cruz Island
Mouse; Santa Cruz Deer Mouse
d Santa-Cruz-Maus

5134 *Peromyscus simulus*
Rodentia - Muridae
e Nayarit Mouse; Nayarit Deer Mouse
d Nayarit-Maus

Peromyscus sitkensis syn. *P. keeni*
q.v.

5135 *Peromyscus slevini*
Rodentia - Muridae
e Slevin's Mouse; Slevin's Deer Mouse
d Slevins Maus

5136 *Peromyscus spicilegus*
Rodentia - Muridae
e Gleaning Deer Mouse; Gleaning
Mouse

Peromyscus stephani syn. *P. boylii
stephani* q.v.

5137 *Peromyscus stirtoni*
Rodentia - Muridae
e Stirton's Deer Mouse
d Stirtons Hirschmaus

Peromyscus thomasi syn.
Megadontomys thomasi q.v.

5138 *Peromyscus truei*
Rodentia - Muridae
e Pinyon Mouse; Piñon Deer Mouse
d Piñon-Maus

5139 *Peromyscus winkelmanni*
Rodentia - Muridae
e Winkelmann's Mouse; Winkelmann's
Deer Mouse
d Winkelmanns Maus

5140 **Peromyscus yucatanicus**
Rodentia - Muridae
e Yucatan Deer Mouse
d Yucatan-Hirschmaus

5141 **Peromyscus zarhynchus**
Rodentia - Muridae
e Chiapan Deer Mouse
d Chiapan-Hirschmaus

Peronymus syn. *Peropteryx* q.v.

5142 **Peropteryx**
Chiroptera - Emballonuridae
e Dog-like Bats; Sac-winged Bats

5143 **Peropteryx kappleri**
Chiroptera - Emballonuridae
e Kappler's Sac-winged Bat; Dusky
Sac-winged Bat; Greater Dog-like
Bat

5144 **Peropteryx leucoptera**
Chiroptera - Emballonuridae
e White-winged Dog-like Bat

5145 **Peropteryx macrotis**
Chiroptera - Emballonuridae
e Lesser Dog-like Bat; Trinidad Sac-
winged Bat

5146 **Peropteryx trinitatis**
Chiroptera - Emballonuridae
e Trinidad Dog-like Bat

5147 **Peroryctes**
Peramelemorphia - Peroryctidae
e New Guinea Bandicoots
d Neuguinea-Nasenbeutler
f Bandicoots de Nouvelle Guinée

5148 **Peroryctes broadbenti**
Peramelemorphia - Peroryctidae
e Giant Bandicoot
d Großer Neuginea-Nasenbeutler;

Groß-Neuguinea-Nasenbeutler
f Bandicoot géant

Peroryctes longicaudata syn.
Microperoryctes longicaudata q.v.

Peroryctes papuensis syn.
Microperoryctes papuensis q.v.

5149 **Peroryctes raffrayana**
Peramelemorphia - Peroryctidae
e New Guinea Bandicoot; Raffray's
Bandicoot
d Raffray-Nasenbeutler
f Bandicoot de Nouvelle Guinée

5150 **Peroryctidae**
Oeramelia
e Rainforest Bandicoots; New Guinea
Bandicoots
d Neuguinea-Nasenbeutler
f Peroryctidés
i Perorictidi

5151 **Petauridae**
Diprotodontia
e Ringtails; Gliding Phalangers;
Ringtail Possums; Gliding Possums
and Striped Possums; Possums and
Gliders
d Gleitbeutler
f Petauridés
i Petauridi

5152 **Petaurillus**
Rodentia - Sciuridae
e Pygmy Flying Squirrels
d Kleinstgleithörnchen
f Écureuils volants pygmées
i Scoattoli pigmei volanti

5153 **Petaurillus emiliae**
Rodentia - Sciuridae
e Lesser Pygmy Flying Squirrel
d Emilias Kleinstgleithörnchen

5154   *Petaurillus hosei*
Rodentia - Sciuridae
e   Hose's Pygmy Flying Squirrel
d   Hoses Kleinstgleithörnchen

5155   *Petaurillus kinlochii*
Rodentia - Sciuridae
e   Selangor Pygmy Flying Squirrel
d   Selangor-Kleinstgleithörnchen

5156   *Petaurista*
Rodentia - Sciuridae
e   Giant Flying Squirrels
d   Riesengleithörnchen
f   Écureuils volants géants
i   Scoattoli pigmei volanti

5157   *Petaurista alborufus*
Rodentia - Sciuridae
e   Red-and-white Flying Squirrel;
Yunnan Giant Flying Squirrel; Red-
and-white Giant Flying Squirrel;
White-faced Giant Flying Squirrel
d   Rotweißes Riesengleithörnchen
f   Écureuil volant géant rouge-et-blanc

5158   *Petaurista alborufus lena*
Rodentia - Sciuridae
e   White-faced Flying Squirrel

5159   *Petaurista elegans*
Rodentia - Sciuridae
e   Lesser Giant Flying Squirrel; Spotted
Giant Flying Squirrel
d   Geflecktes Riesengleithörnchen
i   Scoiattolo volante gigante; Scoiattolo
volante gigante macchiato

5160   *Petaurista leucogenys*
Rodentia - Sciuridae
e   Japanese Giant Flying Squirrel
d   Japanisches Riesengleithörnchen

5161   *Petaurista magnificus*
Rodentia - Sciuridae

e   Hodgson's Flying Squirrel;
Hodgson's Giant Flying Squirrel
d   Hodgson-Riesengleithörnchen

5162   *Petaurista nobilis*
Rodentia - Sciuridae
e   Bhutan Giant Flying Squirrel
d   Bhutan-Riesengleithörnchen

5163   *Petaurista petaurista*
Rodentia - Sciuridae
e   Red Giant Flying Squirrel; Giant
Flying Squirrel; Giant Red Flying
Squirrel
d   Taguan; Rotes Riesengleithörnchen
f   Écureuil volant rouge
i   Taguan

5164   *Petaurista petaurista lanka*
Rodentia - Sciuridae
e   Large Ceylon Flying Squirrel
d   Großes Sri Lanka-Flughörnchen

5165   *Petaurista philippensis*
Rodentia - Sciuridae
e   Indian Giant Flying Squirrel
d   Indisches Riesengleithörnchen

5166   *Petaurista xanthotis*
Rodentia - Sciuridae
e   Chinese Giant Flying Squirrel
d   Chinesisches Riesengleithörnchen
f   Écureuil volant géant de Chine

5167   *Petauroides*
Diprotodontia - Petauridae
e   Greater Gliders (ANZ); Greater
Gliding Possums
d   Großflugbeutler ; Riesenflugbeutler ;
Riesengleitbeutler
f   Grands phalangers volants
i   Petauroide maggiore

5168   *Petauroides volans*
Diprotodontia - Petauridae

e Greater Glider (ANZ); Greater Gliding Possum; Greater Gliding Opossum; Dusky Gliding Opossum; Greater Flying Phalanger; Dusky Glider; Dusky Gliding Possum

d Großflugbeutler; Riesenflugbeutler; Riesengleitbeutler

f Grand phalanger volant

i Petauroide maggiore

5169 ***Petaurus***
Diprotodontia - Petauridae

e Honey Gliders; Sugar Gliders; Squirel Gliders; Flying Possums; Gliding Possums; Lesser Gliding Possums

d Gleithörnchenbeutler ; Flugbeutler

f Phalangers volants; Planeurs

i Petauri; Scoiattoli volanti; Scoiattoli marsupiali

5170 ***Petaurus abidi***
Diprotodontia - Petauridae

e Northern Glider

d Neuguinea Gleitbeutler

f Planeur nordique

5171 ***Petaurus australis***
Diprotodontia - Petauridae

e Yellow-bellied Glider (ANZ); Yellow-bellied Flying Phalanger; Fluffy Glider

d Großer Gleithörnchenbeutler

i Petauro dal ventro giallo

5172 ***Petaurus biacensis***
Diprotodontia - Petauridae

e Biak Glider

d Kiak-Gleitbeutler

5173 ***Petaurus breviceps***
Diprotodontia - Petauridae

e Short-tailed Belideus; Sugar Possum; Lesser Flying Possum; Short-headed Flying Phalanger; Sugar Glider (ANZ)

d Kurzkopfgleitbeutler; Zwergflugbeutler

f Planeur de sucre; Phalanger volant à queue courte

i Petauro del miele

5174 ***Petaurus breviceps papuanus***
Diprotodontia - Petauridae

e Papuan Sugar Glider

d Papua-Gleithörnchenbeutler; Papua-Kurzkopfgleitbeutler

i Petauro dello zucchero

5175 ***Petaurus gracilis***
Diprotodontia - Petauridae

e Mahogony Glider

5176 ***Petaurus norfolcensis***
Diprotodontia - Petauridae

e Lesser Gliding Possum; Squirrel Glider (ANZ)

d Mittlerer Gleithörnchenbeutler; Mittelflugbeutler

f Planeur de l'île Norfolk

i Scoiattolo planatore

5177 ***Petinomys***
Rodentia - Sciuridae

e Bearded Flying Squirrels; Dwarf Flying Squirrels

d Zwerggleithörnchen

5178 ***Petinomys crinitus***
Rodentia - Sciuridae

e Mindanao Flying Squirrel

d Mindanao-Gleithörnchen

5179 ***Petinomys fuscocapillus***
Rodentia - Sciuridae

e Small Travancore Flying Squirrel; Travancore Flying Squirrel

d Travancore-Gleithörnchen

5180 ***Petinomys fuscocapillus layardi***
Rodentia – Sciuridae

e Small Ceylon Flying Squirrel
d Kleines Sri Lanka-Flughörnchen

5181 *Petinomys genibarbis*
Rodentia - Sciuridae
e Whiskered Flying Squirrel
d Schnurrbartgleithörnchen

5182 *Petinomys hageni*
Rodentia - Sciuridae
e Hagen's Flying Squirrel
d Hagen-Gleithörnchen

5183 *Petinomys lugens*
Rodentia - Sciuridae
e Siberut Flying Squirrel
d Siberut-Gleithörnchen

Petinomys morrisi syn. *P. setosus*
q.v.

5184 *Petinomys sagitta*
Rodentia - Sciuridae
e Arrow Flying Squirrel

5185 *Petinomys setosus*
Rodentia - Sciuridae
e White-bellied Flying Squirrel;
Temminck's Pygmy Flying Squirrel;
Temminck's Flying Squirrel
d Temminck-Gleithörnchen

5186 *Petinomys vordermanni*
Rodentia - Sciuridae
e Vorderman's Flying Squirrel
d Vorderman-Gleithörnchen

5187 *Petrodromus*
Macroscelidea - Macroscelididae
e Forest Elephant Shrews; Rock
Elephant Shrews
d Vierzehenrüsselratten; Rüsselratten
f Pétrodromes
i Macroscelidi a quattro dita

Petrodromus sultani syn. *P.
tetradactyla* q.v.

5188 *Petrodromus tetradactylus*
Macroscelidea - Macroscelididae
e Four-toed Elephant Shrew; Four-toed
Sengi
d Vierzehenrüsselratte; Waldrüsselratte
f Pétrodrome
i Macroscelide a quattro dita

Petrodromus tordayi syn. *P.
tetradactyla* q.v.

5189 *Petrogale*
Diprotodontia - Macropodidae
e Rock Wallabies; Ringtailed Rock
Wallabies; Wallabies
d Felskängurus; Bergkängurus;
Steinkängurus
f Pétrogales; Wallabies des rochers
i Wallaby di roccia

5190 *Petrogale assimilis*
Diprotodontia - Macropodidae
e Allied Rock Wallaby
d Verbündeter Felskänguru

5191 *Petrogale brachyotis*
Diprotodontia - Macropodidae
e Short-eared Wallaby; Short-eared
Rock Wallaby
d Kurzohrfelskänguru
f Pétrogale aux oreilles courtes

5192 *Petrogale burbidgei*
Diprotodontia - Macropodidae
e Warabi; Monjon
d Monjon

5193 *Petrogale coenensis*
Diprotodontia - Macropodidae
e Cape York Rock Wallaby
d Kap-York-Felskänguru

Petrogale concinna syn. *Peradorcas concinna* q.v.

5194 *Petrogale godmani*
Diprotodontia - Macropodidae
e Godman's Rock Wallaby
d Godman-Felskänguru

5195 *Petrogale herberti*
Diprotodontia - Macropodidae
e Herbert's Rock Wallaby
d Herbert-Felskänguru

5196 *Petrogale inornata*
Diprotodontia - Macropodidae
e Unadorned Rock Wallaby
d Queensland-Felskänguru
f Pétrogale terne

5197 *Petrogale lateralis*
Diprotodontia - Macropodidae
e Black-footed Rock Wallaby; Black-flanked Rock Wallaby
d Schwarzpfotenfelswallaby
f Pétrogale d'Australie occidentale

5198 *Petrogale lateralis pearsoni*
Diprotodontia - Macropodidae
e Pearson Island Rock Wallaby

5199 *Petrogale mareeba*
Diprotodontia - Macropodidae
e Mareeba Rock Wallaby
d Mariba-Felskänguru

5200 *Petrogale penicillata*
Diprotodontia - Macropodidae
e Brush-tailed Rock Wallaby; Rock Wallaby; Western Rock Wallaby; Godman's Rock Wallaby
d Bürstenfelskänguru; Bürstenschwanzfelskänguru
f Pétrogale à queue touffue; Pétrogale à queue en brosse; Wallaby des rochers à queue en pinceau

i Wallaby delle rocce dalla coda a pennello; Wallaby di Cook

5201 *Petrogale persephone*
Diprotodontia - Macropodidae
e Proserpine Rock Wallaby
d Prosepine-Felskänguru

5202 *Petrogale purpureicollis*
Diprotodontia - Macropodidae
e Purple-necked Rock Wallaby

5203 *Petrogale rothschildi*
Diprotodontia - Macropodidae
e Rothschild's Rock Wallaby
d Rothschild-Felskänguru

5204 *Petrogale sharmani*
Diprotodontia - Macropodidae
e Sharman's Rock Wallaby
d Sharman-Felskänguru

5205 *Petrogale xanthopus*
Diprotodontia - Macropodidae
e Ring-tailed Rock Wallaby; Yellow-footed Rock Wallaby (ANZ)
d Ringschwanzfelskänguru; Gelbfußkänguru
f Pétrogale aux pieds jaunes

5206 **Petromuridae**
Rodentia
e Dassie Rats; Rock Rats
d Felsenratten; Afrikanische Felsenratten
f Petromuridés
i Petromuridi

5207 *Petromus*
Rodentia - Petromuridae
e Rock Mice; Rock Rats; Dassie Rats; African Rock Rats
d Felsenratten
f Rats des rochers

5208   *Petromus typicus*
           Rodentia - Petromuridae
e   African Rock Rat; Rock Rat; Dassie
      Rat
d   Felsenratte
f   Rat typique des rochers

5209   *Petromyscus*
           Rodentia - Muridae
e   Rock Mice; Kopje Mice (CSA)
d   Felsenmäuse

5210   *Petromyscus barbouri*
           Rodentia - Muridae
e   Barbour's Rock Mouse
d   Barbour-Felsenmaus

5211   *Petromyscus collinus*
           Rodentia - Muridae
e   Pygmy Rock Mouse
d   Zwergfelsenmaus

5212   *Petromyscus monticularis*
           Rodentia - Muridae
e   Brukkaros Rock Mouse; Brukkaros
      Pygmy Rock Mouse
d   Brukkaros-Zwergfelsenmaus

5213   *Petromyscus shortridgei*
           Rodentia - Muridae
e   Shortridge's Rock Mouse
d   Shortridge-Felsenmaus

5214   *Petropseudes*
           Diprotodontidae - Petauridae
e   Rock Ringtail Possums (ANZ);
      Rock-haunting Ringtails
d   Felsenringbeutler ;
      Felsenringelschwanzbeutler

5215   *Petropseudes dahli*
           Diprotodontidae - Petauridae
e   Rock-haunting Ringtail; Rock-
      haunting Ring-tailed Phalanger; Rock
      Ringtail Possum (ANZ)

d   Felsenringbeutler;
      Felsenringelschwanzbeutler

5216   *Phacochoerus*
           Artiodactyla - Suidae
e   Warthogs
d   Warzenschweine
f   Phacochères; Sangliers des savanes
i   Facoceri

5217   *Phacochoerus aethiopicus*
           Artiodactyla - Suidae
e   Desert Warthog
d   Wüstenwarzenschwein
f   Sanglier des savanes; Phacochère du
      désert commun; Phacochère de
      Somalie
i   Facocero

5218   *Phacochoerus aethiopicus
           aethiopicus*
           Artiodactyla - Suidae
e   Cape Warthog; Southern Desert
      Warthog; Southern Cape Warthog;
      CommonCape Warthog

d   Kap-Warzenschwein; Südliches
      Warzenschwein; Südliches Kap-
      Warzenschwein; Hartläufer
f   Sanglier des savanes du Sud;
      Phacochère du Cap; Phacochère du
      désert du Sud

5219   *Phacochoerus africanus*
           Artiodactyla - Suidae
e   Warthog; Common Warthog
d   Warzenschwein
f   Phacochère commun
i   Facocero

5220   *Phaenomys*
           Rodentia - Muridae
e   Rio de Janeiro Rice Rats; Rio de
      Janeiro Arboreal Rats
d   Rio-de-Janeiro-Reisratten

5221 **Phaenomys ferrugineus**
Rodentia - Muridae
e Rio de Janeiro Rice Rat; Rio Rice
Rat; Rio de Janeiro Arboreal Rice
Rat; Rio de Janeiro Arboreal Rat
d Rio-Reisratte

5222 **Phalanger**
Diprotodontia - Phalangeridae
e Cuscuses; Australian Cuscuses;
Phalangers
d Kuskuse
f Phalangers; Couscous
i Falangere; Cuschi

5223 **Phalanger alexandrae**
Diprotodontia - Phalangeridae
e Gebe Cuscus

Phalanger atrimaculatus syn.
Spilocuscus rufoniger q.v.

5224 **Phalanger carmelitae**
Diprotodontia - Phalangeridae
e Mountain Cuscus
d Bergkuskus

Phalanger celebensis syn.
Strigocuscus celebensis q.v.

5225 **Phalanger gymnotis**
Diprotodontia - Phalangeridae
e Ground Cuscus; Aru Island Ground
Cuscus
d Bodenkuskus; Gleichfarbkuskus
i Cusco grigio

5226 **Phalanger intercastellanus**
Diprotodontia - Phalangeridae
e Southern Common Cuscus
d Südlicher Wollkuskus

Phalanger interpositus syn. *P.
orientalis* q.v.

5227 **Phalanger lullulae**

Diprotodontia - Phalangeridae
e Woodlark Cuscus
d Woodlark-Kuskus

Phalanger maculatus syn.
Spilocuscus maculatus q.v.

5228 **Phalanger matanim**
Diprotodontia - Phalangeridae
e Telefomin Cuscus
d Telefonim-Kuskus

5229 **Phalanger mimicus**
Diprotodontia - Phalangeridae
e Cryptic Cuscus

5230 **Phalanger orientalis**
Diprotodontia - Phalangeridae
e Grey Cuscus; Common Phalanger;
Stein's Cuscus; Northern Common
Cuscus; Northern Brown Cuscus
d Wollkuskus; Grauer Kuskus
f Couscous gris
i Falangere lanoso; Falangere comune

5231 **Phalanger ornatus**
Diprotodontia - Phalangeridae
e Moluccan Cuscus; Ornate Cuscus

5232 **Phalanger rothschildi**
Diprotodontia - Phalangeridae
e Obi Island Cuscus; Obi Cuscus
d Obi-Kuskus

Phalanger rufoniger syn.
Spilocuscus maculatus q.v.

5233 **Phalanger sericeus**
Diprotodontia - Phalangeridae
e Siky Cuscus
d Seidenkuskus

Phalanger ursinus syn. *Ailurops
ursinus* q.v.

5234   *Phalanger vestitus*
       Diprotodontia - Phalangeridae
e   Stein's Cuscus; Long-haired
       Mountain Cuscus
d   Steins Kuskus

5235   **Phalangeridae**
       Diprotodontia
e   Phalangers; Cuscuses; Possums;
       Brushtails; Cuscuses, Possums and
       Phalangers; Phalangers and Brushtail
       Possums
d   Kletterbeutler ; Possums
f   Phalangeridés
i   Cuschi; Falangeridi

       Phalomys syn. *Microtus* q.v.

5236   *Phaner*
       Primates - Cheirogaleidae
e   Fork-marked Lemurs; Squirrel
       Lemurs; Forked Mouse Lemurs;
       Fork-crowned Lemurs
d   Gabelkatzenmakis; Waluwys;
       Gablestreifige Katzenmakis;
       Gabelstreifenkatzenmakis
f   Phaners
i   Forciferi

5237   *Phaner electromontis*
       Primates - Cheirogaleidae
e   Amber Mountain Fork-marked
       Lemur; Amber Mountain Fork-
       crowned Lemur
d   Nördlicher Gabelstreifenmaki

5238   *Phaner furcifer*
       Primates - Cheirogaleidae
e   Fork-marked Lemur; Fork-marked
       Dwarf Lemur; Fork-marked Mouse
       Lemur; Fork-crowned Lemur;
       Masoala Fork-crowned Lemur;
       Eastern Fork-marked Lemur
d   Gabelstreifiger Katzenmaki
f   Phaner à fourche; Phaner fourché
i   Valuvi forcifero; Valuvi

5239   *Phaner pallescens*
       Primates - Cheirogaleidae
e   Western Fork-marked Lemur
d   Westlicher Gabelstreifenmaki

5240   *Phaner parienti*
       Primates - Cheirogaleidae
e   Sambirano Fork-crowned Lemur
d   Sambirano-Gabelstreifenmaki

5241   *Pharotis*
       Chiroptera - Vespertilionidae
e   New Guinea Big-eared Bats; Papuan
       Big-eared Bats

5242   *Pharotis imogene*
       Chiroptera - Vespertilionidae
e   New Guinea Big-eared Bat; Papuan
       Big-eared Bat; Large-eared
       Nyctophilus

5243   *Phascogale*
       Dasyuromorphia - Dasyuridae
e   Brush-tailed Marsupial Rats;
       Marsupial Rats; Striped Marsupial
       Rats; Brush-tailed Marsupial Mice;
       Phascogales; Wambengers; Brush-
       tailed Rat
d   Pinselschwanz-Beutelmäuse;
       Beutelbilche; Pinselschwanzbeutler
f   Phascogales; Rats marsupiaux
i   Fascogali

5244   *Phascogale calura*
       Dasyuromorphia - Dasyuridae
e   Brush-tailed Marsupial Mouse; Red-
       tailed Phascogale; Wambenger; Red-
       tailed Wambenger; Red-tailed
       Marsupial Mouse

d   Kleinpinselschwanzbeutelmaus;
       Kleiner Pinselschwanzbeutler;
       Rotschwanzphascogale
f   Phascogale rouge
i   Fascogale rosso-unito

5245 *Phascogale tapoatafa*
Dasyuromorphia - Dasyuridae
e Brush-tailed Phascogale (ANZ);
Black-tailed Phascogale; Tuan;
Common Wambenger
d Großpinselschwanzbeutelmaus;
Großer Pinselschwanzbeutler;
Bürstenschwanzphascogale
f Phascogale à queue rousse commune
i Fascogale spazzol-unito

5246 **Phascolarctidae**
Diprotodontia
e Koalas
d Koalas
f Koalas; Phascolartidés
i Fascolartidi

5247 *Phascolarctos*
Diprotodontia - Phascolarctidae
e Koalas; Koala Bears; Native Bears
(ANZ)
d Beutelbären; Koalas
f Koalas
i Koala

5248 *Phascolarctos cinereus*
Diprotodontia - Phascolarctidae
e Koala; Koala Bear; Native Bear
(ANZ)
d Beutelbär; Koala
f Koala
i Koala

5249 *Phascolarctos cinereus adustus*
Diprotodontia - Phascolarctidae
e Queensland Koala
d Queensland-Koala
f Koala de Queensland
i Koala di Queensland

5250 *Phascolarctos cinereus cinereus*
Diprotodontia - Phascolarctidae
e New South Wales Koala
f Koala de New South Wales

5251 *Phascolarctos cinereus victor*
Diprotodontia - Phascolarctidae
e Victoria Koala
f Koala de Victoria

5252 *Phascolosorex*
Dasyuromorphia - Dasyuridae
e Striped Marsupial Rats; Chestnut-
bellied Marsupial Rats; Marsupial
Shrews
d Streifenbeutelmäuse

5253 *Phascolosorex doriae*
Dasyuromorphia - Dasyuridae
e Red-bellied Marsupial Mouse; Red-
bellied Marsupial Shrew
d Orangebauch-Streifenbeutelmaus

5254 *Phascolosorex dorsalis*
Dasyuromorphia - Dasyuridae
e Narrow-striped Marsupial Mouse;
Narrow-striped Marsupial Shrew;
Three-striped Marsupial Mouse
d Braunbauch-Streifenbeutelmaus

5255 *Phataginus*
Cimolesta - Manidae
e Tree Pangolins
d Weißbauchschuppentiere
f Pangolins tricuspides
i Pangolini tricuspide

5256 *Phataginus tricuspis*
Cimolesta - Manidae
e Tree Pangolin; African Tree
Pangolin; Small-scaled Tree
Pangolin; Three-pointed Pangolin;
Three-cusped Pangolin; White-
bellied Pangolin
d Weißbauchschuppentier;
Dreizackschuppentier
f Pangolin tricuspide; Pangolin aux
écailles tricuspides; Pangolin
commun
i Pangolino tricuspide

5257 *Phaulomys*
Rodentia - Muridae
e Japanese Voles
d Japanische Rötelmäuse

5258 *Phaulomys andersoni*
Rodentia - Muridae
e Japanese Red-backed Vole; Smith's Vole

5259 *Phaulomys smithii*
Rodentia - Muridae
e Smith's Vole

5260 *Phenacomys*
Rodentia - Muridae
e Heather Voles; Tree Mice; Spruce Mice
d Tannenmäuse; Heidewühlmäuse; Baumwühlmäuse; Heidekrautwühlmäuse

5261 *Phenacomys albipes*
Rodentia - Muridae
e White-footed Vole
d Weißfußwühlmaus

5262 *Phenacomys intermedius*
Rodentia - Muridae
e Western Heather Vole
d Bergheidewühlmaus

5263 *Phenacomys longicaudus*
Rodentia - Muridae
e Red Tree Mouse; Red Tree Vole; Tree Mouse; Oregon Red Tree Vole
d Baumheidewühlmaus; Rote Baumwühlmaus; Tannenmaus

5264 *Phenacomys longicaudus silvivola*
Rodentia - Muridae
e Northwestern Red Tree Vole

5265 *Phenacomys pomo*
Rodentia - Muridae

e Sonoma Tree Vole; California Red Tree Vole
d Sonora-Baumwühlmaus

5266 *Phenacomys ungava*
Rodentia - Muridae
e Eastern Heather Vole

5267 *Philander*
Didelphimorphia - Didelphidae
e Four-eyed Opossums; Quica Opossums; Grey and Black Four-eyed Opossums
d Vieraugenbeutelratten
f Opossums aux quatre yeux

5268 *Philander andersoni*
Didelphimorphia - Didelphidae
e Black Four-eyed Opossum
d Schwarze Vieraugenbeutelratte

5269 *Philander frenata*
Didelphimorphia - Didelphidae
e Bridled Four-eyed Possum

5270 *Philander mcilhennyi*
Didelphimorphia - Didelphidae
e Mcilhenny's Four-eyed Opossum

Philander nudicaudatus syn. *Metachirus nudicaudatus* q.v.

5271 *Philander opossum*
Didelphimorphia - Didelphidae
e Grey Four-eyed Opossum; Four-eyed Opossum; Grey Four-eyed Quica; Four-eyed Quica
d Graue Vieraugenbeutelratte; Quicka; Graue Vieraugenbeutelmaus
f Opossum aux quatre yeux; Quatre-yeux

Philantomba syn. *Cephalophus* q.v.

5272 *Philetor*
Chiroptera - Vespertilionidae

e New Guinea Brown Bats; Rohu's
Bats

5273 *Philetor brachypterus*
Chiroptera - Vespertilionidae
e New Guinea Brown Bat; Rohu's Bat;
Short-winged Brown Bat

5274 *Phloeomys*
Rodentia - Muridae
e Bushy-tailed Rats; Slender-tailed
Cloud Rats; Philippine Varicoloured
Rats; Giant Cloud Rats
d Gescheckte Riesenborkenratten
f Rats d'écorce tachetés; Rats à queue
mince; Rats des nuages

5275 *Phloeomys cumingi*
Rodentia - Muridae
e Slender-tailed Cloud Rat; Southern
Luzon Giant Cloud Rat
d Gescheckte Riesenborkenratte;
Riesenborkenratte; Buschschwanz-
Borkenratte
f Rat d'écorce tacheté; Rat de Cuming

5276 *Phloeomys pallidus*
Rodentia - Muridae
e Northern Luzon Giant Cloud Rat
d Südliche Riesenborkenratte

5277 *Phoca*
Carnivora - Phocidae
e Hair Seals; Harbour Seals; True
Seals; Common Seals
d Seehunde
f Phoques; Veaux-marins
i Foche

Phoca caspica syn. *Pusa caspica* q.v.

Phoca fasciata syn. *Histriophoca
fasciata* q.v.

Phoca groenlandica syn. *Pagophilus
groenlandicus* q.v.

Phoca hispida syn. *Pusa hispida* q.v.

Phoca kurilensis syn. *P. vitulina* q.v.

5278 *Phoca largha*
Carnivora - Phocidae
e Spotted Seal; Largha Seal
d Largha-Robbe
f Phoque tacheté; Phoque
circumpolaire; Phoque chien de mer
larga
i Foca comune del Pacifico

Phoca richardii syn. *P. vitulina* q.v.

Phoca sibirica syn. *Pusa sibirica* q.v.

5279 *Phoca vitulina*
Carnivora - Phocidae
e Common Seal; Bay Seal; Earless
Seal; Eastern Atlantic Seal; European
Sand Seal; Freshwater Seal; Hair
Seal; Harbour Seal; River Seal;
Ranger Seal
d Europäischer Seehund; Gefleckter
Seehund; Gemeiner Seehund;
Seehund
f Phoque commun; Chien-de-mer;
Veau-marin; Phoque veau-marin
i Foca comune; Foca dei porti; Leone
marino; Vitello marino

5280 *Phoca vitulina concolor*
Carnivora - Phocidae
e Western Atlantic Harbour Seal
d Nordamerikanischer Seehund
i Foca comune dell'Atlantico
occidentale

5281 *Phoca vitulina mellonae*
Carnivora - Phocidae
e Ungava Seal
d Labrador-Seehund; Ungava-Seehund
f Phoque commun d'eau douce

5282   *Phoca vitulina richardsi*
            Carnivora - Phocidae
  e   Pacific Harbour Seal
  d   Ostpazifischer Seehund
  f   Phoque commun du Pacifique
  i   Foca comune orientale

5283   *Phoca vitulina stejnegeri*
            Carnivora - Phocidae
  e   Insular Seal; Kurile Seal
  d   Kurilen-Seehund
  i   Foca comune di Kuril

5284   *Phoca vitulina vitulina*
            Carnivora - Phocidae
  e   Eastern Atlantic Harbour Seal
  d   Osteuropäischer Seehund
  i   Foca comune dell'Atlantico orientale

5285   *Phocarctos*
            Carnivora - Otariidae
  e   New Zealand Sea Lions; Aukland
        Islands Sea Lions
  d   Auckland-Seelöwen;
        Neuseeländische Seelöwen
  f   Lions de mer de Nouvelle Zélande
  i   Leoni marini della Nuova Zelanda

5286   *Phocarctos hookeri*
            Carnivora - Otariidae
  e   New Zealand Sea Lion; Aukland
        Islands Sea Lion; Hooker's Sea Lion
  d   Auckland-Seelöwe;
        Neuseeländischer Seelöwe; Hookers
        Seelöwe
  f   Lion de mer de Nouvelle Zélande
  i   Leone marino della Nuova Zelanda;
        Leone marino di Hooker; Leone
        marino delle Auckland

5287   *Phocidae*
            Carnivora
  e   Earless Seals; True Seals; Hair Seals;
        Seals
  d   Hundsrobben; Seehunde
  f   Phocidés; Phoques; Veaux-marins

  i   Foche; Focidi

5288   *Phocoena*
            Cete - Phocoenidae
  e   Common Porpoises; Harbour
        Porpoises
  d   Braunfische; Meerschweine
  f   Marsouins de mer
  i   Focene

5289   *Phocoena phocoena*
            Cete - Phocoenidae
  e   Common Porpoise; Harbour
        Porpoise; Brownfish
  d   Kleiner Tümmler; Schweinswal;
        Tümmler; Braunfisch; Meerschwein;
        Kleintümmler; Gewöhnlicher
        Schweinswal
  f   Marsouin
  i   Focena; Focena comune; Marsuino;
        Marsovino; Delfino bruno; Porco
        marino

5290   *Phocoena phocoena phocoena*
            Cete - Phocoenidae
  e   North Atlantic Harbour Porpoise
  f   Marsouin commun

5291   *Phocoena sinus*
            Cete - Phocoenidae
  e   Gulf Porpoise; California Porpoise;
        Gulf of California Harbour Porpoise;
        Pacific Harbour Porpoise; Cochito;
        Vaquita; Californian Porpoise
  d   Pazifischer Hafenschweinswal;
        Kalifornischer Schweinswal;
        Kalifornischer Hafenschweinswal
  f   Marsouin du Pacifique; Marsouin du
        golfe de Californie
  i   Focena del Golfo di California;
        Focena del Golfo della California;
        Vaquita

5292   *Phocoena spinipinnis*
            Cete - Phocoenidae
  e   Black Porpoise; Burmeister's
        Porpoise

d Burmeister-Schweinswal;
Spitzflossenschweinswal
f Marsuin spinipenne; Marsouin de
Burmeister
i Focena spinipinne; Focena di
Burmeister

5293 Phocoenidae
Cete
e Porpoises
d Schweinswale
f Phocoenidés; Marsouins
i Focenidi

5294 *Phocoenoides*
Cete - Phocoenidae
e Pacific Porpoises
d Dall-Schweinswale
f Marsouins de Dall
i Focenide

5295 *Phocoenoides dalli*
Cete - Phocoenidae
e Dall's Porpoise; Dall's Harbour
Porpoise; Indian Finnless Porpoise;
White-flanked Porpoise; Spray
Porpoise
d Dall-Hafenschweinswal; Dall-
Schweinswal; Hafen-Schweinswal;
Weißflankenschweinswal
f Marsouin de Dall

5296 *Phocoenoides dalli dalli*
Cete - Phocoenidae
e North Pacific Dall's Porpoise

5297 *Phocoenoides dalli truei*
Cete - Phocoenidae
e West Pacific Dall's Porpoise

5298 *Phodopus*
Rodentia - Muridae
e Dwarf Hamsters; Asian Hamsters;
Small Desert Hamsters; Hairy-footed
Hamsters
d Zwerghamster ; Kurzschwänzige

Zwerghamster ; Kurzschwanz-
Zwerghamster
f Hamsters nains
i Criceti nani

5299 *Phodopus campbelli*
Rodentia - Muridae
e Campbell's Hamster; Campbell's
Dwarf Hamster; Dwarf Campbell's
Russian Hamster; Campell's Hairy-
footed Hamster
d Campbell-Zwerghamster
f Hamster nain de Campbell
i Criceto dalle zampe pelose

5300 *Phodopus roborovskii*
Rodentia - Muridae
e Roborovski's Dwarf Hamster; Desert
Hamster; Roborovski Hamster
d Roborowskis Zwerghamster;
Roborowski-Zwerghamster

f Hamster nain de Roborowski
i Fodopo nano; Criceto nano di
Roborovski; Criceto delle sabbie

5301 *Phodopus sungorus*
Rodentia - Muridae
e Striped Hairy-footed Hamster;
Dzhungarian Hamster
d Dschungarischer Zwerghamster
f Hamster nain de Djoungarie
i Criceto russo nano; Criceto russo

5302 *Phodopus sungorus sungorus*
Rodentia - Muridae
e White Russian Dwarf Hamster;
Winter White Russian Hamster
d Weißer Dschungarischer
Zwerghamster
f Hamster russe

Phoniscus syn. *Kerivoula* q.v.

Phrygetis syn. *Myonicteris* q.v.

5303 *Phylloderma*
Chiroptera - Phyllostomidae
e   Peters's Spear-nosed Bats; Pale-faced Bats

5304 *Phylloderma stenops*
Chiroptera - Phyllostomidae
e   Peters's Spear-nosed Bat; Pale-faced Bat

Phyllodia syn. *Pteronotus* q.v.

5305 *Phyllonycteris*
Chiroptera - Phyllostomidae
e   Smooth-toothed Flower Bats
f   Vampires des fleures

5306 *Phyllonycteris aphylla*
Chiroptera - Phyllostomidae
e   Jamaican Flower Bat; Jamaican Pallid Flower Bat
f   Vampire des fleurs de Jamaïque

Phyllonycteris obtusa syn. *P. poeyi* q.v.

5307 *Phyllonycteris poeyi*
Chiroptera - Phyllostomidae
e   Cuban Flower Bat; Haitian Long-tongued Bat
f   Vampire des fleurs de Cuba

5308 *Phyllops*
Chiroptera - Phyllostomidae
e   Fig-eating Bats; Cuban Fig-eating Bats; Falcate-winged Bats

5309 *Phyllops falcatus*
Chiroptera - Phyllostomidae
e   Falcate-winged Bat; Cuban Fig-eating Bat; Lesser Falcate-winged Bat; White-shouldered Bat

5310 **Phyllostomidae**
Chiroptera
e   American Leaf-nosed Bats; New-

world Leaf-nosed Bats; Neotropical Leaf-nosed Bats
d   Blattnasen; Blattnasenartige; Blattnasenfledermäuse
f   Phyllostomidés
i   Filostomidi

5311 *Phyllostomus*
Chiroptera - Phyllostomidae
e   Spear-nosed Bats
d   Lanzennasen
f   Phyllostomes
i   Pipistrelli frugivori dal naso a lancia

5312 *Phyllostomus discolor*
Chiroptera - Phyllostomidae
e   Coloured Bat; Pale Spear-nosed Bat; Long-tongued Spear-nosed Bat
d   Bunte Lanzennase; Kleiner Lanzennasenfruchtvampir
f   Phyllostome coloré

5313 *Phyllostomus elongatus*
Chiroptera - Phyllostomidae
e   Lesser Spear-nosed Bat

5314 *Phyllostomus hastatus*
Chiroptera - Phyllostomidae
e   Javelin Bat; Greater Spear-nosed Bat; Pallas's Spear-nosed Bat
d   Lanzennase
f   Phyllostome fer-de-lance

5315 *Phyllostomus latifolius*
Chiroptera - Phyllostomidae
e   Guianan Spear-nosed Bat

5316 *Phyllotis*
Rodentia - Muridae
e   Leaf-eared Mice; Pericores
d   Blattohrmäuse
f   Souris aux oreilles en feuille
i   Topi di montagna

5317 *Phyllotis amicus*
Rodentia - Muridae
e Friendly Leaf-eared Mouse

5318 *Phyllotis andium*
Rodentia - Muridae
e Andean Leaf-eared Mouse

Phyllotis boliviensis syn. *Auliscomys boliviensis* q.v.

5319 *Phyllotis bonaeriensis*
Rodentia - Muridae
e Buenos Aires Leaf-eared Mouse

5320 *Phyllotis caprinus*
Rodentia - Muridae
e Capricorn Leaf-eared Mouse

5321 *Phyllotis darwini*
Rodentia - Muridae
e Darwin's Leaf-eared Mouse
d Darwin-Maus
f Souris aux oreilles touffues de Darwin

5322 *Phyllotis definitus*
Rodentia - Muridae
e Definite Leaf-eared Mouse; Definitive Leaf-eared Mouse

Phyllotis domorum syn. *Graomys domorum* q.v.

Phyllotis edithae syn. *Graomys edithae* q.v.

5323 *Phyllotis gerbillus*
Rodentia - Muridae
e Gerbil Leaf-eared Mouse

5324 *Phyllotis haggardi*
Rodentia - Muridae
e Haggard's Leaf-eared Mouse

5325 *Phyllotis limatus*
Rodentia - Muridae
e Narrow-toothed Leaf-eared Mouse

5326 *Phyllotis magister*
Rodentia - Muridae
e Master Leaf-eared Mouse

5327 *Phyllotis osgoodi*
Rodentia - Muridae
e Osgood's Leaf-eared Mouse

5328 *Phyllotis osilae*
Rodentia - Muridae
e Bunchgrass Leaf-eared Mouse

Phyllotis pictus syn. *Auliscomys pictus* q.v.

Phyllotis sublimus syn. *Auliscomys sublimus* q.v.

5329 *Phyllotis wolffsohni*
Rodentia - Muridae
e Wolffson's Leaf-eared Mouse

5330 *Phyllotis xanthopygus*
Rodentia - Muridae
e Yellow-rumped Leaf-eared Mouse

5331 *Physeter*
Cete - Physeteridae
e Sperm Whales
d Pottwale
f Cachalots
i Capodogli

5332 *Physeter catodon*
Cete - Physeteridae
e Cachalot; Sperm Whale; Spermacet Whale; Pot Whale; Great Sperm Whale
d Kaschelot; Großer Pottwal; Pottwal; Spermwal
f Cachalot macrocéphale; Cachalot

   i   Capodoglio; Capodoglio gigante;
       Fisetere; Organante; Cascialotto

     Physeter macrocephalus syn. *P.*
     catodon q.v.

5333 **Physeteridae**
     Cete
   e   Sperm Whales; Pot Whales; Great
       Sperm Whales
   d   Pottwale; Pottwalartige
   f   Cachalots; Physitéridés
   i   Fisiteridi

5334 *Piliocolobus*
     Primates - Cercopithecidae
   e   Red Colobus Monkeys; Red-and-
       olive Colubuses
   d   Rote Stummelaffen
   f   Colobes
   i   Colobi

5335 *Piliocolobus badius*
     Primates - Cercopithecidae
   e   Red Colobus; Bay Colobus; Western
       Red Colobus; West African Red
       Colobus
   d   Roter Stummelaffe; Westlicher Roter
       Stummelaffe
   f   Colobe baie; Colobe baie d'Afrique
       occidentale; Colobe rouge
   i   Colobo ferruginoso

5336 *Piliocolobus badius waldroni*
     Primates - Cercopithecidae
   e   Miss Waldron's Red Colobus
   d   Miss Waldrons Roter Stummelaffe

5337 *Piliocolobus foai*
     Primates - Cercopithecidae
   e   Central African Red Colobus
   d   Zentralafrikanischer Roter
       Stummelaffe

5338 *Piliocolobus gordonorum*
     Primates - Cercopithecidae

   e   Udzungwa Red Colobus
   d   Uhehe-Stummelaffe

5339 *Piliocolobus kirkii*
     Primates - Cercopithecidae
   e   Zanzibar Red Colobus; Kirk's
       Colobus
   d   Sansibar-Stummelaffe
   f   Colobe roux de Zanzibar; Colobe bai
       de Kirk
   i   Colobo dorso rosso; Colobo
       ferruginea; Colobo dorso rosso di
       Zansibar

5340 *Piliocolobus pennantii*
     Primates - Cercopithecidae
   e   West Central African Red Colobus;
       Pennant's Red Colobus; Pennant's
       Colobus; Central Red Colobus
   d   Pennant-Stummelaffe
   f   Colobe bai de Zanzibar; Colobe bai
       de Pennant
   i   Colobo rosso di Pennant

5341 *Piliocolobus pennantii pennantii*
     Primates - Cercopithecidae
   e   Bioko Red Colobus
   f   Colobe de Pennant

5342 *Piliocolobus preussi*
     Primates - Cercopithecidae
   e   Preuss's Red Colobus; Cameroon
       Red Colobus; Preuss's Colobus;
       Korup Red Colobus
   d   Preuss-Stummelaffe; Kamerun-
       Stummelaffe
   f   Colobe roux du Cameroun
   i   Colobo rosso di Preuss

5343 *Piliocolobus rufomitratus*
     Primates - Cercopithecidae
   e   Tana River Red Colobus; East
       Central African Red Colobus
   d   Rotkopfstummelaffe; Roter Colobus
   f   Colobe roux de la Tana; Colobe bai à
       tête rousse
   i   Colobo rosso del Fiume Tana

5344 **Piliocolobus tephrosceles**
　　　Primates - Cercopithecidae
　e　Ugandan Red Colobus
　d　Uganda Roter Stummelaffe
　i　Colobo rosso

5345 **Piliocolobus tholloni**
　　　Primates - Cercopithecidae
　e　Thollon's Red Colobus
　d　Thollons Roter Stummelaffe
　f　Colobe bai aux mains noires

5346 **Pipanacoctomys**
　　　Rodentia - Octodontidae
　e　Viscacha Rats
　d　Viscacha-Ratten

5347 **Pipanacoctomys aureus**
　　　Rodentia - Octodontidae
　e　Golden Viscacha Rat

5348 **Pipistrellus**
　　　Chiroptera - Vespertilionidae
　e　Pipistrelles
　d　Zwergfledermäuse
　f　Pipistrelles
　i　Pipistrelli

5349 **Pipistrellus adamsi**
　　　Chiroptera - Vespertilionidae
　e　Cape York Pipistrelle

5350 **Pipistrellus aegypticus**
　　　Chiroptera - Vespertilionidae
　e　Egyptian Pipistrelle

5351 **Pipistrellus aero**
　　　Chiroptera - Vespertilionidae
　e　Gargues Pipistrelle; Mount Gargues
　　　Pipistrelle

5352 **Pipistrellus affinis**
　　　Chiroptera - Vespertilionidae

　e　Chocolate Pipistrelle

5353 **Pipistrellus anchietae**
　　　Chiroptera - Vespertilionidae
　e　Anchieta's Pipistrelle; d'Anchieta's
　　　Pipistrelle

5354 **Pipistrellus angulatus**
　　　Chiroptera - Vespertilionidae
　e　New Guinea Pipistrelle

5355 **Pipistrellus anthonyi**
　　　Chiroptera - Vespertilionidae
　e　Anthony's Pipistrelle

5356 **Pipistrellus arabicus**
　　　Chiroptera - Vespertilionidae
　e　Arabian Pipistrelle

5357 **Pipistrellus ariel**
　　　Chiroptera - Vespertilionidae
　e　Desert Pipistrelle
　f　Pipistrelle d'Ariel

5358 **Pipistrellus aureocollaris**
　　　Chiroptera - Vespertilionidae
　e　Gold-collared Gilded Pipistrelle

5359 **Pipistrellus babu**
　　　Chiroptera - Vespertilionidae
　e　Himalajan Pipistrelle

5360 **Pipistrellus baverstocki**
　　　Chiroptera - Vespertilionidae
　e　Inland Forest Bat

5361 **Pipistrellus bodenheimeri**
　　　Chiroptera - Vespertilionidae
　e　Bodenheimer's Pipistrelle

5362 **Pipistrellus brunneus**
　　　Chiroptera - Vespertilionidae
　e　Dark-brown Serotine

5363   *Pipistrellus cadornae*
Chiroptera - Vespertilionidae
e   Cadorna's Pipistrelle

5364   *Pipistrellus capensis*
Chiroptera - Vespertilionidae
e   Cape Serotine

5365   *Pipistrellus caurinus*
Chiroptera - Vespertilionidae
e   Northern Cave Bat

5366   *Pipistrellus ceylonicus*
Chiroptera - Vespertilionidae
e   Kelaart's Pipistrelle
d   Sri Lanka Zwergfledermaus

5367   *Pipistrellus circumdatus*
Chiroptera - Vespertilionidae
e   Large Black Pipistrelle; Gilded Black
       Pipistrelle; Black Gilded Pipistrelle

5368   *Pipistrellus collinus*
Chiroptera - Vespertilionidae
e   Mountain Pipistrelle

5369   *Pipistrellus coromandra*
Chiroptera - Vespertilionidae
e   Indian Pipistrelle
d   Indische Zwergfledermaus
f   Pipistrelle de l'Inde

5370   *Pipistrellus crassulus*
Chiroptera - Vespertilionidae
e   Broad-headed Pipistrelle

5371   *Pipistrellus cuprosus*
Chiroptera - Vespertilionidae
e   Coppery Pipistrelle; Coppery Gilded
       Pipistrelle

5372   *Pipistrellus darlingtoni*
Chiroptera - Vespertilionidae
e   Large Forest Bat

5373   *Pipistrellus dormeri*
Chiroptera - Vespertilionidae
e   Dormer's Pipistrelle

5374   *Pipistrellus douglasorum*
Chiroptera - Vespertilionidae
e   Yellow-lipped Bat

5375   *Pipistrellus eisentrauti*
Chiroptera - Vespertilionidae
e   Eisentraut's Pipistrelle

5376   *Pipistrellus endoi*
Chiroptera - Vespertilionidae
e   Endo's Pipistrelle

5377   *Pipistrellus finlaysoni*
Chiroptera - Vespertilionidae
e   Inland Cave Bat

5378   *Pipistrellus flavescens*
Chiroptera - Vespertilionidae
e   Yellow Serotine

5379   *Pipistrellus guineensis*
Chiroptera - Vespertilionidae
e   Tiny Serotine

5380   *Pipistrellus hesperus*
Chiroptera - Vespertilionidae
e   Canyon Bat; Western Pipistrelle;
       American Western Pipistrelle

5381   *Pipistrellus hesperus merriami*
Chiroptera - Vespertilionidae
e   Merriam's Canyon Bat

5382   *Pipistrellus imbricatus*
Chiroptera - Vespertilionidae
e   Brown Pipistrelle

5383   *Pipistrellus inexpectatus*
Chiroptera - Vespertilionidae
e   Aellen's Pipistrelle

5384   *Pipistrellus javanicus*
Chiroptera - Vespertilionidae
e   Javan Pipistrelle; Javanese Pipistrelle

5385   *Pipistrellus joffrei*
Chiroptera - Vespertilionidae
e   Joffre's Pipistrelle

5386   *Pipistrellus kitcheneri*
Chiroptera - Vespertilionidae
e   Red-brown Pipistrelle

5387   *Pipistrellus kuhlii*
Chiroptera - Vespertilionidae
e   Kuhl's Pipistrelle; Kuhl's
Flittermouse
d   Weißrandfledermaus; Weißrandige
Fledermaus
f   Chauve-souris des maisons;
Pipistrelle de Kuhl; Pipistrelle
marginée; Vespérien de Kuhl
i   Pipistrello albolimbato

5388   *Pipistrellus lophurus*
Chiroptera - Vespertilionidae
e   Birma Pipistrelle

5389   *Pipistrellus mackenziei*
Chiroptera - Vespertilionidae
e   Western False Pipistrelle

5390   *Pipistrellus macrotis*
Chiroptera - Vespertilionidae
e   Big-eared Pipistrelle

5391   *Pipistrellus maderensis*
Chiroptera - Vespertilionidae
e   Madeira Pipistrelle
d   Madeira-Fledermaus
f   Pipistrelle de Madère

5392   *Pipistrellus matroka*
Chiroptera - Vespertilionidae
e   Madagascar Serotine

5393   *Pipistrellus melckorum*
Chiroptera - Vespertilionidae
e   Melck's House Bat

5394   *Pipistrellus mimus*
Chiroptera - Vespertilionidae
e   Indian Pygmy Pipistrelle; Pygmy
Pipistrelle
d   Indische Kleine Zwergfledermaus

5395   *Pipistrellus minahassae*
Chiroptera - Vespertilionidae
e   Minehassa Pipistrelle

5396   *Pipistrellus mordax*
Chiroptera - Vespertilionidae
e   Pungent Pipitrelle; Grizzled
Pipistrelle

5397   *Pipistrellus musciculus*
Chiroptera - Vespertilionidae
e   Mouse-like Pipistrelle

5398   *Pipistrellus nanulus*
Chiroptera - Vespertilionidae
e   Tiny Pipistrelle

5399   *Pipistrellus nanus*
Chiroptera - Vespertilionidae
e   Banana Pipistrelle
d   Bananen-Zwergfledermaus
f   Pipistrelle naine aux ailes brunes;
Pipistrelle naine

5400   *Pipistrellus nathusii*
Chiroptera - Vespertilionidae
e   Nathusius's Pipistrelle
d   Rauhhautfledermaus; Rauhhäutige
Fledermaus; Schienenhaarige
Fledermaus
f   Pipistrelle de Nathusius; Vespérien
de Nathusius; Vespérugo de
Nathusius
i   Pipistrello di Nathusius

5401   *Pipistrellus papuanus*
Chiroptera - Vespertilionidae
e   Papuan Pipistrelle

5402   *Pipistrellus paterculus*
Chiroptera - Vespertilionidae
e   Mount Popa Pipistrelle

5403   *Pipistrellus peguensis*
Chiroptera - Vespertilionidae
e   Pegu Pipistrelle

5404   *Pipistrellus permixtus*
Chiroptera - Vespertilionidae
e   Dar es Salam Pipistrelle

5405   *Pipistrellus petersi*
Chiroptera - Vespertilionidae
e   Peters's Pipistrelle

5406   *Pipistrellus pipistrellus*
Chiroptera - Vespertilionidae
e   Common Pipistrelle Bat; Common
Pipistrelle; Pipistrelle
d   Gemeine Zwergfledermaus;
Zwergfledermaus
f   Chauve-souris pipistrelle; Vespérien
pipistrelle; Vespertilion pipistrelle;
Pipistrelle; Pipistrelle commune;
Pipistrelle pygmée
i   Pipistrello nano

5407   *Pipistrellus pulveratus*
Chiroptera - Vespertilionidae
e   Chinese Pipistrelle

5408   *Pipistrellus pumilus*
Chiroptera - Vespertilionidae
e   Eastern Forest Bat; Little Bat

5409   *Pipistrellus pygmaeus*
Chiroptera - Vespertilionidae
e   Soprano Pipistrelle

d   Mückenfledermaus
f   Pipistrelle soprane; Pipistrelle
pygmée; Pipistrelle 55 kHz;
Pipistrelle méditerranéenne; Soprano
i   Pipistrello soprano; Pipistrello
pigmeo

5410   *Pipistrellus regulus*
Chiroptera - Vespertilionidae
e   Southern Forest Bat; King River Bat

5411   *Pipistrellus rendalli*
Chiroptera - Vespertilionidae
e   Rendall's Serotine

5412   *Pipistrellus rueppelli*
Chiroptera - Vespertilionidae
e   Rüppell's Bat; Rüppell's Pipistrelle
f   Pipistrelle de Rüppel

5413   *Pipistrellus rusticus*
Chiroptera - Vespertilionidae
e   Rusty Bat; Rusty Pipistrelle

5414   *Pipistrellus savii*
Chiroptera - Vespertilionidae
e   Savi's Pipistrelle
d   Alpenfledermaus; Bergflatterer
f   Vespérien maure; Pipistrelle de Savi;
Vespère de Savi
i   Pipistrello di Savi

5415   *Pipistrellus societatis*
Chiroptera - Vespertilionidae
e   Social Pipistrelle; Social Gilded
Pipistrelle

5416   *Pipistrellus somalicus*
Chiroptera - Vespertilionidae
e   Somali Pipistrelle

5417   *Pipistrellus stenopterus*
Chiroptera - Vespertilionidae
e   Narrow-winged Pipistrelle

5418 *Pipistrellus sturdeei*
Chiroptera - Vespertilionidae
e Sturdee's Pipistrelle; Bonin
Pipistrelle

5419 *Pipistrellus subflavus*
Chiroptera - Vespertilionidae
e American Eastern Pipistrelle; Eastern
Pipistrelle (NA)
d Ostpipistrelle
f Pipistrelle de l'Est

5420 *Pipistrellus tasmaniensis*
Chiroptera - Vespertilionidae
e False Pipistrelle; Tasmanian
Pipistrelle; Eastern False Pipistrelle
(ANZ)

5421 *Pipistrellus tenuipinnis*
Chiroptera - Vespertilionidae
e White-winged Serotine

5422 *Pipistrellus tenuis*
Chiroptera - Vespertilionidae
e Least Pipistrelle
d Weißflügelfledermaus

5423 *Pipistrellus torquatus*
Chiroptera - Vespertilionidae
e Taiwan Gilded Pipistrelle

5424 *Pipistrellus troughtoni*
Chiroptera - Vespertilionidae
e Eastern Cave Bat

5425 *Pipistrellus vordermanni*
Chiroptera - Vespertilionidae
e White-winged Pipistrelle

5426 *Pipistrellus vulturnus*
Chiroptera - Vespertilionidae
e Little Forest Bat

5427 *Pipistrellus wattsi*
Chiroptera - Vespertilionidae

e Watts's Pipistrelle

5428 *Pipistrellus westralis*
Chiroptera - Vespertilionidae
e Mangrove Pipistrelle

5429 *Pithanotomys*
Rodentia Octodontidae -
Octodontidae
e South American Rock Rat
d Südamerikanische Felsenratten;
Felsenratten
f Octodontes

5430 *Pithanotomys fuscus*
Rodentia Octodontidae -
Octodontidae
e Chilean Rock Rat
d Südamerikanische Felsenratte

5431 *Pithanotomys sagei*
Rodentia Octodontidae -
Octodontidae
e Sage's Rock Rat

5432 *Pithecheir*
Rodentia - Muridae
e Red Tree Rats; Red Bush Rats;
Malayan Tree Rats; Monkey-footed
Rats; Sunda Tree Rats
d Rote Baumratten

5433 *Pithecheir melanurus*
Rodentia - Muridae
e Malayan Red Tree Rat; Red Tree Rat
d Rote Baumratte

5434 *Pithecheir parvus*
Rodentia - Muridae
e Malayan Tree Rat; Monkey-footed
Rat

5435 *Pithecia*
Primates - Atelidae
e Sakis; Saki Monkeys
d Schweifaffen; Sakis

f   Sakis; Pithécies; Sakis moine
i   Pitecie

5436   *Pithecia aequatorialis*
        Primates - Atelidae
e   Eqatorial Saki
f   Saki à perruque
i   Pitecia equatoriale

5437   *Pithecia albicans*
        Primates - Atelidae
e   Buffy Saki; White Saki; Whitish Saki
d   Schwarzrückenmönchsaffe
f   Saki chamois
i   Pitecia bianca

5438   *Pithecia irrorata*
        Primates - Atelidae
e   Grey Monk Saki; Bald-face Saki
i   Pitecia calva

5439   *Pithecia monachus*
        Primates - Atelidae
e   Hairy Saki; Monk Saki; Red-bearded
      Saki
d   Zottelschweifaffe; Mönchsaffe;
      Rotbärtiger Mönchsaffe
f   Saki moine; Saki hirsute; Saki à
      perruque
i   Pitecia monaca; Saki monaco

5440   *Pithecia pithecia*
        Primates - Atelidae
e   Pale-faced Saki; Pale-headed Saki;
      White-masked Saki; Guianan Saki;
      Shaggy Saki; White-headed Saki;
      White-faced Saki
d   Weißkopfsaki; Blasskopfsaki
f   Saki à tête pâle
i   Pitecia dalla faccia bianca; Pitecia
      dalla testa bianca; Saki; Saki dalla
      testa nera

5441   *Pithecia pithecia chrysocephalus*
        Primates - Atelidae
e   Golden-headed Saki

d   Goldkopfsaki
f   Saki à tête dorée

***Pitymys* syn. *Microtus* q.v.**

5442   *Plagiodontia*
        Rodentia - Capromyidae
e   Hispanolian Hutias; Zagoutis
d   Zagutis
f   Plagiodontes

5443   *Plagiodontia aedium*
        Rodentia - Capromyidae
e   Haitian Hutia; Hispaniolan Hutia
d   Cuvier-Zaguti; Cuviers Zaguti
f   Plagiodonte d'Haïti

5444   *Plagiodontia aedium aedium*
        Rodentia - Capromyidae
e   Cuvier's Hutia; Common Cuvier's
      Hutia; Common Cuvier's Zagouti
d   Eigentliches Cuvier-Zaguti;
      Eigentliche Cuvier-Ferkelratte;
      Eigentliche Cuvier-Hutia
f   Rat grimpeur de Cuvier commun;
      Hutia de Cuvier commun;
      Plagiodonte de Cuvier commun;
      Plagiodonte de Haïti commun; Rat
      Cayes commun

5445   *Plagiodontia aedium hylaeum*
        Rodentia - Capromyidae
e   Dominican Hutia
d   Dominikanisches Zaguti

5446   *Planigale*
        Dasyuromorphia - Dasyuridae
e   Planigales (ANZ); Flat-skulled
      Marsupials; Flat-skulled Marsupial
      Mice
d   Flachkopfbeutelmäuse
f   Planigales

5447   *Planigale gilesi*
        Dasyuromorphia - Dasyuridae
e   Paucident Planigale

d   Giles-Flachkopfbeutelmaus
f   Planigale de Giles

5448   *Planigale ingrami*
Dasyuromorphia - Dasyuridae
e   Ingram's Planigale; Northern Planigale; Flat-skulled Marsupial Mouse; Long-tailed Planigale
d   Nördliche Flachkopfbeutelmaus; Nordflachkopfbeutelmaus; Zwergflachkopfbeutelmaus

5449   *Planigale maculata*
Dasyuromorphia - Dasyuridae
e   Pygmy Planigale; Common Planigale (ANZ)
d   Zwergbeutelmaus
f   Planigale commun

5450   *Planigale novaeguineae*
Dasyuromorphia - Dasyuridae
e   New Guinea Planigale; Papua Marsupial Mouse; New Guinean Planigale; Papuan Planigale
d   Neuguinea-Flachkopfbeutelmaus

Planigale subtilissima syn. *P. ingrami* q.v.

5451   *Planigale tenuirostris*
Dasyuromorphia - Dasyuridae
e   Dusky Planigale; Narrow-nosed Planigale; Southern Planigale
d   Südflachkopfbeutelmaus; Südliche Flachkopfbeutelmaus

5452   *Platacanthomys*
Rodentia - Muridae
e   Spiny Dormice; Long-tailed Spiny Mice; Pepper Rats; Malabar Spiny Dormice
d   Südindische Stachelbilche
f   Loirs épineux
i   Ghiri spinosi

5453   *Platacanthomys lasiurus*
Rodentia - Muridae

e   Spiny Dormouse; Malabar Spiny Dormouse; Pepper Rat; Long-tailed Spiny Mouse
d   Südindischer Stachelbilch
f   Loir épineux
i   Ghiro spinoso

5454   *Platalina*
Chiroptera - Phyllostomidae
e   Long-snouted Bats

5455   *Platalina genovensium*
Chiroptera - Phyllostomidae
e   Long-snouted Bat

5456   *Platanista*
Cete - Platanistidae
e   Ganges Dolphins; Ganges River Dolphins; Gangetic Dolphins; Blind Dolphins; Ganges and Indus Dolphins
d   Ganges-Delfine
f   Platanistes; Dauphins du Gange
i   Delfini d'acqua dolce; Plataniste

5457   *Platanista gangetica*
Cete - Platanistidae
e   Ganges Dolphin; Ganges River Dolphin; Gangetic Dolphin; Blind Dolphin; Susu; Indian River Dolphin; South Asian River Dolphin; Ganges Susu
d   Ganges-Delfin; Südasiatischer Flussdelfin; Ganges-Susu
f   Dauphin d'eau douce du Gange; Plataniste du Gange
i   Platanista del Gange

Platanista indi syn. *P. minor* q.v.

5458   *Platanista minor*
Cete - Platanistidae
e   Indus Dolphin; Indus River Dolphin; Indus Susu
d   Indus-Delfin; Indus-Susu

f   Plataniste de l'Indus

i   Platanista dell'Indo

5459   Platanistidae
Cete

e   River Dolphins; Freshwater
Dolphins; Indian River Dolphins

d   Flussdelfine; Flussdelfinartige

f   Dauphins d'eau douce; Dauphins de
rivière; Platanistidés

i   Delfini di fiume; Platanistidi

Platymops syn. *Mormopterus* q.v.

5460   *Platyrrhinus*
Chiroptera - Phyllostomidae

e   White-lined Bats; Broad-nosed Bats

Platyrrhinus aquilus syn. *P.
umbratus* q.v.

5461   *Platyrrhinus aurarius*
Chiroptera - Phyllostomidae

e   Eldorado Broad-nosed Bat

5462   *Platyrrhinus brachycephalus*
Chiroptera - Phyllostomidae

e   Short-headed Broad-nosed Bat;
Dark-lined Bat

5463   *Platyrrhinus chocoensis*
Chiroptera - Phyllostomidae

e   Choco Broad-nosed Bat

5464   *Platyrrhinus dorsalis*
Chiroptera - Phyllostomidae

e   Thomas's Broad-nosed Bat

5465   *Platyrrhinus helleri*
Chiroptera - Phyllostomidae

e   Heller's Bat; Heller's Broad-nosed
Bat; Heller's White-lined Bat

5466   *Platyrrhinus infuscus*
Chiroptera - Phyllostomidae

e   Buffy Broad-nosed Bat

d   Blasse Streifenfruchtfledermaus

5467   *Platyrrhinus lineatus*
Chiroptera - Phyllostomidae

e   White-lined Broad-nosed Bat; White-
lined Bat; Geoffroy's Rayed Bat

f   Sténoderme pseudo-vampire

i   Vampiro dalle strisce bianche

Platyrrhinus nigellus syn. *P.
lineatus* q.v.

Platyrrhinus oratus syn. *V. umbratus*
q.v.

5468   *Platyrrhinus recifinus*
Chiroptera - Phyllostomidae

e   Recife Broad-nosed Bat

5469   *Platyrrhinus umbratus*
Chiroptera - Phyllostomidae

e   Shadowy Broad-nosed Bat

5470   *Platyrrhinus vittatus*
Chiroptera - Phyllostomidae

e   Greater White-lined Bat; White-lined
Bat; Greater Broad-nosed Bat

5471   *Plecotus*
Chiroptera - Vespertilionidae

e   Long-eared Bats; Lump-nosed Bats;
Lapped-eared Bats; Big-eared Bats

d   Großohren; Langohren; Bindeohren;
Ohrenfledermäuse;
Langohrfledermäuse

f   Oreillards

i   Orechioni

5472   *Plecotus alpinus*
Chiroptera - Vespertilionidae

e   Alpine Long-eared Bat

d   Alpenlangohr

f   Oreillard des Alpes

i   Orecchione alpino; Orecchione bruno

5473 *Plecotus auritus*
Chiroptera - Vespertilionidae
- *e* Long-eared Bat; Common Long-eared Bat; European Long-eared Bat; Brown Long-eared Bat; Brown Big-eared Bat
- *d* Langohrige Fledermaus; Langohren; Braunes Langohr; Langohrfledermaus; Gemeine Langohrfledermaus; Großohr; Großohrfledermaus; Großohrenfledermaus; Bindeohr; Ohrenfledermaus
- *f* Oreillard roux; Oreillard settentrional; Oreillard commun; Oreillard brun
- *i* Orecchione; Orecchione comune; Orecchione bruno ; Orecchione settentrionale

5474 *Plecotus austriacus*
Chiroptera - Vespertilionidae
- *e* Grey Long-eared Bat; Grey Big-eared Bat
- *d* Graues Langohr; Graue Langohrfledermaus
- *f* Oreillard gris; Oreillard méridional
- *i* Orecchione meridionale; Orecchione grigio

5475 *Plecotus balensis*
Chiroptera - Vespertilionidae
- *e* Bale Long-eared Bat
- *d* Äthiopisches Langohr

5476 *Plecotus kolombatovici*
Chiroptera - Vespertilionidae
- *e* Kolombatovic's Long-eared Bat
- *d* Balkan-Langohr
- *f* Oreillard des Balkans
- *i* Orecchione di Kolombatovic

5477 *Plecotus macrobullaris*
Chiroptera - Vespertilionidae
- *e* Alpine Long-eared Bat
- *f* Oreillard montagnard; Oreillard de montagne

5478 *Plecotus mexicanus*
Chiroptera - Vespertilionidae
- *e* Mexican Big-eared Bat
- *d* Mexikanische Langohr

Plecotus phyllotis syn. *Idionycteris phyllotis* q.v.

5479 *Plecotus rafinesquei*
Chiroptera - Vespertilionidae
- *e* Eastern Lump-nosed Bat; Eastern Big-eared Bat; Southeastern Big-eared Bat; Rafinesque's Big-eared Bat

5480 *Plecotus sardus*
Chiroptera - Vespertilionidae
- *e* Sardinian Long-eared Bat
- *d* Sardisches Langohr
- *f* Oreillard de Sardaigne
- *i* Orecchione sardo

5481 *Plecotus taivanus*
Chiroptera - Vespertilionidae
- *e* Taiwan Big-eared Bat
- *d* Taiwan-Langohr

5482 *Plecotus teneriffae*
Chiroptera - Vespertilionidae
- *e* Canary Big-eared Bat
- *d* Kanaren-Langohrfledermaus
- *f* Oreillard de Ténérife

5483 *Plecotus townsendi*
Chiroptera - Vespertilionidae
- *e* Townsend's Big-eared Bat; Western Big-eared Bat; Lump-nosed Bat; Townsends Long-eared Bat
- *d* Virginia-Langohrflederenmaus; Townsend-Langohr
- *f* Oreillard de Townsend

5484 *Plecotus townsendi ingens*
Chiroptera - Vespertilionidae
- *e* Ozark Big-eared Bat

5485   *Plerotes*
Chiroptera - Pteropodidae
e   Anchieta's Fruit Bats; d'Anchieta's
Fruit Bats

5486   *Plerotes anchietai*
Chiroptera - Pteropodidae
e   Anchieta's Fruit Bat; d'Anchieta's
Fruit Bat

Podihik syn. *Suncus* q.v.

5487   *Podogymnura*
Ernaceomorpha - Erinaceidae
e   Mindanao Gymnures; Mindanao
Moonrats; Philippine Gymnures;
Woood Shrews
d   Philippinen-Rattenigel
f   Gimnure; Rats de la lune
i   Ricci spinosi; Ratti della luna

5488   *Podogymnura aureospinula*
Ernaceomorpha - Erinaceidae
e   Spiny Moonrat; Dinagat Gymnure;
Golden-spined Hedgehog
d   Dinagat-Rattenigel

5489   *Podogymnura truei*
Ernaceomorpha - Erinaceidae
e   Mindanao Gymnure; Mindanao
Moonrat; Philippine Gymnure;
Woood Shrew
d   Philippinen-Rattenigel

5490   *Podomys*
Rodentia - Muridae
e   Florida Mice
d   Florida-Mäuse
f   Souris de Floride

5491   *Podomys floridanus*
Rodentia - Muridae
e   Florida Mouse
d   Florida-Maus
f   Souris de Floride

5492   *Podoxymys*
Rodentia - Muridae
e   Mount Roraima Mice; Roraima Mice
d   Roraima-Mäuse

5493   *Podoxymys roraimae*
Rodentia - Muridae
e   Mount Roraima Mouse; Roraima
Mouse
d   Roraima-Maus

5494   *Poecilictis*
Carnivora - Mustelidae
e   Saharan Striped Polecats; Libyan
Striped Weasels
d   Libysches Streifenwiesel
f   Zorilles de Libyie; Zorillas de Libyie
i   Zorille libiche

5495   *Poecilictis libyca*
Carnivora - Mustelidae
e   Saharan Striped Polecat; Striped
Weasel; White-naped Weasel; North
African Striped Weasel; Saharan
Striped Weasel; Zoril; Zorilla;
Libyan Striped Weasel
d   Libysches Streifenwiesel
f   Zorille de Libye; Zorilla de Libye
i   Zorilla libica; Zorilla

5496   *Poecilogale*
Carnivora - Mustelidae
e   African Striped Weasels; Cape
Weasels; White-naped Weasels
d   Weißnackenwiesel
f   Poecilogales
i   Donnole striate

5497   *Poecilogale albinucha*
Carnivora - Mustelidae
e   African Striped Weasel; Cape
Weasel; White-naped Weasel; Snake
Mongoose; Striped Weasel; Striped
Polecat
d   Weißnackenwiesel
f   Poecilogale à nuque blanche; Putois

congolais; Chat de brousse
i   Donnola striata; Donnola nucabianca

5498   *Poelagus*
Lagomorpha - Leporidae
e   Scrub Hares (CSA); Uganda Hares; Bunyoro Rabbits
d   Zentralafrikanische Buschkaninchen
f   Lapins sauvages d'Afrique
i   Conigli selvatici africani

5499   *Poelagus marjorita*
Lagomorpha - Leporidae
e   Scrub Hare (CSA); African Grass Hare; Central African Hare; Bunyoro Rabbit; Uganda Grass Hare; Central African Rabbit

d   Zentralafrikanisches Buschkaninchen; Buschkaninchen; Uganda-Grashase
f   Lapin sauvage d'Afrique
i   Coniglio selvatico africano

5500   *Pogonomelomys*
Rodentia - Muridae
e   Rümmler's Mosaic-tailed Rats; Mosaic-tailed Rats; Brush Mice; New Guinean Brush Mice; Pogonomelomyses
d   Rümmler-Mosaikschwanzratten
f   Rats de Rümmler

5501   *Pogonomelomys bruijni*
Rodentia - Muridae
e   Lowland Brush Mouse; Lowland Brushrat; Large Pogonomelomys
d   Salawati-Tieflandratte

5502   *Pogonomelomys mayeri*
Rodentia - Muridae
e   Shaw Mayer's Brush Mouse; Shaw Mayer's Pogonomelomys

5503   *Pogonomelomys sevia*
Rodentia - Muridae

e   Highland Brush Mouse; Menzie's Mouse; Highland Pogonomelomys

5504   *Pogonomys*
Rodentia - Muridae
e   Prehensile-tailed Rats; Prehensile-tailed Mice; Prehensile-tailed Tree Mice
d   Neuguinea-Baummäuse
f   Rats à queue préhensile

5505   *Pogonomys championi*
Rodentia - Muridae
e   Champion's Tree Mouse

Pogonomys fergussoniensis syn. *P. loriae* q.v.

Pogonomys forbesi syn. *Chiruromys forbesi* q.v.

Pogonomys kagi syn. *Chiruromys lamia* q.v.

Pogonomys lamia syn. *Chiruromys lamia* q.v.

5506   *Pogonomys loriae*
Rodentia - Muridae
e   Soft-haired Tree Mouse; Prehensile-tailed Rat (ANZ); Large Tree Mouse
d   Große Baummaus

5507   *Pogonomys macrourus*
Rodentia - Muridae
e   Long-tailed Tree Mouse; Chestnut Tree Mouse
d   Langschwanzbaumratte; Langschwanzbaummaus

Pogonomys mollipilosus syn. *P. loriae* q.v.

Pogonomys shawmayeri syn. *Chiruromys forbesi* q.v.

5508   *Pogonomys sylvestris*
           Rodentia - Muridae
e   Grey-bellied Tree Mouse

           Pogonomys vates syn. *Chiruromys
           vates* q.v.

5509   *Poiana*
           Carnivora - Viverridae
e   African Linsangs
d   Pojanas; Afrikanische Linsangs
f   Poianes
i   Poiane

5510   *Poiana richardsoni*
           Carnivora - Viverridae
e   African Linsang; Richardson's
           Linsang; Oyan
d   Pojana; Afrikanische Linsang
f   Poiane; Poiana
i   Linsang africano; Poiana

           Poliocitellus syn. *Spermophilus* q.v.

5511   *Pongo*
           Primates - Hominidae
e   Orang-Utans; Orang-Utangs; Orang-
           Outangs
d   Orang-Utans
f   Orang-outans
i   Oranghi

5512   *Pongo abelli*
           Primates - Hominidae
e   Sumatran Orang-Utan; Sumatran
           Orang-Utang; Sumatran Orang-
           Outang
d   Sumatra-Orang-Utang
f   Orang-outan de Sumatra
i   Orango di Sumatra

5513   *Pongo pygmaeus*
           Primates - Hominidae
e   Bornean Orang-Utan; Bornean
           Orang-Utang; Bornean Orang-

           Outang
d   Orang-Utan; Borneo-Orang-Utang
f   Orang-outan de Bornéo
i   Orango di Borneo; Orango

5514   *Pontoporia*
           Cete - Pontoporiidae
e   La Plata Dolphins; La Plata River
           Dolphins; Franciscanas
d   La-Plata-Delfine
f   Dauphins de La Plata
i   Pontoporie

5515   *Pontoporia blainvillei*
           Cete - Pontoporiidae
e   La Plata River Dolphin; La Plata
           Dolphin; Franciscana
d   La-Plata-Delfin; Franciscana
f   Dauphin de La Plata; Franciscain
i   Pontoporia

5516   *Pontoporiidae*
           Cete
e   Franciscanas
d   La-Plata-Delfine
f   Pontoporidés
i   Pontoporidi

5517   *Potamochoerus*
           Artiodactyla - Suidae
e   Bush Pigs; African Bush Pigs;
           African Water Hogs; African River
           Hogs
d   Flussschweine; Buschschweine;
           Pinselohrschweine; Pinselschweine
f   Potamochères
i   Potamoceri

5518   *Potamochoerus larvatus*
           Artiodactyla - Suidae
e   Bush Pig
d   Buschschwein
f   Potamochère
i   Potamocero

5519 **Potamochoerus porcus**
Artiodactyla - Suidae
e Red River Hog
d Red River Buschschwein;
Pinselschwein; Pinselohrschwein
f Potamochère de l'Afrique;
Potamochère à pinceaux; Cochon
rouge
i Potamocero; Potamocero dai ciuffetti

5520 **Potamogale**
Soricomorpha - Tenrecidae
e Otter Shrews; Giant African Water
Shrews; Giant Otter Shrews
d Große Otterspitzmäuse
f Potamogales
i Potamogali

5521 **Potamogale velox**
Soricomorpha - Tenrecidae
e Otter Shrew; Giant Otter Shrew;
Giant African Water Shrew
d Große Otterspitzmaus
f Potamogale
i Potamogale gigante

5522 **Potorous**
Diprotodontia - Macropodidae
e Long-nosed Rat Kangaroos; Potoroos
(ANZ)
d Kaninchenkängurus
f Rats kangourous à nez long
i Potoroi

Potorous apicalis syn. *P. tridactylus*
q.v.

5523 **Potorous gilberti**
Diprotodontia - Macropodidae
e Wurrark; Gilbert's Rat Kangaroo;
Gilbert's Potoroo
d Gilbert-Kaninchenkänguru
f Rat kangourou de Gilbert

5524 **Potorous longipes**
Diprotodontia - Macropodidae

e Long-footed Potoroo
d Langfüßiges Potoroo
i Ratto canguro dai piedi lunghi

5525 **Potorous platyops**
Diprotodontia - Macropodidae
e Broad-faced Rat Kangaroo; Broad-
faced Potoroo; Common Broad-faced
Potoroo; Common Broad-faced Rat
Kangaroo
d Breitkopfkänguru

5526 **Potorous tridactylus**
Diprotodontia - Macropodidae
e Long-nosed Rat Kangaroo; Common
Rat Kangaroo; Rat Kangaroo; Long-
nosed Potoroo
d Langschnauzenkaninchenkänguru;
Kängururatte
f Rat kangourou à nez long

5527 **Potorous tridactylus apicalis**
Diprotodontia - Macropodidae
e Tasmanian Rat Kangaroo

5528 **Potos**
Carnivora - Procyonidae
e Kinkajous
d Wickelbären
f Kinkajous; Potos
i Cercolletti

5529 **Potos flavus**
Carnivora - Procyonidae
e American Poto; Kinkajou; Night Ape
d Wickelbär
f Kinkajou; Poto
i Cercoletto

Praesorex syn. *Crocidura* q.v.

5530 **Praomys**
Rodentia - Muridae
e African Soft-furred Rats; Soft-furred
Rats; African Soft-furred Mice;

African Common Rats
d   Große Afrikanische Waldmäuse;
      Afrikanische Waldmäuse
f   Rats à pelage doux

Praomys alleni syn. *Hylomyscus alleni* q.v.

Praomys baeri syn. *Hylomyscus baeri* q.v.

Praomys carillus syn. *Hylomyscus carillus* q.v.

Praomys coucha syn. *Mastomys coucha* q.v.

5531   **Praomys degraafi**
       Rodentia - Muridae
e   De Graaf's Soft-furred Mouse

5532   **Praomys delectorum**
       Rodentia - Muridae
e   Delectable Soft-furred Mouse

5533   **Praomys hartwigi**
       Rodentia - Muridae
e   Hartwig's Soft-furred Mouse

5534   **Praomys jacksoni**
       Rodentia - Muridae
e   Jackson's Soft-furred Mouse
d   Jacksons Waldmaus

5535   **Praomys minor**
       Rodentia - Muridae
e   Least Soft-furred Mouse

5536   **Praomys misonnei**
       Rodentia - Muridae
e   Misonne's Soft-furred Mouse

5537   **Praomys morio**
       Rodentia - Muridae
e   Cameroon Soft-furred Mouse

d   Morio-Waldmaus

5538   **Praomys mutoni**
       Rodentia - Muridae
e   Muton's Soft-furred Mouse

Praomys natalensis syn. *Mastomys natalensis* q.v.

5539   **Praomys rostratus**
       Rodentia - Muridae
e   Forest Soft-furred Mouse

5540   **Praomys tullbergi**
       Rodentia - Muridae
e   Tullberg's Soft-furred Rat
d   Tullbergs Waldmaus; Tullbergs
       Weichfellmaus; Langohrsumpfmaus
f   Rat à pelage doux de Tullberg

Praomys verroxii syn. *Myomys verreauxii* q.v.

5541   **Presbytis**
       Primates - Cercopithecidae
e   Leaf Monkeys; Langurs; Surilis;
       Banded Langurs; Crested Langurs
d   Insellanguren; Languren;
       Mützenlanguren
f   Semnopithèques
i   Presbiti

Presbytis aygula syn. *P. comata* q.v.

5542   **Presbytis chrysomelas**
       Primates - Cercopithecidae
e   Sarawak Surili; Sarawak Leaf
       Monkey
d   Sarawak-Langur

5543   **Presbytis comata**
       Primates - Cercopithecidae
e   Sunda Leaf Monkey; Sunda Island
       Leaf Monkey; Grey Leaf Monkey;
       Grizzled Leaf Monkey; Javan Surili;
       Sunda Langur; Javan Leaf Monkey;

Mitred Leaf Monkey

d Mützenlangur

f Semnopithèque des îles de la Sonde

i Presbite grigio di Giava

Presbytis entellus syn.
Semnopithecus entellus q.v.

5544 Presbytis femoralis
Primates - Cercopithecidae

e Banded Leaf Monkey; Banded Surili

d Bindenlangur

f Semnopithèque malais

i Presbite della Sonda

5545 Presbytis frontata
Primates - Cercopithecidae

e White-fronted Leaf Monkey; White-fronted Surili; White-faced Langur; White-fronted Langur

d Weißstirnlangur

f Semnopithèque à front blanc

i Presbite dalla fronte bianca

5546 Presbytis hosei
Primates - Cercopithecidae

e Grey Leaf Monkey; Hose's Leaf Monkey; Hose's Langur; Everett's Leaf Monkey

d Borneo-Langur

i Prebite di Hose

5547 Presbytis hosei canicrus
Primates - Cercopithecidae

e Miller's Grizzled Surili

Presbytis johnii syn. *Trachypithecus johnii* q.v.

5548 Presbytis melalophos
Primates - Cercopithecidae

e Banded Leaf Monkey; Mitred Leaf Monkey; Sumatran Surili; Banded Langur; Black-crested Leaf Monkey

d Roter Langur

f Semnopithèque de Sumatra;

Semnopithèque melalophe

i Presbite rosso

5549 Presbytis natunae
Primates - Cercopithecidae

e Natuna Islands Surili; Nantuna Banded Leaf Monkey; Natuna Islands Leaf Monkey

Presbytis phayrei syn.
Trachypithecus phayrei q.v.

Presbytis pileatus syn.
Trachypithecus pileatus q.v.

5550 Presbytis potenziani
Primates - Cercopithecidae

e Mentawai Island Leaf Monkey; Mentawai Langur; Joja; Islands Monkey; Long-tailed Langur; Mentawai Leaf Monkey; Mentawai Islands Langur

d Mentawai-Langur

f Semnopithèque de Mentawai

i Presbite delle Mentawai

5551 Presbytis rubicunda
Primates - Cercopithecidae

e Maroon Leaf Monkey; Red Leaf Monkey; Maroon Langur

d Maronenlangur; Kastanienbrauner Langur

f Semnopithèque rubincond

i Presbite marrone

5552 Presbytis siamensis
Primates - Cercopithecidae

e White-thighed Surili; White-thighed Leaf Monkey

d Blassschenkellangur

f Semnopithèque du Siam

5553 Presbytis thomasi
Primates - Cercopithecidae

e North Sumatran Leaf Monkey; Thomas's Leaf Monkey; Thomas's Langur

d   Thomas-Mützenlangur; Thomas-
    Langur
i   Presbite di Thomas

5554   *Priodontes*
    Cingulata - Dasypodidae
e   Giant Armadillos
d   Riesengürteltiere
f   Tatous géants
i   Armadilli giganti

    Priodontes giganteus syn. *P.*
    maximus q.v.

5555   *Priodontes maximus*
    Cingulata - Dasypodidae
e   Giant Armadillo
d   Riesengürteltier
f   Tatou géant
i   Armadillo gigante; Tatù gigante

    Prionailurus syn. *Felis* q.v.

5556   *Prionodon*
    Carnivora - Viverridae
e   Oriental Linsangs; Asiatic Linsanags;
    Banded Linsangs; Spotted Linsangs;
    Linsangs
d   Asiatische Linsangs
f   Linsangs
i   Linsanghi

5557   *Prionodon linsang*
    Carnivora - Viverridae
e   Banded Linsang
d   Bänderlinsang
f   Civette à bandes; Linsang à bandes
i   Linsango fasciato

5558   *Prionodon pardicolor*
    Carnivora - Viverridae
e   Spotted Linsang; Tiger Civet
d   Fleckenlinsang
f   Linsang à taches; Linsang tacheté;
    Civette tigre

i   Linsango macchiato

5559   *Prionomys*
    Rodentia - Muridae
e   Dollman's Tree Mice
d   Dollman-Klettermäuse
f   Souris aboricoles

5560   *Prionomys batesi*
    Rodentia - Muridae
e   Dollman's Tree Mouse; Bates's Tree
    Mouse
d   Dollman-Klettermaus
f   Souris arboricole de Bates

    Procapra syn. *Gazella* q.v.

5561   *Procavia*
    Hyracoidea - Procaviidae
e   Dassies; Rock Dassies; Large-
    toothed Hyraxes; Large-toothed Rock
    Hyraxes; Coneys; Rock Rabbits;
    Rock Hyraxes
d   Klippschliefer; Wüstenschliefer ;
    Shaphans
f   Damans des rochers
i   Iraci delle rocce

5562   *Procavia capensis*
    Hyracoidea - Procaviidae
e   Large-toothed Rock Hyrax; Rock
    Dassie; Large-toothed Hyrax;
    Common Rock Hyrax; Rock Badger;
    Rock Hyrax
d   Felsenklippschliefer
f   Daman des rochers du Cap; Daman
    des rochers; Daman du Cap
i   Procavia del Capo; Irace del Capo

5563   *Procavia capensis capensis*
    Hyracoidea - Procaviidae
e   Cape Hyrax; Klipdas (CSA); Dassie;
    Cape Rock Hyrax; Cape Dassie
d   Kap-Klippschliefer

5564 ***Procavia capensis habessinica***
Hyracoidea - Procaviidae
e   Abyssinian Hyrax
d   Abessinischer Klippspringer

5565 ***Procavia capensis johnstoni***
Hyracoidea - Procaviidae
e   Johnston's Hyrax
d   Johnstons Klippspringer; Sudan-
Klippspringer

5566 ***Procavia capensis ruficeps***
Hyracoidea - Procaviidae
e   Western Hyrax; Cameroon Rock
Hyrax; Red-headed Rock Hyrax;
Central African Hyrax
d   Zentralafrikanischer Klippschliefer
i   Procavia dalla testa rossa

5567 ***Procavia capensis syriaca***
Hyracoidea - Procaviidae
e   Syrian Hyrax; Syrian Rock Hyrax
d   Syrischer Klippschliefer

5568 ***Procavia capensis welwitschii***
Hyracoidea - Procaviidae
e   Kakoveld Hyrax

5569 **Procaviidae**
Hyracoidea
e   Hyraxes; Dassies
d   Kletterschliefer ; Schliefer
f   Procaviidés
i   Procavie; Iraci; Procavidi

5570 ***Procolobus***
Primates - Cercopithecidae
e   Red Colobus Monkeys; Red-and-
olive Colubuses
d   Grüne Stummelaffen
i   Colobi

Procolobus badius syn. *Piliocolobus
badius* q.v.

Procolobus pennantii syn.
Piliocolobus pennantii q.v.

Procolobus preussi syn. *Piliocolobus
preussi* q.v.

Procolobus rufomitratus syn.
Piliocolobus rufomitratus q.v.

5571 ***Procolobus verus***
Primates - Cercopithecidae
e   Olive Colobus; Green Colobus; Van
Benedens Colobus
d   Grüner Stummelaffe;
Schopfstummelaffe
f   Colobus à huppe; Colobe vrai;
Colobe de Van Beneden; Colobe
vert; Colobe vert-olive
i   Colobo verde; Colobo di Van
Beneden

5572 ***Procyon***
Carnivora - Procyonidae
e   Raccoons
d   Waschbären
f   Ratons
i   Procioni

5573 ***Procyon cancrivorus***
Carnivora - Procyonidae
e   Crab-eating Raccoon
d   Krabbenwaschbär
f   Raton crabier
i   Procione granchaiolo

5574 ***Procyon gloveralleni***
Carnivora - Procyonidae
e   Barbados Raccoon; Barbados Island
Raccoon
d   Barbados-Washbär
f   Raton laveur de l'île Barbados; Raton
laveur de Barbados; Raton de l'île
Barbados; Raton de Barbardos;
Raton laveur de la Barbade

5575 ***Procyon insularis***
Carnivora - Procyonidae

e   Tres Marias Raccoon
d   Tres-Marias-Waschbär

5576   *Procyon lotor*
        Carnivora - Procyonidae
e   Common Raccoon; North American
        Raccoon; Raccoon; Northern
        Raccoon
d   Nordamerikanischer Waschbär;
        Waschbär; Schupp; Nördlicher
        Waschbär
f   Raton laveur; Chat sauvage (Qué)
i   Orsetto lavatore; Procione; Orso
        lavatore; Marmotta canadese; Orsetto
        lavatore

5577   *Procyon maynardi*
        Carnivora - Procyonidae
e   Bahaman Raccoon; Nassau Racoon
d   Bahama-Waschbär

5578   *Procyon minor*
        Carnivora - Procyonidae
e   Guadeloupe Raccoon
d   Guadeloupe-Waschbär
f   Raton laveur de Guadeloupe

5579   *Procyon pygmaeus*
        Carnivora - Procyonidae
e   Cozumel Raccoon; Cozumel Island
        Raccoon
d   Cozumel-Waschbär
i   Procione nano

5580   Procyonidae
        Carnivora
e   Raccoons; Raccoons and relatives;
        Raccoons, Coatimundis and
        Ringtails; Procyonids
d   Kleinbären
f   Petits ours; Procyonidés
i   Orsi lavatori; Procionidi

5581   *Proechimys*
        Rodentia – Echimyidae

e   Spiny Rats; Terrestrial Spiny Rats
d   Igelratten
f   Rats terrestres épineux; Rats épineux

5582   *Proechimys albispinus*
        Rodentia - Echimyidae
e   White-spined Spiny Rat

5583   *Proechimys amphichoricus*
        Rodentia - Echimyidae
e   Venezuelan Spiny Rat

5584   *Proechimys barinas*
        Rodentia - Echimyidae
e   Barinas Spiny Rat

5585   *Proechimys bolivianus*
        Rodentia - Echimyidae
e   Bolivian Spiny Rat

5586   *Proechimys brevicauda*
        Rodentia - Echimyidae
e   Huallaga Spiny Rat

5587   *Proechimys canicollis*
        Rodentia - Echimyidae
e   Colombian Spiny Rat
f   Rat épineux de Colombie du Nord

        Proechimys cayennensis syn. *P.*
        guayannensis q.v.

5588   *Proechimys chrysaeolus*
        Rodentia - Echimyidae
e   Boyaca Spiny Rat

5589   *Proechimys cuvieri*
        Rodentia - Echimyidae
e   Cuvier's Spiny Rat

5590   *Proechimys decumanus*
        Rodentia - Echimyidae
e   Pacific Spiny Rat

5591   *Proechimys dimidiatus*
Rodentia - Echimyidae
e   Atlantic Spiny Rat

5592   *Proechimys echinothrix*
Rodentia - Echimyidae
e   Hedgehog Spiny Rat

5593   *Proechimys gardneri*
Rodentia - Echimyidae
e   Garnder's Spiny Rat

5594   *Proechimys goeldii*
Rodentia - Echimyidae
e   Goeldi's Spiny Rat

5595   *Proechimys gorgonae*
Rodentia - Echimyidae
e   Gorgona Spiny Rat

5596   *Proechimys guairae*
Rodentia - Echimyidae
e   Guaira Spiny Rat

5597   *Proechimys gularis*
Rodentia - Echimyidae
e   Ecuadoran Spiny Rat; Ecuadorian
Spiny Rat

5598   *Proechimys guyannensis*
Rodentia - Echimyidae
e   Cayenne Spiny Rat

5599   *Proechimys hendeei*
Rodentia - Echimyidae
e   Hendee's Spiny Rat

5600   *Proechimys hoplomyoides*
Rodentia - Echimyidae
e   Guyanan Spiny Rat

5601   *Proechimys iheringi*
Rodentia – Echimyidae

e   Ihering's Spiny Rat

5602   *Proechimys kulinae*
Rodentia - Echimyidae
e   Kulina's Spiny Rat

5603   *Proechimys longicaudatus*
Rodentia - Echimyidae
e   Long-tailed Spiny Rat

5604   *Proechimys magdalena*
Rodentia - Echimyidae
e   Magdalena Spiny Rat

5605   *Proechimys mincae*
Rodentia - Echimyidae
e   Minca Spiny Rat

5606   *Proechimys mirapatanga*
Rodentia - Echimyidae
e   Mirapatanga Spiny Rat

5607   *Proechimys myosurus*
Rodentia - Echimyidae
e   Mouse-tailed Spiny Rat

5608   *Proechimys oconnellii*
Rodentia - Echimyidae
e   O'Connell's Spiny Rat

Proechimys oris syn. *P. roberti* q.v.

5609   *Proechimys pattoni*
Rodentia - Echimyidae
e   Patton's Spiny Rat

5610   *Proechimys poliopus*
Rodentia - Echimyidae
e   Grey-footed Spiny Rat

5611   *Proechimys quadruplicatus*
Rodentia - Echimyidae
e   Napo Spiny Rat

5612 *Proechimys roberti*
Rodentia - Echimyidae
e Para Spiny Rat

5613 *Proechimys semispinosus*
Rodentia - Echimyidae
e Tome's Spiny Rat

5614 *Proechimys setosus*
Rodentia - Echimyidae
e Hairy Spiny Rat

5615 *Proechimys simonsi*
Rodentia - Echimyidae
e Simons's Spiny Rat

5616 *Proechimys steerei*
Rodentia - Echimyidae
e Steere's Spiny Rat

5617 *Proechimys trinitatis*
Rodentia - Echimyidae
e Trinidad Spiny Rat

5618 *Proechimys urichi*
Rodentia - Echimyidae
e Sucre Spiny Rat

5619 *Proechimys warrreni*
Rodentia - Echimyidae
e Warren's Spiny Rat

5620 *Proechimys yonenagae*
Rodentia - Echimyidae
e Yonenaga-Yassudo's Spiny Rat

5621 *Proedromys*
Rodentia - Muridae
e Duke of Bedford 's Voles
d Bedford-Wühlmäuse

5622 *Proedromys bedfordi*
Rodentia - Muridae
e Duke of Bedford's Vole

d Bedford-Wühlmaus

Profelis syn. *Felis* q.v.

Progerbillurus syn. *Gerbillurus* q.v.

5623 *Prolemur*
Primates - Lemuridae
e Greater Bamboo Lemurs
d Breitschnauzenmakis
f Hapalémurs à nez large
i Apalemuri dal naso largo

5624 *Prolemur simus*
Primates - Lemuridae
e Greater Bamboo Lemur; Broad-nosed Gentle Lemur
d Breitschnauzenhalbmaki; Großer Halbmaki; Breitschnauzenmaki
f Hapalémur à nez large
i Apalemure dal naso largo

5625 *Prometheomys*
Rodentia - Muridae
e Long-clawed Mole Voles; Prometheus Mice
d Prometheus-Mäuse
f Souris prométhées

5626 *Prometheomys schaposchnikowi*
Rodentia - Muridae
e Long-clawed Mole Vole; Prometheus Mouse
d Prometheus-Maus
f Souris prométhée

5627 *Promops*
Chiroptera - Molossidae
e Crested Mastiff Bats
d Bulldoggfledermäuse

5628 *Promops centralis*
Chiroptera - Molossidae
e Mexican Dome-palated Bat; Mexican Dome-palated Mastiff Bat; Thomas's

Dome-palated Bat; Thomas's Dome-
palated Mastiff Bat; Thomas's
Mastiff Bat; Big Crested Mastiff Bat

Promops davisoni syn. *P. centralis*
q.v.

5629   *Promops nasutus*
Chiroptera - Molossidae
e   Brown Mastiff Bat

Promops occultus syn. *P centralis*
q.v.

Promops panama syn. *P. nasutus*
q.v.

5630   *Pronolagus*
Lagomorpha - Leporidae
e   Red Hares; Rock Hares (CSA); Red
Rock Rabbits; Red Rock Hares
d   Rotkaninchen ; Wollschwanzhasen
f   Lièvres à grosse queue d'Afrique
i   Lepre rosse

5631   *Pronolagus crassicaudatus*
Lagomorpha - Leporidae
e   Natal Red Hare; Greater Red Rock
Rabbit; Natal Red Rock Hare
d   Natal-Wollschwanzhase
f   Lièvre à grosse queue d'Afrique;
Lièvre roux de Natal
i   Lepre rossa del Natal

5632   *Pronolagus randensis*
Lagomorpha - Leporidae
e   Rand Red Hare; Jameson's Red Rock
Rabbit; Jameson's Red Rock Hare
d   Rand-Wollschwanzhasen
f   Lapin roux de Jameson

5633   *Pronolagus rupestris*
Lagomorpha - Leporidae
e   Smith's Red Hare; Smith's Red Rock
Rabbit; Smith's Red Rock Hare; Red
Rock Rabbit

d   Rotkaninchen
f   Lapin roux de Smith
i   Lepre rossa di Smith

5634   *Propithecus*
Primates - Indriide
e   Monkey Lemurs; Sifakas
d   Kronenindris; Sifakas; Larvenmakis
f   Prophithèques
i   Propitechi; Sifaka

5635   *Propithecus coquereli*
Primates - Indriide
e   Coquerel's Sifaka
d   Coquerels Sifaka
f   Propithèque de Coquerel

5636   *Propithecus deckenii*
Primates - Indriide
e   Van der Decken's Sifaka
d   Deckens Sifaka
f   Propithèque de Decken

5637   *Propithecus diadema*
Primates - Indriide
e   Monkey Lemur; Diadem Sifaka;
Diademed Sifaka
d   Diademsifaka
f   Prophitèque diadème; Propithèque à
diadème; Sifaka couronné
i   Sifaka coronato; Propiteco coronato;
Sifaka diadema

5638   *Propithecus edwardsi*
Primates - Indriide
e   Milne-Edward's Sifaka
d   Riesenmaki
f   Propithèque de Milne Edwards

5639   *Propithecus perrieri*
Primates - Indriide
e   Perrier's Sifaka
d   Perriers Sifaka
f   Propithèque de Perrier

5640 Propithecus tattersalli
Primates - Indriide
e Golden-crowned Sifaka
d Tattersalls Sifaka; Goldkronensifaka;
Tattersall-Sifaka
f Propithèque de Tattersall
i Sifaka di Tatttersall; Propiteco di
Tattersall

5641 Propithecus verreauxi
Primates - Indriide
e Verreaux's Sifaka
d Larvensifaka; Sifaka
f Prophitèque de Verreaux; Sifaka
i Sifaka di Verreaux; Propiteco di
Verreaux

5642 Propithecus verreauxi coronatus
Primates - Indriide
e Crowned Sifaka
f Propithèque couronné

5643 Propithecus verreauxi major
Primates - Indriide
e Major's Sifaka Lemur
f Propithèque de Forsythe-Major

Prosalpingotus syn. *Salpingotus* q.v.

5644 Prosciurillus
Rodentia - Sciuridae
e Celebes Dwarf Squirrels; Celebes
Pygmy Squirrels; Sulawesi Dwarf
Squirrels; Dwarf Squirrels
d Sulawesi-Zwerghörnchen
f Écureuils des arbres; Écureils de la
brousse; Écureuils nains
i Scoiattoli nani

5645 Prosciurillus abstrusus
Rodentia - Sciuridae
e Secretive Dwarf Squirrel

5646 Prosciurillus leucomus
Rodentia - Sciuridae

e Whitish Dwarf Squirrel
d Weißliches Sulawesi-Zwerghörnchen

5647 Prosciurillus murinus
Rodentia - Sciuridae
e Celebes Dwarf Squirrel; Sulawesi
Dwarf Squirrel
d Gemeines Sulawesi-Zwerghörnchen

5648 Prosciurillus rosenbergii
Rodentia - Sciuridae
e Rosenberg's Dwarf Squirrel
d Sangihe-Zwerghörnchen

5649 Prosciurillus weberi
Rodentia - Sciuridae
e Weber's Dwarf Squirrel
d Weber-Zwerghörnchen

5650 Proteles
Carnivora - Hyaenidae
e Earthwolves; Aardwolves
d Erdwölfe
f Loups fouisseurs; Protèles
i Proteli

5651 Proteles cristatus
Carnivora - Hyaenidae
e Earthwolf; Aardwolf
d Erdwolf
f Loup fouisseur; Protèle
i Protele crestato; Protele

5652 Proteles cristatus cristatus
Carnivora - Hyaenidae
e Southern Afric Earthwolf
d Südlicher Erdwolf

5653 Proteles cristatus septentrionalis
Carnivora - Hyaenidae
e East Africa Earthwolf
d Ostafrikanischer Erdwolf

Protemnodon syn. *Macropus* q.v.

5654 Protoxerus
Rodentia - Sciuridae
e Oil Palm Squirrels; African Giant
Squirrels
d Ölpalmenhörnchen
f Grands écureuils
i Scoiattoli giganti

5655 Protoxerus aubinni
Rodentia - Sciuridae
e Aubinn's Squirrel; Slender-tailed
Giant Squirrel; Slender-tailed
Squirrel; African Giant Squirrel
d Schlankschwanzhörnchen

5656 Protoxerus stangeri
Rodentia - Sciuridae
e Giant Forest Squirrel; Forest Giant
Squirrel
d Ölpalmenhörnchen; Gemeines
Ölpalmenhörnchen
f Grand écureuil de Stanger
i Scoiattolo gigante di Stanger

5657 Psammomys
Rodentia - Muridae
e Sand Rats; Diurnal Sand Rats
d Sandrennmäuse
f Rats des sables
i Ratti della sabbia

5658 Psammomys obesus
Rodentia - Muridae
e Fat Sand Rat; Diurnal Sand Rat
d Sandrennmaus; Dicke Sandratte;
Fette Sandratte; Sandratte
f Rat des sables diurne; Psammomys
obèse; Gros rat du sable

5659 Psammomys obesus dianae
Rodentia - Muridae
e Fat Jird

5660 Psammomys vexillaris
Rodentia – Muridae
e Thin Sand Rat; Pale Sand Rat
d Helle Sandratte

5661 Pseudalopex
Carnivora - Canidae
e South American Foxes
d Anden-Füchse
f Renards de l'Amérique latine;
Renards de l'Amérique du Sud
i Volpi sudamericani

5662 Pseudalopex culpaeus
Carnivora - Canidae
e Colpeo Fox; Culpeo; South
American Fox; Colpeo
d Culpeo-Fuchs; Anden-Fuchs; Anden-
Schakal; Feuerland-Fuchs; Magellan-
Fuchs
f Culpeo; Renard andin; Renard
culpeo; Renard gris d'Argentine;
Renard des Andes
i Volpe delle Ande

Pseudalopex griseus syn.
P.gymnocercus q.v.

5663 Pseudalopex gymnocercus
Carnivora - Canidae
e Azara's Fox; Paraguyan Fox; Pampas
Fox; Azara's Zorro; Argentine Grey
Fox; South American Fox; Argentine
Fox; Darwin's Zorro
d Pampasfuchs; Azara-Fuchs;
Argentinischer Kampfuchs
f Renard d'Azara; Renard gris
i Volpe grigio delle Pampas; Volpe
grigio dell'Argentina; Volpe di
Blanford; Volpe grigio; Volpe
sudamericano

5664 Pseudalopex sechurae
Carnivora - Canidae
e Sechura Fox
d Sechura-Fuchs
f Renard côtier
i Volpe costiera

Pseudalopex vetulus syn. *Lycalopex vetulus* q.v.

Pseudantechinus syn. *Antechinus* q.v.

5665 *Pseudocheirus*
Diprotodontia - Petauride
e Ring-tailed Phalangers; Ring-tailed Possums; Ringtails
d Ringelschwanz-Kletterbeutler ; Ringbeutler
f Queues-zébrées
i Possum coda ad anelli; Opossum coda ad anelli

Pseudocheirus albertisi syn. *Peudochirops albertisi* q.v.

Pseudocheirus archeri syn. *Pseudochirops archeri* q.v.

5666 *Pseudocheirus canescens*
Diprotodontia - Petauride
e Lowland Ringtail; Daintree River Ringtail
d Hundsringbeutler

5667 *Pseudocheirus caroli*
Diprotodontia - Petauride
e Weyland Mountains Ringtail; Weyland Ringtail; Weyland Ringtail Possum
d Bundschwanzringbeutler

Pseudocheirus dahli syn. *Petropseudes dahli* q.v.

5668 *Pseudocheirus forbesi*
Diprotodontia - Petauride
e Moss-forest Ringtail; Painted Ringtail Possum
d Forbes-Ringbeutler

5669 *Pseudocheirus herbertensis*
Diprotodontia - Petauride
e Herbert River Ringtail; Herbert River Ring-tailed Phalanger; Mongan; Herbert River Ringtail Possum (ANZ)
d Herbert-Ringbeutler; Herbert-River-Ringbeutler
f Queue-zebrée du fleuve d'Herbert

Pseudocheirus lemuroides syn. *Hemibelideus lemuroides* q.v.

5670 *Pseudocheirus mayeri*
Diprotodontia - Petauride
e Pygmy Ringtail; Pygmy Ringtail Possum
d Zwergringbeutler
f Queue-zebrée pygmée

5671 *Pseudocheirus occidentalis*
Diprotodontia - Petauride
e Western Ringtail; Western Ringtail Possum
d Ringschwanzopossum

5672 *Pseudocheirus peregrinus*
Diprotodontia - Petauride
e Queensland Ringtail; Common Ringtail; Queensland Ring-tailed Phalanger; Ring-tailed Possum; Common Ringtail Possum (ANZ)
d Wander-Ringelschwanzbeutler; Ostringbeutler; Gewöhnlicher Ringbeutler
f Queue-zebrée de Queensland
i Coda ad anelli comune; Possum coda ad anelli; Opossum coda ad anelli

5673 *Pseudocheirus schlegeli*
Diprotodontia - Petauride
e Arfak Mountains Ringtail; Arfak Ringtail; Arfak Ringtail Possum
d Schlegel-Ringschwanzbeutler; Schlegels Ringbeutler

5674 *Pseudochirops*
Diprotodontia - Petauride
e Ringtail Possums
d Ringbeutler

5675 *Pseudochirops albertisi*
Diprotodontia - Petauride
e D'Albertis's Ringtail; D'Albertis's Ringtail Possum
d Langhaarringbeutler

5676 *Pseudochirops archeri*
Diprotodontia - Petauride
e Green Ringtail (ANZ); Striped Ringtail; Striped Ring-tailed Phalanger; Green Ringtail Possum
d Streifenringelschwanzbeutler; Streifenringbeutler; Grüner Ringbeutler

5677 *Pseudochirops corinnae*
Diprotodontia - Petauride
e Eastern Ringtail; Golden Ringtail Possum; Plush-coated Ringtail Possum
d Glanzringbeutler

5678 *Pseudochirops cupreus*
Diprotodontia - Petauride
e Coppery Ringtail; Copper Ringtail Possum
d Kupferringbeutler

5679 *Pseudohydromys*
Rodentia - Muridae
e New Guinea Water Rats; False Water Rats; New Guinean Shrewmice
d Falsche Neuguinea-Wasserratten

5680 *Pseudohydromys murinus*
Rodentia - Muridae
e Eastern Shrewmouse

5681 *Pseudohydromys occidentalis*
Rodentia - Muridae
e Western Shrewmouse

5682 *Pseudois*
Artiodactyla - Bovidae
e Bharals; Blue Sheep ; Burrhels; Nahurs

d Blauschafe; Nahurs
f Boucs bleus
i Baral; Bahral

5683 *Pseudois nayaur*
Artiodactyla - Bovidae
e Blue Sheep; Bharal; Himalajan Blue Sheep
d Nahur; Blauschaf
f Bouc Bleu; Bharal; Mouton bleu
i Baral; Bahral

5684 *Pseudois nayaur nayaur*
Artiodactyla - Bovidae
e Tibetan Blue Sheep
d Himalaja-Blauschaf

5685 *Pseudois nayaur szechuanensis*
Artiodactyla - Bovidae
e Chinese Blue Sheep
d China-Blauschaf

5686 *Pseudois schaeferi*
Artiodactyla - Bovidae
e Dwarf Blue Sheep; Dwarf Bharal
d Zwergblauschaf
f Bharal sauvage; Bharal pygmée

5687 *Pseudomys*
Rodentia - Muridae
e Pseudorats; Australian False Mice; Australian Mice; Australian Native Mice
d Pseudomys-Mäuse; Australische Kleinmäuse
f Souris d'Australie; Souris de l'Ouest; Fausses souris
i Topi australiani

5688 *Pseudomys albocinereus*
Rodentia - Muridae
e Ashy-grey Mouse

5689 *Pseudomys apodemoides*
Rodentia - Muridae

e   Silky-grey Southern Mouse; Silky
      Mouse (ANZ); Silky Desert Mouse

5690   *Pseudomys australis*
         Rodentia - Muridae
e   Murine Hapalotis; Eastern Pseudorat;
      Eastern Mouse; Plains Rat (ANZ);
      Plains Mouse
f   Souris des plaines

5691   *Pseudomys bolami*
         Rodentia - Muridae
e   Bolam's Mouse

5692   *Pseudomys calabyi*
         Rodentia - Muridae
e   Calaby's Pebble-mound Mouse;
      Kakadu Pebble-mound Mouse

5693   *Pseudomys chapmani*
         Rodentia - Muridae
e   Western Pebble-mound Mouse
      (ANZ)
d   Kieselmaus

5694   *Pseudomys delicatulus*
         Rodentia - Muridae
e   Delicate Mouse; Little Native Mouse

5695   *Pseudomys desertor*
         Rodentia - Muridae
e   Brown Desert Mouse; Desert Mouse
      (ANZ)

5696   *Pseudomys fieldi*
         Rodentia - Muridae
e   Alice Springs Mouse; Djoongari
      Mouse; Shark Bay Mouse;
      Djoongari; Alice Springs Pseudo Rat;
      Field's Australian Native Mouse;
      Field's Mouse
d   Fields Australische Kleinmaus;
      Alice-Springs-Maus
f   Souris d'Australie de Field; Souris de
      Field; Souris d'Alice Springs

         Pseudomys forresti syn. ***Leggadina***

forresti q.v.

5697   *Pseudomys fumeus*
         Rodentia - Muridae
e   Smoky Mouse (ANZ); Smoky
      Pseudo Mouse
d   Rauchige Maus; Russige Maus;
      Russige Falschmaus

5698   *Pseudomys fuscus*
         Rodentia - Muridae
e   Broad-toothed Mouse; Broad-toothed
      Rat (ANZ)

5699   *Pseudomys glaucus*
         Rodentia - Muridae
e   Blue-grey Mouse; Blue-grey
      Australian Native Mouse
d   Blaugraue Australische Kleinmaus;
      Blauegraue Maus
f   Souris d'Australie blue-grise; Souris
      bleue-grise

5700   *Pseudomys gouldi*
         Rodentia - Muridae
e   Gould's Eastern Mouse; White-footed
      Mouse; Gould's Mouse; Gould's
      Australian Native Mouse; Gould's
      Native Mouse
d   Gould-Maus; Goulds Australische
      Kleinmaus
f   Souris d'Australie de Gould; Souris
      de Gould

5701   *Pseudomys gracilicaudatus*
         Rodentia - Muridae
e   Queensland Thetomys; Eastern
      Chestnut Mouse

5702   *Pseudomys hermannsburgensis*
         Rodentia - Muridae
e   Australian Native Mouse; Sandy
      Inland Mouse
d   Hermannsburg-Zwergmaus;
      Hermannsburg-Maus
f   Souris d'Hermannsburg

5703   *Pseudomys higginsi*
Rodentia - Muridae
e   Tasmanian Pseudorat; Tasmanian Mouse; Long-tailed Mouse (ANZ); Tasmanian Long-tailed Mouse

5704   *Pseudomys johnsoni*
Rodentia - Muridae
e   Central Pebble-mound Mouse

5705   *Pseudomys laborifex*
Rodentia - Muridae
e   Kimberley Mouse
d   Kimberley-Maus

Pseudomys lakedownensis syn.
Leggadina lakedownensis q.v.

5706   *Pseudomys nanus*
Rodentia - Muridae
e   Western Chestnut Mouse

5707   *Pseudomys novaehollandiae*
Rodentia - Muridae
e   New Holland Mouse
d   Neuholland-Maus

5708   *Pseudomys occidentalis*
Rodentia - Muridae
e   Western Mouse

5709   *Pseudomys oralis*
Rodentia - Muridae
e   Hastings River Mouse; Smoky Mouse

5710   *Pseudomys patrius*
Rodentia - Muridae
e   Country Mouse

5711   *Pseudomys pilligaensis*
Rodentia - Muridae
e   Pilliga Mouse

5712   *Pseudomys praeconis*

Rodentia - Muridae
e   Shaggy Mouse; Shark Bay False Mouse; Shar Bay Mouse
d   Shark-Bay-Falschmaus
f   Fausse souris de la baie de Shark
i   Falso topo della baia di Shark

5713   *Pseudomys shortridgei*
Rodentia - Muridae
e   Shortridge's Native Mouse; Western Pseudorat; Heath Rat (ANZ)
d   Shortridge-Falschmaus

5714   *Pseudopotto*
Priates - Loridae
e   False Pottos
d   Falsche Pottos
f   Faux Pottos

5715   *Pseudopotto martini*
Priates - Loridae
e   Martin's False Potto
d   Martins Falscher Pottto
f   Faux Potto; Faux Potto de Martin

5716   *Pseudorca*
Cete - Delphinidae
e   Lesser Killer Whales; False Killers
d   Kleine Mörder ; Kleine Schwertwale
f   Pseudorques; Faux-orques
i   Pseudorche

5717   *Pseudorca crassidens*
Cete - Delphinidae
e   False Killer; False Killer Whale
d   Kleiner Mörder; Kleiner Schwertwal; Kleinschwertwal; Unechter Schwertwal; Falscher Schwertwal
f   Pseudorque; Pseudorque aux dents épaisses; Faux orque
i   Pseudorca; Falsa orca

5718   *Pseudoryx*
Artiodactyla - Bovidae
e   Vu Quang Oxen

d   Vietnamesische Waldrinder
f   Boeufs de Vu Quang
i   Saola

5719   Pseudoryx nghetinhensis
Artiodactyla - Bovidae
e   Vu Quang Ox; Saola
d   Vietnamesisches Waldrind; Vu-Quang-Rind
f   Saola; Boeuf de Vu Quang
i   Saola

5720   Pseudoryzomys
Rodentia - Muridae
e   Red-nosed Mice; Brazilian False Rice Rats
d   Brasilianische Reisratten

5721   Pseudoryzomys simplex
Rodentia - Muridae
e   Brazilian False Rice Rat
d   Falsche Reisratte

5722   Ptenochirus
Chiroptera - Pteropodidae
e   Musky Fruit Bats
d   Moschusflughunde

5723   Ptenochirus jagori
Chiroptera - Pteropodidae
e   Greater Musky Fruit Bat; Jagor's Dog-faced Fruit Bat
d   Moschusflughund

5724   Ptenochirus minor
Chiroptera - Pteropodidae
e   Lesser Musky Fruit Bat
d   Kleiner Moschusflughund

5725   Pteralopex
Chiroptera - Pteropodidae
e   Monkey-faced Bats
d   Affengesichtflughunde

5726   Pteralopex acrodonta
Chiroptera - Pteropodidae
e   Fijian Monkey-faced Bat; Monkey-faced Fruit Bat
d   Fidschi-Affengesichtflughund

5727   Pteralopex anceps
Chiroptera - Pteropodidae
e   Bougainville Monkey-faced Bat
d   Bougainville-Affengesichtflughund

5728   Pteralopex atrata
Chiroptera - Pteropodidae
e   Cusp-toothed Flying Fox; Guadalcanal Monkey-faced Bat
d   Guadalcanal-Affengesichtflughund

5729   Pteralopex pulchra
Chiroptera - Pteropodidae
e   Montane Monkey-faced Bat
d   Berg-Affengesichtflughund

5730   Pteralopex taki
Chiroptera - Pteropodidae
e   New Georgia Monkey-faced Bat

5731   Pteromys
Rodentia - Sciuridae
e   Smaller Flying Squirrels; Eurasian Flying Squirrels
d   Flughörnchen ; Gleithörnchen
i   Scoiattoli volanti

5732   Pteromys momonga
Rodentia - Sciuridae
e   Small Japanese Flying Squirrel; Japanese Flying Squirrel
d   Japanisches Gleithörnchen
i   Polatuco

5733   Pteromys volans
Rodentia - Sciuridae
e   Siberian Flying Squirrel; Eurasian Flying Squirrel
d   Fliegendes Eichhorn; Gemeines Flattereichhorn; Gemeines Flattereichhörnchen; Gemeines

Flatterhörnchen; Gemeines
Flughörnchen; Gewöhnliches
Gleithörnchen; Europäisches
Gleithörnchen
f   Écureuil polatouche; Écureuil volant;
Écureuil volant d'Eurasie;
Polatouche; Polatouche volant;
Polatouche de Sibérie
i   Scoiattolo volante russo; Scoiattolo
volante; Petauro; Scoiattolo volante
della Russia

5734   *Pteromys volans orii*
Rodentia - Sciuridae
e   Russian Flying Squirrel
d   Eurasisches Fliegendes Eichhorn

5735   *Pteromyscus*
Rodentia - Sciuridae
e   Günther's Flying Squirrels; Smoky
Flying Squirrels; Small-eared Flying
Squirrels
d   Rauchgraue Gleithörnchen ;
Eurasisches Gleithörnchen

5736   *Pteromyscus pulverulentus*
Rodentia - Sciuridae
e   Günther's Flying Squirrel; Smoky
Flying Squirrel; Small-eared Flying
Squirrel
d   Rauchgraues Gleithörnchen

5737   *Pteronotus*
Chiroptera - Mormoopidae
e   Naked-backed Bats; Moustached
Bats
d   Nacktrükenfledermäuse
f   Vespertillons à dos nus
i   Pipistrelli dai mustacchi

5738   *Pteronotus davyi*
Chiroptera - Mormoopidae
e   Davy's Naked-backed Bat
d   Kleine Nacktrückenfledermaus
f   Vespertillon à dos nu

Pteronotus fuliginosus syn. *P.*

quadridens q.v.

5739   *Pteronotus gymnonotus*
Chiroptera - Mormoopidae
e   Big Naked-backed Bat
d   Große Nacktrückenfledermaus
f   Vespertillon de Suapuré

5740   *Pteronotus macleayi*
Chiroptera - Mormoopidae
e   MacLeays Moustached Bat

5741   *Pteronotus parnelli*
Chiroptera - Mormoopidae
e   Parnell's Moustached Bat; Common
Moustached Bat
d   Schnurrbartfledermaus
i   Pipistrello dai mustacchi

5742   *Pteronotus personatus*
Chiroptera - Mormoopidae
e   Lesser Moustached Bat; Wagner's
Moustached Bat

5743   *Pteronotus quadridens*
Chiroptera - Mormoopidae
e   Sooty Moustached Bat

5744   *Pteronura*
Carnivora - Mustelidae
e   Giant Otters; Giant Brazilian Otters;
Flat-tailed Otters; Wing-tailed Otters;
Margin-tailed Otters
d   Riesenotter
f   Loutres géantes du Brésil
i   Lontre giganti

5745   *Pteronura brasiliensis*
Carnivora - Mustelidae
e   Giant Otter; Giant Brazilian Otter;
Flat-tailed Otter; Wing-tailed Otter;
Margin-tailed Otter; Giant River
Otter; Brazilian Otter
d   Riesenotter; Brasilianischer Otter
f   Loutre géante du Brésil; Loutre
géante

i   Lontra gigante del Brasile;
      Arìraranha; Lontra gigante; Lontra
      Arianna

5746   Pteropodidae
      Chiroptera
e   Fruit Bats; Old World Fruit Bats;
      Large Fruit Bats; Fruit-eating Bats;
      Flying Foxes
d   Flughunde
f   Ptéropidés; Ptéropodidés
i   Volpi volante; Pteropodidi;
      Teropotidi

5747   *Pteropus*
      Chiroptera - Pteropodidae
e   Flying Foxes
d   Flughunde; Langnasenflughunde;
      Eigentliche Flughunde;
      Riesenflughunde; Flederhunde;
      Flugfüchse
f   Roussettes
i   Pteropi ; Volpi volanti; Volpi di volo;
      Teropi

5748   *Pteropus admiralitatum*
      Chiroptera - Pteropodidae
e   Admiralty Flying Fox

5749   *Pteropus aldabrensis*
      Chiroptera - Pteropodidae
e   Aldabra Flying Fox

5750   *Pteropus alecto*
      Chiroptera - Pteropodidae
e   Central Flying Fox; Black Flying Fox
      (ANZ); Funereal Vampire

5751   *Pteropus anetianus*
      Chiroptera - Pteropodidae
e   Vanuatu Flying Fox
d   Vanuata-Flughund

5752   *Pteropus argentatus*
      Chiroptera - Pteropodidae
e   Silvery Flying Fox; Ambon Flying
      Fox
d   Ambon-Flughund

Pteropus balutus syn. *P. pumillus*
q.v.

5753   *Pteropus banakrisi*
      Chiroptera - Pteropodidae
e   Torresian Flying Fox

5754   *Pteropus brunneus*
      Chiroptera - Pteropodidae
e   Dusky Flying Fox; Percy Island
      Flying Fox
d   Percy-Island-Flughund
f   Roussette de l'île Percy; Renard
      volant de l'île Percy

5755   *Pteropus caniceps*
      Chiroptera - Pteropodidae
e   Ashy-headed Flying Fox; North
      Moluccan Flying Fox

5756   *Pteropus chrysoproctus*
      Chiroptera - Pteropodidae
e   Amboina Flying Fox; Moluccan
      Flying Fox

5757   *Pteropus cognatus*
      Chiroptera - Pteropodidae
e   Makira Flying Fox

5758   *Pteropus conspicillatus*
      Chiroptera - Pteropodidae
e   Spectacled Vampire; Spectacled Fruit
      Bat; Spectacled Flying Fox (ANZ)

5759   *Pteropus dasymallus*
      Chiroptera - Pteropodidae
e   Ryukyu Flying Fox; Linkin Islands
      Flying Fox; Ryukyu Islands Flying
      Fox; Ryukyu Islands Fruit Bat;
      Linkin Fruit Bat; Linkin Flying Fox

d   Riukiu-Inselflughund; Riukiu
      Flughund

f Roussette des îles Ryu Kyu;
Roussette des Ryukyu; Chauve-
souris des Ryu Kyu
i Volpe volante delle Ryukyu

5760 *Pteropus dasymallus formosus*
Chiroptera - Pteropodidae
e Formosa Flying Fox; Taiwan Flying
Fox; Taiwan Island Flying Fox;
Taiwan Island Fruit Bat; Taiwan
Fruit Bat
d Formosa-Flughund; Taiwan
Flughund; Taiwan Flederhund
f Roussette des îles Ryu Kyu de
Taiwan; Roussette des Ryukyu de
Taiwan; Chauve-souris des Ryu Kyu
de Taiwan

5761 *Pteropus faunulus*
Chiroptera - Pteropodidae
e Nicobar Flying Fox

5762 *Pteropus fundatus*
Chiroptera - Pteropodidae
e Banks Flying Fox
d Banks-Flughund

5763 *Pteropus giganteus*
Chiroptera - Pteropodidae
e Indian Flying Fox; Great Indian Fruit
Bat; Giant Indian Fruit Bat
d Indischer Flughund; Indischer
Flugfuchs; Indischer Riesenflughund;
Riesenflughund
f Roussette géante

5764 *Pteropus giganteus giganteus*
Chiroptera - Pteropodidae
e Common Flying Fox
d Gemeiner Fliegender Hund

Pteropus gilliardi syn. *P.
gilliardorum* q.v.

5765 *Pteropus gilliardorum*
Chiroptera - Pteropodidae
e Gilliard's Flying Fox

d Neubrittanien-Flughund

5766 *Pteropus griseus*
Chiroptera - Pteropodidae
e Grey Flying Fox

5767 *Pteropus howensis*
Chiroptera - Pteropodidae
e Ontong Java Flying Fox

5768 *Pteropus hypomelanus*
Chiroptera - Pteropodidae
e Small Flying Fox; Island Flying Fox;
Variable Flying Fox
d Inselflughund
f Roussette des îles

5769 *Pteropus hypomelanus condorensis*
Chiroptera - Pteropodidae
e Vietnamese Small Flying Fox
d Vietnam-Inselflughund

5770 *Pteropus insularis*
Chiroptera - Pteropodidae
e Chuuk Flying Fox
d Truk-Flughund
f Roussette des îles Truk
i Pteropo delle isole Truk

5771 *Pteropus leucopterus*
Chiroptera - Pteropodidae
e White-winged Flying Fox

5772 *Pteropus livingstonei*
Chiroptera - Pteropodidae
e Comoro Black Flying Fox
d Komoren-Flughund
f Renard volant de Livingstone
i Pteropo di Livingstone

5773 *Pteropus lombocensis*
Chiroptera - Pteropodidae
e Lombok Flying Fox

5774   *Pteropus lylei*
Chiroptera - Pteropodidae
e   Lyle's Flying Fox
d   Hinterindischer Flughund
f   Renard volant de Lyle

5775   *Pteropus macrotis*
Chiroptera - Pteropodidae
e   Big-eared Flying Fox
f   Roussette aux grands oreilles

5776   *Pteropus mahaganus*
Chiroptera - Pteropodidae
e   Lesser Flying Fox; Sanborn's Flying
Fox

5777   *Pteropus mariannus*
Chiroptera - Pteropodidae
e   Marianas Flying Fox; Marianas Fruit
Bat; Mariana Flying Fox; Mariana
Fruit Bat
d   Marianen-Flughund
f   Roussette d'Okinawa; Renard volant
d'Okinawa
i   Pteropo delle Marianne

5778   *Pteropus mariannus paganensis*
Chiroptera - Pteropodidae
e   Pagan Flying Fox; Pagan Fruit Bat

5779   *Pteropus mariannus pelewensis*
Chiroptera - Pteropodidae
e   Palau Flying Fox

5780   *Pteropus mariannus ulthiensis*
Chiroptera - Pteropodidae
e   Ulithi Flying Fox; Ulithi Fruit Bat

5781   *Pteropus mariannus yapensis*
Chiroptera - Pteropodidae
e   Yap Flying Fox; Yap Fruit Bat

Pteropus mearnsi syn. *P. speciosus
mearnsi*

5782   *Pteropus melanopogon*
Chiroptera - Pteropodidae
e   Black-bearded Flying Fox

5783   *Pteropus melanotus*
Chiroptera - Pteropodidae
e   Black-eared Flying Fox

5784   *Pteropus melanotus natalis*
Chiroptera - Pteropodidae
e   Christmas Island Flying Fox

5785   *Pteropus molossinus*
Chiroptera - Pteropodidae
e   Caroline Flying Fox; Ponape Flying
Fox
d   Ponape-Flughund
f   Roussette de Ponape
i   Pteropo di Ponape; Pterope dell'isola
Ponape

5786   *Pteropus neohibernicus*
Chiroptera - Pteropodidae
e   Bismarck Flying Fox; Great Flying
Fox

5787   *Pteropus neohibernicus hilli*
Chiroptera - Pteropodidae
e   Admiralty Great Flying Fox

5788   *Pteropus niger*
Chiroptera - Pteropodidae
e   Black Flying Fox; Greater Mascarene
Flying Fox
d   Schwarzer Flughund
f   Roussette noire

5789   *Pteropus nitendiensis*
Chiroptera - Pteropodidae
e   Temotu Flying Fox

5790   *Pteropus ocularis*
Chiroptera - Pteropodidae
e   Ceram Flying Fox

5791 *Pteropus ornatus*
Chiroptera - Pteropodidae
e Ornate Flying Fox
f Roussette rousse; Renard volant orné

5792 *Pteropus personatus*
Chiroptera - Pteropodidae
e Masked Flying Fox

5793 *Pteropus phaeocephalus*
Chiroptera - Pteropodidae
e Mortlock Flying Fox
d Mortlock-Flughund
f Roussette de l'île Mortlock
i Pteropo di Mortlock

5794 *Pteropus pilosus*
Chiroptera - Pteropodidae
e Large Palau Flying Fox; Soft-haired Fruit Bat; Palau Fruit Bat; Soft-haired Flying Fox
d Palau-Flughund; Weichhaarflughund; Weichhaarflederhund
f Roussette des îles Palaos; Roussette à poil doux; Chauve-souris à poil doux
i Pteropo di Palau; Pteropo dell'isola Palau

5795 *Pteropus pohlei*
Chiroptera - Pteropodidae
e Geelvink Bay Flying Fox

5796 *Pteropus poliocephalus*
Chiroptera - Pteropodidae
e Grey-headed Vampire Flying Fox; Grey-headed Vampire Fruit Bat; Grey-headed Flying Fox (ANZ)
d Graukopfflughund
f Roussette à tête cendrée
i Pipistrello della frutia a testa grigia

5797 *Pteropus pselaphon*
Chiroptera - Pteropodidae
e Bonin Flying Fox
f Roussette de l'île Bonin

5798 *Pteropus pumilus*
Chiroptera - Pteropodidae
e Little Golden-mantled Flying Fox; Golden-mantled Flying Fox
d Kleiner Goldmantelflughund

5799 *Pteropus rayneri*
Chiroptera - Pteropodidae
e Solomons Flying Fox

5800 *Pteropus rennelli*
Chiroptera - Pteropodidae
e Rennell Flying Fox
d Rennell-Flughund

5801 *Pteropus rodricensis*
Chiroptera - Pteropodidae
e Rodriguez Flying Fox; Rodriguez's Fruit Bat
d Rodriguez-Flughund
f Roussette de Rodriguez
i Pteropo di Rodriques

5802 *Pteropus rufus*
Chiroptera - Pteropodidae
e Rufous Flying Fox; Madagascar Flying Fox; Madagascan Flying Fox
d Roter Flughund
f Roussette rougeâtre

5803 *Pteropus samoensis*
Chiroptera - Pteropodidae
e Samoa Flying Fox; Samoan Flying Fox
d Samoa-Flughund
f Roussette de Samoa
i Pteropo delle Samoa

5804 *Pteropus sanctacrucis*
Chiroptera - Pteropodidae
e Santa Cruz Flying Fox

Pteropus satyrus syn. *P. melanotus* q.v.

5805   *Pteropus scapulatus*
Chiroptera - Pteropodidae
e   Little Red Flying Fox
f   Petit renard volant rouge; Renard volant à collier
i   Pipistrello rosso della fruta

Pteropus sepikensis syn. *P. neohibernicus* q.v.

5806   *Pteropus seychellensis*
Chiroptera - Pteropodidae
e   Seychelles Flying Fox; Seychelles Fruit Bat
d   Seychellen-Flughund
f   Roussette des Seychelles

5807   *Pteropus speciosus*
Chiroptera - Pteropodidae
e   Philippine Grey Flying Fox

5808   *Pteropus speciosus mearnsi*
Chiroptera - Pteropodidae
e   Mearns's Flying Fox

5809   *Pteropus subniger*
Chiroptera - Pteropodidae
e   Dark Flying Fox; Lesser Mascarenes Flying Fox
d   Rauchgrauer Flughund
f   Roussette foncée

Pteropus tablasi syn. *P. pumilus* q.v.

5810   *Pteropus temmincki*
Chiroptera - Pteropodidae
e   Temminck's Flying Fox

5811   *Pteropus tokudae*
Chiroptera - Pteropodidae
e   Guam Flying Fox; Little Mariana Fruit Bat; Fanihi; Guam Island Flying Fox; Guam Fruit Bat; Guam Islasnd Fruit Bat
d   Guam-Flughund; Guam-Flederhund

f   Roussette de l'île Guam; Roussette de Guam; Chauve-souris de Guam; Chauve-souris de l'île Guam

5812   *Pteropus tonganus*
Chiroptera - Pteropodidae
e   Tonga Flying Fox; Insular Flying Fox; Pacific Flying Fox; Tonga Fruit Bat
d   Tonga-Flughund
f   Roussette de Tonga
i   Pterope insulare

5813   *Pteropus tuberculatus*
Chiroptera - Pteropodidae
e   Vanikoro Flying Fox

5814   *Pteropus ualanus*
Chiroptera - Pteropodidae
e   Kosrae Flying Fox; Kosrae Fruit Bat

5815   *Pteropus vampyrus*
Chiroptera - Pteropodidae
e   Red-necked Fruit Bat; Malayan Flying Fox; Malaya Large Flying Fox; Large Flying Fox; Large Fruit Bat
d   Flederhund; Fliegender Hund; Flughund; Kalong; Malaiischer Flughund; Malaiischer Flugfuchs
f   Roussette commune; Roussette à cou rouge; Roussette de Malaisie
i   Cane volante

5816   *Pteropus vampyrus vampyrus*
Chiroptera - Pteropodidae
e   Java Fruit Bat
d   Javanischer Kalong

5817   *Pteropus vetulus*
Chiroptera - Pteropodidae
e   New Caledonia Flying Fox
f   Roussette des roches; Renard volant de Nouvelle-Calédonie

5818 *Pteropus voeltzkowi*
  Chiroptera - Pteropodidae
 e Pemba Flying Fox
 d Pemba-Flughund
 f Roussette de Pemba; Renard volant de Voeltzikow
 i Pteropo di Pemba

5819 *Pteropus woodfordi*
  Chiroptera - Pteropodidae
 e Least Flying Fox; Dwarf Flying Fox

5820 *Pteropus yapensis*
  Chiroptera - Pteropodidae
 e Yap Flying Fox

5821 *Ptilocercus*
  Scandentia - Tupaiidae
 e Pen-tailed Tree Shrews
 d Federschwänze; Federschwanztupajas; Federschwanzspitzhörnchen
 f Ptilocerques

5822 *Ptilocercus lowii*
  Scandentia - Tupaiidae
 e Low's Feather-tailed Tree Shrew; Low's Pen-tailed Tree Shrew; Pen-tailed Tree Shrew; Feather-tailed Tree Shrew
 d Federschwanz; Federschwanzspitzhörnchen; Federschwanztupaja
 f Ptilocerque de Low; Ptilocerque

5823 *Pudu*
  Artiodactyla - Cervidae
 e Pudus
 d Pudus
 f Pudus
 i Pudu

5824 *Pudu mephistophiles*
  Artiodactyla - Cervidae
 e Northern Pudu

 d Nordpudu; Nördlicher Pudu
 f Pudu du Nord
 i Pudu mefistofolo; Pudu del Nord

5825 *Pudu puda*
  Artiodactyla - Cervidae
 e Chilean Pudu; Southern Pudu
 d Südpudu; Südlicher Pudu
 f Pudu du Sud
 i Pudu comune; Pudu del sud

  Puma syn. *Felis* q.v.

5826 *Punomys*
  Rodentia - Muridae
 e Puna Mice
 d Puna-Mäuse

5827 *Punomys kofordi*
  Rodentia - Muridae
 e Koford's Puna Mouse

5828 *Punomys lemminus*
  Rodentia - Muridae
 e Puna Mouse
 d Puna-Maus

5829 *Pusa*
  Carnivora - Phocidae
 e Caspian Seals
 d Ringelrobben
 f Phoques de la Caspienne

5830 *Pusa caspica*
  Carnivora - Phocidae
 e Caspian Seal
 d Kaspische Ringelrobbe; Kaspischer Seehund; Kaspische Robbe
 f Phoque de la Caspienne
 i Foca del Caspio

5831 *Pusa hispida*
  Carnivora - Phocidae
 e Ringed Seal; Marbled Seal; Fjord

Seal; Rough Seal; Jar Seal; Two-
ringed Seal
d   Rauher Seehund; Ringelrobbe;
     Marmorierte Robbe; Geringelter
     Seehund
f   Phoque marbré; Phoque annelé
i   Foca dagli anelli; Foca ring

5832   *Pusa hispida botnica*
Carnivora - Phocidae
e   Baltic Seal
d   Ostsee-Ringelrobbe

5833   *Pusa hispida hispida*
Carnivora - Phocidae
e   Arctic Ringed Seal; Ringed Seal
d   Eismeer-Ringelrobbe

5834   *Pusa hispida krascheninikovi*
Carnivora - Phocidae
e   Bering Sea Ringed Seal
d   Beringmeer-Ringelrobbe

5835   *Pusa hispida ladogensis*
Carnivora - Phocidae
e   Ladoga Seal
d   Lagoda-Ringelrobbe

5836   *Pusa hispida ochotensis*
Carnivora - Phocidae
e   Okhotsk Sea Ringed Seal

5837   *Pusa hispida saimensis*
Carnivora - Phocidae
e   Saimaa Seal
d   Saimaa-Ringelrobbe

5838   *Pusa sibirica*
Carnivora - Phocidae
e   Baikal Seal
d   Sibirische Ringelrobbe; Baikalischer
     Seehund; Baikal-Robbe
f   Phoque du Baikal; Phoque du lac
     Baikal; Phoque de Sibérie
i   Foca dal Baikal; Foca del lago Baikal

Putorius syn. *Mustela* q.v.

5839   *Pygathrix*
Primates - Cercopithecidae
e   Douc Langurs; Langurs; Leaf
     Monkeys; Leaf-eating Monkeys;
     Snub-nosed Monkeys; Snub-nosed
     Langurs; Douc Monkeys
d   Kleideraffen; Strumpfnasenaffen
f   Rhinopithèques; Doucs
i   Langur; Rinopitecchi; Entelli

Pygathrix avunculus syn.
Rhinopithecus avunculus q.v.

Pygathrix bieti syn. *Rhinopithecus
bieti* q.v.

Pygathrix brelichi syn.
Rhinopithecus brelichi q.v.

5840   *Pygathrix cinerea*
Primates - Cercopithecidae
e   Grey-shanked Douc; Grey-shanked
     Douc Langur
d   Grauer Kleideraffe

5841   *Pygathrix nemaeus*
Primates - Cercopithecidae
e   Variegated Langur; Variegated Douc
     Langur; Douc Langur; Red-shanked
     Douc; Red-shanked Douc Langur;
     Douc
d   Rotschenkelkleideraffe; Kleideraffe
f   Douc
i   Langur duca

5842   *Pygathrix nigripes*
Primates - Cercopithecidae
e   Black-shanked Douc; Black-shanked
     Douc Langur
d   Schwarzschenkelkleideraffe
f   Rhinopithèque aux pieds noirs

Pygathrix roxellana syn.
Rhinopithecus roxellana q.v.

5843 *Pygeretmus*
Rodentia - Dipodidae
e Fat-tailed Jerboas
d Plattschwanzspringer ;
Dickschwanzspringer ;
Dickschwänzige Springmäuse;
Fettschwanzspringmäuse; Erdhasen

5844 *Pygeretmus platyurus*
Rodentia - Dipodidae
e Lesser Fat-tailed Jerboa
d Aralischer Dickschwanzspringer;
Winogradows Dickschwanzspringer

5845 *Pygeretmus pumilio*
Rodentia - Dipodidae
e Dwarf Fat-tailed Jerboa; Lesser Five-
toed Jerboa
d Zwergartiges Fettschwanzjerboa;
Erdhäschen

5846 *Pygeretmus shitkovi*
Rodentia - Dipodidae
e Greater Fat-tailed Jerboa
d Balchhasch-Dickschwanzspringer;
Shitkows Dickschwanzspringer;
Größeres Fettschwanzjerboa

5847 *Pygoderma*
Chiroptera - Phyllostomidae
e Ipanema Bats; Isomon Bats

5848 *Pygoderma bilabiatum*
Chiroptera - Phyllostomidae
e Ipanema Bat; Isomon Bat

Pyromys syn. *Mus* q.v.

R

5849 *Rangifer*
         Artiodactyla - Cervidae
e   Reindeer ; Caribou
d   Rener; Rentiere; Karibus
f   Caribous
i   Renne

5850 *Rangifer tarandus*
         Artiodactyla - Cervidae
e   Reindeer; Caribou (NA); Cariboo
d   Ren; Rentier
f   Renne rangifer; Renne
i   Renna

5851 *Rangifer tarandus arcticus*
         Artiodactyla - Cervidae
e   Central Canada Caribou; Barren
         Ground Caribou
d   Barren-Ground-Karibu
f   Caribou des champs

5852 *Rangifer tarandus caribou*
         Artiodactyla - Cervidae
e   Caribou (NA); Woodland Caribou
d   Karibu; Westkanadisches Waldren
f   Caribou
i   Caribù

5853 *Rangifer tarandus dawsoni*
         Artiodactyla - Cervidae
e   Dawson Caribou; Queen Charlotte
         Islands Caribou; Queen Charlotte
         Caribou; Queen Charlotte Islands
         Reindeer; Dawson's Reindeer;
         Dawson's Woodland Caribou; Queen
         Charlotte Woodlands Caribou

d   Queen-Charlotte-Karibu; Queen
         Charlotte-Insel-Karibu; Graham
         Insel-Wildren; Graham-Wildren;
         Graham Insel-Karibu; Dawson-Ren;
         Dawson-Karibu; Dawson-Rentier
f   Renne sauvage des îles Queen
         Charlotte; Caribou des îles Queen
         Charlotte; Renne sauvage de
         Dawson; Renne de Dawson

5854 *Rangifer tarandus fennicus*
         Artiodactyla - Cervidae
e   Finnish Forest Reindeer
d   Waldren
f   Renne des forêts eurasiennes
i   Renna selvatica

5855 *Rangifer tarandus granti*
         Artiodactyla - Cervidae
e   Barren Ground Caribou; Porcupine
         Caribou
d   Alaska-Karibu
f   Caribou de Grant

5856 *Rangifer tarandus groenlandicus*
         Artiodactyla - Cervidae
e   Greenland Reindeer
d   Grönland-Ren; Grönlandisches
         Rentier; Grönlandisches Wildren;
         Grönlandisches Karibu
f   Renne sauvage de Groenland;
         Caribou de Groenland

5857 *Rangifer tarandus pearsoni*
         Artiodactyla - Cervidae
e   Novaya Zemlya Reindeer
d   Nowaja-Semlja Ren

5858 *Rangifer tarandus pearyi*
         Artiodactyla - Cervidae
e   Peary Caribou
d   Peary-Karibu
f   Caribou de Peary

5859 *Rangifer tarandus platyrhynchus*
         Artiodactyla - Cervidae

e Svalbard Reindeer
d Spitzbergen-Ren; Svalbardren
f Renne de Svalbard

5860 *Rangifer tarandus sibiricus*
Artiodactyla - Cervidae
e Siberian Reindeer
d Eurasisches Tundraren; Sibirisches
Tundra-Ren

5861 *Rangifer tarandus sylvestris*
Artiodactyla - Cervidae
e Canadian Reindeer
d Ostkanadisches Waldren

5862 *Rangifer tarandus tarandus*
Artiodactyla - Cervidae
e European Reindeer
d Europäisches Ren; Nordeuropäisches
Ren; Finnisches Ren; Europäisches
Rentier; Fjellren
f Renne d'Europe; Renne

5863 *Rangifer tarandus valentinae*
Artiodactyla - Cervidae
e Siberian Forest Reindeer
d Sibirisches Waldren

5864 *Raphicerus*
Artiodactyla - Bovidae
e Steenboks; Steinbocks; Steinboks;
Steenboks and Grysboks
d Steinantilopen; Steinböckchen
f Raphicères
i Raficeri

5865 *Raphicerus campestris*
Artiodactyla - Bovidae
e Steenbok; Steinbock; Steinbok;
Steinbuck
d Steinantilope; Steinböckchen
f Raphicère champêtre; Raphicère
i Raficero campestre; Steinbock;
Steenbok

5866 *Raphicerus melanotis*
Artiodactyla - Bovidae
e Grysbok; Grysbuck; Cape Grysbok;
Southern Grysbok
d Greisbock; Kap-Greisbock
f Grysbok
i Raficero dalle orecchie nere

5867 *Raphicerus sharpei*
Artiodactyla - Bovidae
e Sharpe's Steinbock; Sharpe's
Steinbok; Sharpe's Grysbok;
Northern Grysbok
d Sharpe-Antilope; Sharpe-Greisbock
f Grysbok de Sharpe

5868 *Rattus*
Rodentia - Muridae
e Rats; Typical Rats
d Ratten; Eigentliche Ratten
f Rats; Rats proprement dits; Rats de
l'ancien monde
i Ratti

Rattus abrae syn. *Apomys abrae* q.v.

5869 *Rattus adustus*
Rodentia - Muridae
e Sunburned Rat

Rattus alticola syn. *Maxomys
alticola* q.v.

Rattus andrewsi syn. *Bunomys
andrewsi* q.v.

5870 *Rattus annandalei*
Rodentia - Muridae
e Annandale's Rat
d Annandales Ratte
f Rat d'Annandale

Rattus apoensis syn. *Tarsomys
apoensis* q.v.

Rattus arcuatus syn. *Taeromys arcuatus* q.v.

5871   Rattus argentiventer
Rodentia - Muridae
- *e*   Ricefield Rat
- *d*   Reisfeldratte
- *f*   Rat des rizières; Rat des champs de riz

Rattus baeodon syn. *Maxomys baeodon* q.v.

Rattus bagobus syn. *Bullimus bagobus* q.v.

5872   Rattus baluensis
Rodentia - Muridae
- *e*   Summit Rat

Rattus bartelsii syn. *Maxomys bartelsii* q.v.

Rattus blanfordi syn. *Cremnomys blanfordi* q.v.

Rattus bocagei syn. *Aethomys bocagei* q.v.

Rattus bocourti syn. *R. exulans* q.v.

5873   Rattus bontanus
Rodentia - Muridae
- *e*   Bonthain Rat

5874   Rattus burrus
Rodentia - Muridae
- *e*   Nonsense Rat

Rattus canus syn. *Lenothrix canus* q.v.

Rattus celebensis syn. *Taeromys celebensis* q.v.

Rattus chrysocomus syn. *Bunomys chrysocomus* q.v.

Rattus chrysophilus syn. *Aethomys chrysophilus* q.v.

Rattus colletti syn. *R. sordidus* q.v.

Rattus cremoriventer syn. *Niviventer cremoriventer* q.v.

Rattus cutchicus syn. *Cremnomys cutchicus* q.v.

Rattus datae syn. *Apomys datae* q.v.

Rattus dollmani syn. *Maxomys dollmani* q.v.

Rattus dominator syn. *Paruromys dominator* q.v.

5875   Rattus elaphinus
Rodentia - Muridae
- *e*   Sula Rat

Rattus elvira syn. *Cremnomys elvira* q.v.

5876   Rattus enganus
Rodentia - Muridae
- *e*   Enggano Rat

5877   Rattus everetti
Rodentia - Muridae
- *e*   Philippine Forest Rat

Rattus exiguus syn. *R. losea* q.v.

5878   Rattus exulans
Rodentia - Muridae
- *e*   Little Rat; Little Polynesian Rat; Polynesian Rat; Pacific Rat
- *d*   Polynesien-Ratte; Polynesische Ratte; Kleine Pazifik-Ratte; Pazifische Ratte
- *f*   Rat des îles du Pacifique
- *i*   Ratto polynesiano

Rattus facetus syn. *R. marmosurus* q.v.

5879   Rattus feliceus
Rodentia - Muridae
e   Spiny Seram Rat; Pepina's Rat; Reddish Rat
d   Seram-Ratte

5880   Rattus foramineus
Rodentia - Muridae
e   Hole Rat

Rattus fratrorum syn. *Bunomys fratrorum* q.v.

Rattus fulvescens syn. *Niviventer fulvescens* q.v.

5881   Rattus fuscipes
Rodentia - Muridae
e   Dusky-footed Rat; Southern Bush Rat; Bush Rat (ANZ); Australian Bush Rat
d   Australische Buschratte
f   Rat brun d'Australie

5882   Rattus giluwensis
Rodentia - Muridae
e   Giluwe Rat

Rattus granti syn. *Aethomys granti* q.v.

5883   Rattus hainaldi
Rodentia - Muridae
e   Hainald's Rat

Rattus hellwaldii syn. *Maxomys hellwaldii* q.v.

Rattus hindei syn. *Aethomys hindei* q.v.

5884   Rattus hoffmanni
Rodentia - Muridae

e   Hoffmann's Rat

5885   Rattus hoogerwerfi
Rodentia - Muridae
e   Hoogerwerf's Rat

Rattus hylocoetes syn. *Apomys hylocoetes* q.v.

Rattus hylomyoides syn. *Maxomys hylomyoides* q.v.

Rattus inas syn. *Maxomys inas* q.v.

Rattus inflatus syn. *Maxomys inflatus* q.v.

Rattus infraluteus syn. *Sundamys infraluteus* q.v.

Rattus insignis syn. *Apomys insignis* q.v.

Rattus jalorensis syn. *R. tiomanicus* q.v.

5886   Rattus jobiensis
Rodentia - Muridae
e   Japen Rat

Rattus kaiseri syn. *Aethomys kaiseri* q.v.

5887   Rattus koopmani
Rodentia - Muridae
e   Koopman's Rat

5888   Rattus korinchi
Rodentia - Muridae
e   Korinch's Rat; Mount Kerinci Rat

Rattus legatus syn. *Diplothrix legatus* q.v.

5889   Rattus leucopus
Rodentia - Muridae

e Mottle-tailed Rat; Mottle-tailed Cape
   York Rat; Cape York Rat; White-
   footed Spiny Rat; Southern Spiny Rat
d Südliche Stachelratte

Rattus longicaudatus syn.
Stochomys longicaudatus q.v.

5890 *Rattus losea*
       Rodentia - Muridae
e   Lesser Ricefield Rat

5891 *Rattus lugens*
       Rodentia - Muridae
e   Mentawai Rat

5892 *Rattus lutreolus*
       Rodentia - Muridae
e   Australian Swamp Rat; Tawny Rat;
       Eastern Swamp Rat; Swamp Rat
       (ANZ)
d   Sumpfratte

Rattus luzonicus syn. *Bullimus
luzonicus* q.v.

5893 *Rattus macleari*
       Rodentia - Muridae
e   MacLear's Rat; Captain MacLear's
       Rat
d   Maclears Ratte; Maclear-Ratte
f   Rat de MacLear

5894 *Rattus marmosurus*
       Rodentia - Muridae
e   Opossum Rat

Rattus maxi syn. *Sundamys maxi* q.v.

Rattus microdon syn. *Apomys
microdon* q.v.

5895 *Rattus mindorensis*
       Rodentia - Muridae
e   Mindoro Black Rat

Rattus moi syn. *Maxomys moi* q.v.

5896 *Rattus mollicomulus*
       Rodentia - Muridae
e   Little Soft-furred Rat; Mount
       Lompobatang Rat

5897 *Rattus montanus*
       Rodentia - Muridae
e   Nillu Rat
d   Nillu-Ratte

5898 *Rattus mordax*
       Rodentia - Muridae
e   Eastern Rat

5899 *Rattus morotaiensis*
       Rodentia - Muridae
e   Moluccan Prehensile-tailed Rat
d   Malaku-Ratte

Rattus muelleri syn. *Sundamys
muelleri* q.v.

Rattus musculus syn. *Apomys
musculus* q.v.

Rattus musschenbroeki syn.
Maxomys musschenbroeki q.v.

Rattus namaquensis syn. *Aethomys
namaquensis* q.v.

5900 *Rattus nativitatis*
       Rodentia - Muridae
e   Bulldog Rat; Burrowing Rat;
       Christmas Island Rat
d   Bulldogratte; Weihnachtsinsel-Ratte
f   Rat de l'île Christmas; Rat de
       Christmas

Rattus neilli syn. *Leopoldamys neilli*
q.v.

5901 *Rattus nitidus*
       Rodentia - Muridae

e Himalajan Rat; Himalajan Field Rat
d Himalaja-Ratte

5902 *Rattus norvegicus*
Rodentia - Muridae
e Common Rat; Norway Rat; Brown Rat; Sewer Rat; Irish Rat
d Siedlungswanderratte; Braune Ratte; Shiffsratte; Graue Wanderratte; Wanderratte
f Surmulot; Rat brun; Rat brun ordinaire; Rat gris des égouts; Rat gris caspien; Rat gris irlandais; Rat gris surmulot; Rat des égouts; Rat surmulot; Rat d'égout
i Surmolotto; Ratto delle chiaviche; Topo di chiavica; Ratto grigio

5903 *Rattus novaeguinae*
Rodentia - Muridae
e New Guinea Rat

Rattus nyikae syn. *Aethomys nyikae* q.v.

Rattus ochraceiventer syn. *Maxomys ochraceiventer* q.v.

Rattus ohiensis syn. *Srilankamys ohiensis* q.v.

5904 *Rattus osgoodi*
Rodentia - Muridae
e Osgood's Rat

Rattus paeduleus syn. *Thallomys paeduleus* q.v.

Rattus pagensis syn. *Maxomys pagensis* q.v.

5905 *Rattus palmarum*
Rodentia - Muridae
e Palm Rat

Rattus panglima syn. *Maxomys panglima* q.v.

5906 *Rattus pelurus*
Rodentia - Muridae
e Peleng Rat

Rattus penitus syn. *Bunomys penitus* q.v.

5907 *Rattus praetor*
Rodentia - Muridae
e Spiny Rat; Large Spiny Rat

Rattus rajah syn. *Maxomys rajah* q.v.

5908 *Rattus ranjiniae*
Rodentia - Muridae
e Kerala Rat

5909 *Rattus rattus*
Rodentia - Muridae
e Roof Rat; Black Rat; House Rat; Ship Rat; Sladen's Rat
d Hausratte; Siedlungshausratte; Schwarze Ratte; Schiffsratte; Dachratte
f Rat commun; Rat noir; Rat noir ordinaire; Rat vulgaire; Rat des greniers; Rat des champs
i Ratto nero; Ratto comune

5910 *Rattus rattus alexandrinus*
Rodentia - Muridae
e Alexandrian Rat; Northern Alexandrine Rat; Alexandrine Rat
d Alexandriner Hausratte; Ägyptische Hausratte
f Rat d'Alexandrie
i Ratto alessandrino

5911 *Rattus rattus brevicaudatus*
Rodentia - Muridae
e Sawah Rat
d Sawah-Ratte
f Rat de Sawah

5912   Rattus rattus diardii
   Rodentia – Muridae
e   Malayan Black Rat
d   Malaiische Hausratte

5913   Rattus rattus frugivorus
   Rodentia - Muridae
e   Fruit Rat
d   Fruchtratte
i   Ratto bruno

5914   Rattus rattus kandianus
   Rodentia - Muridae
e   Common Ceylon House Rat
d   Sri Lanka-Hausratte

5915   Rattus rattus kelaarti
   Rodentia - Muridae
e   Ceylon Highland Rat
d   Sri Lanka-Hochlandratte

5916   Rattus rattus rufescens
   Rodentia - Muridae
e   Indian House Rat
d   Indische Hausratte

   Rattus ruber syn. *R. feliceus* q.v.

   Rattus sabanus syn. *Leopoldamys sabanus* q.v.

   Rattus sacobianus syn. *Apomys sacobianus* q.v.

   Rattus sakeratensis syn. *R. losea* q.v.

   Rattus sibuanus syn. *Limnomys sibuanus* q.v.

5917   Rattus sikkimensis
   Rodentia - Muridae
e   Sikkim Rat

   Rattus silindensis syn. *Aethomys silindensis* q.v.

5918   Rattus simalurensis
   Rodentia - Muridae
e   Simalur Rat; Simeulue Rat

   Rattus siporanus syn. *Leopoldamys siporanus* q.v.

   Rattus sodyi syn. *Kadarsanomys sodyi* q.v.

5919   Rattus sordidus
   Rodentia - Muridae
e   Sordid Rat; Australian Dusky Field Rat; Sombre Downs Rat; Canefield Rat (ANZ); Dusky Field Rat; Dusky Rat

d   Zuckerrohrratte

5920   Rattus steini
   Rodentia - Muridae
e   Stein's Rat; New Guinea Small Spiny Rat

5921   Rattus stoicus
   Rodentia - Muridae
e   Andaman Rat

   Rattus surifer syn. *Maxomys surifer* q.v.

5922   Rattus tanezumi
   Rodentia - Muridae
e   Tanezumi Rat
d   Asiatische Hausratte

   Rattus tatei syn. *R. hoffmanni* q.v.

5923   Rattus tawitawiensis
   Rodentia - Muridae
e   Tawi-Tawi Forest Rat

   Rattus thomasi syn. *Aethomys thomasi* q.v.

5924 **Rattus timorensis**
Rodentia - Muridae
e Timor Rat

5925 **Rattus tiomanicus**
Rodentia - Muridae
e Malaysian Field Rat; Malaysian Wood Rat

5926 **Rattus tunneyi**
Rodentia - Muridae
e Tunney's Rat; Pale Field Rat
d Gelbbraune Ratte

5927 **Rattus turkestanicus**
Rodentia - Muridae
e Turkestan Rat
d Turkestan-Ratte

5928 **Rattus verecundus**
Rodentia - Muridae
e Slender Rat

5929 **Rattus villosissimus**
Rodentia - Muridae
e Long-haired Rat
d Australische Langhaarratte

Rattus whiteheadi syn. *Maxomys whiteheadi* q.v.

5930 **Rattus xanthurus**
Rodentia - Muridae
e Yellow-tailed Rat

5931 **Ratufa**
Rodentia - Sciuridae
e Giant Squirrels; Oriental Giant Squirrels; Oriental Giant Rock Squirrels; Oriental Rock Squirrels
d Riesenhörnchen
f Écureuils géants
i Scoiattoli giganti

5932 **Ratufa affinis**
Rodentia - Sciuridae
e Cream-coloured Giant Squirrel; Pale Giant Squirrel
d Blasses Riesenhörnchen
f Écureuil géant
i Scoiattolo gigante dell'Indonesia

5933 **Ratufa bicolor**
Rodentia - Sciuridae
e Black Giant Squirrel; Malayan Giant Squirrel; Oriental Giant Squirrel
d Riesenhörnchen; Jelarang
f Écureuil d'Inde
i Scoiattolo dell'Asia

5934 **Ratufa indica**
Rodentia - Sciuridae
e Giant Squirrel; Indian Giant Squirrel
d Königsriesenhörnchen
f Écureuil géant de l'Inde
i Scoiattolo indiano gigante; Scoiattolo gigante indiano

5935 **Ratufa indica dealbata**
Rodentia - Sciuridae
e Dang's Giant Squirrel
d Dangs Riesenhörnchen; Dangs-Königriesenhörnchen
f Écureuil des Dangs; Écureuil géant des Dangs; Écureuil géant de l'Inde des Dangs

5936 **Ratufa indica elphinstoni**
Rodentia - Sciuridae
e Malabar Giant Squirrel
d Malabar-Riesenhörnchen

5937 **Ratufa macroura**
Rodentia - Sciuridae
e Grizzled Indian Squirrel; Grizzled Giant Squirrel
d Sri Lanka-Riesenhörnchen
f Écureuil géant gris; Écureuil géant de Ceylan
i Scoiattolo gigante dalla coda lunga

5938   *Ratufa macroura dandolena*
        Rodentia - Sciuridae
e   Common Ceylon Giant Squirrel; Sri
       Lankan Giant Squirrel
d   Gemeines Sri Lanka-Riesenhörnchen

5939   *Ratufa macroura macroura*
        Rodentia - Sciuridae
e   Highland Ceylon Giant Squirrel
d   Sri Lanka-Hochlandriesenhörnchen

5940   *Ratufa macrura melanochra*
        Rodentia - Sciuridae
e   Western Giant Squirrel; Western
       Ceylon Giant Squirrel
d   Westliches Sri Lanka-
       Riesenhörnchen

5941   *Redunca*
        Artiodactyla - Bovidae
e   Reedbucks; Rhinolophe de Mehely
d   Riedböcke

5942   *Redunca arundinum*
        Artiodactyla - Bovidae
e   Reedbuck; Southern Reedbuck;
       Common Reedbuck
d   Großer Riedbock; Großriedbock
f   Antilope des roseaux; Éléotrague;
       Cob des roseaux; Grand cob des
       roseaux
i   Redunca comune

5943   *Redunca fulvorufula*
        Artiodactyla - Bovidae
e   Mountain Reedbuck; Southern
       Mountain Reedbuck
d   Bergriedbock
f   Redunca de montagne
i   Redunca di Lalande; Redunca
       montana

5944   *Redunca fulvorufula adamauae*
        Artiodactyla - Bovidae
e   Western Mountain Reedbuck

d   Westlicher Bergriedbock

5945   *Redunca redunca*
        Artiodactyla - Bovidae
e   Bohor Reedbuck
d   Riedbock; Isabell-Antilope;
       Gemeiner Riedbock
f   Nagor; Nagor de Buffon; Redunca;
       Cob redunca
i   Cervicapra redunca

5946   *Redunca redunca bohor*
        Artiodactyla - Bovidae
e   Abyssinian Bohor Reedbuck
i   Redunca Bohor

5947   *Redunca redunca chanleri*
        Artiodactyla - Bovidae
e   Chanler's Mountain Reedbuck;
       Chanler's Reedbuck

5948   *Redunca redunca cottoni*
        Artiodactyla - Bovidae
e   Sudan Bohor Reedbuck

5949   *Redunca redunca nigeriensis*
        Artiodactyla - Bovidae
e   Nigerian Mountain Reedbuck

5950   *Redunca redunca redunca*
        Artiodactyla - Bovidae
e   Nagor Reedbuck

5951   *Redunca redunca wardi*
        Artiodactyla - Bovidae
e   Eastern Bohor Reedbucl

5952   *Reithrodon*
        Rodentia - Muridae
e   Coney Rats; Rabbit Rats; Pampas
       Gerbils; Bunny Rats
d   Kaninchenratten

5953   *Reithrodon auritus*
        Rodentia - Muridae

e   Coney Rat; Rabbit Rat; Pampas
      Gerbil; Bunny Rat
d   Kaninchenratte

5954   *Reithrodontomys*
       Rodentia - Muridae
e   Harvest Mice; American Harvest
      Mice
d   Amerikanische Erntemäuse;
      Erntemäuse
f   Souris des moissons d'Amérique

5955   *Reithrodontomys brevirostris*
       Rodentia - Muridae
e   Short-nosed Harvest Mouse

5956   *Reithrodontomys burti*
       Rodentia - Muridae
e   Sonoran Harvest Mouse

5957   *Reithrodontomys chrysopsis*
       Rodentia - Muridae
e   Volcano Harvest Mouse

5958   *Reithrodontomys creper*
       Rodentia - Muridae
e   Chiriqui Harvest Mouse
d   Chriqui-Erntemaus

5959   *Reithrodontomys dariensis*
       Rodentia - Muridae
e   Darien Harvest Mouse

5960   *Reithrodontomys fulvescens*
       Rodentia - Muridae
e   Fulvous Harvest Mouse

5961   *Reithrodontomys gracilis*
       Rodentia - Muridae
e   Slender Harvest Mouse

5962   *Reithrodontomys hirsutus*
       Rodentia - Muridae
e   Hairy Harvest Mouse

5963   *Reithrodontomys humulis*
       Rodentia - Muridae
e   American Harvest Mouse; Eastern
      Harvest Mouse
d   Östliche Erntemaus
f   Souris des moissons orientale

5964   *Reithrodontomys megalotis*
       Rodentia - Muridae
e   Western Harvest Mouse
d   Westliche Erntemaus
f   Souris occidentale des moissons;
      Souris des moissons occidentale

5965   *Reithrodontomys megalotis dychei*
       Rodentia - Muridae
e   Prairie Harvest Mouse

5966   *Reithrodontomys mexicanus*
       Rodentia - Muridae
e   Mexican Harvest Mouse

5967   *Reithrodontomys microdon*
       Rodentia - Muridae
e   Small-toothed Harvest Mouse

5968   *Reithrodontomys montanus*
       Rodentia - Muridae
e   Plains Harvest Mouse

5969   *Reithrodontomys paradoxus*
       Rodentia - Muridae
e   Nicaraguan Harvest Mouse

5970   *Reithrodontomys raviventris*
       Rodentia - Muridae
e   Salt-marsh Harvest Mouse
d   Stranderntemaus;
      Salzsumpferntemaus
f   Souris côtière des moissons
i   Topo dei raccolti Americano

5971   *Reithrodontomys rodriguezi*
       Rodentia - Muridae
e   Rodriguez's Harvest Mouse

5972 *Reithrodontomys spectabilis*
Rodentia - Muridae
e   Cozumel Harvest Mouse

5973 *Reithrodontomys sumichrasti*
Rodentia - Muridae
e   Sumichrast's Harvest Mouse

5974 *Reithrodontomys tenuirostris*
Rodentia - Muridae
e   Narrow-nosed Harvest Mouse

5975 *Reithrodontomys zacatecae*
Rodentia - Muridae
e   Zacatecas Harvest Mouse

Reithronycteris syn. *Phyllonycteris* q.v.

5976 *Reithrosciurus*
Rodentia - Sciuridae
e   Groove-toothed Squirrels; Tufted Ground Squirrels; Tufted-ears Ground Squirrels
d   Borneo-Hörnchen
f   Reithosciures

5977 *Reithrosciurus macrotis*
Rodentia - Sciuridae
e   Groove-toothed Squirrel; Tufted Ground Squirrel; Tufted-ears Ground Squirrel; Tassle-eared Ground Squirrel
d   Borneo-Hörnchen
f   Reithosciure de Bornéo

5978 *Rhabdomys*
Rodentia - Muridae
e   Striped Field Mice; Four-striped Grass Mice; Four-striped Rats
d   Afrikanische Striemengrasmäuse; Streifenmäuse; Vierstreifengrasmäuse
f   Rats des champs rayés; Rats des champs

5979 *Rhabdomys pumilio*
Rodentia - Muridae
e   Striped Field Mouse; Four-striped Grass Mouse; Four-striped Rat; Striped Mouse
d   Afrikanische Striemengrasmaus; Streifenmaus; Striemengrasmaus
f   Rat des champs rayé

5980 *Rhagomys*
Rodentia - Muridae
e   Brazilian Arboreal Mice; Brazilian Tree Mice
d   Brasilianische Baummäuse

5981 *Rhagomys rufescens*
Rodentia - Muridae
e   Brazilian Arboreal Mouse; Brazilian Tree Mouse
d   Brasilianische Baummaus

5982 *Rheomys*
Rodentia - Muridae
e   Central American Water Mice; Water Mice; Fishing Rats
d   Zentralamerikanische Wassermäuse; Mittelamerikanische Wassermäuse

Rheomys leander syn. *Anotomys leander* q.v.

5983 *Rheomys mexicanus*
Rodentia - Muridae
e   Mexican Water Mouse

5984 *Rheomys raptor*
Rodentia - Muridae
e   Goldman's Water Mouse

5985 *Rheomys thomasi*
Rodentia - Muridae
e   Thomas's Water Mouse

5986 *Rheomys underwoodi*
Rodentia - Muridae
e   Underwood's Water Mouse

5987 *Rhinoceros*
Perissodactyla - Rhinocerotidae
e Indian Rhinoceroses; Asian One-horned Rhinoceroses
d Nashörner
f Rhinocéros
i Rinoceronti indiani

5988 *Rhinoceros sondaicus*
Perissodactyla - Rhinocerotidae
e Javan Rhinoceros
d Java-Nashorn
f Rhinocéros de la Sonde; Unicorne de la Sonde; Rhinocéros de Java
i Rinoceronte di Giava

5989 *Rhinoceros sondaicus annamaticus*
Perissodactyla - Rhinocerotidae
e Vietnamese Javan Rhinoceros
d Annam-Nashorn
f Rhinocéros de Vietnam

5990 *Rhinoceros unicornis*
Perissodactyla - Rhinocerotidae
e Indian Rhinoceros; Great Indian Rhinoceros; Great Indian One-horned Rhinoceros; Indian Greater One-horned Rhinoceros
d Panzernashorn; Indisches Panzernashorn
f Rhinocéros des Indes; Unicorne de l'Inde
i Rinoceronte indiano

5991 Rhinocerotidae
Perissodactyla
e Rhinoceroses
d Nashörner
f Rhinocérocidés
i Rinoceronti; Rinocerontidi

5992 Rhinolophidae
Chiroptera
e Horseshoe Bats; Horseshoe-nosed Bats; Leaf-nosed Bats; Rhinolophid Bats; Leaf-nosed and Horseshoe Bats

d Dreizehenhufeisennasen; Dreizehige Hufeisennasen; Hufeisennasen; Hufeisennasenartige
f Rhinolophidés; Fers-à-cheval
i Rinolofidi

5993 *Rhinolophus*
Chiroptera - Rhinolophidae
e Horseshoe Bats
d Hufeisennasen; Hufeisennasenfledermäuse
f Fers-à-cheval; Rhinolophes
i Rinolofi

5994 *Rhinolophus acuminatus*
Chiroptera - Rhinolophidae
e Acuminate Horseshoe Bat

5995 *Rhinolophus adami*
Chiroptera - Rhinolophidae
e Adam's Horseshoe Bat
d Adam-Hufeisennase

5996 *Rhinolophus affinis*
Chiroptera - Rhinolophidae
e Intermediate Horseshoe Bat
d Mittlere Hufeisennase

5997 *Rhinolophus alcyone*
Chiroptera - Rhinolophidae
e Halcyon Horseshoe Bat

5998 *Rhinolophus anderseni*
Chiroptera - Rhinolophidae
e Andersen's Horseshoe Bat
d Andersens Hufeisennase

5999 *Rhinolophus arcuatus*
Chiroptera - Rhinolophidae
e Arcuate Horseshoe Bat

6000 *Rhinolophus beddomei*
Chiroptera - Rhinolophidae
e Indian Wooly Horseshoe Bat

6001   ***Rhinolophus blasii***
Chiroptera - Rhinolophidae
e   Peak-saddle Horseshoe Bat; Blasius's Horseshoe Bat
d   Spitzkammhufeisennase; Blasius-Hufeisennase; Spitzkammige Hufeisennase
f   Rhinolophe de Blasius
i   Ferro di cavallo di Blasius; Rinolofo del Blasius

Rhinolophus blythi syn. *R. pusillus* q.v.

6002   ***Rhinolophus borneensis***
Chiroptera - Rhinolophidae
e   Bornean Horseshoe Bat
d   Borneo-Hufeisennase

6003   ***Rhinolophus canuti***
Chiroptera - Rhinolophidae
e   Canut's Horseshoe Bat
d   Canuts Hufeisennase

6004   ***Rhinolophus capensis***
Chiroptera - Rhinolophidae
e   Cape Horseshoe Bat
d   Kap-Hufeisennase

6005   ***Rhinolophus celebensis***
Chiroptera - Rhinolophidae
e   Sulawesi Horseshoe Bat
d   Sulawesi-Hufeisennase

6006   ***Rhinolophus clivosus***
Chiroptera - Rhinolophidae
e   Geoffroy's Horseshoe Bat
d   Geoffroys Hufeisennase
f   Rhinolophe de Geoffroy

6007   ***Rhinolophus coelophyllus***
Chiroptera - Rhinolophidae
e   Peters's Horseshoe Bat; Croslet Horseshoe Bat
d   Croslet-Hufeisennase

6008   ***Rhinolophus cognatus***
Chiroptera - Rhinolophidae
e   Andaman Horseshoe Bat

6009   ***Rhinolophus cornutus***
Chiroptera - Rhinolophidae
e   Little Japanese Horseshoe Bat
d   Kleine Japanische Hufeisennase

6010   ***Rhinolophus creaghi***
Chiroptera - Rhinolophidae
e   Creagh's Horseshoe Bat
d   Creaghs Hufeisennase

6011   ***Rhinolophus darlingi***
Chiroptera - Rhinolophidae
e   Darling's Horseshoe Bat
d   Darlings Hufeisennase

6012   ***Rhinolophus deckenii***
Chiroptera - Rhinolophidae
e   Decken's Horseshoe Bat; Van der Decken's Horseshoe Bat
d   Deckens Hufeisennase

6013   ***Rhinolophus denti***
Chiroptera - Rhinolophidae
e   Dent's Horseshoe Bat
d   Dent-Hufeisennase; Dents Hufeisennase
f   Rhinolophe de Dent

6014   ***Rhinolophus eloquens***
Chiroptera - Rhinolophidae
e   Eloquent Horseshoe Bat

6015   ***Rhinolophus euryale***
Chiroptera - Rhinolophidae
e   Mediterranean Horseshoe Bat
d   Rundkamm-Hufeisennase; Mittelmeer-Hufeisennase; Rundkammige Hufeisennase
f   Rinolophe euryale; Rhinolophe de St. Paterne
i   Ferro di cavallo euriale; Rinolofo

medio; Rinolofo euriale; Rinolofo
mediterraneo; Ferro di cavallo
mediterraneo

6016 *Rhinolophus euryotis*
Chiroptera - Rhinolophidae
e   Broad-eared Horseshoe Bat

6017 *Rhinolophus ferrumequinum*
Chiroptera - Rhinolophidae
e   Greater Horseshoe Bat; Greater
Horseshoe; Greater Horseshoe-nosed
Bat
d   Große Hufeisennase;
Großhufeisennase
f   Grand rinolophe; Grand rinolophe
fer-à-cheval; Grand rinolophe obscur;
Rhinolophe unifer; Grand fer-à-
cheval
i   Ferro di cavallo maggiore; Rinolofo
maggiore

6018 *Rhinolophus fumigatus*
Chiroptera - Rhinolophidae
e   Abyssinian Horseshoe Bat; Rüppell's
Horseshoe Bat
d   Rüppells Hufeisennase

6019 *Rhinolophus guineensis*
Chiroptera - Rhinolophidae
e   Guinenan Horseshoe Bat
d   Guinea-Hufeisennase

6020 *Rhinolophus hildebrandti*
Chiroptera - Rhinolophidae
e   Hildebrandt's Horseshoe Bat
d   Hildebrandt-Hufeisennase
f   Rhinolophe d'Hildebrandt; Fer-à-
cheval du Kenya

Rhinolophus hilli syn. *R. maclaudi*
q.v.

6021 *Rhinolophus hipposideros*
Chiroptera - Rhinolophidae
e   Little Horseshoe Bat; Little
Horseshoe-nosed Bat; Lesser

Horseshoe Bat; Lesser Horseshoe-
nosed Bat
d   Kleine Hufeisennase;
Zwerghufeisennase;
Kleinhufeisennase
f   Petit rhinolophe; Petit rhinolophe fer-
à-cheval; Petit fer-à-cheval; Petite
chauve-souris fer-à-cheval
i   Ferro di cavallo minore; Rinolofo
minore

Rhinolophus hirsutus syn. *R.
macrotis* q.v.

6022 *Rhinolophus imaizumii*
Chiroptera - Rhinolophidae
e   Imaizumi's Horseshoe Bat
d   Imaizumis Hufeisennase

Rhinolophus importunus syn. *R.
borneensis* q.v.

6023 *Rhinolophus inops*
Chiroptera - Rhinolophidae
e   Philippine Forest Horseshoe Bat
d   Philippinen-Hufeisennase

Rhinolophus javanicus syn. *R.
celebensis* q.v.

6024 *Rhinolophus keyensis*
Chiroptera - Rhinolophidae
e   Insular Horseshoe Bat
d   Inselhufeisennase

6025 *Rhinolophus landeri*
Chiroptera - Rhinolophidae
e   Lander's Horseshoe Bat
d   Lander-Hufeisennase; Landers
Hufeisennase
f   Rhinolophe de Lander

6026 *Rhinolophus lepidus*
Chiroptera - Rhinolophidae
e   Blyth's Horseshoe Bat
d   Blyths Hufeisennase

6027   *Rhinolophus luctus*
Chiroptera - Rhinolophidae
e   Great Eastern Horseshoe Bat; Woolly
Horseshoe Bat; Woolly Eastern
Horseshoe Bat
d   Große Östliche Hufeisennase

6028   *Rhinolophus luctus sobrinus*
Chiroptera - Rhinolophidae
e   Ceylon Great Horseshoe Bat
d   Große Sri Lanka
Hufeisennasenfledermaus

6029   *Rhinolophus maclaudi*
Chiroptera - Rhinolophidae
e   Maclaud's Horsehoe Bat
d   Maclaud-Hufeisennase; Maclauds
Hufeisennase
f   Rhinolophe de Maclaud

6030   *Rhinolophus macrotis*
Chiroptera - Rhinolophidae
e   Big-eared Horseshoe Bat; Large-
eared Horseshoe Bat
d   Langohrhufeisennase

Rhinolophus madurensis syn. *R.
celebensis* q.v.

6031   *Rhinolophus malayanus*
Chiroptera - Rhinolophidae
e   Malayan Horseshoe Bat; North
Malayan Horseshoe Bat
d   Malaiische Hufeisennase

6032   *Rhinolophus marshalli*
Chiroptera - Rhinolophidae
e   Marshall's Horseshoe Bat
d   Marschalls Hufeisennase

6033   *Rhinolophus megaphyllus*
Chiroptera - Rhinolophidae
e   Eastern Horseshoe Bat (ANZ);
Great-leaved Horseshoe Bat;
Southern Horseshoe Bat; Smaller

Horseshoe Bat
d   Kleinere Hufeisennase

6034   *Rhinolophus mehelyi*
Chiroptera - Rhinolophidae
e   Mehely's Horseshoe Bat
d   Mehelys Hufeisennase; Mehelys
Rumänische Hufeisennase; Mehely-
Hufeisennase
f   Rhinolophe de Mehely; Rhinolophe
de Roumanie
i   Ferro di cavallo di Mehely; Rinolofo
di Mehely

Rhinolophus minutillus syn. *R.
pusillus* q.v.

6035   *Rhinolophus mitratus*
Chiroptera - Rhinolophidae
e   Mitred Horsehoe Bat; Mitered
Horseshoe Bat (NA)

6036   *Rhinolophus monoceros*
Chiroptera - Rhinolophidae
e   Formosan Lesser Horseshoe Bat
d   Kleinere Formosa-Hufeisennase

6037   *Rhinolophus nereis*
Chiroptera - Rhinolophidae
e   Nereid Horseshoe Bat

6038   *Rhinolophus osgoodi*
Chiroptera - Rhinolophidae
e   Osgood's Horseshoe Bat
d   Osgoods Hufeisennase

6039   *Rhinolophus paradoxolophus*
Chiroptera - Rhinolophidae
e   Bourret's Horseshoe Bat
d   Bourrets Hufeisennase

Rhinolophus parvus syn. *R.
celebensis* q.v.

6040   *Rhinolophus pearsoni*
Chiroptera - Rhinolophidae

e   Pearson's Horseshoe Bat
d   Pearsons Hufeisennase

6041   *Rhinolophus philippinensis*
Chiroptera - Rhinolophidae
e   Philippine Horseshoe Bat
d   Großohrige Hufeisennase

6042   *Rhinolophus pusillus*
Chiroptera - Rhinolophidae
e   Least Horseshoe Bat
d   Kleinste Hufeisennase

Rhinolophus refulgens syn. *R. lepidus* q.v.

6043   *Rhinolophus rex*
Chiroptera - Rhinolophidae
e   King Horseshoe Bat
d   Königshufeisennase

6044   *Rhinolophus robinsoni*
Chiroptera - Rhinolophidae
e   Peninsular Horseshoe Bat
d   Halbinsel-Hufeisennase

6045   *Rhinolophus rouxi*
Chiroptera - Rhinolophidae
e   Rufous Horseshoe Bat
d   Fuchsrote Hufeisennase

6046   *Rhinolophus rufus*
Chiroptera - Rhinolophidae
e   Large Rufous Horseshoe Bat
d   Große Fuchsrote Hufeisennase

Rhinolophus ruwenzorii syn. *R. maclaudi* q.v.

6047   *Rhinolophus sedulus*
Chiroptera - Rhinolophidae
e   Lesser Woolly Horseshoe Bat
d   Kleine Wollige Hufeisennase

6048   *Rhinolophus shameli*
Chiroptera - Rhinolophidae
e   Shamel's Horseshoe Bat
d   Shamels Hufeisennase

6049   *Rhinolophus silvestris*
Chiroptera - Rhinolophidae
e   Forest Horseshoe Bat
d   Waldhufeisennase

6050   *Rhinolophus simplex*
Chiroptera - Rhinolophidae
e   Lombok Horseshoe Bat
d   Lombok-Hufeisennase

6051   *Rhinolophus simulator*
Chiroptera - Rhinolophidae
e   Bushveld Horseshoe Bat
d   Buschveldt-Hufeisennase

6052   *Rhinolophus stheno*
Chiroptera - Rhinolophidae
e   Lesser Brown Horseshoe Bat
d   Kleine Braune Hufeisennase

6053   *Rhinolophus subbadius*
Chiroptera - Rhinolophidae
e   Little Nepalese Horseshoe Bat
d   Kleine Nepal-Hufeisennase

6054   *Rhinolophus subrufus*
Chiroptera - Rhinolophidae
e   Small Rufous Horseshoe Bat
d   Kleine Fuchsrote Hufeisennase

6055   *Rhinolophus swinnyi*
Chiroptera - Rhinolophidae
e   Swinny's Horseshoe Bat
d   Swinnys Hufeisennase

6056   *Rhinolophus thomasi*
Chiroptera - Rhinolophidae
e   Thomas's Horseshoe Bat
d   Thomas-Hufeisennase

6057 *Rhinolophus trifoliatus*
Chiroptera - Rhinolophidae
e Trefoil Horseshoe Bat

6058 *Rhinolophus virgo*
Chiroptera - Rhinolophidae
e Yellow-faced Horseshoe Bat
d Gelbgesichthufeisennase

6059 *Rhinolopus yunanensis*
Chiroptera - Rhinolophidae
e Dobson's Horseshoe Bat
d Dobsons Hufeisennase

6060 *Rhinolopus ziama*
Chiroptera - Rhinolophidae
e Ziama Horseshoe bat

Rhinomegalophus syn. *Rhinolophus* q.v.

6061 *Rhinonicteris*
Chiroptera - Rhinolophide
e Golden Horseshoe Bats
d Goldene Rundblattnasen

6062 *Rhinonycteris aurantia*
Chiroptera - Rhinolophide
e Golden Horseshoe Bat; Orange Horseshoe Bat; Orange Leaf-nosed Bat; Orange Leafnose Bat (ANZ); Australian Orange Leaf-nosed Bat
d Goldene Rundblattnase

6063 *Rhinophylla*
Chiroptera - Phyllostomidae
e Little Fruit Bats

6064 *Rhinophylla alethina*
Chiroptera - Phyllostomidae
e Hairy Little Fruit Bat

6065 *Rhinophylla fischerae*
Chiroptera - Phyllostomidae
e Fischer's Little Fruit Bat

6066 *Rhinophylla pumilio*
Chiroptera - Phyllostomidae
e Dwarf Little Fruit Bat

6067 *Rhinopithecus*
Primates - Cercopithecidae
e Snub-nosed Monkeys
d Stumpfnasenaffen
f Rhinopithèques
i Rinopitechi

6068 *Rhinopithecus avunculus*
Primates - Cercopithecidae
e Tonkin Snub-nosed Monkey; Tonkin Snub-nosed Langur
d Tonkin-Stumpfnase; Tonkin-Stumpfnasenaffe
f Rhinopithèque du Tonkin
i Rinopiteco del Tonchino

6069 *Rhinopithecus bieti*
Primates - Cercopithecidae
e Black Snub-nosed Monkey; Yunnan Snub-nosed Monkey; Brown Snub-nosed Monkey; Black Golden-haired Monkey; Biet's Snub-nosed Monkey
d Braune Stumpfnase
f Rhinopithèque de Mgr Bier; Rhinopithèque brun
i Rinopiteco bruno

6070 *Rhinopithecus brelichi*
Primates - Cercopithecidae
e Brelich's Snub-nosed Monkey; Grey Snub-nosed Monkey; Guizhou Snub-nosed Monkey; Grey Golden-haired Monkey
d Weißmantelstumpfnase
f Rhinopithèque jaune doré
i Rinopiteco dal mantello bianco

6071 *Rhinopithecus roxellana*
Primates - Cercopithecidae
e Sichuan Golden Snub-nosed Monkey; Sichuan Golden Monkey; Snow Monkey; Chinese Snub-nosed

Monkey; Golden Snub-nosed
Monkey
- *d* Goldstumpfnase
- *f* Rhinopithèque de Roxellane
- *i* Rinopiteco dorato; Rinopiteco arancione

6072 *Rhinopoma*
Chiroptera - Rhinopomatidae
- *e* Rat-tailed Bats; Mouse-tailed Bats; Long-tailed Bats
- *d* Mausschwanzfledermäuse; Klappnasen
- *f* Rhinopomes
- *i* Pipistrelli coda di topo

6073 *Rhinopoma hardwickei*
Chiroptera - Rhinopomatidae
- *e* Lesser Rat-tailed Bat; Hardwicke's Mouse-tailed Bat; Lesser Mouse-tailed Bat
- *d* Hardwicke-Klappnase
- *f* Rhinopome d'Hardwicke
- *i* Pipistrello coda di topo minore

6074 *Rhinopoma microphyllum*
Chiroptera - Rhinopomatidae
- *e* Greater Mouse-tailed Bat; Larger Mouse-tailed Bat; Greater Rat-tailed Bat; Larger Rat-tailed Bat
- *d* Ägyptische Klappnase
- *f* Rhinopome microphylle; Grand rhinopome

6075 *Rhinopoma muscatellum*
Chiroptera - Rhinopomatidae
- *e* Small Mouse-tailed Bat; Small Rat-tailed Bat

6076 *Rhinopomatidae*
Chiroptera -
- *e* Rat-tailed Bats; Mouse-tailed Bats; Long-tailed Bats
- *d* Klappnasen; Langschwanzfledermäuse; Mausschwanzfledermäuse
- *f* Rhinopomatidés

- *i* Rinopomatidi

Rhinopterus syn. *Eptesicus* q.v.

6077 *Rhinosciurus*
Rodentia - Sciuridae
- *e* Long-nosed Squirrels; Shrew-faced Squirrels; Long-nosed Ground Squirrels
- *d* Langnasenhörnchen
- *f* Écureuils à nez long
- *i* Scoiattoli dal naso lungo

6078 *Rhinosciurus laticaudatus*
Rodentia - Sciuridae
- *e* Long-nosed Squirrel; Shrew-faced Squirrel; Long-nosed Ground Squirrel; Shrew-faced Ground Squirrel; Ant-eating Squirrel
- *d* Langnasenhörnchen
- *f* Écureuil à nez long
- *i* Scoiattolo dal naso lungo

6079 *Rhipidomys*
Rodentia - Muridae
- *e* Climbing Mice; American Climbing Mice; Long-tailed Climbing Mice
- *d* Südamerikanische Klettermäuse; Klettermäuse

6080 *Rhipidomys austrinus*
Rodentia - Muridae
- *e* Southern Climbing Mouse
- *d* Südliche Klettermaus

6081 *Rhipidomys caucensis*
Rodentia - Muridae
- *e* Cauca Climbing Mouse
- *d* Cauca-Klettermaus

6082 *Rhipidomys couesi*
Rodentia - Muridae
- *e* Coués's Climbing Mouse
- *d* Coués Klettermaus

6083   ***Rhipidomys fulviventer***
Rodentia - Muridae
e   Buff-bellied Climbing Mouse

6084   ***Rhipidomys latimanus***
Rodentia - Muridae
e   Broad-footed Climbing Mouse

6085   ***Rhipidomys leucodactylus***
Rodentia - Muridae
e   White-footed Climbing Mouse
d   Weißfingerklettermaus

6086   ***Rhipidomys macconnelli***
Rodentia - Muridae
e   MacConnell's Climbing Mouse

6087   ***Rhipidomys mastacalis***
Rodentia - Muridae
e   Long-tailed Climbing Mouse

6088   ***Rhipidomys nitela***
Rodentia - Muridae
e   Splendid Climbing Mouse

6089   ***Rhipidomys ochrogaster***
Rodentia - Muridae
e   Yellow-bellied Climbing Mouse

6090   ***Rhipidomys scandens***
Rodentia - Muridae
e   Mount Pirri Climbing Mouse

6091   ***Rhipidomys venezuelae***
Rodentia - Muridae
e   Venezuelan Climbing Mouse

6092   ***Rhipidomys venustus***
Rodentia - Muridae
e   Charming Climbing Mouse

6093   ***Rhipidomys wetzeli***
Rodentia – Muridae

e   Wetzel's Climbing Mouse

6094   ***Rhizomys***
Rodentia - Muridae
e   Bamboo Rats
d   Bambusratten
f   Rats des bambous
i   Ratti del bambú

6095   ***Rhizomys pruinosus***
Rodentia - Muridae
e   Hoary Bamboo Rat
d   Eisgraue Bambusratte

6096   ***Rhizomys sinensis***
Rodentia - Muridae
e   Chinese Bamboo Rat; China Bamboo Rat
d   Chinesische Bambusratte
f   Rat des bambous chinois

6097   ***Rhizomys sumatrensis***
Rodentia - Muridae
e   Sumatran Bamboo Rat; Large Bamboo Rat
d   Sumatra-Bambusratte
f   Rat des bambous de Sumatra

6098   ***Rhogeessa***
Chiroptera - Vespertilionidae
e   Rhogeëssa Bats; Yellow Bats; Little Yellow Bats

6099   ***Rhogeessa alleni***
Chiroptera - Vespertilionidae
e   Allen's Yellow Bat; Allen's Bat

6100   ***Rhogeessa genowaysi***
Chiroptera - Vespertilionidae
e   Genoways's Yellow Bat

6101   ***Rhogeessa gracilis***
Chiroptera - Vespertilionidae
e   Slender Yellow Bat; Big-eared Yellow Bat

6102   *Rhogeessa hussoni*
           Chiroptera - Vespertilionidae
    e   Husson's Yellow Bat

6103   *Rhogeessa minutilla*
           Chiroptera - Vespertilionidae
    e   Tiny Yellow Bat; Little Desert Bat

6104   *Rhogeessa mira*
           Chiroptera - Vespertilionidae
    e   Least Yellow Bat

6105   *Rhogeessa parvula*
           Chiroptera - Vespertilionidae
    e   Little Yellow Bat

6106   *Rhogeessa tumida*
           Chiroptera - Vespertilionidae
    e   Central American Yellow Bat;
           Black-winged Little Yellow Bat

6107   *Rhombomys*
           Rodentia - Muridae
    e   Giant Day Jirds; Great Gerbils
    d   Riesenmäuse; Große Rennmäuse
    f   Gerbilles géantes; Grandes gerbilles

6108   *Rhombomys opimus*
           Rodentia - Muridae
    e   Giant Day Jird; Great Gerbil; Giant
           Day Gerbil; Great Jird
    d   Große Rennmaus; Riesenmaus
    f   Gerbille géante

6109   *Rhynchocyon*
           Macroscelidea - Macroscelididae
    e   Forest Elephant Shrews; Chequered-
           back Elephant Shrews
    d   Rüsselhündchen
    f   Macroscélides
    i   Toporagni elefanti delle foreste;
           Rincioni

6110   *Rhynchocyon chrysopygus*
           Macroscelidea - Macroscelididae

    e   Yellow-rumped Elephant Shrew;
           Golden-rumped Elephant Shrew;
           Golden-rumped Sengi
    d   Goldenes Rüsselhündchen
    i   Toporagno elefante dal dorso dorato

6111   *Rhynchocyon cirnei*
           Macroscelidea - Macroscelididae
    e   Chequered Elephant Shrew; Giant
           Elephant Shrew; Chequered Sengi;
           Checkered Elephant Shrew;
           Stuhlmann's Elephant Shrew
    d   Geflecktes Rüsselhündchen; Dunkles
           Rüsselhündchen
    f   Macroscélide tacheté; Macroscélide
           de Stuhlmann

6112   *Rhynchocyon petersi*
           Macroscelidea - Macroscelididae
    e   Peters's Elephant Shrew; Black-and-
           rufous Elephant Shrew; Black-and-
           rufous Sengi; Zanj Sengi
    d   Rotschulter-Rüsselhündchen
    f   Macroscélide de Peters

           Rhynchocyon stuhlmanni syn. *R.
           cirnei* q.v.

6113   *Rhynchogale*
           Carnivora - Herpestidae
    e   Meller's Mongooses
    d   Langschnauzenmangusten; Meller-
           Mangusten
    f   Mangoustes de Meller
    i   Manguste di Meller

6114   *Rhynchogale melleri*
           Carnivora - Herpestidae
    e   Meller's Mongoose
    d   Langschauzenmanguste; Meller-
           Manguste
    f   Mangouste de Meller
    i   Mangusta di Meller

6115   *Rhyncholestes*
           Paucituberculata - Caenolestidae
    e   Chilean Rat Opossums; Fat-tailed

Rat Opossums; Chilean Shrew
Opossums
d Chile-Opossummäuse; Chilenische
Mausopossums

6116   *Rhyncholestes raphanurus*
Paucituberculata - Caenolestidae
e Chilean Rat Opossum; Fat-tailed Rat
Opossum; Chilean Shrew Opossum
d Chile-Opossummaus; Chilenisches
Mausopossum; Chilenisches
Spitzmausopossum

6117   *Rhynchomeles*
Peromelemorpha - Peroryctidae
e Ceram Long-nosed Bandicoots;
Ceram Island Long-nosed
Bandicoots; Ceram Bandicoots
d Ceram-Nasenbeutler

6118   *Rhynchomeles prattorum*
Peromelemorpha - Peroryctidae
e Ceram Long-nosed Bandicoot;
Ceram Island Long-nosed Bandicoot;
Ceram Bandicoot
d Ceram-Nasenbeutler;
Kaninchennasenbeutler

6119   *Rhynchomys*
Rodentia - Muridae
e Shrewrats; Shrew-like Rats
d Nasenratten
f Rats-musaraignes

6120   *Rhynchomys isarogensis*
Rodentia - Muridae
e Isarog Shrewrat; Shrew-like Rat;
Mount Isarog Shrewrat

6121   *Rhynchomys soricoides*
Rodentia - Muridae
e Philippines Shrewrat; Mount Data
Shrewrat; Long-snouted Shrewrat
d Nasenratte
f Rat au nez longue

Rhynchonax syn. *Uropsilus* q.v.

6122   *Rhynchonycteris*
Chiroptera - Emballonuridae
e Proboscis Bats; Sharp-nosed Bats;
Long-nosed Bats; Tufted Bats
d Nasenfledermäuse
f Rhynchonyctères
i Pipistrelli dalla proboscide

6123   *Rhynchonycteris naso*
Chiroptera - Emballonuridae
e Proboscis Bat; Brazilian Long-nosed
Bat; Long-nosed Bat; Tufted Bat
d Nasenfledermaus
f Chauve-souris à long nez;
Rhynchonyctère à long nez
i Pipistrello dalla proboscide

Rhynchotragus syn. *Madoqua* q.v.

6124   *Romerolagus*
Lagomorpha - Leporidae
e Volcano Rabbits
d Mexikanische Vulkankaninchen
f Lapins de Diaz
i Conigli dei volcani

6125   *Romerolagus diazi*
Lagomorpha - Leporidae
e Volcano Rabbit; Zacatuche; Mexican
Rabbit
d Mexikanisches Vulkankaninchen;
Vulkannaninchen
f Lapin de Diaz; Lapin des volcans
i Coniglio dei volcani

6126   *Rousettus*
Chiroptera - Pteropodidae
e Rousette Fruit Bats; Dog Fruit Bats;
Rousettes
d Höhlenflughunde; Flederhunde
f Roussettes
i Rossetti; Pipistrelli frugivore ;
Pipistrelli della fruta

6127   *Rousettus aegyptiacus*
Chiroptera - Pteropodidae

e Egyptian Rousette; Egyptian
Rousette Bat; Egyptian Fruit Bat;
Southwest Asian Fruit Bat
d Ägyptischer Flughund; Kleiner
Flughund; Nil-Flughund;
Ägyptischer kleiner Flughund
f Roussette d'Égypte; Pipistrelle
frugivore africana
i Rossetto egiziano; Pipistrello
frugivore africano

6128 *Rousettus amplexicaudatus*
Chiroptera - Pteropodidae
e Geoffroy's Rousette; Geoffroy's
Rousette Fruit Bat

6129 *Rousettus angolensis*
Chiroptera - Pteropodidae
e Bocages Fruit Bat; Angolan Rousette
d Angola-Flughund; Angola Flughund

6130 *Rousettus celebensis*
Chiroptera - Pteropodidae
e Celebes Rousette; Sulawesi Rousette

6131 *Rousettus lanosus*
Chiroptera - Pteropodidae
e Ruwenzori Long-haired Rousette;
Long-haired Rousette

6132 *Rousettus lanosus kempi*
Chiroptera - Pteropodidae
e Mountain Fruit Bat

6133 *Rousettus leschenaulti*
Chiroptera - Pteropodidae
e Fulvous Fruit Bat; Leschenault's
Rousette

6134 *Rousettus madagascariensis*
Chiroptera - Pteropodidae
e Madagascar Rousette; Madagascan
Rousette

6135 *Rousettus obliviosus*
Chiroptera - Pteropodidae

e Comoro Rousette
f Petite roussette des Comores

6136 *Rousettus spinalatus*
Chiroptera - Pteropodidae
e Bare-backed Rousette

Rousettus stresemanni syn. *R.
amplexicaudatus* q.v.

6137 *Rubrisciurus*
Rodentia - Sciuridae
e Sulawesi Giant Squirrels
d Sulawesi-Riesenflughunde

6138 *Rubrisciurus rubriventer*
Rodentia - Sciuridae
e Sulawesi Giant Squirrel
d Rotbäuchiges Eichhorn; Sulawesi-
Riesenflughund

Rupestes syn. *Sciurotamias* q.v.

6139 *Rupicapra*
Artiodactyla - Bovidae
e Chamois
d Gemsen
f Chamois
i Camosci

6140 *Rupicapra pyrenaica*
Artiodactyla - Bovidae
e Pyrenean Chamois
d Pyrenäen-Gemse
f Isard
i Camoscio sud-occidentale

6141 *Rupicapra pyrenaica ornata*
Artiodactyla - Bovidae
e Abruzzi Chamois
d Abruzzen-Gemse
f Chamois des Abruzzes
i Camoscio d'Abruzzo; Camoscio
appeninico

6142   ***Rupicapra rupicapra***
        Artiodactyla - Bovidae
    e   Chamois; Alpine Chamois
    d   Alpengemse; Gams; Gemse
    f   Chamois d'Europe; Chamois;
        Chamois des Alpes
    i   Camoscio; Capra selvatica; Muflone;
        Camoscio alpino; Camozza

6143   ***Rupicapra rupicapra parva***
        Artiodactyla - Bovidae
    e   Cantabrian Chamois
    i   Camoscio pirenaica

6144   ***Ruwenzorisorex***
        Soricomorpha - Soricidae
    e   Ruwenzori Shrews

6145   ***Ruwenzorisorex suncoides***
        Soricomorpha - Soricidae
    e   Ruwenzori Shrew

S

6146 *Saccolaimus*
Chiroptera - Emballonuridae
e Pouched Bat

6147 *Saccolaimus flaviventris*
Chiroptera - Emballonuridae
e Yellow-bellied Pouched Bat; Yellow-bellied Tomb Bat; Yellow-bellied Free-tailed Bat; Yellow-bellied Sheathtail Bat (ANZ)
d Größte Australische Fledermaus

6148 *Saccolaimus mixtus*
Chiroptera - Emballonuridae
e Troughton's Pouched Bat; Allied Free-tailed Bat; Papuan Sheath-tailed Bat

6149 *Saccolaimus peli*
Chiroptera - Emballonuridae
e Pel's Pouched Bat

6150 *Saccolaimus pluto*
Chiroptera - Emballonuridae
e Philippine Pouched Bat

6151 *Saccolaimus saccolaimus*
Chiroptera - Emballonuridae
e Naked-rumped Pouched Bat; Bare-rumped Sheathtail Bat (ANZ); Pouch-bearing Tomb Bat; Pouch-bearing Sheath-tailed Bat

6152 *Saccopteryx*
Chiroptera - Emballonuridae
e Sac-winged Bats; Sheath-tailed Bats; Two-lined Bats; White-lined Bats
d Taschenfledermäuse; Glattnasenfreischwänze; Sackflügelfledermäuse
f Saccoptéryx

6153 *Saccopteryx bilineata*
Chiroptera - Emballonuridae
e Temminck's White-lined Bat; Greater White-lined Bat; Greater Sac-winged Bat; El Salvador Sheath-tailed Bat; White-lined Sac-winged Bat
d Zweistreifentaschenfledermaus
f Saccoptéryx à deux raies

6154 *Saccopteryx canescens*
Chiroptera - Emballonuridae
e Frosted Sac-winged Bat; Frosted White-lined Bat

6155 *Saccopteryx gymnura*
Chiroptera - Emballonuridae
e Amazonian Sac-winged Bat; Amazonian White-lined Bat

6156 *Saccopteryx leptura*
Chiroptera - Emballonuridae
e Two-lined Bat; Lesser White-lined Bat; Schreber's White-lined Bat; Lesser Sac-winged Bat; White-lined Bat

6157 *Saccostomus*
Rodentia - Muridae
e African Pouched Mice; Cape Pouched Mice; African Pouched Rats; Short-tailed Pouched Rats; Pouched Mats
d Kurzschwanzhamsterratten
f Saccostomes; Rats à poche

6158 *Saccostomus campestris*
Rodentia - Muridae
e Pouched Mouse; African Pouched Rat; Short-tailed Pouched Mouse
d Backentaschenratte; Kurzschwanzhamsterratte
f Rat fouisseur à queue courte

6159   *Saccostomus mearnsi*
Rodentia - Muridae
e   Mearns's Pouched Mouse

6160   *Saguinus*
Primates - Callitrichidae
e   Tamarins; Long-tusked Marmosets;
Bare-faced Marmosets
d   Tamarins
f   Tamarins
i   Tamarini

6161   *Saguinus bicolor*
Primates - Callitrichidae
e   Pied Tamarin; Martin's Tamarin;
Bare-faced Tamarin; Brazilian Bare-
faced Tamarin; Pied Bare-faced
Tamarin; Brazilian Pied Tamarin
d   Zweifarbenäffchen
f   Tamarin bicolore; Ouistiti bicolore
i   Uistitì calvo; Marikina; Tamarino
calvo

6162   *Saguinus fuscicollis*
Primates - Callitrichidae
e   Brown-headed Tamarin; Saddle-
backed Tamarin; Saddleback
Tamarin; Spix's Saddle-backed
Tamarin

d   Braunrückenäffchen;
Braunrückentamarin
f   Tamarin à tête brune
i   Tamarino dal dorso bruno; Tamarino
di Illiger; Tamarino al dorso bruno;
Tamarino a dorso bruno

6163   *Saguinus fuscicollis avilapiresi*
Primates - Callitrichidae
e   Avila Pires's Saddleback Tamarin

6164   *Saguinus fuscicollis crandalli*
Primates - Callitrichidae
e   Crandall's Saddleback Tamarin

6165   *Saguinus fuscicollis cruzlimai*

Primates - Callitrichidae
e   Cruz Lima's Saddleback Tamarin

6166   *Saguinus fuscicollis fuscicollis*
Primates - Callitrichidae
e   Spix's Saddleback Tamarin

6167   *Saguinus fuscicollis fuscus*
Primates - Callitrichidae
e   Lesson's Saddleback Tamarin

6168   *Saguinus fuscicollis illigeri*
Primates - Callitrichidae
e   Illiger's Saddleback Tamarin
f   Tamarin à manteau rouge

6169   *Saguinus fuscicollis lagonotus*
Primates - Callitrichidae
e   Redmantle Saddleback Tamarin

6170   *Saguinus fuscicollis leucogenys*
Primates - Callitrichidae
e   Andean Saddleback Tamarin

6171   *Saguinus fuscicollis melanoleucus*
Primates - Callitrichidae
e   White Saddleback Tamarin

6172   *Saguinus fuscicollis nigrifrons*
Primates - Callitrichidae
e   Geoffroy's Saddleback Tamarin

6173   *Saguinus fuscicollis weddelli*
Primates - Callitrichidae
e   Weddell's Saddleback Tamarin
f   Tamarin de Weddell

6174   *Saguinus geoffroyi*
Primates - Callitrichidae
e   Geoffroy's Tamarin; Rufous-naped
Tamarin; Panamanian Tamarin; Red-
crested Tamarin; Red-naped Tamarin
d   Geoffroy-Perückenäffchen;
Geoffroy-Perückenaffe

f   Pinché de Geoffroy; Tamarin de Geoffroy

i   Tamarino di Geoffroy

6175   *Saguinus graellsi*
Primates - Callitrichidae

e   Graell's Black-mantled Tamarin; Rio Napo Tamarin

d   Rio-Napa-Tamarin; Schwarzroter Tamarin

f   Tamarin de Rio Napo

i   Tamarino di Grealls

6176   *Saguinus imperator*
Primates - Callitrichidae

e   Emperor Moustached Tamarin; Emperor Tamarin; Emperor Marmoset

d   Kaiserschnurrbart-Tamarin

f   Tamarin empereur

i   Tamarino imperatore

6177   *Saguinus imperator imperator*
Primates - Callitrichidae

e   Black-chinned Emperor Tamarin

6178   *Saguinus imperator subgrisescens*
Primates - Callitrichidae

e   Bearded Emperor Tamarin

6179   *Saguinus inustus*
Primates - Callitrichidae

e   Mottle-faced Tamarin

d   Marmorgesichttamarin

f   Tamarin à face noire

i   Tamarino dalla faccia chiazzata

6180   *Saguinus labiatus*
Primates - Callitrichidae

e   Red-bellied Tamarin; Red-chested Moustached Tamarin; Red-chested Tamarin; White-lipped Tamarin; Geoffroy's Moustached Tamarin

d   Rotbauchtamarin

f   Tamarin labié

i   Tamarino dal ventre rosso

6181   *Saguinus labiatus thomasi*
Primates - Callitrichidae

e   Thomas's Moustached Tamarin

6182   *Saguinus leucopus*
Primates - Callitrichidae

e   White-footed Tamarin; Silvery Brown Bare-faced Tamarin; White-footed Marmoset

d   Weißfußäffchen; Weißfußaffe

f   Ouistiti aux pieds blancs; Pinché aux pieds blancs; Tamarin aux pieds blancs

i   Tamarino dai piedi bianchi

6183   *Saguinus martinsi*
Primates - Callitrichidae

e   Martin's Bare-faced Tamarin

d   Martins-Manteläffchen

6184   *Saguinus martinsi ochraceus*
Primates - Callitrichidae

e   Ochraceous Bare-faced Tamarin

6185   *Saguinus melanoleucus*
Primates - Callitrichidae

e   White-mantled Tamarin

d   Weißmanteltamarin

f   Tamarin blanc

6186   *Saguinus midas*
Primates - Callitrichidae

e   Yellow-handed Marmoset; Red-handed Tamarin; Midas Tamarin; Lacépède's Tamarin; Black Tamarin; Negro Tamarin; Red-backed Tamarin; Golden-handed Tamarin; Rufous-handed Tamarin

d   Rothandtamarin; Mohrentamarin

f   Tamarin aux mains rousses; Tamarin nègre

i   Tamarino dalle mani rosse; Tamarino dalla faccia nera

6187   *Saguinus mystax*
Primates - Callitrichidae

e Moustached Tamarin; Black-chested
    Moustached Tamarin; Spix's
    Moustached Tamarin
d Schnurrbarttamarin
f Tamarin à moustache; Pinché à
    moustaches
i Tamarino dai mustacchi; Tamarino
    mustacchio

6188 *Saguinus niger*
    Primates - Callitrichidae
e Black Tamarin; Black-handed
    Tamarin
d Schwarzhandtamarin
i Tamarino nero

6189 *Saguinus nigricollis*
    Primates - Callitrichidae
e Red-and black Tamarin; Black-and-
    red Tamarin; Black-mantled Tamarin
d Schwarzrückentamarin
f Tamarin rouge-et-noir
i Tamarino dal dorso nero

6190 *Saguinus nigricollis hernandezi*
    Primates - Callitrichidae
e Hernandez-Camacho's Black-mantled
    Tamarin

6191 *Saguinus nigricollis nigricollis*
    Primates - Callitrichidae
e Spix's Black-mantled Tamarin
d Spix-Schwarzrückentamarin

6192 *Saguinus oedipus*
    Primates - Callitrichidae
e Crested Tamarin; Cotton-headed
    Tamarin; Crested Bare-faced
    Tamarin; Cotton-top Tamarin;
    Cotton-top Marmoset; Cotton-top;
    Liszt Monkey
d Liszt-Äffchen; Liszt-Affe
f Tamarin à crête blanche; Pinché à
    crête blanche; Pinché
i Tamarino edipo; Edipomida edipo

6193 *Saguinus pileatus*

    Primates - Callitrichidae
e Red-capped Tamarin; Red-capped
    Moustached Tamarin
d Rotkappentamarin
f Tamarin à calotte rouge

Saguinus tamarin syn. *S. midas* q.v.

6194 *Saguinus tripartitus*
    Primates - Callitrichidae
e Golden-mantled Tanarin; Golden-
    mantled Saddleback Tamarin
d Goldmanteltamarin
f Tamarin à manteau doré
i Tamarino dal mantello dorato

6195 *Saiga*
    Artiodactyla - Bovidae
e Saigas
d Saiga-Antilopen; Saigas;
    Steppenantilopen
f Saïgas
i Saighe; Antilopi delle steppe

6196 *Saiga tatarica*
    Artiodactyla - Bovidae
e Saiga; Saiga Antelope
d Schafsantilope; Saiga-Antilope;
    Steppenantilope; Saiga
f Saïga; Antilope saïga
i Antilope delle steppe; Saiga

6198 *Saiga tatarica mongolica*
    Artiodactyla - Bovidae
e Mongolian Saiga
d Mongolische Saiga

6197 *Saiga tatarica tatarica*
    Artiodactyla - Bovidae
e Russian Saiga
d Russische Saiga

6199 *Saimiri*
    Primates - Atelidae
e Squirrel Monkeys

d   Totenkopfäffchen ; Totenköpchen;
Totenkopfaffen
f   Saïmiris; Sapajous
i   Scimmie scoiattolo; Scimmiette
scoiattolo

6200   *Saimiri boliviensis*
Primates - Atelidae
e   Bolivian Squirrel Monkey; Black-
headed Squirrel
d   Schwarzköpfiges Totenköpchen
f   Singe-écureil de Bolivie; Sapajou à
tète noire
i   Saimiri della Bolivia

6201   *Saimiri boliviensis peruviensis*
Primates - Atelidae
e   Peruvian Squirrel Monkey
d   Peru-Totenkopfäffchen

6202   *Saimiri oerstedi*
Primates - Atelidae
e   Red-backed Squirrel Monkey;
Central American Squirrel Monkey
d   Gelbes Totenköpfchen;
Rotrückentotenkopfaffe; Gelbes
Totenkopfäffchen
f   Sapajou à dos rouge; Saïmiri à tête
noire; Singe écureil à dos rouges
i   Testina di morto; Saimiri del
Centroamerica; Scimmia scoiattolo
dal dorso rosso

6203   *Saimiri sciureus*
Primates - Atelidae
e   Squirrel Monkey; South American
Squirrel Monkey; Common Squirrel
Monkey
d   Totenkopfäffchen; Eigentlicher
Totenkopfaffe; Totenkopfaffe
f   Sapajou jaune; Saïmiri aurore;
Saïmiri commun; Saïmiri à dos brûlé;
Saïmiri sciurin; Singe-écureil
commun
i   Saimiri scoiattolo

6204   *Saimiri sciureus sciureus*
Primates - Atelidae

e   Amazonas Squirrel Monkey
d   Amazonas-Totenkopfäffchen
f   Singe-écureuil

6205   *Saimiri ustus*
Primates - Atelidae
e   Bare-eared Squirrel Monkey;
Geoffroy's Squirrel Monkey; Naked-
ear Squirrel Monkey
d   Nacktohr-Totenkopfäffchen
f   Singe-écureil cinabre
i   Saimiri di Geoffroy

6206   *Saimiri vanzolinii*
Primates - Atelidae
e   Black Squirrel Monkey; Blackish
Squirrel Monkey
d   Dunkler Totenkopfaffe
i   Saimiri di Vanzolini

6207   *Salanoia*
Carnivora - Herpestidae
e   Madagascar Brown-tailed
Mongooses; Brown-tailed
Mongooses
d   Schlichtmungos
f   Mangoustes de Madagascar
i   Manguste coda bruna

6208   *Salanoia concolor*
Carnivora - Herpestidae
e   Madagascar Brown-tailed Mongoose;
Brown-tailed Mongoose; Brown
Mongoose; Plain Mongoose
d   Schlichtmungo
f   Galidie unicolor
i   Mangusta coda bruna

      Salanoia olivacea syn. *S. concolor*
q.v.

6209   *Salinoctomys*
Rodentia Octodontidae -
Octodontidae
e   Chalchalero Viscacha Rats
d   Los Chalchleros-Viscacharatten

6210 *Salinoctomys loschalchalerosum*
Rodentia Octodontidae -
Octodontidae
e Chalchalero Viscacha Rat
d Los Chalchleros-Viscacharatte

6211 *Salinomys*
Rodentia - Muridae
e Salt-flat Mice

6212 *Salinomys delicatus*
Rodentia - Muridae
e Delicate Salt-flat Mouse

Salpingotulus syn. *Salpingotus* q.v.

6213 *Salpingotus*
Rodentia - Dipodidae
e Pygmy Jerboas; Three-toed Pygmy
Jerboas; Three-toed Dwarf Jerboas
d Dreizehige Zwergspringmäuse;
Koslows Zwergspringmäuse;
Zwergspringmäuse
f Gerboises naines à trois doigts

6214 *Salpingotus crassicauda*
Rodentia - Dipodidae
e Thick-tailed Pygmy Jereboa
d Fettschwänzige Zwergspringmaus

6215 *Salpingotus heptneri*
Rodentia - Dipodidae
e Heptner's Pygmy Jerboa
d Heptners Zwergspringmaus

6216 *Salpingotus kozlovi*
Rodentia - Dipodidae
e Kozlov's Pygmy Jerboa; Three-toed
Dwarf Jerboa
d Kozlows Zwergspringmaus

6217 *Salpingotus michaelis*
Rodentia - Dipodidae
e Baluchistan Pygmy Jerboa
d Belutschistan-Zwergspringmaus

6218 *Salpingotus pallidus*
Rodentia - Dipodidae
e Pallid Pygmy Jerboa
d Blasse Zwergspringmaus

6219 *Salpingotus thomasi*
Rodentia - Dipodidae
e Thomas's Pygmy Jerboa
d Thomas-Zwergspringmaus

6220 *Sarcophilus*
Dasyuromorphia - Dasyuridae
e Tasmanian Devils; Ursine
Sarcophiluses
d Beutelteufel ; Tasmanische Teufel
f Sarcophiles; Sarcophiles sataniques;
Diables de Tasmanie; Sarcophiles
oursons
i Diavoli di Tasmania

6221 *Sarcophilus laniarius*
Dasyuromorphia - Dasyuridae
e Ursine Sarcophilus; Tasmanian Devil
d Beutelteufel; Tasmanischer Teufel
f Sarcophile ourson; Sarcophile
satanique; Diable de Tasmanie
i Diavolo di Tasmania

Satanellus syn. *Dasyurus* q.v.

Sauromys syn. *Mormopterus* q.v.

Scaeopus syn. *Bradypus* q.v.

6222 *Scalopus*
Erinaceomorpha - Talpidae
e Eastern American Moles; Eastern
Moles
d Ostamerikanische Maulwürfe
f Taupes

6223 *Scalopus aquaticus*
Erinaceomorpha - Talpidae
e Eastern Mole; Eastern American
Mole; North American Mole; North

American Common Mole

d Ostamerikanischer Maulwurf

f Taupe à queue glabre

6225 *Scalopus aquaticus inflatus*
Erinaceomorpha - Talpidae

e Tamaulipas Mole

d Tamaulipas-Maulwurf

f Taupe de Tamaulipas

6224 *Scalopus aquaticus montanus*
Erinaceomorpha - Talpidae

e Coahuila Mole; Coahuilan Mole

d Coahuila-Maulwurf

f Taupe de Coahuila

6226 *Scapanulus*
Erinaceomorpha - Talpidae

e Kansu Moles; Owen's Moles; Gansu Moles

d Kansu-Maulwürfe; Gansu-Maulwürfe

f Taupes de Kansu; Taupes de Kan-Sou

6227 *Scapanulus oweni*
Erinaceomorpha - Talpidae

e Kansu Mole; Owen's Mole; Gansu Mole

d Kansu-Maulwürf; Gansu-Maulwürf

f Taupe de Kansu; Taupe de Kan-Sou

6228 *Scapanus*
Erinaceomorpha - Talpidae

e Western Moles; Western American Moles

d Westamerikanische Maulwürfe

f Taupes d'Amérique de l'Ouest; Scapanes

6229 *Scapanus latimanus*
Erinaceomorpha – Talpidae

e Broad-footed Mole; Broad-handed Mole

d Kalifornischer Maulwurf;

Breitfußmaulwurf

6230 *Scapanus orarius*
Erinaceomorpha - Talpidae

e Coast Mole; Western American Mole

d Pazifischer Maulwurf

f Taupe de côte; Taupe du Pacifique

6231 *Scapanus townsendi*
Erinaceomorpha - Talpidae

e Townsend's Mole

d Townsends Maulwurf

f Taupe de Townsend

6232 *Scapanus townsendi olympicus*
Erinaceomorpha - Talpidae

e Olympic Snow Mole

6233 *Scapteromys*
Rodentia - Muridae

e South American Water Rats; Water Rats; Swamp Rats

d Südamerikanische Wasserratten

Scapteromys fronto syn. *Kunsia fronto* q.v.

Scapteromys tomentosus syn. *Kunsia tomentosus* q.v.

6234 *Scapteromys tumidus*
Rodentia - Muridae

e South American Water Rat; Water Rat; Argentine Swamp Rat

d Südamerikanische Wasserratte

6235 *Scaptochirus*
Erinaceomorpha - Talpidae

e Short-faced Moles

d Kurzgesichtmaulwürfe

6236 *Scaptochirus moschatus*
Erinaceomorpha - Talpidae

e Short-faced Mole

d Kurzgesichtmaulwurf

6237 *Scaptonyx*
Erinaceomorpha - Talpidae
e Long-tailed Moles; Spindle-tailed
Shrew Moles
d Langschwanzmaulwürfe
f Taupes à longue queue; Taupes
européennes
i Talpa toporagno asiatiche

6238 *Scaptonyx fusicaudus*
Erinaceomorpha - Talpidae
e Long-tailed Mole; Spindle-tailed
Shrew Mole
d Langschwanzmaulwurf
f Taupe à longue queue
i Talpa toporagno asiatica

Scarturus syn. *Allactga* q.v.

Schoinobates syn. *Petauroides* q.v.

6239 *Sciuridae*
Rodentia
e Squirrels; True Squirrels; Typical
Squirrels; Squirrels, Marmots, etc.
d Hörnchen
f Écureuils; Sciuridés
i Sciuridi

6240 *Sciurillus*
Rodentia - Sciuridae
e Neotropical Pygmy Squirrels
d Neuweltliche Kleinsthörnchen
f Écureuils pygmées
i Scoiattoli pigmei tropicali

6241 *Sciurillus pusillus*
Rodentia - Sciuridae
e Neotropical Pygmy Squirrel
d Neuweltliches Kleinsthörnchen
f Écureuil pygmée
i Scoiattolo pigmeo tropicale

6242 *Sciurotamias*
Rodentia - Sciuridae

e Rock Squirrels; Asian Rock Squirrels
d Chinesische Rothörnchen
f Écureuils des rochers
i Scoiattoli delle rocce

6243 *Sciurotamias davidianus*
Rodentia - Sciuridae
e Rock Squirrel; Père David's Rock
Squirrel
d Père Davids Felsenhörnchen; Pater
Davids Felsenhörnchen; Père-David-
Rothörnchen; David-Felsenhörnchen
f Écureuil des rochers; Écureuil de
Père David
i Scoiattolo delle rocce di Padre David

6244 *Sciurotamias forresti*
Rodentia - Sciuridae
e Forrest's Rock Squirrel
d Waldrothörnchen

6245 *Sciurus*
Rodentia - Sciuridae
e Squirrels; European and American
Tree Squirrels; Tree Squirrels
d Taghörnchen ; Eichhörnchen
f Écureuils
i Scoiattoli

6246 *Sciurus aberti*
Rodentia - Sciuridae
e Abert's Squirrel; Tassel-eared
Squirrel
d Abert-Hörnchen; Pinselohrhörnchen
f Écureuil d'Abert

6247 *Sciurus aberti kaibabensis*
Rodentia - Sciuridae
e Kaibab Squirrel
d Kaibab-Hörnchen

6248 *Sciurus aestuans*
Rodentia - Sciuridae
e Brazilian Squirrel

 d   Brasil-Hörnchen; Guyana-Hörnchen
 ƒ   Écureuil de la Guyane

6249   *Sciurus aestuans ingrami*
         Rodentia - Sciuridae
 e   Guianan Squirrel

6250   *Sciurus alleni*
         Rodentia - Sciuridae
 e   Allen's Squirrel
 d   Allens Hörnchen
 ƒ   Écureuil d'Allen

6251   *Sciurus anomalus*
         Rodentia - Sciuridae
 e   Caucasian Squirrel; Persian Squirrel
 d   Kaukasisches Eichhörnchen
 ƒ   Écureuil du Caucase; Ecureuil de Perse
 i   Scoiattolo persiano; Scoiatollo dell'Asia minore

         Sciurus apache syn. *S. nayaritensis* q.v.

6252   *Sciurus argentinius*
         Rodentia - Sciuridae
 e   South Yungas Red Squirrel

6253   *Sciurus arizonensis*
         Rodentia - Sciuridae
 e   Arizona Grey Squirrel
 d   Arizona-Grauhörnchen
 ƒ   Écureuil gris d'Arizona

6254   *Sciurus aureogaster*
         Rodentia - Sciuridae
 e   Red-bellied Squirrel; Guatemalan Grey Squirrel; Mexican Grey Squirrel
 d   Rotbauchhörnchen
 ƒ   Écureuil gris de Mexique

6255   *Sciurus brochus*
         Rodentia - Sciuridae

 e   Neotropical Montane Groove-toothed Squirrel; Montane Squirrel; Panama Mountain Squirrel; Bang's Mountain Squirrel; Central American Montane Squirrel
 d   Mittelamerikanisches Berghörnchen

6256   *Sciurus carolinensis*
         Rodentia - Sciuridae
 e   American Grey Squirrel; Grey Squirrel; Eastern Grey Squirrel
 d   Katzeneichhorn; Graues Eichhörnchen; Grauhörnchen
 ƒ   Écureuil gris; Écureuil gris de l'Est; Écureuil de la Caroline
 i   Scoiattolo grigio; Scoiattolo grigio orientale

6257   *Sciurus carolinensis hypophaeus*
         Rodentia - Sciuridae
 e   Minnesota Grey Squirrel
 d   Minnesota-Grauhörnchen

6258   *Sciurus colliaei*
         Rodentia - Sciuridae
 e   Collie's Squirrel
 d   Collie-Hörnchen
 ƒ   Écureuil de Collie

6259   *Sciurus deppei*
         Rodentia - Sciuridae
 e   Deppe's Squirrel
 d   Deppes Hörnchen
 ƒ   Écureuil de Deppe; Écureuil des pins
 i   Scoiattolo di Depp

6260   *Sciurus flammifer*
         Rodentia - Sciuridae
 e   Fiery Squirrel; Venezuelan Squirrel
 d   Venezolanisches Hörnchen; Venezuelanisches Hörnchen
 ƒ   Écureuil de Venezuela

6261   *Sciurus gilvigularis*
         Rodentia - Sciuridae
 e   Yellow-throated Squirrel

d   Gelbkehlhörnchen
f   Écureuil du Brésil

Sciurus goldmani syn. *S. variegatoides* q.v.

6262   Sciurus granatensis
Rodentia - Sciuridae
e   Tropical Red Squirrel; Red-tailed Squirrel
d   Rotschwanzhörnchen
f   Écureuil à queue rouge

Sciurus griseoflavus syn. *S. aureogaster* q.v.

6263   Sciurus griseus
Rodentia - Sciuridae
e   Western Grey Squirrel
d   Westliches Grauhörnchen
f   Écureuil gris de Californie; Écureuil occidental
i   Scoiattolo americano

6264   Sciurus ignitus
Rodentia - Sciuridae
e   Bolivian Squirrel
d   Bolivianisches Eichhörnchen
f   Écureuil de Bolivie

6265   Sciurus igniventris
Rodentia - Sciuridae
e   Northern Amazon Red Squirrel; North Amazon Red Squirrel
d   Nordamazonisches Rothörnchen
f   Écureuil roux du Nord de l'Amazonie
i   Scoiattolo peruviano

Sciurus langsdorffi syn. *S. spadiceus* q.v.

6266   Sciurus lis
Rodentia - Sciuridae
e   Japanese Squirrel
d   Japanisches Eichhörnchen

f   Écureuil du Japon; Écureuil noir du Japon; Écureuil à ventre blanc
i   Scoiattolo giapponese

6267   Sciurus nayaritensis
Rodentia - Sciuridae
e   Nayarit Squirrel; Mexican Fox Squirrel; Nayarit Fox Squirrel
d   Apache-Fuchshörnchen; Mexikanisches Fuchshörnchen
f   Écureuil de Nayarit

Sciurus nelsoni syn. *S. aureogaster* q.v.

6268   Sciurus niger
Rodentia - Sciuridae
e   Fox Squirrel; Eastern Fox Squirrel; Stump-eared Squirrel; Cat Squirrel
d   Fuchshörnchen; Östliches Fuchshörnchen
f   Écureuil fauve; Écureuil-renard
i   Scoiattolo-volpe orientale

6269   Sciurus niger avicennia
Rodentia - Sciuridae
e   Big Cypress Fox Squirrel

6270   Sciurus niger cinerea
Rodentia - Sciuridae
e   Delmarva Fox Squirrel; Delmarva Peninsula Fox Squirrel; Peninsula Squirrel; Bruant's Squirrel; Delmarva Squirrel

6271   Sciurus niger rufiventris
Rodentia - Sciuridae
e   Western Fox Squirrel

6272   Sciurus niger shermani
Rodentia - Sciuridae
e   Sherman's Fox Squirrel

6273   Sciurus oculatus
Rodentia - Sciuridae
e   Peters's Squirrel

d Peters-Hörnchen
f Écureuil de Peters

Sciurus poliopus syn. *S. aureogaster* q.v.

6274 *Sciurus pucheranii*
Rodentia - Sciuridae
e Andean Squirrel; Colombian Squirrel
d Kolumbianisches Eichhörnchen
f Écureuil de Colombie

6275 *Sciurus pyrrhinus*
Rodentia - Sciuridae
e Junin Red Squirrel
d Junin-Hörnchen
f Écureuil du Junin

Sciurus pyrrhonotus syn. *S. spadiceus* q.v.

6276 *Sciurus richmondi*
Rodentia - Sciuridae
e Richmond's Squirrel
d Richmonds Hörnchen
f Écureuil de Richmond

6277 *Sciurus sanborni*
Rodentia - Sciuridae
e Sanborn's Squirrel
d Sanborns Hörnchen
f Écureuil de Sanborn

6278 *Sciurus spadiceus*
Rodentia - Sciuridae
e Southern Amazon Red Squirrel; Southern Amazonian Red Squirrel
d Südamazonisches Rothörnchen
f Écureuil roux du Sud de l'Amazonie

6279 *Sciurus stramineus*
Rodentia - Sciuridae
e Guayaquil Squirrel
d Guayaquil-Hörnchen
f Écureuil du Guayaquil

i Scoiattolo nuca bianca

6280 *Sciurus variegatoides*
Rodentia - Sciuridae
e Variegated Squirrel
d Bunthörnchen; Veränderliches Hörnchen
f Écureuil multicolore

6281 *Sciurus variegatoides atrirufus*
Rodentia - Sciuridae
e Red Variegated Squirrel
d Rotschwarzes Veränderliches Hörnchen

6282 *Sciurus variegatoides hoffmanni*
Rodentia - Sciuridae
e Hoffmann's Variegated Squirrel
d Hoffmanns Veränderliches Hörnchen

6283 *Sciurus variegatoides rigidus*
Rodentia - Sciuridae
e Central American Variegated Squirrel
d Mittelamerika-Grauhörnchen

6284 *Sciurus vulgaris*
Rodentia - Sciuridae
e Red Squirrel; Eurasian Red Squirrel; Brown Squirrel; European Tree Squirrel; European Red Squirrel; Squirrel
d Europäisches Eichhörnchen; Eichhorn; Eichhörnchen; Gemeines Eichhörnchen; Eichkätzchen; Europäisches Eichhorn
f Écureuil; Écureuil commun; Écureuil vulgaire; Écureuil d'Europe; Écureuil vulgaire d'Europe; Écureil roux
i Scoiattolo; Scoiattolo comune; Scoiattolo rosso europeo

6285 *Sciurus vulgaris exalbidus*
Rodentia - Sciuridae
e Eurasian Red Squirrel; Siberian Red Squirrel

d   Talahutka-Eichhörnchen; Sibirisches
    Eichörnchen

6286   *Sciurus vulgaris fuscoater*
    Rodentia - Sciuridae
e   Central European Squirrel
d   Mitteleuropäisches Eichhörnchen

6287   *Sciurus vulgaris jenissejensis*
    Rodentia - Sciuridae
e   Jenissei Red Squirrel
d   Jenissej-Eichhörnchen

6288   *Sciurus yucatensis*
    Rodentia - Sciuridae
e   Yucatan Squirrel
d   Yucatan-Hörnchen
f   Écureuil du Yucatan

6289   *Scleronycteris*
    Chiroptera - Phyllostomidae
e   Ega Long-tongued Bats

6290   *Scleronycteris ega*
    Chiroptera - Phyllostomidae
e   Ega Long-tongued Bat

6291   *Scolomys*
    Rodentia - Muridae
e   Spiny Mice
d   Stachelmäuse

6292   *Scolomys juraense*
    Rodentia - Muridae
e   Rio Jurá Spiny Mouse

6293   *Scolomys melanops*
    Rodentia - Muridae
e   Ecuador Spiny Mouse; South
    American Spiny Mouse
d   Ecuador-Stachelmaus

6294   *Scolomys ucayalensis*

    Rodentia - Muridae
e   Ucayali Soiny Mouse
d   Ucayali-Stachelmaus

Scoteanax syn. *Nycticeius* q.v.

6295   *Scotinomys*
    Rodentia - Muridae
e   Brown Mice; American Brown Mice
d   Braunmäuse

6296   *Scotinomys teguina*
    Rodentia - Muridae
e   Alston's Brown Mouse; Alston's
    Singing Mouse
d   Ledermaus

6297   *Scotinomys xerampelinus*
    Rodentia - Muridae
e   Chiriqui Brown Mouse; Chiriqui
    Singing Mouse

6298   *Scotoecus*
    Chiroptera - Vespertilionidae
e   House Bats

6299   *Scotoecus albofuscus*
    Chiroptera - Vespertilionidae
e   Gambian Light-winged Bat; Lesser
    House Bat; Light-winged Lesser
    House Bat; Thomas's House Bat

6300   *Scotoecus hirundo*
    Chiroptera - Vespertilionidae
e   Dark-winged Lesser House Bat

6301   *Scotoecus pallidus*
    Chiroptera - Vespertilionidae
e   Desert Yellow Bat

6302   *Scotomanes*
    Chiroptera - Vespertilionidae
e   Harlequin Bats

6303 *Scotomanes emarginatus*
Chiroptera - Vespertilionidae
e   Emarginate Harlequin Bat

6304 *Scotomanes ornatus*
Chiroptera - Vespertilionidae
e   Harlequin Bat

6305 *Scotonycteris*
Chiroptera - Pteropodidae
e   West African Fruit Bats
d   Epaulettenflughunde
f   Scotonyctères

6306 *Scotonycteris ophiodon*
Chiroptera - Pteropodidae
e   Snake-toothed Fruit Bat; Pohle's
Fruit Bat
d   Schlangenzahnflughund
f   Scotonyctères aux dents du serpent

6307 *Scotonycteris zenkeri*
Chiroptera - Pteropodidae
e   Zenker's Fruit Bat
d   Zenker-Flughund
f   Scotonyctère de Zenker

6308 *Scotophilus*
Chiroptera - Vespertilionidae
e   House Bats; Old World Yellow Bats;
Leach's Vesper Bats; Yellow Bats

6309 *Scotophilus borbonicus*
Chiroptera - Vespertilionidae
e   Lesser Yellow Bat; Common
Mascarene Islands House Bat;
Common Mascarene House Bat;
Common Mascarene Islands Yellow
Bat; Common Mascarene Islands
Vesper Bat; Common Mascarene
Islands Bat; Common Mascarene
Bat; Common Mascarene Vesper Bat
d   Eigentliche Maskarenen-
Glattnasenfledermaus; Eigentliche
Maskarenen-Glattnase
f   Chauve-souris commune des

Mascareignes; Vespertillion des
Mascareignes commun; Chauve-
souris des hauts

6310 *Scotophilus celebensis*
Chiroptera - Vespertilionidae
e   Greater Yellow Bat; Celebes Yellow
Bat; Sulawesi House Bat; Sulawesi
Yellow Bat

6311 *Scotophilus collinus*
Chiroptera - Vespertilionidae
e   Small Asiatic Yellow Bat

6312 *Scotophilus dinganii*
Chiroptera - Vespertilionidae
e   African Yellow House Bat; African
Yellow Bat

Scotophilus gigas syn. *S. nigrita* q.v.

6313 *Scotophilus heathii*
Chiroptera - Vespertilionidae
e   Asiatic Greater Yellow Bat; Asiatic
Greater Yellow House Bat; Greater
Yellow House Bat; Common
Yellow-bellied Bat
d   Große Gelbe Fledermaus

6314 *Scotophilus kuhlii*
Chiroptera - Vespertilionidae
e   Asiatic Lesser Yellow House Bat;
Asiatic Lesser Yellow Bat; Lesser
Asiatic Yellow Bat

6315 *Scotophilus leucogaster*
Chiroptera - Vespertilionidae
e   White-bellied Yellow Bat

6316 *Scotophilus nigrita*
Chiroptera - Vespertilionidae
e   Brown Bat; Brown House Bat;
Yellow Bat; Yellow House Bat;
Schreber's Brown Bat; Greater
Brown Bat; Giant Yellow Bat;
Schreber's Brown House Bat;
Schreber's Yellow Bat

6317   *Scotophilus nux*
Chiroptera - Vespertilionidae
e   Nut-coloured Yellow Bat

6318   *Scotophilus robustus*
Chiroptera - Vespertilionidae
e   Robust Yellow Bat

6319   *Scotophilus viridis*
Chiroptera - Vespertilionidae
e   Lesser Yellow House Bat; Greenish
        Yellow Bat

Scotorepens syn. *Nycticeius* q.v.

Scotozous syn. *Pipistrellus* q.v.

6320   *Scutisorex*
Soricomorpha - Soricidae
e   Armoured Shrews; Hero Shrews
d   Panzerspitzmäuse
i   Toporagni armati

Scutisorex congicus syn. *S. somereni*
q.v.

6321   *Scutisorex somereni*
Soricomorpha - Soricidae
e   Uganda Shrew; Hero Shrew; Uganda
        Armoured Shrew; Armoured Shrew
d   Uganda-Panzerspitzmaus;
        Panzerspitzmaus; Kongo-
        Panzerspitzmaus
i   Toporagno armato; Toporagno eroe

6322   *Sekeetamys*
Rodentia - Muridae
e   Bushy-tailed Jirds; Ellerman's
        Gerbils
d   Wollschwanzrennmäuse
f   Gerbilles à queue en plumeau

6323   *Sekeetamys calurus*
Rodentia - Muridae
e   Bushy-tailed Jird; Ellerman's Gerbil;
        Bushy-tailed Gerbil; Bushy-tailed

Dipodil
d   Wollschwanzrennmaus;
        Quastenschwanzrennmaus;
        Schönschweifrennmaus;
        Bilchrennmaus;
        Buschschwanzrennmaus
f   Gerbille à queue en plumeau;
        Gerbille à queue touffue; Mérione à
        queue touffue

Selenarctos syn. *Ursus* q.v.

6324   *Selevinia*
Rodentia - Gliridae
e   Betpakdala Dormice; Selevin's Mice
d   Salzkrutbilche
f   Souris de Selevin; Souris du désert
i   Ghiri del deserto

6325   *Selevinia betpakdalensis*
Rodentia - Gliridae
e   Betpakdala Dormouse; Selevin's
        Mouse
d   Salzkrutbilch
f   Souris de Selevin; Souris du désert
i   Ghiro del deserto

Selysius syn. *Myotis* q.v.

6326   *Semnopithecus*
Primates - Cercopithecidae
e   Hanuman Langurs; Sacred Langurs
d   Hanuman-Languren
i   Semnopitechi

6327   *Semnopithecus ajax*
Primates - Cercopithecidae
e   Kashmir Langur
d   Kaschmir-Langur

6328   *Semnopithecus dussumieri*
Primates - Cercopithecidae
e   Southern Plains Grey Langur
d   Südliches Flachlandlangur

6329   Semnopithecus entellus
Primates - Cercopithecidae
e   Hanuman Langur; Entellus Langur;
Sacred Langur; Grey Langur;
Common Indian Langur; Northern
Plains Langur; Indian Langur;
Northern Plains Grey Langur
d   Hanuman-Langur; Hulman;
Vorderindischer Hulman; Hanuman
f   Entelle; Houleman; Entelle de l'Inde;
Langur sacré
i   Entello; Langur

6330   Semnopithecus hector
Primates - Cercopithecidae
e   Tarai Grey Langur

6331   Semnopithecus hypoleucos
Primates - Cercopithecidae
e   Black-footed Grey Langur
d   Schwarzfußhulman
f   Entelle aux pieds noirs

6332   Semnopithecus priam
Primates - Cercopithecidae
e   Tufted Grey Langur
d   Ceylon-Hulman

6333   Semnopithecus priam thersites
Primates - Cercopithecidae
e   Ceylon Grey Langur

6334   Semnopithecus schistaceus
Primates - Cercopithecidae
e   Nepal Grey Langur
d   Berghulman
i   Langur himalayano

6335   Setifer
Soricomorpha - Tenrecidae
e   Large Madagascar Hedehogs;
Hedgehog Tanrecs
d   Große Igeltanreks
i   Tenrec riccio maggiore

Setifer fontoynonti syn. *S. setosus*
q.v.

6336   Setifer setosus
Soricomorpha - Tenrecidae
e   Greater Hedgehog Tenrec; Hedgehog
Tanrec; Large Madagascar
Hedgehog; Fontoynont's Hedgehog
Tenrec; Madagascar Hedgehog
d   Großer Igeltanrek; Fontoynonts
Igeltanrek
f   Grand hérisson; Sokina
i   Tenrec riccio maggiore

6337   Setonix
Diprotodontia - Macropodidae
e   Short-tailed Pademelons; Scrub
Pademelons; Quokkas (ANZ); Short-
tailed Wallabies; Short-tailed Scrub
Wallabies
d   Kurzschwanzkängurus; Quokkas
f   Kangourous à queue courte
i   Canguri dalla coda corta

6338   Setonix brachyurus
Diprotodontia - Macropodidae
e   Short-tailed Pademelon; Scrub
Pademelon; Quokka (ANZ); Short-
tailed Wallaby; Short-tailed Scrub
Wallaby
d   Kurzschwanzkänguru; Quokka
f   Kangourou à queue courte
i   Quokka

6339   Sicista
Rodentia - Dipodidae
e   Birch Mice; Bush Mice
d   Streifenhüpfmäuse; Birkenmäuse;
Buschmäuse
f   Sicistes
i   Sicista

6340   Sicista armenica
Rodentia - Dipodidae
e   Armenian Birch Mouse

6341   *Sicista betulina*
Rodentia - Dipodidae
e   Northern Birch Mouse
d   Buschmaus; Birkenmaus;
Waldbirkenmaus
f   Sminthe errant; Siciste des bouleaux
i   Sminto betulino

6342   *Sicista caucasica*
Rodentia - Dipodidae
e   Caucasian Birch Mouse
d   Kaukasische Buschmaus

6343   *Sicista caudata*
Rodentia - Dipodidae
e   Long-tailed Birch Mouse

6344   *Sicista concolor*
Rodentia - Dipodidae
e   Chinese Birch Mouse

6345   *Sicista kazbegica*
Rodentia - Dipodidae
e   Kazbeg Birch Mouse

6346   *Sicista kluchorica*
Rodentia - Dipodidae
e   Kuchor Birch Mouse

6347   *Sicista napaea*
Rodentia - Dipodidae
e   Altai Birch Mouse
d   Altaische Buschmaus

6348   *Sicista pseudonapae*
Rodentia - Dipodidae
e   Grey Birch Mouse

6349   *Sicista severtzovi*
Rodentia - Dipodidae
e   Severtzov's Birch Mouse

6350   *Sicista strandi*
Rodentia - Dipodidae

e   Strand's Birch Mouse

6351   *Sicista subtilis*
Rodentia - Dipodidae
e   Southern Birch Mouse; Steppe
Sicista
d   Streifenhüpfmaus;
Steppenstreifenmaus;
Steppenbirkenmaus
f   Siciste des steppes; Sminthe des
steppes
i   Sminto; Sicista delle steppe

6352   *Sicista tianshanica*
Rodentia - Dipodidae
e   Tien Shan Birch Mouse

6353   *Sigmoceros*
Artiodactyla - Bovidae
e   Hartebeests
d   Lichtenstein-Antilopen
f   Bubales de Lichtenstein

6354   *Sigmoceros lichtensteini*
Artiodactyla - Bovidae
e   Lichtenstein's Hartebeest; Konzi's
Hartebeest
d   Lichtenstein-Antilope
f   Bubale de Lichtenstein

6355   *Sigmodon*
Rodentia - Muridae
e   Cotton Rats
d   Baumwollratten
f   Sigmodons; Rats du cotton; Rats des
cottoniers
i   Ratti del cotone

6356   *Sigmodon alleni*
Rodentia - Muridae
e   Allen's Brown Rat; Allen's Cotton
Rat

6357   *Sigmodon alstoni*
Rodentia - Muridae
e   Alston's Cotton Rat

Sigmodon alticola syn. *S. leucotis* q.v.

6358 Sigmodon arizonae
Rodentia - Muridae
e Arizona Cotton Rat

6359 Sigmodon arizonae arizonae
Rodentia - Muridae
e Camp Verde Cotton Rat

6360 Sigmodon arizonae plenus
Rodentia - Muridae
e Colorado River Cotton Rat

6361 Sigmodon fulviventer
Rodentia - Muridae
e Tawny-bellied Cotton Rat

6362 Sigmodon fulviventer goldmani
Rodentia - Muridae
e Hot Springs Cotton Rat

6363 Sigmodon hispidus
Rodentia - Muridae
e Hispid Cotton Rat
d Baumwollrate
f Sigmodon velu
i Ratto ispido del cotone

6364 Sigmodon hispidus eremicus
Rodentia - Muridae
e Yuma Hispid Cotton Rat

6365 Sigmodon inopinatus
Rodentia - Muridae
e Unexpected Cotton Rat

6366 Sigmodon leucotis
Rodentia - Muridae
e White-eared Cotton Rat

6367 Sigmodon mascotensis
Rodentia - Muridae

e Jaliscan Cotton Rat

6368 Sigmodon ochrognathus
Rodentia - Muridae
e Yellow-nosed Cotton Rat

6369 Sigmodon peruanus
Rodentia - Muridae
e Peruvian Cotton Rat

Sigmodon planifrons syn. *S. alleni* q.v.

Sigmodon vulcani syn. *S. alleni* q.v.

6370 Sigmodontomys
Rodentia - Muridae
e Rice Water Rats
d Alfaro-Wasserratten

6371 Sigmodontomys alfari
Rodentia - Muridae
e Alfaro's Rice Water Rat; Cana Rice Rat

6372 Sigmodontomys aphrastus
Rodentia - Muridae
e Harris's Rice Water Rat; Long-tailed Rice Rat

Simias syn. *Nasalis* q.v.

6373 Sminthopsis
Dasyuromorphia - Dasyuridae
e Narrow-footed Marsupial Mice; Pouched Mice; Dunnarts (ANZ); Sminthopsises; Jerboa Marsupials; Jerboa-like Marsupials; Long-legged Jumping Marsupials
d Schmalfußbeutelmäuse; Springbeutelmäuse; Dunnarts
f Souris marsupiales
i Topi marsupiali

6374 Sminthopsis aitkeni
Dasyuromorphia - Dasyuridae

e   Little Mouse Sminthopsis; Kangaroo Island Dunnart; Slender-tailed Pouched Mouse; Mouse Sminthopsis

d   Gemeine Schmalfußbeutelmaus; Känguru-Insel-Dunnart

6375   *Sminthopsis archeri*
        Dasyuromorphia - Dasyuridae

e   Chestnut Dunnart

d   Kastaniendunnart

6376   *Sminthopsis bindi*
        Dasyuromorphia - Dasyuridae

e   Kakadu Dunnart

6377   *Sminthopsis butleri*
        Dasyuromorphia - Dasyuridae

e   Carpentarian Dunnart; Butler's Dunnart

6378   *Sminthopsis crassicaudata*
        Dasyuromorphia - Dasyuridae

e   Crest-tailed Marsupial Mouse; Narrow-footed Marsupial Mouse; Fat-tailed Dunnart (ANZ); Fat-tailed Sminthopsis

d   Dickschwänzige Schmalfußbeutelmaus; Dickschwanzschmalfußbeutelmaus; Fettschwanzdunnart

f   Dunnart

i   Topo marsupiale dalla coda grassa

6379   *Sminthopsis dolichura*
        Dasyuromorphia - Dasyuridae

e   Little Long-tailed Dunnart

d   Kleiner Langschwanzdunnart

6380   *Sminthopsis douglasi*
        Dasyuromorphia - Dasyuridae

e   Julia Creek Dunnart

6381   *Sminthopsis fuliginosus*
        Dasyuromorphia - Dasyuridae

e   Sooty Dunnart

d   Russiger Dunnart

6382   *Sminthopsis gilberti*
        Dasyuromorphia - Dasyuridae

e   Gilbert's Dunnart

d   Gilberts Dunnart; Weißschwanzdunnart

6383   *Sminthopsis granulipes*
        Dasyuromorphia - Dasyuridae

e   White-tailed Dunnart (ANZ); Granule-footed Sminthopsis

d   Kornsohlenschmalfußbeutelmaus

6384   *Sminthopsis griseoventer*
        Dasyuromorphia - Dasyuridae

e   Grey-bellied Dunnart

d   Graubauchdunnart

6385   *Sminthopsis griseoventer boulangerensis*
        Dasyuromorphia - Dasyuridae

e   Boulanger Island Dunnart

6386   *Sminthopsis hirtipes*
        Dasyuromorphia - Dasyuridae

e   Hairy-footed Dunnart; Fringe-footed Sminthopsis; Hairy-footed Sminthopsis

d   Polsterschmalfußbeutelmaus

i   Dunnart

6387   *Sminthopsis laniger*
        Dasyuromorphia - Dasyuridae

e   Kultarr

d   Springbeutelmaus; Kultarr; Östliche Springbeutelmaus

i   Kultarr

6388   *Sminthopsis leucopus*
        Dasyuromorphia - Dasyuridae

e   White-footed Dunnart (ANZ); White-footed Marsupial Mouse; White-footed Pouched Mouse

d   Weißfußdunnart

6389   *Sminthopsis longicaudata*
        Dasyuromorphia - Dasyuridae

e   Long-tailed Dunnart; Long-tailed
     Sminthopsis; Long-tailed Marsupial
     Mouse; Long-tailed Pouched Mouse;
     Long-tailed Narrow-footed
     Marsupial Mouse
d   Langschwanzschmalfußbeutelmaus;
     Langschwänzige
     Dickschwanzbeutelmaus;
     Langschwanzdunnart
f   Souris-marsupiale à longue queue
i   Topo marsupiale dalla coda lunga

6390   *Sminthopsis macroura*
     Dasyuromorphia - Dasyuridae
e   Darling Downs Dunnart; Stripe-faced
     Dunnart; Australian Stripe-faced
     Dunnart

6391   *Sminthopsis murina*
     Dasyuromorphia - Dasyuridae
e   Slender-tailed Dunnart; Slender-
     tailed Pouched Mouse; Common
     Dunnart; Common Pouched Mouse
d   Kleine Schmalfußbeutelmaus;
     Inneraustralische Springbeutelmaus;
     Schlankschwanzdunnart

6392   *Sminthopsis ooldea*
     Dasyuromorphia - Dasyuridae
e   Ooldea Dunnart
d   Ooldea-Dunnart

6393   *Sminthopsis psammophila*
     Dasyuromorphia - Dasyuridae
e   Sandhill Dunnart; Large Desert
     Dunnart; Large Desert Sminthopsis;
     Large Desert Marsupial Mouse
d   Sand-Schmalfußbeutelmaus; Große
     Wüstenschmalfußbeutelmaus
f   Souris-maruspiale du désert
i   Topo marsupiale delle sabbie

Sminthopsis rufigenis syn. *S.
virginiae* q.v.

Sminthopsis spenceri syn. *S. murina*
q.v.

6394   *Sminthopsis virginiae*
     Dasyuromorphia - Dasyuridae
e   Red-cheeked Dunnart; Red-cheeked
     Marsupial Mouse
d   Rotwangendunnart

6395   *Sminthopsis youngsoni*
     Dasyuromorphia - Dasyuridae
e   Lesser Hairy-footed Dunnart
d   Kleines Haarfüßiges Dunnart

6396   *Smutsia*
     Cimolesta - Manidae
e   Pangolins
d   Schuppentiere
f   Pangolins
i   Pangolini

6397   *Smutsia gigantea*
     Cimolesta - Manidae
e   Giant Pangolin; Large African
     Pangolin; Ground Pangolin; Giant
     Ground Pangolin
d   Riesenschuppentier
f   Grand Pangolin; Pangolin géant
i   Pangolino gigante

6398   *Smutsia temminckii*
     Cimolesta - Manidae
e   Ground Pangolin; Cape Pangolin;
     Temminck's Ground Pangolin
d   Steppenschuppentier
f   Pangolin de Temminck
i   Pangolino di Temminck

6399   *Solenodon*
     Soricomorpha - Solenodontidae
e   Solenodons; Haitian Solenodons
d   Schlitzrüssler
f   Solénodons; Solénodontés
i   Solenodonti

6400   *Solenodon cubanus*
     Soricomorpha - Solenodontidae
e   Cuban Solenodon

d Kubanischer Schlitzrüssler; Almiqui;
    Kuba-Schlitzrüssler
f Solénodonte de Cuba; Solénodon de
    Cuba
i Almiqui; Solenodonte di Cuba

6401 *Solenodon paradoxus*
          Soricomorpha - Solenodontidae
e Haitian Solenodon; Hispaniolan
    Solenodon
d Haiti-Schlitzrüssler; Dominkanischer
    Schlitzrüssler
f Almiqui paradoxial
i Solenodonte di Hispaniola;
    Solenodonte di Haiti; Solenodonte
    dei Caraibi

6402 *Solenodontidae*
          Insectivora
e Solenodonts; Solenodons
d Schlitzrüssler
f Solénodontidés
i Solenodontidi

6403 *Solisorex*
          Soricomorpha - Soricidae
e Pearson's Long-clawed Shrews;
    Long-clawed Shrews
d Pearsons Langkrallenmäuse;
    Pearsons Langkrallenspitzmäuse

6404 *Solisorex pearsoni*
          Soricomorpha - Soricidae
e Pearson's Long-clawed Shrew
d Pearsons Langkrallenmaus; Pearsons
    Langkrallenspitzmaus;
    Langkrallenspitzmaus

6405 *Solomys*
          Rodentia - Muridae
e Naked-tailed Rats
d Nacktschwanzratten
f Rat des îles Salomon

6406 *Solomys ponceleti*

Rodentia - Muridae
e Poncelet's Naked-tailed Rat

6407 *Solomys salamonis*
          Rodentia - Muridae
e Florida Naked-tailed Rat

6408 *Solomys salebrosus*
          Rodentia - Muridae
e Bougainville Naked-tailed Rat

6409 *Solomys sapientis*
          Rodentia - Muridae
e Isabel Naked-tailed Rat

6410 *Sommeromys*
          Rodentia - Muridae
e Long-nosed Shrewmice

6411 *Sommeromys macrorhinos*
          Rodentia - Muridae
e Long-nosed Shrewmouse

6412 *Sorex*
          Soricomorpha - Soricidae
e Common Shrews; Long-tailed
    Shrews; Red-toothed Shrews;
    Holarctic Shrews
d Rotzahnspitzmäuse; Spitzmäuse;
    Waldspitzmäuse
f Musaraignes

6413 *Sorex alaskanus*
          Soricomorpha - Soricidae
e Glacier Bay Water Shrew
d Glacier-Bay-Wasserspitzmaus

6414 *Sorex alpinus*
          Soricomorpha - Soricidae
e Alpine Shrew
d Alpenspitzmaus
f Musaraigne alpine; Musaraigne des
    Alpes; Musaraigne des montagnes
i Toporagno alpino

6415 *Sorex araneus*
Soricomorpha - Soricidae
e English Shrew; Eurasian Common
Shrew; Eurasian Shrew;
Shrewmouse; Common Shrew
d Schabrackenwaldspitzmaus;
Gemeine Spitzmaus
f Musaraigne commune; Musaraigne
carrelet; Musaraigne vulgaire
i Toporagno comune

6416 *Sorex arcticus*
Soricomorpha - Soricidae
e Arctic Shrew; Black-backed Shrew
d Arktische Waldspitzmaus
f Musaraigne arctique

6417 *Sorex arizonae*
Soricomorpha - Soricidae
e Arizona Shrew

6418 *Sorex asper*
Soricomorpha - Soricidae
e Tien Shan Shrew

6419 *Sorex bairdii*
Soricomorpha - Soricidae
e Baird's Shrew

6420 *Sorex bedfordiae*
Soricomorpha - Soricidae
e Lesser Striped Shrew; Bedford's
Shrew

6421 *Sorex bendirii*
Soricomorpha - Soricidae
e Bendire's Shrew; Pacific Water
Shrew
d Pazifische Wasserspitzmaus
f Musaraigne des marais; Musaraigne
de Bendire

6422 *Sorex bendirii bendiri*
Soricomorpha - Soricidae
e Marsh Shrew

6423 *Sorex buchariensis*
Soricomorpha - Soricidae
e Pamir Shrew; Bucharan Long-
toothed Shrew
d Bucharische Wasserspitzmaus

6424 *Sorex caecutiens*
Soricomorpha - Soricidae
e Laxmann's Shrew
d Lappenspitzmaus; Maskenspitzmaus;
Lappland-Spitzmaus
f Musaraigne lapone; Musaraigne
masquée
i Toporagno di Laxmann; Toporagno
di Lapponia

6425 *Sorex camschatica*
Soricomorpha - Soricidae
e Kamchatka Shrew

6426 *Sorex cansulus*
Soricomorpha - Soricidae
e Gansu Shrew

6427 *Sorex cinereus*
Soricomorpha - Soricidae
e Masked Shrew (NA); American
Common Shrew; Cinereous Shrew
d Amerikanische Maskenspitzmaus
f Musaraigne cendrée

6428 *Sorex cinereus jacksoni*
Soricomorpha - Soricidae
e St. Lawrence Island Shrew

6429 *Sorex coronatus*
Soricomorpha - Soricidae
e Crowned Shrew; Millet's Shrew;
Jersey Shrew
d Schabrackenspitzmaus
f Musaraigne couronnée; Musette; Rat
aveugle
i Toporagno di Millet

6430 *Sorex cylindricauda*
Soricomorpha - Soricidae

e   Greater Striped Shrew; Stripe-backed Shrew; Sichuan Shrew

6431   *Sorex daphaenodon*
Soricomorpha - Soricidae
e   Large-toothed Shrew; Large-toothed Siberian Shrew; Siberian Large-toothed Shrew
d   Großzähnige Waldspitzmaus

6432   *Sorex dispar*
Soricomorpha - Soricidae
e   Long-tailed Shrew; Rock Shrew; American Long-tiled Shrew
d   Raddes Waldspitzmaus; Langschwanzspitzmaus
f   Musaraigne des Appalaches; Musaraigne longicaude; Musaraigne à longue queue

6433   *Sorex emarginatus*
Soricomorpha - Soricidae
e   Zacatecas Shrew

6434   *Sorex excelsus*
Soricomorpha - Soricidae
e   Lofty Shrew; Allen's Shrew

6435   *Sorex fontinalis*
Soricomorpha - Soricidae
e   Maryland Shrew; Eastern Shrew

6436   *Sorex fumeus*
Soricomorpha - Soricidae
e   Smoky Shrew
f   Musaraigne fuligineuse

6437   *Sorex gaspensis*
Soricomorpha - Soricidae
e   Gaspé Shrew
f   Musaraigne du Gaspé

6438   *Sorex gracillimus*
Soricomorpha - Soricidae
e   Slender Shrew

6439   *Sorex granarius*
Soricomorpha - Soricidae
e   Lagranja Shrew; Granja Shrew; Iberian Shrew; Spanish Shrew
d   Iberische Waldspitzmaus; Spanische Spitzmaus
f   Musaraigne Ibérique
i   Toporagno iberico

6440   *Sorex haydeni*
Soricomorpha - Soricidae
e   Prairie Shrew; Hayden's Shrew; Hayden Shrew

6441   *Sorex hosonoi*
Soricomorpha - Soricidae
e   Azumi Shrew

Sorex hoyi syn. *Microsorex hoyi* q.v.

Sorex hydrodromus syn. *S. pribilofensis* q.v.

6442   *Sorex isodon*
Soricomorpha - Soricidae
e   Even-toothed Shrew; Graves's Shrew; Equal-toothed Shrew; Taiga Shrew
d   Ostibirische Waldspitzmaus
i   Toporagno oscure

Sorex jacksoni syn. *S. cinereus jacksoni* q.v.

Sorex juncensis syn. *S. ornatus* q.v.

6443   *Sorex kozlovi*
Soricomorpha - Soricidae
e   Kozlov's Shrew

6444   *Sorex leucogaster*
Soricomorpha - Soricidae
e   Paramushir Shrew; Paramushir Island Shrew

6445   ***Sorex longirostris***
Soricomorpha - Soricidae
e   Southeastern Shrew

6446   ***Sorex longirostris eionis***
Soricomorpha - Soricidae
e   Homosassa Shrew

6447   ***Sorex longirostris fisheri***
Soricomorpha - Soricidae
e   Dismal Swamp Shrew; Dismal
Swamp Southeastern Shrew

6448   ***Sorex lyelli***
Soricomorpha - Soricidae
e   Mount Lyell Shrew

6449   ***Sorex macrodon***
Soricomorpha - Soricidae
e   Large-toothed Shrew; Mexican
Large-toothed Shrew

6450   ***Sorex merriami***
Soricomorpha - Soricidae
e   Merriam's Shrew

6451   ***Sorex milleri***
Soricomorpha - Soricidae
e   Carmen Mountain Shrew; Miller's
Shrew

6452   ***Sorex minutissimus***
Soricomorpha - Soricidae
e   Least Shrew; Eurasian Least Shrew;
Miniscule Shrew
d   Kleinspitzmaus; Winzige Spitzmaus
f   Musaraigne naine; Musaraigne
minuscule
i   Toporagno minimo

6453   ***Sorex minutissimus tscherskii***
Soricomorpha - Soricidae
e   Least Siberian Shrew

6454   ***Sorex minutus***

Soricomorpha - Soricidae
e   Lesser Shrew; Pygmy Shrew;
Eurasian Pygmy Shrew
d   Zwergspitzmaus; Eurasische
Zwergspitzmaus
f   Musaraigne pygmée
i   Toporagno nano

6455   ***Sorex mirabilis***
Soricomorpha - Soricidae
e   Giant Shrew; Ussuri Shrew

6456   ***Sorex monticolus***
Soricomorpha - Soricidae
e   Dusky Montane Shrew; Montane
Shrew
f   Musaraigne sombre

6457   ***Sorex monticolus monticolus***
Soricomorpha - Soricidae
e   Warren Island Shrew

6458   ***Sorex nanus***
Soricomorpha - Soricidae
e   American Dwarf Shrew

6459   ***Sorex neomexicanus***
Soricomorpha - Soricidae
e   New Mexico Shrew

Sorex obscurus syn. *S. monticolus*
q.v.

6460   ***Sorex oreopolus***
Soricomorpha - Soricidae
e   Mexican Long-tailed Shrew; Volcano
Shrew

6461   ***Sorex ornatus***
Soricomorpha - Soricidae
e   Suisun Shrew; Ornate Shrew

6462   ***Sorex ornatus californicus***
Soricomorpha - Soricidae
e   Californian Ornate Shrew

6463   *Sorex ornatus relictus*
Soricomorpha - Soricidae
e   Buena Vista Lake Shrew

6464   *Sorex ornatus willetti*
Soricomorpha - Soricidae
e   Santa Catalina Island Shrew

6465   *Sorex pacificus*
Soricomorpha - Soricidae
e   Pacific Shrew

6466   *Sorex palustris*
Soricomorpha - Soricidae
e   American Water Shrew
d   Nordische Wasserspitzmaus;
Nördliche Wasserspitzmaus
f   Musaraigne palustre

6467   *Sorex planiceps*
Soricomorpha - Soricidae
e   Kashmir Shrew

6468   *Sorex portenkoi*
Soricomorpha - Soricidae
e   Portenko's Shrew

6469   *Sorex preblei*
Soricomorpha - Soricidae
e   Preble's Shrew; Malheur Shrew

6470   *Sorex pribilofensis*
Soricomorpha - Soricidae
e   Pribilof Shrew; Pribilof Island Shrew
d   Unalaska-Wasserspitzmaus

6471   *Sorex raddei*
Soricomorpha - Soricidae
e   Radde's Shrew

6472   *Sorex roboratus*
Soricomorpha - Soricidae
e   Flat-skulled Shrew

6473   *Sorex sadonis*
Soricomorpha - Soricidae
e   Sado Shrew

6474   *Sorex samniticus*
Soricomorpha - Soricidae
e   Appenine Shrew
d   Italienische Waldspitzmaus;
Apenninen-Spitzmaus
f   Musaraigne des Apennins;
Musaraigne italienne
i   Toporagno apenninico; Toporagno
italico; Toporagno delle Apennini

6475   *Sorex satunini*
Soricomorpha - Soricidae
e   Caucasian Shrew

6476   *Sorex sausssurei*
Soricomorpha - Soricidae
e   Saussure's Shrew

6477   *Sorex sclateri*
Soricomorpha - Soricidae
e   Sclater's Shrew

6478   *Sorex shinto*
Soricomorpha - Soricidae
e   Shinto Shrew

6479   *Sorex sinalis*
Soricomorpha - Soricidae
e   Chinese Shrew; Dusky Shrew

Sorex sinuosus syn. *S. ornatus* q.v.

6480   *Sorex sonomae*
Soricomorpha - Soricidae
e   Fog Shrew

6481   *Sorex stizodon*
Soricomorpha - Soricidae
e   San Cristobal Shrew

6482 Sorex tenellus
Soricomorpha - Soricidae
e Inyo Shrew

6483 Sorex thibetanus
Soricomorpha - Soricidae
e Tibetan Shrew

Sorex trigonirostris syn. *S. vagrans*
q.v.

6484 Sorex trowbridgii
Soricomorpha - Soricidae
e Trowbridge's Shrew

6485 Sorex trowbridgii destructioni
Soricomorpha - Soricidae
e Destruction Island Shrew

6486 Sorex tundrensis
Soricomorpha - Soricidae
e Tundra Shrew

6487 Sorex ugyunak
Soricomorpha - Soricidae
e Barren Ground Shrew

6488 Sorex unguiculatus
Soricomorpha - Soricidae
e Long-clawed Shrew; Large-clawed
Shrew
d Krallenspitzmaus

6489 Sorex vagrans
Soricomorpha - Soricidae
e Wandering Shrew; Ashland Shrew;
Vagrant Shrew
f Musaraigne errante

6490 Sorex vagrans halicoetes
Soricomorpha - Soricidae
e Salt Marsh Wandering Shrew

6491 Sorex ventralis
Soricomorpha - Soricidae

e Chestnut-bellied Shrew; Cerro San
Felipe Shrew

6492 Sorex veraepacis
Soricomorpha - Soricidae
e Verapaz Shrew

Sorex vir syn. *S. roboratus* q.v.

6493 Sorex volnuchini
Soricomorpha - Soricidae
e Caucasian Pygmy Shrew;
Volnuchin's Shrew

6494 Sorex yukonicus
Soricomorpha - Soricidae
e Alaska Tiny Shrew

6495 Soricidae
Soricomorpha
e Shrews
d Spitzmäuse; Spitzrüssler
f Soricidés; Musaraignes
i Soricidi

6496 Soriculus
Soricomorpha - Soricidae
e Asiatic Shrews; Mountain Shrews;
Oriental Shrews

Soriculus baileyi syn. *S. leucops* q.v.

6497 Soriculus caudatus
Soricomorpha - Soricidae
e Hodgson's Brown-toothed Shrew;
Hodgson's Shrew

6498 Soriculus fumidus
Soricomorpha - Soricidae
e Taiwan Brown-toothed Shrew;
Taiwn Mountain Shrew; Taiwan
Shrew

Soriculus gruberi syn. *S. leucops*
q.v.

6499   *Soriculus hypsibius*
Soricomorpha - Soricidae
e   De Winton's Shrew; De Winton's
Mountain Shrew

6500   *Soriculus lamula*
Soricomorpha - Soricidae
e   Lamulate Shrew; Lama Mountain
Shrew

6501   *Soriculus leucops*
Soricomorpha - Soricidae
e   Indian Long-tailed Shrew; Large
Long-tailed Shrew; Long-tailed
Brown-toothed Shrew

Soriculus lowei syn. *S. parca* q.v.

6502   *Soriculus macrurus*
Soricomorpha - Soricidae
e   Long-tailed Mountain Shrew; Small
Long-tailed Shrew

6503   *Soriculus nigrescens*
Soricomorpha - Soricidae
e   Sikkim Large-clawed Shrew;
Himalajan Shrew; Burrowing Forest
Shrew
d   Sikkim-Großklauenspitzmaus
f   Musaraigne de Sikkim

6504   *Soriculus parca*
Soricomorpha - Soricidae
e   Lowe's Shrew

6505   *Soriculus salenskii*
Soricomorpha - Soricidae
e   Salenski's Shrew

6506   *Soriculus smithi*
Soricomorpha - Soricidae
e   Smith's Shrew; Smith's Long-tailed
Shrew

6507   *Sotalia*
Cete - Delphinidae

e   River Dolphins; Ridge-backed
Dolphins; Hump-backed Dolphins;
White Dolphins
d   Brackwasserdelfine; Flussdelfine
f   Sotalies; Dauphins à bosse; Dauphins
blancs
i   Delfini di fiume; sotalia

Sotalia brasiliensis syn. *S. fluviatilis*
q.v.

6508   *Sotalia fluviatilis*
Cete - Delphinidae
e   Tucuxi; Brazilian Dolphin; Guiana
White Dolphin; Guiana River
Dolphin; Estuarine Dolphin
d   Amazonas-Sotalia;
Brackwasserdelfin; Sotalia-Delfin;
Sotalia
f   Sotalie de la Guyane; Sotalie; Tucuxi
i   Sotalia; Bufeo grigio; Bufeo nero;
Delfino di fiume; Delfino tucuxi;
Delfino grigio

6509   *Sotalia fluviatilis fluviatilis*
Cete - Delphinidae
e   Amazon River Tucuxi

6510   *Sotalia fluviatilis guianensis*
Cete - Delphinidae
e   Atlantic Coast Tucuxi
d   Guayana-Delfin
i   Sotalia della Guyana

6511   *Sousa*
Cete - Delphinidae
e   Hump-backed Dolphins; Humpback
Dolphins
d   Buckeldelfine
f   Dauphins à bosse; Sotalies
i   Suse

Sousa borneensis syn. *S. chinensis*
q.v.

6512   *Sousa chinensis*
Cete - Delphinidae

e   Bornean White Dolphin; Borneo White Dolphin; Indo-Pacific Hump-backed Dolphin; Speckled Dolphin ; Freckled Dolphin; Indo-Pacific Sousa; Chinese White Dolphin; White Dolphin; Plumbeous Dolphin; Lead-coloured Dolphin; Malabar Dolphin

d   Chinesischer Weißer Delfin; Weißdelfin; Chinesischer Buckeldelfin

f   Sotalia de Chine; Sousa du Pacifique; Dauphin à bosse du Pacifique; Dauphin a bosse indo-pacifique; Sousa de l'Indo-Pacifique

i   Susa indo-pacifica; Susa pacifica

Sousa lentiginosus syn. *S. chinensis* q.v.

6513   *Sousa teuszii*
       Cete - Delphinidae

e   West African White Dolphin; Atlantic Hump-backed Dolphin; Cameroun River Dolphin; West African Sousa; Cameroon Dolphin

d   Kamerun-Flussdelfin; Kamerun-Delfin; Atlantischer Buckeldelfin

f   Sotalia de Teusz; Sousa de l'Atlantique; Dauphin à bosse de l'Atlantique

i   Susa atlantica

6514   *Spalacopus*
       Rodentia - Octodontidae

e   Coruros
d   Cururos
f   Rats bleus; Octodontes
i   Coruri

6515   *Spalacopus cyanus*
       Rodentia - Octodontidae

e   Coruro
d   Cururo
f   Rat bleu

Spalacopus tabanus syn. *S. cyanus* q.v.

6516   *Spalax*
       Rodentia - Muridae

e   Molerats; Greater Blind Molerats
d   Blindmäuse; Ostblindmäuse
f   Rats-taupes; Spalax ; Rats taupes d'Ukraine
i   Spalaci

6517   *Spalax arenarius*
       Rodentia - Muridae

e   Sandy Molerat

Spalax ehrenbergi syn. *Nannospalax ehrenbrgi* q.v.

6518   *Spalax giganteus*
       Rodentia - Muridae

e   Giant Molerat
d   Riesenblindmaus

6519   *Spalax graecus*
       Rodentia - Muridae

e   Bukovin Molerat; Balkan Molerat
d   Bukowinische Blindmaus
f   Spalax de Bukovine
i   Spalace orientale

Spalax leucodon syn. *Nannospalax leucodon* q.v.

6520   *Spalax micropthalmus*
       Rodentia - Muridae

e   Greater Molerat; Russian Molerat
d   Ostblindmaus; Steppenblindmaus; Südrussische Blindmaus; Große Blindmaus
f   Spalax oriental; Spalax zemni

6521   *Spalax zemni*
       Rodentia - Muridae

e   Podolsk Molerat

6522   *Speothos*
  Carnivora - Canidae
  e   Bush Dogs; Savanna Dogs; Savannah Dogs
  d   Waldhunde
  f   Chiens des buissons
  i   Speoti

6523   *Speothos venaticus*
  Carnivora - Canidae
  e   Bush Dog; Savanna Dog; Vinegar Fox; Savannah Dog
  d   Waldhund
  f   Chien des buissons
  i   Speoto; Itticione

6524   *Spermophilopsis*
  Rodentia - Sciuridae
  e   Long-clawed Ground Squirrels; Thin-toed Ground Squirrels
  d   Langzehige Ziesel ; Zieselmäuse
  f   Écureuils moulu long-griffés

6525   *Spermophilopsis leptodactylus*
  Rodentia - Sciuridae
  e   Long-clawed Ground Squirrel; Thin-toed Ground Squirrel
  d   Langzehiger Ziesel; Zieselmaus
  f   Spermophile leptodactyle; Écureuil moulu long-griffé

6526   *Spermophilus*
  Rodentia - Sciuridae
  e   Ground Squirrels
  d   Ziesel ; Zieselmäuse
  f   Écureuils terrestres; Sousliks; Spermophiles
  i   Spermofili

6527   *Spermophilus adocetus*
  Rodentia - Sciuridae
  e   Tropical Ground Squirrel; Lesser Tropical Ground Squirrel
  d   Tropischer Ziesel

6528   *Spermophilus alaschanicus*
  Rodentia - Sciuridae
  e   Ala Shan Ground Squirrel
  d   Alaschan-Ziesel
  f   Spermophile d'Alashan

6529   *Spermophilus annulatus*
  Rodentia - Sciuridae
  e   Ring-tailed Ground Squirrel
  d   Ringelschwanzziesel

6530   *Spermophilus armatus*
  Rodentia - Sciuridae
  e   Uinta Ground Squirrel
  d   Uinta-Ziesel
  f   Écureuil terrestre; Écureuil gris

6531   *Spermophilus atricapillus*
  Rodentia - Sciuridae
  e   Baja California Rock Squirrel
  d   Baja-California-Ziesel

6532   *Spermophilus beecheyi*
  Rodentia - Sciuridae
  e   California Ground Squirrel
  d   Kalifornische Ziesel
  f   Écureuil terrestre de Californie
  i   Scoiattolo di terra della California; Spermofilo della California

6533   *Spermophilus beldingi*
  Rodentia - Sciuridae
  e   Belding's Ground Squirrel
  d   Belding-Ziesel
  f   Écureuil terrestre de Belding; Spermophile de Belding

6534   *Spermophilus brunneus*
  Rodentia - Sciuridae
  e   Idaho Ground Squirrel
  d   Idaho-Ziesel

6535   *Spermophilus canus*
  Rodentia - Sciuridae

e Merriam's Ground Squirrel
d Merriam-Ziesel

6536 *Spermophilus citellus*
Rodentia - Sciuridae
e European Suslik; European Souslik;
European Ground Squirrel
d Ziesel; Schlichtziesel;
Erdeichhörnchen; Gemeiner Ziesel;
Grauer Ziesel; Westukrainischer
Ziesel; Europäscher Ziesel;
Einfarbige Ziesel
f Souslik; Souslik d'Europe;
Spermophile souslik; Spermophile
d'Europe; Soulik européen;
Spermophile
i Citello comune; Suslik

6537 *Spermophilus columbianus*
Rodentia - Sciuridae
e Columbian Ground Squirrel
d Columbia-Ziesel
f Spermophile du Columbia
i Spermofilo colombiano

6538 *Spermophilus dauricus*
Rodentia - Sciuridae
e Daurian Suslik; Daurian Souslik;
Daurian Ground Squirrel
d Daurischer Ziesel
i Suslik della Mongolia

6539 *Spermophilus elegans*
Rodentia - Sciuridae
e Wyoming Ground Squirrel
d Wyoming-Ziesel
f Spermophile de Wyoming

6540 *Spermophilus erythrogenys*
Rodentia - Sciuridae
e Red-cheeked Suslik; Red-cheeked
Souslik; Red-cheeked Ground
Squirrel
d Rotwangenziesel
f Souslik aux joues rouges

6541 *Spermophilus franklini*
Rodentia - Sciuridae
e Franklin's Ground Squirrel; Grey
Ground Squirrel
d Franklin-Ziesel
f Spermophile de Franklin; Écureuil
terrestre de Franklin

6542 *Spermophilus fulvus*
Rodentia - Sciuridae
e Large-toothed Souslik; Large-toothed
Suslik; Aral Yellow Suslik; Aral
Yellow Souslik; Yellow Ground
Squirrel
d Gelber Ziesel; Falber Ziesel;
Gelbziesel; Sandziesel; Gelbe
Zieselmaus
f Spermophile jaune; Souslik jaune;
Souslik fauve
i Peshanik

6543 *Spermophilus lateralis*
Rodentia - Sciuridae
e Golden-mantled Ground Squirrel
d Goldmantelziesel
f Spermophile à manteau doré;
Écureuil terrestre à manteau doré
i Scoiattolo della California

6544 *Spermophilus madrensis*
Rodentia - Sciuridae
e Sierra Madre Mantled Ground
Squirrel; Sierra Madre Ground
Squirrel
d Sierra-Madre-Ziesel

6545 *Spermophilus major*
Rodentia - Sciuridae
e Russet Suslik; Russet Souslik; Russet
Ground Squirrel
d Rotgelber Ziesel
f Spermophile roussâtre

6546 *Spermophilus mexicanus*
Rodentia - Sciuridae
e Mexican Ground Squirrel
d Mexikanischer Ziesel

f   Spermophile mexicain; Écureuil
      terrestre de Mexique

6547   *Spermophilus mohavensis*
         Rodentia - Sciuridae
e   Mohave Ground Squirrel; Mojave
      Ground Squirrel
d   Mohave-Ziesel; Mojave-Ziesel
f   Spermophile mohave

6548   *Spermophilus mollis*
         Rodentia - Sciuridae
e   Piute Ground Squirrel
d   Piute-Ziesel

6549   *Spermophilus musicus*
         Rodentia - Sciuridae
e   Caucasian Mountain Ground Squirrel
d   Kaukasus-Ziesel

6550   *Spermophilus parryii*
         Rodentia - Sciuridae
e   Arctic Ground Squirrel; Arctic
      Souslik
d   Amerikanischer Langschwanzziesel;
      Parry-Ziesel; Arktischer Ziesel
f   Spermophile arctique; Écureuil
      terrestre arctique

6551   *Spermophilus perotensis*
         Rodentia - Sciuridae
e   Perote Ground Squirrel
d   Perote-Ziesel
f   Spermophile de Perote

6552   *Spermophilus pygmaeus*
         Rodentia - Sciuridae
e   Little Suslik; Little Souslik; Little
      Ground Squirrel
d   Grauziesel; Kleinziesel; Kleiner
      Ziesel; Zwergziesel
f   Souslik nain

6553   *Spermophilus relictus*
         Rodentia - Sciuridae
e   Tien Shan Souslik; Tien Shan Suslik;

Tien Shan Ground Squirrel
d   Tienshan Suslik

6554   *Spermophilus richardsonii*
         Rodentia - Sciuridae
e   Richardson's Ground Squirrel
d   Richardson-Ziesel; Prairieziesel
f   Spermophile de Richardson; Écureuil
      terrestre de Richardson

6555   *Spermophilus saturatus*
         Rodentia - Sciuridae
e   Cascade Golden-mantled Ground
      Squirrel; Cascade Mantled Ground
      Squirrel
d   Kaskaden-Goldmantelziesel
f   Spermophile à manteau doré des
      Cascades

6556   *Spermophilus spilosoma*
         Rodentia - Sciuridae
e   Spotted Ground Squirrel
d   Gefleckter Ziesel; Fleckenziesel
f   Spermophile tacheté

6557   *Spermophilus suslicus*
         Rodentia - Sciuridae
e   Spotted Suslik; Spotted Souslik;
      Speckled Ground Squirrel
d   Suslik; Perlziesel
f   Souslik tacheté; Spermophile tacheté
i   Citello pomellato; Spermofilo
      machiato

6558   *Spermophilus tereticaudus*
         Rodentia - Sciuridae
e   Round-tailed Ground Squirrel;
      Roundtail Ground Squirrel
d   Rundschwanzziesel
f   Écureuil terrestre à queue ronde

6559   *Spermophilus townsendii*
         Rodentia - Sciuridae
e   Townsend's Ground Squirrel
d   Townsend-Ziesel
f   Spermophile de Townsend

6560   *Spermophilus tridecemlineatus*
Rodentia - Sciuridae
e   Thirteen-lined Ground Squirrel
d   Streifenziesel; Leopardenziesel;
Dreizehnstreifenziesel
f   Spermophile rayé; Spermophile à
treize bandes; Souslik à treize bandes

6561   *Spermophilus tridecemlineatus*
pallidus
Rodentia - Sciuridae
e   Pallid Thirteen-lined Ground Squirrel

6562   *Spermophilus undulatus*
Rodentia - Sciuridae
e   Long-tailed Suslik; Long-tailed
Souslik; Long-tailed Siberian Suslik;
Long-tailed Siberian Souslik; Long-
tailed Ground Squirrel
d   Langschwänziger Ziesel;
Langschwanzziesel
f   Souslik de Parry

6563   *Spermophilus variegatus*
Rodentia - Sciuridae
e   Rock Squirrel; Variegated Ground
Squirrel; American Rock Squirrel
d   Felsenziesel; Amerikanisches
Felsenhörnchen
f   Écureuil des rochers

6564   *Spermophilus washingtoni*
Rodentia - Sciuridae
e   Washington Ground Squirrel
d   Washington-Ziesel
f   Écureuil terrestre de Washington

6565   *Spermophilus xanthoprymnus*
Rodentia - Sciuridae
e   Asia Minor Ground Squirrel
d   Kleinasiatischer Ziesel

6566   *Sphaerias*
Chiroptera - Pteropodidae
e   Blanford's Fruit Bats
d   Blanfords Fruchtfledermäuse

6567   *Sphaerias blanfordi*
Chiroptera - Pteropodidae
e   Blanford's Fruit Bat
d   Blanfords Fruchtfledermaus

6568   *Sphaeronycteris*
Chiroptera - Phyllostomidae
e   Visored Bats

6569   *Sphaeronycteris toxophyllum*
Chiroptera - Phyllostomidae
e   Visored Bat

Sphiggurus syn. ***Coendu*** q.v.

6570   *Spilocuscus*
Diprotodontia - Phalangeridae
e   Spotted Cuscuses
d   Tüpfelkuskuse; Fleckenkuskuse
f   Couscous
i   Falangeri

6571   *Spilocuscus kraemeri*
Diprotodontia - Phalangeridae
e   Admiralty Cuscus

6572   *Spilocuscus maculatus*
Diprotodontia - Phalangeridae
e   Spotted New Guinea Cuscus; Spotted
Cuscus; Short-tailed Cuscus; Short-
tailed Spotted Cuscus; Spotted
Phalanger; Common Spotted Cuscus
(ANZ)
d   Tüpfelkuskus;
Kurzschwanzfleckenkuskus
f   Couscous tacheté
i   Falangere macchiato; Cusco
macchiato

6573   *Spilocuscus papuensis*
Diprotodontia - Phalangeridae
e   Waigeo Cuscus

6574   *Spilocuscus rufoniger*
Diprotodontia - Phalangeridae

e   Black-spotted Cuscus
d   Schwarzgefleckter Kuskus

6575   *Spilogale*
Carnivora - Mustelidae
e   Spotted Skunks
d   Fleckenskunks; Gefleckte Stinktiere
f   Spilogales; Moufettes tachetées;
    Mouffettes tachetées(Qué)
i   Moffette; Skunk

Spilogale angustifrons syn. *S.*
putorius q.v.

6576   *Spilogale gracilis*
Carnivora - Mustelidae
e   Western Spotted Skunk
d   Westlicher Fleckenskunk
f   Moufette tachetée occidentale;
    Mouffette tachetée occidentale;
    Mofette tachetée occidentale
i   Puzzola macchiata; Skunk macchiato
    occidentale

6577   *Spilogale putorius*
Carnivora - Mustelidae
e   Spotted Skunk; Civet; North
    American Spotted Skunk
d   Östlicher Fleckenskunk; Südlicher
    Fleckenskunk; Westlicher
    Fleckenskunk; Fleckenskunk
f   Moufette tachetée orientale;
    Mouffette tachetée orientale; Mofette
    tachetée orientale
i   Skunk maculato; Mofetta macchiata;
    Skunk macchiato; Gatto civetta;
    Gatto lyra; Skunk macchiato
    orientale

6578   *Spilogale putorius ambarvalis*
Carnivora - Mustelidae
e   Eastern Spotted Skunk

6579   *Spilogale pygmaea*
Carnivora - Mustelidae
e   Pygmy Spotted Skunk; Little Spotted
    Skunk

d   Zwergfleckenskunk
i   Skunk macchiata pigmea

6580   *Srilankamys*
Rodentia - Muridae
e   Ohiya Rats
d   Sri Lanka-Ratten

6581   *Srilankamys ohiensis*
Rodentia - Muridae
e   Ohiya Rat
d   Sri Lanka-Ratte; Zweifarbige Sri
    Lanka-Ratte

6582   *Steatomys*
Rodentia - Muridae
e   Fat Mice
d   Fettmäuse
f   Rats adipeux; Grosses souris
i   Topi adiposi

6583   *Steatomys caurinus*
Rodentia - Muridae
e   Northwestern Fat Mouse

6584   *Steatomys cuppedius*
Rodentia - Muridae
e   Dainty Fat Mouse

6585   *Steatomys jacksoni*
Rodentia - Muridae
e   Jackson's Fat Mouse

6586   *Steatomys krebsii*
Rodentia - Muridae
e   Krebs's Fat Mouse

Steatomys minutus syn. *S. parvus*
q.v.

6587   *Steatomys parvus*
Rodentia - Muridae
e   Tiny Fat Mouse
d   Fettmaus

6588 *Steatomys pratensis*
Rodentia - Muridae
e Fat Mouse; Common Fat Mouse

6589 *Stenella*
Cete - Delphinidae
e Spotted Dolphins; Stenella Dolphins; Spinner, Spotted and Striped Dolphins
d Fleckendelfine
f Dauphins tachetés
i Stenelle

6590 *Stenella attenuata*
Cete - Delphinidae
e Bridled Slender-beaked Dolphin; Narrow-snouted Dolphin; Malayan Dolphin; Philippine Dolphin; Bridled Dolphin; Bridled Spotted Dolphin; Cuvier's Dolphin; Bridled Porpoise; Cuvier's Porpoise; Pantropical Spotted Dolphin; Pacific Spotted Dolphin
d Schlankdelfin; Fleckendelfin; Pantropischer Fleckendelfin
f Dauphin douteux; Dauphin tacheté du Pacifique; Dauphin tropical; Dauphin tacheté pantropical
i Stenella maculata; Stenella maculata pantropicale

6591 *Stenella attenuata graffmanni*
Cete - Delphinidae
e Eastern Pacific Coastal Spotted Dolphin

6592 *Stenella clymene*
Cete - Delphinidae
e Atlantic Spinner Dolphin; Clymene Dolphin; Short-snouted Spinner Dolphin; Helmet Dolphin; Helmeted Dolphin; Short-snouted Spinner
d Kurzschnauzendelfin; Kurzschnauzenspinner; Clymene-Delfin
f Dauphin clymène
i Stenella climene

6593 *Stenella coeruleoalba*
Cete - Delphinidae
e Blue-white Dolphin; Harnessed Dolphin; Meyen's Dolphin; Striped Dolphin; Euphrosyne Dolphin; Cuvier's Dolphin; Blue-and-white Dolphin
d Blauweißer Delfin; Streifendelfin
f Dauphin bleu-blanc; Dauphin bleu-et-blanc; Dauphin rayé; Dauphin de Thétys
i Delfino bianco ceruleo; Stenella striata; Stenella; Prodelfino bianco-ceruleo; Delfino striato; Delfino eufrosine

Stenella dubia syn. *S. attenuata* q.v.

Stenella euphrosyne syn. *S. coeruleoalba* q.v.

6594 *Stenella frontalis*
Cete - Delphinidae
e Atlantic Spotted Dolphin; Spotted Dolphin
d Zügeldelfin; Antlantischer Gefleckter Delfin
f Dauphin tacheté de l'Atlantique; Dauphin tacheté
i Delfino dalle briglie; Stenella maculata alantica

6595 *Stenella longirostris*
Cete - Delphinidae
e Long-snouted Dolphin; Long-beaked Dolphin; Small-headed Dolphin; Spinner Dolphin; Eastern Pacific Spinner Dolphin; Long-beaked Porpoise; Small-headed Porpoise; Spinner Porpoise; Spinning Porpoise
d Langschnauzendelfin; Ostpazifischer Delfin; Ostpazifischer Spinner; Langschnauzenspinner
f Dauphin à long bec; Dauphin longirostre
i Delfino filatore; Stenella dal lungo rostro

6596   ***Stenella longirostris***
           centroamericana
           Cete - Delphinidae
  e   Central American Spinner Dolphin

6597   ***Stenella longirostris longirostris***
           Cete - Delphinidae
  e   Cosmopolitan Spinner Dolphin

6598   ***Stenella longirostris orientalis***
           Cete - Delphinidae
  e   Tropical Pacific Spinner Dolphin

           Stenella styx syn. *S. coeruleoalba*
           q.v.

6599   ***Steno***
           Cete - Delphinidae
  e   Rough-toothed Dolphins
  d   Langschnauzendelfine;
           Furchenzahndelfine;
           Rauhzahndelfine
  f   Dauphins de Bréda; Sténos
  i   Steni

6600   ***Steno bredanensis***
           Cete - Delphinidae
  e   Rough-toothed Dolphin; Rough-
           toothed Porpoise; Goggle-eyed
           Porpoise; Roughtooth Dolphin
  d   Rauhzahndelfin
  f   Dauphin de Bréda; Sténo rostré;
           Sténo; Dauphin sténo
  i   Steno dal rostro; Steno; Steno
           rostrato; Delfino dai denti rugosi;
           Delfino dai denti minori; Delfino
           comune; Delfino dalla dentatura
           irregolare

           Steno rostratus syn. *S. bredanensis*
           q.v.

6601   ***Stenocephalemys***
           Rodentia - Muridae
  e   Narrow-headed Rats; Ethiopian
           Narrow-headed Rats

  d   Äthiopische Kurzkopfratten

6602   ***Stenocephalemys albocaudata***
           Rodentia - Muridae
  e   Ethiopian Narrow-headed Rat

6603   ***Stenocephalemys griseicauda***
           Rodentia - Muridae
  e   Grey-tailed Narrow-headed Rat

           Stenocranius syn. *Microtus* q.v.

6604   ***Stenoderma***
           Chiroptera - Phyllostomidae
  e   Red Fruit Bats
  d   Fruchtvampire

6605   ***Stenoderma rufum***
           Chiroptera - Phyllostomidae
  e   Red Fruit Bat; Desmarest's Fig-eating
           Bat; Red Fig-eating Bat; Puerto
           Rican Fruit Bat
  d   Fruchtvampir

6606   ***Stenomys***
           Rodentia - Muridae
  e   Slender Rats
  f   Rats de l'île Céram

6607   ***Stenomys ceramicus***
           Rodentia - Muridae
  e   Ceram Rat

6609   ***Stenomys niobe***
           Rodentia - Muridae
  e   Moss-forest Rat

6608   ***Stenomys omichlodes***
           Rodentia - Muridae
  e   Arianus's Rat

6610   ***Stenomys richardsoni***
           Rodentia - Muridae
  e   Glacier Rat

6611 **Stenomys vandeuseni**
Rodentia - Muridae
e Van Deusen's Rat

6612 **Stenomys verecundus**
Rodentia - Muridae
e Slender Rat

Stenonycteris syn. *Rousettus* q.v.

Stictomys syn. *Cuniculus* q.v.

6613 **Stochomys**
Rodentia - Muridae
e Target Rats
d Zielratten

Stochomys defua syn. *Dephomys
defua* q.v.

Stochomys eburnea syn. *Dephomys
eburnea* q.v.

6614 **Stochomys longicaudatus**
Rodentia - Muridae
e Target Rat
d Zielratte

6615 **Strigocuscus**
Diprotodontia - Phalangeridae
e Plain Cuscuses
d Bodenkuskuse

6616 **Strigocuscus celebensis**
Diprotodontia - Phalangeridae
e Celebes Cuscus; Little Celebes
Cuscus; Leaf-eater Bear; Sulawesi
Dwarf Cuscus; Dwarf Cuscus;
Sulawesi Cuscus; Small Sulawesi
Cuscus
d Celebes-Kuskus

Strigocuscus gymnotis syn.
Phalanger gymnotis q.v.

6617 **Strigocuscus pelegensis**
Diprotodontia - Phalangeridae
e Peleng Cuscus

6618 **Sturnira**
Chiroptera - Phyllostomidae
e Yellow-shouldered Bats; Epaulet
Bats; Epauletted Bats; American
Epaulette Bats; American Epauletted
Bats
d Gelbschulterblattnasen
f Sturnires

6619 **Sturnira aratathomasi**
Chiroptera - Phyllostomidae
e Arata and Thomas's Yellow-
shouldered Bat

6620 **Sturnira bidens**
Chiroptera - Phyllostomidae
e Yellow-shouldered Bat; Epaulet Bat;
Epauletted Bat; American Epaulette
Bat; American Epauletted Bat;
Bidentate Yellow-shouldered Bat
f Sturnire

Sturnira bogotensis syn. *S.
oporaphilum* q.v.

6621 **Sturnira erythromos**
Chiroptera - Phyllostomidae
e Hairy Yellow-shouldered Bat

6622 **Sturnira lilium**
Chiroptera - Phyllostomidae
e Common Yellow-shouldered Bat;
Paraguayan Yellow-shouldered Bat;
Little Yellow-shouldered Bat; Little
Yellow-shouldered Fruit Bat
d Gelbschulterblattnase;
Kurznasenvampir
f Sturnire fleur-de-lys

6623 **Sturnira ludovici**
Chiroptera - Phyllostomidae

e   Anthony's Bat; Highland Yellow-
      shouldered Bat
d   Hochlandgelbschulterblattnase

6624   *Sturnira luisi*
Chiroptera - Phyllostomidae
e   Luis's Yellow-shouldered Bat

6625   *Sturnira magna*
Chiroptera - Phyllostomidae
e   Greater Yellow-shouldered Bat

6626   *Sturnira mistratensis*
Chiroptera - Phyllostomidae
e   Mistrato Yellow-shouldered Bat

6627   *Sturnira mordax*
Chiroptera - Phyllostomidae
e   Hairy-footed Bat; Talamancan
      Yellow-shouldered Bat; Talamancan
      Bat

6628   *Sturnira nana*
Chiroptera - Phyllostomidae
e   Lesser Yellow-shouldered Bat

6629   *Sturnira oporaphilum*
Chiroptera - Phyllostomidae
e   Bogota Yellow-shouldered Bat

6630   *Sturnira thomasi*
Chiroptera - Phyllostomidae
e   Sofaian Bat; Thomas's Epauletted
      Bat; Thomas's Yellow-shouldered
      Bat

6631   *Sturnira tildae*
Chiroptera - Phyllostomidae
e   Trinidadian Yellow-shouldered Bat;
      Tilda Yellow-shouldered Bat

6632   *Styloctenium*
Chiroptera - Pteropodidae
e   Stripe-faced Fruit Bats
d   Weißstreifenflughunde

6633   *Styloctenium wallacei*
Chiroptera - Pteropodidae
e   Stripe-faced Fruit Bat
d   Harlekin-Flughund;
      Weißstreifenflughund

6634   *Stylodipus*
Rodentia - Dipodidae
e   Thick-tailed Three-toed Jerboas;
      Feather-tailed Three-toed Jerboas;
      Three-toed Jerboas
d   Wüstenspringmäuse; Zierliche
      Springmäuse;
      Dickschwanzspringmäuse
f   Gerboises à trois doigts

6635   *Stylodipus andrewsi*
Rodentia - Dipodidae
e   Andrews's Three-toed Jerboa
d   Andrews-Dreizehenspringmaus
f   Gerboise à petite queue à trois doigts

6636   *Stylodipus sungorus*
Rodentia - Dipodidae
e   Mongolian Three-toed Jerboa
d   Mongolische Wüstenspringmaus

6637   *Stylodipus telum*
Rodentia - Dipodidae
e   Thick-tailed Three-toed Jerboa;
      Feather-tailed Three-toed Jerboa

d   Zierliche Springmaus;
      Dickschwanzspringmaus

6638   *Suidae*
Artiodactyla
e   True Pigs; Old World Pigs; True
      Hogs; Pigs; Pigs and Hogs
d   Borstentiere; Schweine; Echte
      Schweine; Altweltliche Schweine
f   Suidés; Sangliers; Cochons; Porcs
i   Cinghiali; Suidi

Suillotaxus syn. *Mydaus* q.v.

6639 ***Suncus***
Soricomorpha - Soricidae
e Dwarf Shrews; Musk Shrews; Thick-tailed Shrews; Pygmy and Dwarf Shrews
d Dickschwanzspitzmäuse
f Musaraignes; Pachyures
i Toporagni

6640 ***Suncus ater***
Soricomorpha - Soricidae
e Black Shrew

6641 ***Suncus dayi***
Soricomorpha - Soricidae
e Day's Shrew

6642 ***Suncus etruscus***
Soricomorpha - Soricidae
e Dwarf Shrew; Etruscan Shrew; Mediterranean Pygmy Shrew; Savi's Pygmy Shrew; Pygmy White-toothed Shrew; White-toothed Pygmy Shrew
d Etrusker Spitzmaus; Mittelländische Spitzmaus; Wimperspitzmaus; Toskanische Wimperspitzmaus
f Crocidure étrusque; Musaraigne étrusque; Pachyure étrusque
i Mustiolo; Mustiolo toscano

6643 ***Suncus etruscus madagascariensis***
Soricomorpha - Soricidae
e Madagascan Shrew; Madagascar Pygmy Shrew
d Madagaskar-Spitzmaus
f Pachyure de Madagascar

6644 ***Suncus fellowesgordoni***
Soricomorpha - Soricidae
e Sri Lanka Shrew; Ceylon Pygmy Shrew
d Sri Lanka-Zwergspitzmaus

Suncus granti syn. *Sylvisorex granti* q.v.

6645 ***Suncus hosei***
Soricomorpha - Soricidae
e Hose's Shrew; Borneo Pymy Shrew

Suncus howelli syn. *Sylvisorex howelli* q.v.

6646 ***Suncus infinitesimus***
Soricomorpha - Soricidae
e Least Dwarf Shrew

Suncus johnstoni syn. *Sylvisorex johnstoni* q.v.

6647 ***Suncus lixus***
Soricomorpha - Soricidae
e Greater Dwarf Shrew

Suncus lunaris syn. *Sylvisorex lunaris* q.v.

Suncus madagascariensis syn. *S. etruscus madagascariensis* q.v.

6648 ***Suncus malayanus***
Soricomorpha - Soricidae
e Malayan Pygmy Shrew

Suncus megalura syn. *Sylvisorex megalura* q.v.

6649 ***Suncus mertensi***
Soricomorpha - Soricidae
e Flores Shrew

6650 ***Suncus montanus***
Soricomorpha - Soricidae
e Sri Lanka Highland Shrew

Suncus morio syn. *Sylvisorex morio* q.v.

6651 ***Suncus murinus***
Soricomorpha - Soricidae
e House Shrew; Musk Shrew; Asian House Shrew; Asian Musk Shrew

d   Moschusspitzmaus; Asiatische
     Moschusspitzmaus
f   Musaraigne des maisons

6652 ***Suncus murinus caerulescens***
       Soricomorpha - Soricidae
e   Indian Grey Musk Rat
d   Dickschwanzspitzmaus

6653 ***Suncus murinus kandianus***
       Soricomorpha - Soricidae
e   Kandy Shrew; Kandyan Shrew
d   Kandy-Maus

6654 ***Suncus murinus montanus***
       Soricomorpha - Soricidae
e   Ceylon Highland Shrew
d   Sri Lanka-Hochlandspitzmaus

6655 ***Suncus murinus murinus***
       Soricomorpha - Soricidae
e   Common Indian Musk Shrew; Indian
     Musk Shrew
d   Indische Moschusspitzmaus

      Suncus ollula syn. *Sylvisorex ollula*
      q.v.

6656 ***Suncus remyi***
       Soricomorpha - Soricidae
e   Remy's Shrew; Remy's Pygmy Shrew

6657 ***Suncus stoliczkanus***
       Soricomorpha - Soricidae
e   Anderson's Shrew; Yellow-throated
     Shrew

6658 ***Suncus varilla***
       Soricomorpha - Soricidae
e   Lesser Dwarf Shrew; Termite Shrew

6659 ***Suncus zeylanicus***
       Soricomorpha - Soricidae
e   Jungle Shrew; Ceylon Jungle Shrew

d   Sri Lanka-Dschungelspitzmaus

6660 ***Sundamys***
       Rodentia - Muridae
e   Giant Sunda Rats
d   Sundamys-Ratten

6661 ***Sundamys infraluteus***
       Rodentia - Muridae
e   Mountain Giant Rat; Mountain Giant
     Sund Rat

6662 ***Sundamys maxi***
       Rodentia - Muridae
e   Bartel's Rat; Bartel's Giant Sunda Rat
d   Maxis Ratte

6663 ***Sundamys muelleri***
       Rodentia - Muridae
e   Müller's Rat; Müller's Giant Sunda
     Rat
d   Müllers Sandratte

6664 ***Sundasciurus***
       Rodentia - Sciuridae
e   Sunda Squirrels
d   Sunda-Baumhörnchen
f   Écureuils de l'île de la Sonde

      Sundasciurus albicauda syn. *S.*
      moellendorffi q.v.

6665 ***Sundasciurus brookei***
       Rodentia - Sciuridae
e   Brooke's Squirrel
d   Brooke-Hörnchen

6666 ***Sundasciurus davensis***
       Rodentia - Sciuridae
e   Davao Squirrel
d   Davao-Hörnchen

6667 ***Sundasciurus fraterculus***
       Rodentia - Sciuridae
e   Fraternal Squirrel

6668 *Sundasciurus hippurus*
Rodentia - Sciuridae
e Horse-tailed Squirrel
d Pferdeschwanzhörnchen
f Écureuil à queue de cheval

6669 *Sundasciurus hoogstraali*
Rodentia - Sciuridae
e Busuanga Squirrel; Hoogstraal's
Squirrel
d Busuanga-Hörnchen

6670 *Sundasciurus jentinki*
Rodentia - Sciuridae
e Jentink's Squirrel
d Jentink-Hörnchen

6671 *Sundasciurus juvencus*
Rodentia - Sciuridae
e Northern Palawan Tree Squirrel;
Palawan Squirrel
d Nördliches Palawan-Hörnchen

6672 *Sundasciurus lowii*
Rodentia - Sciuridae
e Low's Squirrel; Sunda Tree Squirrel
d Low-Hörnchen

6673 *Sundasciurus mindanensis*
Rodentia - Sciuridae
e Mindanao Squirrel
d Mindanao-Hörnchen

6674 *Sundasciurus moellendorffi*
Rodentia - Sciuridae
e Culion Tree Squirrel; Möllendorf's
Squirrel
d Culion-Hörnchen

6675 *Sundasciurus philippinensis*
Rodentia - Sciuridae
e Philippine Tree Squirrel; Philippine
Squirrel
d Philippinen-Hörnchen

6676 *Sundasciurus rabori*
Rodentia - Sciuridae
e Palawan Montane Squirrel; Rabor's
Squirrel
d Palawan-Berghörnchen

6677 *Sundasciurus samarensis*
Rodentia - Sciuridae
e Samar Squirrel
d Samar-Hörnchen

6678 *Sundasciurus steerii*
Rodentia - Sciuridae
e Southern Palawan Tree Squirrel;
Steere's Squirrel
d Südliches Palawan-Hörnchen

6679 *Sundasciurus tenuis*
Rodentia - Sciuridae
e Slender Squirrel
d Sunda-Schlankhörnchen

Surdisorex syn. *Myosorex* q.v.

6680 *Suricata*
Carnivora - Herpestidae
e Suricates; Slender-tailed Meerkats;
Meerkats
d Erdmännchen ; Surikaten
f Suricates
i Suricate

6681 *Suricata suricatta*
Carnivora - Herpestidae
e Suricate; Slender-tailed Meerkat;
Meerkat
d Erdmännchen; Surikate
f Suricate
i Suricata

6682 *Sus*
Artiodactyla - Suidae
e Wild Boars; Wild Hogs; Hogs; Pigs;
Eurasian Pigs
d Schweine; Wildschweine

f   Cochons; Sangliers
i   Cinghiali

6683   *Sus barbatus*
Artiodactyla - Suidae
e   Bearded Pig; Bornean Pig
d   Bartschwein; Krausbartschwein
f   Sanglier à barbe
i   Cinghiale barbuto; Maiale barbuto della penisola malese; Maiale barbuta delle isola della Sonda

6684   *Sus barbatus ahoenobarbus*
Artiodactyla - Suidae
e   Palawan Bearded Pig
f   Sanglier géant de Palawan

6685   *Sus barbatus barbatus*
Artiodactyla - Suidae
e   Bornean Bearded Pig
d   Borneo-Bartschwein
f   Sanglier géant de Bornéo

6686   *Sus barbatus oi*
Artiodactyla - Suidae
e   Western Bearded Pig; Curly-bearded Pig
d   Westliches Krausbartschwein
f   Sanglier à barbe; Sanglier à moustaches

6687   *Sus bucculentus*
Artiodactyla - Suidae
e   Viet Nam Warty Pig; Vietnamese Warty Pig
d   Anamitisches Pustelschwein
f   Sanglier du Vietnam

6688   *Sus cebifrons*
Artiodactyla - Suidae
e   Visayan Warty Pig; Visayan Islands Warty Pig; Visayas Warty Pig
d   Visayas-Mähnenschwein; Visayas-Pustelwchwein
f   Sanglier des Visayas

6689   *Sus cebifrons cebifrons*
Artiodactyla - Suidae
e   Cebu Warty Pig; Cebu Island Warty Pig
d   Cebu-Mähnenschwein; Cebu-Pustelschwein
f   Sanglier pustule des îles Visayas; Sanglier pustule des Visayas

6690   *Sus celebensis*
Artiodactyla - Suidae
e   Celebes Wild Boar; Celebes Warty Pig; Sulawesi Warty Pig; Sulawesi Wild Pig
d   Nordcelebes-Pustelschwein; Sulawesi-Pustelschwein
f   Sanglier des Célèbes; Sanglier pustule de l'île Cebu
i   Cinghiale delle Celebes

6691   *Sus domesticus*
Artiodactyla - Suidae
e   Pig; Domestic Pig
d   Hausschwein
f   Cochon domestique
i   Maiale domestico

6692   *Sus heureni*
Artiodactyla - Suidae
e   Flores Warty Pig
d   Flores-Pustelschwein
f   Sanglier de l'île de Flores

6693   *Sus philippensis*
Artiodactyla - Suidae
e   Philippine Warty Pig
d   Philippinisches Pustelschwein
f   Sanglier des Philippines

6694   *Sus salvanius*
Artiodactyla - Suidae
e   Pygmy Hog
d   Zwergwildschwein
f   Sanglier pygmée; Sanglier nain
i   Cinghiale nano; Cinghiale pigmeo; Maiale selvatico nano del Nepal

6695 *Sus scrofa*
Artiodactyla - Suidae

e Wild Hog; Eurasian Wild Pig; Wild Boar; Eurasian Wild Boar; Wild Pig; Feral Hog (NA)

d Wildschwein; Wildes Schwein; Eurasisches Wildschwein; Schwarzwild

f Sanglier; Sanglier d'Europe; Porc marron

i Cinghiale; Cignàlo; Porco cignàlo; Maiale selvatico eurasiatico

6696 *Sus scrofa barbarus*
Artiodactyla - Suidae

e Barbary Wild Boar; African Wild Boar

d Berber-Schwein

i Cinghiale berbero

6697 *Sus scrofa cristatus*
Artiodactyla - Suidae

e Indian Wild Pig

d Kammwildschwein

f Sanglier à crinière

i Cinghiale crestato

6698 *Sus scrofa domestica*
Artiodactyla - Suidae

e Vietnamese Potbellied Pig; Vietnamese Potbelly Pig; Potbelly

d Vietnamesisches Hausschwein

i Maiale domestico di Vietnam

6699 *Sus scrofa falzfeini*
Artiodactyla - Suidae

e Polish Wild Boar; Poland Wild Boar

6700 *Sus scrofa leucomystax*
Artiodactyla - Suidae

e Japanese Wild Boar; Japan Wild Boar

d Japanisches Wildschwein

f Sanglier du Japon

6701 *Sus scrofa meridionalis*

Artiodactyla - Suidae

e Iberian Pig

d Sardenschwein; Sardisches Wildschwein

f Sanglier de Corse

i Cinghiale della Sardegna

6702 *Sus scrofa moupinensis*
Artiodactyla - Suidae

e Chinese Wild Boar

f Sanglier de Chine

i Cinghiale di Moupin

6703 *Sus scrofa scrofa*
Artiodactyla - Suidae

e European Wild Hog; European Wild Boar

d Mitteleuropäisches Wildschwein

f Sanglier européen

i Cinghiale eurasiatico

6704 *Sus scrofa taivanus*
Artiodactyla - Suidae

e Formosan Wild Boar; Formosan Wild Pig; Taiwan Wild Boar

f Sanglier de Taiwan

6705 *Sus scrofa ussuricus*
Artiodactyla - Suidae

e Manchurian Wild Boar; Ussuriysk Wild Boar

d Ussurisches Wildschwein

f Sanglier de Mandchourie

6706 *Sus scrofa vittatus*
Artiodactyla - Suidae

e Miniature Vietnamese Pot-bellied Pig; Chinese Pot-bellied Pig; Oriental Pot-bellied Pig

d Bindenschwein

f Sanglier d'Asie

i Cinghiale dalle bande; Maiale fasciato dell'Asia sudorientale

6707 *Sus timoriensis*
Artiodactyla - Suidae

e   Timor Wild Boar; Timor Warty Pig

d   Timor-Pustelschwein

f   Sanglier de Timor

6708   *Sus verrucosus*
Artiodactyla - Suidae

e   Javan Pig; Javan Warty Pig

d   Pustelschwein

f   Sanglier de Java

i   Cinghiale dalle verruche

6709   *Syconycteris*
Chiroptera - Pteropodidae

e   Blossom Bats

d   Langzungenflughunde

6710   *Syconycteris australis*
Chiroptera - Pteropodidae

e   Queensland Blossom Bat; Southern
Blossom Bat; Common Blossom Bat
(ANZ)

d   Queensland-Langzungenflughund;
Australische Blütenfledermaus

6711   *Syconycteris carolinae*
Chiroptera - Pteropodidae

e   Halmahera Blossom Bat

***Syconycteris crassa* syn.** *S. australis*
q.v.

6712   *Syconycteris hobbit*
Chiroptera - Pteropodidae

e   Moss-forest Blossom Bat

***Syconycteris naias* syn.** *S. australis*
q.v.

***Sylvaemus* syn.** *Apodemus* q.v.

6713   *Sylvicapra*
Artiodactyla - Bovidae

e   Forest Duikers; Grey Duikers;
Duikerboks; Bush Duikers

d   Ducker ; Kronenducker

f   Céphalophes de Grimm; Sylvicapres;
Biches cochons

i   Silvicapre

6714   *Sylvicapra grimmia*
Artiodactyla - Bovidae

e   Common Duiker; Grey Duiker; Bush
Duiker; Grimm's Duiker; Crowned
Duiker

d   Echter Ducker; Kronenducker

f   Céphalophe de Grimm; Sylvicapre de
Grimm

i   Cefalofo di Grimm; Silvicapra

6715   *Sylvicapra grimmia coronata*
Artiodactyla - Bovidae

e   Western Bush Duiker

d   Westlicher Kronenducker

f   Céphalophe couronné

6716   *Sylvicapra grimmia grimmia*
Artiodactyla - Bovidae

e   Southern Bush Duiker

d   Südlicher Kronenducker

6717   *Sylvicapra grimmia hindei*
Artiodactyla - Bovidae

e   East African Bush Duiker

d   Ostafrikanischer Kronenducker

6718   *Sylvicapra grimmia splendidula*
Artiodactyla - Bovidae

e   Angolan Bush Duiker

d   Angola-Kronenducker

f   Céphalophe grise

6719   *Sylvilagus*
Lagomorpha - Leporidae

e   Cottontail Rabbits; Cottontails (NA);
American Rabbits

d   Baumwollschwanzkaninchen

f   Lapins américains

i   Minilepri; Silvilagi

6720 *Sylvilagus aquaticus*
Lagomorpha - Leporidae
e Swamp Rabbit; Water Rabbit
d Wasserkaninchen
f Lapin aquatique
i Coniglio delle palude; Silvilago acquatico

6721 *Sylvilagus auduboni*
Lagomorpha - Leporidae
e Audubon's Cottontail; Desert Cottontail; Audubon Cottontail
d Audubon-Kaninchen
i Silvilago del deserto

6722 *Sylvilagus auduboni arizonae*
Lagomorpha - Leporidae
e Desert Cottontail

6723 *Sylvilagus bachmani*
Lagomorpha - Leporidae
e Brush Rabbit
d Strauchkaninchen
i Silvilago di Bachman

6724 *Sylvilagus bachmani mariposae*
Lagomorpha - Leporidae
e Mariposa Brush Rabbit

6725 *Sylvilagus bachmani riparius*
Lagomorpha - Leporidae
e Riparian Brush Rabbit

6726 *Sylvilagus brasiliensis*
Lagomorpha - Leporidae
e Forest Rabbit; Tapeti; Brazilian Rabbit; Tropical Cottontail Rabbit; Tapiti
d Brasilien-Waldkaninchen; Baumwollschwanzkaninchen
f Lapin du Brésil

6727 *Sylvilagus cunicularius*
Lagomorpha – Leporidae

e Mexican Cottontail
d Mexikanisches Baumwollschwanzkaninchen

6728 *Sylvilagus dicei*
Lagomorpha - Leporidae
e Dice's Cottontail; Dice's Rabbit; Dice's Mountain Rabbit
d Dice-Baumwollschwanzkaninchen

6729 *Sylvilagus floridanus*
Lagomorpha - Leporidae
e Cottontail Rabbit; Cottontail (NA); Eastern Cottontail
d Florida-Waldkaninchen
f Lapin de Floride; Lapin à queue blanche
i Coda di cotone orientale; Coda di cotone de l'Est

6730 *Sylvilagus floridanus floridanus*
Lagomorpha - Leporidae
e Eastern Cottontail Rabbit; Florida Rabbit

6731 *Sylvilagus floridanus hutchensii*
Lagomorpha - Leporidae
e Hutchen's Eastern Cottontail

6732 *Sylvilagus floridanus margaritae*
Lagomorpha - Leporidae
e Margarita's Cottontail Rabbit

6733 *Sylvilagus floridanus similis*
Lagomorpha - Leporidae
e Nebraska Cottontail

6734 *Sylvilagus graysoni*
Lagomorpha - Leporidae
e Tres Marias Cottontail
d Tres-Marias- Baumwollschwanzkaninchen

Sylvilagus idahoensis syn.
Brachylagus idahoensis q.v.

6735 *Sylvilagus insonus*
Lagomorpha - Leporidae
e Omilteme Cottontail
d Omiteme-
Baumwollschwanzkaninchen

6736 *Sylvilagus mansuetus*
Lagomorpha - Leporidae
e San José Brush Rabbit
d San-José-Bürstenkaninchen

6737 *Sylvilagus nuttallii*
Lagomorpha - Leporidae
e Mountain Cottontail
d Nuttals-Baumwollschwanzkaninchen

6738 *Sylvilagus nuttallii grangerii*
Lagomorpha - Leporidae
e Black Hills Cotttontail

6739 *Sylvilagus nuttallii nuttallii*
Lagomorpha - Leporidae
e Nuttall's Cottontail
f Lapin de Nuttall

6740 *Sylvilagus obscurus*
Lagomorpha - Leporidae
e Appalachian Cottontail

6741 *Sylvilagus palustris*
Lagomorpha - Leporidae
e Marsh Rabbit
d Sumpfkaninchen; Marschkaninchen
i Silvilago palustre

6742 *Sylvilagus palustris heffneri*
Lagomorpha - Leporidae
e Lower Keys Rabbit

6743 *Sylvilagus transitionalis*
Lagomorpha - Leporidae
e New England Cottontail
d Neuengland-
Baumwollschwanzkaninchen

f Lapin de Nouvelle-Angleterre

6744 *Sylvilagus varynaensis*
Lagomorpha - Leporidae
e Barinas Rabbit

6745 *Sylvisorex*
Soricomorpha - Soricidae
e Forest Musk Shrews

6746 *Sylvisorex granti*
Soricomorpha - Soricidae
e Grant's Shrew

6747 *Sylvisorex howelli*
Soricomorpha - Soricidae
e Howell's Shrew

6748 *Sylvisorex isabellae*
Soricomorpha - Soricidae
e Isabella Shrew

6749 *Sylvisorex johnstoni*
Soricomorpha - Soricidae
e Johnston's Shrew

6750 *Sylvisorex konganensis*
Soricomorpha - Soricidae
e Centrl African Shrew

6751 *Sylvisorex lunaris*
Soricomorpha - Soricidae
e Crescent Shrew

6752 *Sylvisorex megalura*
Soricomorpha - Soricidae
e Climbing Shrew

6753 *Sylvisorex morio*
Soricomorpha - Soricidae
e Arrogant Shrew; Mount Cameroon
Forest Shrew

6754	***Sylvisorex ollula***		*e*	African Buffaloes
	Soricomorpha - Soricidae		*d*	Afrikanische Büffel ; Steppenbüffel ; Kaffernbüffel
e	Forest Musk Shrew		*f*	Buffles africains; Buffles d'Afrique
f	Grande musaraigne grise		*i*	Bufali

6755 ***Sylvisorex oriundus***
Soricomorpha - Soricidae
e Mountain Shrew

6756 ***Sylvisorex pluvialis***
Soricomorpha - Soricidae
e Rainforest Shrew

6757 ***Sylvisorex vulcanorum***
Soricomorpha - Soricidae
e Volcano Shrew

Symphalangus syn. *Hylobates* q.v.

6758 ***Synaptomys***
Rodentia - Muridae
e Bog Lemmings; Bog Mice; Lemming Mice
d Lemmingmäuse; Moorlemminge
f Campagnols-lemmings
i Lemming delle paludi

6759 ***Synaptomys borealis***
Rodentia - Muridae
e Northern Bog Lemming
d Nördliche Lemmingmaus
f Campagnol-lemming boréal; Lemming du Nord
i Lemming del Nord

6760 ***Synaptomys cooperi***
Rodentia - Muridae
e Southern Bog Lemming
d Südliche Lemmingmaus
f Campagnol-lemming de Cooper; Lemming du Sud
i Lemming del Sud

6761 ***Syncerus***
Artiodactyla - Bovidae

6762 ***Syncerus caffer***
Artiodactyla - Bovidae
e African Buffalo; Cape Buffalo
d Afrikanischer Büffel; Steppenbüffel; Kaffernbüffel
f Buffle africain; Buffle d'Afrique
i Bufalo cafro; Bufalo nero

6763 ***Syncerus caffer aequinoctialis***
Artiodactyla - Bovidae
e Northeastern Buffalo; Nile Buffalo
d Nil-Büffel
f Buffle équinoxial

6764 ***Syncerus caffer brachyceros***
Artiodactyla - Bovidae
e West African Buffalo; Northwestern Buffalo
d Grasbüffel; Sudan-Büffel
f Buffle de savane

6765 ***Syncerus caffer caffer***
Artiodactyla - Bovidae
e Southern Buffalo; Savanna Buffalo; Savannah Buffalo
d Eigentlicher Kaffernbüffel; Kaffernbüffel; Schwarzbüffel
f Buffle noir; Buffle de Cafrerie
i Bufalo del Capo; Bufalo rosso

6766 ***Syncerus caffer nanus***
Artiodactyla - Bovidae
e Red Buffalo; Forest Buffalo; Dwarf Buffalo; Congo Buffalo; Dwarf Forest Buffalo; Red Forest Buffalo
d Rotbüffel; Waldbüffel
f Bufle de forêt; Buffle rouge
i Bufalo rosso di foresta; Bufalo nano

Syndesmotis syn. *Hipposideros* q.v.

Syntheosciurus syn. *Sciurus* q.v.

T

6767 Tachyglossidae
Tachyglossa
e Spiny Anteaters; Echidnas; Spiny Echidnas
d Ameisenigel ; Kurzschnabeligel ; Schnabeligel
f Échidnés; Tachyglossidés
i Tachyglossidi; Echidne

6768 *Tachyglossus*
Tachyglossa - Tachyglossidae
e Porcupine Anteaters; Short-beaked Spiny Anteaters; Short-nosed Echidnas; Short-beaked Echidnas
d Australien-Kurzschnabeligel ; Australische Kurzschnabeligel ; Australische Stacheligel
f Échidnés; Échidnés à bec droit
i Echidne

6769 *Tachyglossus aculeatus*
Tachyglossa - Tachyglossidae
e Australian Echidna; Australian Short-nosed Echidna; Spiny Anteater; Australian Spiny Anteater; Short-nosed Echidna; Short-beaked Echidna (ANZ)
d Australischer Kurzschnabeligel; Kurzschnabeligel
f Échidné d'Australie
i Echidna; Echidna istrice

6770 *Tachyglossus aculeatus multiaculeatus*
Tachyglossa - Tachyglossidae
e Kangaroo Island Echidna
f Échidne à nez court épineux

6771 *Tachyglossus aculeatus setosus*
Tachyglossa - Tachyglossidae
e Tasmanian Echidna
d Tasmanien-Kurzschnabeligel
f Échidnéde Tasmanie
i Echidna della Tasmania

6772 *Tachyoryctes*
Rodentia - Muridae
e African Molerats; East African Molerats; Root Rats
d Maulwurfsratten
f Rats-taupes africains
i Ratti talpa

6773 *Tachyoryctes ankoliae*
Rodentia - Muridae
e Ankole Molerat

6774 *Tachyoryctes annectens*
Rodentia - Muridae
e Mianzini Molerat

6775 *Tachyoryctes audax*
Rodentia - Muridae
e Audacious Molerat

6776 *Tachyoryctes daemon*
Rodentia - Muridae
e Demon Molerat; Tanzanian Mole Rat
d Afrikanische Maulwurfsratte

6777 *Tachyoryctes macrocephalus*
Rodentia - Muridae
e Big-headed Molerat
d Riesenmaulwurfsratte
i Ratto talpa africano gigante; Ratto talpa gigante

6778 *Tachyoryctes naivashae*
Rodentia - Muridae
e Naivasha Molerat

6779   ***Tachyoryctes rex***
     Rodentia - Muridae
e   King Molerat

6780   ***Tachyoryctes ruandae***
     Rodentia - Muridae
e   Ruanda Molerat; Rwanda Molerat

6781   ***Tachyoryctes ruddi***
     Rodentia - Muridae
e   Rudd's Molerat

6782   ***Tachyoryctes spalacinus***
     Rodentia - Muridae
e   Embi Molerat

6783   ***Tachyoryctes splendens***
     Rodentia - Muridae
e   East African Molerat
d   Ostafrikanische Maulwurfsratte

6784   ***Tadarida***
     Chiroptera - Molossidae
e   Tadarine Bats; Free-tailed Bats;
     Wrinkle-lipped Bats; Mastiff Bats
d   Faltlippenfledermäuse;
     Faltenlippenfledermäuse;
     Bulldoggfledermäuse
f   Tadarides; Molosses

6785   ***Tadarida aegyptiaca***
     Chiroptera - Molossidae
e   Egyptian Free-tailed Bat; Egyptian
     Wrinkle-lipped Bat; Indian Wrinkle-
     lipped Bat
d   Indische Krauslippenfledermaus
f   Molosse d'Égypte; Oreillard
     d'Égypte; Molosse du désert

Tadarida aloysiiabaudiae syn.
Chaerephon aloysiisabaudiae q.v.

Tadarida ansorgei syn. *Chaerephon
ansorgei* q.v.

6786   ***Tadarida australis***

     Chiroptera - Molossidae
e   Australian Molossus; Southern
     Mastiff Bat; White-striped Mastiff
     Bat

Tadarida bemmeleni syn.
Chaerephon bemmeleni q.v.

Tadarida bivittata syn. *Chaerephon
bivittata* q.v.

6787   ***Tadarida brasiliensis***
     Chiroptera - Molossidae
e   Geoffroy's Little Free-tailed Bat
d   Mexikanische Bulldogfledermaus
f   Tadaride du Brésil

6788   ***Tadarida brasiliensis cynocephala***
     Chiroptera - Molossidae
e   Brazilian Free-tailed Bat

6789   ***Tadarida brasiliensis mexicana***
     Chiroptera - Molossidae
e   Guano Bat; Mexican Free-tailed Bat
d   Guano-Fledermaus

Tadarida chapini syn. *Chaerephon
chapini* q.v.

Tadarida espiritosantensis syn.
Nyctinomops laticaudatus q.v.

6790   ***Tadarida fulminans***
     Chiroptera - Molossidae
e   Madagascan Large Free-tailed Bat;
     Large Free-tailed Bat

Tadarida gallagheri syn.
Chaerephon galagheri q.v.

Tadarida jobensis syn. *Chaerephon
jobensis* q.v.

Tadarida johorensis syn.
Chaerephon johorensis q.v.

6791   *Tadarida kuboriensis*
           Chiroptera - Molossidae
  e   New Guinea Mastiff Bat

6792   *Tadarida latouchei*
           Chiroptera - Molossidae
  e   Latouche's Free-tailed Bat

6793   *Tadarida lobata*
           Chiroptera - Molossidae
  e   Big-eared Kenya Free-tailed Bat;
           Kenyan Big-eared Free-tailed Bat

           Tadarida major syn. *Chaerephon
           major* q.v.

           Tadarida nigeriae syn. *Chaerephon
           nigeriae* q.v.

           Tadarida plicata syn. *Chaerephon
           plicata* q.v.

           Tadarida pumila syn. *Chaerephon
           pumila* q.v.

           Tadarida russata syn. *Chaerephon
           russata* q.v.

6794   *Tadarida teniotis*
           Chiroptera - Molossidae
  e   European Free-tailed Bat
  d   Bulldoggfledermaus;
           Faltlippenfledermaus;
           Faltenlippenfledermaus; Europäische
           Bulldoggfledermaus
  f   Molosse de Cestoni; Tadaride
           bouledogue
  i   Molosso del Cestoni; Molosso di
           Cestoni

6795   *Tadarida ventralis*
           Chiroptera - Molossidae
  e   African Giant Free-tailed Bat

6796   *Taeromys*
           Rodentia – Muridae

  e   Sulawesi Rats

6797   *Taeromys arcuatus*
           Rodentia - Muridae
  e   Salokko Rat

6798   *Taeromys callitrichus*
           Rodentia - Muridae
  e   Lovely-haired Rat

6799   *Taeromys celebensis*
           Rodentia - Muridae
  e   Celebes Rat; Sulawesi Rat

6800   *Taeromys hamatus*
           Rodentia - Muridae
  e   Sulawesi Montane Rat

6801   *Taeromys punicans*
           Rodentia - Muridae
  e   Sulawesi Forest Rat

6802   *Taeromys taerae*
           Rodentia - Muridae
  e   Tondano Rat

6803   *Talpa*
           Erinaceomorpha - Talpidae
  e   Moles; Eurasian Moles; Old World
           Moles; Common Old World Moles
  d   Maulwürfe
  f   Taupes
  i   Talpe

6804   *Talpa altaica*
           Erinaceomorpha - Talpidae
  e   Siberian Mole
  d   Sibirischer Maulwurf; Altai-
           Maulwurf
  f   Taupe de Sibérie

6805   *Talpa caeca*
           Erinaceomorpha - Talpidae
  e   Blind Mole; Mediterranean Mole

d	Blindmaulwurf; Blinder Maulwurf
f	Taupe aveugle
i	Talpa cieca

6806   *Talpa caucasica*
Erinaceomorpha - Talpidae
e   Caucasian Mole
d   Kaukasischer Maulwurf
f   Taupe du Caucase

6807   *Talpa etigo*
Erinaceomorpha - Talpidae
e   Echigo Mole; Echigo Plain Mole

6808   *Talpa europaea*
Erinaceomorpha - Talpidae
e   Common Mole; Mole; Common Eurasian Mole; European Mole; Northern Mole

d   Europäischer Maulwurf; Eurasischer Maulwurf; Gemeiner Maulwurf; Maulwurf
f   Taupe; Taupe commune; Taupe d'Europe; Taupe européenne; Taupe ordinaire; Taupe vulgaire
i   Talpa europea; Talpa; Talpa comune; Talpa illuminata

6809   *Talpa insularis*
Erinaceomorpha - Talpidae
e   Insular Mole

Talpa klossi syn. *Euroscaptor klossi* q.v.

6810   *Talpa kobeae*
Erinaceomorpha - Talpidae
e   Kobe Mole; Large Japanese Mole

6811   *Talpa leucura*
Erinaceomorpha - Talpidae
e   White-tailed Mole
d   Weißschwanzmaulwurf

6812   *Talpa levantis*

Erinaceomorpha - Talpidae
e   Levantine Mole; Black Sea Mole
d   Levantinischer Maulwurf; Schwarzmeer-Maulwurf
f   Taupe du Levant

Talpa longirostris syn. *Euroscaptor longirostris* q.v.

6813   *Talpa minor*
Erinaceomorpha - Talpidae
e   Small Japanese Mole

Talpa mizura syn. *Euroscaptor mizura* q.v.

Talpa moschatus syn. *Scaptochirus moschatus* q.v.

6814   *Talpa occidentalis*
Erinaceomorpha - Talpidae
e   Iberian Mole; Iberian Blind Mole
d   Iberischer Maulwurf
f   Taupe ibérique; Taupe d'Espagne
i   Talpa occidentale

6815   *Talpa robusta*
Erinaceomorpha - Talpidae
e   Greater Japanese Mole; Large Mole

6816   *Talpa romana*
Erinaceomorpha - Talpidae
e   Roman Mole
d   Römischer Maulwurf
f   Taupe romaine
i   Talpa romana

6817   *Talpa stankovici*
Erinaceomorpha - Talpidae
e   Stankovic's Mole; Balkan Mole
d   Balkan-Maulwurf
f   Taupe orientale; Taupe des Balkans; Taupe de Stankovic
i   Talpa dei Balcani

6818 *Talpa streeti*
Erinaceomorpha - Talpidae
e Persian Mole
d Persischer Maulwurf
f Taupe de Perse

6819 *Talpa tokudae*
Erinaceomorpha - Talpidae
e Tokuda's Mole; Large Sadu Island Mole

6820 *Talpa wogura*
Erinaceomorpha - Talpidae
e Japanese Mole; Lesser Japanese Mole
d Japanischer Maulwurf

6821 **Talpidae**
Erinaceomorpha
e Moles; Moles, Desmans and Shrew Moles
d Maulwürfe; Landmaulwürfe
f Talpidés
i Talpidi

6822 *Tamandua*
Pilosa - Myrmecophagidae
e Tamanduas; Collared Anteaters
d Tamanduas
f Tamanduas; Tamanduas à quatre doigts
i Tamandua

Tamandua longicaudata syn. *T. tetradactyla* q.v.

6823 *Tamandua mexicana*
Pilosa - Myrmecophagidae
e Northern Tamandua; Collared Anteater; Tamandua; Banded Anteater; Vested Anteater
d Nördlicher Tamandua
f Fourmillier à collier
i Tamandua del Messico

6824 *Tamandua tetradactyla*
Pilosa - Myrmecophagidae
e Tamandua; Lesser Anteater; Southern Tamandua
d Tamandua; Caguare; Kleiner Ameisenbär; Südlicher Tamandua
f Tamandua à quatre doigts
i Tamandua tetradattilo; Formichiere minore; Tamandua

6825 *Tamandua tetradactyla chapadensis*
Pilosa - Myrmecophagidae
e Matto GrossoTamandua

6826 *Tamias*
Rodentia - Sciuridae
e Eastern Chipmunks (NA); Eastern American Chipmunks
d Chipmunks; Streifenhörnchen ; Backenhörnchen
f Tamias; Néotamias; Chipmunks
i Tamia

6827 *Tamias alpinus*
Rodentia - Sciuridae
e Alpine Chipmunk
d Gebirgschipmunk
f Néotamia de montagne

6828 *Tamias amoenus*
Rodentia - Sciuridae
e Yellow-pine Chipmunk
d Gelber Fichtenchipmunk
f Néotamia jaune

6829 *Tamias bulleri*
Rodentia - Sciuridae
e Buller's Chipmunk
d Buller-Chipmunk

6830 *Tamias canipes*
Rodentia - Sciuridae
e Grey-footed Chipmunk
d Graufußchipmunk
f Tamia aux pieds grises

6831   *Tamias cinereicollis*
           Rodentia - Sciuridae
  e   Grey-collared Chipmunk
  d   Grauhalschipmunk
  f   Tamia à col gris

6832   *Tamias dorsalis*
           Rodentia - Sciuridae
  e   Cliff Chipmunk
  d   Felsenchipmunk

6833   *Tamias durangae*
           Rodentia - Sciuridae
  e   Durango Chipmunk
  d   Durango-Chipmunk

6834   *Tamias merriami*
           Rodentia - Sciuridae
  e   Merriam's Chipmunk
  d   Merriam-Chipmunk

6835   *Tamias minimus*
           Rodentia - Sciuridae
  e   Least Chipmunk; Western Chipmunk
  d   Kleiner Chipmunk; Kleines
        Streifenhörnchen; Kleines
        Backenhörnchen
  f   Tamia mineur; Néotamia nain;
        Chipmunk nain

6836   *Tamias minimus astristriatus*
           Rodentia - Sciuridae
  e   Penasco Chipmunk; Black-striped
        Least Chipmunk; Black-striped
        Western Chipmunk
  d   Schwarzgestreiftes Kleines
        Streifenhörnchen; Schwarzgestreiftes
        Kleines Backenhörnchen;
        Schwarzgestreifter Kleiner
        Chipmunk
  f   Chipmunk nain à raie noir; Néotamia
        nain à raie noir

6837   *Tamias obscurus*
           Rodentia - Sciuridae
  e   Baja California Chipmunk;

       California Chipmunk
  d   Kalifornischer Chipmunk

6838   *Tamias obscurus davisi*
           Rodentia - Sciuridae
  e   Chaparral Chipmunk

6839   *Tamias ochrogenys*
           Rodentia - Sciuridae
  e   Yellow-cheeked Chipmunk
  d   Gelbwangenchipmunk

6840   *Tamias palmeri*
           Rodentia - Sciuridae
  e   Palmer's Chipmunk
  d   Palmer-Chipmunk

6841   *Tamias panamintinus*
           Rodentia - Sciuridae
  e   Panamint Chipmunk
  d   Panamint-Chipmunk

6842   *Tamias quadrimaculatus*
           Rodentia - Sciuridae
  e   Long-eared Chipmunk
  d   Langohrchipmunk

6843   *Tamias quadrivittatus*
           Rodentia - Sciuridae
  e   Colorado Chipmunk
  d   Colorado-Chipmunk
  f   Néotamia du Colorado

6844   *Tamias ruficaudus*
           Rodentia - Sciuridae
  e   Red-tailed Chipmunk
  d   Rotschwanzchipmunk
  f   Tamia à queue rousse

6845   *Tamias rufus*
           Rodentia - Sciuridae
  e   Hopi Chipmunk
  d   Hopi-Chipmunk

6846 *Tamias senex*
Rodentia - Sciuridae
e Allen's Chipmunk; Shadow
Chipmunk
d Allen-Chipmunk

6847 *Tamias sibiricus*
Rodentia - Sciuridae
e Siberian Chipmunk; Siberian Ground
Squirrel
d Burunduk; Sibirisches
Streifenhörnchen; Asiatisches
Streifenhörnchen; Edles
Backenhörnchen; Eurasisches
Erdhörnchen
f Écureuil terrestre de Sibérie; Tamia
de Sibérie; Scoiattolo giapponese
i Tamia siberiano; Topo di Russia;
Topo di Siberia; Burunduki

6848 *Tamias siskiyou*
Rodentia - Sciuridae
e Siskiyou Chipmuink
d Siskiyou-Chipmunk

6849 *Tamias sonomae*
Rodentia - Sciuridae
e Sonoma Chipmunk
d Sonoma-Chipmunk

6850 *Tamias speciosus*
Rodentia - Sciuridae
e Lodgepole Chipmunk

6851 *Tamias speciosus frater*
Rodentia - Sciuridae
e Mount Pinos Cchipmunk

6852 *Tamias striatus*
Rodentia - Sciuridae
e Eastern Chipmunk (NA); Eastern
American Chipmunk
d Streifenbackenhörnchen; Östlicher
Chipmunk; Backenhörnchen;
Östliches Streifenhörnchen

f Tamia strié; Tamia strié de
l'Amérique de l'Est
i Tamia striato; Cipmunk

6853 *Tamias striatus griseus*
Rodentia - Sciuridae
e Grey Chipmunk

6854 *Tamias townsendi*
Rodentia - Sciuridae
e Townsend's Chipmunk; Siskiyou
Chipmunk; Redwood Chipmunk
d Townsend-Chipmunk
f Tamia de Townsend; Néotamia de
Townsend

6855 *Tamias umbrinus*
Rodentia - Sciuridae
e Uinta Chipmunk; Colorado
Chipmunk
d Uinta-Chipmunk

6856 *Tamiasciurus*
Rodentia - Sciuridae
e Chickarees (NA); American Red
Squirrels; Red Squirrels (NA)
d Nordamerikanische Rothörnchen ;
Chickarees; Rothörnchen
f Écureuils roux d'Amérique

6857 *Tamiasciurus douglasii*
Rodentia - Sciuridae
e Douglas's Squirrel; Chickaree (NA)
d Chickaree; Douglas-Hörnchen
f Écureuil de Douglas

Tamiasciurus fremonti syn. *T.*
hudsonicus q.v.

6858 *Tamiasciurus hudsonicus*
Rodentia - Sciuridae
e American Red Squirrel; Eastern Red
Squirrel (NA); Red Squirrel (NA)
d Rothörnchen; Hudson-Hörnchen;
Gemeines Rothörnchen;
Amerikanisches Rothörnchen

f   Écureuil roux; Écureuil de l'Hudson;
      Écureuil du Canada
i   Scoiattolo rosso del Nordamerica

**6859   *Tamiasciurus hudsonicus
          grahamensis***
          Rodentia - Sciuridae
e   Mount Graham Red Squirrel
d   Mount-Graham-Rothörnchen

6860   *Tamiasciurus mearnsi*
          Rodentia - Sciuridae
e   Mearns's Squirrel
d   Mearns-Rothörnchen

6861   *Tamiops*
          Rodentia - Sciuridae
e   Striped Himalajan and Burmese
      Squirrels; Asiatic Striped Squirrels;
      Himalajan Striped Squirrels
d   Baumstreifenhörnchen
f   Écureuils rayés

6862   *Tamiops macclellandi*
          Rodentia - Sciuridae
e   Himalayan Striped Squirrel
d   Himalaja-Streifenhörnchen;
      Himalaja-Zwergstreifenhörnchen

6863   *Tamiops maritimus*
          Rodentia - Sciuridae
e   Maritime Striped Squirrel; Coastal
      Striped Squirrel
d   Küstenstreifenhörnchen

6864   *Tamiops rodolphei*
          Rodentia - Sciuridae
e   Cambodian Striped Squirrel
d   Kambodscha-Streifenhörnchen

6865   *Tamiops swinhoei*
          Rodentia - Sciuridae
e   Swinhoe's Striped Squirrel
d   Swinhoe-Schönhörnchen
f   Écureuil de tibet

6866   *Tapecomys*
          Rodentia - Muridae
e   Tapecua Rats

6867   *Tapecomys primus*
          Rodentia - Muridae
e   Tapecua Rat

      Tapeti syn. *Sylvilagus* q.v.

6868   *Taphozous*
          Chiroptera - Emballonuridae
e   Tomb Bats; Pouched Bats;
      Taphozouses
d   Grabflatterer ; Grabfledermäuse
f   Taphiens; Chauve-souris des tombes
i   Pipistrelli delle tombe

6869   *Taphozous achates*
          Chiroptera - Emballonuridae
e   Brown-bearded Tomb Bat

6870   *Taphozous australis*
          Chiroptera - Emballonuridae
e   Australian Taphozous; Little Free-
      tailed Bat (ANZ); Gould's Pouched
      Bat; Coastal Tomb Bat; Coastal
      Sheathtail Bat

      Taphozous capito syn. *Saccolaimus
      pluto* q.v.

      Taphozous flaviventris syn.
      Saccolaimus flaviventris q.v.

6871   *Taphozous georgianus*
          Chiroptera - Emballonuridae
e   Sharp-nosed Pouched Bat; Sharp-
      nosed Tomb Bat; Common Sheathtail
      Bat (ANZ)

6872   *Taphozous hamiltoni*
          Chiroptera - Emballonuridae
e   Hamilton's Tomb Bat

6873 **Taphozous hildegardeae**
Chiroptera - Emballonuridae
e Hidegarde's Tomb Bat

6874 **Taphozous hilli**
Chiroptera - Emballonuridae
e Hill's Pouched Bat; Hill's Tomb Bat

6875 **Taphozous kapalgensis**
Chiroptera - Emballonuridae
e White-striped Sheath-tailed Bat;
Arnhem Tomb Bat

6876 **Taphozous longimanus**
Chiroptera - Emballonuridae
e Long-winged Tomb Bat; Long-armed
Sheath-tailed Bat

6877 **Taphozous mauritianus**
Chiroptera - Emballonuridae
e Mauritian Tomb Bat
d Mauritianischer Grabflatterer
f Taphien de Maurice; Taphien de l'île
Maurice; Chauve-souris banane
i Pipistrello delle tombe delle
Mauritius

6878 **Taphozous melanopogon**
Chiroptera - Emballonuridae
e Black-bearded Tomb Bat; Black-
bearded Sheath-tailed Bat

Taphozous mixtus syn. *Saccolaimus
mixtus* q.v.

Taphozous nudicluniatus syn.
Saccolaimus saccolaimus q.v.

6879 **Taphozous nudiventris**
Chiroptera - Emballonuridae
e Naked-rumped Bat; Naked-rumped
Tomb Bat; Naked-bellied Bat;
Naked-bellied Tomb Bat; Sheath-
tailed Bat
d Nacktbauchiger Grabflatterer
f Taphien à ventre nu

Taphozous peli syn. *Saccolaimus
peli* q.v.

6880 **Taphozous perforatus**
Chiroptera - Emballonuridae
e Egyptian Tomb Bat
f Taphien peforé

6881 **Taphozous philippinensis**
Chiroptera - Emballonuridae
e Philippine Tomb Bat

Taphozous pluto syn. *Saccolaimus
pluto* q.v.

6882 **Taphozous theobaldi**
Chiroptera - Emballonuridae
e Theobalds Tomb Bat

6883 **Taphozous troughtoni**
Chiroptera - Emballonuridae
e Troughton's Tomb Bat

6884 **Tapiridae**
Perissodactyla
e Tapirs
d Tapire
f Tapirs; Tapiridés
i Tapiridi

6885 **Tapirus**
Perissodactyla - Tapiridae
e Tapirs
d Tapire
f Tapirs
i Tapiri

6886 **Tapirus bairdii**
Perissodactyla - Tapiridae
e Central American Tapir; Baird's
Tapir
d Mittelamerikanischer Tapir
f Tapir de Baird
i Tapiro di Baird

6887   Tapirus indicus
Perissodactyla - Tapiridae
e   Malayan Tapir; Indian Tapir
d   Schabrackentapir
f   Tapir de l'Inde; Tapir à chabraque
i   Tapiro dalla Gualdrappa

6888   Tapirus pinchaque
Perissodactyla - Tapiridae
e   Mountain Tapir; Woolly Tapir;
Woolly Mountain Tapir; Andean
Tapir
d   Bergtapir
f   Tapir des Indes; Tapir de Roulin
i   Tapiro di montagna

Tapirus roulini syn. *T. pinchaque*
q.v.

6889   Tapirus terrestris
Perissodactyla - Tapiridae
e   Lowland Tapir; Brazilian Tapir;
South American Tapir; Amazonian
Tapir; Amazon Tapir
d   Flachlandtapir; Südamerikanischer
Tapir
f   Tapir terrestre; Tapir d'Amérique
i   Tapiro sudamericano; Tapiro
americano; Tapiro del Brasile; Tapiro
comuune

6890   Tapirus terrestris spegazzinii
Perissodactyla - Tapiridae
e   Spegazzini's Tapir
i   Tapiro di Spegazzini

6891   Tarsiidae
Primates
e   Tarsiers
d   Koboldmakis; Fußwurzeltiere
f   Tarsiidés
i   Tarsidi; Tarsi

6892   Tarsipedidae
Diprotodontia
e   Honey Phalangers; Honey Possums;

Noolbenders (ANZ); Noolbengers
(ANZ)
d   Rüsselbeutler
f   Souris à miel
i   Tarsipedidi

6893   Tarsipes
Diprotodontia - Tarsipedidae
e   Honey Possums (ANZ); Honey
Phalangers; Long-snouted
Phalangers; Honey Opossums
d   Honigbeutler ; Rüsselbeutler
f   Souris à miel
i   Possum del miele; Topi del miele

6894   Tarsipes rostratus
Diprotodontia - Tarsipedidae
e   Honey Phalanger; Honey Possum
(ANZ); Long-snouted Phalanger;
Honey Opossum; Slender-nosed
Honey Possum; Honey Mouse;
Noolbenger (ANZ); Noolbender
(ANZ)

d   Honigbeutler; Rüsselbeutler
f   Souris à miel
i   Possum del miele; Topo del miele

Tarsipes spencerae syn. *T. rostratus*
q.v.

6895   Tarsius
Primates - Tarsiidae
e   Tarsiers
d   Fußwurzeltiere; Koboldmakis
f   Tarsiers
i   Tarsi

6896   Tarsius bancanus
Primates - Tarsiidae
e   Malayan Tarsier; Western Tarsier;
Horsfield's Tarsier
d   Sunda-Koboldmaki
f   Tarsier de Horsfield
i   Tarsio malese; Tarsio delle Filippine

6897   *Tarsius bancanus borneanus*
Primates - Tarsiidae
e   Bornean Tarsier
d   Borneo-Koboldmaki

6898   *Tarsius dianae*
Primates - Tarsiidae
e   Diana Tarsier; Dian's Tarsier
d   Diana-Koboldmaki
f   Tarsier pygmée des montagnes
i   Tarsio di Diana

6899   *Tarsius pelegensis*
Primates - Tarsiidae
e   Peleng Tarsier

6900   *Tarsius pumilus*
Primates - Tarsiidae
e   Pygmy Tarsier; Lesser Spectral Tarsier
d   Zwergkoboldmaki
f   Tarsier pygmée
i   Tarsio nano

6901   *Tarsius sangirensis*
Primates - Tarsiidae
e   Sangihe Islands Tarsier; Sangihe Tarsier
d   Sangihe-Koboldmaki
i   Tarsio spettro delle isole Sangihe

6902   *Tarsius spectrum*
Primates - Tarsiidae
e   Eastern Tarsier; Celebes Tarsier; Dusky-handed Tarsier; Spectral Tarsier; Yellow-bearded Tarsier; Celebesian Tarsier; Spectacled Tarsier
d   Celebes-Koboldmaki
f   Tarsier spectre
i   Tarsio spettro

6903   *Tarsius syrichta*
Primates - Tarsiidae
e   Philippine Tarsier

d   Philippinen-Koboldmaki
f   Tarsier des Philippines
i   Tarsio delle Filippine; Tarsio del Borneo

6904   *Tarsius syrichta carbonarius*
Primates - Tarsiidae
e   Mindanao Tarsier
d   Mindanao-Koboldmaki

6905   *Tarsomys*
Rodentia - Muridae
e   Long-footed Rats

6906   *Tarsomys apoensis*
Rodentia - Muridae
e   Long-footed Rat

6907   *Tarsomys echinatus*
Rodentia - Muridae
e   Spiny Long-footed Rat

6908   *Tasmacetus*
Cete - Hyperoodontidae
e   Tasmanian Beaked Whales; Shepherd's Beaked Whales
d   Shepherd-Wale
f   Tasmacètes
i   Tamasceti

6909   *Tasmacetus shepherdi*
Cete - Hyperoodontidae
e   Tasmanian Beaked Whale; Shepherd's Beaked Whale; Tasman Whale; Tasman Beaked Whale; Shepherd's Whale
d   Shepherd-Wal
f   Tasmacète de Shepherd
i   Tasmaceto

6910   *Tateomys*
Rodentia - Muridae
e   Tate's Rats; Greater Sulawesian Shrewrats
d   Tateomys-Spitzmausratten

6911   *Tateomys macrocercus*
           Rodentia - Muridae
e   Long-tailed Shrewrat

6912   *Tateomys rhinogradoides*
           Rodentia - Muridae
e   Tate's Rat; Tate's Shrewrat

6913   *Tatera*
           Rodentia - Muridae
e   Large Gerbils; Naked-soled Gerbils;
           Taterine Gerbils; Rat-like Gerbils;
           Large Naked-soled Gerbils
d   Nacktsohlenrennmäuse
f   Gerbilles africaines

6914   *Tatera afra*
           Rodentia - Muridae
e   Cape Gerbil; Large Naked-sole
           Gerbil
d   Kap-Renmaus
f   Gerbille du Cap

6915   *Tatera boehmi*
           Rodentia - Muridae
e   Böhm's Gerbil; King Gerbil
f   Gerbille de Boehm

6916   *Tatera brantsii*
           Rodentia - Muridae
e   Highveld Gerbil; Brant's Gerbil
d   Hochlandrennmaus
f   Gerbille de Brant

           Tatera gambiana syn. *T. valida* q.v.

6917   *Tatera guineae*
           Rodentia - Muridae
e   Guinea Gerbil
f   Gerbille du Guinée

6918   *Tatera inclusa*
           Rodentia - Muridae
e   Gorogonza Gerbil
d   Gorogonza-Rennmaus

f   Gerbille de Gorogoza

6919   *Tatera indica*
           Rodentia - Muridae
e   Indian Gerbil; Large Gerbil;
           Antelope Rat; Kangaroo Rat; Sand
           Rat
d   Indische Nacktsohlenrennmaus
f   Gerbille des Indes
i   Gerbillo indiano

6920   *Tatera indica ceylonica*
           Rodentia - Muridae
e   Ceylon Gerbil
f   Gerbille de Ceylan

6921   *Tatera kempi*
           Rodentia - Muridae
e   Kemp's Gerbil
f   Gerbille de Kemp

6922   *Tatera leucogaster*
           Rodentia - Muridae
e   Bushveld Gerbil
d   Buschveldt-Rennmaus
f   Gerbille du bushveld

6923   *Tatera nigricauda*
           Rodentia - Muridae
e   Black-tailed Gerbil
f   Gerbille à queue noire

6924   *Tatera phillipsi*
           Rodentia - Muridae
e   Phillips's Gerbil

6925   *Tatera robusta*
           Rodentia - Muridae
e   Fringe-tailed Gerbil
d   Nacktsohlenrennmaus

6926   *Tatera valida*
           Rodentia - Muridae
e   Savanna Gerbil; Bocage's Gerbil;
           Gambian Gerbil; Savannah Gerbil;

Togo Savanna Gerbil
d Kleine Togo-Nacktsohlenrennmaus
f Gerbille de savane

6927 Taterillus
Rodentia - Muridae
e Naked-sole Gerbils; Small Naked-
sole Gerbils
d Kleine Nacktsohlenrennmäuse
f Gerbilles

6928 Taterillus arenarius
Rodentia - Muridae
e Sahel Gerbil

6929 Taterillus congicus
Rodentia - Muridae
e Congo Gerbil

6930 Taterillus emini
Rodentia - Muridae
e Emin's Gerbil
d Kleine Nacktsohlenrennmaus;
Antilopenmaus
f Gerbille d'Emin

6931 Taterillus gracilis
Rodentia - Muridae
e Slender Gerbil

6932 Taterillus harringtoni
Rodentia - Muridae
e Harrington's Gerbil
d Harrington-Nacktsohlenrennmaus

6933 Taterillus lacustris
Rodentia - Muridae
e Lake Chad Gerbil

6934 Taterillus petteri
Rodentia - Muridae
e Petter's Gerbil
f Gerbille du Burkina; Gerbille du
Burkina Faso

6935 Taterillus pygargus
Rodentia - Muridae
e Senegal Gerbil
d Senegal-Nacktsohlenrennmaus

Taterona syn. *Tatera* q.v.

6936 Taurotragus
Artiodactyla - Bovidae
e Elands
d Elenantilopen
f Élands

6937 Taurotragus derbianus
Artiodactyla - Bovidae
e Giant Eland; Lord Derby's Eland;
Lord Derby's Giant Eland; Derby
Eland
d Riesenelen; Riesenelenantilope
f Éland de Derby; Éland géant
i Eland gigante

6938 Taurotragus derbianus derbianus
Artiodactyla - Bovidae
e Western Giant Eland
d Westliches Riesenelen; Westsudan-
Riesenelen
f Éland de Derby
i Eland di Lord Derby

6939 Taurotragus derbianus gigas
Artiodactyla - Bovidae
e Eastern Giant Eland
d Ostsudan-Riesenelen; Östliches
Riesenelen

6940 Taurotragus oryx
Artiodactyla - Bovidae
e Eland; Common Eland
d Elenantilope
f Antilope canna; Éland du Cap
i Antilope alcina; Eland comune;
Eland

6941   *Taurotragus oryx livingstoni*
Artiodactyla - Bovidae
e   Livingstone's Eland

6942   *Taurotragus oryx oryx*
Artiodactyla - Bovidae
e   Southern Eland; Cape Eland

6943   *Taurotragus oryx pattersonianus*
Artiodactyla - Bovidae
e   East African Eland

6944   *Taxidea*
Carnivora - Mustelidae
e   American Badgers
d   Amerikanische Dachse;
        Amerikanische Silberdachse;
        Silberdachse
f   Blaireaux nord-américains
i   Tassi; Tassi americani

6945   *Taxidea taxus*
Carnivora - Mustelidae
e   American Badger; Badger (NA)
d   Amerikanischer Dachs;
        Amerikanischer Silberdachs;
        Nordamerikanischer Dachs;
        Silberdachs
f   Blaireau nord-américain; Blaireau
        d'Amérique
i   Tasso americano; Tasso argentato;
        Tasso del Canada

6946   *Taxidea taxus jeffersoni*
Carnivora - Mustelidae
e   British Columbia Badger

6947   *Tayassu*
Artiodactyla - Tayassuidae
e   Jabalinas; Javelinas; White-lipped
        Peccaries
d   Nabelschweine; Pekaris;
        Weißbartpekaris
f   Pécaris; Cochons d'Amérique
i   Pecari

Tayassu albirostris syn. *Dicotyles tacaju* q.v.

6948   *Tayassu pecari*
Artiodactyla - Tayassuidae
e   White-lipped Peccary; Collared
        Peccary
d   Bisamschwein; Weißbartpekari
f   Pécari à barbe blanche
i   Pecari labiato

Tayassu tajacu syn. *Dicotyles tajacu* q.v.

6949   *Tayassuidae*
Artiodactyla
e   Peccaries; Javelinas
d   Nabelschweine; Pekaris
f   Pécaris; Tayassuidés
i   Pecari; Taiassuidi

Tayra syn. *Eira* q.v.

Teanopus syn. *Neotoma* q.v.

Tenes syn. *Sciurus* q.v.

6950   *Tenrec*
Soricomorpha - Tenrecidae
e   Common Tenrec; Common Tanrecs;
        Tanrecs; Tailless Tanrecs
d   Große Tanreks; Tanreks
f   Tanrecs
i   Tenrec

6951   *Tenrec ecaudatus*
Soricomorpha - Tenrecidae
e   Common Tenrec; Tanrec; Tailless
        Tenrec; Madagascar Hedgehog;
        Common Tanrec; Tanrec; Tailless
        Tanrec
d   Großer Tanrek
f   Tanrec
i   Tenrec comune

6952 **Tenrecidae**
Soricomorpha
e Tenrecs; Otter Shrews; Tanrecs
d Tanreks; Borstenigel;
Tenrekverwandte
f Tenrecidés
i Tenrecidi

Teonoma syn. *Neatoma* q.v.

Terricola syn. *Microtus* q.v.

6953 *Tetracerus*
Artiodactyla - Bovidae
e Four-horned Antelopes
d Vierhornantilopen
f Tétracères; Antilopes à quatre cornes
i Antilopi quadricorni

6954 *Tetracerus quadricornis*
Artiodactyla - Bovidae
e Four-horned Antelope; Nilgai
Antelope; Chousingha
d Vierhornantilope
f Antilope aux quatre cornes; Tétracère
i Antilope quadricorne; Chousingha

6955 *Thallomys*
Rodentia - Muridae
e Acacia Rats
d Akazienratten
f Rats des acacias

6956 *Thallomys loringi*
Rodentia - Muridae
e Loring's Rat

6957 *Thallomys nigricauda*
Rodentia - Muridae
e Black-tailed Tree Rat
d Gescheckte Akazienratte

6958 *Thallomys paedulcus*
Rodentia - Muridae

e Acacia Rat
d Afrikanische Akazienratte
f Petit rat des acacias

6959 *Thallomys shortridgei*
Rodentia - Muridae
e Shortridge's Rat

6960 *Thalpomys*
Rodentia - Muridae
e Cerrado Mice
d Cerrado-Mäuse

6961 *Thalpomys cerradensis*
Rodentia - Muridae
e Cerrado Mouse

6962 *Thalpomys lasiotis*
Rodentia - Muridae
e Hairy-eared Cerrado Mouse

6963 *Thamnomys*
Rodentia - Muridae
e Thicket Rats

Thamnomys cometes syn.
Grammomys cometes q.v.

Thamnomys dolichurus syn.
Grammomys dolichurus q.v.

6964 *Thamnomys kempi*
Rodentia - Muridae
e Kemp's Thicket Rat; Kemp's Forest
Rat

Thamnomys rutilans syn.
Grammomys rutilans q.v.

6965 *Thamnomys venustus*
Rodentia - Muridae
e African Forest Mouse; Charming
Thicket Rat

Thaptomys syn. *Akodon* q.v.

Thecurus syn. *Hystrix* q.v.

6966   Thelarctos
Carnivora - Ursidae
e   Polar Bears
d   Polarbären
f   Ours polaires
i   Orsi polari

6967   Thelarctos maritimus
Carnivora - Ursidae
e   Polar Bear; White Bear
d   Eisbär; Polarbär
f   Ours blanc; Ours polaire; Ours blanc
        polaire; Ours maritime
i   Orso polare; Orso bianco

6968   Theropithecus
Primates - Cercopithecidae
e   Geladas; Gelada Baboons
d   Hundsaffen; Dscheladas
f   Géladas; Théropithèques
i   Geladi

6969   Theropithecus gelada
Primates - Cercopithecidae
e   Gelada; Gelada Baboon
d   Hundsaffe; Blutbrustpavian;
        Dschelada
f   Théropithèque gélada; Gélada
i   Gelada

Thetamys syn. *Pseudomys* q.v.

6970   Thomasomys
Rodentia - Muridae
e   Thomas's Paramo Mice; Thomas's
        Oldfield Mice; Oldfield Mice
d   Thomas-Paramo-Mäuse

6971   Thomasomys apeco
Rodentia - Muridae
e   Apeco Oldfield Mouse

6972   Thomasomys aureus
Rodentia - Muridae
e   Golden Oldfield Mouse

6973   Thomasomys baeops
Rodentia - Muridae
e   Beady-eyed Mouse; Beady-eyed
        Oldfield Mouse

6974   Thomasomys bombycinus
Rodentia - Muridae
e   Silky Oldfield Mouse

6975   Thomasomys cinereiventer
Rodentia - Muridae
e   Ashy-bellied Oldfield Mouse

6976   Thomasomys cinereus
Rodentia - Muridae
e   Ash-coloured Oldfield Mouse

6977   Thomasomys daphne
Rodentia - Muridae
e   Daphne's Oldfield Mouse

6978   Thomasomys eleusis
Rodentia - Muridae
e   Peruvian Oldfield Mouse

Thomasomys fuscatus syn. *Aepomys
fuscatus* q.v.

6979   Thomasomys gracilis
Rodentia - Muridae
e   Slender Oldfield Mouse

6980   Thomasomys hylophilus
Rodentia - Muridae
e   Woodland Oldfield Mouse

6981   Thomasomys incanus
Rodentia - Muridae
e   Inca Oldfield Mouse

6982 *Thomasomys ischyurus*
Rodentia - Muridae
e   Strong-tailed Oldfield Mouse

6983 *Thomasomys kalinowskii*
Rodentia - Muridae
e   Kalinowski's Oldfield Mouse

6984 *Thomasomys ladewi*
Rodentia - Muridae
e   Ladew's Oldfield Mouse

6985 *Thomasomys laniger*
Rodentia - Muridae
e   Butcher Oldfield Mouse

Thomasomys lugens syn. *Aepomys lugens* q.v.

6986 *Thomasomys macrotis*
Rodentia - Muridae
e   Large-eared Oldfield Mouse

6987 *Thomasomys monochromos*
Rodentia - Muridae
e   Unicoloured Oldfield Mouse

6988 *Thomasomys niveipes*
Rodentia - Muridae
e   Snow-footed Oldfield Mouse

6989 *Thomasomys notatus*
Rodentia - Muridae
e   Distinguished Oldfield Mouse

6990 *Thomasomys onkiro*
Rodentia - Muridae
e   Onkiro Oldfield Mouse

6991 *Thomasomys oreas*
Rodentia - Muridae
e   Montane Oldfield Mouse

6992 *Thomasomys paramorum*

Rodentia - Muridae
e   Paramo Oldfield Mouse

6993 *Thomasomys pyrrhonotus*
Rodentia - Muridae
e   Thomas's Oldfield Mouse

Thomasomys pyrrhorhinos syn.
Wiedomys pyrrhorhinos q.v.

6994 *Thomasomys rhoadsi*
Rodentia - Muridae
e   Rhoads's Oldfield Mouse

6995 *Thomasomys rosalinda*
Rodentia - Muridae
e   Rosalinda Oldfield Mouse

6996 *Thomasomys silvestris*
Rodentia - Muridae
e   Forest Oldfield Mouse

6997 *Thomasomys taczanowskii*
Rodentia - Muridae
e   Taczanowski's Oldfield Mouse

6998 *Thomasomys vestitus*
Rodentia - Muridae
e   Dressy Oldfield Mouse

6999 *Thomomys*
Rodentia - Geomyidae
e   Western Pocket Gophers (NA);
Smooth-toothed Pocket Gophers
d   Gebirgstaschenratten
f   Gaufres à poche des montagnes;
Geomys

7000 *Thomomys bottae*
Rodentia - Geomyidae
e   Botta's Pocket Gopher; Valley Pocket
Gopher
d   Gebirgstaschenratte
f   Gaufre de Botta
i   Gopher di Botta

7001 *Thomomys bottae angularis*
Rodentia - Geomyidae
e Buena Vista Lake Pocket Gopher

7002 *Thomomys bulbivorus*
Rodentia - Geomyidae
e Camas Pocket Gopher
d Camas-Taschenratte
f Gaufre bulbivore
i Gopher gigante

7003 *Thomomys clusius*
Rodentia - Geomyidae
e Wyoming Pocket Gopher

7004 *Thomomys idahoensis*
Rodentia - Geomyidae
e Idaho Pocket Gopher

7005 *Thomomys mazama*
Rodentia - Geomyidae
e Western Pocket Gopher; Mazama
Pocket Gopher
d Mazama-Tachenratte

7006 *Thomomys monticola*
Rodentia - Geomyidae
e Mountain Pocket Gopher
d Sierra-Taschenratte

7007 *Thomomys talpoides*
Rodentia - Geomyidae
e Northern Pocket Gopher
d Nördliche Taschenratte
f Gaufre gris

7008 *Thomomys talpoides bullatus*
Rodentia - Geomyidae
e Sagebrush Pocket Gopher

7009 *Thomomys talpoides rufescens*
Rodentia - Geomyidae
e Dakota Pocket Gopher

7010 *Thomomys townsendii*
Rodentia - Geomyidae
e Townsend's Pocket Gopher

7011 *Thomomys umbrinus*
Rodentia - Geomyidae
e Southern Pocket Gopher
d Zwergtaschenratte

7012 *Thoopterus*
Chiroptera - Pteropodidae
e Short-nosed Fruit Bats; Swift Fruit
Bats
d Schwarzflügelflughunde

7013 *Thoopterus nigrescens*
Chiroptera - Pteropodidae
e Short-nosed Fruit Bat; Swift Fruit
Bat
d Schwarzflügelflughund

7014 *Thrichomys*
Rodentia - Echimyidae
e Punares
d Punare

7015 *Thrichomys apereoides*
Rodentia - Echimyidae
e Punare
d Punare

Thrichomys cunicularis syn. *T.
aperoides* q.v.

7016 *Thryonomys*
Rodentia - Thryonomyidae
e Ground Hogs; African Cane Rats;
African Spiny Rats; Cane Rats
d Borstenferkel; Rohrratten
f Aulacodes
i Trionomi

7017 *Thryonomys gregorianus*
Rodentia - Thryonomyidae
e Lesser Cane Rat

   d   Kleine Rohrratte
   f   Petit Aulacode; Aulacode; Rat des roseaux
   i   Trionomio gregoriano

7018   ***Thryonomys swinderianus***
        Rodentia - Thryonomyidae
   e   Great Cane Rat; Greater Cane Rat
   d   Große Rohrratte
   f   Grand aulacode; Rat-taupe
   i   Trionomio swinderiano

7019   **Thylacinidae**
        Dasyuromorphia
   e   Thylacines; Tasmanian Wolves
   d   Beutelwölfe
   f   Thylacinidés
   i   Tilacinidi

7020   ***Thylacinus***
        Dasyuromorphia - Thylacinidae
   e   Thylacines; Tasmanian Wolves; Tasmanian Pouched Wolves; Zebra Wolf
   d   Beutelwölfe
   f   Thylacins; Loups marsupiaux
   i   Lupi marsupiali

7021   ***Thylacinus cynocephalus***
        Dasyuromorphia - Thylacinidae
   e   Thylacine; Tasmanian Wolf; Tasmanian Pouched Wolf; Tasman Tiger
   d   Beutelwolf
   f   Loup marsupial; Thylacin
   i   Tilacino ; Lupo marsupiale

7022   ***Thylamys***
        Didelphimorphia - Didelphidae
   e   Fat-tailed Opossum
   d   Fettschwanzbeutelratten; Fettschwanzopossums

7023   ***Thylamys cinderella***
        Didelphimorphia - Didelphidae

   e   Cinderella Fat-tailed Opossum
   d   Cinderella-Fettschwanzbeutelratte

7024   ***Thylamys elegans***
        Didelphimorphia - Didelphidae
   e   Elegant Fat-tailed Opossum; Mouse Opossum
   d   Fettschwanzopossum

7025   ***Thylamys macrura***
        Didelphimorphia - Didelphidae
   e   Long-tailed Fat-tailed Opossum
   d   Langschwanzfettschwanzopossum

7026   ***Thylamys pallidior***
        Didelphimorphia - Didelphidae
   e   Pallid Fat-tailed Opossum
   d   Blasses Fettschwanzopossum

7027   ***Thylamys pusillus***
        Didelphimorphia - Didelphidae
   e   Small Fat-tailed Opossum
   d   Kleines Fettschwanzopossum

7028   ***Thylamys velutinus***
        Didelphimorphia - Didelphidae
   e   Velvety Fat-tailed Opossum

7029   ***Thylamys venustus***
        Didelphimorphia - Didelphidae
   e   Pretty Fat-tailed Opossum

7030   ***Thylogale***
        Marsupialia - Macropodidae
   e   Pademelon Wallabies; Scrub Wallabies; Pademelons; Tasmanian Wallaby
   d   Filander
   f   Thylogales; Wallabies de Tasmanie
   i   Tilogale

7031   ***Thylogale billardieri***
        Marsupialia - Macropodidae
   e   Tasmanian Pademelon (ANZ); Red-bellied Pademelon; Rufous-bellied

Pademelon; Red-bellied Wallaby
d   Rotbauchfilander
f   Wallaby de Billardier; Padmelon à
     ventre rouge

7032   *Thylogale brunii*
     Marsupialia - Macropodidae
e   Bruijn's Pademelon; Dusky Wallaby;
     Dusky Padmelon; Branded Wallaby;
     Common Bush Wallaby
d   Neuguinea-Filander
f   Wallaby de Bruijn; Padmelon à
     queue courte
i   Tilogale della Nova Guinea

7033   *Thylogale calabyi*
     Marsupialia - Macropodidae
e   Calaby's Padmelon

7034   *Thylogale stigmatica*
     Marsupialia - Macropodidae
e   Red-legged Pademelon; Northern
     Red-legged Pademelon (ANZ); Red-
     legged Scrub Wallaby
d   Rotbeinfilander
f   Padmelon aux pattes rouges
i   Tilogale dal ventre rosso

7035   *Thylogale thetis*
     Marsupialia - Macropodidae
e   Pademelon Wallaby; Red-necked
     Pademelon (ANZ)
d   Rothalsfilander; Rotnackenfilander
f   Padmelon à cou rouge

7036   *Thyroptera*
     Chiroptera - Thyropteridae
e   Disk-winged Bats; Sucker-footed
     Bats; New-world Sucker-footed Bats
d   Haftscheibenfledermäuse;
     Amerikanische
     Haftscheibenfledermäuse

7037   *Thyroptera discifera*
     Chiroptera – Thyropteridae

e   Honduran Disk-winged Bat; Peters's
     Disk-winged Bat; Brazilian Sucker-
     winged Bat
d   Honduras-Haftscheibenfledermaus
f   Vespertilion d'Honduras

7038   *Thyroptera lavali*
     Chiroptera - Thyropteridae
e   La Val's Disk-winged Bat

7039   *Thyroptera tricolor*
     Chiroptera - Thyropteridae
e   Spix's Disk-winged Bat; Brazilian
     Disk-winged Bat; Tri-coloured Bat;
     Brazilian Sucker-footed Bat
d   Dreifarbige Haftscheibenfledermaus
f   Vespertilion tricolore; Chauve-souris
     tricolore

7040   Thyropteridae
     Chiroptera
e   Disk-winged Bats; American Disk-
     winged Bats; New-world Sucker-
     footed Bats
d   Amerikanische
     Haftscheibenfledermäuse
f   Thyroptéridés
i   Tiropteridi

Tigris syn. *Panthera* q.v.

7041   *Tokudaia*
     Rodentia - Muridae
e   Ryukyu Spiny Rats; Riu Kiu Spiny
     Rats; Liukiu Spiny Rats
d   Riukiu-Stachelratten

7042   *Tokudaia muenninki*
     Rodentia - Muridae
e   Muennink's Spiny Rat
d   Riukiu-Stachelratten

7043   *Tokudaia osimensis*
     Rodentia - Muridae
e   Ryukyu Spiny Rat; Amami Spiny Rat

7044 *Tolypeutes*
Cingulata - Dasypodidae
e Three-banded Armadillos
d Kugelgürteltiere
f Tatous à trois bandes
i Armadilli dalle tre fasce

7045 *Tolypeutes matacus*
Cingulata - Dasypodidae
e La Plata Three-banded Armadillo;
Southern Three-banded Armadillo
d Südliches Kugelgürteltier;
Kugelgürteltier
f Tatou à trois bandes du Sud
i Armadillo dalle tre fasce del sud

7046 *Tolypeutes tricinctus*
Cingulata - Dasypodidae
e Brazilian Three-banded Armadillo
d Dreibindenkugelgürteltier
f Tatou à trois bandes
i Armadillo dalle tre fasce

7047 *Tomopeas*
Chiroptera - Molossidae
e Blunt-eared Bats

7048 *Tomopeas ravus*
Chiroptera - Molossidae
e Blunt-eared Bat

7049 *Tonatia*
Chiroptera - Phyllostomidae
e Round-eared Bats
d Rundohrblattnasen
f Chauve-souris aux oreilles rondes

7050 *Tonatia bidens*
Chiroptera - Phyllostomidae
e Spix's Greater Round-eared Bat;
Greater Round-eared Bat; Spix's
Round-eared Bat
d Große Rundohrblattnase

7051 *Tonatia brasiliensis*

Chiroptera - Phyllostomidae
e Pygmy Round-eared Bat

7052 *Tonatia carrikeri*
Chiroptera - Phyllostomidae
e Allen's Round-eared Bat; Carriker's
Round-eared Bat

7053 *Tonatia evotis*
Chiroptera - Phyllostomidae
e Davis's Round-eared Bat

7054 *Tonatia saurophila*
Chiroptera - Phyllostomidae
e Stripe-headed Round-eared Bat

7055 *Tonatia schultzi*
Chiroptera - Phyllostomidae
e Schultz's Round-eared Bat

7056 *Tonatia silvicola*
Chiroptera - Phyllostomidae
e D'Orbigny's Round-eared Bat;
White-throated Round-eared Bat
d D'Orbignys Rundohrfledermaus

7057 *Trachops*
Chiroptera - Phyllostomidae
e Fringe-lipped Bats
d Fransenlippenfledermäuse

7058 *Trachops cirrhosus*
Chiroptera - Phyllostomidae
e Lizard-eating Bat; Fringe-lipped Bat
d Fransenlippenfledermaus

7059 *Trachypithecus*
Primates - Cercopicecidae
e Brown-ridged Langurs; Dusky
Langurs; Dusky Leaf Monkeys;
Lutungs
d Kappenlanguren; Haubenlanguren
f Semnopithèques; Langurs
i Presbiti

7060   ***Trachypithecus auratus***
Primates - Cercopicecidae

e   Javan Langur; Ebony Leaf Monkey;
Javan Lutung; Budeng; Moor Leaf
Monkey; Negro Leaf Monkey
d   Mohrenlangur; Schwarzlangur
f   Semnopithèque noir
i   Budeng

7061   ***Trachypithecus auratus auratus***
Primates - Cercopicecidae

e   Spangled Ebony Langur
f   Semnopithèque étoilé; Langur de
Java

7062   ***Trachypithecus auratus mauritius***
Primates - Cercopicecidae

e   West Javan Ebony Langur

7063   ***Trachypithecus barbei***
Primates - Cercopicecidae

e   Tenasserim Lutung; Tenasserim Leaf
Monkey
d   Tenasserim-Lutung
f   Semnopithèque de Barbe

7064   ***Trachypithecus cristatus***
Primates - Cercopicecidae

e   Silvered Leaf Monkey; Silvery Leaf
Monkey; Indochinese Lutung;
Silvered Langur
d   Haubenlangur; Silberlangur
f   Semnopithèque à coiffe
i   Presbite dalla cresta; Entello crestato;
Prebite moro

7065   ***Trachypithecus delacouri***
Primates - Cercopicecidae

e   Delacour's Langur; Delacour's Leaf
Monkey
d   Delacour-Langur; Pandalangur
f   Langur de Delacour
i   Langur di Delacour

7066   ***Trachypithecus ebenus***
Primates - Cercopicecidae

e   Indochinese Black Lemur; Black
Leaf Monkey
d   Schwarzer Langur
f   Langur noir

7067   ***Trachypithecus francoisi***
Primates - Cercopicecidae

e   Francois's Leaf Monkey; Francois's
Langur; Indochinese Langur
d   Tonkin-Langur; Francois-Langur
f   Semnopithèque de François
i   Presbite di Francois; Presbite del
Tonchino

7068   ***Trachypithecus geei***
Primates - Cercopicecidae

e   Golden Leaf Monkey; Gee's Golden
Langur; Golden Langur; Gee's Leaf
Monkey
d   Goldlangur; Gees Langur
f   Semnopithèque de Gee; Entelle dorée
i   Presbite d'oro; Presbite dorato

7069   ***Trachypithecus germaini***
Primates - Cercopicecidae

e   Indochinese Leaf Monkey
f   Gibbon blanc

7070   ***Trachypithecus hatinhensis***
Primates - Cercopicecidae

e   Hatinh Langur; Haitinh Leaf Monkey
d   Hatih-Langur
f   Semnopithèque de Hà Tin

7071   ***Trachypithecus johnii***
Primates - Cercopicecidae

e   Hooded Leaf Monkey; Nilgiri Leaf
Monkey; Nilgiri Langur; John's Leaf
Monkey
d   Nilgiri-Langur
f   Semnopithèque du Nilgiri
i   Presbite del Nilgiri

7072   ***Trachypithecus laotum***
Primates - Cercopicecidae

e Laotian Langur; Laotian Leaf
Monkey
d Laos-Langur
f Semnopithèque de Laos

7073 *Trachypithecus obscurus*
Primates - Cercopicecidae
e Dusky Leaf Monkey; Spectacled
Leaf Monkey; Dusky Langur;
Spectacled Langur
d Brillenlangur
f Semnopithèque obscur
i Presbite dagli occhiali

7074 *Trachypithecus obscurus flavicauda*
Primates - Cercopicecidae
e Blond-tailed Dusky Leaf Monkey
d Gelbschwanzbrillenlangur

7075 *Trachypithecus phayrei*
Primates - Cercopicecidae
e Phayre's Leaf Monkey; Phayre's
Langur
d Phayres Langur; Phayre-Langur
f Semnopithèque de Phayre
i Presbite di Phayre

7076 *Trachypithecus pileatus*
Primates - Cercopicecidae
e Capped Leaf Monkey; Capped
Langur; Southern Capped Langur
d Schopflangur
f Entelle pileuse; Semnopithèque à
bonnet
i Prebite dal ciuffo

7077 *Trachypithecus poliocephalus*
Primates - Cercopicecidae
e Golden-headed Langur; White-
headed Leaf Monkey
d Goldkopflangur

7078 *Trachypithecus shortridgei*
Primates - Cercopicecidae
e Shortridge's Langur; Shortridge's
Leaf Monkey

d Shortridges Langur

7079 *Trachypithecus vetulus*
Primates - Cercopicecidae
e Purple-faced Leaf Monkey; Purple-
faced Langur
d Weißbartlangur;
Purpurgesichtslangur
f Semnopithèque blanchâtre
i Presbite dalla barba bianca

7080 *Trachypithecus vetulus monticola*
Primates - Cercopicecidae
e Bear Monkey
d Nördlicher Purpurgesichtslangur

7081 *Trachypithecus vetulus nestor*
Primates - Cercopicecidae
e Western Purple-faced Langur
d Westlicher Purpurgesichtslangur

7082 *Tragelaphus*
Artiodactyla - Bovidae
e Bushbucks; Spiral-horned Bovines;
African Antelopes; Kudus
d Drehornantilopen; Waldböcke;
Drehhörner
f Koudous
i Tragelafi

7083 *Tragelaphus angasii*
Artiodactyla - Bovidae
e Nyala; Lowland Nyala
d Nyala; Tiefland-Nyala
f Nyala

7084 *Tragelaphus buxtonii*
Artiodactyla - Bovidae
e Mountain Nyala
d Bergnyala; Bergnyala-Antilope
f Nyala de Montagne
i Nyala di monte; Nyala di montagna

Tragelaphus derbianus syn.
Taurotragus derbianus q.v.

Tragelaphus eurycerus syn.
Boocercus eurycerus q.v.

7085   *Tragelaphus imberbis*
  Artiodactyla - Bovidae
e Lesser Kudu; Lesser Koodoo
d Kleiner Kudu; Kleinkudu;
  Wenigkudu
f Petit Kudu; Petit Koudou
i Tragelafo minore

Tragelaphus oryx syn. *Taurotragus
oryx* q.v.

7086   *Tragelaphus scriptus*
  Artiodactyla - Bovidae
e Bushbuck; Harnessed Antelope
d Buschbock
f Guib; Guib harnaché; Antilope
  harnachée
i Tragelafo striato; Bushbuck;
  Antilope nera

7087   *Tragelaphus scriptus bor*
  Artiodactyla - Bovidae
e Nile Bushbuck
d Nil-Buschbock

7088   *Tragelaphus scriptus decula*
  Artiodactyla - Bovidae
e Abyssinian Bushbuck
d Abessinischer Buschbock

7089   *Tragelaphus scriptus delamerei*
  Artiodactyla - Bovidae
e East Africa Bushbuck; Masai
  Bushbuck
d Delameres Buschbock

7090   *Tragelaphus scriptus meneliki*
  Artiodactyla - Bovidae
e Arusi Bushbuck; Menilik Bushbuck
d Menelik-Buschbock; Schwarzer
  Buschbock

f Guib de Menelik

i Tragelafo di Menelik

7091   *Tragelaphus scriptus ornatus*
  Artiodactyla - Bovidae
e Chobe Bushbuck
d Chobe-Buschbock

7092   *Tragelaphus scriptus powelli*
  Artiodactyla - Bovidae
e Shoan Bushbuck

7093   *Tragelaphus scriptus scriptus*
  Artiodactyla - Bovidae
e Harnessed Bushbuck
d Schirrantilope

7094   *Tragelaphus scriptus sylvaticus*
  Artiodactyla - Bovidae
e South African Bushbuck
d Kap-Buschbock

7095   *Tragelaphus spekii*
  Artiodactyla - Bovidae
e Sitatunga; Marshbuck
d Sitatunga; Wasserkudu; Sumpfbock;
  Wasserantilope; Sumpfantilope
f Sitatunga; Guib d'eau
i Sitatunga

7096   *Tragelaphus spekii gratus*
  Artiodactyla - Bovidae
e Forest Sitatunga
d Waldsitatunga

7097   *Tragelaphus spekii selousi*
  Artiodactyla - Bovidae
e Zambesi Sitatunga
d Zambezi-Sitatunga

7098   *Tragelaphus spekii spekii*
  Artiodactyla - Bovidae
e East African Sitatunga

7099 ***Tragelaphus spekii sylvestris***
Artiodactyla - Bovidae
e Sesse Islands Sitatunga

7100 ***Tragelaphus strepsiceros***
Artiodactyla - Bovidae
e Greater Kudu; Greater Koodoo; Kudu
d Kudu; Großes Kudu; Großkudu; Größeres Kudu
f Grand Koudou
i Cudù maggiore

7101 ***Tragelaphus strepsiceros bea***
Artiodactyla - Bovidae
e East African Greater Kudu

7102 ***Tragelaphus strepsiceros chora***
Artiodactyla - Bovidae
e Northern Greater Kudu; Abyssinian Greater Koodoo

7103 ***Tragulidae***
Artiodactyla
e Chevrotains; Mouse Deer
d Hirschferkel; Zwergböckchen
f Tragulidés
i Tragulidi

7104 ***Tragulus***
Artiodactyla - Tragulidae
e Chevrotains; Mouse Deer ; Asian Chevrotains; Asiatic Mouse Deer .; Asiatic Chevrotain
d Kantschile
f Chevrotains d'Asie
i Traguli

7105 ***Tragulus javanicus***
Artiodactyla - Tragulidae
e Lesser Malay Mouse Deer; Lesser Malay Chevrotain; Spotted Mouse Deer; Lesser Mouse Deer; Lesser Malayan Mouse Deer
d Kleinkantschil

f Chevrotain malais mineur
i Tragulo pigmeo

7106 ***Tragulus meminna***
Artiodactyla - Tragulidae
e Indian Spotted Chevrotain; Spotted Chevrotain; Indian Spotted Mouse Deer; Greater Malay Chevrotain; Memmina; Indian Chevrotain; Indian Mouse Deer
d Fleckenkantschil; Indien-Kantschil
f Chevrotain tacheté
i Tragulo memina; Cervo topo maculato; Tragulo macchiato

7107 ***Tragulus napu***
Artiodactyla - Tragulidae
e Larger Malay Mouse Deer; Greater Malay Mouse Deer; Larger Malay Chevrotain; Greater Mouse Deer; Large Mouse Deer; Asiatic Mouse Deer; Asiatic Chevrotains
d Großkantschil
f Chevrotain malais majeur; Napu
i Tragulo grande

7108 ***Tragulus napu nigricans***
Artiodactyla - Tragulidae
e Balabac Chevrotain
d Balabac-Zwergböckchen

7109 ***Tremarctos***
Carnivora - Ursidae
e Spectacled Bears
d Anden-Bären; Brillenbären
f Ours à lunettes
i Orsi delle Ande

7110 ***Tremarctos ornatus***
Carnivora - Ursidae
e Spectacled Bear; Andean Bear; Ucumari
d Anden-Bär; Brillenbär
f Ours à lunettes
i Orso dagli occhiali; Orso andino

7111 *Triaenops*
Chiroptera - Rhinolophidae
e Triple Leaf-nosed Bats; Trident Bats;
Triple Nose-leaf Bats
d Dreiblattfledermäuse

7112 *Triaenops furculus*
Chiroptera - Rhinolophidae
e Trouessart's Trident Bat

7113 *Triaenops persicus*
Chiroptera - Rhinolophidae
e Persian Leaf-nosed Bat; Persian
Trident Bat
d Dreiblattfledermaus
f Triaenops de Perse

7114 *Trichechidae*
Uranotheria -
e Manatees; Amazon Oxen
d Rundschwanzseekühe;
Rundschwanzsirenen; Manatis;
Lamantine
f Lamantins; Trichéchidés
i Lamantini; Manati; Trichechidi

7115 *Trichechus*
Uranotheria - Trichechidae
e Manatees; Amazon Oxen;
Amazonian Manatees
d Flussmanatis; Manatis
f Lamantins; Lamantins de l'Amazone
i Lamantini

7116 *Trichechus inunguis*
Uranotheria - Trichechidae
e Amazon Manatee; South American
Manatee; South American River
Manatee; Amazonian Manatee;
Amazon Ox Manatee
d Flussmanati; Südamerikanischer
Manati
f Lamantin de l'Amazone

7117 *Trichechus manatus*
Uranotheria - Trichechidae

e American Manatee; North American
Manatee; Caribean Manatee; West
Indian Manatee; Caribbean Manatee;
Antillean Manatee
d Nagel-Manati; Karribik-Manati;
Westindien-Manati
f Lamantin d'Amérique du Nord;
Lamantin des Antilles
i Manato comune; Lamantino dei
Caraibi

7118 *Trichechus manatus latirostris*
Uranotheria - Trichechidae
e Florida Manatee
d Florida-Manati; Florida-Nagelmanati
f Lamantin de Floride

7119 *Trichechus senegalensis*
Uranotheria - Trichechidae
e African Manatee; West African
Manatee
d Afrikanischer Manati; Senegal-
Manati
f Lamantin d'Afrique; Lamantin du
Sénégal; Lamantin
i Lamantino africano; Manato

7120 *Trichosurus*
Diprotodontia - Phalangeridae
e Common Phalangers; Brush-tailed
Phalangers; Vulpine Phalangers;
Brush-tailed Possums; Brushtail
Possums
d Hundskusus; Kusus; Eigentliche
Kusus
f Opossums d'Australie; Phalangers
i Tricosuri

Trichosurus arnhemensis syn. *T.*
vulpecula q.v.

7121 *Trichosurus caninus*
Diprotodontia - Phalangeridae
e Short-eared Phalangista; Short-tailed
Brush-eared Phalanger; Short-tailed
Brush-eared Possum; Brush Possum;
Mountain Possum; Short-eared
Possum; Bobuck

d   Kusu; Hundskusu; Bergkusu
f   Phalanger de montagne

7122   Trichosurus cunninghami
Diprotodontia - Phalangeridae
e   Mountain Brushtail Possum
i   Opossum di montagna dalla coda a spazzola

7123   Trichosurus vulpecula
Diprotodontia - Phalangeridae
e   Brush-tailed Phalanger; Vulpine Phalanger; Brush-tailed Possum; Common Brushtail Possum; Bushy-tailed Possum; Silver-grey Brushtail; Tasmanian Brushtail Possum; Northern Brush-tailed Possum
d   Fuchskusu; Nördlicher Fuchskusu; Gewöhlicher Fuchskusu
f   Phalanger-renard
i   Tricosuro volpino; Possum comune dalla coda a spazzola; Opossum australiano; Opossum d'Australia; Opossum di Tasmania

7124   Trichosurus vulpecula fuliginosus
Diprotodontia - Phalangeridae
e   Opossum Ringtail
i   Opossum di Tasmania

7125   Trichys
Rodentia - Hystricidae
e   Long-tailed Porcupines
d   Pinselstachler
f   Porcs-épics
i   Trichide del Borneo

7126   Trichys fasciculata
Rodentia - Hystricidae
e   Long-tailed Porcupine
d   Malaiischer Pinselstachler; Borneo-Pinselstachler
f   Porc-épic à longue queue
i   Trichide del Borneo

Trichys lipura syn. *T. fasciculata*

q.v.

Trinomys syn. *Proechimys* q.v.

Trinycteris syn. *Micronycteris* q.v.

7127   Trogopterus
Rodentia - Sciuridae
e   Complex-toothed Flying Squirrels
d   Komplexzahngleithörnchen
i   Scoiattoli volanti cinese

7128   Trogopterus xanthipes
Rodentia - Sciuridae
e   Complex-toothed Flying Squirrel
d   Komplexzahnfleithörnchen
i   Scoiattolo volante cinese

7129   Tryphomys
Rodentia - Muridae
e   Mearn's Luzon Rats; Luzon Short-nosed Rats
d   Mearns-Luzon-Ratten

7130   Tryphomys adustus
Rodentia - Muridae
e   Mearn's Luzon Rat; Luzon Short-nosed Rat
d   Mearns-Luzon-Ratte

7131   Tscherskia
Rodentia - Muridae
e   Greater Long-tailed Hamsters
d   Rattenhamster
f   Hamsters-rats nain
i   Cricet coreani

7132   Tscherskia triton
Rodentia - Muridae
e   Greater Long-tailed Hamster; Korean Grey Hamster
d   Rattenhamster; Rattenartiger Zwerghamster
f   Hamster-rat nain
i   Criceto coreano

7133 *Tupaia*
Scandentia - Tupaiidae
e Tree Shrews
d Spitzhörnchen ; Eigentliche Tupaias
f Tupaïas; Tupajas
i Tupaie

7134 *Tupaia belangeri*
Scandentia - Tupaiidae
e Northern Tree Shrew; Belanger's
Tree Shrew
d Belanger-Tupaja; Nördliches
Spitzhörnchen; Belangers
Spitzhörnchen
f Tupaïa de Belanger; Tupaja de
Belanger

Tupaia chinensis syn. *T. glis* q.v.

7135 *Tupaia chrysogaster*
Scandentia - Tupaiidae
e Golden-bellied Tree Shrew
d Mentawai-Spitzhörnchen
f Tupaïa de Mentawai; Tupaja de
Mentawai

7136 *Tupaia dorsalis*
Scandentia - Tupaiidae
e Striped Tree Shrew
d Gestreiftes Sitzhörnchen;
Streifenspitzhörnchen
f Tupaïa rayé; Tupaja rayé

7137 *Tupaia glis*
Scandentia - Tupaiidae
e Tree Shrew
d Gewöhnliches Spitzhörnchen;
Spitzhörnchen; Eigentliches Tupaia
f Tupaïa commun; Tupaja commun
i Tupaia comune

7138 *Tupaia glis ferruginea*
Scandentia - Tupaiidae
e Common Tree Shrew
d Rostrotes Spitzhörnchen

7139 *Tupaia gracilis*
Scandentia - Tupaiidae
e Slender Tree Shrew
d Schlanktupaja; Schlankspitzhörnchen
f Tupaïa grêle; Tupaja grêle

7140 *Tupaia javanica*
Scandentia - Tupaiidae
e Javan Tree Shrew; Small Tree
Shrew; Indonesian Tree Shrew
d Java-Spitzhörnchen
f Tupaïa d'Indonésie; Tupaja
d'Indonésie
i Tupaia di Giava

7141 *Tupaia longipes*
Scandentia - Tupaiidae
e Long-footed Tree Shrew
d Großfüßiges Spitzhörnchen;
Langfußspitzhörnchen
f Tupaïa aux pattes longues; Tupaja
aux pattes longues

7142 *Tupaia minor*
Scandentia - Tupaiidae
e Günther's Tree Shrew; Pygmy Tree
Shrew; Lesser Tree Shrew
d Zwergtupaja; Zwergspitzhörnchen
f Tupaïa nain; Tupaja nain
i Tupaia minore

7143 *Tupaia montana*
Scandentia - Tupaiidae
e Mountain Tree Shrew
d Hochlandtupaja;
Hochlandspitzhörnchen
f Tupaïa des montagnes; Tupaja des
montagnes

7144 *Tupaia nicobarica*
Scandentia - Tupaiidae
e Nicobar Tree Shrew
d Philippinen-Tupaja; Nikobaren-
Spitzhörnchen
f Tupaïa de Nicobar; Toupaye de
Nicobar

7145 *Tupaia palawensis*
Scandentia - Tupaiidae
e Palawan Tree Shrew
d Palawan-Spitzhörnchen
f Tupaïa de Palawan; Tupaja de Palawan

7146 *Tupaia picta*
Scandentia - Tupaiidae
e Painted Tree Shrew
d Tiefland-Tupaja; Tiefland-Spitzhörnchen
f Tupaïa peint; Tupaja peint

7147 *Tupaia splendidula*
Scandentia - Tupaiidae
e Red-tailed Tree Shrew; Ruddy Tree Shrew; Rufous-tailed Tree Shrew
d Rotschwanztupaja; Rotschwanzspitzhörnchen
f Tupaïa à queue rousse; Tupaja à queue rousse

7148 *Tupaia tana*
Scandentia - Tupaiidae
e Large Tree Shrew; Terrestrial Tree Shrew
d Tana; Malaiisches Tupaja
f Tupaïa de Malaisie; Tupaja de Malaisie; Tana
i Tupaia della Malesia

7149 **Tupaiidae**
Scandentia
e Tree Shrews
d Tupajas; Spitzhörnchen
f Tupaïdés; Tupaïas; Tupajas
i Tupaidi; Toporagni arboricoli

7150 *Tursiops*
Cete - Delphinidae
e Bottle-nosed Dolphins
d Tümmler ; Große Tümmler
f Souffleurs; Dauphins souffleurs; Souffleurs à gros nez; Tursiops
i Tursiopi

7151 *Tursiops aduncus*
Cete - Delphinidae
e Indo-Pacific Bottlenosed Dolphin; Indian Ocean Bottle-nosed Dolphin
d Indopazifischer Großer Tümmler; Rotmeer-Tümmler
f Grand dauphin de l'Indo-Pacifique
i Tursiope indopacifico

Tursiops gilli syn. *T. truncatus* q.v.

7152 *Tursiops truncatus*
Cete - Delphinidae
e Bottle-nosed Dolphin; Common Bottle-nosed Dolphin; Bottle-nosed Porpoise; Bottlenose Dolphin (ANZ); Coastal Porpoise; Grey Porpoise; Common Bottlenose Dolphin; Pacific Bottle-nosed Dolphin
d Großer Tümmler; Großtümmler; Tümmler; Gill-Tümmler
f Souffleur à gros nez; Dauphin souffleur; Grand dauphin; Tursiops tronqué; Souffleur
i Tursiope troncato; Tursiope; Delfino maggiore; Delfino soffiatore; Tursio; Tursione

7153 *Tylomys*
Rodentia - Muridae
e Climbing Rats; American Climbing Rats; Naked-tailed Climbing Rats
d Kletterratten
f Rats grimpants

7154 *Tylomys bullaris*
Rodentia - Muridae
e Chiapan Climbing Rat

7155 *Tylomys fulviventer*
Rodentia - Muridae
e Fulvous Bellied Climbing Rat

7156 *Tylomys mirae*
Rodentia - Muridae
e Mira Climbing Rat

7157   *Tylomys nudicaudus*
Rodentia - Muridae
e   Peters's Climbing Rat; Naked-tailed
Climbing Rat; Northern Climbing
Rat
d   Nacktschwanzkletterratten

7158   *Tylomys panamensis*
Rodentia - Muridae
e   Panama Climbing Rat; Panamanian
Climbing Rat

7159   *Tylomys tumbalensis*
Rodentia - Muridae
e   Tumbala Climbing Rat

7160   *Tylomys watsoni*
Rodentia - Muridae
e   Watson's Climbing Rat; Watson's
Tree-climbing Rat

7161   *Tylonycteris*
Chiroptera - Vespertilionidae
e   Club-footed Bats; Flat-headed Bats;
Bamboo Bats
d   Bambusfledermäuse

7162   *Tylonycteris pachypus*
Chiroptera - Vespertilionidae
e   Lesser Club-footed Bat; Lesser Flat-
headed Bat; Bamboo Bat; Lesser
Bamboo Bat
d   Bambusfledermaus

7163   *Tylonycteris robustella*
Chiroptera - Vespertilionidae
e   Greater Club-footed Bat; Greater
Flat-headed Bat; Flat-headed Bat;
Greater Bamboo Bat

7164   *Tympanoctomys*
Rodentia - Octodontidae
e   Plains Vicacha Rats
d   Wüstenratten
f   Octodontes

7165   *Tympanoctomys barrerae*
Rodentia - Octodontidae
e   Plains Viscacha Rat; Red Viscacha
Rat
d   Rote Viscacha-Ratte
f   Octodonte

7166   *Typhlomys*
Rodentia - Muridae
e   Chinese Pygmy Dormice; Blind
Dormice
d   Chinesische Zwergschlafmäuse;
Chinesische Zwergbilche
f   Souris naines de Chine

7167   *Typhlomys chapensis*
Rodentia - Muridae
e   Chapa Pygmy Dormouse

7168   *Typhlomys cinereus*
Rodentia - Muridae
e   Chinese Pygmy Dormouse; Blind
Dormouse
d   Chinesische Zwerschlafmaus;
Chinesischer Zwergbilch
f   Souris naine de Chine; Loir pygmée
de Chine

U

Uncia syn. *Panthera* q.v.

7169 Uranomys
Rodentia - Muridae
e African Big-toothed Mice; Giant Naked-tailed Rats; White-bellied Brush-furred Rats; Rudd's Mice
d Weißbauchbürstenhaarratten

7170 Uranomys ruddi
Rodentia - Muridae
e African Big-toothed Mouse; Rudd's Brush-furred Rat; Rudd's Mouse
d Weißbauchbürstenhaarratte

Urocitellus syn. *Spermophilus* q.v.

7171 Urocyon
Carnivora - Canidae
e Grey Foxes
d Graufüchse
f Remards gris
i Volpi grige

7172 Urocyon cinereoargenteus
Carnivora - Canidae
e Grey Fox; Tree Fox; Mainland Grey Fox
d Graufuchs; Festland-Graufuchs
f Renard gris
i Volpe grigia nordamericana; Volpe arboricola

7173 Urocyon littoralis
Carnivora - Canidae
e Island Grey Fox (NA); Island Fox (NA); Coast Fox (NA); Short-tailed Fox; Insular Grey Fox (NA); Channel Islands Fox (NA); Channel Islands Grey Fox (NA); California Channel Islands Fox
d Inselgraufuchs
f Renard gris insulaire; Renard insulaire

7174 Urocyon littoralis catalinae
Carnivora - Canidae
e Santa Catalina Island Fox
d Santa Catalina Insel-Graufuchs

7175 Urocyon littoralis clementae
Carnivora - Canidae
e San Clemente Island Fox
d San Clemente Insel-Graufuchs

7176 Urocyon littoralis dickeyi
Carnivora - Canidae
e San Nicolas Island Fox
d San Nicolas Insel-Graufuchs

7177 Urocyon littoralis littoralis
Carnivora - Canidae
e San Miguel Island Fox
d San Miguel Insel-Graufuchs

7178 Urocyon littoralis santacruzae
Carnivora - Canidae
e Santa Cruz Island Fox
d Santa Cruz Insel-Graufuchs

7179 Urocyon littoralis santarosae
Carnivora - Canidae
e Santa Rosa Island Fox
d Santa Rosa Insel-Graufuchs

7180 Uroderma
Chiroptera - Phyllostomidae
e Tent-making Bats; Tent-building Bats
d Gelbohrfledermäuse
i Pipistrelli costruttori di tende

7181   Uroderma bilobatum
Chiroptera - Phyllostomidae
e   Tent-making Bat; Tent-building Bat;
Common Tent-making Bat; Peters's
Tentmaking Bat
d   Gelbohrfledermaus; Zeltbauende
Fledermaus
f   Vespertilion bilobé

7182   Uroderma magnirostrum
Chiroptera - Phyllostomidae
e   Davis's Bat; Brown Tent-making Bat

7183   Urogale
Scandentia - Tupaiidae
e   Philippine Tree Shrews
d   Philippinen-Spitzhörnchen
f   Tupaïas des Philippines; Tupajas des
Philippines

7184   Urogale everetti
Scandentia - Tupaiidae
e   Philippine Tree Shrew; Mindanao
Tree Shrew; Everett's Tupaia
d   Everett-Spitzhörnchen ; Everetts
Spitzhörnchen ; Philippinen-Tupaja;
Philippinen-Spitzhörnchen
f   Tupaïa des Philippines; Tupaja des
Philippines

7185   Uromanis
Cimolesta - Manidae
e   Ground Pangolins
d   Langschwanzschuppentiere
f   Pangolins à longue queue du Congo
i   Pangolini dalla coda lunga

7186   Uromanis tetradactyla
Cimolesta - Manidae
e   Long-tailed Pangolin; Asiatic
Pangolin; Black-bellied Pangolin;
Long-tailed Tree Pangolin
d   Schwarzbauchschuppentier;
Langschwanzschuppentier
f   Pangolin à longue queue du Congo;
Pangolin à longue queue
i   Pangolino a coda lunga; Pangolino

dalla coda lunga

7187   Uromys
Rodentia - Muridae
e   Mosaic-tailed Rats; Giant Naked-
tailed Rats
d   Neuguinea-Riesenratten;
Riesenratten
f   Rats à queue en mosaique; Rats
géants à queue nue

7188   Uromys anak
Rodentia - Muridae
e   Giant Naked-tailed Rat; Black-tailed
Giant Rat
d   Mosaikschwanz-Riesenratte;
Uromys-Riesenratte;
Schwarzschwanzriesenratte
f   Rat à queue en mosaique des
montagnes

7189   Uromys boeadii
Rodentia - Muridae
e   Black Giant Rat

7190   Uromys caudimaculatus
Rodentia - Muridae
e   Cape York Uromys; Giant White-
tailed Rat (ANZ); Mottle-tailed Tree
Rat; Mottle-tailed Giant Rat
d   Gebirgs-Mosaikschwanzriesenratte;
Weißschwanzriesenratte

7191   Uromys emmae
Rodentia - Muridae
e   Emma's Giant Rat

7192   Uromys hadrourus
Rodentia - Muridae
e   Masked White-tailed Rat

7193   Uromys imperator
Rodentia - Muridae
e   Emperor Rat
d   Riesenratte

7194 *Uromys neobrittannicus*
Rodentia - Muridae
e Bismarck Giant Rat

7195 *Uromys porculus*
Rodentia - Muridae
e Guadalcanal Rat

7196 *Uromys rex*
Rodentia - Muridae
e King Rat

7197 *Uropsilus*
Erinaceomorpha - Talpidae
e Asiatic Shrew Moles; Shrew Moles; Chinese Shrew Moles
d Ohrenspitzmausmaulwürfe
f Musaraignes-taupes

7198 *Uropsilus andersoni*
Erinaceomorpha - Talpidae
e Anderson's Shrew Mole
d Andersons-Spitzmausmaulwurf

7199 *Uropsilus gracilis*
Erinaceomorpha - Talpidae
e Gracile Shrew Mole; Chinese Shrew Mole
d Chinesischer Spitzmausmaulwurf

7200 *Uropsilus investigator*
Erinaceomorpha - Talpidae
e Inquisitive Shrew Mole; Yunnan Shrew Mole
d Yunnan-Spitzmausmaulwurf

7201 *Uropsilus soriceps*
Erinaceomorpha - Talpidae
e Chinese Shrew Mole; Asiatic Shrew Mole; Sichuan Shrew Mole
d Ohrenspitzmausmaulwurf; Sichuan-Spitzmausmaulwurf
f Musaraigne-taupe

Urosciurus syn. *Sciurus* q.v.

7202 *Urotrichus*
Erinaceomorpha - Talpidae
e Japanese Shrew Moles
d Japanische Spitzmulle; Spitzmausmaulwürfe
f Taupes du Japon

7203 *Urotrichus pilirostris*
Erinaceomorpha - Talpidae
e True's Shrew Mole; Lesser Japanese Shrew Mole
d Trues Spitzmull
f Taupe de True

7204 *Urotrichus talpoides*
Erinaceomorpha - Talpidae
e Japanese Shrewmouse; Greater Japanese Shrew Mole; Japanese Shrew Mole
d Japanischer Spitzmull
f Taupe des montagnes du Japon

7205 **Ursidae**
Carnivora
e Bears
d Großbären; Bären; Bärenartige Raubtiere
f Ursidés; Ours
i Orsi; Ursidi

7206 *Ursus*
Carnivora - Ursidae
e True Bears; Black and Brown Bears; Brown and Grizzly Bears; Black, Brown and Polar Bears
d Bären; Echte Bären
f Ours
i Orsi

7207 *Ursus americanus*
Carnivora - Ursidae
e American Black Bear; Black Bear
d Baribal; Schwarzbär
f Ours noir; Ours noir américain; Ours noir d'Amérique

i   Orso nero americano; Orso nero;
     Orso nero del america del Nord

7208   *Ursus americanus altifrontalis*
         Carnivora - Ursidae
e   British Columbia Bear

7209   *Ursus americanus americanus*
         Carnivora - Ursidae
e   Minnesota Black Bear
f   Ours noir

7210   *Ursus americanus carlottae*
         Carnivora - Ursidae
e   Queen Charlotte Bear

7211   *Ursus americanus cinnamomum*
         Carnivora - Ursidae
e   Cinnamon Bear
d   Zimtbär

7212   *Ursus americanus emmonsi*
         Carnivora - Ursidae
e   Glacier Bear; Silver Bear
d   Silberbär; Inselbaribal
f   Ours bleu

7213   *Ursus americanus floridanus*
         Carnivora - Ursidae
e   Florida Bear; Florida Black Bear

7214   *Ursus americanus hamiltoni*
         Carnivora - Ursidae
e   Newfoundland Bear

7215   *Ursus americanus kermodei*
         Carnivora - Ursidae
e   Kermode Bear; Spirit Bear
d   Kermode-Bär
f   Ours de Kermode; Ours kermode

7216   *Ursus americanus luteolus*
         Carnivora – Ursidae

e   Louisiana Black Bear; Texas Bear
f   Ours noir de Louisiane

7217   *Ursus americanus vancouveri*
         Carnivora - Ursidae
e   Vancouver Black Bear

7218   *Ursus arctos*
         Carnivora - Ursidae
e   Grizzly Bear
d   Braunbär; Gemeiner Bär; Brauner
     Bär; Landbär
f   Ours brun; Ours vulgaire
i   Orso bruno

7219   *Ursus arctos arctos*
         Carnivora - Ursidae
e   Eurasian Brown Bear; Brown Bear
d   Europäischer Braunbär
f   Ours d'Europe; Ours des Pyrénées
i   Orso bruno alpino

7220   *Ursus arctos beringianus*
         Carnivora - Ursidae
e   Siberian Bear; Kamchatka Bear
d   Sibirischer Braunbär; Kamschatka-
     Bär
f   Ours de Kamchatka

7221   *Ursus arctos californicus*
         Carnivora - Ursidae
e   Golden Bear; Bajan Bear; California
     Brown Bear; California Grizzly Bear;
     California Grizzly; California Coast
     Brown Bear; California Coast
     Grizzly Bear; California Coast
     Grizzly
d   Niederkalifornischer Grizzlybär;
     Kalifornischer Braunbär;
     Kalifornischer Grizzly;
     Kalifornischer Graubär

f   Ours brun de Californie; Ours
     vulgaire de Californie; Grizzly de
     Californie; Grizzli de Californie

7222 **Ursus arctos dalli**
Carnivora - Ursidae
e Alaskan Bear; Alaskan Brown Bear

7223 **Ursus arctos gyas**
Carnivora - Ursidae
e Peninsula Giant Bear

7225 **Ursus arctos horriaeus**
Carnivora - Ursidae
e New Mexican Brown Bear; New Mexican Grizzly Bear; New Mexican Grizzly
f Ours vulgaire du Nouveau Mexique; Grizzly du Nouveau Mexique; Grizzli du Nouveau Mexique

7224 **Ursus arctos horribilis**
Carnivora - Ursidae
e Grizzly Bear; Grizzly; American Grizzly Bear; North American Grizzly Bear; Grizzly Giant Bear
d Grizzlybär
f Grizzli; Ours grizzli; Grizzly; Ours grizzly
i Orso grizzly; Grizzly

7226 **Ursus arctos isabellinus**
Carnivora - Ursidae
e Himalajan Brown Bear; Red Bear; Red Giant Bear
d Isabell-Braunbär; Isabell-Bär
f Ours brun
i Orso rosso tibetano

7227 **Ursus arctos jeniseensis**
Carnivora - Ursidae
e Upper Jenesei Bear

7228 **Ursus arctos lasiotus**
Carnivora - Ursidae
e Manchurian Grizzly Bear
d Mongolischer Braunbär

7229 **Ursus arctos manchuricus**
Carnivora - Ursidae
e Manchurian Bear
d Mandschurischer Braunbär

7230 **Ursus arctos meridionalis**
Carnivora - Ursidae
e Caucasian Wild Bear

7231 **Ursus arctos middendorfi**
Carnivora - Ursidae
e Kodiak Bear; Kodiak
d Kodiak-Bär
f Kodiak; Ours kodiak
i Orso kodiak; Kodiak

7232 **Ursus arctos nelsoni**
Carnivora - Ursidae
e Mexican Grizzly Bear; Mexican Grizzly; Mexican Brown Bear; Mexican Big Brown Bear; Silver Grizzly; Silver Grizzly Bear; Mexican Silver Grizzly Bear; Mexican Silver Grizzly
d Mexikanischer Grizzlybär; Mexikanischer Braunbär; Mexikanischer Graubär; Mexikanischer Grizzly
f Grizzli mexicain; Grizzly mexicain

7233 **Ursus arctos pruinosus**
Carnivora - Ursidae
e Horse Bear; Tibetan Horse Bear
d Tibet-Bär; Tibetanischer Braunbär; Tibetischer Braunbär
f Ours bleu de Tibet; Ours bleu

7234 **Ursus arctos richardsoni**
Carnivora - Ursidae
e Barren Ground Grizzly Bear

7235 **Ursus arctos yesoensis**
Carnivora - Ursidae
e Hokkaido Bear
d Hokkaido-Braunbär
f Ours de Hokkaido

Ursus maritimus syn. *Thelarctos maritimus* q.v.

7236   Ursus thibetanus
Carnivora - Ursidae
  e   Asiatic Black Bear; Himalajan Black
        Bear; Asian Black Bear; Tibetan
        Black Bear; White-breasted Bear
  d   Kragenbär; Weißbrustbär;
        Asiatischer Schwarzbär
  f   Ours à collier; Ours noir du Tibet
  i   Orso tibetano; Orso dal collare; Orso
        nero asiatico

7237   Ursus thibetanus gedrosianus
Carnivora - Ursidae
  e   Pakistan Black Bear

V

7238   Vampyressa
Chiroptera - Phyllostomidae
e   Yellow-eared Bats

7239   Vampyressa bidens
Chiroptera - Phyllostomidae
e   Bidentate Yellow-eared Bat

7240   Vampyressa brocki
Chiroptera - Phyllostomidae
e   Brock's Yellow-eared Bat

7241   Vampyressa melissa
Chiroptera - Phyllostomidae
e   Melissa's Yellow-eared Bat
f   Chauve-souris melisa

7242   Vampyressa nymphaea
Chiroptera - Phyllostomidae
e   Big Yellow-eared Bat; Striped
     Yellow-eared Bat

7243   Vampyressa pusilla
Chiroptera - Phyllostomidae
e   Little Yellow-eared Bat

Vampyriscus syn. *Vampyressa* q.v.

7244   Vampyrodes
Chiroptera - Phyllostomidae
e   Trinidad White-lined Tailless Bats;
     Trinidadian White-lined Tailless
     Bats; Great Stripe-faced Bats
d   Caracciolo-Fledermäuse

7245   Vampyrodes caraccioli
Chiroptera - Phyllostomidae
e   Trinidad White-lined Tailless Bat;

Trinidadian White-lined Tailless Bat;
Great Stripe-faced Bat
d   Spießblattnase; Caracciolo-
     Fledermaus

7246   Vampyrum
Chiroptera - Phyllostomidae
e   Linnaeus's False Vampire Bats;
     Neotropical False Vampire Bats;
     Tropical American False Vampire
     Bats; Giant Spear-nosed Bats;
     Spectral Vampire Bats; Spectral Bats
d   Falsche Vampire
f   Vampires faux
i   Vampiri

7247   Vampyrum spectrum
Chiroptera - Phyllostomidae
e   Linnaeus's False Vampire Bat;
     Neotropical False Vampire Bat;
     Tropical American False Vampire
     Bat; Giant Spear-nosed Bat; Spectral
     Vampire Bat; American False
     Vampire Bat; Spectral Bat; False
     Vampire Bat; Great False Vampire
     Bat; Spectrum Bat
d   Große Spießblattnase; Großer
     Vampir; Südamerikanischer Vampir;
     Großer Südamerikanischer Vampir;
     Falscher Vampir
f   Faux vampire; Faux vampire
     commun
i   Vampiro

7248   Vandeleuria
Rodentia - Muridae
e   Palm Mice; Long-tailed Climbing
     Mice
d   Langschwänzige Indische
     Baummäuse; Indische Baummäuse
f   Souris à longue queue

7249   Vandeleuria nolthenii
Rodentia - Muridae
e   Nolthenius's Long-tailed Climbing
     Mouse; Ceylon Highland Long-tailed
     Tree Mouse

d   Ceylon Langschwanzbaummaus
f   Souris à longue queue

7250   ***Vandeleuria oleracea***
Rodentia - Muridae
e   Palm Mouse; Asiatic Long-tailed
Climbing Mouse
d   Langschwänzige Indische
Baummaus; Indische Baummaus

7251   ***Vandeleuria oleracea nilagirica***
Rodentia - Muridae
e   Indian Long-tailed Tree Mouse
d   Indische Langschwanzbaummaus

7252   ***Varecia***
Primates - Lemuridae
e   Ruffed Lemurs
d   Varis
f   Lémurs variés
i   Lemuri

7253   ***Varecia rubra***
Primates - Lemuridae
e   Red Ruffed Lemur
d   Roter Vari
f   Lémur varié roux

7254   ***Varecia variegata***
Primates - Lemuridae
e   Black-and-white Ruffed Lemur;
Black-and-white Lemur
d   Schwarzweißer Vari; Vari
f   Lémur vari; Lémur varié
i   Vari; Lemure variegato

7255   ***Vernaya***
Rodentia - Muridae
e   Vernay's Climbing Mice; Oriental
Climbing Mice
d   Vernay-Klettermäuse

7256   ***Vernaya fulva***
Rodentia - Muridae
e   Vernay's Climbing Mouse; Red

Climbing Mouse
d   Vernay-Klettermaus

Vespadelus syn. *Pipistrellus* q.v.

7257   ***Vespertilio***
Chiroptera - Vespertilionidae
e   Particoloured Red Bats;
Particoloured Bats
d   Zweifarbige Fledermäuse;
Zweifarbfledermäuse;
Nachtschwirrer
f   Sérotines bicolor
i   Vespertili

7258   ***Vespertilio murinus***
Chiroptera - Vespertilionidae
e   Particoloured Red Bat; Particoloured
Frosted Bat
d   Zweifarbige Fledermaus;
Zweifarbfledermaus; Nachtschwirrer;
Großer Nachtschwirrer;
Weißscheckige Fledermaus;
Gemeine Fledermaus
f   Sérotine bicolor; Petite chauve-souris
murine; Chauve-souris à deux
couleurs; Vespertilion murin;
Vespérion discolor
i   Serotino bicolore; Serotino discolore

7259   ***Vespertilio sinensis***
Chiroptera - Vespertilionidae
e   Asian Particolored Bat
d   Östliche Fledermaus

Vespertilio superans syn. *V. sinensis*
q.v.

7260   **Vespertilionidae**
Chiroptera
e   Vesper Bats; Vespertilionid Bats;
Typical Insect-eating Bats; Common
Bats
d   Glattnasenfledermäuse; Gemeine
Fledermäuse; Glattnasen;
Abendflatterer ; Nachtschwirrer ;
Dämmerungsmäuse;

Mausohrfledermäuse
f Vespertilionidés
i Vespertilionidi

7261 *Vicugna*
Artiodactyla - Camelidae
e Vicuñas; Vicugnas
d Vikunjas; Vikugnas; Vikunas
f Vigognes
i Vicugne

7262 *Vicugna pacos*
Artiodactyla - Camelidae
e Alpaca
d Alpaka; Paco
f Alpaca; Alpaga
i Alpaca

7263 *Vicugna vicugna*
Artiodactyla - Camelidae
e Vicuña; Vicugna
d Vikunja; Vikugna; Vikuna
f Vigogne
i Vigogna

7264 *Viverra*
Carnivora - Viverridae
e True Civets; Oriental Civets
d Echte Zibetkatzen; Zibetkatzen
f Civettes
i Civette

7265 *Viverra civettina*
Carnivora - Viverridae
e Malabar Civet
d Malabar-Zibetkatze
f Civette de Malabar; Civette d'Afrique
i Civetta grande di Malabar; Zibetto; Gatto Zibetto

7266 *Viverra megaspila*
Carnivora - Viverridae
e Large-spotted Civet
d Großfleckzibetkatze

f Civette aux grandes taches

7267 *Viverra tainguensis*
Carnivora - Viverridae
e Tainguen Civet

7268 *Viverra tangalunga*
Carnivora - Viverridae
e Tangalunga; Oriental Civet; Malay Civet; Malayan Civet; Ground Civet
d Tangalunga
f Civette tangalunga; Civette de Malaisie
i Viverra tagalunga

7269 *Viverra zibetha*
Carnivora - Viverridae
e Large Indian Civet
d Asiatische Zibetkatze; Indien-Zibetkatze; Indische Zibetkatze
f Grande civette de l'inde
i Civetta indiana maggiore; Gatto civetta asiatico

7270 *Viverricula*
Carnivora - Viverridae
e Small Indian Civets; Lesser Oriental Civets
d Kleinzibetkatzen
f Civettes de l'Inde

7271 *Viverricula indica*
Carnivora - Viverridae
e Small Indian Civet; Lesser Oriental Civet; Rasse; Lesser Civet
d Kleine Indische Zibetkatze
f Civette de l'Inde
i Civetta indiana minore; Piccola civetta indana

7272 *Viverricula indica mayori*
Carnivora - Viverridae
e Ceylon Small Civet Cat

Viverricula malaccensis syn. *V. indica* q.v.

7273 Viverridae
  Carnivora
  e Viverids; Civets, etc.; Genets, Civets
       and allies; Genets and Civets
  d Schleichkatzen; Zibetkatzen;
       Zibettiere; Viverren
  f Viverridés; Genettes; Civettes
  i Viverridi

7274 *Voalavo*
  Rodentia - Muridae
  e Voalavos

7275 *Voalavo gymnocaudus*
  Rodentia - Muridae
  e Naked-tailed Voalavo

7276 *Volemys*
  Rodentia - Muridae
  e Musser's Voles

7277 *Volemys clarkei*
  Rodentia - Muridae
  e Clarke's Vole

7278 *Volemys kikuchii*
  Rodentia - Muridae
  e Taiwan Vole

7279 *Volemys millicens*
  Rodentia - Muridae
  e Szechuan Vole

7280 *Volemys musseri*
  Rodentia - Muridae
  e Marie's Vole

7281 *Vombatidae*
  Diprodotodontia -
  e Wombats
  d Wombats; Plumpbeutler
  f Wombats; Wombatidés; Vombats;
       Vombatidés
  i Vombati; Vombatidi

7282 *Vombatus*
  Diprodotodontia - Vombatidae
  e Wombats; Coarse-haired Wombats;
       Naked-nose Wombats; Eastern
       Wombat
  d Nacktnasenwombats; Wombats
  f Wombats; Vombats; Wombats aux
       narines dénudées; Vombats aux
       narines dénudées
  i Vombati

7283 *Vombatus ursinus*
  Diprodotodontia - Vombatidae
  e Wombat; Common Wombat (ANZ);
       Coarse-haired Wombat; Ursine
       Wombat; Forest Wombat; Naked-
       nosed Wombat
  d Wombat; Nacktnasenwombat
  f Wombat aux narines dénudées;
       Vombat aux narines dénudées
  i Vombato; Vombato comune;
       Vombato dal naso nudo

7284 *Vombatus ursinus platyrrhinus*
  Diprodotodontia - Vombatidae
  e Australian Wombat
  d Australischer Nacktnasenwombat

7285 *Vombatus ursinus tasmaniensis*
  Diprodotodontia - Vombatidae
  e Tasmanian Wombat
  d Tasmanischer Nacktnasenwombat

7286 *Vombatus ursinus ursinus*
  Diprodotodontia - Vombatidae
  e Flinders Island Wombat

7287 *Vormela*
  Carnivora - Mustelidae
  e Marbled Polecats; Mottled Polecats
  d Tigeriltisse; Tigermarder
  f Poutois marbrés
  i Puzzole marmorizzate

7288 *Vormela peregusna*
  Carnivora - Mustelidae

e Marbled Polecat; Mottled Polecat

d Osteuropäischer Tigeriltis; Tigeriltis; Tigermarder; Fleckeniltis; Pantheriltis; Perwitzky

f Putois marbré; Putois marbré de Pologne

i Puzzola marmorizzata; Puzzola striata; Puzzola tigrata

7289   *Vulpes*
Carnivora - Canidae

e Foxes

d Füchse; Echte Füchse

f Renards

i Volpi

7290   *Vulpes bengalensis*
Carnivora - Canidae

e Bengal Fox; Indian Fox

d Bengal-Fuchs

f Renard du Bengal

i Volpe del Bengala

7291   *Vulpes cana*
Carnivora - Canidae

e Afghan Fox; Blanford's Fox; King Fox

d Cana-Fuchs; Afghan-Fuchs

f Renard de Blanford; Renard de l'Afghanistan

i Volpe di Blanford

7292   *Vulpes chama*
Carnivora - Canidae

e Cape Fox; Silver Jackal; Silver-backed Fox; Gray Fox (NA)

d Kap-Fuchs; Silberrückenfuchs; Kama-Fuchs; Chama-Fuchs

f Renard du Cap

i Chama; Volpe del Capo; Volpe cama

7293   *Vulpes corsac*
Carnivora - Canidae

e Corsac Fox

d Steppenfuchs; Korsak-Fuchs; Korsak

f Renard corsac; Corsac

i Volpe corsac; Volpe delle steppe

7294   *Vulpes ferrilatta*
Carnivora - Canidae

e Tibetan Fox; Tibetan Sand Fox

d Tibet-Fuchs

f Renard des sables du Tibet

i Volpe delle sabbie tibetane

Vulpes fulva syn. *V. vulpes* q.v.

7295   *Vulpes lagopus*
Carnivora - Canidae

e Polar Fox; White Fox; Blue Fox; Arctic Fox

d Eisfuchs; Polarfuchs; Steinfuchs; Weißfuchs; Blaufuchs

f Renard arctique; Renard blanc; Renard polaire; Renard bleu; Isatis

i Volpe artica; Volpe polare; Volpe azzurra

7296   *Vulpes pallida*
Carnivora - Canidae

e Pale Fox; Sand Fox

d Blassfuchs

f Renard pâle des sables; Renard pâle

i Volpe delle sabbie; Volpe pallida

7297   *Vulpes pallida edwardsi*
Carnivora - Canidae

e Edwards's Pale Fox

d Edwards-Blassfuchs

7298   *Vulpes rueppelli*
Carnivora - Canidae

e Rüppell's Fox; Rüppell's Sand Fox

d Rüppel-Fuchs

f Renard famélique

i Volpe di Rüppel

7299   *Vulpes velox*
Carnivora - Canidae

e Kit Fox; Swift Fox

d Kit-Fuchs

f   Renard veloce
i   Volpe veloce; Volpe kit; Volpe
      pigmca

7300   *Vulpes velox macrotis*
        Carnivora - Canidae
e   San Joaquin Kit Fox; South
      California Kit Fox; Long-eared Kit
      Fox
d   Großohr-Kit-Fuchs
f   Renard véloce de Californie du Sud;
      Renard véloce aux longues oreilles;
      Renard véloce oreillard

7301   *Vulpes velox velox*
        Carnivora - Canidae
e   Southern Swift Fox
d   Swift-Fuchs
f   Renard véloce du Sud

7302   *Vulpes vulpes*
        Carnivora - Canidae
e   Common Fox; Common Red Fox;
      European Fox; Red Fox; Fox
d   Fuchs; Nordischer Fuchs; Rotfuchs
f   Renard; Renard commun; Renard du
      Nord; Renard d'Europe; Renard
      vulgaire
i   Volpe; Volpe comune; Vòlopa;
      Volpe rossa

7303   *Vulpes vulpes argenteus*
        Carnivora - Canidae
e   Silver Fox; Black Fox; Silver-grey
      Fox
d   Silberner Fuchs
f   Renard argenté
i   Volpe argentata

7304   *Vulpes vulpes cascadensis*
        Carnivora - Canidae
e   Red Fox (NA)

7305   *Vulpes vulpes crucigera*
        Carnivora - Canidae
e   European Red Fox

d   Europäischer Rotfuchs
f   Renard roux
i   Volpe europea

7306   *Vulpes vulpes fulva*
        Carnivora - Canidae
e   American Red Fox
d   Nordamerikanischer Rotfuchs
f   Renard d'Amérique

7307   *Vulpes vulpes griffithi*
        Carnivora - Canidae
e   Griffith's Fox
d   Griffith-Rotfuchs
f   Renard roux de Griffith
i   Volpe rossa dell'Afghanistan

7308   *Vulpes vulpes macroura*
        Carnivora - Canidae
e   Utah Red Fox

7309   *Vulpes vulpes montana*
        Carnivora - Canidae
e   Himalajan Fox
d   Himalaja-Rotfuchs
f   Renard roux de l'Himalaja
i   Volpe rossa delle Himalaja

7310   *Vulpes vulpes nevadensis*
        Carnivora - Canidae
e   Nevada Kit Fox

7311   *Vulpes vulpes pusilla*
        Carnivora - Canidae
e   Little Red Fox
f   Renard roux; Renard fauve
i   Volpe rossa del Punjab

7312   *Vulpes vulpes regalis*
        Carnivora - Canidae
e   Yellow-red Fox

7313   *Vulpes vulpes vulpes*
        Carnivora - Canidae

  e Scandinavian Red Fox
  d Rotfuchs
  i Volpe rossa

7314 ***Vulpes zerda***
   Carnivora - Canidae
  e Fennec; Fennec Fox; Desert Fox
  d Wüstenfuchs; Fennek
  f Fennec
  i Fennec; Volpe del deserto

W

Rodentia - Muridae
e   Wilfred's Mouse
d   Wilfred-Maus

7321   *Wyulda*
Diprotodontia - Phalangeridae
e   Scaly-tailed Possums (ANZ); Scaly-
tailed Phalangers
d   Schuppenkusus;
Schuppenschwanzkusus

7322   *Wyulda squamicaudata*
Diprotodontia - Phalangeridae
e   Scaly-tailed Possum (ANZ); Scaly-
tailed Phalanger
d   Schuppenkusu;
Schuppenschwanzkusu

7315   *Wallabia*
Diprotodonia - Macropodidae
e   Black-tailed Swamp Wallabies;
Swamp Wallabies (ANZ)
d   Sumpfwallabys
f   Wallabies bicolores
i   Wallaby delle paludi

7316   *Wallabia bicolor*
Diprotodonia - Macropodidae
e   Black-tailed Wallaby; Black-tailed
Swamp Wallaby; Swamp Wallaby
(ANZ); Black Wallaby
d   Sumpfwallaby
f   Wallaby bicolore
i   Wallaby delle paludi; Puzzone;
Wallaby nero

7317   *Wiedomys*
Rodentia - Muridae
e   Wied's Red-nosed Mice
d   Rotnasenmäuse
i   Topi di Wied

7318   *Wiedomys pyrrhorhinos*
Rodentia - Muridae
e   Wied's Red-nosed Mouse; Coatinga
Mouse
d   Rotnasenmaus
i   Topo di Wied

7319   *Wilfredomys*
Rodentia - Muridae
e   Wilfred's Mice
d   Wilfred-Mäuse

7320   *Wilfredomys oenax*

X

7323 Xenomys
Rodentia - Muridae
e Magdalena Rats; Colima Woodrats
d Magdalena-Ratten

7324 Xenomys nelsoni
Rodentia - Muridae
e Magdalena Rat
d Magdalena-Ratte

7325 Xenuromys
Rodentia - Muridae
e White-tailed Rats; White-tailed New Guinea Rats; Giant Rats; Rock-dwelling Giant Rats
d Neuguinea-Weißschwanzratten

7326 Xenuromys barbatus
Rodentia - Muridae
e White-tailed Rat; White-tailed New Guinea Rat; Mimic Tree Rat; Rock-dwelling Giant Rat
d Falsche Baumratte

7327 Xeromys
Rodentia - Muridae
e False Water Rats (ANZ); False Swamp Rats
d Falsche Schwimmratten
f Faux rats d'eau; Xeromys
i Falsi ratti d'acqua

7328 Xeromys myoides
Rodentia - Muridae
e False Water Rat (ANZ); False Swamp Rat
d Falsche Schwimmratte; Australische Landmaus

f Faux rat d'eau; Xeromys myoides
i Falso ratto d'acqua

Xerospermophilus syn.
Spermophilus q.v.

7329 Xerus
Rodentia - Sciuridae
e African Ground Squirrels; African Bristly Ground Squirrels; Spiny Squirrels
d Afrikanische Borstenhörnchen ; Zieselhörnchen ; Borstenhörnchen
f Écureuils fouisseurs; Écureuils terrestres; Écureuils de la brousse
i Xeri

7330 Xerus erythropus
Rodentia - Sciuridae
e Western Ground Squirrel; Geoffrey's Ground Squirrel; Striped Ground Squirrel; Chad Ground Squirrel
d Gestreiftes Eichhörnchen; Westhörnchen; Gestreiftes Borstenhörnchen
f Écureuil fouisseur; Rat palmiste

7331 Xerus inauris
Rodentia - Sciuridae
e Cape Ground Squirrel; South African Ground Squirrel
d Kap-Borstenhörnchen
f Écureuil fouisseur du Cap
i Xero del Capo; Scoiattolo di terra; Scoiattolo terrestre del Capo

7332 Xerus princeps
Rodentia - Sciuridae
e Kaokoveld Ground Squirrel (CSA); Mountain Ground Squirrel; Damara Ground Squirrel
d Kaokoveld-Borstenhörnchen

7333 Xerus rutilus
Rodentia - Sciuridae
e Unstriped Ground Squirrel; East African Ground Squirrel

d   Schlichtborstenhörnchen; Schilu
i   Xero orientale

Xiphonycteris syn. *Mops* q.v.

Xylomys syn. *Heteromys* q.v.

Z

7334 Zaedyus
Cingulata - Dasypodidae
e Pichis
d Zwerggürteltiere
i Armadilli pigmei

7335 Zaedyus pichiy
Cingulata - Dasypodidae
e Small Armadillo; Pichi; Little
Armadillo
d Zwerggürteltier
i Armadillo pigmeo; Pichy

7336 Zaglossus
Tachyglossa - Tachyglossidae
e Long-beaked Spiny Anteaters; New
Guinea Long-nosed Echidnas; Long-
nosed Echidnas; Three-toed Spiny
Anteater; Long-nosed Spiny
Anteaters
d Langschnabeligel ;
Langschnabelameisenigel
f Échidnés à bec courbé
i Zaglossi della Nuova Guinea;
Echidne della Nuova Guinea

7337 Zaglossus attenboroughi
Tachyglossa - Tachyglossidae
e Attenborough's Echidna
d Attenborough-Langschnabeligel
f Échidné de Attenborough

7338 Zaglossus bartoni
Tachyglossa - Tachyglossidae
e Barton's Echidna
d Barton-Langschnabeligel
f Échidné à long bec; Échidné de
Barton

i Echidna a becco lungo

7339 Zaglossus bruijni
Tachyglossa - Tachyglossidae
e Bruijn's Echidna; Bruijn's Long-
nosed Echidna; New Guinea Spiny
Anteater; Long-nosed Echidna;
Long-beaked Echidna; Long-nosed
Spiny Anteater; New Guinea Echidna
d Bruijn-Langschnabeligel;
Langschnabelameisenigel;
Langschnabeligel
f Échidné de Bruijn; Échidné à longue
trompe; Échidné à long nez
i Zaglosso di Bruijn

7340 Zalophus
Carnivora - Otariidae
e Black Sea Lions; Californian Sea
Lions
d Kalifornische Seelöwen
f Lions de mer de Californie; Otaries
i Leoni marini

7341 Zalophus californianus
Carnivora - Otariidae
e Black Sea Lion
d Kalifornischer Seelöwe
f Lion de mer de Californie; Otarie de
California
i Leone marino della California; Otaria
della California

**7342 Zalophus californianus
californianus**
Carnivora - Otariidae
e Californian Sea Lion
d Kalifornischer Seelöwe
f Lion de mer de Californie
i Leone marino della California

7343 Zalophus californianus japonicus
Carnivora - Otariidae
e Japanese Sea Lion; Japanese Black
Sea Lion
d Japanischer Seelöwe

f   Lion de mer du Japon; Lion marin du Japon; Otarie du Japon

7344   ***Zalophus californianus wollebacki***
Carnivora - Otariidae
e   Galapagos Sea Lion
d   Galapagos-Seelöwe
i   Leone marino delle Galapagos

7345   ***Zapus***
Rodentia - Dipodidae
e   Jumping Mice; Meadow Jumping Mice; American Jumping Mice
d   Hüpfmäuse; Feldhüpfmäuse
f   Zapodes des prés; Souris sauteuses

7346   ***Zapus hudsonius***
Rodentia - Dipodidae
e   Meadow Jumping Mouse; North American Meadow Jumping Mouse; Common Jumping Mouse
d   Wiesenhüpfmaus
f   Zapode du Canada; Souris sauteuse des champs

7347   ***Zapus hudsonius campestris***
Rodentia - Dipodidae
e   Prairie Jumping Mouse

7348   ***Zapus princeps***
Rodentia - Dipodidae
e   Western Jumping Mouse
d   Westliche Hüpfmaus
f   Souris sauteuse de l'Ouest

7349   ***Zapus trinotatus***
Rodentia - Dipodidae
e   Pacific Jumping Mouse
d   Pazifik-Hüpfmaus
f   Souris sauteuse du Pacifique

7350   ***Zapus trinotatus orarinus***
Rodentia - Dipodidae
e   Point Reyes Jumping Mouse

7351   ***Zelotomys***
Rodentia - Muridae
e   Broad-headed Mice; Broad-headed Rats
d   Breitkopfmäuse

7352   ***Zelotomys hildegardeae***
Rodentia - Muridae
e   Hildegard's Broad-headed Mouse

7353   ***Zelotomys woosnami***
Rodentia - Muridae
e   Pale Rat; Woosnam's Desert Rat; Woosnam's Broad-headed Mouse

7354   ***Zenkerella***
Rodentia - Anomaluridae
e   Flightless Scaly-tailed Squirrels; Non-flying Scaly-tailed Squirrels; Scalytails; Flightless Scalytails; Cameroon Scaly-tails
d   Dornschwanzhörnchen ; Dornschwanzbilche
f   Zenkerelles

7355   ***Zenkerella insignis***
Rodentia - Anomaluridae
e   Flightless Scaly-tailed Squirrel; Non-flying Scaly-tailed Squirrel; Scalytail; Flightless Scalytail; Cameroon Scaly-tail
d   Stachelbilch; Dornschwanzbilch
f   Zenkerelle; Anomalure aptère

7356   ***Ziphius***
Cete - Hyperoodontidae
e   Goose-beaked Whales; Cuvier's Beaked Whales
d   Cuvier-Schnabelwale
f   Baleines à bec de Cuvier; Baleines à bec des oies
i   Zifi di Cuvier

7357   ***Ziphius cavirostris***
Cete - Hyperoodontidae
e   Goose-beaked Whale; Cuvier's Beaked Whale; Two-toothed Whale;

Cuvier's Whale
d Cuviers Schnabelwal; Cuvier-
     Schnabelwal
f Baleine à bec de Cuvier; Baleine à
     bec des oies; Ziphius; Baleine de
     Cuvier
i Zifio di Cuvier; Zifio; Balaeno a
     becco di Cuvier

Zorilla syn. *Ictonyx* q.v.

7358   Zygodontomys
         Rodentia - Muridae
e Cane Mice; Cane Rats; American
     Cane Rats
d Rohrmäuse

7359   Zygodontomys brevicauda
         Rodentia - Muridae
e Short-tailed Cane Mouse; Common
     Cane Rat; Common Cane Mouse
d Rohrmaus

7360   Zygodontomys brunneus
         Rodentia - Muridae
e Brown Cane Mouse; Colombian
     Cane Mouse

Zygodontomys cherriei syn. *Z.*
brevicauda q.v.

Zygodontomys reigi syn. *Z.*
brevicauda q.v.

7361   Zygogeomys
         Rodentia - Geomyidae
e Michoacan Pocket Gophers; Taltuzas
f Geomys

7362   Zygogeomys trichopus
         Rodentia - Geomyidae
e Michoacan Pocket Gopher; Taltuza;
     Tuza

7363   Zyzomys
         Rodentia - Muridae

e Thick-tailed Rats; Australian Rock
     Rats
d Dickschwanzratten; Felsenratten
f Rats des rochers
i Ratti dalla grossa coda

7364   Zyzomys argurus
         Rodentia - Muridae
e White-tailed Rat; Common
     Australian Rock Rat; Silver-tailed
     Rock Rat; Common Rock Rat
d Gemeine Felsenratte
f Rat à queue blanche

7365   Zyzomys maini
         Rodentia - Muridae
e Arnhem Land Rock Rat
d Arnhem-Land Felsenratte

7366   Zyzomys palatilis
         Rodentia - Muridae
e Carpentarian Rock Rat
d Carpentaria-Felsenratte

7367   Zyzomys pedunculatus
         Rodentia - Muridae
e Macdonnell's Rock Rat; Thick-tailed
     Rat; Macdonnell Range Rock Rat;
     Central Rock Rat; Central Thick-
     tailed Rat
d Zentrale Felsenratte
f Rat à grosse queue
i Ratto di roccia dalla grossa coda

7368   Zyzomys woodwardi
         Rodentia - Muridae
e Western Thick-tailed Rat;
     Woodward's Rock Rat; Large Rock
     Rat; Kimberly Rock Rat
d Kimberley-Felsenratte

English Index

Australian Dusky Field Rat 5919
Australian Echidna 6769
Australian False Mice 5687
Australian False Vampire 3409
Australian False Vampire Bat 3409
Australian False Vampire Bats 3408
Australian False Vampires 3408
Australian Ghost Bat 3409
Australian Giant False Vampire Bat 3409
Australian Giant False Vampire Bats 3408
Australian Hopping Mice 4497
Australian Jerboa Rat 4499
Australian Kangaroo Mice 4497
Australian Meat-and-Insect-eating Marsupials 1761
Australian Mice 5687
Australian Molossus 6786
Australian Myotis 4227
Australian Native Dog 846
Australian Native Mice 3157, 5687
Australian Native Mouse 5702
Australian Orange Leaf-nosed Bat 6062
Australian Rock Rats 7363
Australian Sea Lion 4397
Australian Sea Lions 4396
Australian Short-nosed Echidna 6769
Australian Spiny Anteater 6769
Australian Sticknest Rats 3209
Australian Stripe-faced Dunnart 6390
Australian Swamp Rat 5892
Australian Taphozous 6870
Australian Water Rat 2875
Australian Water Rats 2874

Australian Wombat 7284
Austro-Hungarian Wolf 862
Avahi 467
Avahi Lemur 467
Avahis 2990
Avahis 466
Avian Vampire Bats 1834
Avila Pires's Saddleback Tamarin 6163
Awash Multimammate Mouse 3555
Axis Deer 470, 471
Aye-aye 1772
Aye-ayes 1771, 1773
Azara's Agouti 1740
Azara's Fox 5663
Azara's Grass Mouse 107
Azara's Night Monkey 319
Azara's Opossum 1863
Azara's Tuco-tuco 1625
Azara's Tucutucu 1625
Azara's Zorro 5663
Azores Noctule 4516
Aztec Deer Mouse 5073
Aztec Fruit-eating Bat 404
Aztec Mouse 5073
Azumi Shrew 6441
Babakoto 2989
Babakotos 2988
Babault's Mouse Shrew 4191
Babiroussa 480
Babiroussas 479
Babirussa 480
Babirussas 479
Baboons 1037, 4960
Back-striped Mice 2864
Back-striped Weasel 4151
Bactrian Camel 817
Bactrian Deer 1081
Bactrian Red Deer 1081
Badger (NA) 6945
Badger 3629
Badlands American Bighorn 4849

Badlands Bighorn Sheep (NA) 4849
Badlands White-footed Mouse 5098
Baer's Wood Mouse 2924
Baffin Island Tundra Wolf 861
Baffin Island Wolf 861
Bagobo Rat 676
Bahama Hutia 2460
Bahama Island Hutia 2460
Bahaman Funnel-eared Bat 4362
Bahaman Hutias 2458
Bahaman Raccoon 5577
Bahamonde's Beaked Whale 3744
Bahia Hairy Dwarf Porcupine 1332
Baiji 3283
Baijitun 3283
Baikal Seal 5838
Bailey's Pocket Mouse 1154
Bailey's Shrew 1416
Baird's Pocket Gopher 2468
Baird's Beaked Whale 560
Baird's Shrew 6419
Baird's Tapir 6886
Baird's Whale 560
Baja California Chipmunk 6837
Baja California Rock Squirrel 6531
Baja Lynx 2318
Bajan Bear 7221
Balabac Chevrotain 7108
Bald Uakari 698
Bald-face Saki 5438
Bald-headed Uakari 698
Bale Long-eared Bat 5475
Bale Monkey 1261
Bale Mountains Vervet 1261
Bale Shrew 1421
Bali Cattle 622

Borneo Gibbon 2906
Borneo Pygmy Elephant 2030
Borneo Pymy Shrew 6645
Borneo Roundleaf Bat 2792
Borneo Smooth-tailed Tree Shrew 1787
Borneo Water Shrew 1214
Borneo White Dolphin 6512
Bororo 3590
Bosman's Potto 5053
Bosman's Pottos 5052
Boto 2992
Botswanan Long-eared Bat 3087
Botta's Gerbil 2494
Botta's Pocket Gopher 7000
Botta's Serotine 2112
Bottego's Pygmy Shrew 1420
Bottego's Shrew 1420
Bottlenose 2951
Bottlenose Dolphin (ANZ) 7152
Bottlenose Whales 2950
Bottle-nosed Dolphin 7152
Bottle-nosed Dolphins 7150
Bottle-nosed Porpoise 7152
Bottle-nosed Whales 2950
Botttle-nose Dolphin 2951
Bougainville Melomys 3649
Bougainville Monkey-faced Bat 5727
Bougainville Mosaic-tailed Rat 3649
Bougainville Naked-tailed Rat 6408
Boulanger Island Dunnart 6385
Bourret's Horseshoe Bat 6039
Bourton's Genet 2448
Boutu 2992
Bowdoin's Beaked Whale 3732

Bower's White-toothed Rat 563
Bowers's Rat 563
Bowhead 490
Bowhead Whale 490
Bow-headed Greenland Whales 486
Bow-headed Whale 490
Bow-headed Whales 491
Boyaca Spiny Rat 5588
Brahma White-bellied Rat 4472
Bramble Cay Melomys 3673
Bramble Cay Mosaic-tailed Rat 3673
Branded Wallaby 7032
Brandt's Bat 4234
Brandt's Hamster 3719
Brandt's Hedgehog 2706
Brandt's Vole 3838
Brandt's Whistling Rat 5015
Brandt's Yellow-toothed Cavy 2389
Branick's Rat 1872
Branick's Rats 1870, 1871
Brant's Climbing Mouse 1811
Brant's Gerbil 6916
Brauer's Dwarf Gerbil 1827
Brauer's Gerbil 1827
Brazenor's Hopping Mouse 4499
Brazilian Agouti 1746
Brazilian Arboreal Mice 5980
Brazilian Arboreal Mouse 5981
Brazilian Arboreal Rice Rat 4650
Brazilian Bare-faced Tamarin 6161
Brazilian Big-eared Bat 3804
Brazilian Big-eyed Bat 1230
Brazilian Brown Bat 2113

Brazilian Burrowing Mouse 3025
Brazilian Disk-winged Bat 7039
Brazilian Dolphin 6508
Brazilian False Rice Rat 5721
Brazilian False Rice Rats 5720
Brazilian Free-tailed Bat 6788
Brazilian Gracile Mouse Opossum 2601
Brazilian Guinea-pig 968
Brazilian Lesser Long-nosed Armadillo 1759
Brazilian Long-nosed Armadillo 1759
Brazilian Long-nosed Bat 6123
Brazilian Long-tongued Bat 3299
Brazilian Otter 5745
Brazilian Pied Tamarin 6161
Brazilian Porcupine 1337
Brazilian Puma 2280
Brazilian Pygmy Rice Rat 4672
Brazilian Rabbit 6726
Brazilian River Otter 3361
Brazilian Shrewmice 596
Brazilian Shrewmouse 597
Brazilian Spiny Tree Rat 3466
Brazilian Squirrel 6248
Brazilian Sucker-footed Bat 7039
Brazilian Sucker-winged Bat 7037
Brazilian Tapir 6889
Brazilian Three-banded Armadillo 7046
Brazilian Tree Mice 5980
Brazilian Tree Mouse 5981
Brazilian Tree Porcupine 1337
Brazilian Tuco-tuco 1628
Brazilian Tucutucu 1628

Dato Mastif Bat 1139

Dato Meldrum's Bat 1139

Daubenton's Bat 4243

Daubenton's Free-tailed Bat 4184

Daurian Ground Squirrel 6538

Daurian Hedgehog 2176

Daurian Pika 4588

Daurian Souslik 6538

Daurian Suslik 6538

Davao Squirrel 6666

Davies's Big-eared Bat 3800

Davies's Long-eared Bat 3800

Davis's Bat 7182

Davis's Long-tongued Bat 2580

Davis's Round-eared Bat 7053

Davy's Naked-backed Bat 5738

Dawn Bat 2083

Dawn Bats 2081

Dawn Fruit Bats 2081

Dawn Meadow Mice 2084

Dawson Caribou 5853

Dawson's Reindeer 5853

Dawson's Woodland Caribou 5853

Dayak Leaf-nosed Bat 2793

Day's Grass Mouse 112

Day's Shrew 6641

De Blainville's Beaked Whale 3734

De Brazza's Guenon 1056

De Brazza's Monkey 1056

De Graaf's Soft-furred Mouse 5531

De Vies's Bare-backed Fruit Bat 1937

De Vies's Woolly Rat 3477

De Winton's Golden Mole 1583

De Winton's Long-eared Bat 3089

De Winton's Mountain Shrew 6499

De Winton's Shrew 6499

De Winton's Tree Squirrel 2367

Decken's Horseshoe Bat 6012

Deep-crested Whale 3732

Deep-socketed Whale 3738

Deer (pl) 1070

Deer Hog 480

Deer Mice 5071

Deer Mouse 5102

Deer Tiger 2274

Deer Wallaroo 3432

Deer, Elk. Moose and allies 1070

Defassa Waterbuck 3061

Definite Leaf-eared Mouse 5322

Definitive Leaf-eared Mouse 5322

Defua Rat 1822

Defua Rats 1821

Degu 4619

Degus 4617, 4622

Dekeyser's Nectar Bat 3296

Dekhan Leaf-nosed Bat 2798

Delacour's Langur 7065

Delacour's Leaf Monkey 7065

Delacour's Marmoset Rat 2668

Delandi's Fox 4792

Delany's Mouse 1775

Delany's Swamp Mice 1774

Delany's Swamp Mouse 1775

Delectable Soft-furred Mouse 5532

Delicate Mouse 5694

Delicate Salt-flat Mouse 6212

Delicate Slender Mouse Opossum 3510

Delicate Vesper Mouse 798

Delmarva Peninsula Fox Squirrel 6270

Delmarva Fox Squirrel 6270

Delmarva Squirrel 6270

Delta Pygmy Rice Rat 4670

Demidoff's Bushbaby 2374

Demidoff's Dwarf Galago 2374

Demidoff's Galago 2374

Demidoff's Pymy Galago 2374

Demon Molerat 6776

Demonic Tube-nosed Fruit Bat 4553

Denliquin Soft-furred Wombat 3137

Dense-beaked Whale 3734

Dent's Guenon 1044

Dent's Horseshoe Bat 6013

Dent's Mona Monkey 1044

Dent's Shrew 1432

Dent's Vlei Rat 4810

Deppe's Squirrel 6259

Derby Eland 6937

Derby's Woolly Opossum 810

Dero 2438

Deroo's Mouse 4171

Desert Bandicoot 5048

Desert Bat 313

Desert Bats 312

Desert Bighorn 4848

Desert Cavy 1943

Desert Cottontail 6721, 6722

Desert Dormouse 2043

Desert Dwarf Mongoose 2692

Desert Fox 7314

Desert Grey Shrew 4494

Desert Gundi 1621

Desert Hamster 5300

Desert Hedgehog 2700

Desert Hedgehogs 2699

Desert Jerboa 3020

Flores Giant Rat 4959
Flores Giant Rats 4958
Flores Giant Tree Rat 4959
Flores Island Giant Tree
     Rat 4959
Flores Island Giant Tree
     Rats 4958
Flores Island Mouse 5021
Flores Long-nosed Rat
     5021
Flores Long-nosed Rats
     5020
Flores Shrew 6649
Flores Tube-nosed Bat 4062
Flores Warty Pig 6692
Flores Woolly Bat 3040
Florida Bear 7213
Florida Black Bear 7213
Florida Black wolf 875
Florida Bobcat 2315
Florida Cougar 2281
Florida Deer Tiger 2281
Florida Lynx 2315
Florida Manatee 7118
Florida Mastiff Bat 2223
Florida Mice 5490
Florida Mountain Lion
     2281
Florida Mouse 5491
Florida Naked-tailed Rat
     6407
Florida Otter 3355
Florida Pack Rat 4415
Florida Panther 2281
Florida Puma 2281
Florida Rabbit 6730
Florida Red Tiger 2281
Florida Red Wolf 875
Florida Water Rat 4387
Florida Water Rats 4386
Florida Wolf 875
Flower-faced Bat 301
Flower-faced Bats 300
Flower's Bottle-nosed
     Whale 2952
Flower's Gerbil 2506

Flower's Shrew 1444
Fluffy Glider 5171
Fly River Leaf-nosed Bat
     2814
Fly River Leptomys 3215
Fly River Roundleaf Bat
     2814
Fly River Trumpet-eared
     Bat 3047
Fly River Water Rat 3215
Flying Foxes 22
Flying Foxes 261
Flying Foxes 5746, 5747
Flying Lemur 1683
Flying Lemurs 1682, 2397
Flying Mouse 60
Flying Mouse Squirrels
     2983
Flying Possums 5169
Fog Shrew 6480
Fontaine's Cat 2333
Fontoynont's Hedgehog
     Tenrec 6336
Foothill Arboreal Rice Rat
     4656
Forbes's Tree Mouse 1252
Forest Baboons 3481
Forest Bat 3041
Forest Buffalo 6766
Forest Dormice 1962
Forest Dormouse 1965
Forest Duikers 6713
Forest Duikers 991
Forest Elephant 3349
Forest Elephant Shrews
     5187, 6109
Forest Fox 1069
Forest Genet 2452
Forest Giant Squirrel 5656
Forest Grass Mouse 143
Forest Horseshoe Bat 6049
Forest Mice 2603, 3441
Forest Musk Deer 4028
Forest Musk Shrew 6754
Forest Musk Shrews 6745
Forest Oldfield Mouse 6996

Forest Rabbit 6726
Forest Shrew 4206
Forest Shrews 4190
Forest Sitatunga 7096
Forest Small Rice Rat 3823
Forest Soft-furred Mouse
     5539
Forest Spiny Pocket Mice
     2750
Forest Spiny Pocket Mouse
     2753
Forest Thicket Rat 2609
Forest Tube-nosed Bat 4070
Forest Tuco-tuco 1634
Forest Tucutucu 1634
Forest Wallabies 1946,
     1951
Forest Wild Cat 2331
Forest Wombat 7283
Forester 3425
Foresters 3418
Fork-crowned Lemur 5238
Fork-crowned Lemurs 5236
Forked Mouse Lemurs 5236
Fork-marked Dwarf Lemur
     5238
Fork-marked Lemur 5238
Fork-marked Lemurs 5236
Fork-marked Mouse Lemur
     5238
Formosa Flying Fox 5760
Formosa Macaque 3381
Formosa Serow 916
Formosan Clouded Leopard
     4384
Formosan Field Mouse 341
Formosan Gem-faced Cat
     4884
Formosan Lesser Horseshoe
     Bat 6036
Formosan Rock Macaquw
     3381
Formosan Sika 1105
Formosan Sika Deer 1116
Formosan Wild Boar 6704
Formosan Wild Pig 6704
Forrest Mouse 3158

Gray's Long-tongued Bat
2580
Gray's Molerats 1585
Gray's Spear-nosed Bats
3917
Gray's Whale 3737
Gray's White-sided Dolphin
3097
Great Polar Whale 490
Great Anteater 4318
Great Anteaters 4317
Great Apes 2856
Great Bare-backed Fruit Bat
1934
Great Basin Kangaroo Rat
1907
Great Basin Pocket Mouse
5070
Great Bat 4520
Great Bent-winged Bat
3936
Great Blue Whale 501
Great Cane Rat 7018
Great Ceylon Leaf-nosed
Bat 2805
Great Eastern Horseshoe
Bat 6027
Great Evening Bat 2971
Great Evening Bats 2970
Great False Vampire Bat
7247
Great Flying Fox 5786
Great Fruit-eating Bat 417
Great Gerbil 6108
Great Gerbils 6107
Great Grey Kangaroo 3425
Great Grey Kangaroos 3418
Great Himalajan Leaf-
nosed Bat 2769
Great Indian Fruit Bat 5763
Great Indian One-horned
Rhinoceros 5990
Great Indian Rhinoceros
5990
Great Jerboa 176
Great Jird 6108
Great Killer Whale 4702

Great Long-nosed
Armadillo 1754
Great Northern Rorqual 501
Great Pamir Sheep 4843
Great Panda 95
Great Pipistrelle 2971
Great Pipistrelles 2970
Great Plains Grey Wolf 865
Great Plains Lobo Wolf 865
Great Plains Wolf 865
Great Right Whale 489
Great Roundleaf Bat 2769
Great Roundleaf Horseshoe
Bat 2769
Great Seal 2166
Great Seals 2165
Great Slit-faced Bat 4528
Great Sperm Whale 5332
Great Sperm Whales 5333
Great Stripe-faced Bat 7245
Great Stripe-faced Bats
7244
Great Tibetan Sheep 4841
Great Tube-nosed Bat 4066
Great Whale 501
Great White-toothed Shrew
1532
Greater Antillean Fruit-
eating Bat 639
Greater Antillean Long-
tongued Bat 3989
Greater Bamboo Bat 7163
Greater Bamboo Lemur
5624
Greater Bamboo Lemurs
5623
Greater Bandicoot 520
Greater Bandicoot Rat 520
Greater Big-footed Mouse
3443
Greater Bilby 3445
Greater Blind Molerats
6516
Greater Broad-nosed Bat
(ANZ) 4548
Greater Broad-nosed Bat
5470

Greater Brown Bat 6316
Greater Bulldog Bat 4488
Greater Bushbabies 4795
Greater Bushbaby 4796
Greater Cane Rat 7018
Greater Cavy 971
Greater Chinese Mole 2245
Greater Club-footed Bat
7163
Greater Cyclops Bat 2778
Greater Dawn Bat 2082
Greater Dog-faced Fruit Bat
1703
Greater Dog-like Bat 5143
Greater Dwarf Lemur 1196
Greater Dwarf Shrew 6647
Greater Egyptian Gerbil
2534
Greater Egyptian Jerboa
3020
Greater Fairy Armadillo
1256
Greater False Vampire 3596
Greater False Vampire Bat
3596
Greater Fat-tailed Jerboa
5846
Greater Fishing Bat 4488
Greater Flat-headed Bat
7163
Greater Flying Phalanger
5168
Greater Forest Wallaby
1948
Greater Free-tailed Bats
3992
Greater Galago 4796
Greater Galagos 4795
Greater Gerbil 2534
Greater Ghost Bat 1849
Greater Glider (ANZ) 5168
Greater Gliders (ANZ)
5167
Greater Gliding Opossum
5168
Greater Gliding Possum
5168

Indian Mouse Deer 7106
Indian Muntjac 4046
Indian Muntjak 4046
Indian Musk Shrew 6655
Indian Ocean Blue Whale 503
Indian Ocean Bottle-nosed Dolphin 7151
Indian Otter 216
Indian Palm Civet 4986
Indian Palm Squirrel 2349
Indian Pangolin 3486
Indian Pika 4607
Indian Pilot Whale 2576
Indian Pipistrelle 5369
Indian Pygmy Pipistrelle 5394
Indian Rats 1391
Indian Red Wolf 1670
Indian Rhinoceros 5990
Indian Rhinoceroses 5987
Indian River Dolphin 5457
Indian River Dolphins 5459
Indian Roundleaf Bat 2804
Indian Sambar 1123
Indian Sambar Deer 1126
Indian Sambur 1123
Indian Sambur Deer 1126
Indian Short-eared Fruit Bat 1695
Indian Short-nosed Fruit Bat 1702
Indian Sloth Bear 3683
Indian Smooth-coated Otter 3371
Indian Soft-furred Rat 3914
Indian Soft-furred Rats 3910
Indian Spotted Chevrotain 7106
Indian Spotted Deer 471
Indian Spotted Mouse Deer 7106
Indian Striped Squirrels 2346
Indian Tapir 6887
Indian Tiger 4953

Indian Tree Shrew 251
Indian Tree Shrews 250
Indian Water Buffalo 666
Indian Wild Ass 2143
Indian Wild Dog 1668
Indian Wild Pig 6697
Indian Wolf 868
Indian Wooly Horseshoe Bat 6000
Indian Wrinkle-lipped Bat 6785
Indiana Bat 4302
Indiana Myotis 4302
Indochina Leopard 4924
Indochina Tiger 4950
Indochinese Black Lemur 7066
Indochinese Flying Squirrel 2937
Indochinese Gaur 618
Indochinese Ground Squirrel 3685
Indochinese Ground Squirrels 3684
Indochinese Hog Deer 477
Indochinese Langur 7067
Indochinese Leaf Monkey 7069
Indochinese Lutung 7064
Indochinese Shrew 1414
Indo-Malayan Flying Squirrel 2935
Indo-Malayan Flying Squirrels 2931
Indonesian Mountain Weasel 4140
Indonesian Short-nosed Fruit Bat 1703
Indonesian Stink Badger 4156
Indonesian Tree Shrew 7140
Indo-Pacific Beaked Whale 2987
Indo-Pacific Bottlenosed Dolphin 7151
Indo-Pacific Finless Porpoise 4401

Indo-Pacific Hump-backed Dolphin 6512
Indo-Pacific Sousa 6512
Indri 2989
Indri Lemurs 2988
Indris 2988, 2990
Indus Dolphin 5458
Indus Dolphins 5456
Indus River Dolphin 5458
Indus Susu 5458
Ingram's Hutia 2460
Ingram's Planigale 5448
Inia 2992
Inland Broad-nosed Bat 4544
Inland Cave Bat 5377
Inland Forest Bat 5360
Inland Hill Rat 687
Inornate Squirrel 775
Inquisitive Shrew Mole 7200
Insectivorous Bats 4057
Insular Flying Fox 5812
Insular Grey Fox (NA) 7173
Insular Horseshoe Bat 6024
Insular Long-tongued Bat 3988
Insular Mole 6809
Insular Myotis 4260
Insular Seal 5283
Insular Vole 3832
Intelligent Grass Mouse 116
Interior Alaskan Wolf 869
Intermediate Fruit-eating Bat 414
Intermediate Guinea-pig 970
Intermediate Horseshoe Bat 5996
Intermediate Leaf-nosed Bat 2806
Intermediate Lesser Grass Mice 3753
Intermediate Lesser Grass Mouse 3754
Intermediate Rice Rat 4755

Madagascar Straw-coloured Fruit Bat 2006

Madagascar Sucker-footed Bat 4327

Madeira Pipistrelle 5391

Madras Macaque 3401

Madras Tree Shrew 251

Maduran Roundleaf Bat 2810

Magdalena Rat 7324

Magdalena Rats 7323

Magdalena Spiny Rat 5604

Magellanic Pygmy Rice Rat 4677

Magellanic Tuco-tuco 1641

Magellanic Tucutucu 1641

Maggie Taylor's Roundleaf Bat 2811

Maggie's Leaf-nosed Bat 2811

Maggie's Roundleaf Bat 2811

Maghreb Gerbil 2520

Maghreb Ground Squirrel 459

Magistrate Colobus 1351

Magot 3405

Mahogony Glider 5175

Mahomet Mouse 4094

Maine Lynx 2316

Mainland Babiroussa 481

Mainland Golden Bandicoot 3005

Mainland Grey Fox 7172

Mainland Serow 915

Mainland Slender-tailed Tree Shrew 1788

Major's Long-fingered Bat 3928

Major's Pine Vole 3871

Major's Sifaka Lemur 5643

Major's Tufted-tail Rat 2047

Makira Flying Fox 5757

Makira Roundleaf Bat 2788

Malabar Civet 7265

Malabar Dolphin 6512

Malabar Giant Squirrel 5936

Malabar Spiny Dormice 5452

Malabar Spiny Dormouse 5453

Malacca Prevost's Squirrel 784

Malagas Free-tailed Bats 3946

Malagasy Civet 2345

Malagasy Civets 2344

Malagasy Giant Rat 2955

Malagasy Giant Rats 2954

Malagasy Least Long-fingered Bat 3929

Malagasy Mountain Mice 3990

Malagasy Mountain Mouse 3991

Malagasy Mouse-eared Bat 4253

Malagasy Ring-tailed Mongoose 2402

Malagasy Voles 645

Malaita Island Tube-nosed Bat 4555

Malaita Tube-nosed Bat 4555

Malaita Tube-nosed Fruit Bat 4555

Malawi Galago 2380

Malay Civet 7268

Malay Leaf-nosed Bat 2815

Malay Pipistrelle 4291

Malaya Large Flying Fox 5815

Malayan Bear 2680

Malayan Black Rat 5912

Malayan Black-striped Squirrel 3131

Malayan Black-striped Squirrels 3129

Malayan Brush-tailed Porcupine 455

Malayan Civet 7268

Malayan Dolphin 6590

Malayan Elephant 2031

Malayan False Vampire 3598

Malayan False Vampire Bat 3598

Malayan Flying Fox 5815

Malayan Flying Lemur 1683

Malayan Free-tailed Bat 3999

Malayan Gaur 617

Malayan Giant Squirrel 5933

Malayan Gliding Lemur 1683

Malayan Gymnure 1993

Malayan Horseshoe Bat 6031

Malayan Leaf-nosed Bat 2815

Malayan Mountain Spiny Rat 3570

Malayan Pangolin 3487

Malayan Porcupine 2963

Malayan Pygmy Shrew 6648

Malayan Rat Shrew 1993

Malayan Red Tree Rat 5433

Malayan Roundleaf Bat 2815

Malayan Roundleaf Horshoe Bat 2815

Malayan Sambar Deer 1125

Malayan Sambur Deer 1125

Malayan Shrew 1489

Malayan Slit-faced Bat 4537

Malayan Squirrel 777

Malayan Stink Badger 4156

Malayan Sun Bear 2680

Malayan Sun Bears 2679

Malayan Tailless Horseshoe Bat 1328

Malayan Tailless Leaf-nosed Bat 1328

Malayan Tapir 6887

Malayan Tarsier 6896

Malayan Tree Rat 5434

Malayan Tree Rats 5432

Malayan Water Shrew 1212

Malayan Weasel 4145

Malayan White-toothed
Shrew 1489

Malaysian Field Rat 5925

Malaysian Fruit Bat 1695

Malaysian Hollow-faced
Bat 4531

Malaysian Weasel 4145

Malaysian Wood Rat 5925

Malbrouk Monkey 1260

Malheur Shrew 6469

Malpas's Bat 3042

Man 2858

Manacou 1865

Manado Fruit Bat 608

Manado Fruit Bats 607

Manatees 7114, 7115

Manchurian Bear 7229

Manchurian Goral 4332

Manchurian Grizzly Bear
7228

Manchurian Hare 3242

Manchurian Hedgehog
2174

Manchurian Red Deer 1098

Manchurian Sika Deer 1114

Manchurian Tiger 4947

Manchurian Weasel 4149

Manchurian Wild Boar
6705

Manchurian Zokor 4209

Mandarin Sika 1113

Mandarin Sika Deer 1113

Mandarin Vole 3872

Mandelli's Mouse-eared Bat
4299

Mandrill 3483

Mandrills 3481

Maned Hamster 3311

Maned Hamsters 3310

Maned Mouflon 239

Maned Rat 3311

Maned Rats 3310

Maned Sloth 651

Maned Sloths 649

Maned Tamarins 3190

Maned Three-toed Sloth
651

Maned Wolf 1297

Maned Wolves 1296

Manenguba Shrew 1490

Mangabeys 1029

Mangrove Pipistrelle 5428

Mangues 1571

Manicore Marmoset 755

Manipur Brown-antlered
Deer 1101

Manipur Bush Rat 2654

Manipur Bush Rats 2653

Manipur Mice 1873

Manipur Mouse 1874

Manipur Rat 565

Manipur White-toothed Rat
565

Manitoba Moose 162

Manitoba Wolf 848

Manso Grass Mouse 19

Mantbuls 3713

Mantled African Bush
Squirrel 5008

Mantled Black Howler
Monkey 197

Mantled Colobus 1351

Mantled Guereza 1351

Mantled Howler 197

Mantled Howler Monkey
197

Mantled Mastiff Bat 4805

Manul 2297

Manus Melomys 3666

Manusela Melomys 3656

Manusela Mosaic-tailed Rat
3656

Many-toothed Blackfish
5042

Maquassie Musk Shrew
1491

Mara 1942

Marajo Short-tailed
Opossum 3976

Maral Deer 1090

Maras 1941

Marbled Cat 2300

Marbled Polecat 7288

Marbled Polecats 7287

Marbled Seal 5831

Marca's Marmoset 756

Marco Polo Sheep 4843

Mareeba Rock Wallaby
5199

Margareta Rats 3489

Margarita Island Kangaroo
Rat 1903

Margarita's Cottontail
Rabbit 6732

Margay 2336

Margay Cat 2336

Margin-tailed Otter 5745

Margin-tailed Otters 5744

Mari 5046

Maria Madre Island Rice
Rat 4763

Maria Madre Rice Rat 4763

Mariana Flying Fox 5777

Mariana Fruit Bat 5777

Marianas Flying Fox 5777

Marianas Fruit Bat 5777

Marico Shrew 1492

Marie-Galante Tree Bat 398

Marie's Vole 7280

Marimonda 440

Marine Dolphins 1781

Marine Otter 3359

Marine Wolf 1971

Marinkelle's Sword-nosed
Bat 3306

Mariposa Brush Rabbit
6724

Mariput 2980

Maritime Striped Squirrel
6863

Markham's Grass Mouse
122

Markhor 893

Marmore Arboreal Rice Rat
4649

Marmoset Mice 2667

Marmoset Mouse 2669

Marmoset Rat 2669
Marmoset Rats 2667
Marmosets 741, 765
Marmots 3517, 6239
Maroon Langur 5551
Maroon Leaf Monkey 5551
Marsh Deer (pl) 598
Marsh Deer 599
Marsh Deer Mouse 5120
Marsh Mongoose 457
Marsh Mongooses 456
Marsh Mouse 5120
Marsh Otter 4139
Marsh Rabbit 6741
Marsh Rat 1734, 2855, 4434
Marsh Rats 1732, 2852, 4433
Marsh Rice Rat 4766
Marsh Shrew 6422
Marsh Tenrec 3271
Marsh Tenrecs 3270
Marshall's Horseshoe Bat 6032
Marshbuck 7095
Marsupial Anteater 4314
Marsupial Anteaters 4313
Marsupial Cats 1764
Marsupial Mice 1761
Marsupial Mice 278
Marsupial Mole 4511
Marsupial Moles 4510, 4513
Marsupial Rats 5243
Marsupial Shrews 5252
Marten 3538, 3546
Martens 3535
Martienssen's Free-tailed Bat 4803
Martinique Island Giant Rice Rat 3615
Martinique Island Swamp Rice Rat 3615
Martinique Musk Rat 3615
Martinique Rice Rat 3615
Martinique Swamp Rice

Rat 3615
Martino's Snow Vole 1869
Martino's Snow Voles 1868
Martino's Vole 1869
Martin's Bare-faced Tamarin 6183
Martin's False Potto 5715
Martin's Tamarin 6161
Maryland Shrew 6435
Maryland Sika Deer 1117
Ma's Night Monkey 325
Masai Bushbuck 7089
Masai Giraffe 2557
Masked Flying Fox 5792
Masked Palm Civet 4883
Masked Palm Civets 4882
Masked Shrew (NA) 6427
Masked Titi 734
Masked Titi Monkey 734
Masked White-tailed Rat 7192
Masoala Fork-crowned Lemur 5238
Massai Lion 4903
Master Leaf-eared Mouse 5326
Mastiff Bats 2218, 3956, 4489, 6784
Mato Grosso Dog-faced Bat 3950
Matschie's Bushbaby 2378
Matschie's Tree Kangaroo 1800
Matses's Big-eared Bat 3803
Matthey's Mouse 4095
Matto GrossoTamandua 6825
Matundu Dwarf Galago 2386
Matundu Galago 2386
Maues Marmoset 757
Maui's Dolphin 1019
Maule Tuco-tuco 1642
Maule Tucutucu 1642
Mauritanian Gerbil 2521
Mauritanian Shrew 1485

Mauritian Tomb Bat 6877
Maximilian Pocket Mouse 5058
Maximilian's Bat 988
Maximilians Shaggy-haired Bat 988
Maximovicz's Vole 3873
Max's Shrew 1497
Maxwell's Duiker 998
Maxwell's Duikers 991
Maya Deer Mouse 5106
Maya Mouse 5106
Maya Small-eared Shrew 1606
Mayor's Mouse 4096
Mayotte Lemur 2209
Mazama Pocket Gopher 7005
Mbarapi Antelope 2837
McConnell's Bat 3729
McConnell's Bats 3728
Mcilhenny's Four-eyed Opossum 5270
Meadow Jumping Mice 7345
Meadow Jumping Mouse 7346
Meadow Mice 3831
Meadow Mouse 3891
Meadow Vole 3891
Meadow Voles 3831
Mearn's Luzon Rat 7130
Mearn's Luzon Rats 7129
Mearns's Flying Fox 5808
Mearns's Grasshopper Mouse 4695
Mearns's Pouched Mouse 6159
Mearns's Squirrel 6860
Mechow's Molerat 1593
Medem's Titi 726
Mediterranean Hare 3231
Mediterranean Horseshoe Bat 6015
Mediterranean Mole 6805
Mediterranean Monk Seal 3964

Mountain Brushtail Possum 7122
Mountain Cat 2292
Mountain Cavies 3756
Mountain Chinchilla 3103
Mountain Chinchillas 3101
Mountain Coati 4356
Mountain Cottontail 6737
Mountain Cuscus 5224
Mountain Degu 4624
Mountain Degus 4623
Mountain Forest Tree Hyrax 1792
Mountain Fruit Bat 6132
Mountain Gazelle 2410, 2418
Mountain Giant Rat 6661
Mountain Giant Sund Rat 6661
Mountain Goat (NA) 4704
Mountain Goats (NA) 4703
Mountain Gorilla 2589
Mountain Ground Squirrel 7332
Mountain Hare 3252
Mountain Lion 2274
Mountain Marmot 3525
Mountain Melomys 3672
Mountain Monkey 1052
Mountain Mosaic-tailed Rat 3672
Mountain Noctule 4519
Mountain Nyala 7084
Mountain Paca 1666
Mountain Pipistrelle 5368
Mountain Pocket Gopher 7006
Mountain Possum 7121
Mountain Pygmy Possum 691
Mountain Pygmy Possums 690
Mountain Rats 4405
Mountain Reedbuck 5943
Mountain Sheep 4848
Mountain Shrew 6755

Mountain Shrews 6496
Mountain Spiny Pocket Mouse 2757
Mountain Spiny Rat 3564
Mountain Tapir 6888
Mountain Tree Mouse 1811
Mountain Tree Shrew 7143
Mountain Tree Squirrel 2361
Mountain Tube-nosed Fruit Bat 4557
Mountain Viscacha 3103
Mountain Viscacha Rat 4626
Mountain Viscachas 3101
Mountain Voles 201
Mountain Water Rat 2876
Mountain Weasel 4130
Mountain Zebra 2153
Mountains Goat 4704
Moupin Pika 4610
Mourning Bat 3140
Mourning Titi 725
Mouse Bandicoot 3812
Mouse Bandicoots 3811
Mouse Deer (pl) 7103, 7104
Mouse Dwarf Lemurs 3760
Mouse Gerbil 3475
Mouse Hare 4602
Mouse Hares 4583, 4612
Mouse Lemurs 1192, 1193, 3760
Mouse Opossum 1961, 7024
Mouse Opossums 1960, 3493
Mouse Shrews 4190
Mouse Sminthopsis 6374
Mouse-eared Bat 4277
Mouse-eared Bats 4215
Mouse-like Dormice 4164
Mouse-like Dormouse 4165
Mouse-like Hamsters 800
Mouse-like Pipistrelle 5397
Mouse-tailed Bats 6072, 6076

Mouse-tailed Dormice 4164
Mouse-tailed Dormouse 4165
Mouse-tailed Shrew 1499
Mouse-tailed Spiny Rat 5607
Moustached Bats 4008, 5737
Moustached Guenon 1043
Moustached Monkey 1043
Moustached Tamarin 6187
Moutain Sheep (pl) 4835
Mozambique Genet 2447
Mozambique Sheath-tailed Bat 1347
Mozambique Thicket Rat 2607
Mozambique Woodland Mouse 2607
Mrs. Gray's Kob 3074
Mrs. Gray's Lechwe 3074
Muennink's Spiny Rat 7042
Muenster Guinea-pig 2390
Muisk Vole 3880
Mule Deer 4632, 4633
Mulgara (ANZ) 1730
Mulgaras (ANZ) 1729
Muli Pika 4599
Mulita Armadillo 1753
Mullah Spiny Mouse 51
Müller's Bornean Gibbon 2906
Müller's Giant Sunda Rat 6663
Müller's Gibbon 2906
Müller's Grey Gibbon 2909
Müller's Rat 6663
Multimammate Mice 3553
Multimammate Rats 3553
Muna-Butung Macaque 3399
Mundarda (ANZ) 1026
Munning 3116
Muntjac 4044
Muntjacs 4039
Muntjak 4044

Pygmy Hippotamus 2759
Pygmy Hippotamuses 2758
Pygmy Hog 6694
Pygmy Jerboas 6213
Pygmy Kangaroo Rats 3770
Pygmy Kashmir Flying
    Squirrel 2933
Pygmy Killer 2341
Pygmy Killer Whale 2341
Pygmy Killer Whales 2340
Pygmy Long-eared Bat
    4582
Pygmy Loris 4542
Pygmy Marmoset 761
Pygmy Mice 483
Pygmy Mouse 4099
Pygmy Mouse Lemur 3763
Pygmy Nyctophilus 4582
Pygmy Phalanger 1028
Pygmy Phalangers 1024
Pygmy Pipistrelle 5394
Pygmy Planigale 5449
Pygmy Possum (ANZ) 691
Pygmy Possums 1024
Pygmy Possums 689
Pygmy Rabbit 636
Pygmy Rabbits 635
Pygmy Rice Rats 4666
Pygmy Right Whale 884
Pygmy Right Whales 883
Pygmy Ringtail 5670
Pygmy Ringtail Possum
    5670
Pygmy Rock Mouse 5211
Pygmy Rock Wallabies
    5043
Pygmy Rock Wallaby 5044
Pygmy Round-eared Bat
    7051
Pygmy Scaly-tailed Flying
    Squirrel 2985
Pygmy Scaly-tailed Flying
    Squirrels 2983
Pygmy Scaly-tailed Squirrel
    265
Pygmy Short-tailed

Opossum 3975
Pygmy Shrew (NA) 3830
Pygmy Shrew 6454
Pygmy Shrew Tanrec 3784
Pygmy Shrew Tenrec 3784
Pygmy Shrews 6639
Pygmy Slow Loris 4542
Pygmy Sperm Whale 3077
Pygmy Sperm Whales 3076
Pygmy Spotted Skunk 6579
Pygmy Squirrel 4189
Pygmy Squirrels 2253
Pygmy Tarsier 6900
Pygmy Three-toed Sloth
    650
Pygmy Tree Mice 2655
Pygmy Tree Rats 2655
Pygmy Tree Shrew 7142
Pygmy Weasel 4143
Pygmy White-toothed
    Shrew 6642
Pyramid Gerbil 2534
Pyrenean Chamois 6140
Pyrenean Desman 2394
Pyrenean Desmans 2393
Pyrenean Ibex 905
Pyrenean Pine Vole 3857
Quagga 2148
Quechuan Hocicudo 4867
Queen Charlotte Bear 7210
Queen Charlotte Caribou
    5853
Queen Charlotte Islands
    Caribou 5853
Queen Charlotte Islands
    Reindeer 5853
Queen Charlotte
    Woodlands Caribou
    5853
Queen of Sheba's Gazelle
    2408
Queensland Blossom Bat
    6710
Queensland Hairy-nosed
    Wombat 3137
Queensland Koala 5249

Queensland Ringtail 5672
Queensland Ring-tailed
    Phalanger 5672
Queensland Thetomys 5701
Queensland Tube-nosed Bat
    4563
Queensland Tube-nosed
    Fruit Bat 4563
Quelpart Shrew 1437
Quenda 3008
Querétaro Pocket Gopher
    4974
Quetta Mole Vole 2056
Quica Opossums 5267
Quichua Hairy Dwarf
    Porcupine 1340
Quirquincho 1171
Quirquinchos 1170
Quokka (ANZ) 6338
Quokkas (ANZ) 6337
Quoll (ANZ) 1770
Quolls 1764
Rabbit 4728
Rabbit Bandicoot 3445
Rabbit Bandicoots 3444
Rabbit Rat (ANZ) 1371
Rabbit Rat 5953
Rabbit Rats (ANZ) 1370
Rabbit Rats 3713, 5952
Rabbit-eared Bandicoot
    3445
Rabbit-eared Bandicoots
    3444
Rabbit-eared Rats 3713
Rabbit-eared Tree Rat 1371
Rabbits 3208, 4727
Rabor's Squirrel 6676
Rabor's Tube-nosed Bat
    4562
Raccoon 5576
Raccoon Dog 4522
Raccoon Dogs 4521
Raccoon-like Dog 4522
Raccoon-like Dogs 4521
Raccoons 5572, 5580

Spotted Bats 2198
Spotted Bolo Mouse 605
Spotted Cavy 1665
Spotted Chevrotain 7106
Spotted Cuscus 6572
Spotted Cuscuses 6570
Spotted Deer 471
Spotted Dolphin 6594
Spotted Dolphins 6589
Spotted Free-tailed Bat
    1134
Spotted Giant Flying
    Squirrel 5159
Spotted Grass Mice 3162
Spotted Grass Mouse 3174
Spotted Ground Squirrel
    6556
Spotted Hyena 1570
Spotted Hyenas 1569
Spotted Linsang 5558
Spotted Linsangs 5556
Spotted Marten 1768
Spotted Mongoose 2729
Spotted Mouse Deer 7105
Spotted Native Cat 1768
Spotted New Guinea
    Cuscus 6572
Spotted Phalanger 6572
Spotted Seal 5278
Spotted Skunk 6577
Spotted Skunks 6575
Spotted Souslik 6557
Spotted Suslik 6557
Spotted Vampire Bat 1835
Spotted-necked Otter 3365
Spotted-tailed Dasyure
    1768
Spotted-tailed Quoll (ANZ)
    1768
Spotted-wings Fruit Bat 515
Spotted-wings Fruit Bats
    514
Spot-winged Fruit Bat 515
Spot-winged Fruit Bats 514
Spray Porpoise 5295
Sprightly Pygmy Rice Rat

    4681
Spring Hares 5028
Spring Tamarin 740
Spring Tamarins 739
Springboc 303
Springbock 303
Springbocs 302
Springboks 302
Springhaas (CSA) 5027
Springhaases 5026, 5028
Springhare 5027
Springhares 5026
Spruce Mice 5260
Spurred Leaf-nosed Bat
    2777
Spurred Roundleaf Bat
    2777
Spurrell's Free-tailed Bat
    4005
Spurrell's Woolly Bat 3052
Spy Hocicudo 4864
Squareflipper 2166
Squareflippers 2165
Square-lipped Rhinoceros
    1021
Square-lipped Rhinoceroses
    1020
Squirel Gliders 5169
Squirrel 6284
Squirrel Glider (ANZ) 5176
Squirrel Lemurs 5236
Squirrel Monkey 6203
Squirrel Monkeys 6199
Squirrels 6239, 6245
Squirrel-tailed Dormouse
    2571
Squirrel-toothed Rat 255
Squirrel-toothed Rats 254
Sri Lanka Highland Shrew
    6650
Sri Lanka Shrew 6644
Sri Lanka Water Buffalo
    669
Sri Lankan Elephant 2032
Sri Lankan Giant Squirrel
    5938

Sri Lankan Leopard 4929
Sri Lankan Long-tailed
    Shrew 1495
Sri Lankan Sloth Bear 3682
Sri Lankan Spiny Mouse
    4088
St. Aignan's Trumpet-eared
    Bat 3035
St. Andrews Beach Mouse
    5127
St. Kilda Field Mouse 356
St. Kilda House Mouse
    4104
St. Kilda Island House
    Mouse 4104
St. Kilda Island Mouse
    4104
St. Kilda Mouse 4104
St. Lawrence Island
    Collared Lemming
    1856
St. Lawrence Island Shrew
    6428
St. Lucia Giant Rice Rat
    3616
St. Lucia Island Giant Rice
    Rat 3616
St. Lucia Island Muskrat
    3616
St. Lucia Island Rice Rat
    3616
St. Lucia Island Swamp
    Rice Rat 3616
St. Lucia Musk Rat 3616
St. Lucia Muskrat 3616
St. Lucia Rice Rat 3616
St. Lucia Swamp Rice Rat
    3616
St. Vincent Fruit-eating Bat
    638
St. Vincent Pygmy Rice Rat
    4682
Stampfl's Putty-nosed
    Guenon 1059
Stankovic's Mole 6817
Star-nosed Mole 1362
Star-nosed Moles 1361
Steenbok 5865
Steenboks 5864

Twilight Bat 4546

Twilight Bats 4543

Two-coloured Leaf-nosed Bat 2774

Two-humped Camel 817

Two-lined Bat 6156

Two-lined Bats 6152

Two-ringed Seal 5831

Two-spotted Palm Civet 4335

Two-toed Anteater 1678

Two-toed Anteaters 1677

Two-toed Sloths 1277, 3617

Two-toothed Whale 7357

Tyler's Mouse Opossum 3515

Typical Gerbils 2485

Typical Insect-eating Bats 7260

Typical Mongooses 2715

Typical Otters 3354

Typical Rats 5868

Typical Squirrels 6239

Typical Striped Grass Mouse 3174

Typical Vlei Rat 4817

Uakaris 697

Ucayali Soiny Mouse 6294

Ucumari 7110

Udzungwa Mouse Shrew 4196

Udzungwa Red Colobus 5338

Ugadan Shrew 1547

Uganda Armoured Shrew 6321

Uganda Elephant Shrew 2019

Uganda Grass Hare 5499

Uganda Hares 5498

Uganda Kob 3068

Uganda Large-toothed Shrew 1500

Uganda Leopard 4921

Uganda Shrew 6321

Ugandan Bushbaby 2384

Ugandan Defassa Waterbuck 3063

Ugandan Red Colobus 5344

Ugrug 2166

Ugruk 2166

Uinta Chipmunk 6855

Uinta Ground Squirrel 6530

Ulithi Flying Fox 5780

Ulithi Fruit Bat 5780

Ultimate Shrew 1555

Uluguru Forest Shrew 4195

Uluguru Galago 2381

Umboi Tube-nosed Fruit Bat 4553

Unadorned Rock Wallaby 5196

Unalaska Collared Lemming 1856

Unau 1278

Underwood's Bonneted Bat 2229

Underwood's Long-nosed Bat 2930

Underwood's Long-nosed Bats 2929

Underwood's Long-tongued Bat 2930

Underwood's Long-tongued Bats 2929

Underwood's Mastiff Bat 2229

Underwood's Pocket Gopher 4723

Underwood's Water Mouse 5986

Uneven-toothed Rat 255

Unexpected Cotton Rat 6365

Ungava Collared Lemming 1857

Ungava Lemming 1857

Ungava Seal 5281

Unicoloured Arboreal Rice Rat 4647

Unicoloured Avahi 469

Unicoloured Oldfield Mouse 6987

Unicoloured Tree Kangaroo 1795

Unicoloured Tree Rat 1989

Unicorn Whale 3985

Unstriped Ground Squirrel 7333

Unstriped Tube-nose Bats 4995

Unstriped Tube-nosed Bat 4996

Upemba Shrew 1567

Upper Amazonian Porcupine 1343

Upper Jenesei Bear 7227

Ural Field Mouse 357

Urartsk Mouse-like Hamster 806

Urial 4857

Urical Mouflon 4838

Ursine Colobus 1354, 1358

Ursine Howler 194

Ursine Howler Monkey 194

Ursine Sarcophilus 6221

Ursine Sarcophiluses 6220

Ursine Wombat 7283

Usambara Shrew 1556

Uspallata Chinchilla Rat 12

Ussuri Large White-toothed Shrew 1475

Ussuri Moose 163

Ussuri Shrew 6455

Ussuri Tiger 4947

Ussuri Tube-nosed Bat 4074

Ussuri White-toothed Shrew 1475

Ussuriysk Wild Boar 6705

Ustuyrt Mountain Sheep 4839

Utah Prairie Dog 1693

Utah Red Fox 7308

Utah White-tailed Jackrabbit 3259

Vaal Rhebok 5030

Vaal Rhebuck 5030

Vaal Ribbok 5030

Vacas Chinchilla Rat 13

Vagrant Hedgehog 2173

Vagrant Shrew 6489

Deutsches Register

Blattnasen 4977, 5310

Blattnasenartige 5310

Blattnasen-Federmäuse 935

Blattnasenfledermäuse 5310

Blattohrmäuse 2617, 5316

Blauaugenlemur 2212

Blauäugiger Maki 2212

Blaubock 631

Blauböckchen 999

Blauducker 999

Blauegraue Maus 5699

Blauer Affe 1054

Blaues Gnu 1375

Blaufuchs 7295

Blaugraue Australische
    Kleinmaus 5699

Blaumäulige Meerkatze
    1043

Blaumaulmeerkatze 1043

Blaurobbe 2166

Blaurückenducker 1008

Blauschaf 5683

Blauschafe 5682

Blauwal 501

Blauweißer Delfin 6593

Bleichböckchen (pl) 4827

Bleichböckchen 4828

Blessbock 1727

Blessmull 2477

Blessmulle 2476

Blinder Maulwurf 6805

Blindmaulwurf 6805

Blindmäuse 6516

Blindmull 4211

Blindmulle 4207

Bloßschwanz Opossum 812

Blumenfledermaus 2579

Blumennasenfledermaus
    301

Blumennasenfledermäuse
    300

Blutbrustpavian 6969

Blyths Hufeisennase 6026

Bobak 3519

Bobrinskis Fledermaus
    2111

Bobrinskis Pferdespringer
    182

Bobrinskis Springer 182

Bobrinski-Sprimgmäuse
    181

Bobrinski-Springmaus 182

Bocage-Buschratte 83

Bocages Graumull 1587

Böcke 238

Bodenkuskus 5225

Bodenkuskuse 6615

Boehm-Hörnchen 5002

Bogota-Grasmaus 108

Böhm-Zebra 2150

Bokhara-Wildschaf 4838

Bolivianische Feldmaus
    109

Bolivianischer Roter
    Brüllaffe 195

Bolivianisches
    Dreizehenfaultier 653

Bolivianisches
    Eichhörnchen 6264

Bolivien-Brüllaffe 199

Bolivische Chinchillaratte 7

Bolomäuse 600

Bonda-Samtfledermaus
    3959

Bongo 610

Bongos 609

Bonobo 4890

Bori 4624

Borkenkletterer 1386

Borkenratten 1386

Borneo-Banteng 626

Borneo-Bartschwein 6685

Borneo-Berghörnchen 1955

Borneo-Bergtupaja 1787

Borneo-Delfin 3091

Borneo-Delfine 3090

Borneo-Gibbon 2906

Borneo-Goldkatze 2260

Borneo-Hörnchen (pl) 5976

Borneo-Hörnchen 5977

Borneo-Hufeisennase 6002

Borneo-Koboldmaki 6897

Borneo-Langur 5546

Borneo-Muntjak 4040

Borneo-Orang-Utang 5513

Borneo-
    Pinselschwanzbaumma
    us 1245

Borneo-Pinselstachler 7126

Borneo-
    Schwarzbindenhörnche
    n 779

Borneo-Stachelschwein
    2964

Borneo-Wasserbüffel 668

Borneo-Wasserspitzmaus
    1214

Borneo-Zwergelefant 2030

Borneo-Zwerghörnchen (pl)
    2583

Borneo-Zwerghörnchen
    2584

Borstenbaumstachler (pl)
    1168

Borstenbaumstachler 1169

Borstenferkel 7016

Borstengürteltier 1173

Borstengürteltiere 1170,
    2240

Borstenhörnchen 458

Borstenhörnchen 7329

Borstenigel 6952

Borstenkaninchen (pl) 917

Borstenkaninchen 918

Borstenreisratte 4371

Borstenschwanzrennmaus
    2484

Borstentiere 6638

Borstige Taschenmaus 1160

Boto 2992

Bottas Fledermaus 2112

Bougainville-
    Affengesichtflughund
    5727

Bougainville-
    Langnasenbeutler 5046

Bourrets Hufeisennase 6039

Bowdoin-Schnabelwal 3732

Bowers Ratte 563
Brackwasserdelfin 6508
Brackwasserdelfine 6507
Brandmaus 336
Brandmäuse 335
Brandt-Fledermaus 4234
Brandts Hamster 3719
Brandts Igel 2706
Brandts Steppenwühlmaus
    3838
Brandt-Steppenwühlmaus
    3838
Brants Pfeifratte 5015
Brasil-Hörnchen 6248
Brasilianische Baummaus
    5981
Brasilianische Baummäuse
    5980
Brasilianische Reisratten
    5720
Brasilianische
    Spitzmausratte 597
Brasilianische
    Spitzmausratten 596
Brasilianische Vespermaus
    796
Brasilianischer Flussotter
    3361
Brasilianischer Kampfuchs
    3373
Brasilianischer Kampfüchse
    3372
Brasilianischer Otter 5745
Brasilianischer Puma 2280
Brasilianisches Zartes
    Mausopossum 2601
Brasilien-Sumpfratte 2853
Brasilien-Waldkaninchen
    6726
Brauer-Rennmaus 1827
Brauers Rennmäuse 1826
Braunbär 7218
Braunbauch-
    Streifenbeutelmaus
    5254
Braunborstengürteltier
    1171, 1173
Braunbrustigel 2178

Braunbrustiger Igel 2178
Braune Hirschmaus 5107
Braune Hyäne 4876
Braune Ratte 5902
Braune Stumpfnase 6069
Braune Zwerghörnchen (pl)
    4337
Brauner Bär 7218
Brauner Brüllaffe 194
Brauner Kapuziner 983
Brauner Maki 2208
Brauner Nord-
    Kurznasenbeutler 3007
Brauner Sri Lanka
    Hochland Mungo 2723
Brauner Süd-
    Kurznasenbeutler 3008
Brauner Wollaffe 3121
Braunes Aguti 1751
Braunes Langohr 5473
Braunes Lemur 2208
Braunes Opossum 3746
Braunes Zwerghörnchen
    4338
Braunfisch 5289
Braunfische 5288
Braunhaargürteltier 1173
Braunkehldreifingerfaultier
    653
Braunkopfklammeraffe 442
Braunmäuse 6295
Braunmazama 3589
Braunrückenäffchen 6162
Braunrückentamarin 6162
Braunzottiges
    Borstengürteltier 1173
Braunzottiges Gürteltier
    1173
Brazza-Meerkatze 1056
Breitflügelfledermaus 2128
Breitflügelfledermäuse
    2110
Breitfußbeutelmäuse 278
Breitfußmaulwurf 6229
Breitkopfbaumratte 1253
Breitkopfkänguru 5525

Breitkopfmäuse 7351
Breitmaulnashorn 1021
Breitmaulnashörner 1020
Breitohr 525
Breitohrfledermaus 525
Breitohrfledermäuse 524
Breitohrige Fledermaus 525
Breitschnabeldelfin 5042
Breitschnabeldelfine 5041
Breitschnauzenhalbmaki
    5624
Breitschnauzenmaki 5624
Breitschnauzenmakis 5623
Breitschwanz 4846
Breitstirnwombat 3138
Breitstreifenmungo 2404
Breitstreifenmungos 2403
Brillenbär 7110
Brillenbären 7109
Brillenblattnase 938
Brillenhasenkänguru 3107
Brillenlangur 7073
Brillenschweinswal 465
Brillenschweinswale 464
Brillentümmler 465
British-Columbia-Wolf 841
Bronzequoll 1769
Brooke-Hörnchen 6665
Brucies 654
Bruijn-Langschnabeligel
    7339
Brukkaros-
    Zwergfelsenmaus 5212
Brüllaffe 740
Brüllaffen 190
Brydes Wal 500
Bryde-Wal 500
Buchara-Hirsch 1081
Bucharische
    Schraubenziege 897
Bucharische
    Wasserspitzmaus 6423
Buckeldelfine 6511
Buckelrind 621
Buckelwal 3619

Buckelwale 3618
Budin-Grasmaus 110
Büffel (pl) 663
Büffelwolf 865
Bukowinische Blindmaus 6519
Bulldoggfledermaus 2227, 6794
Bulldoggfledermäuse 2218, 3945, 5627, 6784
Bulldogratte 5900
Buller-Chipmunk 6829
Bulmers Fruchtfledermaus 370
Bulmers Fruchtfledermäuse 369
Bundschwanzringbeutler 5667
Buntbock 1726, 1728
Bunte Fledermaus 3053
Bunte Lanzennase 5312
Bunthamster 1406
Bunthörnchen 6280
Buntmarder 3540
Burchell-Zebra 2136
Burmeister-Gürtelmull 1256
Burmeister-Schweinswal 5292
Bürstenfelskänguru 5200
Bürstenkängurus 566
Bürstenmaulwurf 4999
Bürstenmaulwürfe 4998
Bürstenschwanzfelskänguru 5200
Bürstenschwanzkaninchenr atte 1372
Bürstenschwanzphascogale 5245
Bürstenschwanz-Rattenkänguru 569
Burts Hirschmaus 5083
Burunduk 6847
Buschbabys 2371
Buschbock 7086
Büschelaffen 741
Büschelohrige Katzenmakis

185
Büschelohriger Katzenmaki 186
Büschelohrkatzenmaki 186
Büschelohr-Katzenmakis 185
Büschelohrspießbock 4735
Buschfelsenratten 82
Buschhase 3249
Buschkängurus 1946, 1951
Buschkaninchen 5499
Buschkatze 2322
Buschmannhase 680
Buschmannhasen 679
Buschmaus 6341
Buschmäuse 6339
Buschratten 4405
Buschschliefer (pl) 2747
Buschschwanzbeutelratte 2567
Buschschwanzbeutelratten 2566
Buschschwanzborkenratte 1390
Buschschwanz-Borkenratte 5275
Buschschwanzgundi 5025
Buschschwanzmanguste 549
Buschschwanzratte 4412
Buschschwanzrennmaus 6323
Buschschwein 5518
Buschschweine 5517
Buschveldt-Hufeisennase 6051
Buschveldt-Rennmaus 6922
Buschwaldgalago 2372, 4796
Busuanga-Hörnchen 6669
Büttikofer-Epaulettenflughund 2107
Büttikofers Epaulettenflughund 2107
Büttner-Baummaus 3161

Büttners Waldmaus 3161
Büttners Waldmäuse 3160
Butzkopf 2951, 4702
Cabrera-Maus 3840
Caguare 6824
Calabar-Bärenmaki 378
Calamian-Hirsch 474
Calamian-Schweinshirsch 474
Camas-Taschenratte 7002
Campbell-Meerkatze 1042
Campbell-Mona 1042
Campbells Meerkatze 1042
Campbell-Zwerghamster 5299
Camperdown-Wal 3737
Cana-Fuchs 7291
Canuts Hufeisennase 6003
Canyon-Maus 5079
Capybara 2873
Caracciolo-Fledermaus 7245
Caracciolo-Fledermäuse 7244
Carolina-Gleithörnchen 2561
Carpentaria-Felsenratte 7366
Carpenter-Weißhandgibbon 2898
Carruthers-Rotschenkelhörnchen 2361
Cascade Mountains-Wolf 847
Cauca-Klettermaus 6081
Cebu-Mähnenschwein 6689
Cebu-Pustelschwein 6689
Celebes-Flugfuchs 23
Celebes-Koboldmaki 6902
Celebes-Kuskus 6616
Celebes-Roller (pl) 3410
Celebes-Roller 3411
Ceram-Nasenbeutler (pl) 6117
Ceram-Nasenbeutler 6118
Cerrado-Mäuse 6960

Chriqui-Erntemaus 5958
Chriqui-Olingo 533
Cinderella-
    Fettschwanzbeutelratte
    7023
Cinderella-Spitzmaus 1426
Clara-Stachelnasenbeutler
    1997
Clymene-Delfin 6592
Coahuila-Maulwurf 6224
Coati 4351
Coatis 4350
Cobaya 972
Cochabamba-Grasmaus 138
Coiba-Aguti 1741
Coiba-Brüllaffe 193
Collie-Hörnchen 6258
Colorado-Chipmunk 6843
Columbia-Weißwedelhirsch
    4640
Columbia-Ziesel 6537
Columbische Maus 5095
Columbischer Bergwolf
    3120
Commerson-Delfin 1014
Cook-Maus 4083
Cookson-Gnu 1377
Cooper-Buschhörnchen
    5004
Cooper-Hörnchen 5004
Coquerels Sifaka 5635
Coquerels Zwergmaki 3944
Coquerels Zwergmakis
    3943
Costa-Rica-Otter 3360
Costa-Rica-Ozelot 2304
Costa-Rica-Puma 2282
Coucha-Vielzitzenmaus
    3556
Coués Klettermaus 6082
Coués-Reisratte 4747
Couguar 2274
Cozumel-Waschbär 5579
Crayshaw-Zebra 2151
Creaghs Hufeisennase 6010
Croslet-Hufeisennase 6007

Crump-Maus 1874
Crump-Mäuse 1873
Cuandu 1337
Culion-Hörnchen 6674
Culpeo-Fuchs 5662
Cursor-Grasmaus 111
Cururo 6515
Cururos 6514
Cuvier-Gazelle 2410
Cuvier-Hasenmaus 3103
Cuvier-
    Rotschenkelhörnchen
    2366
Cuviers Schnabelwal 7357
Cuviers Stummeldaumen
    2370
Cuviers Zaguti 5443
Cuvier-Schnabelwal 7357
Cuvier-Schnabelwale 7356
Cuvier-Zaguti 5443
Dachratte 5909
Dachs 3629
Dachsbär 3629
Dachse 3628
Dagestanischer Hamster
    3721
Dagestanischer Steinbock
    892
Dagestanischer Tur 892
Dall-Hafenschweinswal
    5295
Dall-Schaf 4853
Dall-Schweinswal 5295
Dall-Schweinswale 5294
Daltons Maus 4170
Dama-Gazelle 2411
Damara-Dikdik 3457
Damara-Graumull 1588
Damara-Zebra 2149
Damhirsch 1720
Damhirsche 1719
Dämmerungsmäuse 7260
Dangs Riesenhörnchen
    5935
Dangs-
    Königriesenhörnchen

5935
Darling-Downs-Hüpfmaus
    4506
Darling-Downs-
    Kängurumaus 4506
Darlings Hufeisennase 6011
Darwin-Maus 5321
Darwin-Reisratte 4450
Daurischer Blindmull 4208
Daurischer Igel 2176
Daurischer Pfeifhase 4588
Daurischer Ziesel 6538
Daurischer Zwerghamster
    1400
Davao-Hörnchen 6666
David-Felsenhörnchen 6243
David-Hirsch 2014
David-Hirsche 2013
Davids Hirsch 2014
Davids Hirsche 2013
David-Wühlmaus 2089
David-Wühlmäuse 2084
Dawson-Karibu 5853
Dawson-Ren 5853
Dawson-Rentier 5853
Day-Grasmaus 112
De Winton-Goldmull 1583
De Winton-Goldmulle 1582
Deckens Hufeisennase 6012
Deckens Sifaka 5636
Defassa-Wasserbock 3061
Defua-Ratte 1822
Defua-Ratten 1821
Degu 4619
Degus 4622
Dekhan-Rothund 1670
Dekhan-
    Rundblattnasenflederma
    us 2798
Delacour-Langur 7065
Delameres Buschbock 7089
Delanys Sumpfklettermaus
    1775
Delanys Sumpfklettermäuse
    1774
Delfin 1784

Gabelkatzenmakis 5236
Gabelkrall-Lemming 1859
Gabelkrall-Lemminge 1855
Gabelschwanzseekühe 1969
Gabelstreifenkatzenmakis
    5236
Gabelstreifiger Katzenmaki
    5238
Gablestreifige Katzenmakis
    5236
Gabun-Ducker 997
Gabun-Meerkatze 1066
Gaimards
    Bürstenrattenkänguru
    567
Galagos 2371
Galapagos-Inseln Reisratte
    4751
Galapagos-Mäuse 4449
Galapagos-Reisratte 4751
Galapagos-Seebär 384
Galapagos-Seelöwe 7344
Gallaghers
    Freischwanzfledermaus
    1137
Gambia-
    Epaulettenflughund
    2101
Gambia-Manguste 4035
Gambia-Mungo 4035
Gambia-Riesenhamsterratte
    1397
Gambia-Schlitznase 4527
Gams 6142
Ganges-Delfin 5457
Ganges-Delfine 5456
Ganges-Susu 5457
Gansu-Maulwürf 6227
Gansu-Maulwürfe 6226
Gansu-Pika 4585
Gaoligong-Pika 4591
Gappers Rötelmaus 1307
Garlepp-Maus 2396
Garlepp-Mäuse 2395
Garnetts Galago 4799
Garridos Hutia 923

Gartenmaus 4101
Gartenschläfer (pl) 2040
Gartenschläfer 2042
Gartenspitzmaus 1542
Gartenwaldmaus 354
Gaumer-Zwergbeutelratte
    3496
Gaur 614
Gayal 615
Gazellen 2406
Gebänderte Hasenkängurus
    3115
Gebändertes Hasenkänguru
    3116
Gebändertes
    Rotschenkelhörnchen
    2364
Gebirge Bongo 612
Gebirgsbachspitzmaus 4374
Gebirgsbachspitzmäuse
    4373
Gebirgschipmunk 6827
Gebirgshase 3252
Gebirgsmaus 213
Gebirgsmäuse 201
Gebirgs-
    Mosaikschwanzriesenra
    tte 7190
Gebirgsnasenbär 4356
Gebirgsnasenbären 4355
Gebirgsschwimmratte 2876
Gebirgsschwimmratten
    4990
Gebirgstaschenratte 7000
Gebirgstaschenratten 4967,
    6999
Gebirgswasserratte 4991
Gebirgswühlmäuse 201
Gees Langur 7068
Gefleckte Fledermaus 2199
Gefleckte Fledermäuse
    2198
Gefleckte Ginsterkatze
    2451
Gefleckte Stinktiere 6575
Gefleckter Seehund 5279
Gefleckter Ziesel 6556

Geflecktes
    Riesengleithörnchen
    5159
Geflecktes Rüsselhündchen
    6111
Gefranste Fledermaus 4279
Gehaupter Kapuziner 977
Gehufte Eisfuchsmaus 1859
Gekröntes Lemur 2207
Gelbbäuchiges Murmeltier
    3525
Gelbbauchmaus 4101
Gelbbauchmurmeltier 3525
Gelbbauchwiesel 4138
Gelbbraune Maus 354
Gelbbraune Ratte 5926
Gelbbraune Rötelmaus
    1315
Gelbbrustkapuziner 984
Gelbbrustmaus 343
Gelbe Kängurumaus 5061
Gelbe Taschenratten 4967
Gelbe Wollbeutelratte 812
Gelbe Zieselmaus 6542
Gelber Babuin 4962
Gelber Fichtenchipmunk
    6828
Gelber Pavian 4962
Gelber Steppenlemming
    2079
Gelber Ziesel 6542
Gelbes Totenkopfäffchen
    6202
Gelbes Totenköpfchen 6202
Gelbflügelige Großblattnase
    3156
Gelbflügelige
    Großblattnasen 3155
Gelbfußäffchen 749
Gelbfußbeutelmaus 284
Gelbfüssiges Antechinus
    284
Gelbfußkänguru 5205
Gelbgesichthufeisennase
    6058
Gelbgrüne Meerkatze 1263
Gelbhalsmaus 342

Goldkopflöwenäffchen 3192

Goldkopfsaki 5441

Goldkronenflughund 25

Goldkronensifaka 5640

Goldlangur 7068

Goldman-Stacheltaschenmaus 2755

Goldmanteltamarin 6194

Goldmantelziesel 6543

Goldmaulwürfe 1291

Goldmaus 4614

Goldmäuse 4613

Goldmullartige 1291

Goldmulle 1291, 1292

Goldmusang 4989

Goldrückenaguti 1746

Goldschakal 824

Goldstachelmaus 54

Goldstaubmanguste 2716

Goldsteißlöwenäffchen 3193

Goldstirnklammeraffe 440, 446

Goldstumpfnase 6071

Golduakari 701

Goldwolf 824

Golfküsten-Kängururatte 1892

Goliathratte 2943

Gongshan-Muntjak 4043

Goodenough-Buschkänguru 1947

Goodfellow-Baumkänguru 1796

Goral 4332

Gorale 4329

Gorilla 2591

Gorillas 2588

Gorkhar 2143

Gorogonza-Rennmaus 6918

Gould-Maus 5700

Goulds Australische Kleinmaus 5700

Grabflatterer (pl) 6868

Grabfledermäuse 6868

Graham Insel-Karibu 5853

Graham Insel-Wildren 5853

Graham-Wildren 5853

Gramper 2616

Granada-Hase 3239

Grant-Buschratte 85

Grant-Gazelle 2419

Grant-Goldmull 2157

Grants Bushbaby 2377

Grants Waldratte 543

Grant-Zebra 2150

Grasbüffel 6764

Grashüpfermäuse 4694

Graslandmelomysratte 3650

Grasmäuse 3162

Grasratten 423

Grauarmmakak 3399

Graubauchdunnart 6384

Graubauchgrasmaus 139

Graubauchhörnchen 772

Graubäuchige Opossummaus 705

Graubauchmausopossum 3499

Graue Baumratte 3187

Graue Baumratten 3186

Graue Kap-Manguste 2730

Graue Langohrfledermaus 5474

Graue Mausohrfledermaus 4254

Graue Riesenkängurus 3418

Graue Rötelmaus 1314

Graue Stachelmaus 46

Graue Sulawesi-Ratte 2187

Graue Vieraugenbeutelmaus 5271

Graue Vieraugenbeutelratte 5271

Graue Wanderratte 5902

Graue Wüstenspitzmaus 4494

Graue Zwerghamster (pl) 1398

Grauer Bambuslemur 2665

Grauer Gibbon 2906

Grauer Gorilla 2590

Grauer Halbmaki 2665

Grauer Kleideraffe 5840

Grauer Kuskus 5230

Grauer Mausmaki 3762

Grauer Seehund 2660

Grauer Springaffe 729

Grauer Steppenlemming 3123

Grauer Wollaffe 3118

Grauer Ziesel 6536

Grauer Zwerghamster 1403

Graues Baumkänguru 1798

Graues Eichhörnchen 6256

Graues Großkänguru 3425

Graues Langohr 5474

Graues Murmeltier 3518

Graues Riesenkänguru 3425

Graues Schlankes Mausopossum 3502

Graues Springäffchen 729

Graues Wieselmeerschweinchen 2391

Graufuchs 7172

Graufüchse 7171

Graufußchipmunk 6830

Graufußhörnchen (pl) 2683

Graufußhörnchen 2684

Graugelbe Großohrmaus 2620

Graugrüne Meerkatze 1259

Grauhalschipmunk 6831

Grauhörnchen 6256

Graukatze 2265

Graukopfbaumstachelratte 1882

Graukopfflughund 5796

Graulemming 3123

Graulemminge 3122

Grauliches Mausopossum 3496

Graumakak 3399

Graumazama 3589

Insektenfressende
  Waldmaus 1820
Insektenfressende
  Waldmäuse 1819
Inselbandikutratte 519
Inselbaribal 7212
Inselflughund 5768
Inselgraufuchs 7173
Inselhufeisennase 6024
Insellanguren 5541
Inselmäuse 3441
Inselratte 4448
Inselratten 3441, 4447
Iranischer Pferdespringer
  174
Iran-Maushamster 801
Iran-Waldmaus 339
Irawadi-Hörnchen 786
Irbis 4955
Iriomote-Katze 2264
Irma-Wallaby 3428
Irrawaddi-Delfin 4700
Irrawaddi-Delfine 4699
Isabella-Gazelle 2415
Isabell-Antilope 5945
Isabell-Bär 7226
Isabell-Braunbär 7226
Isarog-Streifenratte 1286
Italienische Kleinwühlmaus
  3899
Italienische Waldspitzmaus
  6474
Italienischer Igel 2180
Italienischer Wolf 854
Jackson-Manguste 550
Jacksons Waldmaus 5534
Jacobita 1014
Jagdhyäne 3375
Jagdleopard 29
Jagdleoparden 28
Jaguar 4909
Jaguarundi 2337
Jaik-Hamster 1403
Jamaika-Ferkelratte 2459
Jamaika-Fruchtfledermaus

415
Jamaika-Fruchtvampir 415
Jangtse-Delfin 3283
Japaniches
  Glattschweinswal 4402
Japanische Feldmaus 3879
Japanische Rötelmäuse
  5257
Japanische Schläfer (pl)
  2568
Japanische Spitzmulle 7202
Japanischer Dachs 3631
Japanischer Hase 3227
Japanischer Hirsch 1105
Japanischer Marder 3547
Japanischer Maulwurf 6820
Japanischer Pfeifhase 4594
Japanischer Schläfer 2569
Japanischer Schnabelwal
  3736
Japanischer Seelöwe 7343
Japanischer Serau 914
Japanischer Spitzmull 7204
Japanischer Wolf 851
Japanisches Eichhörnchen
  6266
Japanisches Gleithörnchen
  5732
Japanisches
  Riesengleithörnchen
  5160
Japanisches Wildschwein
  6700
Japan-Makak 3390
Japan-Sika 1105
Japan-Stachelnasenbeutler
  1997
Jarkand-Hase 3261
Jarkand-Hirsch 1099
Järv 2640
Ja-Schlitznase 4534
Java-Banteng 625
Java-Gibbon 2903
Java-Hausmaus 4123
Java-Hohlnase 4531
Java-Leopard 4932

Java-Mähnenhirsch 1122
Java-Muntjak 4045
Java-Nashorn 5988
Javaner Affe 3382
Javanischer Kalong 5816
Javanischer Sonnendachs
  3644
Javanisches Gleithörnchen
  2935
Javanisches Schuppentier
  3487
Javanisch-Malaiisches
  Schuppentier 3487
Java-
  Pinselschwanzbaumma
  us 1243
Java-Rothund 1672
Java-Sonnendachs 3644
Java-Spitzhörnchen 7140
Java-Stachelschwein 2967
Java-Stinkdachs 4156
Java-Tiger 4951
Jelarang 5933
Jemen-Gazelle 2408
Jenissej-Eichhörnchen 6287
Jentink-Ducker 996
Jentink-Hörnchen 6670
Jerboa 176
Jerdon-Musang 4988
Jiangtzhu-Schweinswal
  4400
Johnston-Ginsterkatze 2450
Johnstons Klippspringer
  5565
Jordanische Hausmaus
  4086
Jordan-Stachelmaus 55
Juan-Fernandez-Seebär 386
Jugoslawische Schermaus
  1869
Junin-Grasmaus 117
Junin-Hörnchen 6275
Kaama 151
Kaberu 878
Kabul-Markhor 899
Kaffeeratten 2585

Koboldmakis 6891, 6895
Kodiak-Bär 7231
Kodkod 2291
Koford-Grasmaus 119
Kojote 828
Kolinsky 4150
Kolonok 4147
Kolumbia-Maultierhirsch 4635
Kolumbianische Grasmaus 104
Kolumbianische Waldmas 1210
Kolumbianische Waldmäuse 1209
Kolumbianisches Eichhörnchen 6274
Kolumbianisches Wiesel 4134
Kolumbien-Spitzmausbeutelratte 3968
Komodo-Ratte 3081
Komodo-Ratten 3080
Komoren-Flughund 5772
Komplexzahnfleithörnchen 7128
Komplexzahngleithörnchen (pl) 7127
Kongo-Fingerotter 316
Kongo-Kleinkrallenotter 316
Kongo-Kusimanse 1572
Kongo-Leopard 4926
Kongo-Löwe 4898
Kongo-Manguste 2728
Kongoni 152
Kongo-Otter 3365
Kongo-Panzerspitzmaus 6321
Kongo-Rotschenkelhörnchen 2362
Kongo-Weißwangenotter 316
Kongo-Wimperspitzmaus 4982
Kongo-Wimperspitzmäuse

4979
Königscolobus 1354
Königsgepard 35
Königshufeisennase 6043
Königsriesenhörnchen 5934
Königstiger 4953
Königswüstenmaus 3703
Kontsal 878
Kordofan-Giraffe 2550
Kordofan-Mähnenspringer 241
Korea-Hase 3234
Korea-Igel 2181
Koreanische Waldmaus 349
Koreanischer Hase 3234
Koreanisches Wasserreh 2882
Korea-Wasserreh 2882
Kornsohlenschmalfußbeutel maus 6383
Korrigum 1725
Korsak 7293
Korsak-Fuchs 7293
Korsika-Hase 3235
Koslows Pika 4596
Koslows Zwergspringmäuse 6213
Kouprey 628
Kowari 1763
Koyote 828
Kozlows Zwergspringmaus 6216
Krabbenfresser (pl) 3292
Krabbenfuchs 1069
Krabbenmanguste 2736
Krabbenwaschbär 5573
Krabenesser (pl) 3292
Krabenesser 3293
Krabenfresser 3293
Kräftige Neuseeland-Fledermaus 4321
Kragenbär 7236
Kragenfaultier 651
Kragenflughunde 4176
Krallenäffchen (pl) 765

Krallenäffchen 750
Krallenaffen 765
Krallenrennmaus 3710
Krallenspitzmaus 6488
Krausbartschwein 6683
Krebsotter 4139
Kreishornschaf 4839
Kreta-Stachelmaus 50
Kretische Bezoarziege 889
Kretische Wildziege 889
Kreuzbanddelfin 3096
Kronenducker (pl) 6713
Kronenducker 6714
Kronenindris 5634
Kronenmaki 2207
Kronenmeerkatze 1061
Kropfgazelle 2439
Krummnasige Kegelrobbe 2660
Kuba-Baumratte 928
Kuba-Baumratten 920
Kubanischer Schlitzrüssler 6400
Kubanischer Tur 891
Kuban-Tur 891
Kuba-Schlitzrüssler 6400
Kudu 7100
Kufermulle 217
Kugelgürteltier 7045
Kugelgürteltiere 7044
Kuguar 2274
Kuhantilope 149
Kuhantilopen 148
Kuhl-Hirsch 475
Kulan 2144
Kultarr 6387
Kuns-Spitzmausbeutelratte 3975
Kupferfarbener Springaffe 718
Kupfergoldmulle 217
Kupferringbeutler 5678
Kurilen-Seehund 5283
Kurzohrrüsselspringer (pl) 3438

Moschustiere 4025, 4026
Mount-Data-Ratte 361
Mount-Graham-
    Rothörnchen 6859
Mount-Isarog-
    Spitzmausratte 372
Moupin-Pika 4610
Mückenfledermaus 5409
Muffelwild 4855
Mufflon 4855
Mulgara 1730
Muli-Pika 4599
Mullah-Stachelmaus 51
Müllers Borneo-Gibbon
    2906
Müllers Sandratte 6663
Mull-Lemminge 2054
Mullmäuse 4207
Mungo 2725
Mungos 4034
Münstersches
    Meerschweinchen 2390
Muntjak 4044
Muntjaks 4039
Mura-Nasenbeutler 3814
Mura-Neuguinea-
    Nasenbeutler 3814
Muriki 643
Murmele 3527
Murmeltier 3527
Murmeltiere 3517
Musangs 4985
Musmuqui 326
Mussos Fischratte 4460
Mützenlangur 5543
Mützenlanguren 5541
Mützenrobbe 1705
Mützenrobben 1704
Myers Grasmaus 134
Nabelschweine 6947, 6949
Nachtaffe 327
Nachtaffen 318
Nachtfledermäuse 4215
Nachtkatze 2291
Nachtschwirrer (pl) 7257,

7260
Nachtschwirrer 7258
Nacht-Zwergbeutelratte
    3509
Nacktbauchigel 2708
Nacktbauchiger
    Grabflatterer 6879
Nacktbrustkänguru 808
Nacktbrustkängurus 807
Nacktbrustrattenkänguru
    808
Nacktfledermaus 1202
Nacktfußwiesel 4145
Nacktmull 2746
Nacktmulle 2745
Nacktnasenwombat 7283
Nacktnasenwombats 7282
Nacktohrige Hirschmaus
    5093
Nacktohr-Totenkopfäffchen
    6205
Nacktrückenflughunde
    1926
Nacktrükenfledermäuse
    5737
Nacktschwanzbeutelratte
    3746
Nacktschwanzbeutelratten
    3745
Nacktschwanzgürteltier 696
Nacktschwanzgürteltiere
    692
Nacktschwanzkletterratten
    7157
Nacktschwanzratten 6405
Nacktsohlenrennmaus 6925
Nacktsohlenrennmäuse
    6913
Nagelkängurus 4690
Nagel-Manati 7117
Nahur 5683
Nahurs 5682
Namaqua-Buschratte 88
Namaqua-Felsenmaus 88
Namaqua-Manguste 2735
Namaqua-Strandgräber 539
Namdapha-Gleithörnchen

(pl) 583
Namdapha-Gleithörnchen
    584
Namib-Goldmull 2158
Namib-Rennmäuse 2480
Narwal 3985
Narwale 3984
Nasenaffe 4349
Nasenaffen 4347
Nasenbär 4351
Nasenbären 4350
Nasenbeuteldachs 5050
Nasenbeutler (pl) 5051
Nasenfledermaus 6123
Nasenfledermäuse 6122
Nasenratte 6121
Nasenratten 6119
Nashörner 5987, 5991
Natal-Ducker 1001
Natal-Vielzitzenmaus 3559
Natal-Wollschwanzhase
    5631
Natterers Fledermaus 4279
Nayarit-Maus 5134
Nebelparder (pl) 4382
Nebelparder 4383
Nebraska-Wolf 865
Negev-Rennmaus 3704
Nehrings Bergmaus 1869
Nelson-Antilopenziesel 237
Nelson-Nasenbär 4352
Nelson-Reisratte 4763
Nelsons Kängururatte 1908
Neotropische Wasserratten
    4375
Nepal-Leopard 4939
Nepal-Stachelschwein 2963
Nesokia-Ratten 4441
Netzgiraffe 2554
Neubrittanien-Flughund
    5765
Neuengland-
    Baumwollschwanzkani
    nchen 6743
Neufundland Biber 951

Riesenbaummäuse 3593
Riesenbeutelmarder 1768
Riesenblindmaus 6518
Riesenborkenratte 5275
Riesendachs 395
Riesenducker 1009
Riesenelen 6937
Riesenelenantilope 6937
Riesenfledermaus 4277
Riesenflugbeutler (pl) 5167
Riesenflugbeutler 5168
Riesenflughund 5763
Riesenflughunde 22
Riesenflughunde 5747
Riesengalago 4796
Riesengalagos 4795
Riesenginsterkatze 2457
Riesenglattwale 486
Riesengleitbeutler (pl) 5167
Riesengleitbeutler 5168
Riesengleitflieger (pl) 1682
Riesengleitflieger 2397
Riesengleithörnchen (pl)
    5156
Riesengoldmull 1299
Riesengoldmulle 1298
Riesengraumull 1593
Riesengürteltier 5555
Riesengürteltiere 5554
Riesenhamsterratten 1395
Riesenhohlnase 4528
Riesenhörnchen (pl) 5931
Riesenhörnchen 5933
Riesenkängururatte 1901
Riesenkängurus 3418
Riesenleierantilope 1725
Riesenmaki 5638
Riesenmaulwurfsratte 6777
Riesenmaus 6108
Riesenmäuse 6107
Riesenmulle 1298
Riesenmutjak 4050
Riesennager (pl) 2871
Riesennager 2873

Riesenohrspringmaus 2197
Riesenohrspringmäuse 2196
Riesenotter (pl) 5744
Riesenotter 5745
Riesenpanda 95
Riesenrappenantilope 2841
Riesenratte 7193
Riesenratten 7187
Riesenrundblattnase 2781
Riesenschuppentier 6397
Riesentaschenratte 4717
Riesentaschenratten 4712
Riesenwal 501
Riesenwaldschwein 2914
Riesenwaldschweine 2913
Riesenwildschaf 4836
Riesenwimperspitzmaus
    1443
Rinder 632
Rinderartige 632
Rindergemsen 670
Ringbeutler (pl) 5665, 5674
Ringelmungo 2402
Ringelmungos 2401
Ringelrennmaus 3708
Ringelrobbe 5831
Ringelrobben 5829
Ringelschwanz-
    Kletterbeutler (pl) 5665
Ringelschwanzmaki 3182
Ringelschwanzmungo 2402
Ringelschwanzziesel 6529
Ringschwanzfelskänguru
    5205
Ringschwanzopossum 5671
Rio-de-Janeiro-Reisratten
    5220
Rio-Grande-Biber 953
Rio-Napa-Tamarin 6175
Rio-Reisratte 5221
Risso-Delfin 2616
Riukiu Flughund 5759
Riukiu-Inselflughund 5759
Riukiu-Kaninchen (pl) 5037
Riukiu-Kaninchen 5038

Riukiu-Maulwurf 4455
Riukiu-Maulwürfe 4454
Riukiu-Ratte 1886
Riukiu-Ratten 1885
Riukiu-Stachelratten 7041,
    7042
Roberts Letschwe-
    Wasserbock 3072
Roberts Litschi-Wasserbock
    3072
Roberts Nacktschwanzmaus
    2647
Roberts-Letschwe
    Wasserbock 3072
Roberts-Litschi 3072
Roberts-Wasserbock 3072
Robinson-
    Röhrennasenflughund
    4563
Robinson-Zwergbeutelratte
    3513
Roborowskis Zwerghamster
    5300
Roborowski-Zwerghamster
    5300
Rocky-Mountains-
    Wühlmaus 3878
Rodriguez-Flughund 5801
Röhrennasen 4057
Röhrennasenflughunde
    4551
Rohrkatze 2271
Rohrmaus 7359
Rohrmäuse 7358
Rohrratten 7016
Rohrwolf 862
Rollmarder (pl) 4985
Roloway-Affe 1064
Roloway-Meerkatze 1064
Römischer Maulwurf 6816
Rondo-Bushbaby 2382
Roraima-Maus 5493
Roraima-Mäuse 5492
Rossantilopen 2831
Rossmeer-Robbe 4686
Ross-Robbe 4686
Ross-Robben 4685

Taiwan Flughund 5760

Taiwanische Waldmaus 352

Taiwanischer Nebelparder 4384

Taiwanisches Hochgebirgs-Zwergwiesel 4135

Taiwan-Langohr 5481

Taiwan-Serau 916

Taiwan-Sika 1116

Takin 671

Takine 670

Talahutka-Eichhörnchen 6285

Tamandua 6824

Tamanduas 6822

Tamarau 258

Tamarins 6160

Tamariskenrennmaus 3708

Tamaulipas-Maulwurf 6225

Tammar 3423

Tammar-Wallaby 3423

Tana 7148

Tana Fluss-Mangabe 1034

Tana-Haubenmangabe 1034

Tanala-Bilchschwanz 2052

Tangalunga 7268

Tannenmaus 5263

Tannenmäuse 5260

Tannenzapfenschupper 3484

Tanreks 6950, 6952

Tansania-Gepard 33

Tantalus-Affe 1264

Tantalus-Meerkatze 1264

Tapirböckchen 3454

Tapire 6884, 6885

Tarbagan 3533

Tarpan 2137

Taschenfledermäuse 6152

Taschenmäuse 1151

Taschennager 2464

Taschenratten 2464

Taschenspringer (pl) 1888

Taschenstachelmaus 2751

Taschenstachelmäuse 2750

Tasmanien-Bilchbeutler 1027

Tasmanien-Kurzschnabeligel 6771

Tasmanien-Langnasenbeutler 5049

Tasmanische Teufel (pl) 6220

Tasmanischer Langnasenbeutler 5049

Tasmanischer Nacktnasenwombat 7285

Tasmanischer Teufel 6221

Tasmanisches Bennet-Känguru 3437

Tasmanisches Bettong 567

Tasmanisches Riesenkänguru 3426

Tateomys-Spitzmausratten 6910

Tatra-Kleinwühlmaus 3904

Tattersalls Sifaka 5640

Tattersall-Sifaka 5640

Tatu 1755

Tayra 2009

Tayras 2008

Tchiru 4957

Tehuantepec-Eselhase 3238

Teichfledermaus 4242

Teledu 4156

Telefonim-Kuskus 5228

Temminck-Gleitflieger 1683

Temminck-Gleithörnchen 5185

Temminck-Zwergmaus 4100

Tenasserim-Lutung 7063

Tenasserim-Muntjak 4042

Tenrekverwandte 6952

Texanische Höhlenfledermaus 4306

Texas-Antilopenziesel 235

Texas-Grauwolf 864

Texas-Kängururatte 1895

Texas-Maus 5072

Texas-Puma 2289

Texas-Taschenratte 2471

Texas-Wolf 864

Theresas Kurzschwanz Opossum 3982

Thollons Roter Stummelaffe 5345

Thomas-Buschratte 92

Thomas-Bushbaby 2385

Thomas-Gleithörnchen 79

Thomas-Hirschmaus 3604

Thomas-Hirschmäuse 3601

Thomas-Hufeisennase 6056

Thomas-Langschwanztanrek 3790

Thomas-Langur 5553

Thomas-Mützenlangur 5553

Thomas-Paramo-Mäuse 6970

Thomas-Pika 4611

Thomas-Rotschenkelhörnchen 2359

Thomas-Zwergspringmaus 6219

Thomson-Gazelle 2443

Tibesti-Stachelmaus 44

Tibetanischer Bärenmakak 3406

Tibetanischer Braunbär 7233

Tibetanischer Halbesel 2146

Tibetanischer Rothirsch 1097

Tibetanischer Wolf 856

Tibetanischer Wollhase 3233

Tibet-Antilope 4957

Tibet-Antilopen 4956

Tibet-Bär 7233

Tibet-Fuchs 7294

Tibet-Gazelle 2428

Tibetischer Braunbär 7233

Tibetischer Pika 4610

Tibetischer Zwerghamster 1399

Index Français

Grand cob des roseaux 5942
Grand cobaye 971
Grand dauphin 7152
Grand dauphin de l'Indo-Pacifique 7151
Grand écureuil de Stanger 5656
Grand fer-à-cheval 6017
Grand fourmillier 4318
Grand galago 4799
Grand Hamster 1406
Grand hérisson 6336
Grand Kangourou rouge 3611
Grand Koudou 7100
Grand murin 4277
Grand panda 95
Grand Pangolin 6397
Grand phalanger volant 5168
Grand phoque 2166
Grand phoquie 2660
Grand polatouche 2560
Grand rhinopome 6074
Grand rinolophe 6017
Grand rinolophe fer-à-cheval 6017
Grand rinolophe obscur 6017
Grand Rorqual 501
Grande taupe dorée 1299
Grande Artibée 417
Grande baleine bleue 501
Grande chauve-souris à long nez de Mexique 3220
Grande chauve-souris brune 2119
Grande chauve-souris de Nouvelle-Zélande 4321
Grande chauve-souris murine 4277
Grande civette de l'inde 7269
Grande gerbille 2534
Grande gerbille à queue courte 2520

Grande gerbille de l'Aden 2530
Grande gerboise 3020
Grande musaraigne 589
Grande musaraigne à queue courte 589
Grande musaraigne du désert 3621
Grande musaraigne grise 6754
Grande mystacine 4321
Grande Noctule 4517
Grande Nyctère 4528
Grande sérotine 2128
Grande taupe marsupial 4512
Grandes gerbilles 6107
Grandes musaraignes du désert 3620
Grandes taupes dorées 1298
Grands écureuils 5654
Grands fourmilliers 4317
Grands kangourous 3418
Grands phalangers volants 5167
Grands wallabies 3418
Grans baleines à bec 558
Grinde 2577
Grison 2400
Grison d'Allamand 2400
Grisons 2398
Grivet 1259
Grizzli 7224
Grizzli de Californie 7221
Grizzli du Nouveau Mexique 7225
Grizzli mexicain 7232
Grizzly 7224
Grizzly de Californie 7221
Grizzly du Nouveau Mexique 7225
Grizzly mexicain 7232
Gros cerf 1077
Gros rat du sable 5658
Grosses souris 6582
Grysbok 5866

Grysbok de Sharpe 5867
Guanaco 3126
Guémal 2762
Guémal de Pérou 2762
Guémal du Chili 2763
Guenons 1038
Guépard 29
Guépard commun 32
Guépard d'Iran 38
Guépard royal 35
Guépards 28
Guérénouk 3289
Guérénouks 3288
Guéreza d'Angola 1350
Guéreza du Kilimandjaro 1351
Guéreza noir 1357
Guib 7086
Guib de Menelik 7090
Guib d'eau 7095
Guib harnaché 7086
Guimbo 638
Gymnure 1993
Gymnures 1992
Halichère gris 2660
Halichères 2659
Hamadryas 4964
Hamster 1406
Hamster à longue queue 1402
Hamster à queue blanche 4325
Hamster commun 1406
Hamster de Chine 1400
Hamster de la Dobroudja 3720
Hamster de Ladakh 1399
Hamster de Mongolie 188
Hamster de Newton 3720
Hamster de Roumanie 3720
Hamster de Turquie 3719
Hamster d'Europe 1406
Hamster d'Eversmann 189
Hamster d'Imhaus 3311
Hamster doré 3717

Hamster doré de Roumanie 3720
Hamster du Daghestan 3721
Hamster géorgien 3719
Hamster kazakh 189
Hamster migrateur 1403
Hamster nain de Campbell 5299
Hamster nain de Djoungarie 5301
Hamster nain de Roborowski 5300
Hamster nain de Sokolov 1404
Hamster roumain 3720
Hamster russe 5302
Hamster souriciforme 800
Hamster taupe sibérien 4211
Hamster tibétain 1401
Hamster-rat nain 7132
Hamsters 1405
Hamsters 881
Hamsters à queue blanche 4324
Hamsters communs 1405
Hamsters d'Europe 1405
Hamsters d'Imhaus 3310
Hamsters dorés 3716
Hamsters nains 187
Hamsters nains 5298
Hamsters nains gris 1398
Hamsters-rats nain 7131
Hamsters-taupes 4207
Hamster-taupe chinois 4212
Hapalémur à nez large 5624
Hapalémur doré 2664
Hapalémur gris 2665
Hapalémur gris d'Alaotra 2663
Hapalémur gris occidental 2666
Hapalémurs 2662
Hapalémurs à nez large 5623
Harpionyctère de Whitehead 2676

Harpionyctères 2674
Hay-Hay 1772
Helarctos (pl) 2679
Hélicres 3640
Hélioscure aux pattes rousses 2687
Hélioscure de Gambie 2684
Hélioscure du Ruwenzori 2688
Hemicentètes 2696
Hémione 2141
Hémippe 2141
Hérisson 2177
Hérisson à ventre blanc 2171
Hérisson aux longues oreilles d'Éthiopie 2700
Hérisson commun 2177
Hérisson d'Afrique du Sud 2183
Hérisson d'Algérie 2172
Hérisson daurien 2176
Hérisson de Brandt 2706
Hérisson de l'Amur 2174
Hérisson de l'Europe de l'Est 2175
Hérisson de l'Europe de l'Ouest 2178
Hérisson de Portugal 2179
Hérisson d'Europe 2177
Hérisson d'Europe orientale 2175
Hérisson du désert 2700
Hérisson du désert d'Éthiope 2700
Hérisson européen 2177
Hérisson ordinaire 2177
Hérisson oreillard 2701
Hérisson oreillard d'Éthiopie 2700
Hérisson oriental 2175
Hérisson sarde 2180
Hérissons 2169
Hérissons communs 2170
Hérissons oreillards 2699
Hermine 4131
Herminette 4131

Herpestidés 2738
Hippopotame 2766
Hippopotame amphibie 2766
Hippopotame nain 2759
Hippopotame nain de l'Est 2760
Hippopotames 2765
Hippopotames nains 2758
Hippopotamidés 2764
Hippotrague 2832
Hippotrague noir 2837
Hippotrague noir géant 2841
Hippotragues 2831
Hirola 1724
Hocheur 1057
Hocheur à nez rouge 1049
Hocheur à ventre rouge 1048
Hocheur à ventre roux 1048
Hocheur blanc-nez 1057, 1060
Hocheur blanc-nez de Bénin 1060
Hocheur de Martin 1058
Hocheur de Stampfl 1059
Hominidés 2856
Homme 2858
Hommes 2857
Hoolock 2895
Houleman 6329
Houlock 2895
Huemul 2763
Huillin 3366
Hurleur à manteau 197
Hurleur aux mains rousses 191
Hurleur de Bolivie 199
Hurleur de Guatemala 198
Hurleur moir 192
Hurleur roux 200
Hurleurs 190
Hutia d'Allen 3002
Hutia de Cuvier commun 5444

Zorilles communs 2979
Zorilles de Libyie 5494

Indice Italiano